Reactive Transport Modeling

Reactive Transport Modeling

Applications in Subsurface Energy and Environmental Problems

Edited by

Yitian Xiao
ExxonMobil Upstream Research Company, USA

Fiona Whitaker
School of Earth Sciences, University of Bristol, UK

Tianfu Xu
Key Laboratory of Groundwater Resources and Environment, Ministry of Education, Jilin University, Changchun, China

Consulting Editor:

Carl Steefel
Lawrence Berkeley National Laboratory, USA

WILEY

This edition first published 2018

Registered Office(s)
John Wiley & Sons, Inc., 111 River Street, Hoboken, NJ 07030, USA
John Wiley & Sons Ltd, The Atrium, Southern Gate, Chichester, West Sussex, PO19 8SQ, UK

Editorial Office
9600 Garsington Road, Oxford, OX4 2DQ, UK

For details of our global editorial offices, customer services, and more information about Wiley products visit us at www.wiley.com.

Wiley also publishes its books in a variety of electronic formats and by print-on-demand. Some content that appears in standard print versions of this book may not be available in other formats.

Library of Congress Cataloging-in-Publication Data

Names: Xiao, Yitian, editor.
Title: Reactive transport modeling : applications in subsurface energy and environmental problems / edited by Yitian Xiao (ExxonMobil Upstream Research Company, Spring, TX, USA) [and three others].
Description: First edition. | Hoboken, NJ : Wiley, [2018] | Includes bibliographical references and index. |
Identifiers: LCCN 2017054860 (print) | LCCN 2018001750 (ebook) | ISBN 9781119060017 (pdf) | ISBN 9781119060024 (epub) | ISBN 9781119060000 (cloth)
Subjects: LCSH: Transport theory. | Porous materials–Permeability–Mathematical models. | Groundwater flow–Mathematical models. | Geological carbon sequestration. | Geochemistry.
Classification: LCC QC175.2 (ebook) | LCC QC175.2 .R4245 2018 (print) | DDC 628.4–dc23
LC record available at https://lccn.loc.gov/2017054860

Cover Design: Wiley
Cover Image: Courtesy of CSIRO Publishing

Set in 10/12pt Warnock by SPi Global, Pondicherry, India

Printed in the UK by Bell & Bain Ltd, Glasgow

10 9 8 7 6 5 4 3 2 1

The manufacturer's authorized representative according to the EU General Product Safety Regulation is Wiley-VCH GmbH, Boschstr. 12, 69469 Weinheim, Germany, e-mail: Product_Safety@wiley.com.

Contents

List of Contributors

Pascal Audigane
BRGM (Bureau de Recherches Géologiques et Minières), Water Environment and Ecotechnologies Division (D3E), D3E/GDR Groundwater Resources Unit, Orleans Cedex 2
France

Zhangxin (John) Chen
University of Calgary
Canada

Francis Claret, PhD
BRGM
FRANCE

Xiaohu Dong
China University of Petroleum (Beijing), China
and
University of Calgary
Canada

Miles Frazer
Chevron Energy Technology Company
Houston
USA

Irina Gaus
NAGRA

Fabrizio Gherardi
Istituto di Geoscienze e Georisorse (IGG) – Consiglio Nazionale delle Ricerche
Pisa
Italy

Esra Inan-Villegas
Shell Global Solutions (UK)
London
UK

Xinfeng Jia
China University of Petroleum (Beijing), China and
University of Calgary
Canada

Gareth D. Jones
ExxonMobil Upstream Research Company
Spring
USA

Günter Langergraber
Institute of Sanitary Engineering and Water Pollution Control
University of Natural Resources and Life Sciences Vienna (BOKU)
Austria

Nicolas N. Marty, PhD
BRGM Orléans Cedex 2
FRANCE

Christof Meile
Department of Marine Sciences
University of Georgia
Athens
USA

Jin Na
Key Laboratory of Groundwater
Resources and Environment
Ministry of Education
Jilin University
Changchun
China

Timothy D. Scheibe, PhD
Pacific Northwest National
Laboratory
Richland
USA

Jirka Šimůnek
Department of Environmental Sciences
University of California Riverside,
USA

Jeroen Snippe
Shell Global Solutions International B.V.
Amsterdam
The Netherlands

Nicolas Spycher
Earth Sciences Division
Lawrence Berkeley National Laboratory
USA

Carl Steefel
Earth Science Division
Lawrence Berkeley National Laboratory
USA

Alexandre M. Tartakovsky
Pacific Northwest National Laboratory
Richland
USA

Paul Taylor
Shell Global Solutions (UK)
London
UK

Hailong Tian
Key Laboratory of Groundwater
Resources and Environment
Ministry of Education
Jilin University
Changchun
China

Christophe Tournassat, PhD
BRGM Orléans Cedex 2
FRANCE

Fiona Whitaker
School of Earth Sciences
University of Bristol
UK

Yitian Xiao
ExxonMobil Upstream Research
Company
Spring
USA

Jinze Xu
University of Calgary
Canada

Tianfu Xu
Key Laboratory of Groundwater
Resources and Environment
Ministry of Education
Jilin University
Changchun
China

Guoxiang Zhang
Shell Exploration and Production Inc.
USA

Liange Zheng
Earth Sciences Division
Lawrence Berkeley National Laboratory
USA

Preface

Reactive transport modeling is a highly multidisciplinary area of research that has seen rapid advances in the past 20 years. The field encompasses a number of diverse disciplines including hydrology, geochemistry, geology, biology, engineering, environmental and computational sciences. Reactive transport in the subsurface can result in a wide variety of geochemical processes, including rock/mineral alteration in natural diagenetic systems and in response to injection of CO_2 and acid gases, enhanced oil and gas recovery, transport and storage of radiogenic and toxic waste products in geological formations, as well as biogeochemical. Reactive transport models are thus important engineering tools which, as this summary of the state of the art demonstrates, have already had significant impact in some disciplines and offer promise in many more. However, at a higher level they also offer the potential to provide the 'dynamic glue' with which to integrate fundamental process-based research into the study of complex natural Earth systems.

This book assimilates contributions from leading experts from universities and national and industrial research institutes around the world, who authoritatively discuss recent applications of reactive transport modeling to a wide range of subsurface and environmental problems. As such, the book is a timely reference and can be used as an advanced text book with relevance to a broad geoscience and engineering audience from both academia and industry. It spans the application of reactive transport models to a range of engineering and environmental fields, from understanding interactions between fluids, solutes and minerals to better manage extraction of energy resources and subsurface storage of waste, to incorporation of geobiology and addressing the challenges of modeling at a range of spatial and temporal scales.

CO_2 sequestration and geothermal energy: Large-scale carbon capture and sequestration (CCS) and geothermal energy are potential solutions to reduce CO_2 emissions and provide alternative and renewable energy. These processes involve a complex interplay of multiphase flow, capillary trapping, diffusion, convection, and chemical reactions that may have significant impacts on both injection performance and storage security. In Chapters 1 and 2, Xu et al. from Jilin University, China, and Audigane et al. from the French Geological Survey and other institutions in the EU, evaluate the applications of reactive transport modeling to CO_2 geological sequestration and the development of geothermal energy. These discussions elucidate the basic theory of reactive transport modeling and its application in understanding interactions among gases, liquids, solutes and minerals, and the implications for CO_2 geological sequestration and geothermal energy development. The modeling results help to identify the short- and long-term storage capacities of sedimentary formations and the geochemical processes associated with CO_2 leakage during storage, including near-well phenomena and impacts on groundwater, as well as performance issues associated with enhanced geothermal energy. The authors also discuss the current capabilities and limitations of reactive transport models for simulating the geochemical processes associated with CO_2 geological storage.

Diagenesis and reservoir quality: The key challenge in reservoir characterization is predicting the spatial distribution of diagenesis, which is often a critical control on reservoir quality heterogeneity and thus producibility of oil and gas. In Chapters 3 and 4, Whitaker from the University of Bristol, UK, Frazer from Chevron, USA, and Xiao and Jones from ExxonMobil, USA, review and discuss a wide range of geoscience applications pertinent to reservoir quality prediction. These include simulation of syn-depositional diagenesis of carbonate rocks controlled by interactions between episodes of submarine deposition of reactive sediment, and rapid changes in hydrological and biogeochemical conditions in response to changes in sea-level. Diagenesis continues to alter reservoir quality during burial, when fault-controlled flows play an increasingly important role, for example in hydrothermal fluid flow and illitization. Finally, at an operational time-scale, reservoirs can be altered by acid gas and steam injection. Reactive transport models are used to examine how key natural variables impact different styles of diagenesis and reservoir porosity evolution, and provide insight on a range of important issues, such as the occurrence and distribution of karstic and dolomite geobodies which can give rise to high permeability 'super-k' zones, calcite and anhydrite cements, and the impact on reservoir quality of fault-controlled fluid flow during both early and burial diagenesis. They are also used to provide rule sets in developing a coupling forward stratigraphic and diagenetic model for carbonate platform development. The authors discuss frameworks and examples to link fundamental geochemical processes, integrate with traditional methods, and calibrate with field and production data that have the potential to significantly improve the ability to predict carbonate and siliciclastic reservoir quality.

Enhanced oil recovery: During petroleum production, enhanced oil recovery (EOR) methods such as CO_2 flooding, acid stimulation, steam injection and in situ combustion can cause chemical reactions between the injected fluids and the reservoir rocks (artificial diagenesis) that may be either beneficial or detrimental to enhanced oil recovery. In Chapter 5, Zhang et al. from Shell, USA, discuss enhanced production from conventional hydrocarbon reservoirs which are characterized by high temperature, high pressure, high salinity and highly reducing conditions (4-high). Injection to and/or production from these reservoirs can induce significant geochemical reactions due to the high contrast between geochemical conditions in the reservoir and those of the invading fluids, or between the reservoir and the surface conditions, introducing risks for hydrocarbon (HC) recovery. Risk evaluation and management require thorough understanding and modeling of these geochemical systems. Potential geochemical reactions taking place in 4-high reservoirs during HC recovery are discussed, based on the characterization of the reservoirs. Quantitative description, modeling approaches, tool and examples of real cases are presented in this chapter.

Of the approximately 9 – 11 trillion barrels of crude oil resources in the world, about two-thirds are unconventional heavy oil and bitumen. Thermal processes, using heat to reduce oil viscosity in situ, have been the most successful methods of recovery. In Chapter 6, Jia et al. from the University of Calgary, Canada, cover current modeling approaches to dynamic heat and mass transfer in several heavy oil recovery processes, such as SAGD, hybrid steam-solvent, and steam-solvent-gas co-injection. The respective contribution of conduction and convection to the viscosity reduction of heavy oil is analysed. In addition, the effects of heterogeneous reservoir properties and the impact on heavy oil recovery efficiency are discussed.

Acid gas injection and groundwater contamination: One potential risk of geological storage of CO_2 is the leakage of CO_2 and deep brines from a deep storage formation into overlying shallow, potable groundwater aquifers. In Chapter 7, Zheng and Spycher from Lawrence Berkeley National Laboratory, USA, discuss results from reactive transport models used to investigate the potential impact of acid gas injection and brine leakage on shallow groundwater, particularly the fate of trace metals and organics and the environmental consequences. Their simulation results provide insight into acid gas injection site selection and operational feasibility to ensure the safety of drinking-water resources near these locations.

Nuclear waste disposal in geological formations: Large amounts of nuclear waste around the world need to be disposed of in repository sites that can ensure the safety and security of the waste for millennia. In Chapter 8, Claret et al. from the French Geological Survey discuss the grand challenge of long-term safe storage of nuclear waste in geological formations. They cover the framework of modeling the long-term stability of multi-barrier systems, which consist of waste overpacks (e.g. metal canisters, concrete), engineered barriers such as bentonite, and natural barriers such as clay rocks. They summarize recent improvements and discuss future challenges in the application of reactive transport modeling to nuclear repository systems in order to understand how a repository system will evolve due to thermal, hydraulic, mechanical, chemical and radiological processes.

Sustainable constructed wetlands: Constructed wetlands are sustainable, environmentally friendly solutions to engineering problems such as wastewater treatment, which involve a large number of physical, chemical, and biological processes. In Chapter 9, Langergraber from the University of Natural Resources and Life Sciences Vienna (BOKU), Austria, and Simunek from the University of California Riverside, USA, discuss their mechanistic modeling approach for saturated/unsaturated water flow and the application of the convection-dispersion equation for heat and solute transport, transformation and degradation processes of the pollutants, plants taking up nutrients and releasing organic matter and nutrients, and clogging that would impact the long-term performance of constructed wetlands. They also discuss the limitations with the existing approach and remaining challenges, and summarize the need for further development and research.

Biogeochemical cycles: Biogeochemical transformations have significant impacts on many of Earth's processes and environments at the surface and in the subsurface of the

Earth. In Chapter 10, Meile from the University of Georgia and Scheibe from the Pacific Northwest National Laboratory, USA, review challenges in modeling microbially-driven subsurface systems. They summarize recent reactive transport modeling approaches, including the representation of microbes in the models and how microbial community interactions can be described. They also discuss the challenges associated with the integration of data, and ways to tackle the vast range of scales relevant to models of biogeochemical cycling.

Meso-scale model: Most reactive transport models rely on the traditional Darcy-scale advection–dispersion model with constant dispersion coefficient that, under certain conditions, may overestimate the extent of reactions due to overestimating mixing by treating it as an effective Fickian process, and lead to inaccurate predictions of concentration-gradient driven flows. In Chapter 11, Tartakovsky from the Pacific Northwest National Laboratory, USA, reviews an alternative Langevin advection–diffusion model. His study shows that separate treatment of the advective and diffusive mixing in the Langevin model results in a more accurate prediction of the multicomponent reactive transport processes.

Acknowledgements

The editors would like to acknowledge the dedication and commitment from all chapter authors whose contributions made the book possible. The editors would also like to acknowledge the collaborations and support from colleagues at their corresponding institutes. Significant support and guidance from the Wiley editorial team, in particular Sarah Higginbotham, Emma Strickland, Elsie Merlin, Samanaa Srinivas, Rebecca Ralf, Sathishwaran, Samantha Jones, and Shalisha Sukanya, has been greatly appreciated.

January 2018

Yitian Xiao
Houston, USA

Fiona Whitaker
Bristol, UK

Tianfu Xu
Jilin, China

Carl Steefel
Berkeley, USA

1

Application of Reactive Transport Modeling to CO_2 Geological Sequestration and Chemical Stimulation of an Enhanced Geothermal Reservoir

Tianfu Xu, Hailong Tian and Jin Na

Key Laboratory of Groundwater Resources and Environment, Ministry of Education, Jilin University, Changchun, 130021, China

This chapter is devoted to giving a deep insight into the applications of reactive transport modeling to two problems where geochemistry plays an important role. The two problems involved here are CO_2 geological sequestration and the development of geothermal energy. These applications elucidate the use of reactive transport modeling in understanding the reactions among water, solute and minerals occurring in the areas of CO_2 geological sequestration and geothermal energy development.

1.1 Introduction

Coupled modeling of subsurface multiphase fluid and heat flow, solute transport, and chemical reactions can be applied to many geological systems and environmental problems, including geothermal systems, diagenetic and weathering processes, subsurface waste disposal, acid mine drainage remediation, contaminant transport, and groundwater quality.

The processes involved in coupling include geochemical, thermal and hydrological interactions. These processes are strongly coupled through linear and non-linear relationships, and make up a significant part of the subject area of 'reactive transport', in which we focus on the development of models that quantitatively represent the evolution of subsurface geological environments.

Reactive transport models that consider conductive and convective heat transport, dissolution and precipitation kinetics, aqueous speciation and coupled porosity evolution within saturated and unsaturated geological materials are very demanding in terms of computing time and memory, and results might either be difficult to obtain or limited in terms of precision. Very recently, such modeling work has achieved a high degree of sophistication as access to high-performance computer increases.

Recent numerical modeling for subsurface environment applications can be split into three categories, including hydrodynamic modeling, geochemical modeling (batch

Reactive Transport Modeling: Applications in Subsurface Energy and Environmental Problems, First Edition. Edited by Yitian Xiao, Fiona Whitaker, Tianfu Xu and Carl Steefel.

reaction) and reactive transport modeling. Reactive transport modeling is the most realistic modeling technique to quantify the complex of coupling processes over a long period of time.

For each category there are special modeling packages for performing specific simulations. For example, ECLIPSE, TOUGH2, SIMED, and so on, were developed for hydrodynamic modeling and associated issues. For batch geochemical modeling, usages of PHREEQC and TOUGHREACT are reported in previous works (Gaus et al., 2004; Xu et al., 2004a). Batch geochemical modeling is ideal when modeling laboratory batch experiments. However, it becomes limited when the hydrodynamics plays a dominant role in the whole process, and then TOUGHREACT, SIMUSCOPP, STOMP, HYTEC, UNFT, SUPCRT92 and GEMBOCHS were developed for reactive transport modeling.

Among these codes, TOUGHREACT was developed by introducing geochemistry into the existing TOUGH2, a framework of a non-isothermal, multicomponent fluid and heat in a porous and fractured media simulator. This program has been widely used for many geological systems and environmental problems. For the worldwide recognition of TOUGHREACT, this chapter will briefly introduce the fundamental theories of reactive transport modeling, and illustrate its basic applications to two problems: CO_2 geological sequestration and geothermal energy development.

1.2 Fundamental Theories

1.2.1 Governing Equations for Flow and Transport

The primary governing equations for multiphase fluid and heat flow and chemical transport have the same structure, derived from the principle of mass (or energy) conservation. These equations are presented in Table 1.1.

Aqueous (dissolved) species are subject to transport in the liquid phase as well as to local chemical interactions with the solid and gas phases. The transport equations are written in terms of total dissolved concentrations of chemical components, which are concentrations of the basis species plus their associated aqueous

Table 1.1 Primary governing equations for fluid and heat flow, and chemical transport. Symbol meanings are given in Table 1.2.

General governing equation: $\frac{\partial M_\kappa}{\partial t} = -\nabla F_\kappa + q_\kappa$

Water: $M_w = \phi\left(S_l \rho_l X_{wl} + S_g \rho_g X_{wg}\right) F_w = X_{wl}\rho_l u_l + X_{wg}\rho_g u_g q_w = q_{wl} + q_{wg}$

Air: $M_c = \phi\left(S_l \rho_l X_{cl} + S_g \rho_g X_{cg}\right) F_c = X_{cl}\rho_l u_l + X_{cg}\rho_g u_g q_c = q_{cl} + q_{cg}$

Heat: $M_h = \phi\left(S_l \rho_l U_l + S_g \rho_g U_g\right) + (1-\phi)\rho_s U_s F_h = \sum_{\beta=l,g} h_\beta \rho_\beta u_\beta - \lambda \nabla T q_h$

where $u_\beta = -k\frac{k_{r\beta}}{\mu_\beta}\left(\nabla P_\beta - \rho_{\beta g}\right)\beta = l, g$ (Darcy's law)

Chemical components in the liquid phase ($j = 1, 2, \ldots, N_l$):

$M_j = \phi S_l C_{jl}\ F_j = u_l C_{jl} - \left(\tau\phi S_l D_l\right)\nabla C_{jl}\ q_j = q_{jl} + q_{js} + q_{jg}\ \tau_\beta = \phi^{1/3} S_\beta^{7/3}$

Table 1.2 Symbols used in Table 1.1.

C	component concentration, $mol\,l^{-1}$	ρ	density, $kg\,m^{-3}$
D	diffusion coefficient, $m^2\,s^{-1}$	μ	viscosity, $kg\,m^{-1}\,s^{-1}$
F	mass flux, $kg\,m^{-2}\,s^{-1}$	λ	heat conductivity, $W\,m^{-1}\,K^{-1}$
k	permeability, m^2		
k_r	relative permeability	Subscripts:	
g	gravitational acceleration, $m\,s^{-2}$	c	air
M	mass accumulation, $kg\,m^{-3}$	g	gas phase
N	number of chemical components	h	heat
P	pressure, Pa	j	aqueous chemical component
q	source/sink	κ	governing equation index
S	saturation	l	liquid phase
T	temperature, °C	r	reaction
U	internal energy, $J\,kg^{-1}$	s	solid phase
u	Darcy velocity, $m\,s^{-1}$	w	water
X	mass fraction	β	phase index
φ	porosity	τ	medium tortuosity

secondary species (Yeh and Tripathi, 1991; Steefel and Lasaga, 1994; Xu and Pruess, 2001). If kinetically controlled reactions occur between aqueous species, then additional ordinary differential equations need to be solved to link the total concentrations of the primary species with the evolving concentrations of the secondary species (Steefel and MacQuarrie, 1996). Advection and diffusion processes are considered for both the aqueous and gaseous species. Aqueous species diffusion coefficients are assumed to be the same. Gaseous species, having a neutral valence, can have differing diffusion coefficients calculated as a function of T, P, molecular weight, and molecular diameter.

The local chemical interactions in the transport equations are represented by reaction source/sink terms.

1.2.2 Equations for Chemical Reactions

To represent a geochemical system, it is convenient to select a subset of N_C aqueous species as basis species (or component or primary species). All other species are called secondary species that include aqueous complexes, precipitated (mineral) and gaseous species (Reed, 1982; Yeh and Tripathi, 1991; Steefel and Lasaga, 1994). The number of secondary species must be equal to the number of independent reactions. Any of these secondary species can be represented as a linear combination of the set of basis species, such as

$$S_i = \sum_{j=1}^{N_c} \nu_{ij} S_j \quad i = 1, \ldots, N_R \tag{1.1}$$

where S_i represents chemical species, j is the basis species index, i is the secondary species index, N_R is the number of reactions (or secondary species), and ν_{ij} is the stoichiometric coefficient of the jth basis species in the ith reaction.

1) Dissolution/precipitation for equilibrium minerals

The equations for chemical equilibrium are similar to those by Parkhurst et al. (1980), Reed (1982), Yeh and Tripathi (1991), Wolery (1992), and Steefel and Lasaga (1994), and are presented in Table 1.3.

Aqueous species activity coefficients are calculated from the extended Debye-Hückel equation (Helgeson et al., 1981). The calculations of gas fugacity coefficients are given in Spycher and Pruess (2005).

The mineral saturation ratio can be expressed as:

$$\Omega_m = K_m^{-1} \times \prod_{j=1}^{N_c} C_j^{\nu_{mj}} \gamma_j^{\nu_{mj}} \quad \left(m = 1, \ \ldots, \ N_P\right) \tag{1.2}$$

where m is the equilibrium mineral index and K_m is the corresponding equilibrium constant. At equilibrium, we have:

$$SI_m = \log_{10} \Omega_m = 0 \tag{1.3}$$

where SI_m is the mineral saturation index.

Table 1.3 List of equations governing chemical equilibrium (illustrated by specific examples; in fact the model is valid for general geochemistry). Symbol meanings are given in Table 1.4.

General dissociation reactions	$S_i^s = \sum_{j=1}^{N_C} \nu_{ij} S_j^p$
General mass action equations	$K_i a_{S_i^s} = \sum_{j=1}^{N_C} \left(a_{S_j^s}\right)^{\nu_{ij}}$
Aqueousdissociation:	$HCO_3^- = CO_3^{2-} + H^+ K_{HCO_3^-} \gamma_{HCO_3^-} c_{HCO_3^-} = \gamma_{CO_3^{2-}} c_{CO_3^{2-}} \gamma_{H^+} c_{H^+}$
Mineral dissolution:	$CaCO_3(s) = CO_3^{2-} + Ca^{2+} K_{CaCO_3(s)} = \gamma_{Ca^{2+}} c_{Ca^{2+}} \gamma_{CO_3^{2-}} c_{CO_3^{2-}}$
Gas dissolution:	$CO_2(g) = CO_2(aq) K_{CO_2(g)} f_{CO_2(g)} = \gamma_{CO_2(aq)} c_{CO_2(aq)}$

Table 1.4 Symbols used in Table 1.3.

a	thermodynamic activity	γ	thermodynamic activity coefficient
f	gaseous species partial pressure, bar	ν	stoichiometric coefficient
K	equilibrium constant		
N_C	number of component (basis species)	Subscript:	
S^p	basis (primary) species	i	secondary species index
S^s	secondary species	j	basis (primary) species index

2) **Kinetic mineral dissolution/precipitation**
Kinetic rates can also be functions of non-basis species. Usually the species appearing in rate laws happen to be basis species. In this model, we use a rate expression given by Lasaga et al. (1994):

$$r_n = f\left(c_1, c_1, \ldots, c_{NC}\right) = \pm k_n A_n \left|1 - \Omega_n^{\theta}\right|^{\eta} \left(n = 1, \ldots, N_q\right) \tag{1.4}$$

where positive values of r_n indicate dissolution, and negative values precipitation, k_n is the rate constant (moles per unit mineral surface area and unit time) which is temperature dependent, A_n is the specific reactive surface area per kg H_2O, and n is the kinetic mineral saturation ratio. The parameters must be determined from experiments; usually, but not always, they are taken equal to one.

The temperature dependence of the reaction rate constant can be expressed reasonably well via an Arrhenius equation (Lasaga, 1984; Steefel and Lasaga, 1994). Because many rate constants are reported at 25 °C, it is convenient to approximate rate constant dependency as a function of temperature thus:

$$k = k_{25} \exp\left[\frac{-E_a}{R}\left(\frac{1}{T} - \frac{1}{298.15}\right)\right] \tag{1.5}$$

where E_a is the activation energy, k_{25} is the rate constant at 25 °C, R is the gas constant, and T is the absolute temperature.

For many minerals, the kinetic rate constant k can be summed from three mechanisms (Palandri and Kharaka, 2004):

$$\begin{aligned} k = {} & k_{25}^{nu} \exp\left[\frac{-E_a^{nu}}{R}\left(\frac{1}{T} - \frac{1}{298.15}\right)\right] + k_{25}^{H} \exp\left[\frac{-E_a^{H}}{R}\left(\frac{1}{T} - \frac{1}{298.15}\right)\right] a_H^{n_H} \\ & + k_{25}^{OH} \exp\left[\frac{-E_a^{OH}}{R}\left(\frac{1}{T} - \frac{1}{298.15}\right)\right] a_{OH}^{n_{OH}} \end{aligned} \tag{1.6}$$

where superscripts or subscripts *nu*, *H*, and *OH* indicate neutral, acid, and base mechanisms, respectively, E_a is the activation energy, k_{25} is the rate constant at 25 °C, R is the gas constant, T is absolute temperature, a is the activity of the species, and n is an exponent (constant). The rate constant k can be also dependent on other species such as Al^{3+} and Fe^{3+}. Two or more species may be involved in one mechanism by:

$$k = k_{25}^{nu} \exp\left[\frac{-E_a^{nu}}{R}\left(\frac{1}{T} - \frac{1}{298.15}\right)\right] + \sum_i k_{25}^{i} \exp\left[\frac{-E_a^{i}}{R}\left(\frac{1}{T} - \frac{1}{298.15}\right)\right] \prod_j a_{ij}^{n_{ij}} \tag{1.7}$$

where superscript or subscript i is the additional mechanism index, and j is the species index involved in one mechanism that can be primary or secondary species.

1.2.3 Solution Method for Transport Equations

Most chemical species are only subject to transport in the liquid phase. A few species can be transported in both liquid and gas phases, such as O_2 and CO_2. We first derive the numerical formulation of reactive transport in the liquid phase. This will then be extended to transport in the gas phases for some gaseous species.

1) **Transport in the liquid phase**
   In the sequential iteration approach (SIA), the mass transport equations and chemical reaction equations are considered as two relatively independent subsystems. They are solved separately in a sequential manner following an iterative procedure. If reactions taking place in the liquid phase are assumed to be at local equilibrium, mass transport equations can be written in terms of total dissolved component concentrations. By lumping all mass accumulation terms due to mass transfer between aqueous and solid phases including precipitated (kinetics and equilibrium), exchanged and sorbed species, we can write equations for multicomponent chemical transport in the liquid phase as:

$$\frac{\Delta t}{V_n}\sum_m A_{nm}\left[u_{nm}^{k+1}C_{nm}^{(j),k+1,s+1/2}+D_{nm}\frac{C_m^{(j),k+1,s+1/2}-C_n^{(j),k+1,s+1/2}}{d_{nm}}\right]$$
$$=\Delta M_n^{(j),k+1}-q_n^{(j),k+1}\Delta t-R_n^{(j),k+1,s}\Delta t\ (j=1,2,\ \ldots,N_C) \quad (1.8)$$

   where n labels the grid block, m labels the adjacent grid blocks connected to n, j labels the chemical component, N_C is the total number of chemical components, l labels the liquid phase, k labels the number of the time step, s labels the number of the transport-chemistry iteration, u_{nm} is the liquid volumetric flux or the Darcy velocity (m/s), D_{nm} is the effective diffusion coefficient (including effects of porosity, phase saturation, tortuosity and weighting factors between the two grid blocks), d_{nm} is the nodal distance, and $R_n^{(j),k+1}$ are the overall chemical reaction source/sink terms.

2) **Transport in the gas phase**
   Gaseous species concentrations can be related to partial pressures by

$$C^g=\frac{10^2}{RT}P^g \quad (1.9)$$

   where C^g are gaseous species concentrations (in mol/l), P^g is the gaseous species partial pressure (in bar), R is the gas constant (8.314 J mol^{-1} K^{-1}) and T is the absolute temperature. By following the same principle as used for transport in the liquid phase and by considering Equation (1.9), the numerical formulation of gaseous transport in the gas phases can be expressed as:

$$\left[\frac{10^2}{RT}S_{g,n}^{k+1}\phi_n^{k+1}+\frac{\Delta t}{V_n}\sum_m A_{nm}\frac{10^2}{RT}-\left(u_{nm}^{g,k+1}\varepsilon_{nm}^{g}+\frac{D_{nm}^{g}}{d_{nm}}\right)\right]P_n^{g,(j),k+1,s+\frac{1}{2}}$$
$$+\frac{\Delta t}{V_n}\sum_m A_{nm}\frac{10^2}{RT}\left[u_{nm}^{g,k+1}(\varepsilon_{nm}^{g}-1)-\frac{D_{nm}^{g}}{d_{nm}}\right]P_n^{g,(j),k+1,s+\frac{1}{2}} \tag{1.10}$$
$$=\frac{10^2}{RT}S_{g,n}^{k}\phi_n^{k}P_n^{g,(j),k}+q_n^{(j)k+1}\Delta t+R_n^{(j),k+1,s}\Delta t\quad (j=1,2,\ldots,N_g)$$

where N_g is the number of gaseous species.

1.2.4 Solution Method for Mixed Equilibrium-Kinetics Chemical System

Aqueous complexation and gas dissolution/exsolution proceed according to local equilibrium, while mineral dissolution/precipitation is subject to equilibrium and/or kinetic conditions. Gas dissolution/exsolution is included in the model and treated in a similar way as equilibrium mineral dissolution/precipitation, but with fugacity correction. The formulation is based on mass balances in terms of basis species as used by Parkhurst et al. (1980) and Reed (1982) for the equilibrium chemical system. The kinetic rate expressions for mineral dissolution/precipitation are included in the equations along with the mass balances of basis species. At time zero (initial), the total concentrations of basis species j in the system are assumed to be known, and are given by

$$T_j^0=c_j^0+\sum_{k=1}^{N_x}v_{kj}c_k^0+\sum_{m=1}^{N_p}v_{mj}c_m^0+\sum_{n=1}^{N_q}v_{nj}c_n^0+\sum_{z=1}^{N_z}v_{zj}c_z^0+\sum_{s=1}^{N_s}v_{sj}c_s^0\; j=1\ldots N_C \tag{1.11}$$

where superscript 0 represents time zero; c denotes concentration; subscripts j, k, m, n, z and s are the indices of the primary species, aqueous complexes, minerals at equilibrium, minerals under kinetic constraints, and exchanged and surface complexes, respectively; N_c, N_x, N_p, N_q, N_z and N_s are the number of corresponding species and minerals; and v_{kj}, v_{mj}, v_{nj}, v_{zj} and v_{sj} are stoichiometric coefficients of the primary species in the aqueous complexes, equilibrium, kinetic minerals, and exchanged and surface complexes, respectively.

After a time step Δt, the total concentration of primary species j (T_j) is given by

$$T_j=c_j+\sum_{k=1}^{N_x}v_{kj}c_k+\sum_{m=1}^{N_p}v_{mj}c_m+\sum_{n=1}^{N_q}v_{nj}\left(c_n^0-r_n\Delta t\right)+\sum_{z=1}^{N_z}v_{zj}c_z+\sum_{s=1}^{N_s}v_{sj}c_s\quad\left(j=1,\;\ldots,\;N_C\right) \tag{1.12}$$

where r_n is the kinetic rate of mineral dissolution (negative for precipitation, units used here are moles per kilogram of water per time), for which a general multi-mechanism rate law can be used. Δt and T_j are related through generation of j among primary species as follows:

$$T_j-T_j^0=\sum_{l=1}^{N_a}v_{lj}r_l\Delta t\quad\left(j=1,\;\ldots,\;N_C\right) \tag{1.13}$$

where l is the aqueous kinetic reaction (including biodegradation) index, N_a is the total number of kinetic reactions among primary species, and r_l is the kinetic rate, which is in terms of one mole of product species.

By substituting Equations (1.11) and (1.12) into Equation (1.13), and denoting residuals as F^c_j (which are zero in the limit of convergence), we have:

$$
\begin{aligned}
F_j^c = & (c_j - c_j^0) && \text{primary species}\\
& + \sum_{k=1}^{N_x} \nu_{kj}(c_k - c_k^0) && \text{equilibrium aqueous complexes}\\
& + \sum_{z=1}^{N_z} \nu_{zj}(c_z - c_z^0) && \text{exchanged complexes}\\
& + \sum_{s=1}^{N_s} \nu_{sj}(c_s - c_s^0) && \text{surface complexes}\\
& + \sum_{m=1}^{N_p} \nu_{mj}(c_m - c_m^0) && \text{equilibrium minerals}\\
& - \sum_{n=1}^{N_q} \nu_{nj} r_n \Delta t && \text{kinetic minerals}\\
& - \sum_{l=1}^{N_a} \nu_{lj} r_l \Delta t && \text{kinetics among primary species}\\
= & \, 0 && (j = 1, \ldots, N_C)
\end{aligned}
\tag{1.14}
$$

According to mass-action equations, the concentrations of aqueous and exchanged complexes c_k and c_z can be expressed as functions of concentrations of the primary species c_j. Kinetic rates r_n and r_l are functions of c_j. Surface complexes are expressed as the product of primary species and an additional potential term Ψ. Additional equilibrium equations depending on the surface complexation model have to be solved together with Equation (1.14). No explicit expressions relate equilibrium mineral concentrations c_m to c_j. Therefore, N_P additional mass action equations (one per mineral) are needed. At equilibrium we have the residual functions

$$F_m^P = SI_m = 0 \quad \left(m = 1, \ldots, N_P\right) \tag{1.15}$$

where SI_m is the mth equilibrium mineral saturation index, which is a function of c_j. N_C equations in (1.14) and N_p in (1.15) constitute a complete set of equations needed for $N_C + N_P$ primary variables ($c_1, c_2, \ldots, c_{Nc}; c_1, c_2, \ldots, c_{Np}$).

Equations (1.14) and (1.15) can be solved together with the robust Newton-Raphson iterative method.

1.3 Application to CO_2 Geological Storage (CGS)

1.3.1 Overview of Applications in CGS

Carbon capture and storage (CCS) has been validated by the IPCC (2005) as an important option to mitigate the increasing atmospheric concentrations of anthropogenic carbon dioxide (CO_2). In order for the technology to be deployed on a large scale, its viability in terms of injectivity, the containment of the injected CO_2 and the long-term safety with respect to humans and the environment needs to be guaranteed. Carbon dioxide geological storage (hereafter abbreviated to CGS) has been conducted on a commercial scale (e.g. a CO_2 injection rate more than 1.0 Mt/a) at several sites including the Weyburn-Midale field, Canada (Petroleum Technology Research Center, 2011). Due to the corrosive character of CO_2, certainly once it is dissolved, geochemical reactions play an important role and might affect the chemical and physical properties of the wells, the reservoir and its surroundings.

Geochemical interactions caused by the presence of CO_2 in geological sequences where CO_2 occurs naturally are particularly valuable, since this illustrates the long-term impact of CO_2 on natural rocks that cannot be reproduced during experiments or field tests. Chemical equilibrium seems not to be reached in some natural analogues, even when subjected to very long (geological) contact time, suggesting that chemical equilibrium might not be the natural state of injected CO_2 during the whole duration of the storage period (thousands to hundreds of thousands of years).

Numerical modeling of a storage site requires estimates for both the short- and long-term fate of the injected CO_2. Such modeling work can be very demanding in terms of CPU time and memory, and results might either be difficult to obtain or limited in terms of precision. Recent numerical modeling for CO_2 storage applications can be split into three categories: (i) hydrodynamic modeling simulating structural, residual gas and dissolution trapping processes; (ii) geochemical modeling simulating batch geochemical reactivity (closed system without any fluid flow); and (iii) reactive transport modeling combining the two previous types of simulations, and therefore providing a complete calculation over time of the amount of CO_2 trapped through structural, dissolution or mineral trapping. Reactive transport modeling is the most realistic modeling technique to quantify the long-term fate of CO_2 as well as other aspects during geological storage, but also the most challenging to perform.

Geochemical and solute transport modeling has many application domains when assessing the geochemical impact of CO_2 storage, and each has its own spatial scale and timeframe of interest and therefore requires an adapted modeling approach.

One application of reactive transport modeling is to assess the ultimate fate of the injected CO_2 and its impacts on physical properties. There are four distinct processes during sequestering the injected CO_2, including structural trapping, residual CO_2 trapping, dissolution trapping and mineral trapping. Mineral trapping kinetics is controlled by both dissolution and precipitation kinetics. While the dissolution kinetics for carbonates and sulfate reactions are generally fairly rapid, the kinetics of alumino-silicate mineral reactions are much slower (total dissolution of these minerals can take up to thousands of years at ambient temperatures), making mineral trapping a slow process since mainly the latter are involved in CO_2 trapping processes.

Another application domain of interest for researchers is to predict the risk potentially occurring in CGS through reactive transport modeling, such as caprock integrity and damage, CO_2 leakage, or groundwater quality reduction. However, the caprock as the first natural barrier of the reservoir particularly concerns scientists. Caprock is generally defined as a low to very low permeability formation, and composed mainly of clay-rich shale or mudstone that consists of aluminates or aluminosilicates. Because aluminosilicate mineral alteration is very slow under ambient deep-formation conditions, and is not amenable to experimental studies, the numerical modeling of hydrogeochemical processes is necessary to investigate long-term evolution of caprock sealing efficiency.

1.3.2 Long-Term Fate of Injected CO_2 in Deep Saline Aquifers

As the negative effects of global warming are widely discernible, the need for the deployment of CO_2 capture and storage (CCS) incorporating CO_2 geological storage (IPPC, 2005) has been increased to confront this worldwide environmental issue. Geological CO_2 sequestration offers a most promising solution for reducing net emissions of greenhouse gases into the atmosphere. Injection of CO_2 produced by different human activities into depleted oil and gas reservoirs has been considered, but aquifers containing non-drinking water offer the largest storage volume. CO_2 is injected into these aquifers in the supercritical state with a higher density, and therefore occupying less volume underground. The supercritical state can be obtained at pressures greater than 7.4 MPa and temperatures higher than 31.1 °C. These conditions correspond to a reservoir at a depth of 800 m beneath the Earth's surface. There are three interdependent yet conceptually distinct processes storing CO_2 in deep saline aquifers, which are distinguished as hydrodynamic trapping of the gas (or supercritical) phase, dissolution trapping in the liquid (groundwater) phase (dissolution of CO_2), and mineral trapping of the solid phase (mineral alteration, leading to precipitation of secondary mineral phases).

In terms of migration, injected CO_2 moves by volumetric displacement of formation waters, with which it is largely immiscible; by gravity segregation, which cause the immiscible plume to rise owing to its relatively low density; and by viscous fingering, which causes it to migrate preferentially into local high-permeability zones owing to its relatively low viscosity compared with water. In terms of sequestration, some fraction of the rising plume will dissolve into formation waters (solubility trapping); some fraction will react with formation minerals to precipitate carbonates (mineral trapping); and the remaining fraction will reach and become isolated beneath the caprock (hydrodynamic trapping), migrate up-dip along this interface, and accumulate in any local topographic highs (structural trapping). Numerical simulation of these interdependent migration and sequestration processes requires a computational capability that explicitly represents and couples multiphase flow and kinetically controlled geochemical processes within porous media characterized by physical and compositional heterogeneity.

The ultimate fate of CO_2 injected into saline aquifers for environmental isolation is governed by the above-mentioned three interdependent yet conceptually distinct processes. The first process is directly linked to hydrodynamic trapping, the second to solubility trapping and pH evolution, and the third to mineral trapping (and pH evolution). In this section, using the reactive transport modeling approach, we will quantify and compare the relative effectiveness of these trapping mechanisms during propagation of injected CO_2 within a specified deep saline aquifer.

1.3.2.1 Brief Description of CO_2 Storage Site in the Songliao Basin

In order to ensure the effectiveness and safety of any geological storage project, site selection is crucial to the long-term containment of injected CO_2 in the deep reservoir and a very important component in the life-cycle assessment of CCS, and also plays a key role during the implementation process of CCS (Bachu, 2008; Li et al., 2012). Sites are chosen under different conditions, such as performance or risk. Integrated scales of site assessment are divided into four stages: country/state-scale screening, basin-scale assessment, site characterization, and site deployment (CO_2CRC, 2008; CSLF, 2008). Screening criteria for CO_2 sequestration have been proposed for reservoir properties and surface facilities, and different value ranges of indicators have been given (Kovscek, 2002; LBNL, 2004). Chadwick et al. (2008) defined the criteria as reservoir efficacy, reservoir properties and caprock.

The site used in this study, Songliao Basin, is located in the northeast of China, which spans Heilong Jiang, Jilin and Liaoning provinces and the Inner Mongolia Autonomous Region. The basin was formed first by rifting and then by subsidence and inversion. The Songliao Basin is one of the most important productive oil basins in China, as well as one of the most important heavy industrial bases. Every year a large amount of CO_2 is generated in this basin. Therefore, it is essential to implement CGS in this basin for mitigating these CO_2 emissions. The target layer used in this study is the sandstone formation located in the third and fourth Members of the Lower Cretaceous Quantou Formation (K1q3 and K1q4), which is a deep saline formation and oil-producing unit, with a thickness of about 50 m. This formation is expected to be a potential candidate for CO_2 geological storage.

In this study, based on the geological, hydrogeological and geochemical conditions of K1q3 and K1q4, the ultimate fate of injected CO_2 and efficiencies of three trapping processes will be simulated with a multiphase, reactive transport simulator, the TOUGHREACT program.

1.3.2.2 Conceptual Model

Geometry and boundary conditions

A 2-dimensional (2D) radial model was developed to investigate the transformation of injected CO_2 within the target reservoir (Figure 1.1). The 2D model is a homogeneous sandstone formation of 50 m thickness with a cylindrical geometry. Ten layers were used vertically with a constant spacing of 5 m. In the horizontal direction, a distance of 100 km was modeled with a radial spacing that increases gradually away from the injection well. This distance was discretized into 45 radial grid elements. CO_2 is injected through two elements at the bottom of the well (Figure 1.1) at a constant rate of 20 kg/s, and the total amount of CO_2 is calculated to be 0.64 Mt/a. The injection lasts for 100 years, while the simulation time is 10,000 years.

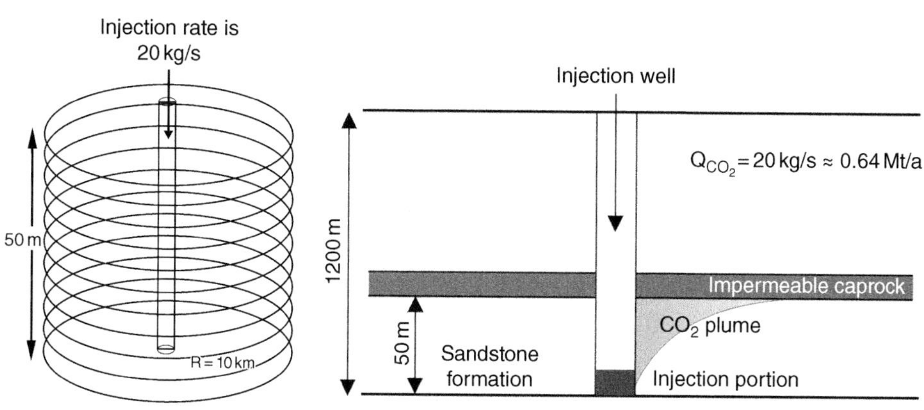

Figure 1.1 Two-dimensional radial schematic representation and discretization for CO_2 injection model in the formation.

Physical parameters

Without consideration of the caprock for this reservoir, the model was treated as impermeable at the top and bottom boundaries. Considering that the model length of up to 100 km is out of pressure-affecting range, lateral boundaries were assigned with fixed pressure. Hydrostatic pressure was imposed along the vertical layers.

The hydrogeological parameters used in the simulations are listed in Table 1.5. The whole formation was assumed to be homogeneous and isotropic. The petrophysical characteristics such as porosity and permeability were taken from the averaged experimental measurements and data collection (Zhang et al., 2009). Other parameters such as those used in the capillary pressure and relative permeability models were taken from Zhang et al. (2009) and are summarized in Table 1.5.

Table 1.5 Hydrogeological parameters used in the simulations.

Parameter	Sandstone saline aquifer
Porosity	0.2
Permeability (m^2)	1.0×10^{-13}
Ratio of vertical to horizontal permeability	0.1
Compressibility (Pa^{-1})	4.5×10^{-10}
Rock grain density (kg/m^3)	2600
Formation heat conductivity (W/(m °C))	2.51
Rock grain specific heat (J/(kg °C))	920
Temperature (°C)	50
Pressure of bottom (bar)	120
Salinity (wt.%)	0.01
Relative permeability model	
Liquid (Van Genuchten, 1980):	
$k_{rl} = \sqrt{S^*}\left\{1-\left[1-\left(S^*\right)^{1/m}\right]^m\right\}^2$	$S^* = (S_l - S_{lr})/(1 - S_{lr})$
S_{lr}: residual water saturation	$S_{lr} = 0.30$
m: exponent	$m = 0.457$
Gas (Corey, 1954):	
$k_{rg} = (1-\hat{S})^2(1-\hat{S}^2)$	$\hat{S} = (S_l - S_{lr})/(S_l - S_{lr} - S_{gr})$
S_{gr}: residual gas saturation	$S_{gr} = 0.05$
Capillary pressure model (Van Genuchten, 1980):	
$P_{cap} = -P_0\left[\left(S^*\right)^{-1/m} - 1\right]^{1-m}$	$S^* = (S_l - S_{lr})/(1 - S_{lr})$
S_{lr}: residual water saturation	$S_{lr} = 0.00$
m: exponent	$m = 0.457$
P_0: strength coefficient	$P_0 = 19.61$ kPa

Initial mineral composition

According to X-ray diffraction (XRD) mineralogical analysis of the clastic rocks from the Songliao Basin, they are mainly composed of quartz, illite, chlorite, calcite, plagioclase and K-feldspar. The composition of plagioclase is uncertain, because the ratio between albite and anorthite may affect the simulation results. In the work of Gaus et al. (2005), anorthite did have an impact on the mineral trapping of CO_2. However, in the modeling of Audigane et al. (2007), all plagioclase is treated as pure albite because albite is much more resistant to weathering than anorthite, and Na will be the dominant cation in plagioclase. For the simplification in this study, plagioclase was represented using pure albite. The choice of secondary minerals may also affect the simulation results significantly. Almost all possible secondary carbonate and clay minerals are considered in the modeling, which refers to previous simulation studies (Xu et al., 2006; Gaus et al., 2005), laboratory research (Wolf et al., 2004), and field observations (Moore et al., 2005). The resulting mineralogical composition is presented in Table 1.6.

Water chemistry

Prior to the simulations, a batch geochemical modeling of water–rock interaction was performed to obtain a quasi-stable aqueous chemistry, where a 0.171 M (mol/kg H_2O) solution of NaCl was used to react with the initially present minerals (listed in Table 1.6) at an initial CO_2 partial pressure of 0.01 bar. After about 10 years the whole geochemical system reached an approximate steady state, resulting in a quasi-stable solution composition (Table 1.7). This composition is used as the starting-point water for the subsequent reactive transport modeling.

Table 1.6 Initial volume fractions of primary and secondary minerals used in the simulations.

Mineral	Chemical formula	Initial volume fraction
Primary mineral		
Calcite	$CaCO_3$	0.030
Quartz	SiO_2	0.258
Kaolinite	$Al_2Si_2O_5(OH)_4$	0.009
Illite	$K_{0.6}Mg_{0.25}Al_{1.8}(Al_{0.5}Si_{3.5}O_{10})(OH)_2$	0.028
Albite	$NaAlSi_3O_8$	0.415
K-feldspar	$KAlSi_3O_8$	0.233
Chlorite	$Mg_{2.5}Fe_{2.5}Al_2Si_3O_{10}(OH)_8$	0.027
Secondary mineral		
Smectite-Na	$Na_{0.290}Mg_{0.26}Al_{1.77}Si_{3.97}O_{10}(OH)_2$	0.0
Smectite-Ca	$Ca_{0.145}Mg_{0.26}Al_{1.77}Si_{3.97}O_{10}(OH)_2$	0.0
Magnesite	$MgCO_3$	0.0
Siderite	$FeCO_3$	0.0
Dawsonite	$NaAlCO_3(OH)_2$	0.0
Dolomite	$CaMg(CO_3)_2$	0.0
Ankerite	$CaMg_{0.3}Fe_{0.7}(CO_3)_2$	0.0

Table 1.7 Initial total dissolved component concentrations of the formation water at reservoir conditions of 50 °C and 120 bars.

Parameter	Value	Component	Concentration (mol/kg H_2O)
Temperature (°C)	50.0	Na^+	1.978×10^{-1}
pH	8.28	K^+	0.7339×10^{-3}
Pressure (bars)	120	Ca^{2+}	0.3431×10^{-4}
		Mg^{2+}	0.1576×10^{-7}
		Fe^{2+}	0.4172×10^{-6}
		SiO_2 (aq)	0.5717×10^{-3}
		AlO_2^-	0.1125×10^{-6}
		HCO_3^-	0.2762×10^{-1}
		Cl^-	1.6970×10^{-1}
		SO_4^{2-}	0.9983×10^{-10}
		O_2 (aq)	0.4879×10^{-68}

Thermodynamic and kinetic parameters

The mineralogy in this study is described in terms of 16 minerals including primary minerals and secondary minerals, as shown in Table 1.6. The thermodynamic data for minerals, gases and aqueous species are mostly taken from the EQ3/6 V7.2b database of Wolery (1992). For kinetically controlled mineral dissolution and precipitation, a general form of rate law (Lasaga, 1984; Steefel and Lasaga, 1994) is used (Equation 1.4). The kinetic rate constants $k(T)$ of concerned minerals are calculated using Equation (1.6). The values of the most relevant kinetic parameters are summarized in Appendix A to this chapter (Xu et al., 2011; Gherardi et al., 2007).

1.3.2.3 Results and Discussion

The injection of CO_2 lasts 100 years with a rate of 20 kg/s (0.64 Mt/a), and the total reactive transport simulation was performed for a period of 10,000 years.

Changes in the distribution of gas saturation (Sg) with time

Once carbon dioxide is injected into the reservoir in a supercritical state (hereafter referred to as gas CO_2), it will displace the ambient porewater, which will cause different regions with varying combinations of CO_2 and brine. For about 100 m distance around the injection well, the pore spaces are completely filled with gas CO_2 after 100 years. These zones are called the 'dry area', within which the CO_2 gas saturations (abbreviated to Sg) are higher or at least not lower than 0.95. Beyond 100 m the Sg lies in the range from 0.4 to 0.6, where gas CO_2 and porewater exist together. In the outermost zones Sg declines to almost zero.

Due to the relatively lower density of supercritical CO_2 compared with brine, the buoyancy force moves supercritical CO_2 upwards until it reaches the caprock. With the return of groundwater, a very small amount of the CO_2 is immobilized and permanently trapped; this trapping process is referred to as residual sequestration. However, a large amount of CO_2 will move towards and accumulate beneath the caprock, which gives

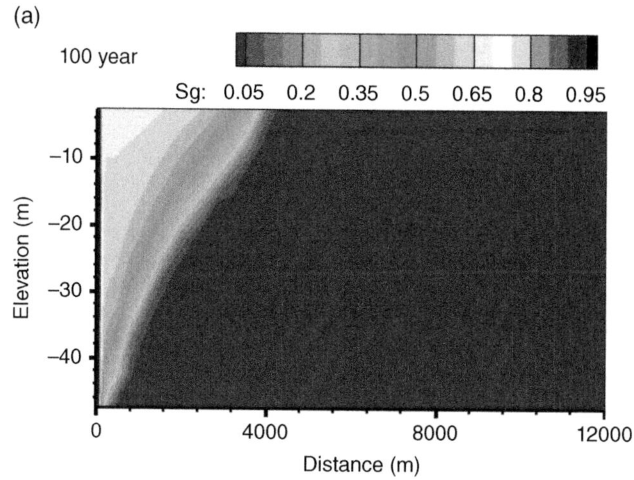

Figure 1.2 Spatial distributions of CO_2 saturation after (a) 100, (b) 1000, (c) 5000 and (d) 10,000 years.

rise to pressure under the base of the caprock. Locally higher pressure forces CO_2 to migrate laterally.

At the beginning of injection, CO_2 mainly distributes around the injection well, displacing groundwater. Sg of CO_2 is able to reach 0.95. After injection (100 years), along with vertical and lateral propagation of CO_2, the displaced groundwater returns, which causes a mutual displacement between supercritical CO_2 and groundwater. At the time point of 5000 years, the space occupied by CO_2 is small due to upward movement (Figure 1.2). At the end of the simulation, most of the injected CO_2 accumulates beneath the caprock, and supercritical CO_2 gradually dissolves into groundwater.

Solubility trapping

Part of the injected supercritical CO_2 dissolved into the formation water. This dissolved CO_2 initiates the increase of aqueous solution density (density change depends on the

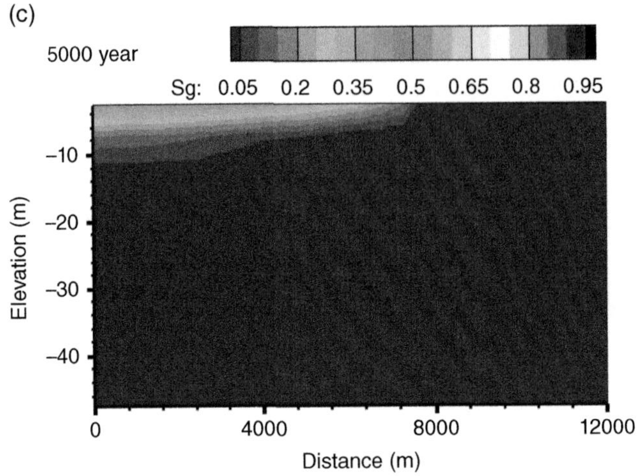

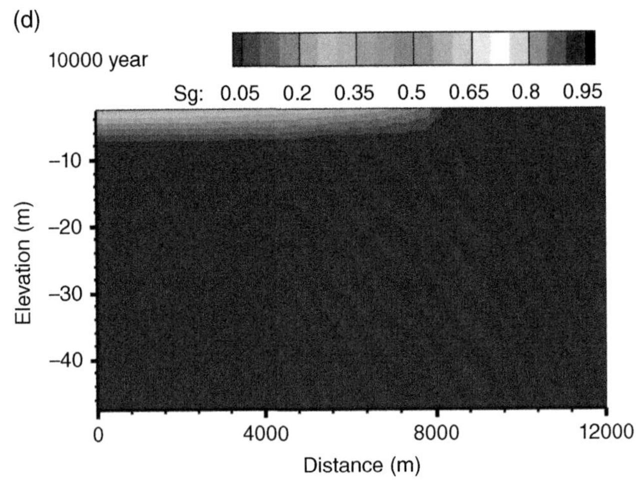

Figure 1.2 (Cont'd)

salinity, temperature and pressure), consequently causing gravitational instability of the 'gas'-water interface. The dissolved CO_2 migrates through molecular diffusion, diffusion, and convection. As the large amount of dissolved CO_2 accumulates beneath the caprock, a high-density solution saturated with dissolved CO_2 moved downwards under gravity (Figure 1.3a, b and c), which was accompanied by upward migration of gaseous CO_2. This phenomenon is so-called 'convective mixing'. At the end of the simulation at 10,000 years (Figure 1.3d), an obvious fingering phenomenon can be observed, and the maximum distance of dissolved CO_2 migration reaches 9000 m beneath the bottom of the caprock.

These above-mentioned processes triggered the solubility trapping mechanism of injected CO_2. Subsequently, the dissolved CO_2 could react with formation water forming carbonic acid (H_2CO_3), and then this acid would dissociate into proton (H^+) and bicarbonate ion (HCO_3^-), resulting in the acidification of brine.

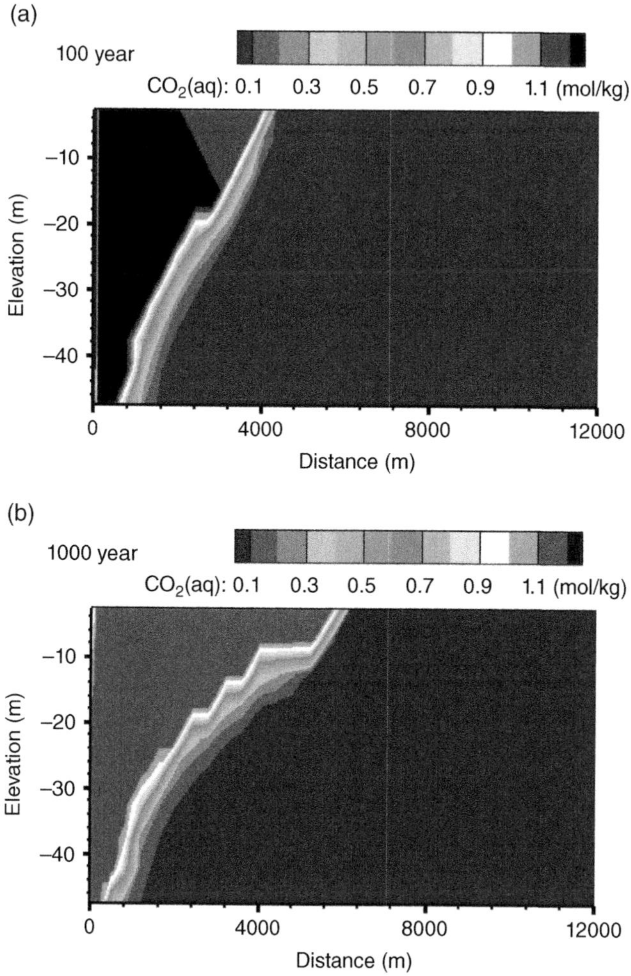

Figure 1.3 Spatial distribution of dissolved CO_2 at (a) 100, (b) 1000, (c) 5000 and (d) 10,000 years. (See colour plate section for the colour representation of this figure.)

Variation of pH

As mentioned above, the injected CO_2 acidified the formation water, and this process is reflected by the spatial and temporal distribution of pH. As depicted in Figure 1.4, during CO_2 injection pH continued dropping. At the end of injection (Figure 1.4a), pH declined to about 4.5. Along with the increment of distance from injection well, owing to the consumption of dissolved CO_2, pH of the formation water decreased. After injection stopped, CO_2 was trapped in three forms: gaseous, aqueous and solid (trapped during mineral alteration). The zones with low pH also move upwards and laterally (Figure 1.4c,d). Spatial and temporal distribution of pH is subject to the dissolution of CO_2. Therefore, the patterns of gas CO_2 and dissolved CO_2 are consistent.

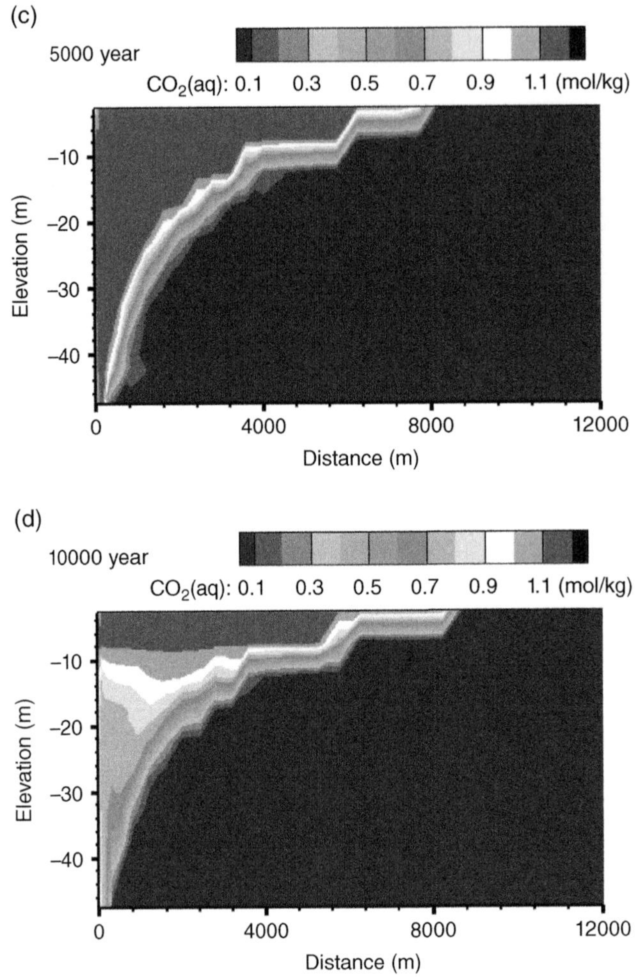

Figure 1.3 (Cont'd)

Mineral alterations

The increased acidity (decreased pH) induces mineral dissolution/precipitation. In this study (the end point of simulation at 10,000 years was chosen to conduct discussion), spatial distributions of the volume fraction changes in minerals, which experienced dissolution/precipitation, are presented in Figures 1.5 and 1.6, respectively.

From Figure 1.5 we can see that albite, calcite, K-feldspar and chlorite dissolved significantly. A large amount of dissolved CO_2 acidified the ambient water, reducing the pH. The equilibrium between calcite and formation water was broken, which resulted in the dissolution of calcite (Figure 1.5a). At the same time, plagioclase (albite) and K-feldspar also underwent dramatic dissolutions (Figure 1.5b,c). Under acidic conditions, chlorite is vulnerable to dissolution. After injection, the chlorite began to dissolve, and the dissolution extent gradually decreased with the radial distance from the injection well (Figure 1.5d).

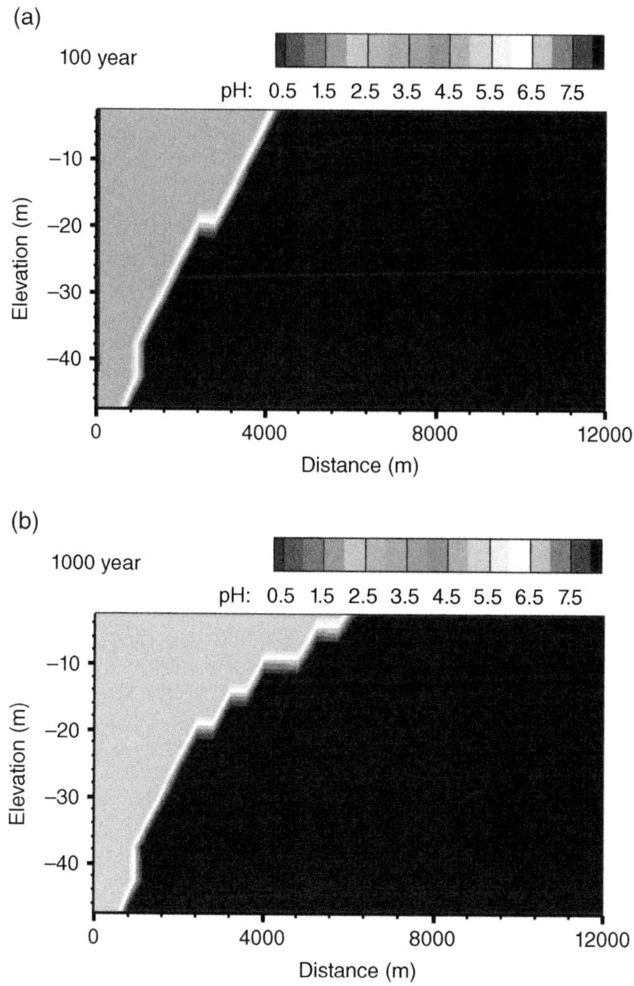

Figure 1.4 Spatial distribution of pH at (a) 100, (b) 1000, (c) 5000 and (d) 10,000 years.

At the end of the simulation, maximum volume fraction change for albite was 0.07%, calcite 0.022%, K-feldspar 0.015%, and chlorite 0.013%, respectively.

In contrast, due to release of ions from particular minerals, concentrations will increase until saturation of a specific ion-bearing mineral is reached. As chlorite, calcite and plagioclase dissolve, sufficient Mg^{2+}, Na^{+} and Ca^{2+} are released into the formation water to make the precipitation of smectite-Na (Figure 1.6a), illite (Figure 1.6b), ankerite (Figure 1.6c), and smectite-Ca (Figure 1.6e) possible

The redundant Mg^{2+} and concentrated HCO_3^- formed small amounts of magnesite (Figure 1.6 h). Additionally, sufficient Fe^{2+} and HCO_3^- facilitate the formation of siderite (Figure 1.6 g).

Due to the significant dissolution within the two-phase mixing region, abundant ions required for precipitation of particular minerals were provided, which caused precipitation of minerals mainly within these regions. Magnesite, siderite and dawsonite deposit

(c)

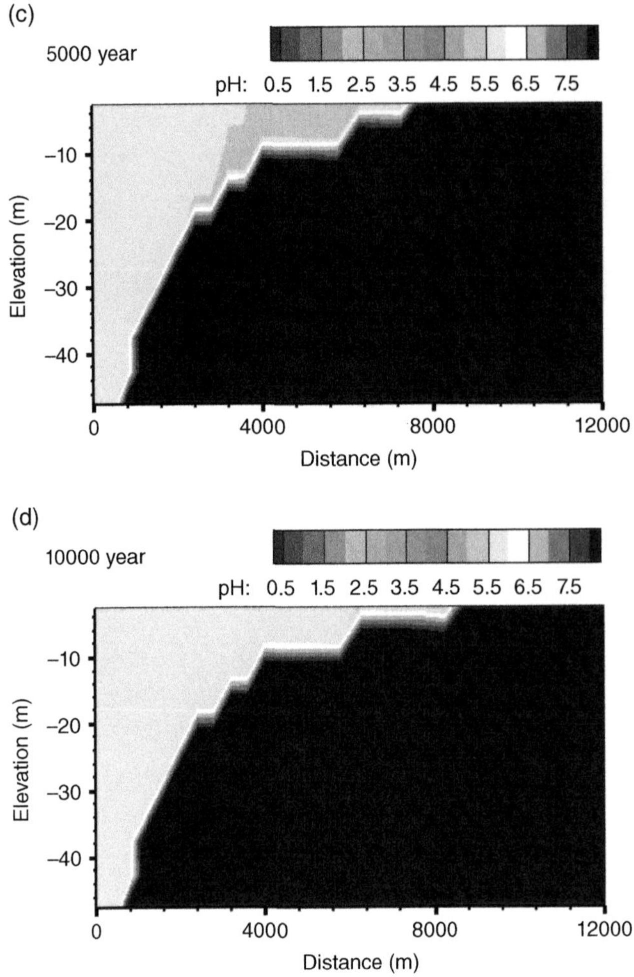

Figure 1.4 (Cont'd)

in the case of ample cations. This is the reason why these minerals precipitated later. By 10,000 years, the maximum precipitation of smectite-Ca is 0.065% in volume fraction, illite 0.02%, ankerite 0.015%, quartz 0.0018%, and magnesite 0.0016%, respectively. Ankerite and dawsonite are the main minerals to solidify CO_2, followed by magnesite and siderite.

The amount of CO_2 trapped by minerals rose from the bottom to top of reservoir, which was caused by a large accumulation of CO_2 at the top. At the end of the simulation the amount of CO_2 trapped by minerals reached $8\,kg/m^3$.

Transformation among three CO_2 trapping forms over time

The injected CO_2 was sequestered through three mechanisms in gaseous, liquid and solid forms, respectively. Figure 1.7 shows the amount of transformation of trapped CO_2 between the three phases with time. At the beginning of the simulation, the injected

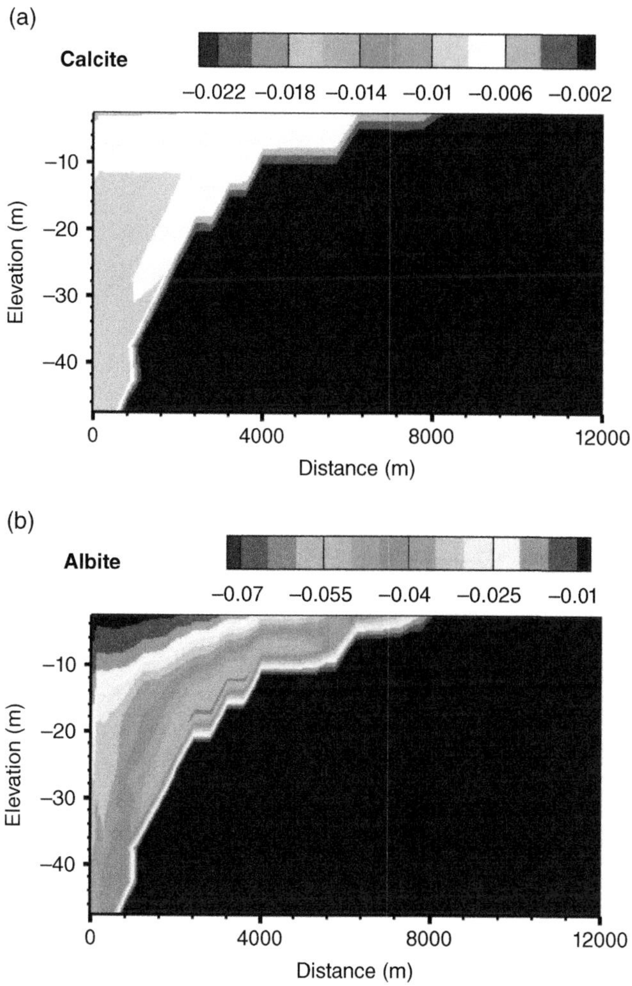

Figure 1.5 Spatial distribution of volume fraction change for dissolved minerals at 10,000 years (negative values indicate dissolution, positive precipitation). (See colour plate section for the colour representation of this figure.)

CO_2 was sequestered mainly by hydrodynamic trapping. With increasing simulation time, the CO_2 dissolved into the formation water gradually. Due to the formation of carbonic acid, some susceptible minerals dissolved, supplying ions for the formation of carbonate minerals (dawsonite, magnesite, ankerite, siderite, etc.), resulting in mineral trapping of CO_2. The amounts of CO_2 trapped in different phases by the end of the simulation are summarized in Table 1.8.

1.3.2.4 Summary and Conclusions

When CO_2 is injected into saline aquifers, it will undertake a series of physical and chemical processes for hundreds or thousands of years, or even longer. This processing mode of CO_2 can affect the global carbon cycle. However, the deep saline aquifers in

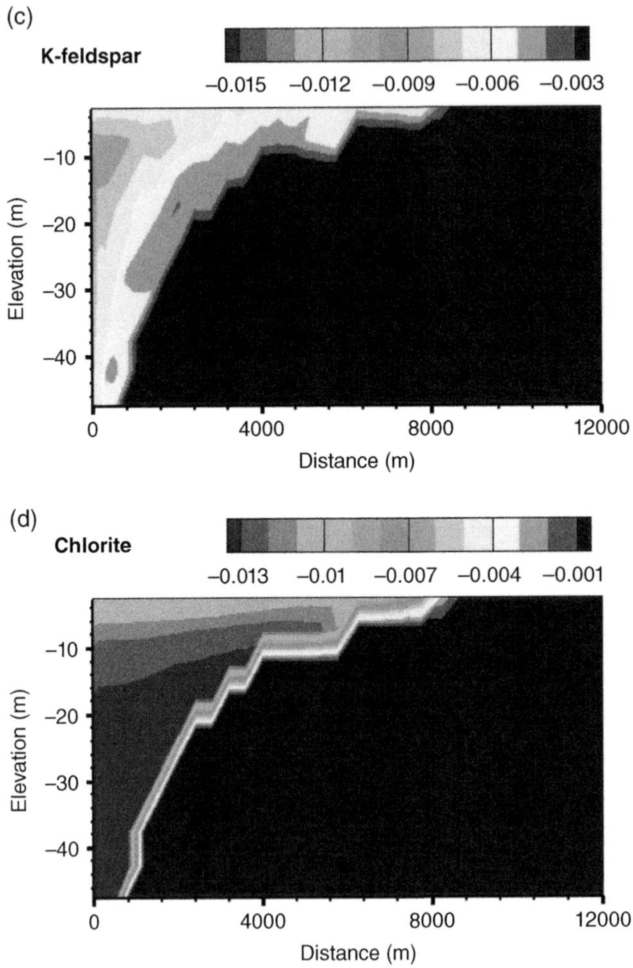

Figure 1.5 (Cont'd)

sedimentary basins are very complex, and processes involved in CO_2 geological sequestration are complicated as well. Therefore, reactive fluid flow and geochemical transport numerical simulation is necessary.

It can be seen from the simulation results that during the period of CO_2 injection, a large amount of CO_2 remained as a free supercritical phase, then moved upwards due to buoyancy and gathered beneath the caprock. In the post-injection stage, the content of supercritical CO_2 gradually declined, and the dissolved CO_2 increased significantly, then reduced due to the formation of secondary carbonate minerals.

Mineral alteration and CO_2 trapping capability depend on the primary mineral composition. Precipitation of siderite and ankerite requires Fe^{2+}, which can be supplied by the dissolution of iron-bearing minerals, such as chlorite, or by reduction of Fe^{3+} in small amounts of hematite. Variation in Ca content in plagioclase significantly affects

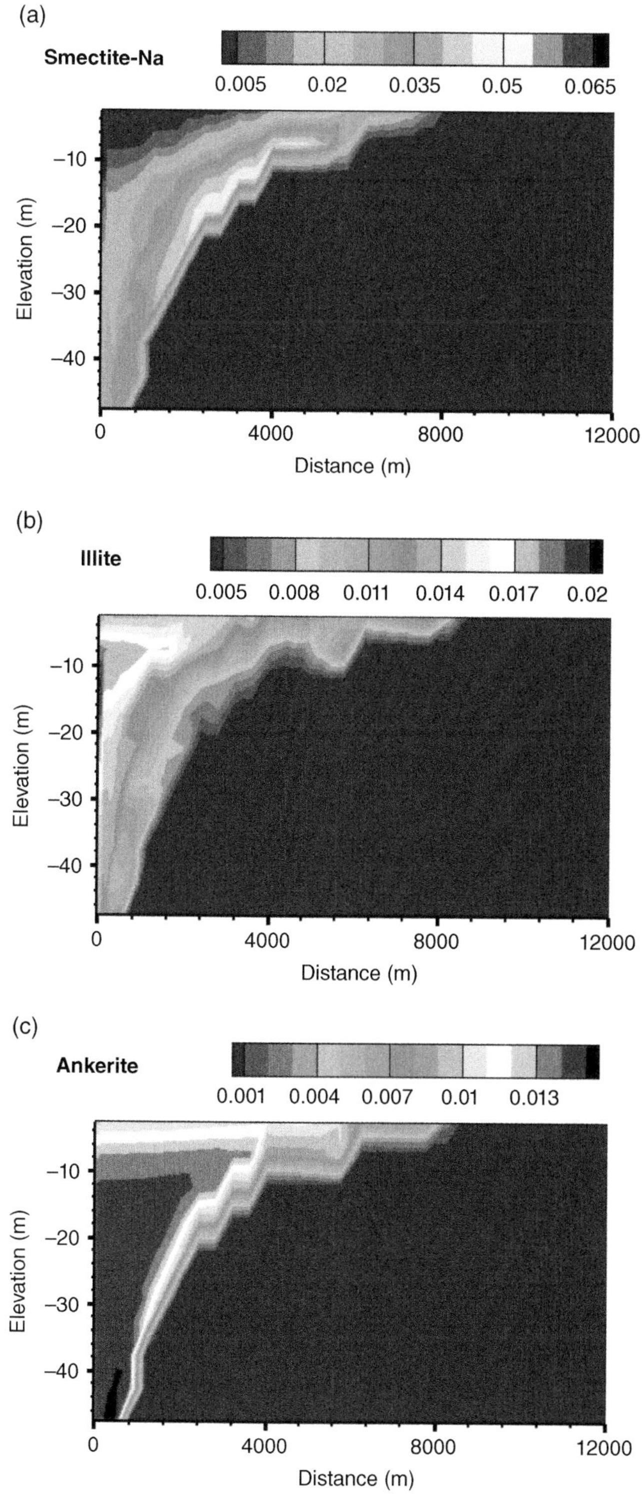

Figure 1.6 Spatial distribution of volume fraction change for precipitated minerals at 10,000 years. (See colour plate section for the colour representation of this figure.)

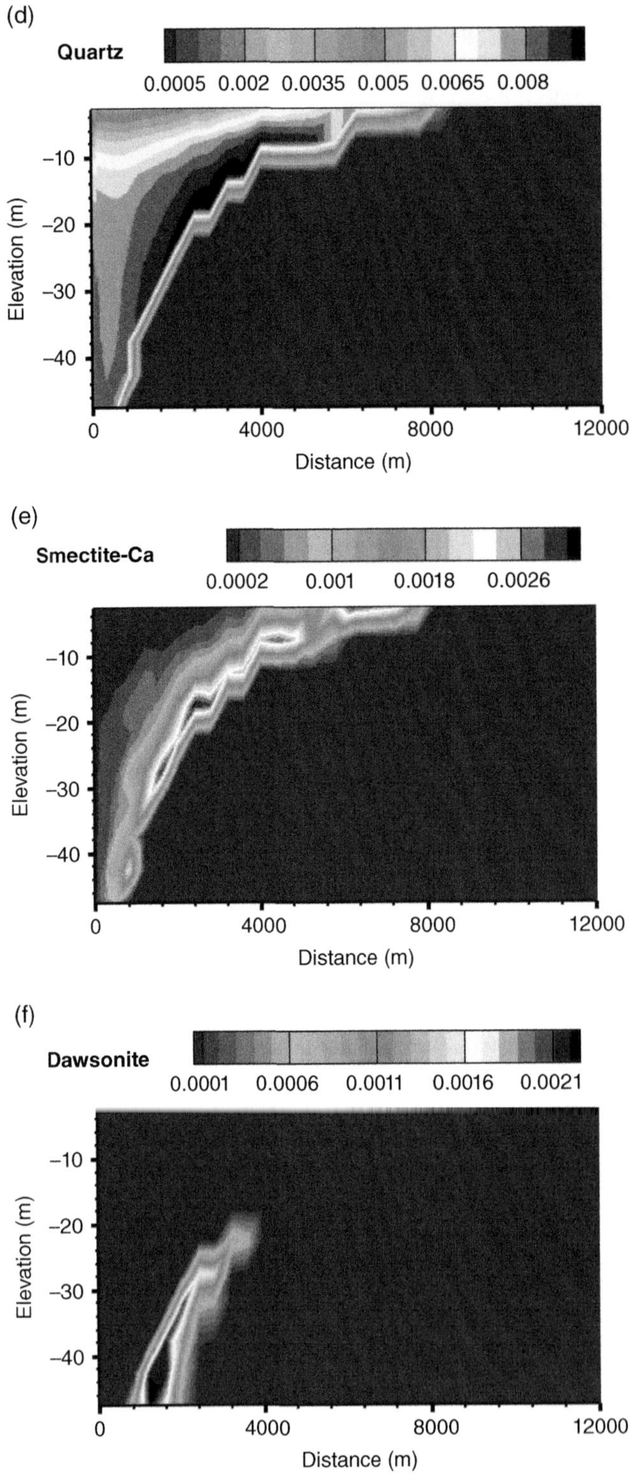

Figure 1.6 (Cont'd)

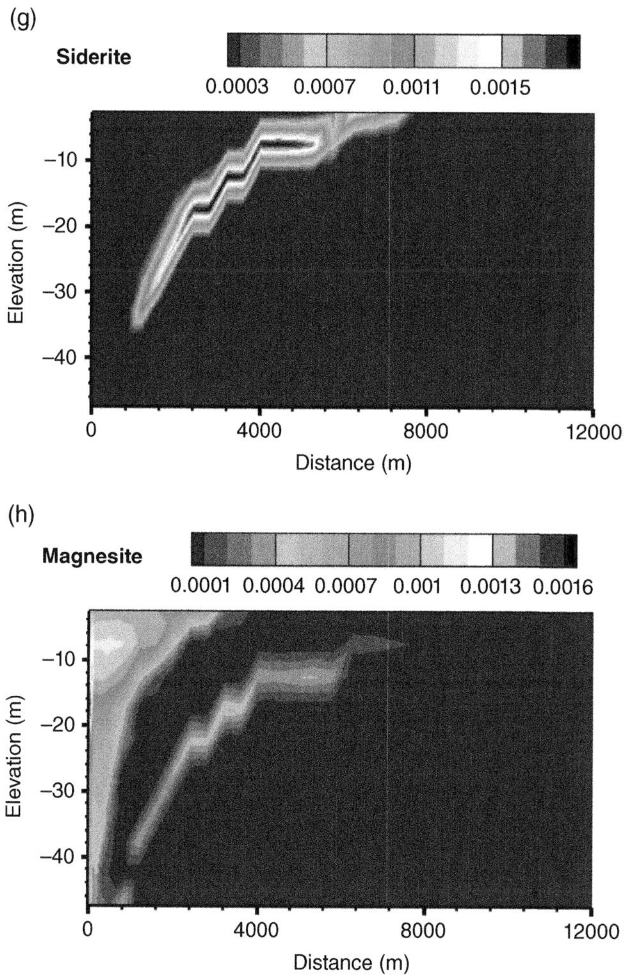

Figure 1.6 (Cont'd)

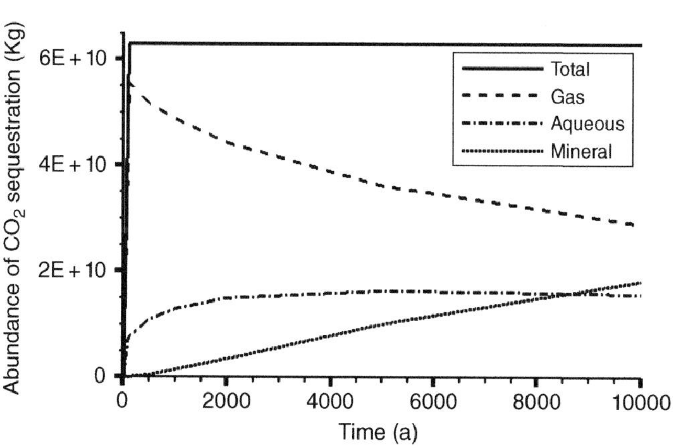

Figure 1.7 Changes in amounts of CO_2 in three phases with time.

Table 1.8 Amount of CO_2 trapped in the three phases at the end of the simulation.

Phase	Gaseous	Aqueous	Solid	Total
Amount (Mt)	0.29	0.16	0.19	0.64
Percentage	41.02%	25.06%	28.92%	100%

carbonate mineral precipitation, and thus CO_2 mineral trapping. The time required for mineral alteration and CO_2 sequestration depends on the rates of mineral dissolution and precipitation.

1.3.3 Evolution of Caprock Sealing Efficiency after the Intrusion of CO_2

1.3.3.1 Introduction

Although carbon dioxide geological storage (CGS) appears to be the most promising option to mitigate CO_2 emission to the atmosphere, by isolating the emissions in the subsurface for a long time period, risks may nevertheless be present, such as caprock integrity damage, CO_2 leakage, or groundwater quality reduction due to acidic fluid intrusion. As an integral part of a CGS system, a caprock with low to very low permeability is considered to be the first natural barrier, needs to be at the desired depth for keeping CO_2 in a supercritical state ($scCO_2$), and should prevent leakage of CO_2 over a long time period.

Leakage of $scCO_2$ may occur through fractures, faults, or caprock weakness, which may provide preferential pathways for deep-seated gases. Several researchers have studied potential leakages through wells, fractures and faults (e.g., Celia and Bachu, 2003; Lindeberg and Bergmo, 2003). Studies cover issues such as gas–brine–rock interactions, multiphase flow, reservoir engineering and modeling, monitoring, and verification. A common aspect of all CO_2 capture and storage options is the sealing efficiency of the caprocks overlying the potential CO_2 storage reservoir.

Leakage through the caprock is reported to occur mainly in three ways: (1) rapid (catastrophic) leakage by seal-breaching (mechanical failure) or damage of well casing (corrosion of pipes and cements), resulting in gas flow through a (micro-) fracture network; (2) long-term leakage controlled by capillary sealing efficiency and permeability (after the capillary break-through pressure is exceeded); and (3) diffusive loss of dissolved gas through brine-saturated pore spaces.

The intruded $scCO_2$ may lead to the acidity of resident brine, of which pH reaches about 3–5. This low-pH brine might invade into the pore space, inducing interactions with the minerals. For a long time period, this geochemical process may inhibit or facilitate the reaction, significantly contributing to the evolution of permeability/porosity, thereby enhancing or weakening the sealing efficiency of the caprock, which depends upon the varying geochemical settings (Alemu et al., 2011). In a geological time-frame the resistance of caprock to geochemical processes seems to be a site-specific phenomenon. Therefore, reactive transport modeling is an efficient and effective way to predict the evolution of caprock sealing efficiency on a large timescale.

In China, the government makes much effort to reduce total greenhouse gas emissions. The first full-chain CGS demonstration at a pilot scale by China Shenhua Group is currently underway. In terms of the safety of CGS on a large timescale, precisely assessing the evolution of caprock sealing efficiency with geochemical processes is essential. In this section, based upon the measurement data from Shiqianfeng Formation in Ordos Basin, we evaluated the evolution of caprock sealing efficiency due to alteration of minerals, and identified the minerals contributing to the changes in sealing efficiency of the caprock. All simulations were performed using the comprehensive reactive transport simulator TOUGHREACT.

1.3.3.2 Geological Setting

The Shiqianfeng Formation in the Ordos Basin is screened as one of the CO_2 storage reservoirs with the best potential, and is roughly divided into many sub-reservoirs and sub-caprocks, in terms of lithological characteristics. Employing the approach reported by Zhang and Zhou (2010), Tian (2014) evaluated the sealing capacities of sub-caprocks. The results suggested that the uppermost mudstone sub-caprock, with a thickness of 52.4 m, possesses the greatest sealing capacity. In this section we selected this 52.4 m thick sublayer as the local caprock to evaluate the evolution of caprock sealing efficiency due to mineral alteration on a large timescale after the intrusion of CO_2.

1.3.3.3 Conceptual Model

Model configuration and boundary condition

Our main attention here is paid to the changes in sealing capacity of the uppermost sub-caprock of Shiqianfeng Formation in the Ordos Basin, after intrusion of supercritical CO_2 from the underlying reservoir. Therefore, the conceptual model consists at least of two parts: the reservoir and the caprock.

Based on the measurement data, the thickness of the uppermost and most effective sub-caprock layer is 52.4 m. For simplification, a one-dimensional, 52.4 m long, vertical column geometric model (Figure 1.8) with a cross-sectional area of 1.0 m^2 was adopted to represent the target caprock formation. Simulations aiming to investigate the effects of geochemical processes on caprock sealing ability evolution were performed with the comprehensive simulator TOUGHREACT.

At the bottom of the column, an element with a large volume (10^{10} m^3) is set to act as the reservoir supplying supercritical CO_2 for the caprock. The caprock is discretized into a number of small grids with a spacing of the order of a few mm, near the interface between caprock and reservoir, to improve the resolution of chemical concentration gradients near the caprock boundary.

At the top of the column, an infinite volume (10^{30} m^3) element is imposed to behave like a constant pressure boundary for the model. Along the lateral boundaries, no-flow and no-transport boundary conditions are assigned, representing a compartmentalized section of caprock, thus guaranteeing the intruded CO_2 will not flow out laterally.

Initial physical parameters

The data utilized here for configuring the model are restricted to basic physical parameters (temperature, pressure, permeability and porosity as listed in Table 1.9), as well as caprock mineralogy (which will be further detailed later).

A pressure of 170 bars was used initially in the simulation, which is close to the hydrostatic pressure at the top of the Shiqianfeng Formation. Nevertheless, the pressure of the reservoir would be reassigned later for ensuring that the $scCO_2$ in the reservoir is

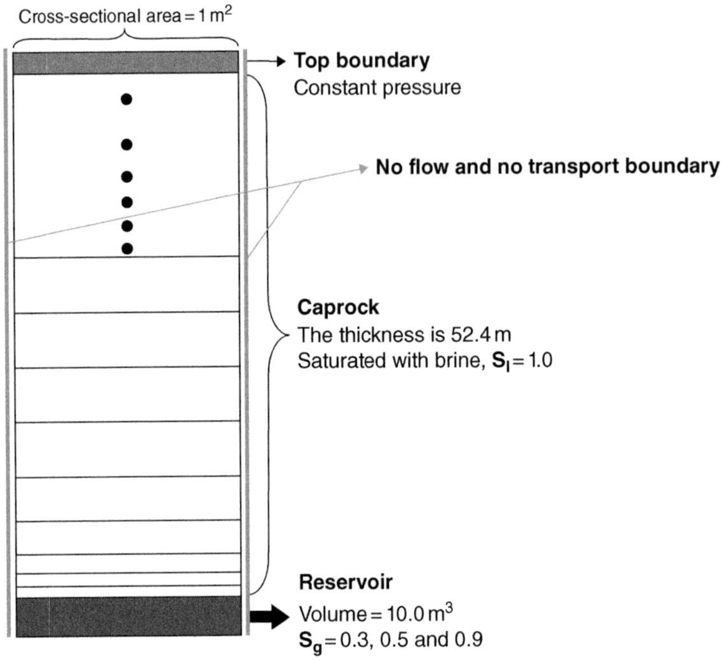

Figure 1.8 Geometric model used for representing the caprock.

able to invade the caprock and would not overflow the top boundary of the model. A temperature of 56 °C was used, based on a land surface temperature of 10.5 °C and a geothermal gradient of 21.8 °C/1000 km. The target sub-caprock consists mainly of clay minerals and has a lower porosity and permeability. The permeability of the clay-rich caprock was set an average of 8.0×10^{-17} m^2, accompanied by a porosity of 0.022. Directly underlying the target caprock is a subreservoir consisting of sandstones with an average porosity of 0.124 and permeability of 5.47×10^{-15} m^2.

The relative permeability function for liquid refers to the van Genuchten-Mualem function (van Genuchten, 1980), and for the gas phase we adopt the model of Corey (1954) to calculate its relative permeability. The Van Genuchten (1980) calculation method is used in this study for obtaining capillary pressure. All parameters involved in these two models are cited from previous works (e.g. Xu et al., 2006; Tian, 2014), and summarized in Table 1.9.

Initial mineralogy

In this study, although the focus is on the geochemical processes occurred within the caprock, because the reservoir designed in the model acts as the $scCO_2$ resource, all physical and geochemical conditions should be assigned initially.

The mineral composition of the caprock is characterized on the basis of X-ray diffraction (XRD) analysis conducted on samples taken from the uppermost mudstone caprock of the Shiqianfeng Formation in Ordos Basin. We treated the averages of data as the initial composition for each mineral presented in all simulations. To meet the format of TOUGHREACT for defining mineral abundance, the mass fractions of minerals

Table 1.9 Hydrological properties of the base case model.

Parameters	Reservoir*	Caprock
Temperature (°C)	51.0	51.0
Pressure (bar)	177.9	170.0
Rock grain density (kg/m^3)	2100.0	2300.0
Porosity (%)	12.4	2.20
$k_z = k_x$, vertical and horizontal absolute permeabilities (m^2)	5.47×10^{-15}	8.00×10^{-17}
Liquid relative permeability, parameters for the van Genuchten-Mualem function (van Genuchten, 1980): $K_{rl} = \sqrt{S^*}\left\{1-\left[1-\left(S^*\right)^{1/\lambda}\right]^{\lambda}\right\}^2$, here $S^* = (S_l - S_{lr})/(S_{ls} - S_{lr})$		
λ	0.457	0.457
S_{lr}	0.30	0.30
S_{ls}	1.0	1.0
Gas relative permeability, parameters for Corey (1954) model:		
$k_{rg} = (1-\hat{S})^2(1-\hat{S}^2)$, here $S^* = (S_l - S_{lr})/(S_{ls} - S_{lr})$		
S_{gr}	0.05	0.05
Capillary pressure, parameters for van Genuchten (1980) function:		
$P_{cap} = -P_0\left[\left(S^*\right)^{-1/\lambda} - 1\right]^{1-\lambda}$, here $S^* = (S_l - S_{lr})/(S_{ls} - S_{lr})$		
λ	0.457	0.457
S_{lr}	0.20	0.20
S_{ls}	0.999	0.999
P_{max} (Pa)	1.0×10^7	1.0×10^8
P_0 (Pa)	2.4×10^5	2.1×10^6

* All physical parameters were taken from the subreservoir directly underlying the target caprock.

directly taken from XRD were converted to volume fractions. The analysis showed that the caprock formation mainly consists of mudstones with a significant component of clay minerals accounting for 61.71% by volume (see Table 1.10; chlorite, kaolinite, illite, and smectite). The remaining volume fraction is composed of quartz, 11.28% by volume, K-feldspar 13.81%, feldspar 7.2%, and calcite 1%.

For the reservoir formation, the mineral composition was drawn from a previous work (Wan, 2012) consisting of quartz-rich sandstone, represented by quartz, which amounts to 53.70%. The remaining minerals are represented in order of decreasing abundance by feldspar, illite, smectite, kaolinite and hematite (Table 1.10). This mineralogical assemblage was assigned to the aquifer directly underlying the target caprock.

Selection of secondary minerals

The selection of the secondary mineral assemblage can markedly affect the modeling results. Therefore enough attention should be paid to the choice of secondary minerals for the modeling system. To guarantee the reasonable selection of secondary minerals

Table 1.10 Initial values of mineral volume fractions for caprock and reservoir (minerals with zero volume fraction are secondary minerals, which could be formed with the presence of CO_2 during the simulation).

Mineral	Chemical formula	Caprock[a]	Reservoir[b]
Quartz	SiO_2	0.1128	0.5370
K-feldspar	$KAlSi_3O_8$	0.1381	
Oligoclase	$CaNa_4Al_6Si_{14}O_{40}$	0.0720	0.2764
Calcite	$CaCO_3$	0.0010	
Smectite-Ca	$Ca_{0.145}Mg_{0.26}Al_{1.77}Si_{3.97}O_{10}(OH)_2$	0.1921	0.0271
Smectite-Na	$Na_{0.29}Mg_{0.26}Al_{1.77}Si_{3.97}O_{10}(OH)_2$	0.1916	0.0166
Illite	$K_{0.6}Mg_{0.25}Al_{1.8}(Al_{0.5}Si_{3.5}O_{10})(OH)_2$	0.1770	0.1120
Kaolinite	$Al_2Si_2O_5(OH)_4$	0.0524	0.0207
Chlorite	$Mg_{2.5}Fe_{2.5}Al_2Si_3O_{10}(OH)_8$	0.0540	
Hematite	Fe_2O_3		0.0102
Dolomite	$CaMg(CO_3)_2$		
Albite	$NaAlSi_3O_8$		
Anorthite	$CaAl_2Si_2O_8$		
Siderite	$FeCO_3$		
Magnesite	$MgCO_3$		
Ankerite	$CaMg_{0.3}Fe_{0.7}(CO_3)_2$		
Dawsonite	$NaAlCO_3(OH)_2$		
Halite	$NaCl$		

[a] Initial mineral composition for caprock formation from an average of core samples.
[b] Initial mineral volume fractions assigned to the subreservoir directly underlying the target caprock.

possibly formed, in the current study, an analysis based on equilibrium batch modeling (Gaus et al., 2005; Tian et al., 2014, 2015) was carried out. Supercritical CO_2 was added to the initial formation brine in contact with the primary mineral assemblage, and then the saturation indices of all minerals contained in the built-in database were examined. The minerals that become supersaturated have the potential to form under the given conditions and are included in the secondary mineral list. In addition, in this study, all carbonate minerals are of special interest and selected as secondary minerals (Table 1.10), because they are expected reaction products of the interaction with $scCO_2$.

Aqueous chemistry

To obtain an early equilibrated water for the caprock, due to the lack of measured data, a batch geochemical model of brine–rock interaction was performed using a pure 0.53 M NaCl solution corresponding to the salinity of reservoir water, and primary minerals listed in Table 1.10. After a 1000-year batch reaction simulation, the initial aqueous components of the caprock (listed in Table 1.12) were obtained. In order to check whether the estimated initial aqueous solution is reasonably stable or not, saturation indices of the presenting minerals were checked (see Table 1.11) at the end of the batch

Table 1.11 Saturation indices (SIs) of primary minerals before and after batch reaction.

Mineral	Chemical formula	Caprock	Reservoir	
		AB[a]	BB[b]	AB[a]
Quartz	SiO_2	0.0038	2.46	0.012
K-feldspar	$KAlSi_3O_8$	−0.0004		
Oligoclase	$CaNa_4Al_6Si_{14}O_{40}$	0.00047	1.68	−0.094
Calcite	$CaCO_3$	−0.000237		
Smectite-Ca	$Ca_{0.145}Mg_{0.26}Al_{1.77}Si_{3.97}O_{10}(OH)_2$	−0.00028	5.39	0.0321
Smectite-Na	$Na_{0.29}Mg_{0.26}Al_{1.77}Si_{3.97}O_{10}(OH)_2$	−0.0014	5.266	0.0054
Illite	$K_{0.6}Mg_{0.25}Al_{1.8}(Al_{0.5}Si_{3.5}O_{10})(OH)_2$	−0.00111	4.90	−0.013
Kaolinite	$Al_2Si_2O_5(OH)_4$	−0.00101	3.2	0.0689
Chlorite	$Mg_{2.5}Fe_{2.5}Al_2Si_3O_{10}(OH)_8$	−0.00161		
Hematite	Fe_2O_3		5.7	−0.0223

[a] After the batch reaction simulation.
[b] Before the batch reaction simulation.

Table 1.12 Initial aqueous chemistry used in all simulations.

Component	Reservoir*	Caprock
	C_{TOT} (mol/kg H_2O)	C_{TOT} (mol/kg H_2O)
Ca^{2+}	0.0287	0.076
Mg^{2+}	0.452×10^{-5}	0.582×10^{-5}
Na^+	0.498	0.477
K^+	0.213×10^{-4}	0.356×10^{-4}
Fe^{2+}	0.733×10^{-4}	0.111×10^{-4}
$SiO_2(aq)$	0.544×10^{-3}	0.213×10^{-2}
AlO_2^-	0.174×10^{-7}	0.890×10^{-6}
Cl^-	0.5160	0.521
HCO_3^-	0.360×10^{-2}	0.364×10^{-2}
SO_4^{2-}	0.204×10^{-6}	0.183×10^{-6}
Tracer	1.000	1.000×10^{-7}
pH	6.52	7.25

* Water chemistry of the reservoir was derived from a batch reaction on the basis of measured data.

simulation. The calculations imply that the water is very close to an equilibrium state with respect to the initially existing minerals. The resulting water was cast into the model as the initial water composition for the caprock.

For tracking the front of fluid invading from the reservoir, a non-reactive tracer concentration of 1 mol/kg H_2O was set initially in the inventory of reservoir water chemistry.

Thermodynamic and kinetic parameters

The mineralogy in this study is described in terms of 18 minerals including primary minerals and secondary minerals, as shown in Table 1.10. The thermodynamic data for minerals, gases and aqueous species are mostly taken from the EQ3/6 V7.2b database of Wolery (1992). For kinetically controlled mineral dissolution and precipitation, a general form of rate law (Lasaga, 1984; Steefel and Lasaga, 1994) is used (Equation 1.7). The values of the most relevant kinetic parameters are summarized in Appendix A (Gherardi et al., 2007; Xu et al., 2011).

1.3.3.4 Results and Discussion

One-dimensional (1D) simulations, performed over a period of 1000 years, were carried out to evaluate the effects of geochemical processes on the self-sealing and self-enhancement efficiencies induced by CO_2 intrusion, including occurrence time and position, and the extent of changes in porosity used to represent the evolution of caprock sealing efficiency. Initial conditions with varying gas saturations (Sg) are set to mimic different scenarios of CO_2–brine two-phase flow. Gas CO_2 saturations of 0.3, 0.5 and 0.9 are used to represent conditions that are likely to occur in the zones far away from the injection well, within the scCO_2 plume and near the injection well, respectively. The caprock grids are treated initially as fully saturated with brine.

The dynamics of CO_2–brine fluids have been represented by applying an overpressure (by trial and error) at the reservoir grid and a constant underpressure (the value with respect to numerical values obtained at the end of an initial gravity equilibration calculation) at the top grid of the caprock. The extent of reservoir overpressure is a critical parameter that determines the propagation velocity of the gas front. Reservoir overpressure has been determined by gradually increasing the reservoir pressure until the two-phase fluid penetrates into the caprock grids. After a number of trials, when the pressure difference between reservoir and caprock reaches 2.5 bars, the scCO_2 can penetrate the caprock and not overflow the top boundary of model.

Migration of scCO_2 within the caprock

Due to the pressure gradient between the reservoir and caprock, scCO_2 intrudes into the caprock with brine. After intrusion the scCO_2 still migrates under the pressure gradient and buoyancy force. Figure 1.9 plots the percolated distance by scCO_2 within the caprock in the three scenarios.

From Figure 1.9 we can see that, in the case with $Sg = 0.3$ (Figure 1.9a), after 10 years of propagation, the migration of the CO_2 front reaches about 0.06 m, and after 100 years reaches 0.2 m, and 1000 years 0.4 m. For the case with $Sg = 0.5$ (Figure 1.9b), the migration of scCO_2 in the caprock is similar to that for $Sg = 0.3$. When the reservoir initial Sg rises to 0.9 (Figure 1.9c), the front of scCO_2 in the caprock reaches 6.0 m after 10 years. At the end of the simulation, the scCO_2 front almost reaches the top boundary of the model. This phenomenon is explained as due to the lower viscosity of scCO_2 than that

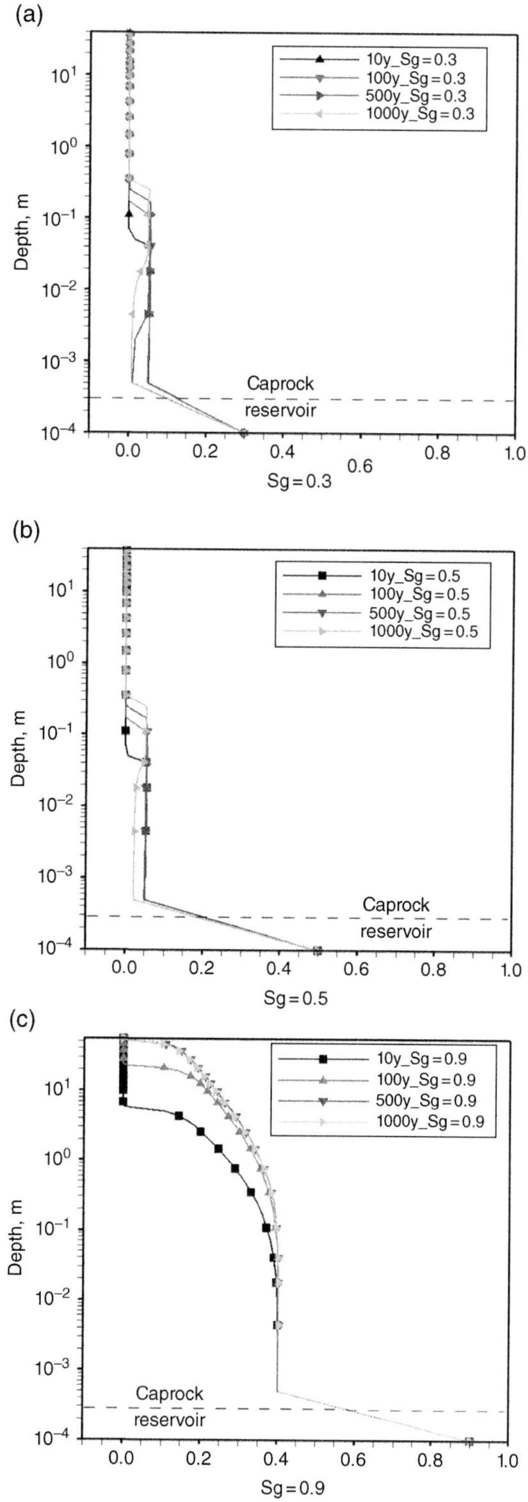

Figure 1.9 Temporal and spatial distribution of $scCO_2$ under different reservoir initial gas saturations.

of brine, so once the gas capillary pressure is exceeded, $scCO_2$ migrates faster than brine under pressure and buoyancy. The distance penetrated by $scCO_2$ depends on the pressure difference between the reservoir and caprock.

Migration of tracer

Considering the migration of $scCO_2$ with brine from the reservoir, the effects of reservoir initial water chemistry should be taken into account. In order to detect the scope influenced by the reservoir water, we modeled the spatial distribution of tracer over time (Figure 1.10). From Figure 1.10 we can see that, in cases with reservoir initial Sg = 0.3 (Figure 1.10a) and 0.5 (Figure 1.10b), the tracers are higher than in the case with Sg = 0.9 (Figure 1.10c). For example, tracer penetrates 4 m after 10 years in cases where Sg = 0.3 and 0.5, compared with 0.3 m in the case with Sg = 0.9. After 500 years, the tracers almost reached the top boundary in the cases where gas saturations of reservoirs are initially 0.3 and 0.5, respectively. In the third case with Sg = 0.9, the tracer only migrated about 3.0 m. We interpret this as due to the difference in initial Sg of reservoirs causing different capillary pressures within the unsaturated reservoir grids, which leads to various pressure gradients in the liquid phase. The higher the reservoir Sg, the smaller the pressure gradient in the liquid phase between the first grid of caprock and reservoir.

Brine pH change

After intrusion, the $scCO_2$ dissolves into the resident porewater, which acidifies the formation water. Figure 1.11 shows the distribution of pH within the caprock for the three cases.

As delineated by Figure 1.11, the extents of acidification in the three cases are different. Before 10 years, pH in the cases of Sg = 0.3 (Figure 1.11a), 0.5 (Figure 1.11b) and 0.9 (Figure 1.11c) all dropped to 3.6 at the bottom of the caprock. Moving up along the column, no CO_2 is available to dissolve into native brine, so with distance, pH becomes higher. Consumption of H^+ associated with mineral alteration also contributes to the increment of pH. The combination of these two mechanisms caused a higher pH in the upper parts of the caprock.

Comparison of the three cases indicates that in the case with reservoir initial Sg = 0.9, the lower pH locates at the front of $scCO_2$. In contrast, the lower pH zones are mostly within the $scCO_2$. This is because, in the case with reservoir initial Sg = 0.9, the process of $scCO_2$ penetration is dominated by displacement of liquid by gas, where only the front of supercritical CO_2 ($scCO_2$) contacts with formation water, which causes a lower pH at the front of $scCO_2$. At the end of displacement, because of the decrease in the pressure gradient, the displaced water flows back due to the negative capillary pressure. The $scCO_2$ dissolves into the returned fluid. Within the reflux, consumption of H^+ due to mineral alteration accounts for the increment of pH.

Evolution of caprock sealing efficiency

The acidified porewater reacts with the ambient minerals. The alteration of minerals, including precipitation and dissolution, modifies the porosity and permeability of the caprock. Figure 1.12 shows the spatial and temporal change in caprock porosity for the three cases. Based on Figure 1.12 we can see that at the bottom of the caprock, the porosities increase because mineral dissolution dominates during $scCO_2$ penetration under the lower pH condition. In response to the evolution of pH, at the end of the

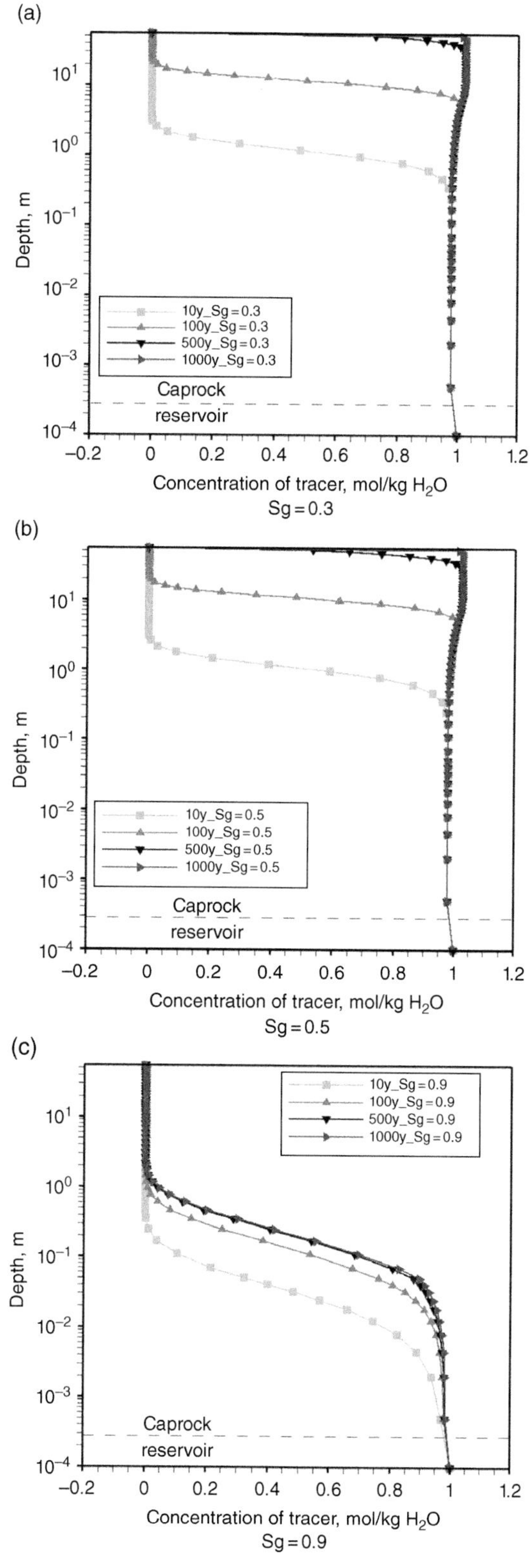

Figure 1.10 Concentration of tracer in the different cases.

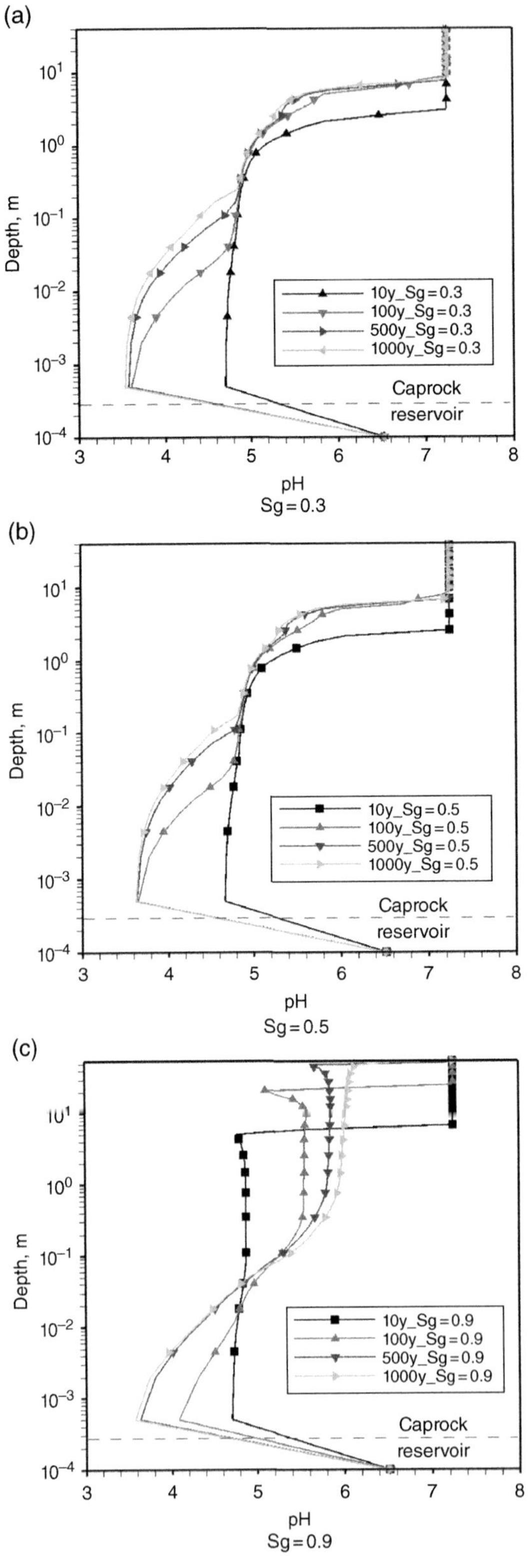

Figure 1.11 Spatial distribution of pH over simulation time in the three cases.

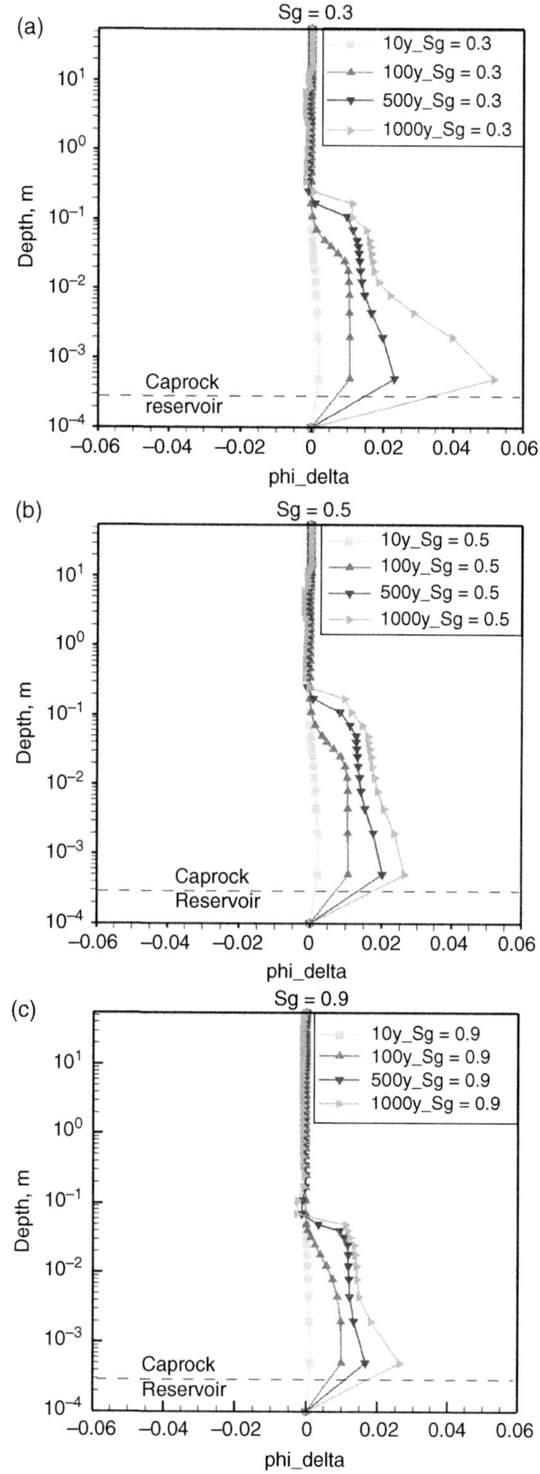

Figure 1.12 Snapshots of spatial and temporal distribution of change in porosity in the three cases.

simulation, the maximum change occurring in the case with Sg = 0.3 (Figure 1.12a) reaches +0.052. In the case of reservoir initial Sg = 0.5 (Figure 1.12b), the maximum change is +0.027, which is slightly higher than +0.026 taking place in the case with Sg = 0.9 (Figure 1.12c). After 10 years the porosity change occurs for a distance of 0.002 m from the interface between reservoir and caprock in all three cases. As the time goes on, after 100 years, the distances with porosity change are 0.08 m in the case of Sg = 0.9, but 0.1 m in other two cases. When time reaches 500 years, the distance in the Sg = 0.9 case rises to 0.1 m, and rises to 0.2 m for the Sg = 0.3 and 0.5 cases. At the end of the simulations, the porosity changes reach 0.3 m in the Sg = 0.3 and 0.5 cases, but 0.1 m for the Sg = 0.9 case. The different porosity effects are due to the different extents of acidification after $scCO_2$ intrusion.

Mineral alteration involved in $scCO_2$ penetration

The change in porosity results derive only from the imbalance of mineral volume, without considering the effects of compression of fluid. Figures 1.13, 1.14 and 1.15 depict the changes in mineral volume fraction and porosity for Sg = 0.3, 0.5 and 0.9, after 10, 100, 500 and 1000 years, respectively.

After 10 years, we can see from Figures 1.13a, 1.14a and 1.15a that changes in porosity match those in the volume fractions of calcite. This means that, before 10 years, the change in porosity is mainly caused by dissolution of calcite. At the bottom of the caprock, the extent of change in calcite volume fraction is larger than that in the upper portion. Changes in porosity in both cases with reservoir initial Sg = 0.3 (Figure 1.13a) and 0.5 (Figure 1.14a) are +0.0016, which is slightly higher than +0.0014 in the case of Sg = 0.9 (Figure 1.15a). At the same time, minerals with obvious changes in volume fraction include k-feldspar, oligoclase, illite, smectite-Ca, smectite-Na and chlorite. The minerals precipitated in the process include smectite-Ca and smectite-Na. A small amount of dawsonite precipitation was predicted.

Qu et al. (2006) reported that dawsonite only deposits under alkalescent and high CO_2 partial pressure conditions. The consumption of H^+ due to dissolution of minerals increased the pH of resident water, and initial Sg = 0.9 (Figure 1.15a) in the reservoir provides high partial pressure, resulting in the formation of dawsonite.

At the 10 years point, the Fe^{2+} dissociated from chlorite dissolution is redundant. Even though the Fe^{2+} in solution is abundant, the shortage of dissolved CO_2 makes it impossible to deposit minerals bearing Fe. As $scCO_2$ increments into the reservoir, the dissolved CO_2 rises. After 100 years, in the upper part of the caprock, calcite no longer dissolves under a higher pH condition, and illite, siderite and ankerite precipitate (Figures 1.13b, 1.14b and 1.15b).

In the lower part of the caprock, the porosity still increases. Dissolution of calcite dominantly contributes to the change in porosity. The maximum increment in porosity, +0.0096, occurs in the case of Sg = 0.3 (Figure 1.13b), and the minimum increment (+0.006) is in the case of Sg = 0.9 (Figure 1.15b).

When time reaches 500 years, besides the alterations of minerals occurring at the 100 years point, smectite-Ca starts to dissolve. The increment in ankerite volume fraction is enhanced locally. Zones where increases in smectite-Na volume fraction take place correspond to zones where reductions in smectite-Ca volume fraction occur. This means that, due to the precipitation rate of ankerite being higher than that of smectite-Ca, Ca^{2+} dissociated from smectite-Ca is deposited with ankerite. Under the condition

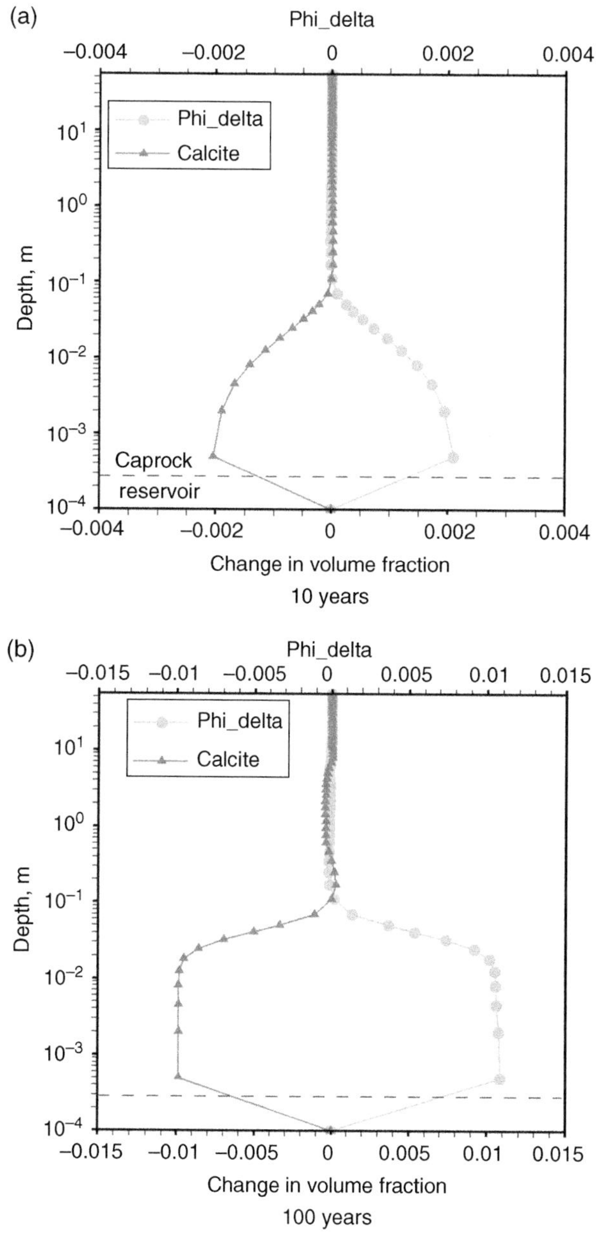

Figure 1.13 Spatial changes in mineral volume fraction and porosity after 10, 100, 500 and 1000 years in the case with Sg = 0.3.

where there is abundant Mg^{2+} and Na^{+} supplied by chlorite and oligoclase, smectite-Na keeps precipitating, and the extent is gently enhanced.

The maximum change in porosity occurs in the case of Sg = 0.3 (Figure 1.13c), when the value reaches +0.0225, followed by +0.02 in the case of Sg = 0.5 (Figure 1.14c). The minimum value appears in the case of Sg = 0.9 (Figure 1.15c), at +0.016. These

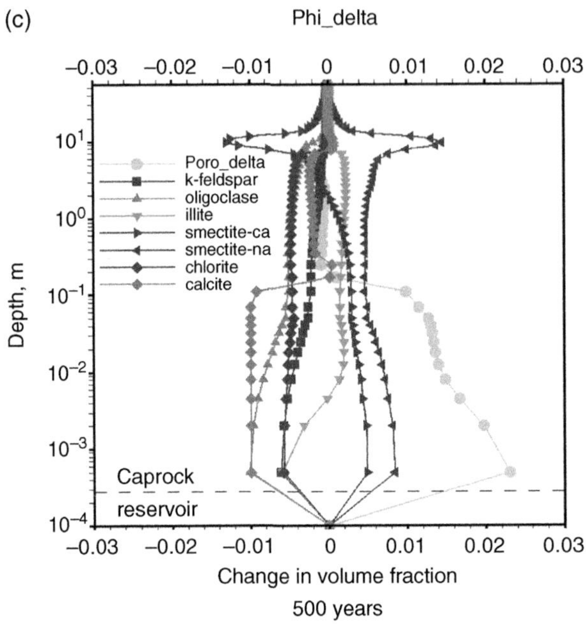

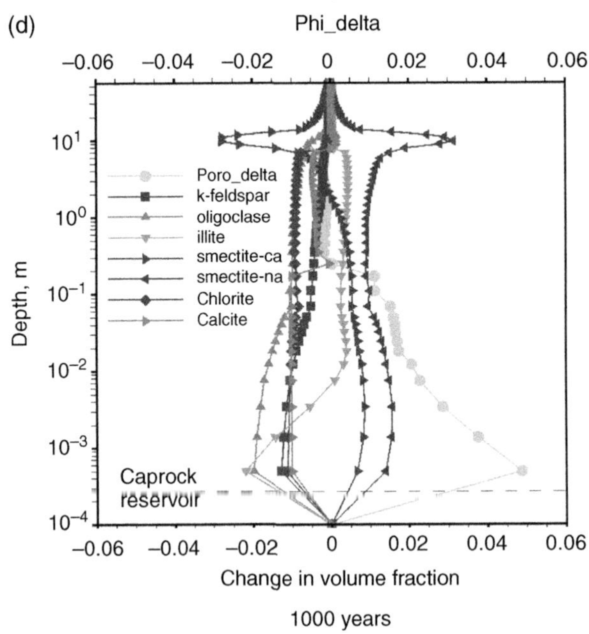

Figure 1.13 (Cont'd)

differences are attributed to the varying pH resulting from different extents of $scCO_2$ penetration. In the cases with initial Sg = 0.3 and 0.5, at the bottom of the caprock, pH is lower than in the case of Sg = 0.9, which leads to different acidifications and thereby different reaction extents. The differences lead to varying degrees of mineral alteration, and then changes in porosity.

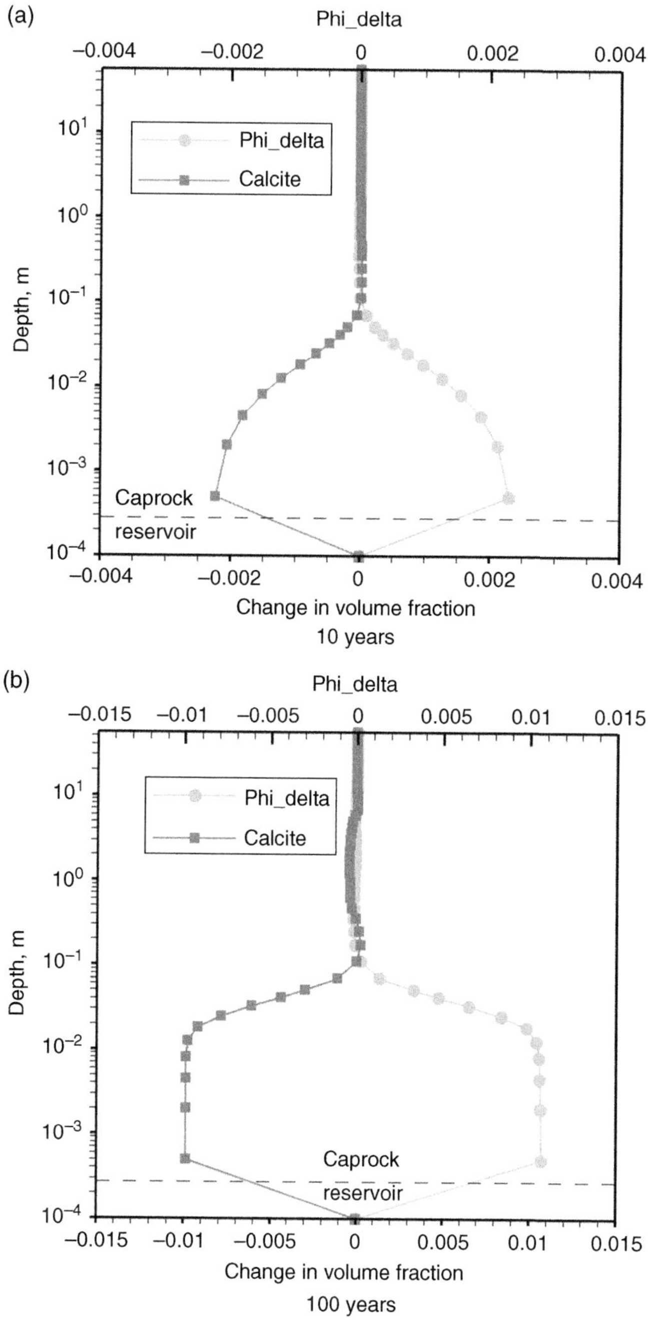

Figure 1.14 Spatial changes in mineral volume fraction and porosity after 10, 100, 500 and 1000 years in the case with Sg = 0.5.

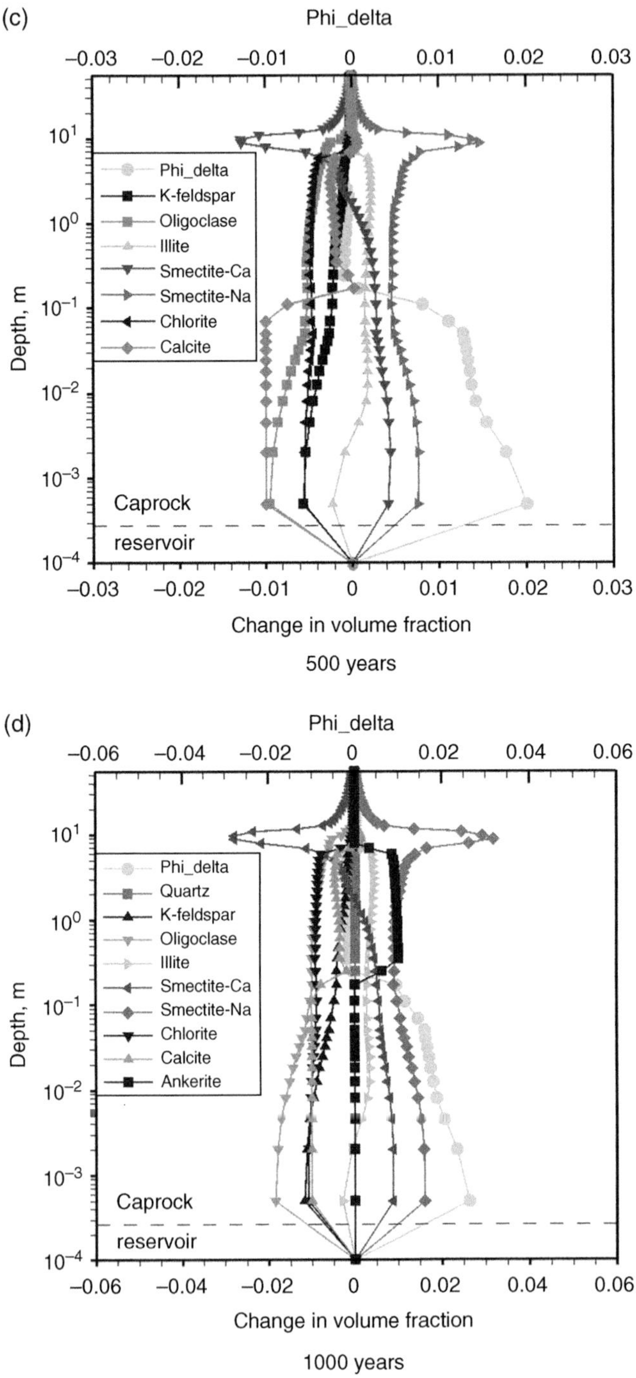

Figure 1.14 (Cont'd)

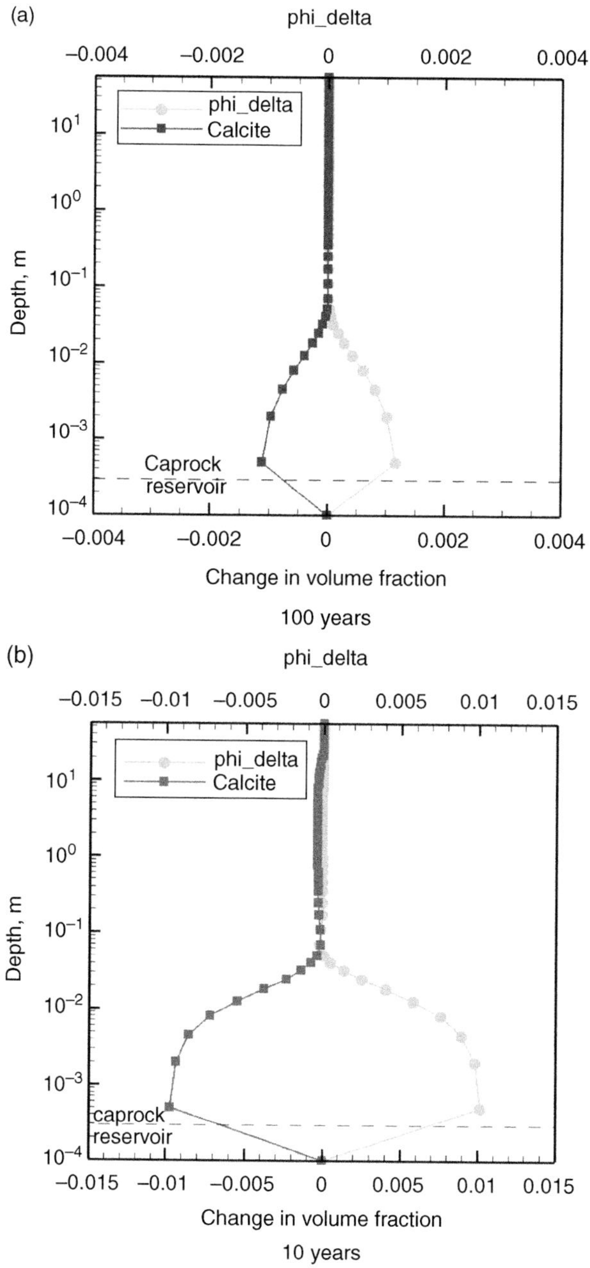

Figure 1.15 Spatial changes in mineral volume fraction and porosity after 10, 100, 500 and 1000 years in the case with Sg = 0.9.

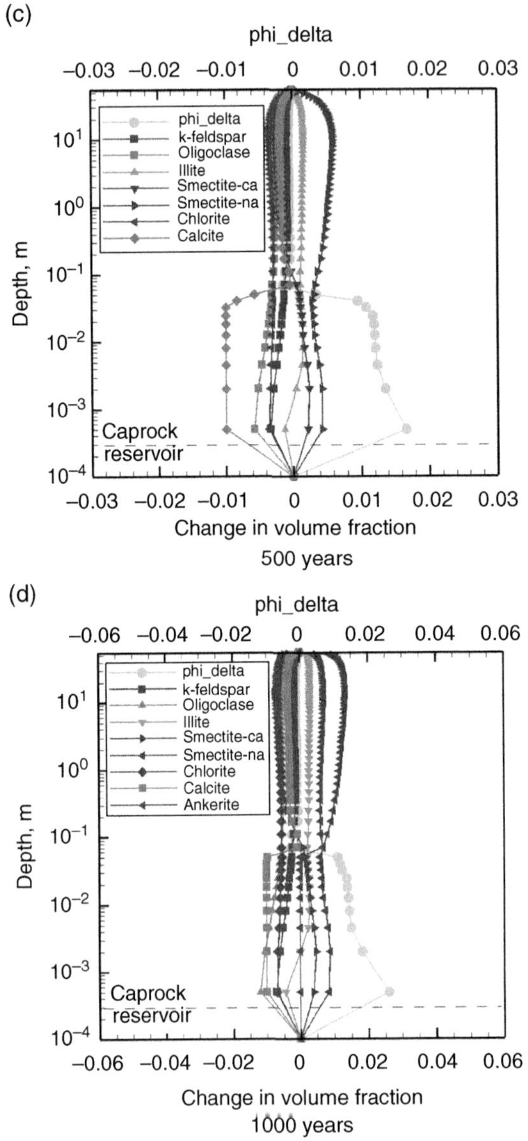

Figure 1.15 (Cont'd)

At the end of the simulation (1000 years), patterns of change in mineral volume fraction are similar to those of 500 years. The changes in porosity are still enhanced with time. At the 1000 years point, the porosity change is mainly caused by alteration of oligoclase and illite (Figures 1.13d, 1.14d and 1.15d).

1.3.3.5 Concluding Remarks

In this section, we evaluated the evolution of caprock sealing efficiency by employing the reactive transport simulation approach. A simplified 1D column model was adopted to investigate the change in porosity due to mineral alteration after $scCO_2$ penetration. Geological, hydrogeological and geochemical conditions were assigned based on measurement data from the uppermost sub-caprock of the Shiqianfeng Formation in Ordos Basin. On the given data basis, by comparing the simulation results, we can conclude:

1) When the supercritical CO_2 invades the caprock, dissolution of minerals dominates the whole geochemical processes, which causes an increment in porosity, weakening the sealing efficiency of the caprock.
2) The CO_2 penetrating position (far away from the injection well, within the $scCO_2$ plume, and near the injection well, with gas CO_2 saturations of the reservoir of 0.3, 0.5 and 0.9, respectively) affects the evolution of the caprock sealing capability. Near the injection well, in the 'dry area', $scCO_2$ barely reacts with the surrounding rocks in the short term, which means the porosity hardly changes; with the increment of distance from the injection well (Sg drops from 0.9 to 0.5, and finally reaches 0.3 in this study), the geochemical reaction becomes stronger, and the sealing efficiency of the caprock was weakened more severely, especially after 500 years. This means that the leakage of CO_2 through the caprock is predicted to take place potentially at the front of and within the CO_2 plume.
3) Minerals involved in this study displayed different behaviours after intrusion of $scCO_2$. Minerals dissolved throughout the simulation include calcite, chlorite, oligoclase and k-feldspar. Deposited minerals predicted by simulation are siderite, ankerite, smectite-Na and a little dawsonite. Minerals undergoing both dissolution and precipitation are illite and smectite-Ca.

As the first natural barrier for a CO_2 geological storage site, a caprock should prevent the $scCO_2$ from leaking from the deep saline reservoir. Therefore, the sealing efficiency of the caprock is the necessary prerequisite for screening a site as a potential reservoir for CO_2 storage. An evaluation of the caprock's sealing capability is imperative to guarantee the security of CO_2 geological storage for the long term. The simulation results of this study showed that the geochemical process plays an important role during the evolution of caprock sealing efficiency after CO_2 intrusion. Penetration position, mineral composition of the caprock and time-span all affect the change in caprock sealing capability, which is a complex process coupled with fluid flow, heat conduction and chemical reaction. Therefore, reactive transport modeling is an effective tool to give deep insight into the coupled processes. However, geomechanical processes induced by the injection of CO_2 are ignored in this study. Mechanical damage of the caprock should be carefully considered in future work, and the coupling of mechanics with geochemistry still faces a challenge.

1.4 Reactive Transport Modeling for Chemical Stimulation of an Enhanced Geothermal Reservoir

1.4.1 General Description

The concept of an enhanced geothermal system (EGS) involves mining heat from hot dry rocks by pumping cold fluid into a hydrofractured reservoir through the injection well, and bringing hot water from the production well, then utilizing the hot water to generate electricity. One major consideration in the development of EGS reservoirs is achieving and maintaining adequate injection and production. A reactive transport model can be used to examine the rock–fluid interactions and associated mineral dissolution and precipitation, which modify the porosity and permeability of the EGS reservoirs. Xu et al. (2006) have illustrated how the chemical composition of re-injected waters can be regulated to improve reservoir performance by maintaining or even enhancing injection, by using the non-isothermal multiphase reactive geochemical transport code TOUGHREACT. Hydrological and geochemical parameters used in the simulation were taken from 3500 m depth in the Soultz Well GPK1, which is located in the Deep Geothermal research site situated at Soultz-sous-Forêts, France. The thermo–hydraulic–chemical coupled computer code FRACHEM has been developed specifically to predict changes in the EGS reservoir of the Soultz (Durst, 2002; Bachler, 2003). André et al. (2006) predicted changes in reservoir porosity and permeability during water–rock interactions, and mineral behaviours have been investigated during a long-term production/injection operation using FRACHEM. Their study showed that calcite dissolved significantly, while silicates and alumino-silicates tended to precipitate in small quantities near the injection well.

As part of an effort to reduce atmospheric emissions of carbon dioxide (CO_2), a novel concept of operating an enhanced geothermal system (EGS) using CO_2 instead of water as a working fluid (CO_2-EGS), and achieving simultaneous geological sequestration of CO_2, has been proposed and evaluated (Brown, 2000; Pruess, 2006). Reactive transport modeling could also be applied to the study of CO_2-induced mineral alteration for a CO_2-based geothermal system and its impact on fluid circulation. Field experiments have been conducted at the Ogachi EGS site in order to study CO_2 sequestration in solid minerals (Kaieda et al., 2009). Xu et al. (2011) developed a reactive transport model using TOUGHREACT for aqueous chemical evolution and mineral alteration induced by mixtures of CO_2 and water in the field experiments, to investigate the development and operation of a CO_2-based EGS for actual reservoir conditions. Results revealed that some measured concentrations of aqueous components can be reproduced by the initial simulations, such as pH and $SiO_{2(aq)}$. Given the data from the quartz monzodiorite unit at the Desert Peak EGS site (Nevada), Wan et al. (2011) described the impact of fluid–rock interactions on CO_2-EGS. Xu et al. (2014) performed reactive transport simulations to study the impact of fluid–rock interactions in CO_2 geothermal systems.

The major problems observed in the exploration of geothermal reservoirs are the poor hydraulic connection between the production/injection wells, and the fractured reservoir restricting the extraction of the thermal energy from the reservoir. Injecting aqueous chemical agents, such as mineral acids, could be effective for mineral dissolution and porosity enhancement at distances of several metres around a well. The chemical interactions between rocks and chemical agents could be understood and reliably predicted through reactive transport modeling, in order that stimulation treatments can be properly planned and executed. Xu et al. (2004b) have performed a 1D reactive transport model to investigate injectivity recovery due to acid (HF) injection in the Tiwi geothermal field (Philippines), and the predicted amount of amorphous silica dissolved by acid was consistent with the estimated amount. Xu et al. (2009) presented the results of reactive transport simulations to investigate the impact of a chelating agent (NTA) at a high pH on the quartz monzodiorite unit at the EGS site at Desert Peak (Nevada). Portier and Vuataz (2010) performed reactive transport modeling to reproduce chemical interactions between host rocks and a mixture of HCl and HF, as well as its potential effects on the Soultz EGS reservoir, using the FRACHEM code. Results indicated that the predicted dissolution of fracture minerals by injection of a mixture of acids, Regular Mud Acid (RMA), was consistent with the test performed on the GPK4 well at the Soultz EGS site. Xu et al. (2010) carried out a series of reactive transport simulations in which the supercritical CO_2 was treated as a stimulation agent for EGS.

The current study is not intended to represent any particular sit, but some geological and geochemical conditions and parameters were taken from the China EGS research site in Songliao Basin.

1.4.2 Brief Description of the EGS Site in Songliao Basin

The Songliao Basin, located in northeastern China, is a large Mesozoic–Cenozoic sedimentary basin developed by the initial rifting of the Eurasian continent during the late Jurassic. Deep volcanic rocks are widely distributed in this region. The central depression of Songliao Basin is the location of a Chinese EGS research project, with an average geothermal gradient of about 44.5 °C/km.

The potential EGS geothermal reservoir is the Yingcheng Formation, interbedded with volcanic rocks in the central depression. The burial depth of the reservoir is approximately 3500 m and the formation temperature corresponding to the depth reaches 150 °C, which meets the requirements for geothermal development (Huang et al., 2004; Hu et al., 1998). Moreover, the naturally existing fractures are favourable for generating artificial fractures for the EGS reservoir.

In this study, we illustrate how reactive transport models are used to evaluate the impact of chemical stimulation on the enhanced geothermal system (EGS) reservoir when developing such models. Chemical reaction processes in the whole system will be predicted with reactive transport simulator TOUGHREACT. Geophysical and geochemical parameters of the Yingcheng Formation in Songliao Basin were used.

1.4.3 Conceptual Model

1.4.3.1 Geometry and Boundary Conditions

We considered an idealized fractured porous medium with two perpendicular sets of plane-parallel fractures of equal aperture and spacing. Because of the assumed symmetry, we employed a one-dimensional (1D) flow tube between injection and production wells for our current modeling (Figure 1.16). From the injection well to the production well, the model consists of 96 grid blocks representing a distance of 600 m. The method of 'multiple interacting continua' (MINC; Pruess, 2006) was used to represent matrix-to-fracture fluid transfer (Figure 1.17). Conductive heat exchange with low permeability rocks is treated with a semi-analytical technique according to Vinsome and Westerveld (1980).

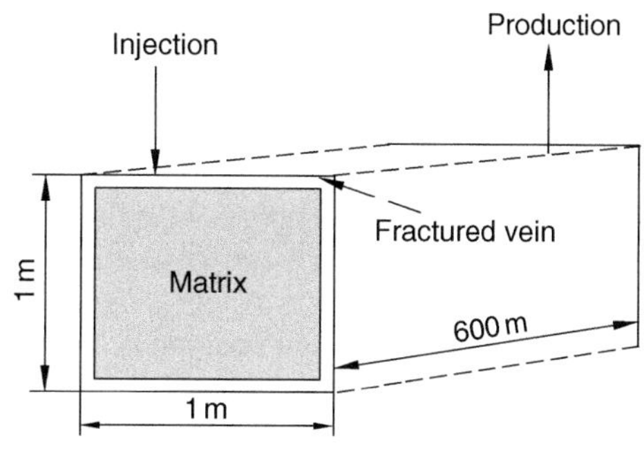

Figure 1.16 Schematic representation of the 1D model.

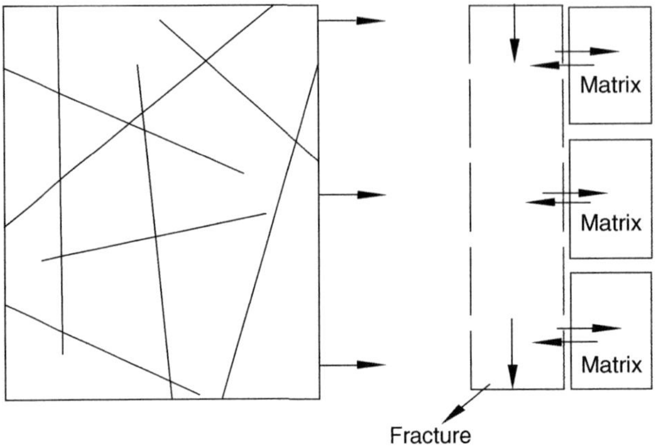

Figure 1.17 Schematic representation of the 'multiple interacting continua' (MINC) method.

Table 1.13 Basic parameters of the artificial geothermal reservoir.

Reservoir parameters				Chemical agent injection	
Porosity in matrix	3.45%	Permeability in matrix	81.0×10^{-17} m^2	System uptime	2.5 h
Permeable volume fraction	10%	Permeability in fracture	1.0×10^{-11} m^2	Injection temperature	30 °C
Porosity in fracture	50%	Rock grain density	2500 kg/m^3	Injection rate	0.08 kg/s
Rock specific heat	1000 J/kg·°C	Rock thermal conductivity	2.9 W/m·°C	Chemical agents	15% HCl
Reservoir pressure	35 MPa	Reservoir temperature	150 °C		6% HCl + 0.5% HF

1.4.3.2 Physical Parameters

Hydrological parameters used in the present simulations are listed in Table 1.13, and the formation is assumed to be homogeneous. The initial fracture permeability and porosity were assumed to be 1.0×10^{-11} m^2 and 0.3, respectively. We considered the injection of chemical agent into the reservoir under the condition with temperature and pressure being 150 °C and 35 MPa, respectively. A scenario was conducted with a maximum test period of 2.5 hours, a preflush of 15 wt% HCl solution for 1 hour and a main flush of mud acid (6% HCl + 0.5% HF) for 1.5 hours at a flow rate of 0.08 kg/s.

1.4.3.3 Initial Mineral Composition

The mineralogical composition is listed in Table 1.14, according to mineralogical analysis of core samples from the Yingcheng Formation; the volcanic rock in the reservoir mainly consists of calcite, quartz, alkali-feldspar and plagioclase. All plagioclase was represented using pure albite, because anorthite is rare in the Yingcheng Formation,

Table 1.14 Initial mineral volume fractions.

Mineral	Volume fraction (%)
Calcite	2
Quartz	71
K-feldspar	17
Albite	10

Table 1.15 Initial total dissolved component concentrations in mol/l for the geothermal reservoir water.

Component	Concentration	Component	Concentration
K	0.718×10^{-2}	HCO_3^-	0.529×10^{-3}
Al	0.883×10^{-6}	Ca	0.530×10^{-3}
Na	0.213	Cl^-	0.220
Si	0.268×10^{-2}	pH	7.63

according to the literature (Zhao, 2005). Alkali-feldspar was represented by K-feldspar because it is the main mineral in the feldspar group of the reservoir, according to geological data (Zhao, 2005).

1.4.3.4 Water Chemistry

The aqueous composition of the formation water was obtained by an initial equilibrium simulation conducted using the mineralogical composition to establish a quasi-steady-state chemical composition of the formation water. A pure 0.22 M NaCl solution was used to react with the primary minerals listed in Table 1.14 at a temperature of 150 °C, and the resulting solution composition listed in Table 1.15 is treated as the initial aqueous composition for the simulation.

1.4.3.5 Thermodynamic and Kinetic Parameters

The thermodynamic data for minerals and aqueous species are mostly taken from the EQ3/6 V7.2b database of Wolery (1992). The reaction rates between chemical agents and minerals in the reservoir are affected by kinetics. The kinetics of the reaction between HCl and carbonate minerals have been calibrated from a lab experiment, and the results were summarized as follows (Fogler et al., 1976):

$$r = A_m k_m \left(C_{\mathrm{HCl}}\right)^{\alpha} \tag{1.16}$$

$$k_m = k_{25} \exp\left[\frac{-E_a}{R}\left(\frac{1}{T} - \frac{1}{298.15}\right)\right] \tag{1.17}$$

where r is the effective reaction rate (mol/m^2/s), A_m is the specific surface area (cm^2/g), k_m is the kinetic rate constant, C_{HCl} is HCl activity (mol/kg H_2O), α is a dimensionless

Table 1.16 Parameters for kinetic rate constants of minerals.

Mineral	A (cm²/g)	α	β	K	k_{25} (mol/m²/s)	E_a (kJ/mol)
Calcite	9.8	0.63	–	–	7.0×10^{-2}	61
Quartz	9.8	–	1	–	2.32×10^{-8}	9.56
K-feldspar	9.8	0.4	1.2	1.4 exp[956(1/T – 1/298.15)]	1.9×10^{-5}	38.9
Albite	9.8	1	1	0.4 exp[554(1/T – 1/298.15)]	1.8×10^{-5}	32.7

parameter, E_a is the activation energy, k_{25} is the rate constant at 25 °C, R is the gas constant (8.31 J/(mol K)), and T is the absolute temperature in kelvin.

HF reacts with quartz and feldspars in the reservoir, and all these kinetic processes can be expressed by Equation (1.18) (Fogler et al., 1975).

$$r = A_m k_m \left[1 + K\left(C_{\mathrm{HCl}}\right)^{\alpha}\right] \cdot \left(C_{\mathrm{HF}}\right)^{\beta} \tag{1.18}$$

where r is the effective reaction rate (mol/m²/s), A_m is the specific surface area (cm²/g), k_m is the kinetic rate constant, C_{HCl} is the HCl activity (mol/kg H_2O), C_{HF} is the HF activity (mol/kg H_2O), K, α and β are dimensionless parameters, E_a is the activation energy, k_{25} is the rate constant at 25 °C, R is the gas constant (8.31 J/(mol K)), and T is the absolute temperature.

Table 1.16 lists parameters for the kinetics of mineral reactions involved in the model, which were taken from previous studies (Fogler et al., 1975, 1976; Portier et al., 2010; Xu et al., 2014).

1.4.4 Results and Discussion

1.4.4.1 HCl Preflush

The presence of calcium (Ca^{2+}) inherited from the initial reaction solution and produced by the dissolution of calcite can cause CaF_2 to precipitate in the mud acid injection stage of chemical stimulation. In view of this, the reservoir was treated using HCl in the first hour of operation prior to the main stage of acid injection. The HCl preflush produces calcite dissolution within a short distance of the well. Moreover, the HCl injection could also decrease pH of the formation water, and then inhibit the precipitation of CaF_2. During the simulation, HCl preflush induces a significant calcite dissolution (1.4%) close to the injection well, and the alteration zone extends to almost 8 m from the injection point (Figure 1.18). The small effective distance is mainly due to the high rate of reaction HCl–$CaCO_3$. The pH drops to −0.14 close to the injection point (Figure 1.18), and the zone with lowered pH extends to 17 m after the HCl preflush stage.

1.4.4.2 Mud Acid Main Flush

Figure 1.19 shows predicted changes in the volume fractions of calcite, K-feldspar and albite after 0.5, 1.0 and 1.5 hours in the reservoir during the main flush of mud acid. More calcite dissolves close to the injection well, with a maximum volume of 1.4%

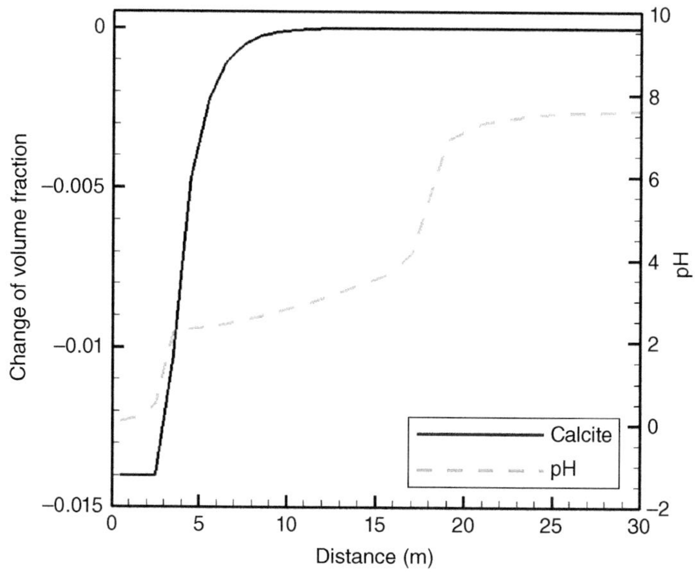

Figure 1.18 Calcite volume fraction change and pH in the fracture after HCl injection.

because of the high rate for the HCl–$CaCO_3$ reaction. K-feldspar in the fracture adjacent to the wellbore significantly dissolves in the mud acid after 1.5 hours. The maximum dissolved volume fraction reaches 0.4%, and the zone where K-feldspar dissolves extends to about 14 m. Significant dissolution of albite also occurs, and its maximum changed value is 0.15% in volume fraction, and its dissolution pattern is similar to that of K-feldspar. Quartz is not included in this figure, because the smallest amount (less than 1.0×10^{-6}) of quartz was exhausted during chemical stimulation. The reaction rates of mud acid with carbonate minerals and feldspars are orders of magnitude higher than that of the reaction rate of HF with quartz. The quartz reaction becomes significant in regions where calcite and feldspars have been completely dissolved. Dissolutions of calcite, K-feldspar, albite and quartz can be written as:

$$CaCO_3 + 2HCl \longrightarrow 2Cl^- + Ca^{2+} + H_2O + CO_{2(aq)} \tag{1.19}$$

$$KAlSi_3O_8 + 22HF \longrightarrow KF + AlF_3 + 3H_2SiF_6 + 8H_2O \tag{1.20}$$

$$NaAlSi_3O_8 + 22HF \longrightarrow NaF + AlF_3 + 3H_2SiF_6 + 8H_2O \tag{1.21}$$

$$4HF + SiO_2 \longrightarrow SiF_4 + 2H_2O; \quad SiF_4 + 2HF \rightarrow H_2SiF_6 \tag{1.22}$$

We assumed that the change in porosity is ideally caused by the imbalance of mineral volume fractions. An enhancement in porosity is revealed in Figure 1.19d. Compared with the initial porosity of 0.3, the porosity rises to 0.32 near the injection point. The improvement in porosity extends to a distance of about 20 m from the injection well. Increases in porosity are mainly caused by dissolution of calcite, K-feldspar and albite, as mentioned above.

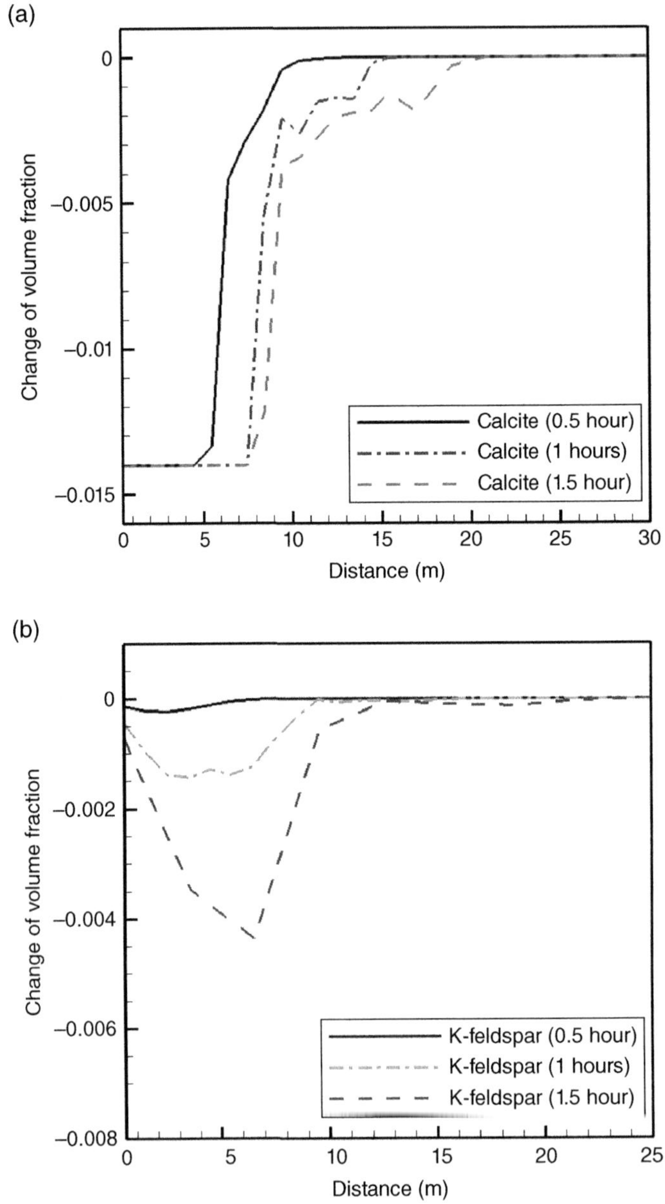

Figure 1.19 Changes of mineral volume fraction (negative values indicate dissolution and positive precipitation) and porosity of the fracture during mud acid injection.

1.4.5 Concluding Remarks

Reactive transport modeling is an important tool for the study of geothermal development, such as chemical stimulation for an enhanced geothermal system (EGS). In this section, the numerical simulations using TOUGHREACT were performed to

(c)

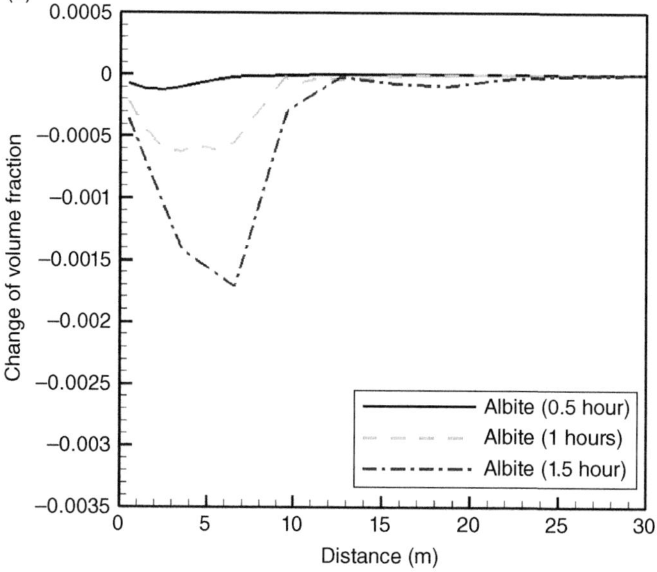

(d)

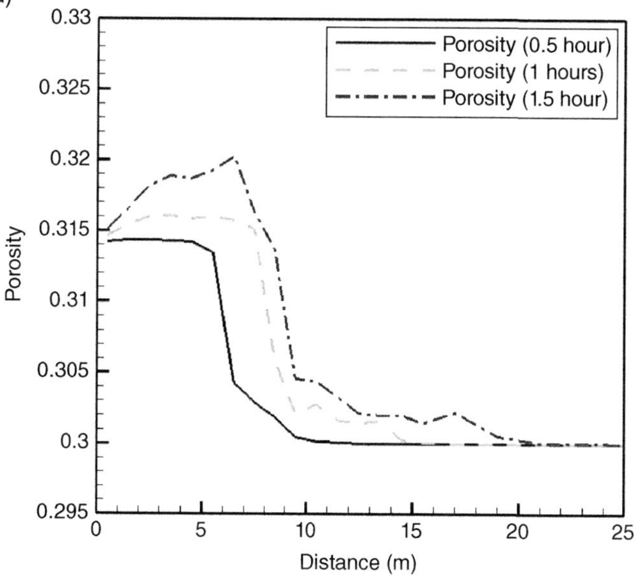

Figure 1.19 (Cont'd)

investigate the behaviour of chemical stimulation within the EGS reservoir in Songliao Basin, China. The following conclusions can be drawn based on this study:

1) The injections of HCl and mud acid result in dissolution of minerals in the EGS reservoir. Reservoir porosity and permeability can be enhanced around the injection well.

2) Increases in porosity are mainly caused by dissolutions of calcite, K-feldspar and albite during chemical stimulation.

It is shown by reactive transport modeling that chemical processes play a significant role in the development of porosity around injection wells. Reactivity between mud acid and rock in high-temperature reservoirs is not well understood, especially the precipitation resulting from the secondary and tertiary reactions between the spent acid and the rock. Further experimental studies are required to describe the reaction kinetics quantitatively. Investigations on different operation modes in chemical stimulation should also be conducted in the future.

1.5 Conclusions and Outlook

Reactive transport modeling is an essential tool for the analysis of coupled physical, chemical and biological processes in earth systems. In this chapter, we mainly introduced two application domains: carbon dioxide geological storage (CGS) and enhanced geothermal system (EGS). For CGS the impact of geochemical reactions induced by CO_2 and its feedback on physical properties has to be assessed for long-term CO_2 storage. One major consideration in the development of EGS reservoirs is achieving and maintaining adequate injection and production. Injecting aqueous chemical agents, such as mineral acids, could enhance the hydraulic connection between the production/injection wells and the fractured reservoir. Reactive transport modeling is able to examine the rock–fluid interactions and associated mineral dissolution and precipitation involved in the above-mentioned domains, which modify the porosity and permeability of reservoirs.

During the last decade, significant progress has been made in geochemical and coupled geochemical and flow modeling, evolving from simple geochemical bath models to three-dimensional fully coupled models. Although many application domains are identified with respect to different purposes, each domain faces its own challenges during analysis of the geochemical impacts. For example, in long-term modeling prediction, weaknesses remain in the identification of dominant geochemical reactions (initial mineral selection) and poor knowledge of reaction kinetics, making it difficult to attribute an adequate timescale to the occurrence of the trapping processes. Another major challenge when applying coupled modeling is that, for the relevant temperature and pressure conditions and brine compositions, calculations have to be made outside the validity ranges of certain concepts that are generally integrated in geochemical models (e.g. ion association models, solubility calculations) and that are based on parameter values that were determined for different conditions. This creates large uncertainties in model outcomes, especially when compared with uncertainty ranges generally encountered in purely physical (reservoir) modeling. Therefore, experimental studies are further required to quantitatively describe the reaction kinetics.

Coupling of fluid flow, geochemical reactions and mechanical deformation in porous media plays a critical role. One of the key issues in simulation of complicated system is the ability to describe the mechanical, hydraulic and geochemical behaviours. However, our knowledge of the geomechanical and geochemical processes in geological formations at high pressure and high temperature is still incomplete, and should be investigated as a key focus of future study.

Appendix A

Parameters used in kinetic rate law (Equation 1.6) (Xu et al., 2011, 2014; Gherardi et al., 2007) for the three mechanisms specific to minerals involved in simulations.

		Neutral mechanism		Acid mechanism			Base mechanism		
Mineral	**A (cm^2/g)**	**k_{25} ($mol/m^2/s$)**	**E_a (kJ/mol)**	**k_{25} ($mol/m^2/s$)**	**E_a (kJ/mol)**	**n (H^+)**	**k_{25} ($mol/m^2/s$)**	**E_a (kJ/mol)**	**n (H^+)**
Quartz	9.8	1.023×10^{-14}	87.7						
K-feldspar	9.8	3.890×10^{-13}	38	8.71×10^{-11}	51.7	0.5	6.31×10^{-22}	94.1	−0.823
Oligoclase	9.8	1.445×10^{-12}	69.8	2.13×10^{-10}	65.0	0.457			
Calcite	9.8	1.55×10^{-9}	23.5	1.55×10^{-6}	14.4	1.0			
Smectite-Ca	151.6	1.66×10^{-13}	35.0	1.05×10^{-11}	23.6	0.34	3.02×10^{-17}	58.9	−0.40
Smectite-Na	151.6	1.66×10^{-13}	35.0	1.05×10^{-11}	23.6	0.34	3.02×10^{-17}	58.9	−0.40
Illite	151.6	1.66×10^{-13}	35	1.05×10^{-11}	23.6	0.10	3.02×10^{-17}	58.9	−0.400
Kaolinite	151.6	6.92×10^{-14}	22.2	4.90×10^{-12}	65.9	0.78	8.91×10^{-18}	17.9	−0.472
Chlorite	9.8	3.02×10^{-13}	88.0	7.76×10^{-12}	88.0	0.5			
Hematite	12.9	2.51×10^{-15}	66.2	4.07×10^{-10}	66.2	1.0			
Dolomite	12.9	2.52×10^{-12}	62.76	2.34×10^{-7}	43.54	1.0			
Albite	9.8	2.75×10^{-13}	69.8	6.92×10^{-11}	65.0	0.457	2.51×10^{-16}	71.0	−0.572
Anorthite	9.8	7.60×10^{-13}	17.8	7.60×10^{-4}	16.6	1.41			
Siderite	9.8	1.26×10^{-9}	62.76	6.46×10^{-4}	36.1	0.5			
Magnesite	9.8	4.57×10^{-10}	23.5	4.17×10^{-7}	14.4	0.5			
Ankerite	9.8	1.26×10^{-9}	62.76	6.46×10^{-4}	36.1	0.2			
Dawsonite	9.8	1.26×10^{-9}	62.76	6.46×10^{-6}	36.1	0.5			
Halite					Equilibrium				

Acknowledgements

This work was supported by the National Natural Science Foundation of China (Grant No. 41572215), and by the China Geological Survey working project (Grant No. 12120113006300).

References

Alemu, B.L., Aagaard, P., Munzi, A., and Skurtveit, E. (2011). Caprock interaction with CO_2: A laboratory study of reactivity of shale with supercritical CO_2 and brine. *Applied Geochemistry* 26 (12): 1975–1989.

André, L., Rabemanana, V., Vuataz, F.D. (2006). Influence of water–rock interactions on fracture permeability of the deep reservoir at Soultz-sous-Forêts, France. *Geothermics* 35 (5): 507–531.

Audigane, P., Gaus, I., Czernichowski-Lauriol, I., et al. (2007). Two-dimensional reactive transport modeling of CO_2 injection in a saline aquifer at the Sleipner site, *North Sea. Am. J. Sci.* 307: 974–1008.

Bachler, D. (2003). Coupled thermal–hydraulic–chemical modeling at the Soultz-sous-Forêts HDR reservoir (France). PhD dissertation, Swiss Federal Institute of Technology, Zurich, Switzerland.

Bachu, S. (2008). CO_2 storage in geological media: role, means, status and barriers to deployment. *Progress in Energy and Combustion Science* 34: 254–273.

Brown, D.W. (2000). A hot dry rock geothermal energy concept utilizing supercritical CO_2 instead of water. *Proceedings of the Twenty-Fifth Workshop on Geothermal Reservoir Engineering*, Stanford University, 233–238.

Celia, M.A. and Bachu, S. (2003). Geological sequestration of CO_2: is leakage unavoidable and acceptable? In: *Proceedings of the Sixth International Greenhouse Gas Technologies Conference, Kyoto* (ed. J. Gale and Y. Kaya), Volume I, 477–482.

Chadwick, A., Arts, R., Bernstone, C., et al. (2008). *Best Practice for the Storage of CO_2 in Saline Aquifers – Observations and Guidelines from the SACS and CO_2STORE Projects.* Nottingham: British Geological Survey.

CO2CRC. (2008). *Storage Capacity Estimation, Site Selection and Characterisation for CO_2 Storage Projects.* Cooperative Research Centre for Greenhouse Gas Technologies, Canberra, Australia: Cooperative Research Centre for Greenhouse Gas Technologies.

Corey, A.T. (1954). The interrelation between gas and oil relative permeabilities. *Producers Monthly* 19 (1): 38–41.

CSLF. (2008). *Comparison between Methodologies Recommended for Estimation of CO_2 Storage Capacity in Geological Media – Phase III Report.* Washington, DC: CSLF.

Durst, D. (2002). Geochemical modeling of the Soultz-sous-Forêts hot dry rock test site: coupled fluid–rock interaction to heat and fluid transport. PhD dissertation, Université de Neuchâtel, Switzerland.

Fogler, H.S., Lund, K., and Mccune, C.C. (1975a). Acidization III – The kinetics of the dissolution of sodium and potassium feldspar in HF/HCl acid mixtures. *Chemical Engineering Science* 30 (75): 1325–1332.

Fogler, H.S., Lund, K., and Mccune, C.C. (1975b). Predicting the flow and reaction of HCl/HF acid mixtures in porous sandstone cores. *Society of Petroleum Engineers Journal* 16 (5): 248–260.

Gaus, I., Azaroual, M., and Czernichowski-Lauriol. (2005). Reactive transport modelling of the impact of CO_2 injection on the clayey cap rock at Sleipner (North Sea). *Chemical Geology* 217 (3): 319–337.
Gaus, I., Le Guern, C., Pearce, J., et al. (2004). Comparison of long term geochemical interactions at two natural CO_2-analogues: Montmiral (Southeast basin, France) and Messoampos (Florina basin, Greece) case studies. In: *Proceedings of 7th International Conference on Greenhouse Gas Control Technologies* (ed. E.S. Rubin, D.W. Keith and C.F. Giboy), Volume 1, Peer-Reviewed Papers and Plenary Presentations. Cheltenham: IEA Greenhouse Gas Program.
Gherardi, F., Xu, T., and Pruess, K. (2007). Numerical modeling of self-limiting and self-enhancing caprock alteration induced by CO_2 storage in a depleted gas reservoir. *Chemical Geology* 244 (1): 103–129.
Helgeson, H.C., Kirkham, D.H., and Flowers, D.C. (1981). Theoretical prediction of the thermodynamic behavior of aqueous electrolytes at high pressures and temperatures: IV. Calculation of activity coefficients, osmotic coefficients, and apparent molal and standard and relative partial molal properties to 600 °C and 5 kb. *American Journal of Science* 281: 1249–1516.
Hu, W., Cai, C., Wu, Z., et al. (1998). Structural style and its relation to hydrocarbon exploration in the Songliao basin, northeast China. *Marine and Petroleum Geology* 15 (1): 41–55.
Huang, H., Yang, J., Yang, Y., et al. (2004). Geochemistry of natural gases in deep strata of the Songliao Basin, NE China. *International Journal of Coal Geology* 58 (4): 231–244.
IPCC. (2005). *Carbon Dioxide Capture and Storage*. Cambridge: Cambridge University Press.
Kaieda, H., Ueda, A., Kubota, K., et al. (2009). Field experiments for studying on CO_2 sequestration in solid minerals at the Ogachihdr geothermal site, Japan. *Proceedings, Thirty-Fourth Workshop on Geothermal Reservoir Engineering*, Stanford University, California (February 9–11, 2009).
Kovscek, A.R. (2002). Screening criteria for CO_2 storage in oil reservoirs. *Petroleum Science and Technology* 20: 841–866.
Lasaga, A.C. (1984). Chemical kinetics of water-rock interactions. *Journal of Geophysical Research* 89: 4009–4025.
Lasaga, A.C., Soler, J.M., Ganor, J., et al. (1994). Chemical weathering rate laws and global geochemical cycles. *Geochimica et Cosmochimica Acta* 58: 2361–2386.
LBNL. (2004). *GEO-SEQ Best Practices Manual – Geologic Carbon Dioxide Sequestration: Site Evaluation to Implementation*. Ernest Orlando Lawrence Berkeley National Laboratory.
Li, Q., Li, X., Du, L., et al. (2012). Potential sites and early opportunities of acid gas re-injection in China. In: *Sour Gas and Related Technologies* (ed. Y. Wu, J.J. Carroll, and W. Zhu), 131–140. New York: Wiley Scrivener.
Lindeberg, E. and Bergmo, P. (2003). The long-term fate of CO_2 injected into an aquifer. In: *Proceeding of the 6th International Greenhouse Gas Control Technologies* (ed. J. Gale and Y. Kaya), Volume I, 489–494. Elsevier Science Ltd.
Moore, J., Adams, M., Allis, R., et al. (2005). Mineralogical and geochemical consequences of the long-term presence of CO_2 in natural reservoirs: an example from the Springerville-St. Johns Field, Arizona, and New Mexico, USA. *Chemical Geology* 217: 365–385.
Palandri, J., and Kharaka, Y.K. (2004). A compilation of rate parameters of water-mineral interaction kinetics for application to geochemical modeling. US Geological Survey Open File Report 2004-1068, 64.

Parkhurst, D.L., Thorstenson, D.C., and Plummer, L.N. (1980). PHREEQE: A computer program for geochemical calculations. *US Geological Survey Water Resources Invest.* 80–96, 174.

Petroleum Technology Research Center. (2011). *Weyburn-Midale CO2 Project*. Available at http://www.ptrc.ca/weyburn_overview.php

Portier, S. and Vuataz, F.D. (2010). Developing the ability to model acid–rock interactions and mineral dissolution during the RMA stimulation test performed at the Soultz-sous-Forêts EGS site, France. *Comptes Rendus Geosciences* 342 (7–8): 668–675.

Pruess, K. (2006). Enhanced geothermal systems (EGS) using CO_2 as working fluid – A novel approach for generating renewable energy with simultaneous sequestration of carbon. *Geothermics* 35 (4): 351–367.

Qu, X., Liu, L., Wang, Z., et al. (2006). Chemical features and origin of dawsonite-bearing stratum water in Wuerxun sag. *Journal of Daqing Petroleum Institute* 30 (5): 7–10.

Reed, M.H. (1982). Calculation of simultaneous chemical equilibria and reactions processes in system involving minerals, gases and aqueous phase. *Geochimica et Cosmochimica Acta* 46: 513–528.

Spycher, N. and Pruess, K. (2005). CO_2–H_2O mixtures in the geological sequestration of CO_2: II. Partitioning in chloride brines at 12–100 °C and up to 600 bar. *Geochimica et Cosmochimica Acta* 69: 3309–3320.

Steefel, C.I. and Lasaga, A.C. (1994). A coupled model for transport of multiple chemical species and kinetic precipitation/dissolution reactions with applications to reactive flow in single phase hydrothermal system. *American Journal of Science* 294: 529–592.

Steefel, C.I. and MacQuarrie, K.T.B. (1996). Approaches to modeling of reactive transport in porous media. In: Reactive Transport in Porous Media (ed. P.C. Lichtner, C.I. Steefel, and E.H. Oelkers). Reviews in Mineralogy, Mineral Society of America 34, 83–19.

Tian, H. (2014). Impacts of CO_2-brine-rock interaction on the caprock sealing efficiency: A case study of Shiqianfeng Formation mudstone caprock in Ordos Basin. PhD Dissertation, Jilin University.

Tian, H., Pan, F., Xu, T., et al. (2014). Impacts of hydrological heterogeneities on the caprock mineral alteration and containment of CO_2 geological storage sites. *International Journal of Greenhouse Gas Control* 24: 30–42.

Tian, H., Xu, T., Li, Y., et al. (2015). Evolution of sealing efficiency for CO_2 geological storage due to mineral alteration within a hydrogeologically heterogeneous caprock. *Applied Geochemistry* 63: 380–397.

Van Genuchten, M.T. (1980). A closed-form equation for predicting the hydraulic conductivity of unsaturated soils. *Soil Science Society of America Journal* 44 (5): 892–898.

Vinsome, P.K.W. and Westerveld, J. (1980). A simple method for predicting cap and base rock heat losses in thermal reservoir simulators. *Journal of Canadian Petroleum Technology* 19 (3): 87–90.

Wan, Y. (2012). Migration and transformation of CO_2 in CO_2 geological sequestration process of Shiqianfeng saline aquifers in Ordos Basin. PhD dissertation, Jilin University.

Wan, Y., Xu, T., and Pruess, K. (2011). Impact of fluid–rock interactions on enhanced geothermal systems with CO_2 as heat transmission fluid. *Proceedings, Thirty-Sixth Workshop on Geothermal Reservoir Engineering,* Stanford University, Stanford, CA, January 31 – February 2.

Wolery, T.J. (1992). *EQ3/6: Software package for geochemical modeling of aqueous systems: Package overview and installation guide (version 8.0).* Lawrence Livermore National Laboratory Report UCRL-MA-110662 PT I, Livermore, California.
Wolf, G.H., Chizmeshya, A.V.G., Diefenbacher, J., and Mckelvy, M.J. (2004). In situ observation of CO_2 sequestration reactions using a novel microreaction system. *Environmental Science Technology* 38: 932–936.
Xu, T. and Pruess, K. (2001). Modeling multiphase fluid flow and reactive geochemical transport in variably saturated fractured rocks: 1. *Methodology. American Journal of Science* 301: 16–33.
Xu, T., Apps, J., and Pruess, K. (2004a). Numerical simulation of CO_2 disposal by mineral trapping in deep aquifers. *Applied Geochemistry* 19: 917–936.
Xu, T., Ontoy, Y., Molling, P., et al. (2004b). Reactive transport modeling of injection well scaling and acidizing at Tiwi field, Philippines. *Geothermics* 33 (4): 477–491.
Xu, T., Sonnenthal, E., Spycher, N., and Pruess, K. (2006). TOUGHREACT – A simulation program for non-isothermal multiphase reactive geochemical transport in variably saturated geologic media: Applications to geothermal injectivity and CO_2 geological sequestration. *Computers & Geosciences* 32 (2): 145–165.
Xu, T., Rose, P., Fayer, S., and Pruess, K. (2009). On modeling of chemical stimulation of an enhanced geothermal system using a high pH solution with chelating agent. *Geofluids* 9 (2): 167–177.
Xu, T., Kaieda, H., Ueda, A., et al. (2010). Fluid-rock interactions in enhanced geothermal systems with CO_2 as working fluid: modeling of geochemical changes induced by CO_2 injection into the Ogachi (Japan) EGS Site. *Transactions of the Geothermal Resources Council* 34 (1): 459–464.
Xu, T., Zheng, L., and Tian, H. (2011). Reactive transport modeling for CO_2 geological sequestration. *Journal of Petroleum Science and Engineering* 78: 765–777.
Xu, T., Feng, G., and Shi, Y. (2014). On fluid–rock chemical interaction in CO_2-based geothermal systems. *Journal of Geochemical Exploration* 144: 179–193.
Yeh, G.T. and Tripathi, V.S. (1991). A model for simulating transport of reactive multispecies components: model development and demonstration. *Water Resources Research* 27: 3075–3094.
Zhang, L. and Zhou, G. (2010). Improvement and application of the methods of gas reservoir cap sealing ability. *Acta Sedimentologica Sinica* 28 (2): 388–394.
Zhang, W., Li, Y.L., Xu, T.F., et al. (2009). Long-term variations of CO_2 trapped in different mechanisms in deep saline formations: A case study of the Songliao Basin, China. *International Journal of Greenhouse Gas Control* 3: 161–180.
Zhao, G.Q. (2005). Study of Petrological Characteristics and Thermodynamic Mechanism of Secondary Pores in Deep Reservoirs, Songliao Basin. Beijing: China University of Geosciences.

2

Modeling Reactive Transport in CO_2 Geological Storage: Applications at the Site Scale and Near-Well Effects

Pascal Audigane[1], Irina Gaus[2] and Fabrizio Gherardi[3]

[1] *BRGM (Bureau de Recherches Géologiques et Minières), Water Environment and Ecotechnologies Division (D3E), D3E/GDR Groundwater Resources Unit, 3 Av. Claude Guillemin, BP 36009, 45060, Orleans, Cedex 2, France*
[2] *NAGRA, Hardstrasse 73, 5430, Wettingen, Switzerland*
[3] *Istituto di Geoscienze e Georisorse (IGG) – Consiglio Nazionale delle Ricerche (CNR), Area della Ricerca CNR, Via G. Moruzzi 1, 56124, Pisa, Italy*

2.1 Introduction

The development of subsurface reactive transport modeling enables us to understand the behaviour of hydrogeological and hydrogeochemical systems under natural conditions or during various industrial operations (hydrocarbon extraction, gas storage, geothermal wells, etc.). As concerns CO_2 geological storage, a considerable amount of modeling has been done since the concept was first developed in the 1990s. CO_2 storage is considered to be one of the possible means of reducing greenhouse gas emissions and thus combatting climate change, much of which is caused by the atmospheric emission of carbon dioxide by industrial activity, increasingly fueled by coal and hydrocarbons. After several years of research, pilot sites were created in North America, Europe, Australia, Asia, and Africa. Of the many lines of research, the role of chemical reactivity on the behaviour, efficiency, and safety of a storage site has raised many questions. Indeed, carbon dioxide, once it has dissolved in water, produces carbonic acid, which might dissolve some of the minerals making up the reservoir rock into which the CO_2 has been injected, the caprock that blocks all vertical buoyancy of the CO_2 gas (supercritical), and the cement used to plug the injection wells or present in the storage complex. The EU's CCS Directive (2009) mentions the need to model the reactive transport of CO_2 in order to understand, when characterizing the dynamic behaviour of storage sites, the reactive processes occurring notably with the minerals in the storage reservoir and the surrounding formations, the evolution with time of the chemical composition of the formation fluids, and any precipitation of secondary phases, all of these for timescales ranging from decades to millennia (EC Directive, 2009).

Injection and storage conditions must allow the maintenance of the CO_2 as a dense (or supercritical) gas phase in order to maximize the storage capacity. Indeed, beyond 31 °C

Reactive Transport Modeling: Applications in Subsurface Energy and Environmental Problems, First Edition. Edited by Yitian Xiao, Fiona Whitaker, Tianfu Xu and Carl Steefel.

and 75 bars, the thermodynamic properties of carbon dioxide become 'supercritical' with a density that is similar to that of a liquid (between 600 and 800 kg/m^3) and a viscosity that is similar to that of a gas (0.5 cp). As such, 1000 m^3 of CO_2 at the surface occupies only 3 m^3 deep underground. These pressure and temperature conditions are reached at a depth of about 800 m, given a natural geothermal and hydrostatic gradient with no thermal or hydrogeological perturbation. Once in contact with the water, the CO_2 dissolves to produce a weak bicarbonic acid (pH around 3), and acid–base reactions conducive to the dissolution and precipitation of minerals can occur over the short and long terms. Dissolution has been identified as being one of the more effective and rapid trapping mechanisms, whereas the efficiency – or even the occurrence – of so-called 'mineral' trapping, which corresponds to the formation of secondary carbonate phases containing the carbon coming from the injected CO_2, is still in doubt (Hellevang et al., 2005).

In addition to mineral dissolution and precipitation, other reactive processes have been identified. These include aqueous phase reactions, multisite diffuse double- or triple-layer surface complexation reactions, multisite ion exchange reactions, gas–aqueous phase exchange reactions, and micro-bacterial activity (Steefel et al., 2014a). These reactive processes interact with transport processes in porous media and can directly affect the advective, dispersive and diffusive phenomena that govern fluid flow. Coupling these processes is still a numerical challenge, and Steefel et al. (2014a) describe the principal equations governing subsurface reactive fluid flow in a continuous medium and the ability of existing research codes to model these phenomena.

Gaus et al. (2008) discussed the importance of studying the geochemical impacts of CO_2 storage in sedimentary formations and notably over long periods (>1000 years), because these are indeed governed by reaction kinetics. Among the major processes identified, there are four types of *geological trapping*: structural, solubility, residual and mineral. Recognizing that all of these are very strongly coupled with a temporal sequential process should enable an increase in the security of the storage site (Figure 2.1) (IPCC, 2005).

Gaus et al. (2008) also emphasized the importance of modeling geochemical impacts on *reservoir injectivity*, with all of the near-well phenomena of drying and precipitation that might modify, in some cases drastically, the system's permeability, and then its injectivity (André et al., 2007). It is also considered crucial that *well integrity* be assessed through modeling studies in order to help risk-based approaches at defining the security of the site (also for long timescales). Casing corrosion and the properties of cement–clay–reservoir seals in acid environments like those of CO_2 storage are still poorly understood and require additional research to be conclusive (Manceau et al., 2013, 2015). A final important step is the modeling of laboratory experiments (temperature-controlled batch or percolation experiments) that make it possible to determine the geochemical reactions associated with acidification caused by CO_2 on reservoir rock or caprock samples, and at the various pressures and temperatures that might be reached under storage conditions. In many cases, these experiments make it possible to study only short-term phenomena (several weeks or months), which limits the interpretation of the results and our knowledge of other, slower reaction pathways. Natural analogues, on the other hand, provide a long-term view that is difficult to interpret because the history of the deposition of CO_2 in these natural reservoirs is sometimes complex and poorly understood (Gaus et al., 2008). Nevertheless, by modeling experiments, we can calibrate models in closed or open systems and make predictions for longer timescales.

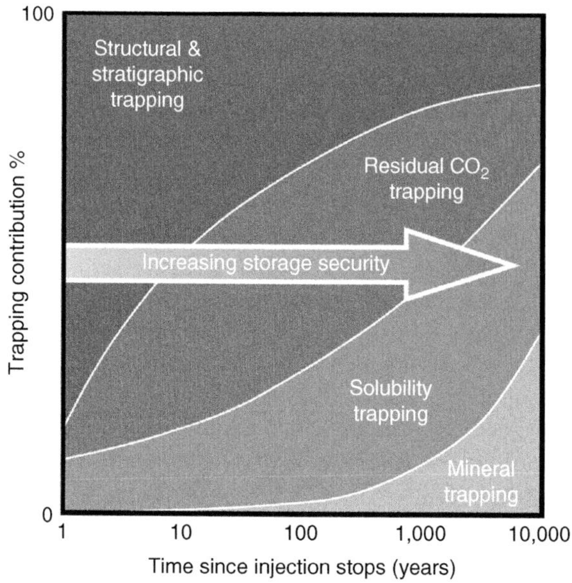

Figure 2.1 CO_2 trapping mechanisms vs time (IPCC, 2005).

Geochemical and transport processes have been extensively modeled in the last 15 years for CO_2 storage, and the most recent studies describe many applications. Site studies (pilot or large-scale), like the one in Wabamun Lake, Alberta, Canada, use radial multilayer 2D models to determine long-term gravity effects and estimate the different CO_2 trapping processes (Dalkhaa et al., 2013). The pilot site in Ketzin, Germany, created to evaluate various monitoring, well-sealing and modeling techniques, has furnished a large amount of data to calibrate reservoir flow models. Klein et al. (2013) developed a coupling method that enabled them to exploit 3D models using porosity and gas saturation as coupling parameters between the geochemical model and the reservoir simulations. This made it possible to decrease calculation times and estimate the mineral trapping at the reservoir scale by scaling up.

Trémosa et al. (2014) studied the limits of reactive transport models for simulating reactive flow in highly saline environments (>30 g/l, i.e. seawater). Basing their study on fluid and rock composition data from existing sites (Ketzin, In Salah, and Snøvith), they compared mineral solubility calculations done using various thermodynamic databases with experimental data, and observed notable differences in behaviour as a function of the increase in salinity of the aqueous phase. They concluded that it is difficult to estimate the mineral trapping of sandstone formations for high ionic strengths.

Jacquemet et al. (2009) were among the first to determine the ability of calculation codes to simulate the presence of impurities or co-injected gases associated with the technologies used for CO_2 capture that might have a greater impact on the subsurface due to reactivity, notably SO_x and NO_x. Progress has been made since then, and several models are now able to handle the composition of the gas phase with appropriate equations of state (see Steefel et al., 2014a and references therein; Nghiem et al., 2004; Corvisier et al., 2013; Tambach et al., 2013). These impacts have also been determined, notably for H_2S and SO_2 (Xu et al., 2007; Ziabakhsh-Ganji and Kooi, 2014).

In terms of sensitivity studies, Zhang et al. (2011) shed new light on the long-term effects of the convection of dissolved CO_2-enriched brines. They identified the role of mineral dissolution kinetics on the positioning of convection cells and their circulation rates. Balashov et al. (2013) studied the role of mineral reaction kinetics in the mineral trapping of CO_2 in sandstone. They described mineral dissolution and precipitation sequences over time and their variability as a function of the kinetics of the minerals in question. Tian et al. (2014) studied the impact of reservoir property heterogeneities on the gravity migration of CO_2 in a clay caprock, and compared their results with those of similar homogeneous systems. They showed how the change in porosity associated with the dissolution of clay by the acid plume modifies the gas penetration distances in the rock and therefore impacts site safety.

Another crucial point concerning CO_2 storage site security is the potential impact on groundwater used for drinking water supplies (e.g. review article by Lions et al., 2014). Studies involving modeling of a French geological context (Paris Basin) are described in greater detail in the following sections of this article. One of these studies describes a novel approach that consists of using isotopic tools to track changes in major- and trace-element concentrations induced by fluid–rock interactions with the dissolved CO_2 (Humez et al., 2013). The multi-isotopic approach, calibrated by batch experiments, uses carbon and strontium isotopes ($\delta^{13}C_{DIC}$ and $^{87}Sr/^{86}Sr$) to detect all of the changes in concentrations in a glauconite-rich sandstone system. Another approach involves studying natural analogues to analyse possible leakage scenarios. The most recent work of Keating et al. (2013) concerns water samples from a shallow aquifer (Chimayo, New Mexico, USA) with a high CO_2 content. The study aimed to determine the ability of models to simulate the mobilization of a trace element (uranium) contained in a mineral (calcite), which is dissolved by the acid brine. After calibrating the geochemical models using batch experiments studying uranium, arsenic, and lead (Viswanathan et al., 2012), 3D reactive transport models enabled them to reproduce the trends observed on the site, despite a limited knowledge of the hydrodynamic system.

Among the many possible pathways for CO_2 leakage from a CO_2 storage site, boreholes are a principal target of preliminary studies because they are the most direct pathways from the reservoir to the surface. Many studies have been done on the integrity of both drilling cement and barrier interfaces: casing-cement-rock (Carey et al., 2007; Fabbri et al., 2009, 2012; Duguid and Sherer, 2010). One of the essential questions is how to evaluate long-term well tightness in order to ensure permanent and safe storage. Here again, reactive transport modeling plays a crucial role since the slow dissolution kinetics of some mineral phases can influence the changes in permeability of the system. Calculation codes are constantly being improved in order to include more processes such as diffusive ion transport (Shen et al., 2013), and models are being calibrated using experimental results (Luquot and Gouze, 2009; Huet et al., 2010; Hermanrud et al., 2010). Zhang et al. (2013) have recently attempted to predict the carbonation front from experiments done on cements with Pozzolan additives with CO_2 and H_2S.

Other areas of investigation for modeling applications of CO_2 storage in more 'reactive' geological systems have also been studied in recent years. The mineralogy of basaltic rocks is favourable to mineral trapping of CO_2, and the potential of these formations is under investigation at two pilot sites (Gislason and Oelkers, 2014). One of these, the Big Sky Carbon Sequestration Partnership site, is in Wallula, Washington (USA), and the other is in Iceland (Project CarbFix). Aradóttir et al. (2012) estimated the potential trapping capacity of the basalt formation in project CarbFix to be between 1200 and 22,000 tons per year, and Aradóttir et al. (2014), using the gas emitted by geothermal plants in Iceland, have predicted rapid CO_2 (and H_2S) mineralization in the form of calcite or magnesite. Paukert et al. (2012) used peridotite rocks from the mantle as a reactive potential based on data from a natural analogue in Oman. This involves using the high olivine (and pyroxene) concentration as the reactive mineral with water and CO_2 to form serpentine, magnesite, calcite, and quartz, as observed naturally with meteoric water and carbonate or travertine veins (Kelemen and Matter, 2008).

We describe here, in greater detail, simulations of reactive transport done to identify the short- and long-term storage capacities of sedimentary formations and the geochemical processes associated with CO_2 leakage during storage. We studied, in particular, near-well phenomena and any impacts on groundwater. We conclude with a discussion of the current capabilities and limitations of reactive transport models for simulating the geochemical processes associated with CO_2 geological storage.

2.2 Short- and Long-term Predictive Simulations of Trapping Mechanisms

The current theory on the fate of CO_2, initially a gas, injected into deep porous sedimentary formations (i.e. > 800 m to ensure a sufficiently high density and supercritical conditions) is a four-step sequence (Figure 2.1). The CO_2 gas percolates in the porous rock system until it reaches a low permeable geological structure (clay, sealed fault, unconformity, facies change, etc.). Some of the percolating CO_2 gas is trapped by capillary forces in the porous network. Some of the gas dissolves in the available aqueous phase (brine in saline aquifers) and forms a weakly carbonic acid. Lastly, the acidification of the aqueous system results in a dissolution of sequences of the minerals making up the rock (sandstone or carbonate) that can lead to the precipitation of secondary carbonated phases and ultimately turn the CO_2 into an immobile and permanently secure solid form. The classic representation of these four types of trapping is illustrated by the IPCC (2005) sketch in Figure 2.1. This sketch aims to illustrate how, over a long timescale (millennia), the gas phase decreases while CO_2 mineralization increases. However, when a realistic data-set and a reactive transport model capable of handling multiphase flow are used to simulate trapping, the fate of the CO_2 is somewhat different.

An initial qualitative response was proposed by the Australian CO2CRC team (CO2CRC, 2011) based on various geological concepts (Figure 2.2). Indeed, it is easy to imagine that, in a thin silica sand anticline reservoir in which the CO_2 does not react with the quartz sand, there will be little mineral trapping and the injected CO_2 will move to the caprock and spread out somewhat (therefore, with minor residual trapping) and some will dissolve. In this system, structural trapping predominates. On the other hand, if the CO_2 is injected into an extensive, highly permeable, saline aquifer formation, it will spread out and dissolve and, once again, there will be minor residual mineral trapping. In this system, solubility (dissolution) trapping predominates. For a very heterogeneous sedimentary reservoir with low vertical permeability and a marked dip, the CO_2 will spread horizontally, resulting in major residual trapping together with major dissolution and, eventually, the disappearance of the gas phase. In this system, residual trapping predominates. There are certainly many other geological and hydrodynamic

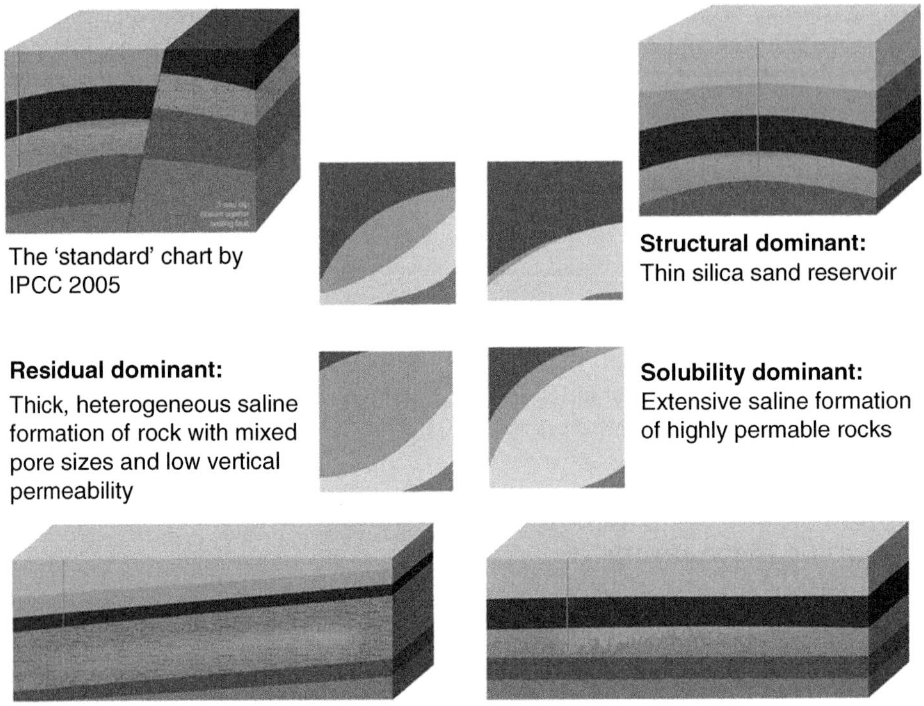

Figure 2.2 Various types of trapping depending on the type of geological reservoir considered (CO2CRC, 2011).

contexts that could illustrate cases in which the four main trapping mechanisms are much different from those shown in Figure 2.1.

Case studies of sites that include modeling have enabled researchers to estimate trapping more quantitatively. Indeed, only reactive transport models can provide long-term predictions of the processes involved and notably slow mineral dissolution kinetics: Sleipner in the North Sea (Johnson et al., 2001; Audigane et al., 2007), Ketzin in Germany (Kempka et al., 2010), Frio, Texas (Xu et al., 2010), and Mount Simon, Illinois (Liu et al., 2011) in the United States, Wabamun Lake Area, Alberta in Canada (Dalkhaa et al., 2013), Nagaoka in Japan (Mito et al., 2008, 2013; Sato et al., 2011), and many others.

In the literature, we find relatively few comprehensive analyses that are able to identify the magnitude of trapping over time as a function of geological context. Xu et al. (2004) gave an initial estimation of the mineral trapping potential of minerals in the storage reservoir rocks whose weathering might lead to the precipitation of carbonates (Table 2.1). The calculations overestimate the results by considering the total dissolution of the primary mineral. Moreover, the results vary as a function of pressure and temperature. However, we can easily identify the high potential of minerals making up magmatic and basaltic rocks compared with those in sedimentary formations. A more complete list of minerals can be found in Xu et al. (2000).

Frykman et al. (2011; Frykman, 2014) used the results of simulations done by other authors, some working on the same site (Sleipner, North Sea), to draw diagrams of trapping as a function of time (Figure 2.3). These differ greatly from the trends proposed by

Table 2.1 Mineral trapping potential of some minerals making up rocks under consideration for CO_2 geological storage (Xu et al., 2004).

Mineral name	Mineral formula	Potential CO_2 fixed, kg/m^3 mineral
Mica group-biotite	$K_2(Mg,Fe(II))_{6-4}(Fe(III),Al)_{0-2}[Si_{6-5}Al_{2-3}O_{20}](OH)_4$	2671.0
Olivine (forsterite-fayalite)	$Mg_2SiO_4–Fe_2SiO_4$	2014.7–1896.3
Pyroxene group-enstatite	$(Mg,Fe)_2Si_2O_6$	1404.2
Augite	$(Ca,Mg,Fe(II),Al)_2(Si,Al)_2O_6$	1306.3
Serpentine	$Mg_6Si_4O_{10}(OH)_8$	1232.7
Calcium amphiboles – tremolite	$Ca_2Na_{0-1}(Mg,Fe(II))_{3-5}(Al,Fe(III))_{2-0}[Si_6–8Al_{2-0}O_{22}]$ $(O,OH)_2$	1119.3
Amphibole group – anthophyllite–cummingtonite	$(Mg,Fe(II),Fe(III))_{5-7}Al_{0-2}[Si_{6-8}Al_{2-0}O_{22}](OH)_2$	1169.5–1041.8
Common hornblende	$Ca_2Na_{0-1}(Mg,Fe(II))_{3-5}(Al,Fe(III))_{2-0}[Si_{6-8}Al_{2-0}O_{22}]$ $(O,OH)_2$	1000.4
Chlorite group	$(Mg,Al,Fe(II))_{12}[(Si,Al)_8O_{20}](OH)_{16}$	923.4
Mica group – phlogopite	$K_2(Mg,Fe(II))_6[Si_6Al_2O_{20}](OH)_4$	881.8
Plagioclase (anorthite)	$Ca[Al_2Si_2O_8]$	436.4
Clay minerals – smectite	$(1/2Ca,Na)_{0.7}(Al,Mg,Fe)_4(Si,Al)_8O_{20}(OH)_4.nH_2O$	161.2
Clay minerals – illite	$K_{1-1.5}(Fe(III),Al,Fe(II),Mg)_4[Si_{7-6.5}Al_{1-1.5}](OH)_4$	78.42
Mica group – glauconite	$(K,Na,Ca)_{1.2-2.0}(Fe(III),Al,Fe(II),Mg)_4[Si_{7-7.6}Al_{0.4}O_{20}]$ $(OH)_4.nH_2O$	61.97
Plagioclase (anorthite)	$Ca[Al_2Si_2O_8]$	436.4

the IPCC reference diagram. Most of the simulations predict both more structural trapping (>60% of the mass of injected CO_2) and more stable trapping during the first millennia, and less mineral trapping, which is almost entirely absent. Dissolution seems to be the second type of major trapping, sometimes able to dissolve all of the free-phase supercritical CO_2 in 10,000 years. The role of mineral trapping is predicted to be less than expected, rarely exceeding 10% after 10,000 years. The level of uncertainty of these calculations is high and is related to many factors. We will return to this point in the discussion section below.

In the next section we present three case studies using classic simulation methods and the TOUGHREACT multiphase reactive transport model (Xu and Pruess, 2001), a member of the TOUGH2 (Pruess et al., 1999) family of codes. The first study concerns an unconsolidated aquifer formation at the Sleipner site in the North Sea. The second is a hypothetical case of injection in a carbonate formation in the Paris Basin, and the third involves enhanced recovery of a nearly depleted gas deposit in a sandstone formation in the North Sea, off the Netherlands.

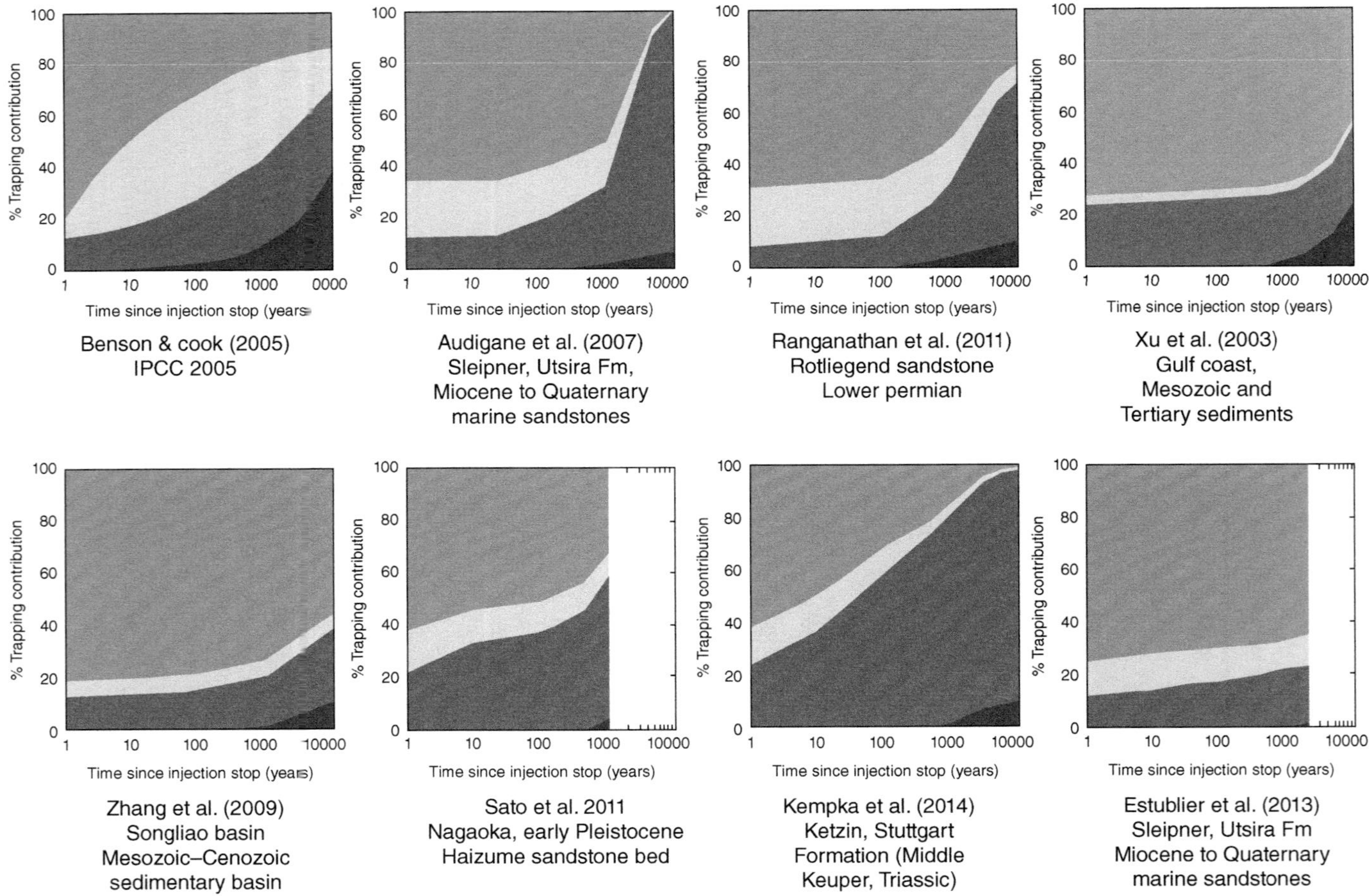

Figure 2.3 Trapping diagrams: structural (orange), residual (green), dissolution (blue), and mineral (purple) trapping of CO_2 as a function of time (Frykman et al., 2011; Frykman, 2014). *Source:* Reproduced with kind permission of Peter Frykman.

2.2.1 Sandy Aquifer: Predictions of Long-term Effects of Storage in Sleipner, North Sea, Norway

The Sleipner site was the first site used to demonstrate the possibility of injecting and storing CO_2 in a 1-km-deep sedimentary formation in the North Sea, and at commercial-scale injection rates of over 1 million tons per year (SACS, 2002). STATOIL, which operates the site, initially pumped natural gas from a deeper deposit, 3 km deep. A platform dedicated to CO_2 capture and separation from the gases produced was purpose-built for the site, and since 1996, the CO_2, once separated, has been compressed and reinjected into the Utsira unconsolidated silica formation that extends over more than 200 km to the north and several tens of kilometres east–west (Figure 2.4).

In compliance with recommendations of the European Directive, several geophysical surveillance methods (3D seismic, gravimetry, seabed imaging) have been used to monitor the site. The results revealed the movement of a CO_2 plume and were therefore used to calibrate the flow model. According to 3D seismic data, the CO_2 seems to move upward and reach the top of the formation after 3 years of injection, after which it accumulates under the shale layer identified by a horizon 160 m above the injection point (Figure 2.5). The seismic data also revealed stratified accumulations of CO_2 that seem to be related to the presence of more or less permeable shale lenses.

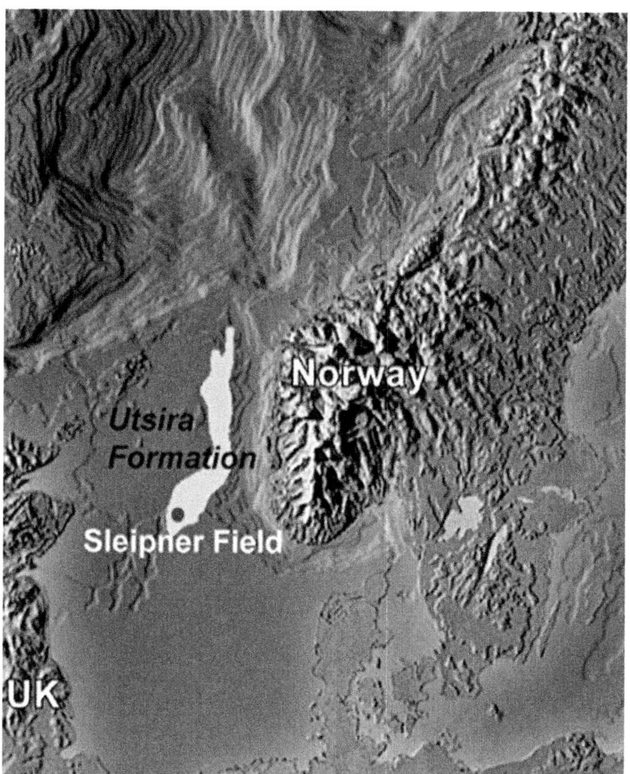

Figure 2.4 Location of the Sleipner site and the sandy Utsira formation used for CO_2 storage (Arts et al., 2008).

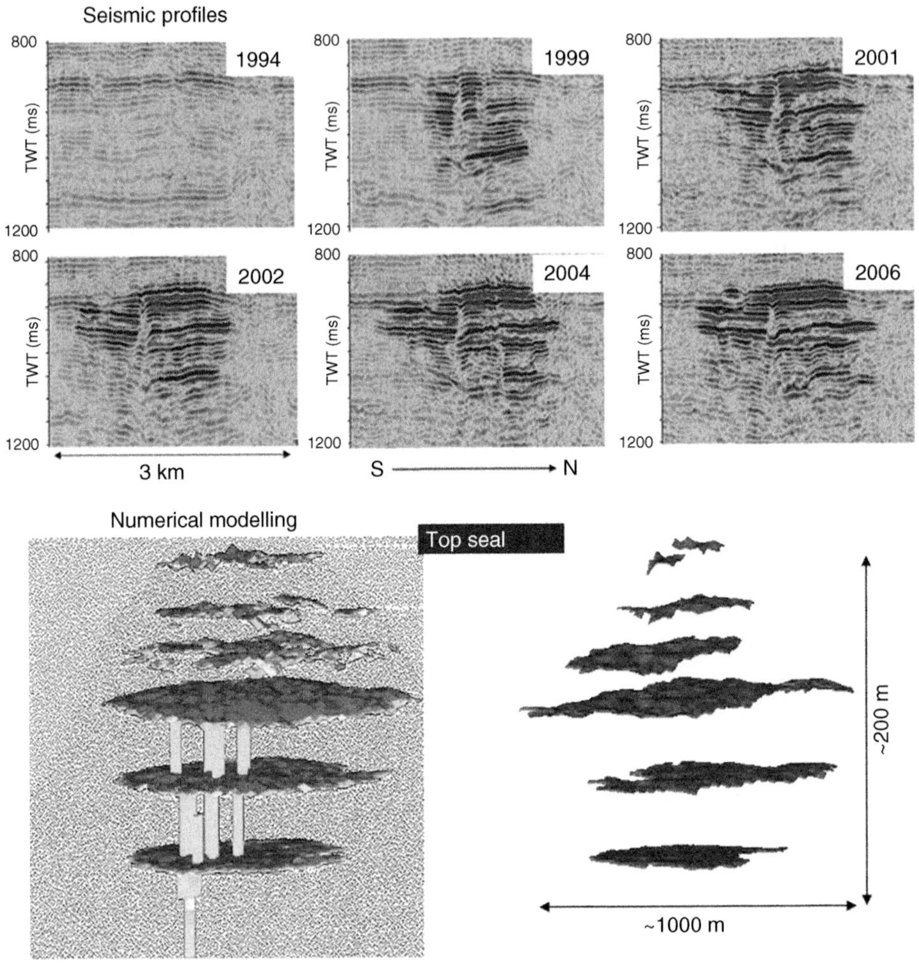

Figure 2.5 Seismic profiles before and during injection (Arts et al., 2008) and example of modeling of the related CO_2 plume illustrating CO_2 accumulation beneath semi-permeable shale lenses (Lindeberg et al., 2000). *Source:* Reproduced from Lindeberg et al. (2000), with permission from CSIRO Publishing. (See colour plate section for the colour representation of this figure.)

Audigane et al. (2007) used this information to build a 2D reactive transport model with cylindrical multilayer symmetry (Figure 2.6). This model makes it possible to include semi-permeable shale lenses with a mineralogy different from that of the Utsira silica sand (Table 2.2).

For the clay, some minerals, such as pyrite, were disregarded and others were added as secondary phases, such as dolomite, magnesite, and notably dawsonite, which plays a significant role here (see below). Preliminary batch calculations (0D, i.e. equivalent to one beaker) made it possible to identify the important phases of the reactivity of dissolved CO_2 in the sand and clay under consideration. For the clays, in contact with the CO_2, the pH decreases rapidly, causing dissolution of the carbonates, which stabilizes

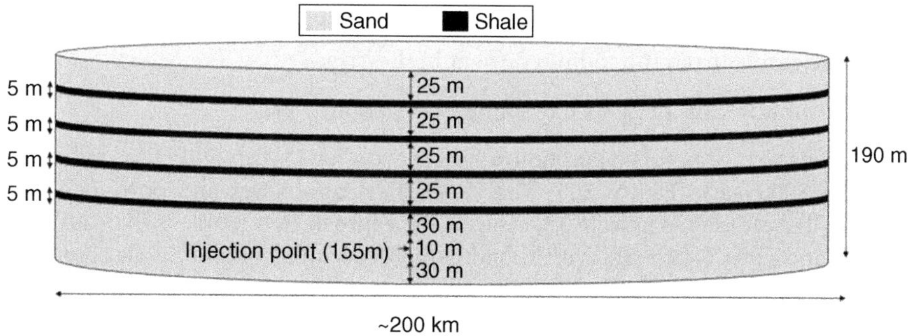

Figure 2.6 The cylindrical multilayer 2D model used to simulate the CO_2 plume in Sleipner (Audigane et al., 2007) as a function of the 3D seismicity observed between 1996 and 1999. The caprock is not represented in the model.

Table 2.2 Mineral composition of the shale and sand of the Utsira formation used in the models by Audigane et al. (2007).

Minerals	Sand (vol. %)	Clay (vol. %)
Chalcedony	77.0	33.4
Albite	3.0	13.2
Calcite	6.7	1.0
Chlorite	1.3	4.4
Muscovite	5.2	25.1
K-feldspar	6.9	2.3
Kaolinite		19.5
Siderite		1.1

Source: Audigane et al. (2007).

the pH at 5.13. For long-term reactions, the alumino-silicate minerals dominate the geochemical interactions. Three major reactions were identified:

1) Weathering of chlorite, which consumes the calcite and leads to the trapping of CO_2:

$$\text{chlorite} + \text{calcite} \rightarrow \text{siderite} + \text{dolomite} + \text{kaolinite} + \text{chalcedony} + H_2O$$

2) Weathering of albite, leading to the formation of dawsonite and chalcedony:

$$\text{albite} + CO_2 + H_2O \rightarrow \text{dawsonite} + \text{chalcedony}$$

3) The last reaction is indirectly related to the two previous reactions, which produce a large quantity of SiO_2, catalysing muscovite dissolution and causing a large transfer of moles in the system and a change in porosity.

$$\text{muscovite} + \text{chalcedony} + H_2O \rightarrow \text{K-feldspar} + \text{kaolinite}$$

A slightly different reactivity was observed for the sand. The two predominant reactions are chlorite and albite weathering, but less for the latter. The newly-formed dawsonite also results from the sodium present in the brine:

$$2Na^+ + albite + CO_2 + H_2O \rightarrow dawsonite + chalcedony + 2H^+$$

2D simulations were done for an injection of 1 Mt per year for 25 years and a post-injection duration of 10,000 years (Figure 2.7). The system's permeabilities and porosity were calibrated so that the supercritical CO_2 rises to the caprock in 3 years, as indicated by seismic data, and extends laterally over around 1 km. An accumulation of CO_2 is observed under the shale layers, accompanied by an acidification of the aqueous phase enriched in dissolved CO_2, and convective phenomena over the long term (500 to 10,000 years) related to the difference in density between the initial brine (pH 7) and the acidified brine (pH 5.13), slightly weighted down by the dissolved CO_2, and therefore moving to the bottom of the formation. This mixing leads, after several millennia, to the total disappearance of the supercritical CO_2 and a return of the brine to a neutral state. As with the batch simulations, albite, muscovite and chlorite weathering, together with the dissolution of the calcite, was observed. Fluid convection also causes exchanges of reactivity between the sand and clay formations that can lead to processes such as the precipitation of calcite in the clays, which are not identified in batch experiments. Indeed, the calcium coming from the dissolution of the calcite in the sandy medium is transported to the clays, where the pH is slightly higher, and triggers a new precipitation of calcite.

Considering dawsonite as a secondary phase in the simulated system increases the mineral trapping of CO_2. Dawsonite precipitation in deep saline aquifers is currently very much a subject of debate because numerical simulations have suggested large-scale formation, whereas natural occurrences in present-day or previously CO_2-charged reservoirs are scarce (Hellevang et al., 2005, 2010, 2011; Bénézeth et al., 2007). The exact conditions that suggest dawsonite can form under CO_2 storage conditions (CO_2 pressure, alkalinity, temperature, ionic strength) are highly uncertain. In this study, although dawsonite precipitation is permitted, the mineral trapping of CO_2 turns out to be rather limited (<3% in 10,000 years) (Figure 2.3). This can be explained by the predominance of a sand facies that is weakly reactive with CO_2. The geological formation here is siliceous and unconsolidated, so highly permeable. This therefore creates conditions favourable to dissolution trapping over the long term. Using a multilayer 2D model makes it possible to include fluid mixing processes and therefore the reactivity related to convective phenomena, which is impossible to estimate using batch or 1D models.

2.2.2 Near-well Effects in Saline Aquifers in Carbonate Formations: Carbonate Dissolution, Drying, and Salt Crystallization in the Dogger, Paris Basin

Carbonate formations are susceptible to being much more reactive to the injection of CO_2, which, after dissolving in the aquifer brine, acidifies the aqueous phase, fosters the dissolution of carbonate minerals (calcite, dolomite, and siderite), and releases a certain number of chemical elements such as calcium, magnesium, and iron. The rock's pore structure might, therefore, be altered, which would modify well injectivity and storage performance either positively, by increasing the system's porosity, or negatively, by clogging caused by the precipitation of secondary phases.

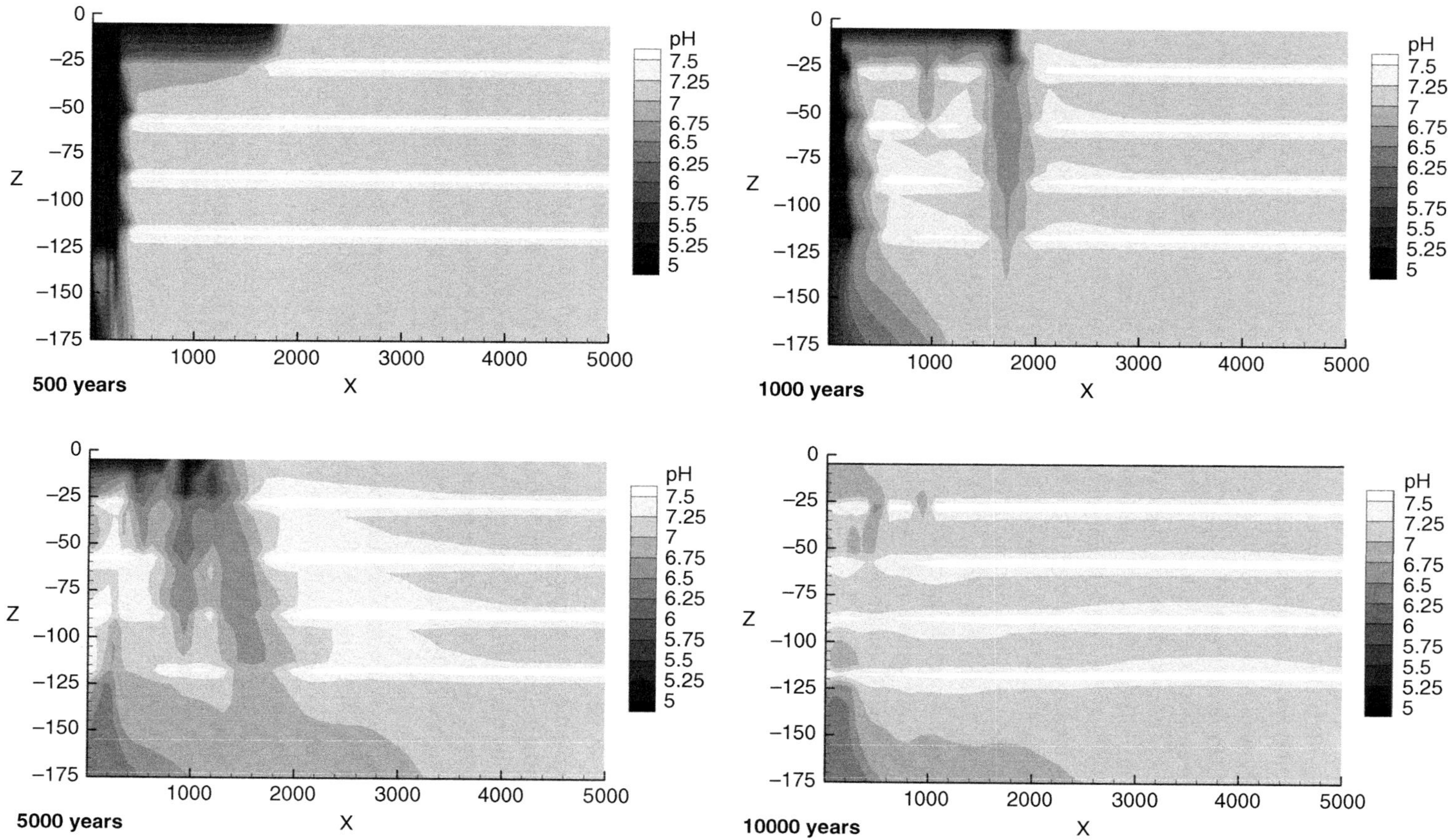

Figure 2.7 Simulations of the CO_2 plume in Sleipner in a cylindrical, multilayer 2D model (only half the symmetrical model is shown here, the CO_2 is injected at 155 m). Notice the difference in pH between the shale layers and the sand, which is related to the difference in mineralogy. The CO_2 plume, once dissolved, acidifies the brine, which tends to fall to the bottom of the formation, whereas the 'fresh' brine rises towards the plume to dissolve more supercritical CO_2. After 10,000 years, all of the supercritical CO_2 has dissolved (Audigane et al., 2007).

Injecting 'dry' CO_2 (no water vapour in the gas phase) in a porous water- or brine-saturated medium can lead to drying that only calculation codes able to handle at least two phases (supercritical CO_2 and water) can estimate. In addition to the thermodynamic and capillary processes occurring around the injection point, there are also geochemical phenomena like the precipitation of salts from the brine 'dried' by the CO_2.

André et al. (2007) estimated these near-well carbonate-dissolution and drying phenomena using the geological context of the Dogger carbonate formation in the Paris Basin (France). They demonstrated the importance of considering two distinct phases in the reactive transport calculations by comparing their results to a single-phase approach with CO_2-enriched water. Their model was a cylinder centred around an injection well with a single layer (1D) 100 m thick and 100 km wide. Both scenarios (single- and two-phase) considered an injection of 0.3 Mt of CO_2 for 10 years at 75 °C and 100 bars, based on existing data for the Paris Basin, which has been studied for its geothermal properties (Rojas et al., 1989; Lopez et al., 2010). The mineralogical composition of the Jurassic oolitic limestone is (by volume) 70% calcite, 10% dolomite, and 5% siderite, illite, albite and K-feldspar. The formation water is considered to have a moderate salinity of 5 g per kilogram of water. In order to compare the results in terms of reactivity of the single-phase and two-phase calculations, André et al. (2007) used the same CO_2 mass injection rate of 1 kg s^{-1} in both cases. In single-phase, the injected water was in equilibrium with CO_2 at 75 °C and for a pressure of 200 bars, equivalent to a CO_2 partial pressure of about 100 bars. After 10 years of injection, the results of the two simulations diverge. There is a massive increase in porosity (from 12% to 90%) for the single-phase scenario, but an increase to only 12.8% for the two-phase scenario (Figure 2.8).

These calculations show how a single-phase approach can overestimate the impacts associated with dry supercritical CO_2 storage. This large difference can be explained mainly by the availability of the aqueous phase for dissolving the CO_2. Indeed, in the single-phase scenario, dissolution is total because the injected CO_2 is already dissolved, whereas in the two-phase scenario, a large part of the CO_2 remains in the aqueous phase (Figure 2.9). Furthermore, in the single-phase scenario, the continuous injection of water in disequilibrium with the formation water creates a strong, increasing dissolution front (Lagneau et al., 2005). On the other hand, in the two-phase scenario, only the water at the front of the CO_2 plume (containing the dissolved CO_2) causes a dissolution reaction on the limited edge.

Another process that cannot be simulated with a single-phase scenario, but which is very important for well injectivity, is the drying caused by the injection of a dry gas into a water-saturated formation for many years. This was simulated by TOUGHREACT for 10 years of uninterrupted injection of 1 kg s^{-1} of pure dry CO_2 at 75 °C and 100 bars (Figure 2.9). The calculations predicted a dried zone 5 m wide around the injection well – a zone in which pH cannot be calculated. A second zone, called the acid zone, extending up to 650 m around the well, contains an aqueous phase enriched with dissolved CO_2 and a pH of 4.8 (initially 3.6, then buffered by carbonates) and a gas phase (supercritical CO_2 and water vapour). There is also a third zone that contains the water initially in the formation with an initial pH of around 7. Around the well, the injected CO_2 first pushes the so-called mobile water from the porous reservoir. The immobile (residual) water remains and is, in a second phase, vapourized by the dry CO_2, which results in the gas saturation of the system at values of 1. If the aqueous phase is salty, the

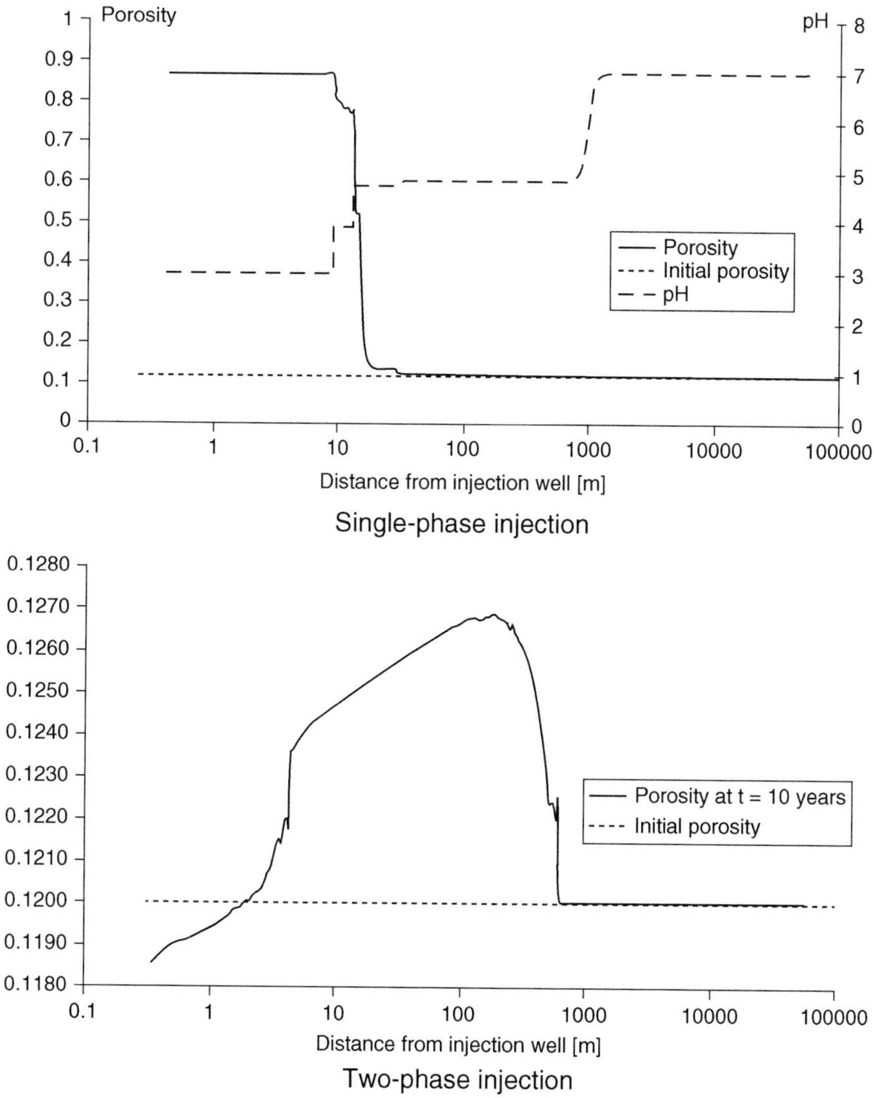

Figure 2.8 Comparison of an injection of 1 kg s^{-1} of CO_2 in single-phase (CO_2-enriched water) and two-phase injection (only supercritical CO_2) at 75 °C and 100 bars in the Jurassic limestone of the Dogger, Paris Basin (André et al., 2007).

brine precipitates salts while drying – in this case, dolomite, anhydrite, and, to a lesser extent, calcite. Each of these mineral phases leads to an overall change in porosity calculated using a mass and volume balance that depends on the quantities precipitated and dissolved through chemical reactions (Xu and Pruess, 2001).

It is important to point out here the limitations of current reactive transport models for simulating geochemical interactions for high-salinity brines. In order to correctly estimate mixtures of species in the aqueous phase under conditions that are very different from those of ideal solutions, activity coefficients are calculated using

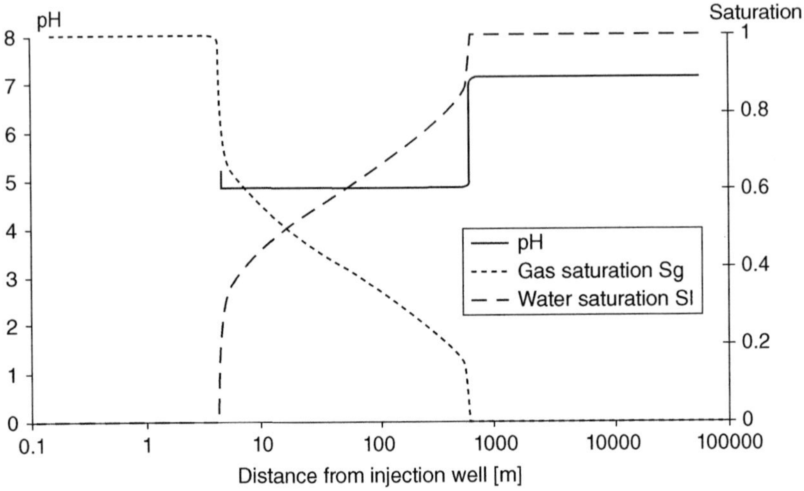

Figure 2.9 Two-phase injection of CO_2 in carbonates, with pH and near-well drying (André et al., 2007).

specially-adapted models. The Debye-Hückel model (Debye and Hückel, 1923), the extended Debye-Hückel model, or the Davies model are usually used and included in codes for ionic strengths of solutions below 0.5–0.7 M (Trémosa et al., 2014). For more concentrated solutions, the B-dot model (Helgeson et al., 1969), an extension of the Debye-Hückel model, can be used for solutions with high NaCl concentrations. However, the most suitable model for considering non-ideal solutions is still the Pitzer formalism (Pitzer, 1973, 1991), which includes the interaction of ions between doublets and triplets. However, determining the parameters of Pitzer equations is difficult and limits their range of application with regard to available databases. Trémosa et al. (2014) show that few species are considered (for instance Al, Si and C), and very often data are available at standard (25 °C) temperature conditions. Incoherencies can also be observed in the database, in particular for redox issues. As a consequence, species for which behaviour is highly dependent on the redox state, in particular Fe and S systems, cannot be considered in models when the Pitzer formalism is used.

To return to the case study in question, it therefore seems inappropriate to predict the precipitation of salts in a drying system near the borehole using an unsuitable formalism – since TOUGHREACT does not include Pitzer's model. As an example, André et al. (2007) compared, for two ionic strengths (0.1 M and 10 M), the saturation indices for calcite, dolomite, siderite, magnesite, chalcedony, and halite (which should have been identified as a secondary phase) using the SCALE 2000 geochemical calculation code (Azaroual et al., 2004) suitable for highly saline solutions – but for a limited number of mineral phases (i.e. carbonates, but not alumino-silicates). For indices calculated for a solution with a low ionic strength, the results coincide perfectly for all of the mineral phases except magnesite. On the other hand, for a high ionic strength of 10 M, the calculations show notable discrepancies for all of the mineral phases and notably for halite carbonates. We will come back to these limitations in the discussion section below.

2.2.3 Depleted Offshore Gas Field: Mixing with Methane K12B Field

Among the various possible solutions under consideration for CO_2 geological storage, depleted oil reservoirs are one of the most advanced due to the existence of suitable transport and injection infrastructures. We distinguish between enhanced recovery done in an oil reservoir and the gas reservoirs. The former is a three-phase, liquid (oil and immiscible water) and gas (CO_2 + …) system, whereas the latter is a two-phase, gas (CO_2 + CH_4 + …) and liquid (water) system. The simulation of reactive transport for three-phase fluid systems and several components is still in its infancy because it is extremely complicated and difficult. Nghiem et al. (2004) were among the first to develop a commercial tool (GEM-GHG) able to handle a compositional formulation for the equation of state of fluids with an implicit coupling of the fluid–rock interaction component. More recently, Tambach et al. (2015) have attempted to simulate reactive transport associated with a generic depleted gas site using existing reservoir conditions in the North Sea. Using TOUGH2/EOS7C (Oldenburg et al., 2004), TOUGHREACT, and PHREEQC (Parkhurst and Appelo, 1999), they have evaluated several processes associated with storage in a methane reservoir – CO_2 injectivity and CH_4 production, near-well dry-out, mineral trapping, dissolution and diffusion in the caprock, and weathering of borehole cement, all for the very long-term. Their use of several different models shows the current lack of a model that is able to handle a multi-compositional formulation of equation of state of fluids with the calculation of fluid–rock interactions (Tambach et al., 2013).

In a similar manner, Audigane et al. (2008) used TOUGH2/EOS7C and TOUGHREACT to evaluate geochemical reactions for a CO_2 geological storage site in the K12B depleted reservoir in the North Sea, the Netherlands. Operated by Gaz de France, this pilot site demonstrated the feasibility of storing CO_2 in a depleted natural gas deposit using a reservoir 3800 m deep. The fractured geology in this part of the North Sea breaks the K12B field up into several reservoirs that are separated by sealed faults and are not, therefore, hydraulically connected (Figure 2.10). Compartment 3 is equipped with an injector well, K12-B6, and two producing wells, K12-B1 and K12-B5, used for enhanced recovery of methane coupled with CO_2 storage. The authors first converted a 3D Petrel grid (©Sclumberger) of compartment 3 into a grid in the TOUGH2 and TOUGHREACT format using BRGM tools (Audigane et al., 2011). Two types of simulations were then run separately: (i) methane injection and production calculations using the EOS7C module developed for the CO_2-CH_4-H_2O system and (ii) reactive transport calculations related to CO_2 and disregarding methane (inert) while using the same injector- and producer-well flow rates (10 kg/s and 1 kg/s per well, respectively), an initial reservoir pressure of 47 bars, and a temperature of 130 °C. The porosity and permeability values were mean scaled using the fine-grid reference model. A mineralogy was identified for four facies: the Ten Boer, representing the caprock, the Upper and Lower Slochteren making up the sandstone part of the reservoir, and the Ameland representing a shale interlayer located about halfway down the reservoir.

The notable difference with a saline aquifer is that the K12B site initially contains a natural gas rich in CO_2 (around 13%), signifying a geochemical equilibrium between the fluids and the rock with an initial pH of around 4.5. Moreover, injecting CO_2 into methane consists essentially of compressing the system's gas phase without really making the gas circulate or percolate in an aqueous-phase-saturated porous medium. Relative

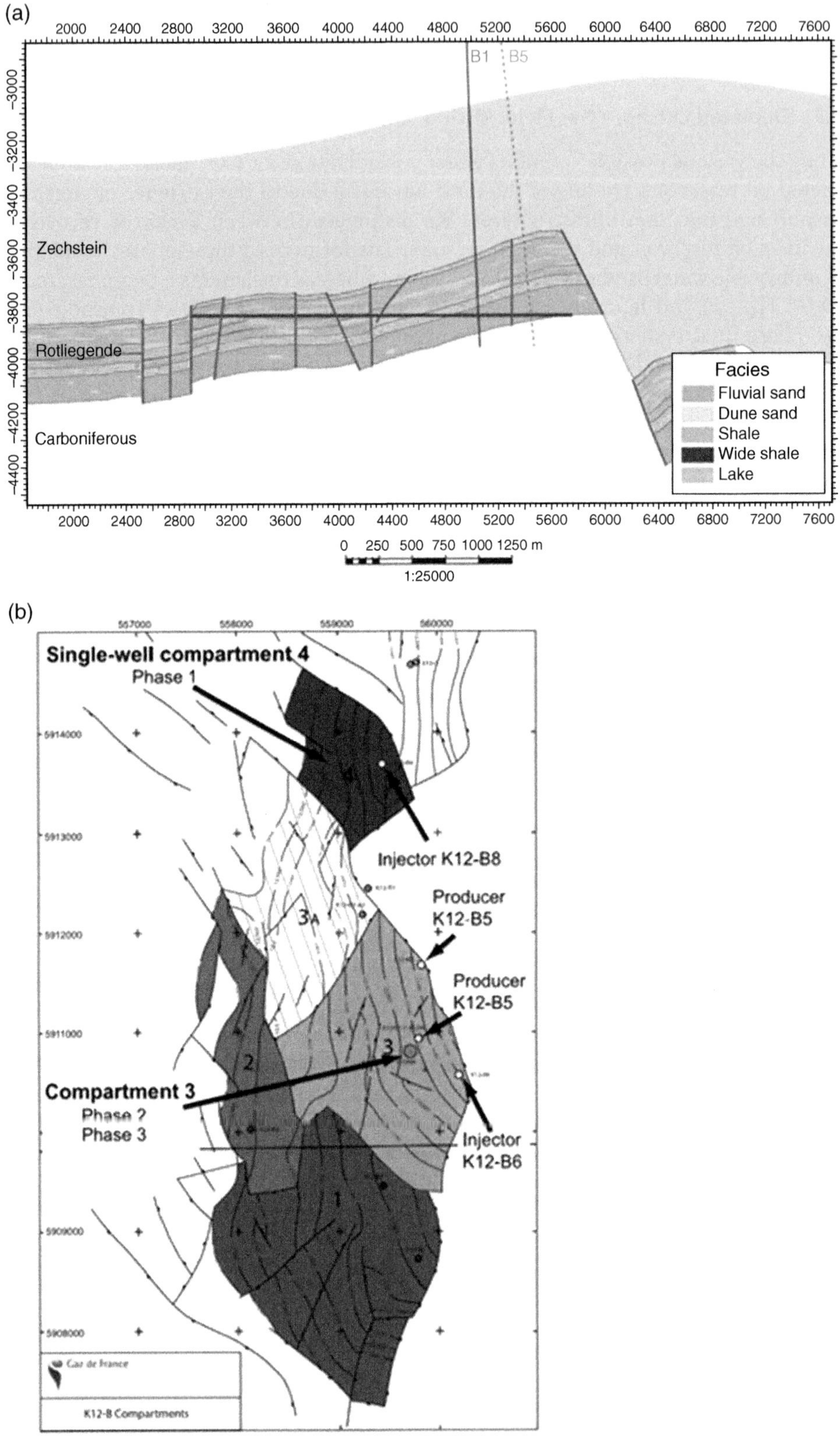

Figure 2.10 The K12B field (North Sea, the Netherlands): (a) geological cross-section, (b) plan of the field and the compartments, (c) gridded numerical model of compartment 3 used for reactive transport calculations with TOUGH2 from the Petrel model developed by SINTEF (CASTOR project) (Audigane et al., 2008, *Source:* used with permission of AAPG).

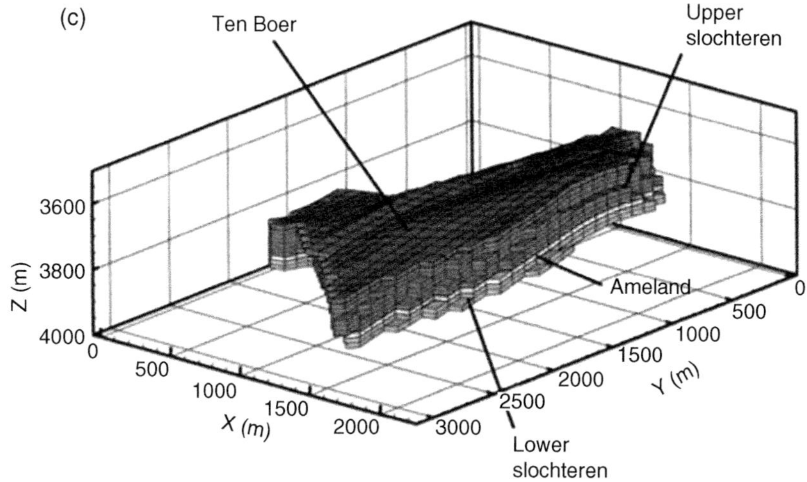

Figure 2.10 (Cont'd)

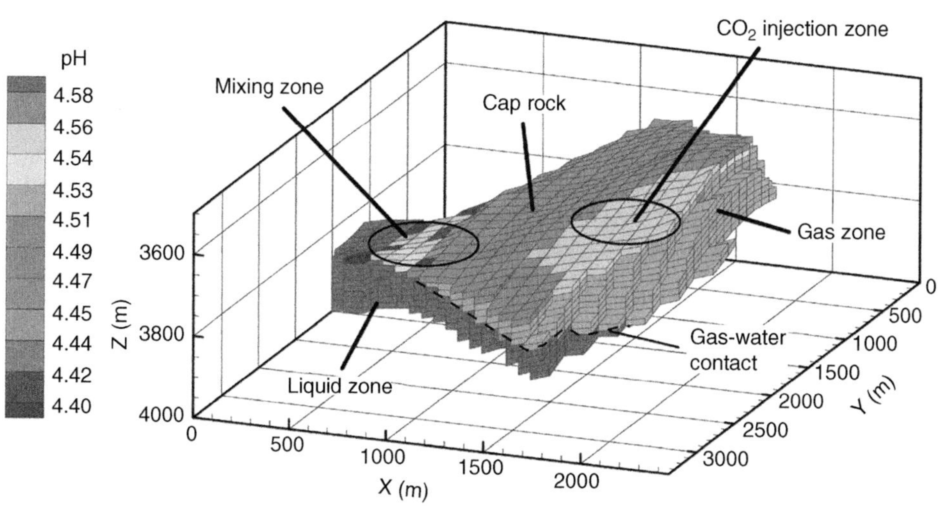

Figure 2.11 pH simulated in the K12B reservoir after 10 years of injection (Audigane et al., 2008, *Source:* used with permission of AAPG). (See colour plate section for the colour representation of this figure.)

permeability and capillary pressure laws are therefore relatively unimportant in this type of model. The results show almost no variation of the geochemistry of the reservoir's aqueous phase (Figure 2.11), and only the cells located at the reservoir's gas–water interface are slightly impacted by a change in gas saturation ('mixing zone') that might cause a weak reactivity that is considered to be negligible. A drying-up accompanied by precipitation of anhydrite and carbonates (calcite and siderite) is also calculated around the injector well. The proximity of the injector well to producer wells (450 m for K12-B1 and 850 m for K12-B5) results in rather rapid CO_2 arrival times of 60 days and 1 year, respectively, for an injection of 310,000 tons per year and a producer flow-rate ten times less for each well. The potential to recover more methane is therefore rather limited with a rather moderate storage of CO_2, and the CO_2 trapping is mainly structural since it is confined to the gas phase of the reservoir.

2.3 Studying CO_2 Leakage and Well Integrity by Reactive Transport Modeling

We describe here reactive transport simulations of CO_2 leakage during CO_2 geological storage. Both the *geochemical phenomena* that might cause leakage and the *hydrogeochemical impacts* associated with leakage are studied.

The weathering of minerals by dissolution affects the integrity of the well cement used to seal wells, and it is notably at the interfaces between the rock formation (clay, sandstone or carbonate), the well cement, and the casing that possible seal failures are identified. Problems of weathering within the matrix are usually considered to be negligible for clays (Gaus et al., 2005) and for cement (Carey et al., 2007). In a different context, the studies on nuclear waste geological disposal also address the question of modeling geochemical interactions between iron, bentonite and clay interfaces (Savage et al., 2011 and references therein; Wilson et al., 2015). The difficulty in reproducing these interfaces and simulating the geochemical impacts on the quality of the bonding has led researchers to carry out experiments that are able to reconstruct pseudo wells at various scales. Sterpenich et al. (2014) installed an injection well at laboratory scale (1:20) using a limestone rock reservoir, an argillite caprock, a class G cement, a pseudo casing, and pressure and temperature conditions similar to those of a deep reservoir. Manceau et al. (2013, 2015) used the underground laboratory in Mont Terri (Switzerland) to reconstruct a full-size (1:1) pseudo well with casing and cement in the formation clay of the rock laboratory's galleries. These experiments provide valuable data-sets for calibrating geochemical models and benchmarking codes.

Once some understanding of the geochemical processes that may lead to CO_2 leakage from storage reservoirs has been gained, the related impacts must also be determined. As with storage, the leaking fluid (supercritical CO_2 or dissolved-CO_2 enriched brine) causes the minerals in the rock with which it comes into contact to dissolve, and frees chemical species that can modify the chemistry of the groundwater in an impacted aquifer, increasing their concentrations to undesirable or even non-regulatory levels (Lions et al., 2014). Wang and Jaffe (2004) studied the release of lead sulfide from a rock with a simplified mineralogy made up mainly of galena (as well as calcite and quartz). Their calculations predicted an exceedance of the US Environmental Protection Agency's acceptability threshold for lead. Kharaka et al. (2009) and Xu et al. (2010) simulated the geochemical impacts of the mobilization of elements (notably iron hydroxides) observed on the Frio Brine Pilot site in Texas (USA). The drop in pH can also lead to the release of hazardous trace elements by desorption/absorption, mobilizing toxic trace elements or organic components (Kharaka et al., 2006, 2010).

Simulating CO_2 leakage and related hydrogeochemical impacts is a necessary prerequisite for the creation of a CO_2 storage site. Indeed, the European Directive states the importance of being able to predict the consequences of leakage and the possible remedial solutions in the event that leakage is observed. Below, we describe two French case studies in the Paris Basin for which fictional scenarios of short- and long-term CO_2 leakage were simulated.

2.3.1 Near-well Problem in the Paris Basin

2.3.1.1 Weathering of Drilling Cement Prior to Injection

Gherardi and Audigane (2013) modeled the ageing of the cement in an abandoned well prior to contact with CO_2. This involved studying the mineralogical transformations in a cement, taking into consideration the life span of a well during its working life and after closure, in order to determine the initial state of the geochemistry of a cement–caprock system. Field data showed that Portland cement was able to maintain well integrity, even in the oldest wells (55 years old), after 30 years of exposure to CO_2 (Carey et al., 2007; Crow et al., 2010; Scherer et al., 2011). Gherardi and Audigane (2013) used the geological context of the Paris Basin (Callovo-Oxfordian (COx) clay) and an ordinary Portland cement to define their mineralogical system and the chemical composition of the water. Their model was an axisymetric radial 1D cylinder centred on the well (Figure 2.12). Only the interface between the cement and the caprock was studied. The casing was considered geochemically inert. This assumption conceptualizes the condition of a wellbore properly completed, with tight sealing of the cement–casing interface. In fact, as observed in the SACROC Unit, USA (Carey et al., 2007), and in the URL[1] of Mt. Terri, Switzerland (Nussbaum, personal communication 2014), the steel of the casing may passivate in the relatively short term if proper isolation from pH-acid fluids is guaranteed over time. The calculations were done with TOUGHREACT (Xu et al., 2003) and diffusion alone governed the migration of chemical elements. The cement had a radial thickness of about 10 cm, compared with 2 m for the caprock.

The cement domain was modeled both as a homogeneous and a dual-porosity system ('heterogeneous model') to account for the poorly connected nature of the pores in the porous/fractured media. The 'homogeneous' model considered a single cement domain with relatively high porosity. The 'heterogeneous' model considered a high-porosity volume ('accessible cement': AC) representing the highly connected, macro-porous portion of the cement directly connected to the caprock, and a low-porosity volume ('cement matrix': CM), directly connected only to the accessible cement, representing the micro-porous, poorly-connected portion of the cement. The AC:CM volume ratio was arbitrarily fixed to 20:80. The simulated cement has the composition of a hydrated and hardened Class G cement, mostly made of calcium-silicate-hydrate (CSH:1.6; 38.1% by volume), portlandite (26.1%) and ettringite (13.8%), with relatively small amounts of hydrogarnets (katoite-Si, 10.5%, and C_3FH_6, 5.5%) and hydrotalcite, and traces of calcite (0.6%). The gravel was not taken into consideration. For the caprock, the mineralogy of COx has been studied in detail for French nuclear projects using experimental and numerical methods (Gaucher et al., 2004). An average composition by Marty et al. (2009), Gherardi et al. (2012) and Wertz et al. (2013) was determined, with, by volume, 42% illite-Mg, 24% quartz, 14% calcite, 6% montmorillonite-Mg-Na, 1% daphnite and pyrite, and 0.7% celestite. The mineralogical system (including secondary phases) contained over 45 minerals. The chemical compositions of the porewater in the cement and caprock were reconstructed based on previous simulations. Gaucher et al. (2009) developed a reference model that was adapted by Gherardi et al. (2012) and used here. The cement porewater was alkaline (pH around 11 at 75 °C) whereas the COx porewater has near-neutral pH (around 6.33) and higher solute contents (except for aluminium). The thermodynamic database *Thermoddem*, developed at BRGM mainly for nuclear projects (Blanc et al., 2012), was used.

---

1 Underground Rock Laboratory.

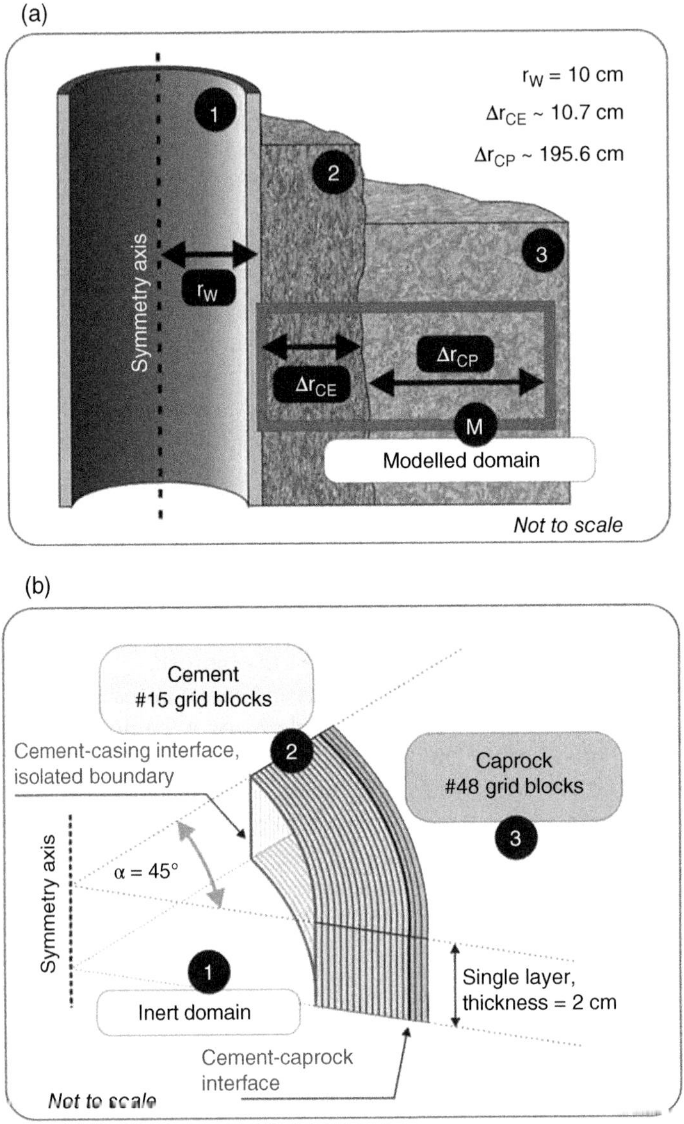

Figure 2.12 (a) Conceptual model of a well and its (1) casing, (2) cement, and (3) caprock interfaces. The cement–casing interface has been assumed to act as an isolated boundary (outlined in red in (b). (b) Axisymmetric geometry and dimension of grid used by Gherardi and Audigane (2013).

Since the migration processes are solely diffusive, the concentration gradients govern the migration of chemical elements. The simulations illustrate well the competition between the alkaline plume flowing from the cement towards the caprock and, inversely, the slightly acidic plume flowing from the caprock towards the cement. The calculations predict a high capacity of the cement to buffer the acidity of the caprock by stabilizing at a pH of 10.99, whereas an alkaline plume diffuses a pH value of 8 at more than 1 metre from the interface with the cement (Figure 2.13). However, simulated tracer tests show that the caprock's porewater controls the fluid mixing rate, such that most of

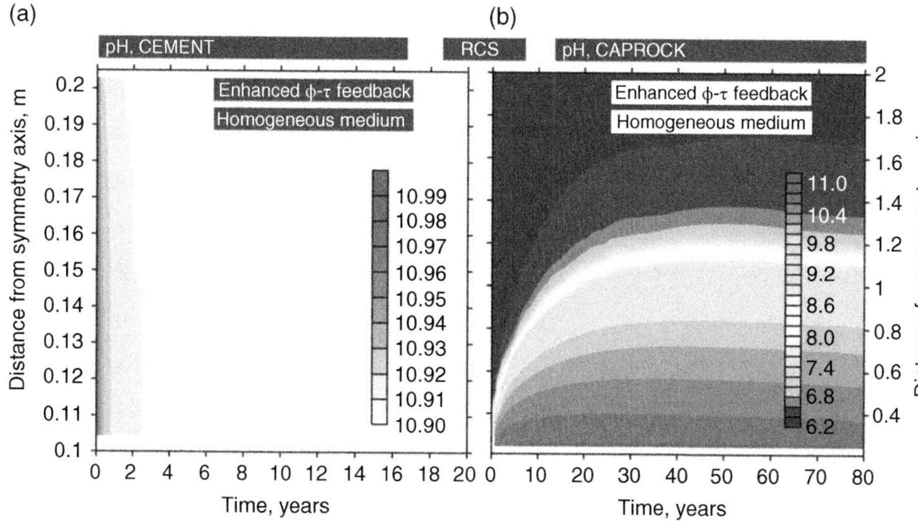

Figure 2.13 Diffusion of alkaline and acid plumes in the cement and caprock domains after 80 years in contact (Gherardi and Audigane, 2013).

the chemical elements in the caprock water diffuse towards the cement, showing once again the strong pH buffering capacity of cement.

The mineralogical transformations in the cement are controlled by the evolution of the calcium Ca, carbon HCO_3 (total dissolved inorganic compounds, TDIC), aluminium Al, silica SiO_2, magnesium Mg, and sulfate SO_4 near the cement–clay interface, with notably a large increase in calcium in the two concomitant domains and a decrease in total carbon (TDIC) in the caprock. These trends coincide with a strong dissolution of portlandite inducing a precipitation of calcite at the interface, a carbonation that has been observed experimentally by many authors (Figure 2.14). However, the model does not predict a massive precipitation of calcite. In addition to the dissolution of portlandite, the simulations also predict the partial dissolution of the hydrogarnets (katoite-Si, C_3FH_6) and the precipitation of CSH:1.6, ettringite, and to a letter extent, of hydrotalcite and brucite.

This 'homogeneous' model reproduces rather well the observations made on samples by Carey et al. (2007) who observed (in areas that were not weathered by CO_2) the presence of portlandite, CSH, ettringite, katoite, brucite, and a small amount of calcite. Although the models presented show, for the most part, that cement is well able to retain its integrity for several decades under the conditions of the Paris Basin, the variations in porosity near the interface do not confirm this trend. Indeed, the models predict an increase in porosity near the interface due to the absence of the massive carbonation that is normally observed.

The lack of extensive carbonation within the cement mirrors the complex interplay between cement and caprock porewaters near the contact between the two materials. The limiting factor is the low concentration of dissolved aqueous carbon in the cement, which is buffered at low levels by the precipitation of calcite in the caprock, a few centimetres further from the contact. This feature suggests that reliable predictive numerical models need to account for the fact that the carbonation potential of the caprock porewater is an essential but not sufficient condition to guarantee enhanced carbonation of the cement. The setup of realistic initial (IC) and boundary (BC) geochemical conditions is of paramount importance to allow numerical models to capture

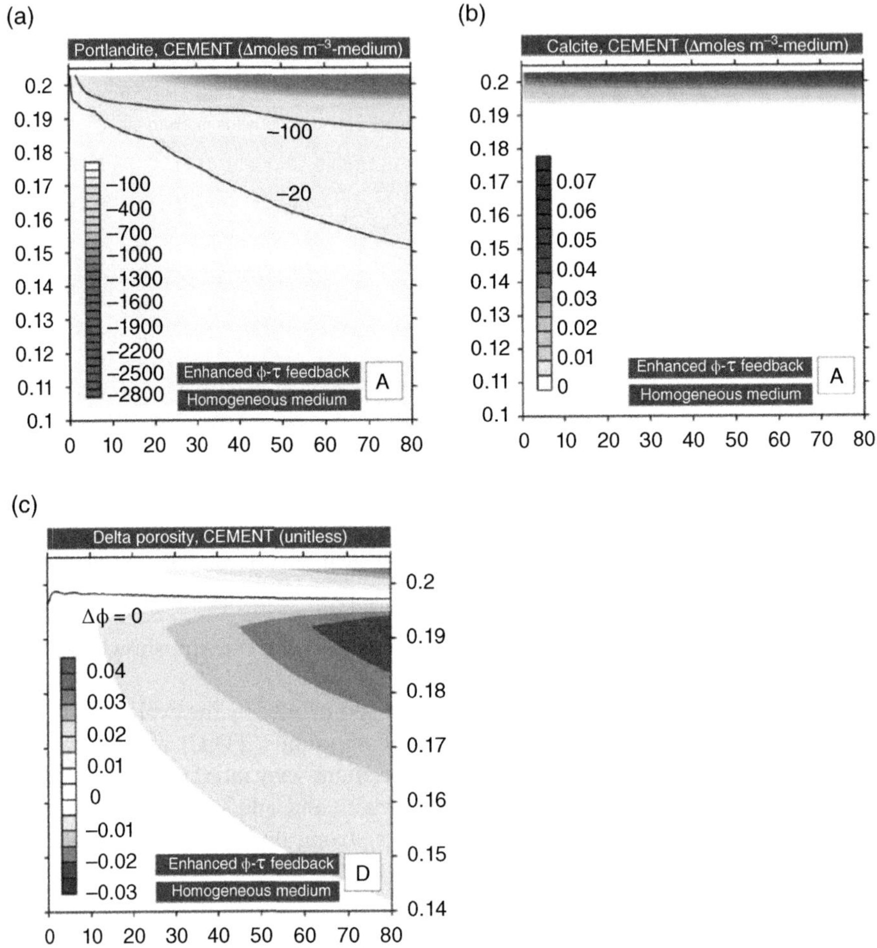

Figure 2.14 (a) Portlandite, (b) calcite, and (c) porosity calculated for evolution over 80 years in cement (Gherardi and Audigane, 2013).

this aspect, by accounting for the reciprocal geochemical feedback between caprock and cement fluids and minerals. The use of unappropriated IC and BC (such as the occurrence of carbonated fluids with time-invariant, pH-acid, chemical composition at the cement interface, as widely assumed in laboratory experiments) would lead to an unrealistic description of the processes expected at the cement/caprock interface, and possibly to erratic predictions about carbonation depths, integrity and sealing efficiency of the cement over time. The moderate mineralogical rearrangement predicted by the code reinforces a number of field observations that indicate slow rates of Portland cement deterioration under diffusion-controlled field conditions, and over the time interval (up to 80 years) considered by the numerical model.

2.3.1.2 Cement–Reservoir–Caprock Interface

As part of a study of the geochemical impacts of a CO_2 storage site where there is an abandoned well near the injection point (Figure 2.15), Gherardi et al. (2012) simulated the long-term resistance of well cement to CO_2 dissolved in water. In their model,

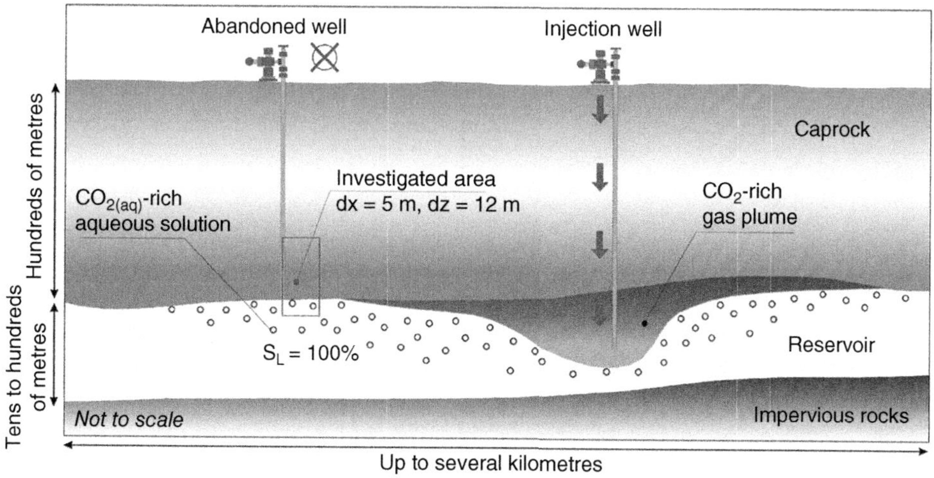

Figure 2.15 Diagrammatic representation of the context of the study by Gherardi et al. (2012) showing the supercritical CO_2 plume and the saturated part of the aquifer. The bubbles represent the part of the aquifer where the groundwater is enriched in dissolved CO_2. The area studied includes caprock, reservoir and well cement.

diffusion alone controls the transport of chemical elements since it reproduces a post-injection scenario in which a CO_2 plume is steadily dissolving in an aquifer with no flow, and therefore for an immobile aqueous phase.

In order to decrease calculation time, the domain is limited to several metres around the cement–clay–reservoir interfaces. A 2D Cartesian grid was used in the model, with finer mesh at the interfaces (Figure 2.16). The geological context is, once again, that of the Paris Basin, using the carbonate Dogger formation as the reservoir and the COx argillite for the caprock. The mineralogy of an ordinary Portland cement was used for the well (Table 2.3). Degradation of the casing caused by corrosion was not taken into consideration. The temperature and pressure conditions for a reservoir located between 1500 and 2000 m deep were 75 °C and 200 bars, respectively. The initial chemical composition of the reservoir porewater was that of groundwater in Fontainbleau (Michard and Bastide, 1988) with a higher salinity (0.1 mol $kg^{-1}{}_{H2O}$) similar to the salinity encountered in the Dogger (5–35 g $kg^{-1}{}_{H2O}$). The water was then saturated in dissolved CO_2 and reached a pH of 3.7. The porewater in the COx has been studied at length in nuclear waste storage contexts. In order to obtain, at 75 °C, the most stable mineralogical system with water almost in equilibrium, the work of Marty et al. (2009) was adapted for this study, giving a very slightly acidic water (pH 6.6). Using a standard mineralogical assemblage for Portland cement, an ideal alkaline chemical composition (pH 11.2) was calculated.

Different diffusivity conditions have been explored, and two limiting reference cases have been defined, based on the values of the effective diffusion coefficient calculated at the reservoir–cement interface (low diffusivity (LD) $=1.65 \times 10^{-11}\,m^2\,s^{-1}$; high diffusivity (HD) $= 1.79 \times 10^{-10}\,m^2\,s^{-1}$). Diffusion is expressed with an average diffusion coefficient applied to all dissolved species, which is an approximation of that used in TOUGHREACT.

The results of simulation after 1000 years of contact between the carbonate Dogger reservoir saturated in dissolved CO_2, the COx caprock, and the Portland cement show that there is competition between the acid and alkaline fronts. After 1000 years, the acid

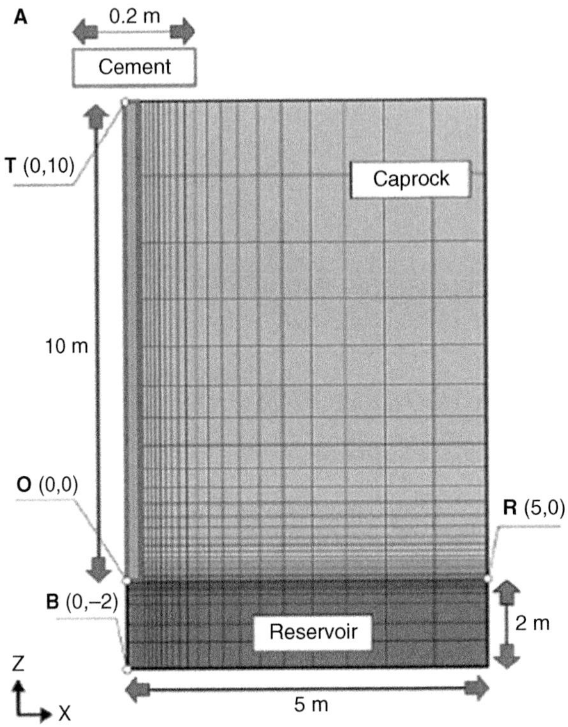

Figure 2.16 Gridding and dimensions of simulated domain that encompasses the cement, caprock, and reservoir media (Gherardi et al., 2012).

Table 2.3 Mineralogical composition of the reservoir domains, caprock and cement considered by Gherardi et al. (2012).

Primary minerals	Reservoir	Caprock	Primary minerals	Cement
Calcite	0.775	0.14	Portlandite	0.40
Dolomite	0.102	0.11	Tobermorite-14 Å	0.35
Quartz	0.058	0.24	Katoite-Si	0.18
Mg-illite	0.001	0.423	Hydrotalcite	0.03
Mg-montm-Na	–	0.06	Ettringite	0.02
Daphnite-14 Å	–	0.01	C_3FH_6	0.02
Pyrite	–	0.01		
Celestite	–	0.007		
Kaolinite	0.02	–		
Siderite	0.041	–		
Magnesite	0.003	–		

front has penetrated 2 m and 8 m into the caprock for the LD and HD scenarios, respectively (Figure 2.17a). Due to the fact that there is much less cement than there is caprock or reservoir rock, the cement porewater is less able to diffuse into the other domains, whereas the fluid coming from the caprock seems able to totally invade the cement. Based on the particular geometry of this system in which the three domains have relatively different volumes, the alkaline plume seems to be less efficient than the acid front in controlling the system's geochemical reactivity. We observe therefore mainly a decrease in pH diffusing in the caprock outside of the zone farthest from the reservoir, at the caprock–cement interface where the alkaline plume is able to migrate over a distance of, at most, 2 m (Figure 2.17a).

The mineralogical transformations of the system therefore occur mainly in the lower part of the cement (near the reservoir) where the porosity is most heavily impacted

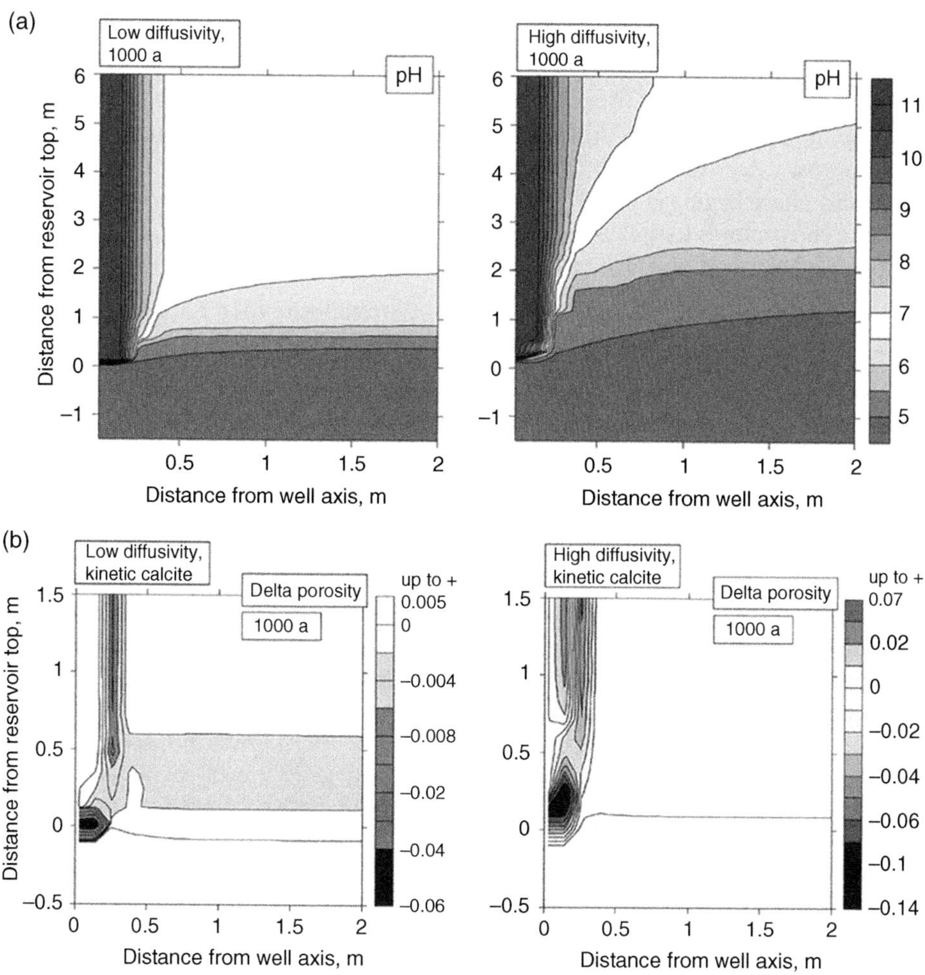

Figure 2.17 Simulated spatial distribution of (a) pH and (b) porosity after 1000 years (Gherardi et al., 2012). (See colour plate section for the colour representation of this figure.)

(clogging) after 1000 years (Figure 2.17b). There is a massive dissolution of primary phases accompanied by an equally massive precipitation of calcite. Three reaction plateaus can be identified by changes in porosity (Figure 2.18). The first phase corresponds to the total dissolution of portlandite and the precipitation of calcite according to the following reaction:

$$\text{portlandite} + CO_2 \rightarrow \text{calcite} + H_2O$$

The second phase starts after 20 and 180 years for the HD and LD scenarios, respectively, coincides with the almost total disappearance of all other phases, and also contributes to a massive precipitation of calcite according to the following reactions:

$$\text{tobermorite} + CO_2 \rightarrow \text{calcite} + H_2O + SiO_{2(aq)}$$

$$\text{katoite} + CO_{2(aq)} + H_2O \rightarrow \text{calcite} + \text{straetlingite}$$

$$\text{hydrotalcite} + OH^- + SiO_{2(aq)} \rightarrow \text{clinochlore} + AlO^- + H_2O$$

$$\text{ettringite} + CO_{2(aq)} + OH^- \rightarrow \text{calcite} + AlO^- + H_2O + SO_4^{2-}$$

$$\text{Fe-katoite} + 4H^+ + 3CO_{2(aq)} + H_{2(aq)} \rightarrow 3\ \text{calcite} + 9H_2O + 2Fe^{2+}$$

At this point, the alkaline conditions persist but with a decrease in pH (to about 10.2) and the porosity decreases to 0.12 for phase 1 and to 0.03 for phase 2.

The third phase begins at about 25 and 250 years for scenarios HD and LD, respectively, and corresponds to the disappearance of straetlingite, an intermediate secondary phase, totally dissolved and also turning into calcite and Ca-phillipsite by the reaction:

$$\text{straetlingite} + CO_{2(aq)} + SiO_{2(aq)} \rightarrow \text{calcite} + \text{Ca-phillipsite} + H_2O$$

This is accompanied by the total disappearance of tobermorite, the precipitation of magnesite, and a drop in pH to 4.8. The magnesium released into solution by the dissolution of Mg-chlorite precipitates as Mg-illite and magnesite. After the total weathering of the minerals initially in the cement, the fluids from the reservoir control the geochemistry of the system. This intensifies with time and leads to the total disappearance of the Ca-phillipsite and the precipitation of kaolinite, dawsonite, and quartz. The porosity increases to 0.10. In this very localized area, therefore, there are large variations in porosity. These might be associated with possible preferential leakage pathways (if the porosity increases) or with clogged areas (if the porosity decreases), knowing that this type of simplified model does not take into account petrophysical or mechanical variations in the porous structure of the rock or cement under consideration. The accuracy of the numerical predictions should be improved when more detailed site-specific physical and chemical data are available, and the presence of physical discontinuities, together with the spatial variability of hydraulic and geochemical parameters, can be considered. Further modeling could improve on some of the simplifications made, also by considering the presence of amorphous and/or not completely hydrated components in the starting cementitious material and addressing the simulation of their alteration mechanisms by means of a fully kinetic treatment of precipitation/dissolution reactions, and by accounting for possible in situ ageing effects induced by the interaction of the cement with host formation porewaters before the beginning of the confinement of CO_2 at depth.

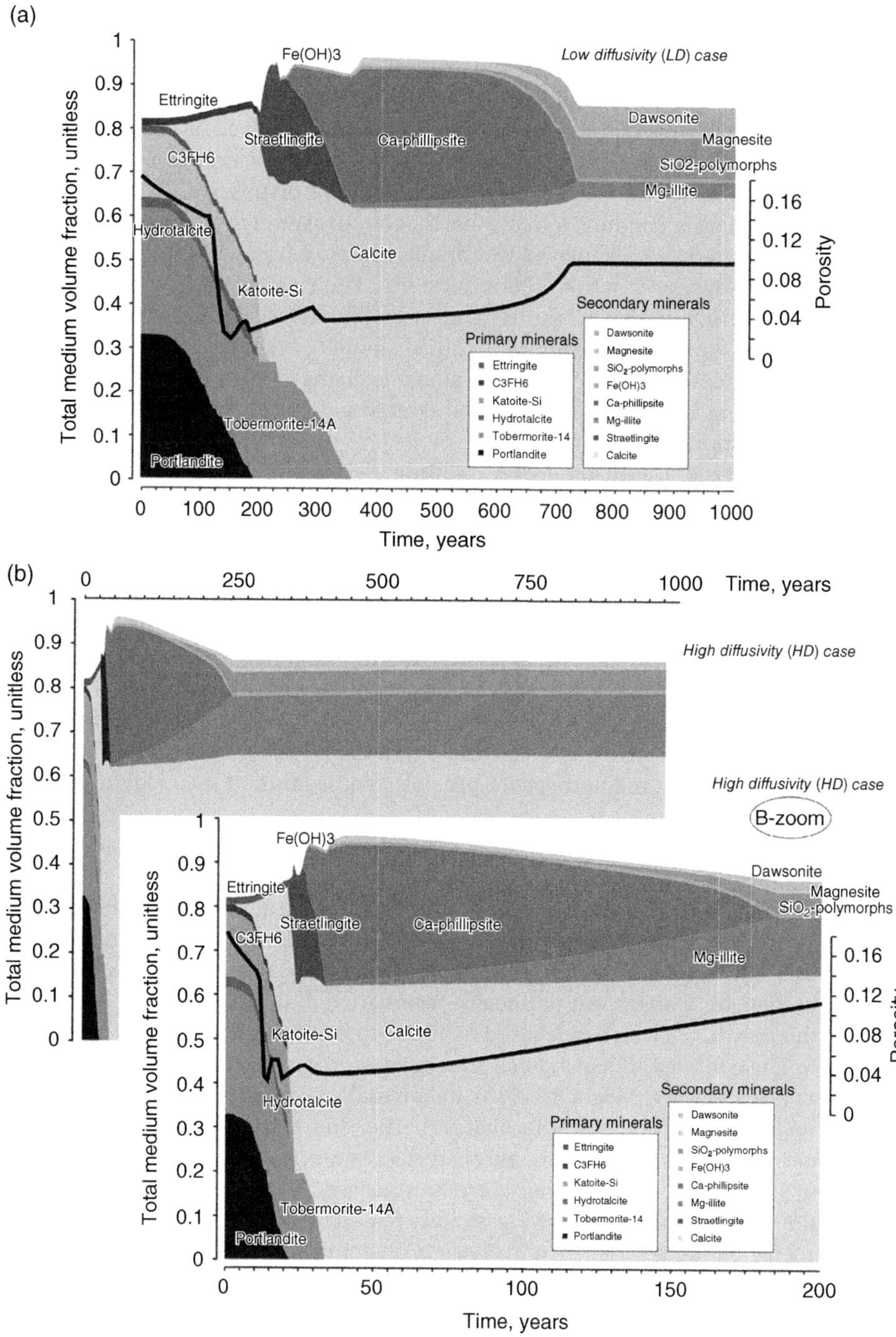

Figure 2.18 Sequence of dissolution and precipitation of the mineral phases of the cement at the interface with the reservoir and the caprock (Gherardi et al., 2012). (See colour plate section for the colour representation of this figure.)

2.3.2 The Impact of CO_2 Leakage on Groundwater

CO_2 leakage can have potential impacts on groundwater in a shallow aquifer, which is often a strategic target for drinking water supplies. Natural analogues can be very useful for understanding the geochemical mechanisms involved in the release of harmful chemical elements. Viswanathan et al. (2012) have, for example, studied the Chimayo site in New Mexico (USA) where high levels of dissolved CO_2, rising with groundwater through a conductive fault, have been measured in many wells. These authors demonstrated the role played by surface complexation with illite in the desorption of arsenic observed in the water on the site. Their model predicted the major role of kaolinite precipitation in the reduction of the release of As in longer-term processes. On the same site, Keating et al. (2013) simulated the observations made of several major elements in 3D in order to evaluate exchanges or adsorption of cations and dissolution and precipitation of calcite containing traces of uranium possibly released and transported in the site.

Gherardi et al. (2007) used modeling to study chemical reactivity in a fractured caprock overlying a CO_2 storage reservoir. The simulated caprock contained both a fault zone (with high permeability and porosity) and a fractured area (simulated as a dual medium). These authors showed that, for a sandstone reservoir with a mineralogy rich in calcite and dolomite (16% by volume) and a fractured clay caprock, leakage varied drastically depending on the type of fluid that was leaking. For a fluid that was mainly gas (or supercritical) and moving by advection within the fractured zone and the fault, a sharp drop in pH was predicted, along with a major dissolution of calcite and a substantial increase in porosity. On the other hand, for a fluid made up solely of water or brine enriched with dissolved CO_2, moving by diffusion in the fractured system, the pH was buffered and calcite precipitated, leading to the clogging of the porous medium.

Humez et al. (2011) simulated a leak from the carbonate formation of the Paris Basin towards an Albian aquifer, which is a strategic drinking water resource for the Paris region (Figure 2.19). The reservoir proposed for CO_2 storage in this study is located in the Mid-Jurassic aquifer (Dogger) in the centre of the Paris Basin at a depth of around 1500–2000 m. This aquifer is tapped for its geothermal and oil resources and is made up mainly of carbonates whose most permeable facies are oolitic limestone, the high permeability and porosity of which is believed to be due to fracturing and dissolution (Rojas et al., 1989). The confined Albian aquifer extends over 75 km^2 and is made up of several sandy layers (Megnien and Megnien, 1980). Its mineralogy is mainly quartz, glauconite, kaolinite, muscovite, carbonates, phosphates, barite, and pyrite (Raoult et al., 1997). The formation water has a highly variable chemical composition depending on the area studied, but has several common characteristics as well. It is anoxic, has a high iron concentration, a pH of around 7, and low salinity (0.3–0.6 g/l).

The geometry of the modeled system is based on a numerical benchmark study carried out by the University of Stuttgart that aimed to compare the ability of calculation codes to predict leakage to an overlying aquifer from a leaky well in a storage reservoir (Class et al., 2009). The injector well and the leaky well were 100 m apart and the geological context of the Paris Basin was used to define the depths of the two aquifers and their pressure and temperature ranges. TOUGHREACT (Xu et al., 2003) was used for the calculations and could therefore handle the variations in the properties of the

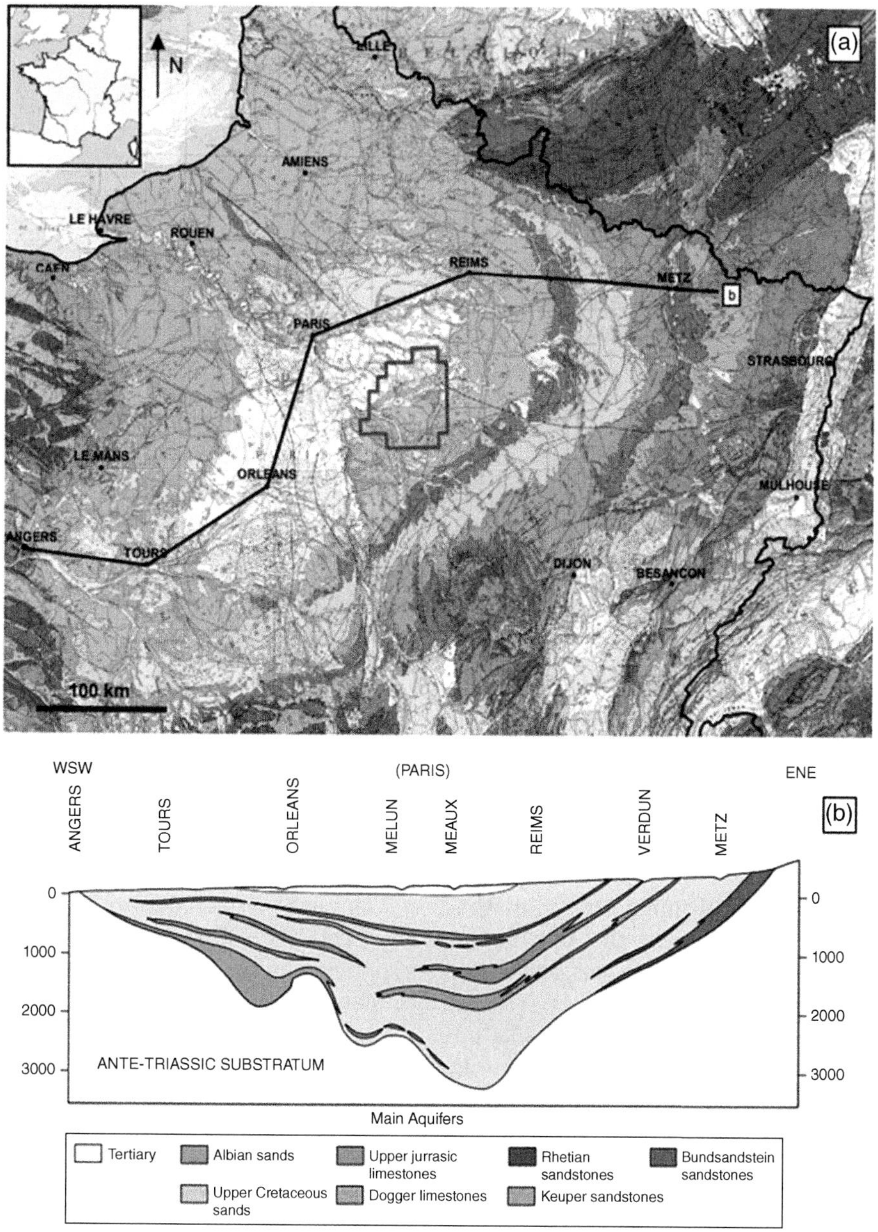

Figure 2.19 Geological map of the Paris Basin and cross-section identifying the Dogger carbonate formations and the Albian sandstone formations.

system's gas and liquid phases associated with these pressure and temperature ranges. The difficulty when using this model is in gathering of all of the information needed to characterize the hydrogeological, geochemical, and 'reservoir' (i.e. two-phase flow) properties of the entire model, and notably those for the leaky well. A sensitivity study was carried out in order to compensate for the high level of uncertainty associated with these data. The objective of this study was to show the feasibility of developing a reactive transport model capable of taking into account fluid–rock interactions occurring in the reservoir during storage, along the leakage pathway (here, a well represented by a 1D geometric system) and in the impacted aquifer, while considering a two-phase system and a realistic geological context.

A data-set was compiled from previous studies of the two aquifers (Vernoux et al., 1997; Brosse et al., 2007; André et al., 2007) in order to identify the petrophysical and hydrogeological properties that are coherent with this simplified system. The properties of the leaky well were approximated by an equivalent porous medium, although the identified possible leakage pathways are distributed in a complex manner at the cement, clay, and casing interfaces (Gasda et al., 2004). Using the average chemical composition measured in several wells in the two aquifers, an initial calculation of speciation at equilibrium with the mineral phases was conducted to initiate the system. The Dogger was considered to be 100% calcite and the Albian to be 49% quartz, 19% calcite, 15% kaolinite, 14% glauconite, 3% siderite, and 1% pyrite. Once the initial chemical compositions had been calculated, CO_2 injection was simulated using reaction kinetics from Palandri and Kharaka (2004) for the dissolution and precipitation of the mineral phases, except for calcite and siderite, which remained in equilibrium. One of the key issues in these simulations consisted of determining the role of glauconite in the release of iron previously observed in the Albian's green sands (green due to its high glauconite content). We have here not only a coupling of the intrusion of an equilibrated fluid with the mineralogy of a carbonate storage reservoir, but also the ability to supply a leakage rate and a spatiotemporal progression of the impacts, all for fluid volumes that are consistent with the aquifer dimensions. The calculated leakage rates and plume extents are linearly dependent on the injection flow rates, the distances separating the wells, and the average permeability of the leaky well. A sensitivity study was done of these aspects related solely to flow. Concerning the reactivity of the system, the intrusion of an acidified fluid into the Albian caused the pH to drop, buffered by the carbonates. The calcite and glauconite dissolved to release calcium, iron, and magnesium. The calcite dissolved more strongly at the point of intrusion into the Albian (Figure 2.20). Potassium also increased due to the dissolution of alumino-silicates, and the sulfate content increased due to the dissolution of pyrite, but to a lesser extent. The calcium increased significantly from 1.3 to 33.6 mmol/kgw, and iron from 13 to 2500 μmol/kgw. Although iron is not harmful to human health, it modifies the taste and appearance of water when concentrations exceed French drinking water standards (3.6 μmol/l). In the Albian aquifer, this value is often exceeded and the groundwater is therefore always treated. The dissolution of siderite plays a relatively small role here in the variability of the iron concentration. The impacts upon porosity and permeability were considered to be minor (<1% variation).

2.4 Discussion and Conclusion

Reactive transport models unquestionably improve our understanding of the hydrogeochemical behaviour of storage reservoirs at different spatial scales – from pore-scale to basin-scale – and temporal scales – from seconds to millennia, or even several hundreds of thousands of years in the context of nuclear waste disposal (Gaus, 2010). Predicting these physicochemical phenomena over the long term cannot be done in a

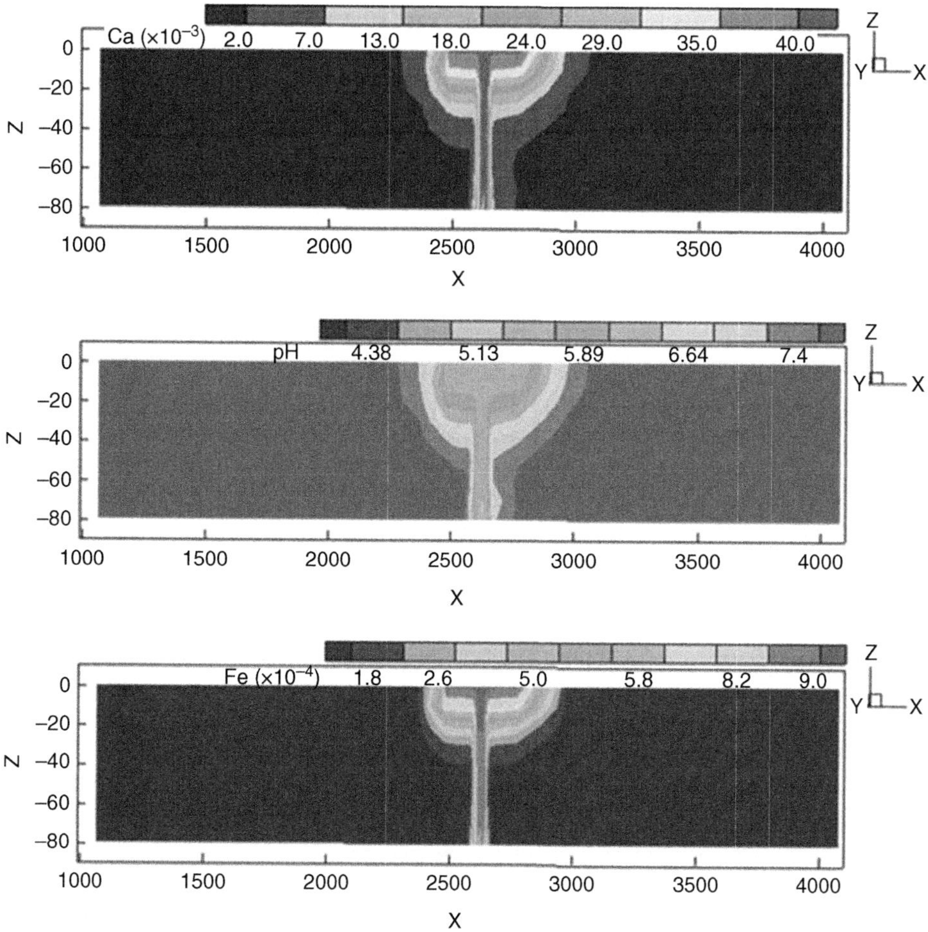

Figure 2.20 Calcium concentration, pH and iron concentration in the sandy Albian aquifer after 5 years of leakage of CO_2 from a storage site in the Dogger carbonate reservoir (Humez et al., 2011).

quantitative fashion using experiments or natural analogues, as experiments are usually conducted on short timescales (weeks to years) and the origin of the CO_2 nature and flow in natural analogues remains often complex and difficult to interpret. The European Commission's Directive requires the use of models to confront the modeling predictions with monitored data (migration of the CO_2 plume, reservoir pressure and temperature, changes in fluid composition, etc.) in order to evaluate the behaviour of the reservoir and also to have a picture of the reservoir behaviour at longer timescales. Operators have the responsibility that they control the evolution of the storage reservoir and they need numerical modeling to support their arguments. Very often, these models are usually fluid flow simulators, commonly used in petroleum reservoir engineering, to match production data. In the context of CO_2 storage, reactive transport models can never become predictive because of the reasons mentioned above. It is too ambitious to aim at fitting monitored chemical composition data from observations wells in a quantitative manner. However, these models help to improve understanding and support simpler decision-making or engineering models that can claim to be sufficiently predictive, as these focus on the impacts without having the ambition to accurately describe the detailed processes. These simpler models feed into risk and safety assessments and thus respond to requirements written in the legislation. Therefore constant developments of such tools are required.

Modeling now makes it possible to investigate both long-term behaviours that will result from changes in the chemical composition of fluids in a CO_2 storage site, and the impacts associated with potential leakage to overlying formations such as drinking water aquifers (Lions et al., 2014). Models are also technical tools for risk assessments and remediation or mitigation studies that might be necessary in the event that problems, such as the mobilization of undesirable elements (Apps et al., 2010) or the shifting of brines (Oldenburg and Rinaldi, 2011; Lions et al., 2014), are identified.

Reactive transport models are often used to interpret field observations from research pilot sites used notably to characterize the possible release of trace elements associated with surface complexation phenomena and ion exchange (Kharaka et al., 2006; Hovorka et al., 2006). However, researchers have not been able to develop models that can correctly simulate many physicochemical processes so as to produce results that are in full agreement with data measured on site, in laboratory experiments, or in samples taken from natural analogues. Indeed, the equivalent porous medium that constitutes the sedimentary rocks being studied for CO_2 geological storage is made up of grains deposited following complex depositional processes, and subjected to multiple diageneses, resulting in a highly heterogeneous distribution of the properties of the medium on different scales. It is not surprising, therefore, that researchers are unable to sufficiently define the often very heterogeneous initial geochemical state of an aquifer (and its chemical composition) based on a few water samples and mineralogical assemblages resulting from observations of drilling samples. The simplification or homogenization of data often makes it difficult to calculate the chemical composition of water at equilibrium with the primary mineral phases that the model considers are representative of the rock under consideration. This difficulty is related to heterogeneity at the scale of the microstructure of the porous medium, which modifies the geochemical properties at the macroscopic scale. Indeed, the upscaling of the reactive surface of mineral grains to the kilometre-wide scale, and the upscaling from the mineralogy at the core sample scale to the whole aquifer, has not been achieved. It is uncertain whether this will ever be the case. All of these questions are essential for subsurface modeling. One of the direct impacts of such a high degree of uncertainty in the initial conditions is that it is also reflected in the prediction of secondary phases that might precipitate, over the short and long term. In the field of CO_2 storage, dawsonite is subject to much debate due to the major role it might play as secondary carbonate for mineral trapping of CO_2 (Bénézeth et al., 2007; Hellevang et al., 2014 and references therein). Trémosa et al. (2014) gathered compositional data from natural analogues to study the occurrences of dawsonite and concluded that storage sites such as Snovith, Sleipner, In Salah, and Ketzin were favourable contexts for dawsonite precipitation and so, to a less extent, for mineral trapping (Figure 2.21). This situation is a perfect illustration of the limits of modeling of the predictive behaviour of a chemical system to react, and the need for expert interpretation backed by field observations to guide the selection of reactional pathways to simulate.

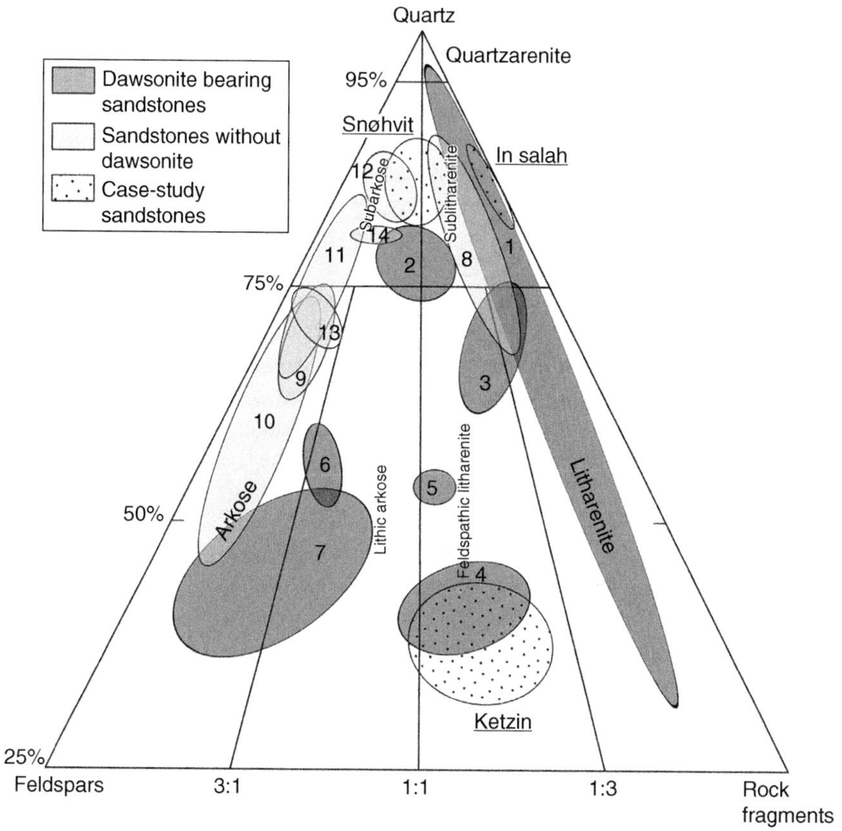

Figure 2.21 Ternary diagram of the mineralogical composition of natural analogue sites for CO_2 storage showing the presence or absence of dawsonite. The numbers correspond to the following sites: (1) Bowen-Gunnedah-Sydney, (2) Lam, (3) Fizzy, (4) Songliao, (5) Cerro Barcino, (6) Springerville-St Johns, (7) Hailer, (8) Miller, (9) Magnus, (10) Bravo Dome, (11) Vert le Grand, (12) Ladbroke Grove, (13) Montmiral, and (14) Paradox. Compositions in Ketzin, In Salah, and Snøhvit are also given (Tremosa et al., 2014).

Reactive transport modeling also enables us to study the long-term mechanical resistance and integrity of fault or cement seals exposed to the percolation of acidic fluids (CO_2-enriched brine) and can improve our assessment of the risk of leakage through a fractured caprock, a faulted reservoir formation, or an abandoned well. Coupling with the mechanical behaviour of the rocks in a multi-phase hydrodynamic context is therefore necessary. To date, however, few studies have coupled reactive transport and rock mechanics. Rutqvist (2012) listed several attempts that have been made in recent years to couple reactive multiphase fluid flow with geomechanical numerical models, with various levels of sophistication (Rutqvist et al., 2002; Taron et al., 2009). Kim and Santamarina (2014) studied the impacts of mineral dissolution on reservoir compaction and the mechanical resistance of caprock using an analytical solution for the mechanics, coupled with the results of a PHREEQC simulation. Wojtacki et al. (2015) used homogenization techniques to determine the influence of the evolving microstructure on the macroscopic properties of the rocks and to estimate the degradation of the stiffness tensor. Additional work is needed to couple multiphase reactive flow processes and deep reservoir geomechanics.

Reactive transport models use thermodynamic databases containing information about the equilibrium constants and the stoichiometry of a number of reactions involving aqueous species in solution, gases, and solid phases. Thermodynamic databases are generally based on sparse data, only in part derived from experimental determinations, and experimental data are generally obtained under specific temperature and pressure conditions. Estimation methods are also largely employed to 'create' thermodynamic data for conditions/species not explored in lab experiments. Quite often, experimental data, measured and/or estimated, are extrapolated/regressed over relatively large temperature and pressure intervals to allow geochemical codes to perform calculations/predictions in a wide range of geological/engineered systems. It follows that the currently available thermodynamic databases are far from being complete and accurate. Thermodynamic databases are constantly being updated and adapted to the conditions of the investigated systems. Concerted validation efforts and the traceability of changes in the compiled data have now become a necessity, guided notably by nomenclatures and technical protocols (Blanc et al., 2012). Some of the most commonly used databases include CTDP (van der Lee and Lomenech, 2004), Thermoddem (Blanc et al., 2012), and the databases associated with EQ3/6 (Wolery, 1992), PHREEQC (Parkhurst and Appelo, 1999), and MINTEQA2 (Allison et al., 1991) geochemical codes.

The validity range of these databases depends not only on pressure and temperature, but also on the salinity (or ionic strength) of the solutions studied and, for CO_2, storage in a saline aquifer requires the use of appropriate formalisms. The Ketzin storage site, for example, located under an evaporitic formation, has an extremely high salinity (over 300 g/l) that cannot be reliably modeled with existing geochemical models. Conventional activity models like Debye-Hückel (Debye and Hückel, 1923), extended Debye-Hückel, or Davies are often used to simulate CO_2 storage, but these models are valid only for diluted systems (0.5–0.7 M). Only the Pitzer formalism (Pitzer, 1973, 1991) is suited to non-ideal solutions with high salinities (>1 M). This has already been pointed out by Gaus et al. (2008) for calcite solubility. Trémosa et al. (2014) have studied the problem in greater depth by comparing several databases using the Pitzer formalism for scenarios with no transport (batch model) and using the chemical compositions of the Ketzin, In Salah, and Snovith sites based on available mineralogy and fluid data. The Pitzer model is also compared with the B-dot formalism developed for solutions with high NaCl concentrations (Helgeson et al., 1969). Trémosa et al. (2014) used the (i) Lawrence Livermore National Laboratory (Johnson et al., 1992); (ii) Thermoddem (Blanc et al., 2012); (iii) Pitzer.dat (Parkhurst and Appelo, 1999); (iv) THEREDA (Altmaier et al., 2011); (v) Sandia National Laboratory (Mariner, 2004; reformatted for PHREEQC by Quintessa (Benbow et al., 2008)); and (vi) SCALE2000 (Azaroual et al., 2004) databases.

These databases are often incomplete, notably for elements such as Al and Si that are essential for the characterization of the alumino-silicates in the mineralogical composition of most of the formations studied. Redox processes are also difficult to evaluate using the Pitzer formalism and the behaviour of Fe and Si is almost impossible to evaluate in the case of high salinity. Trémosa et al. (2014) first compared databases using laboratory measurements of mineral solubility at ambient temperature (Figure 2.22). They then determined, case-by-case and depending on the data available and the validity ranges of the available databases, the compositions of the Ketzin, In Salah, and

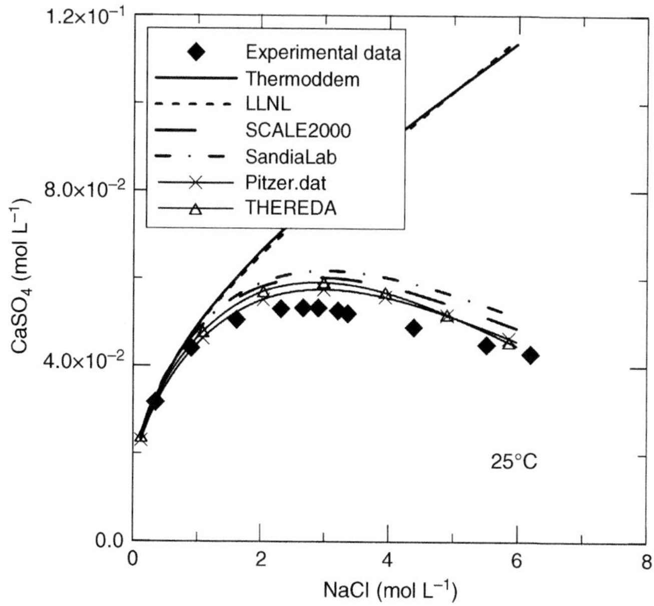

Figure 2.22 Gypsum solubility, measured and calculated, in an NaCl solution at 25 °C from Trémosa et al. (2014) (using experimental data from Linke and Seidel, 1965).

Snovith sites. Although none of the databases seems to be perfectly suited, various trends are clearly visible, depending on the case. The B-dot model seems poorly suited to reproducing the simple mineral solubilities for solutions with salinities over 1 mol/l and the calculations done with the Pitzer formalism are much nearer to the experimental data. The Ketzin site seems to have the highest mineral trapping potential (although this could not be quantified). However, for many ion species, few thermodynamic databases contain enough interaction coefficients (SCALE2000 and SandiaLab seem to have the most) and few are able to calculate energy interactions in the aqueous phase and reproduce mineral solubilities under the conditions encountered in the deep saline aquifer formations foreseen for CO_2 storage.

When reactive transport is modeled, the geometry and fluid dynamics of a system must be carefully adapted so that the geochemical reactivity can be simulated correctly. André et al. (2007) showed that using a one-dimensional, single-phase approach to determine the geochemical impacts of a CO_2 storage site tends to overestimate mineral dissolution (and therefore the related mineral trapping) due to the fact that water that is constantly being injected is not in equilibrium with the chemical composition of the aqueous reservoir. Using a one-dimensional, two-phase approach disregards gravitational processes and those related to the convection of the CO_2-enriched brines, and therefore underestimates the long-term dissolution of gas (Liu et al., 2011). Each approach therefore has its limitations, and it is important that one is aware of them before drawing conclusions concerning the processes being studied.

Depleted reservoirs are a preferential target for CO_2 storage because the coupling of storage with oil or gas production is economically advantageous. However, within the framework of enhanced recovery (of natural gas or oil), currently available models are still poorly suited and few are able to assess the geochemical impacts of storage. Modeling three-phase flow is still at the leading edge of model development and, to date, little coupling with geochemical reactivity has been done (Nghiem et al., 2004).

There are many ways in which models developed to study CO_2 storage might be improved. Retroactive coupling of chemistry and flow is a huge area of research. Changes in microstructure due to chemical reactivity (the opening up of porosity or the creation of channels in the event of dissolution, or a loss of permeability or near-well clogging in the event of precipitation) are still hard to simulate because researchers have not yet been able to estimate and include in models the change of scale of pore-structure-changing processes (Steefel et al., 2013; Bracco et al., 2013).

Remarkable progress is, however, being made thanks to constant improvement in computational resources that make it possible to increase the model resolution or include heterogeneity (Lichtner and Hammond 2012; Trebotich et al., 2014; Hammond et al., 2014; Tian et al., 2014). Other techniques such as streamlines also make it possible to increase the calculation speed considerably – this is often a limiting factor for reactive models (Atchley et al., 2013). Imaging and visualisation at the pore scale also contributes to providing basis data to back up solute transport modelling (Andrew et al., 2014, 2015). Other innovative approaches like the Lattice-Boltzmann theory, which describes the movement of a fluid at the macroscopic scale by resolving microscopic particle movement, provide promising results (Tian et al., 2014). Another new area of research is the integration of isotopic fractionation in models (notably of carbon and oxygen) to detect reactive mechanisms and therefore provide indicators of CO_2 migration or leakage for site monitoring (Humez et al., 2014; Steefel et al., 2014b).

References

Allison, J.D., Brown, D.S., and Novo-Gradac, K.J. (1991). *MINTEQA2/PRODEFA2, a geochemical assessment model for environmental systems, version 3.0. User's manual.* USEPA Report EPA/600/3-91/021.

Altmaier, M., Brendler, V., Bube, C., et al. (2011). THEREDA. *Thermodynamic Reference Database. Final Report, GRS-265*. Available at www.thereda.de/

André, L., Audigane, P., Azaroual, M., and Menjoz, A. (2007). Numerical modeling of fluid-rock chemical interactions at the supercritical CO_2-liquid interface during supercritical carbon dioxide injection into a carbonated reservoir, the Dogger aquifer (Paris Basin, France). *Energy Conversion Management* 48 (6): 1782–1797.

Andrew, M., Bijeljic, B., and Blunt, M.J. (2014). Pore-scale imaging of trapped supercritical carbon dioxide in sandstones and carbonates. *International Journal of Greenhouse Gas Control* 22: 1–14.

Andrew, M., Menke, H., Blunt, M.J., and Bijeljic, B. (2015). The imaging of dynamic multiphase fluid flow using synchrotron-based X-ray microtomography at reservoir conditions. *Transport in Porous Media* 110: 1–24.

Apps, J.A., Zheng, L., Zhang, Y., and Xu, T. (2010). Evaluation of potential changes in groundwater quality in response to CO_2 leakage from deep geologic storage. *Transport in Porous Media* 82: 215–246.

Aradóttir, E.S.P., Sonnenthal, E.L., Björnsson, G., and Jónsson, H. (2012). Multidimensional reactive transport modeling of CO_2 mineral sequestration in basalts at the Hellisheidi geothermal field, Iceland. *International Journal of Greenhouse Gas Control* 9: 24–40.

Aradóttir E., Gunnarsson I., Sigfússon B., et al. (2014). Toward cleaner geothermal energy utilization: Capturing and sequestering CO_2 and H_2S emissions from geothermal power plants. *Transport in Porous Media* 108 (1): 61–84.

Arts, R., Chadwick, A., Eiken, O., et al. (2008). Ten years' experience of monitoring CO_2 injection in the Utsira Sand at Sleipner, offshore Norway. *First Break* 26: 65–72.

Atchley, A., Maxwell, R.M., and Navarre-Sitchler, A.K. (2013). Using streamlines to simulate stochastic reactive transport in heterogeneous aquifers: Kinetic metal release and transport in CO_2 impacted drinking water aquifers. *Advances in Water Resources* 52: 93–106.

Audigane, P., Gaus, I., Czernichowski-Lauriol, I., et al. (2007). Two dimensional reactive transport modeling of CO_2 injection in a saline aquifer at the Sleipner site, North Sea. *American Journal of Science* 307: 974–1008.

Audigane, P., Lions, J., Gaus, I., et al. (2008). Geochemical modeling of CO_2 injection into a methane gas reservoir at the K12-B field, North Sea. In: *Carbon Dioxide Sequestration in Geological Media – State of the Science* (ed. M. Grobe, J C. Pashin, and R.L. Dodge), 1–20. AAPG Studies 59.

Audigane, P., Chiaberge, C., Mathurin, F., et al. (2011). A workflow for handling heterogeneous 3D models with the TOUGH2 family of codes: Applications to numerical modelling of CO_2 geological storage. *Computers and Geosciences* 37: 610–620.

Azaroual, M., Kervévan, C., Durance, M.V., et al. (2004). *SCALE2000 (v3.1): Logiciel de calculs thermodynamiques et cinétiques applicables aux saumures pétrolières, hydrothermales et industrielles* [User's Manual in French]. BRGM.

Balashov, V., Guthrie, G., Hakala, J., et al. (2013). Predictive modeling of CO_2 sequestration in deep saline sandstone reservoirs: Impacts of geochemical kinetics. *Applied Geochemistry* 30: 41–56.

Benbow, S.J., Metcalfe, R., and Wilson, J.C. (2008). *Pitzer Databases for Use in Thermodynamic Modeling*. Quintessa.

Bénézeth, P., Palmer, D.A., Anovitz, L.M., and Horita, J. (2007). Dawsonite synthesis and reevaluation of its thermodynamic properties from solubility measurements: implications for mineral trapping of CO_2. *Geochimica et Cosmochimica Acta* 71: 4438–4455.

Blanc, P., Lassin, A., Piantone, P., et al. (2012). Thermoddem: a geochemical database focused on low temperature water/rock interactions and waste materials. *Applied Geochemistry* 27: 2107–2116.

Bracco, J.N., Stack, A.G., and Steefel, C.I. (2013). Upscaling calcite growth rates from the mesoscale to the macroscale. *Environmental Science Technology* 47 (13): 7555–7562.

Brosse, É., Hasanov, V., Bonijoly, D., and Garcia, D. (2007). The PICOREF project: selection of geological sites for pilot CO_2 injection and storage in the Paris Basin. *1st French-German Symposium on Geological Storage of CO_2*, Potsdam, June 21–22.

Carey, J., Wigand, M., Chipera, S., et al. (2007). Analysis and performance of oil well cement with 30 years of CO_2 exposure from the SACROC Unit, West Texas, USA. *International Journal of Greenhouse Gas Control* 1: 75–85.

Class, H., Ebigbo, A., and Helmig, R. (2009) A benchmark study on problems related to CO_2 storage in geologic formations. *Computers & Geosciences* 13: 409–434.

CO2CRC. (2011). Storage trapping mechanisms according to the geology of the considered reservoir. Available at http://www.co2crc.com.au/aboutccs/stor_trapping.html#trapping

Corvisier, J., Bonvalot, A-F., Lagneau, V., et al. (2013). Impact of co-injected gases on CO_2 storage sites: Geochemical modeling of experimental results. *Energy Procedia* 37: 3699–3710.

Crow, W., Carey, J.W., Gasda, S., et al. (2010). Wellbore integrity analysis of a natural CO_2 producer. *International Journal of Greenhouse Gas Control* 4: 186–197.

Dalkhaa, C., Shevalier, M., Nightingale, M., and Mayer, B. (2013). 2D reactive transport modeling of the fate of CO_2 injected into a saline aquifer in the Wabamun Lake Area, Alberta, Canada. *Applied Geochemistry* 38: 10–23.

Debye, P. and Hückel, E. (1923). The theory of electrolytes. I. Lowering of freezing point and related phenomena. *Physikalische Zeitschrift* 24: 185–206.

Duguid, A. and Scherer, G. (2010). Degradation of oil well cement due to exposure to carbonated brine. *International Journal of Greenhouse Gas Control* 4: 546–560.

EC Directive (2009) Directive 2009/31/EC of the European Parliament and of the Council. *Official Journal of the European Union* L140/114, 29 April 2009. Available at http://eur-lex.europa.eu/LexUriServ/LexUriServ.do?uri=OJ:L:2009:140:0114:0135:EN:PDF

Fabbri, A., Corvisier, J., Schubnel, A., et al. (2009). Effect of carbonation on the hydro-mechanical properties of Portland cements. *Cement and Concrete Research* 39: 1156–1163.

Fabbri, A., Jacquemet, N., and Seyedi, D.M. (2012). A chemo-mechanical model of oil well cement carbonation under CO_2 geological storage conditions. *Cement and Concrete Research* 42: 8–19.

Frykman, P. (2014). The long-term time-trapping-security – a meta-study of published simulations (GEUS). *9th CO2GEONET Open Forum*, Venice, Italy, 20–22 May.

Frykman, P., Nielsen, C.M., Dalhoff, F., et al. (2011). Geological modelling for site evaluation at the Vedsted structure, NW Denmark. *Energy Procedia* 4: 4711–4718.

Gasda, S., Bachu, S., and Celia, M.A. (2004). Spatial characterization of the location of potentially leaky wells penetrating a deep saline aquifer in a mature sedimentary basin. *Environmental Geology* 46: 707–720. doi:10.1007/s00254-004-1073-5

Gaucher, E.C., Robelin, C., Matray, J.M., et al. (2004). ANDRA underground research laboratory: interpretation of the mineralogical and geochemical data acquired in the Callovian-Oxfordian formation by investigative drilling. *Physics and Chemistry of the Earth* 29: 55–77.

Gaucher, E.C., Tournassat, C., Pearson, F.J., et al. (2009). A robust model for pore-water chemistry of clayrock. *Geochimica et Cosmochimica Acta* 73: 6470–6487.

Gaus, I. (2010). Role and impact of CO_2–rock interactions during CO_2 storage in sedimentary rocks. *International Journal of Greenhouse Gas Control* 4: 73–89.

Gaus, I., Azaroual, M., and Czernichowski-Lauriol, I. (2005). Reactive transport modeling of the impact of CO_2 injection on the clayey cap rock at Sleipner (North Sea). *Chemical Geology* 217: 319–337.

Gaus, I., Audigane, P., André, L., et al. (2008). Geochemical and solute transport modelling for CO_2 storage, what to expect from it? *International Journal of Greenhouse Gas Control* 2 (4): 605–625.

Gherardi, F. and Audigane, P. (2013). Modeling geochemical reactions in wellbore cement: assessing pre-injection integrity in a site for CO_2 geological storage. *Greenhouse Gas Science and Technology* 3: 447–474.

Gherardi, F., Xu, T., and Pruess, K. (2007). Numerical modeling of self-limiting and self-enhancing caprock alteration induced by CO_2 storage in a depleted gas reservoir. *Chemical Geology* 244: 103–129.

Gherardi, F., Audigane, P., and Gaucher, E.C. (2012). Predicting long-term geochemical alteration of wellbore cement in a generic geological CO_2 confinement site: tackling a difficult reactive transport modeling challenge. *Journal of Hydrology* 420–421: 340–359.

Gislason, S.R. and Oelkers, E.H. (2014). Carbon storage in basalt. *Science* 344: 373.

Hammond, G.E., Lichtner, P.C., and Mills, R.T. (2014). Evaluating the performance of parallel subsurface simulators: An illustrative example with PFLOTRAN. *Water Resources Research* 50: 208–228.

Helgeson, H.C., Brown, T.H., and Leeper, R.H. (1969). *Handbook of Theoretical Activity Diagrams Depicting Chemical Equilibria in Geologic Systems Involving an Aqueous Phase at One Atm. and 0 to 300 °C*. Cooper & Company, Freeman.

Hellevang, H., Aagaard, P., Oelkers, E.H., and Kvamme, B. (2005). Can dawsonite permanently trap CO_2? *Environmental Science Technology* 39: 8281–8287.

Hellevang, H., Declercq, J., Kvamme, B., and Aagaard, P. (2010). The dissolution rates of dawsonite at pH 0.9 to 5 and temperatures of 22, 60 and 77 °C. *Applied Geochemistry* 25: 1575–1586.

Hellevang, H., Declercq, J., and Aagaard, P. (2011). Why is dawsonite absent in CO_2 charged reservoirs? *Oil & Gas Science and Technology – Revue de l'IFP* 66: 119–135.

Hellevang, H., Aagaard, P., and Jahren, J. (2014). Will dawsonite form during CO_2 storage? *Greenhouse Gases Science and Technology* 4 (2): 191–199.

Hermanrud, C., Huet, B., Prevost, J., and Scherer, G. (2010). Quantitative reactive transport modeling of Portland cement in CO_2-saturated water. *International Journal of Greenhouse Gas Control* 4: 561–574.

Hovorka, S.D., Benson, S.M., Doughty, C.K., et al. (2006). Measuring permanence of CO_2 storage in saline formations – the Frio experiment. *Environmental Geoscience* 13: 105–121.

Huet, B.M., Prevost, J.H., and Scherer, G.W. (2010). Quantitative reactive transport modeling of Portland cement in CO_2-saturated water. *International Journal of Greenhouse Gas Control* 4: 561–574.

Humez, P., Audigane, P., Lions, J., et al. (2011). Modeling of CO_2 leakage up through an abandoned well from deep saline aquifer to shallow fresh groundwaters. *Transport in Porous Media* 90: 153–181.

Humez, P., Lagneau, V., Lions, J., and Negrel, Ph. (2013). Assessing the potential consequences of CO_2 leakage to freshwater resources: A batch-reaction experiment towards an isotopic tracing tool. *Applied Geochemistry* 30: 178–190.

Humez, P., Lions, J., Negrel, Ph., and Lagneau, V. (2014). CO_2 intrusion in freshwater aquifers: Review of geochemical tracers and monitoring tools, classical uses and innovative approaches. *Applied Geochemistry* 46: 95–108.

IPCC (2005). *IPCC Special Report on Carbon Dioxide Capture and Storage*. Prepared by Working Group III of the Intergovernmental Panel on Climate Change (ed. B. Metz, O. Davidson, H.C. de Coninck, et al.). Cambridge: Cambridge University Press.

Jacquemet, N., Le Gallo, Y., Estublier, A., et al. (2009). CO_2 streams containing associated components – a review of the thermodynamic and geochemical properties and assessment of some reactive transport codes. *Energy Procedia* 1: 3739–3746.

Johnson, J.W., Oelkers, E.H., and Helgeson, H.C. (1992). SUPCRT92: a software package for calculating the standard molal thermodynamic properties of minerals, gases, aqueous species, and reactions from 1 to 5000 bar and 0 to 1000 °C. *Computers & Geosciences* 18: 899–947.

Johnson, J.W., Nitao, J.J., Steefel, C., and Knaus, K.G. (2001). Reactive transport modeling of CO_2 storage in saline aquifers to elucidate fundamental processes; trapping mechanisms, and sequestration partitioning. Geological Society Special Publication on Carbon Sequestration Technologies, UCRL-JRNL-205627. London: Geological Society.

Keating, E.H., Hakala, J.A., Viswanathan, H., et al. (2013). CO_2 leakage impacts on shallow groundwater: Field-scale reactive-transport simulations informed by observations at a natural analog site. *Applied Geochemistry* 30: 136–147.

Kelemen, P. and Matter, J. (2008). In situ carbonation of peridotite for CO_2 storage. *Proceedings of the National Academy of Sciences* 11 (45): 17295–17300.

Kempka, T., Kühna, M., Class, H., et al. (2010). Modelling of CO_2 arrival time at Ketzin – Part I. *International Journal of Greenhouse Gas Control* 4: 1007–1015.

Kharaka, Y.K., Cole, D.R., Hovorka, S.D., et al. (2006). Gas–water–rock interactions in Frio Formation following CO_2 injection: implications to the storage of greenhouse gases in sedimentary basins. *Geology* 34: 577–580.

Kharaka, Y.K., Thordsen, J.J., Hovorka, S.D., et al. (2009). Potential environmental issues of CO_2 storage in deep saline aquifers: geochemical results from the Friot-I brine pilot test, Texas, USA. *Applied Geochemistry* 24: 1106–1112.

Kharaka, Y.K., Thordsen, J.J., Kakouros, E., et al. (2010). Changes in the chemistry of shallow groundwater related to the 2008 injection of CO_2 at the ZERT field site, Bozeman, Montana. *Environmental Earth Sciences* 60 (2): 273–284.

Kim, S. and Santamarina, J.C. (2014). CO_2 geological storage: hydrochemo-mechanical analyses and implications. *Greenhouse Gas Science and Technology* 4: 528–543.

Klein, E., De Lucia, M., Kempka, T., and Kühn, M. (2013). Evaluation of long-term mineral trapping at the Ketzin pilot site for CO_2 storage: An integrative approach using geochemical modelling and reservoir simulation. *International Journal of Greenhouse Gas Control* 19: 720–730.

Lagneau, V., Pipart, A., and Catalette, H. (2005). Reactive transport modelling of CO_2 sequestration in deep saline aquifers. *Oil & Gas Science and Technology – Revue de l'IFP* 60 (2): 231–247.

Lichtner, P.C. and Hammond, G.E. (2012). Using high performance computing to understand roles of labile and nonlabile U(VI) on Hanford 300 Area plume longevity. *Vadoze Zone Journal* 120–121: 115–128.

Lindeberg, E., Zweigel, P., Bergmo, P., et al. (2000). Prediction of CO_2 dispersal pattern improved by geology and reservoir simulation and verified by time lapse seismic. *Proceedings of the 5th International Conference on Greenhouse Gas Control Technologies, Cairns,* Australia.

Linke, W.F. and Seidell, A. (1965). *Solubilities of Inorganic and Metalorganic Compounds* (4th ed.). Washington, DC: American Chemical Society.

Lions, J., Devau, N., de Lary, L., et al. (2014). Potential impacts of leakage from CO_2 geological storage on geochemical processes controlling fresh groundwater quality: A review. *International Journal of Greenhouse Gas Control* 22: 165–175.

Liu, F., Lua, P., Zhua, C., and Xiao, Y. (2011). Coupled reactive flow and transport modeling of CO_2 sequestration in the Mt. Simon sandstone formation, Midwest USA. *International Journal of Greenhouse Gas Control* 5: 294–307.

Lopez, S., Hamm, V., Le Brun, M., et al. (2010). 40 years of Dogger aquifer management in Ile-de-France, Paris Basin, France. *Geothermics* 39: 339–356.

Luquot, L. and Gouze, Ph. (2009). Experimental determination of porosity and permeability changes induced by massive injection of CO_2 into carbonate reservoirs. *Chemical Geology*. doi:10.1016/j.chemgeo.2009.03.028

Manceau, J-C., Audigane, P., Claret, F., et al. (2013). ULTimate CO_2 project: Field experiment in an underground rock laboratory to study the well integrity in the context of CO_2 geological storage. *Energy Procedia* 37: 5722–5729.

Manceau, J.C., Tremosa, J., Audigane, P., et al. (2015). 1:1 scale wellbore experiment for a better understanding of well integrity in the context of CO_2 geological storage: design and year 1 observations and modeling. *Water Resources Research* 51 (8): 6093–6109. doi:10.1002/2014WR016786

Mariner, P. (2004). *In-Drift Precipitates/Salts Model*. USDOE Technical Report number ANL-EBS-MD-000045, REV 02, DOC.20041111.0002, DC42152; TRN: US0502943, United States Department of Energy.

Marty, N.C.M., Tournassat, C., Burnol, A., et al. (2009). Influence of reaction kinetics and mesh refinement on the numerical modelling of concrete/clay interactions. *Journal of Hydrology* 364: 58–72.

Mégnien, C. and Mégnien, L. (1980). Synthèse géologique du Bassin de Paris. Stratigraphie et paléogéographie. *Mémoire BRGM no. 101-102-103*. Orléans: BRGM.

Michard, G. and Bastide, J.-P. (1988). Etude géochimique de la nappe du Dogger du Bassin de Paris. *Journal of Volcanology and Geothermal Research* 35: 151–163 [in French].

Mito, S., Xue, Z., and Ohsumi, T. (2008). Case study of geochemical reactions at the Nagaoka CO_2 injection site, Japan. *International Journal of Greenhouse Gas Control* 2: 309–318.

Mito, S., Xue, Z., and Sato, T. (2013). Effect of formation water composition on predicting CO_2 behavior: A case study at the Nagaoka post-injection monitoring site. *Applied Geochemistry* 30: 33–40.

Nghiem, L., Sammon, P., Grabenstetter, J., and Ohkuma, H. (2004). Modeling CO_2 storage in aquifers with a fully-coupled geochemical EOS compositional simulator. *Proceedings of the SPE/DOE 14th Symposium on Improved Oil Recovery, Paper SPE-89474*. Tulsa, Oklahoma, 17–21 April 2004.

Oldenburg, C.M., Moridis, G.J., Spycher, N., and Pruess, K. (2004). EOS7C Version 1.0: TOUGH2 Module for Carbon Dioxide or Nitrogen in Natural Gas (Methane) Reservoirs. Earth Sciences Division, Lawrence Berkeley National Laboratory, University of California, Berkeley, CA.

Oldenburg, C.M. and Rinaldi, A.P. (2011). Buoyancy effects on upward brine displacement caused by CO_2 injection. *Transport in Porous Media* 87: 525–540.

Palandri, J. and Kharaka, Y.K. (2004). A compilation of rate parameters of water–mineral interaction kinetics for application to geochemical modeling. *US Geological Survey Open File Report* 2004-1068.

Parkhurst, D.L. and Appelo, C.A.J. (1999). *User's Guide to PHREEQC (Version 2) – A Computer Program for Speciation, Batch Reaction, One-Dimensional Transport, and Inverse Geochemical Calculations*. Report 99-4259, US Geological Survey Water-Resources Investigations.

Paukert, A.N., Matter, J.M., Kelemen, P.B., et al. (2012). Reaction path modeling of enhanced in situ CO_2 mineralization for carbon sequestration in the peridotite of the Samail Ophiolite, Sultanate of Oman. *Chemical Geology* 330–331: 86–100.

Pitzer, K.S. (1973). Thermodynamics of electrolytes. 1. Theoretical basis and general equations. *Journal of Physical Chemistry* 77: 268–277.

Pitzer, K.S. (1991). Activity Coefficients in Electrolyte Solutions. Boca Raton, FL: CRC Press.

Pruess K., Oldenburg, C., and Moridis, G. (1999). *TOUGH2 User's Guide, Version 2.0.* Lawrence Berkeley National Laboratory, Berkeley, CA. Rep LBNL-43134.

Raoult, Y., Boulègue, J., Lauverjat, J., and Olive, P. (1997) Geochemistry of the Albian aquifer in the Paris Basin area contributes to understanding complex hydrogeological behaviour. *Comptes Rendus de l'Académie des Sciences Série II Fascicule A-Sciences de la Terre et des Planètes* 325 (6): 419–425.

Rojas, J., Giot, D., Le Nindre, Y.M., et al. (1989). *Caractérisation et modélisation du réservoir géothermique du Dogger, Bassin Parisien, France.* Rapport final CCE, EN 3G-0046-F(CD), BRGM R 30 IRG SGN 89.

Rutqvist, J. (2012). The geomechanics of CO_2 storage in deep sedimentary formations. *Geotechnical and Geological Engineering* 30: 525–551.

Rutqvist, J., Wu, Y-S., Tsang, C-F., and Bodvarsson, G. (2002). A modeling approach for analysis of coupled multiphase fluid flow, heat transfer, and deformation in fractured porous rock. *International Journal of Rock Mechanics and Mining Sciences* 39: 429–442.

SACS2 (2002). *Final Technical Report: EU-contract ENK6-CT-1999-00014.* Available at http://carbonsequestration.us/News&Projects/htm/IEAGreen-sacshome.htm

Sato, K., Mito, S., Horie, T., et al. (2011). Monitoring and simulation studies for assessing macro- and meso-scale migration of CO_2 sequestered in an onshore aquifer: Experiences from the Nagaoka pilot site, Japan. *International Journal of Greenhouse Gas Control* 5: 125–137.

Savage, D., Soler, J.M., Yamaguchi, K., et al. (2011). A comparative study of the modelling of cement hydration and cement–rock laboratory experiments. *Applied Geochemistry* 26 (7): 1138–1152.

Scherer, G.W., Kutchko, B., Thaulow, N., et al. (2011). Characterization of cement from a well at Teapot Dome Oil Field: Implications for geological sequestration. *International Journal of Greenhouse Gas Control* 5: 115–124.

Shen, J., Dangla, P., and Thiery, M. (2013). Reactive transport modeling of CO_2 through cementitious materials under CO_2 geological storage conditions. *International Journal of Greenhouse Gas Control* 18: 75–87.

Steefel, C., Molins, S., and Trebotich, D. (2013). Pore scale processes associated with subsurface CO_2 injection and sequestration. In: *Geochemistry of Geologic CO_2 Sequestration* (ed. D.J. DePaolo, D.R. Cole, A. Navrotsky, and I.C. Bourg), 259–303. Available at https://pubs.geoscienceworld.org/msa/rimg/article/77/1/259/140970/Pore-Scale-Processes-Associated-with-Subsurface

Steefel, C., Appelo, C., Arora, B., et al. (2014a). Reactive transport codes for subsurface environmental simulation. *Original Papers in Computational Geosciences.* doi:10.1007/s10596-014-9443-x

Steefel, C., Druhan, J., and Maher, K. (2014b). Modeling coupled chemical and isotopic equilibration rates. *Procedia Earth and Planetary Science* 10: 208–217.

Sterpenich, J., Jobard, E., El Hajj, H., et al. (2014). Experimental study of CO_2 injection in a simulated injection well: the MIRAGES experiment. *Greenhouse Gas Science and Technology* 4: 210–224. doi:10.1002/ghg

Tambach, T.J., Pentland, C.H., Zhang, G., et al. (2013). A comparative study of reactive transport modeling using Toughreact and Mores for modeling CO_2 sequestration. *Second EAGE Sustainable Earth Sciences (SES) Conference and Exhibition*, Pau, France, 30 September 2013.

Tambach, T., Koenen, M., Wasch, L., and van Bergen, F. (2015). Geochemical evaluation of CO_2 injection and containment in a depleted gas field. *International Journal of Greenhouse Gas Control* 32: 61–80.

Taron, J., Elsworth, D., and Min, K-B. (2009). Numerical simulation of thermal-hydrologic-mechanical-chemical processes in deformable fractured porous media. *International Journal of Rock Mechanics and Mining Sciences* 46: 842–854.

Tian, H., Pan, F., Xu, T., et al. (2014). Impacts of hydrological heterogeneities on caprock mineral alteration and containment of CO_2 in geological storage sites. *International Journal of Greenhouse Gas Control* 24: 30–42.

Trebotich, D., Adams, M.F., Molins, S., et al. (2014). High-resolution simulation of pore-scale reactive transport processes associated with carbon sequestration. *Computing in Science & Engineering* 16 (6): 22–31.

Trémosa, J., Castillo, C., Vong, C.Q., et al. (2014). Long-term assessment of geochemical reactivity of CO_2 storage in highly saline aquifers: Application to Ketzin, In Salah and Snøhvit storage sites. *International Journal of Greenhouse Gas Control* 20: 2–26.

van der Lee, J. and Lomenech, C. (2004). Towards a common thermodynamic database for speciation models. *Radiochimica Acta* 92: 811–818.

Vernoux, J.F., Maget, P., Afzali, H., et al. (1997). Synthèse hydrogéologique du Crétacé inférieur du Bassin de Paris. Report BRGM DSGR/IDF R39702. Orleans: BRGM.

Viswanathan, H., Dai, Z., Lopano, C., et al. (2012). Developing a robust geochemical and reactive transport model to evaluate possible sources of arsenic at the CO_2 sequestration natural analog site in Chimayo, New Mexico. *International Journal of Greenhouse Gas Control* 10: 199–214.

Wang, S. and Jaffe, P.R. (2004). Dissolution of a mineral phase in potable aquifers due to CO_2 releases from deep formations; effect of dissolution kinetics. *Energy Conversion Management* 45: 2833–2848.

Wertz, F., Gherardi, F., Blanc, P., et al. (2013). Cement CO_2-alteration propagation at the well–caprock–reservoir interface and influence of diffusion. *International Journal of Greenhouse Gas Control* 12: 9–17.

Wilson, J.C., Benbow, S., Sasamoto, H., et al. (2015). Thermodynamic and fully-coupled reactive transport models of a steel–bentonite interface. *Applied Geochemistry* 61: 10–28.

Wojtacki, K., Lewandowska, J., Gouze, Ph., and Lipkowski, A. (2015). Numerical computations of rock dissolution and geomechanical effects for CO_2 geological storage. *International Journal for Numerical and Analytical Methods in Geomechanics* 39: 482–506.

Wolery, T.J. (1992). EQ3/6: *Software Package for Geochemical Modeling of Aqueous Systems: Package Overview and Installation Guide (Version 7.2).* Lawrence Livermore National Laboratory Report UCRL-MA-10662 PT I, Livermore, California.

Xu, T. and Pruess, K. (2001). Modeling multiphase non-isothermal fluid flow and reactive geochemical transport in variably saturated fractured rocks: 1. *Methodology. American Journal of Science* 301: 16–33.

Xu, T., Apps, J.A., and Pruess, K. (2000). Analysis of mineral trapping for CO_2 disposal in deep aquifers. *Lawrence Berkeley National Laboratory Report LBNL-46992*, Berkeley, CA.

Xu, T., Apps, J.A., and Pruess, K. (2003). Reactive geochemical transport simulation to study mineral trapping for CO_2 disposal in deep arenaceous formations. *Journal of Geophysical Research* 108: 2071. doi:10.1029/2002JB001979

Xu, T., Apps, J.A., and Pruess, K. (2004). Numerical simulation of CO_2 disposal by mineral trapping in deep aquifers. *Applied Geochemistry* 19: 917–936.

Xu, T., Apps, J.A., Pruess, K., and Yamamoto, H. (2007). Numerical modeling of injection and mineral trapping of CO_2 with H_2S and SO_2 in a sandstone formation. *Chemical Geology* 242: 319–346.

Xu, T., Kharaka, Y., Doughty, Ch., et al. (2010). Reactive transport modeling to study changes in water chemistry induced by CO_2 injection at the Frio-I Brine Pilot. *Chemical Geology* 271: 153–164.

Zhang, L., Dzombak, D., Nakles, D., et al. (2013). Reactive transport modeling of interactions between acid gas (CO2 + H2S) and Pozzolan-amended wellbore cement under geologic carbon sequestration conditions. *Energy Fuels* 27: 6921 – 6937.

Zhang, W., Li, Y., and Nyatichi Omambia, A. (2011). Reactive transport modeling of effects of convective mixing on long-term CO_2 geological storage in deep saline formations. *International Journal of Greenhouse Gas Control* 5: 241–256.

Ziabakhsh-Ganji, Z. and Kooi, H. (2014). Sensitivity of the CO_2 storage capacity of underground geological structures to the presence of SO_2 and other impurities. *Applied Energy* 135: 43–52.

3

Process-based Modelling of Syn-depositional Diagenesis

Fiona Whitaker[1] *and Miles Frazer*[2]

[1] *School of Earth Sciences, University of Bristol, Queen's Road, Bristol, BS8 1RJ, UK*
[2] *Chevron Energy Technology Company, 1500 Louisiana, Houston, Texas, 77002, USA*

3.1 Introduction

Carbonate rocks are of global importance both as economically significant reservoirs for oil and gas and as aquifers for water supply. On deposition, most carbonate sediments transmit fluids through matrix porosity, and thus they act as single-permeability systems. However, the reactivity of their constituent minerals makes them especially prone to chemical diagenesis – the alteration of minerals and of pore volume and geometry that occurs largely as a result of water:rock interaction. This can involve precipitation of pore-occluding cements or generation of secondary mouldic, linked-vug and channel porosity, and even cave formation (Moore, 1989). Diagenesis can strongly impact the performance of these rocks as reservoirs and aquifers by modifying the spatial distribution of primary (depositional) porosity and permeability that together determine key properties of storage and flow (Skalinski et al., 2013). Even minor diagenetic changes in porosity can result in major shifts in permeability at core scale, and significantly enhance both the heterogeneity and scale dependence of permeability (Whitaker and Smart, 2000). In addition, diagenesis can affect rock strength and ductability, wettability and absorption characteristics, and control their response to acidization and measurement of porosity via wireline logs (Roehl and Choquette, 2012). Developing a predictive understanding of carbonate diagenesis is thus of critical economic importance.

Reservoir geologists are often called upon to predict diagenetic modification of rock quality across a carbonate platform based on core samples and downhole data for a limited number of sites, and seismic data that are rarely of sufficient resolution to image most diagenetic features. This relies on a workflow that involves inferring diagenetic process from product, and predicting distributions based on an understanding of controls on process (e.g. Dickson and Kenter, 2014). For example, petrographic and geochemical evidence may suggest a given phase of cement was precipitated close to

Reactive Transport Modeling: Applications in Subsurface Energy and Environmental Problems,
First Edition. Edited by Yitian Xiao, Fiona Whitaker, Tianfu Xu and Carl Steefel.

the water table, and thus, by reference to models of the spatial configuration of modern water tables in carbonate environments, we may predict a near-horizontal zone of cementation controlled by the paleo-position of sea-level. This approach has enabled geologists to effectively highlight potential mechanisms and conditions for carbonate diagenesis within the context of a geological history (e.g. Hollis and Walkden, 1996; Hollis, 1998). However, it is only relatively recently that the application of process-based models has allowed us to quantitatively assess the relative merits and shortcomings of mechanisms hypothesized by these more inferential methods (e.g. Frazer et al., 2014).

This contribution provides a review of the status of process-based forward modelling of syn-depositional diagenesis. The unstable mineralogy and high matrix porosity and permeability means that carbonate sediments are at their most vulnerable to diagenesis soon after deposition. We review the processes of syn-depositional diagenesis and the application of reactive transport models (RTMs) to understand controls on the rates, distribution and nature of alteration. The limitations of simulating reactions within a static sedimentological framework are then explored, and an alternative approach is presented that involves explicit coupling of stratigraphic and diagenetic forward models. Applications of this approach to understanding fundamental controls on early meteoric digenesis within a sequence stratigraphic framework are explored using examples of simulations of modern, Cenozoic, Mesozoic and Paleozoic platform carbonates. Findings for outcrop studies are compared with those for reservoirs where constraints on system behaviour are less well known, but where forward stratigraphic–diagenetic modelling offers the potential to generate fully 3D models to help reduce uncertainty in the prediction of those reservoir properties key to the efficient extraction of hydrocarbons.

3.2 Fundamentals of Syn-depositional Carbonate Diagenesis

Syn-depositional (also termed 'eogenetic') diagenesis involves alteration within the context of carbonate deposition and platform growth, close to the surface, involving oceanic and/or meteorically-derived waters. The reactivity of carbonate minerals makes them prone to rapid alteration whenever they encounter fluid chemistries with which they are not at equilibrium. This can lead to a complex diagenetic record of changing environmental conditions. This is particularly the case where primary minerals are dominated by metastable minerals such as high-Mg calcite and aragonite. Most carbonate sediments are precipitated from seawater, either directly or by biotic mediation. The mineralogy of biogenic sediments is determined primarily by the type of calcifying organism, but that of primary precipitates reflects the Mg/Ca of the fluid from which they precipitated. As such, aragonite is the dominant primary marine precipitate at present and during periods in the past when seawater had a similarly high Mg/Ca ratio ('aragonite seas': Sandberg, 1983), whilst low Mg/Ca ratios in seawater ('calcitic seas') promote the precipitation of calcite. This initial mineralogy can have profound implications for the diagenetic potential and thus the evolution of porosity and permeability within the sediment. High-Mg calcite is most reactive and is more prone to neomorphic replacement by low-Mg calcite, accompanied by minimal pore-structure rearrangement. Aragonite tends to stabilize to low-Mg calcite via dissolution–precipitation. Typically, this leads to extensive porosity rearrangement as aragonitic grains are dissolved and pore-filling calcite cements are precipitated either locally or down the flow path. Alteration of low-Mg calcite to form dolomite can further modify the porosity and permeability, and increase mechanical strength, and is discussed by Xiao et al. (this volume).

Active circulation of seawater through the accumulating sediments can lead to precipitation of marine cements (James and Jones, 2015). Platform margins, where permeability and hydrodynamic energy are high, may cement rapidly, whilst in quiescent environments a slow sedimentation rate can permit marine hardground formation (Tucker, 1993). Excessive marine cementation can totally occlude porosity, but moderate cementation at grain-grain contacts can help preserve porosity to depth by strengthening the rock against compaction (Choquette and James, 1987). However, within centimetres of the sediment surface, organic-matter oxidation can significantly modify the carbonate chemistry of the porewaters, leading to dissolution or precipitation depending on the reactivity of the organics and availability of sulfate and iron (Mackenzie, 2005). Marine hardgrounds have been widely recognized as an important control on later fluid flow (Read and Horbury, 1993; Christ et al., 2012; Shekdar et al., 2014), but the impact of early marine dissolution or stabilization of aragonite has received less attention in the geological literature (Christ et al., 2015).

In areas of restricted seawater circulation, evaporation can increase fluid density and generate the potential for downward 'reflux' under the influence of gravity through the underlying sediments. If temperature and fluid chemistry are favourable, this can lead to the replacement of limestone by dolomite and the precipitation of diagenetic gypsum (Machel and Mountjoy, 1986; Machel, 2004). RTM of this variant of syn-depositional diagenesis is discussed in detail by Xiao et al. in Chapter 4 of this volume.

Carbonate units predominantly build during the transgressive stages of a sea-level cycle, with the rate of accommodation generation being closely linked with the rate of relative sea-level rise (eustatic sea-level rise plus subsidence). Because carbonate platform sediments tend to accumulate to sea-level when the platform is flooded, even a relatively minor fall in relative sea-level will expose sediments deposited in a shallow marine environment to meteoric fluids. Exposure-driven diagenesis plays an integral role in the sequence stratigraphic development of a carbonate platform, and importantly, this has the potential to modify sediments across sequence boundaries (Matthews and Frohlich, 1987; Tucker, 1993). In fact, while many sequence stratigraphic studies focus on understanding the depositional phases of a platform, the resulting sediment packages are separated by unconformities that represent extended periods of subaerial exposure. These periods of exposure, and the meteoric diagenesis that occurs during these periods, represent the default state of many carbonate platforms, with phases of deposition occurring only briefly throughout their history. This is not was true during icehouse periods, when high-amplitude variations in sea-level means those platforms that were not drowned by the rapid sea-level rise are left high and dry for much of their early history, interrupted only by relatively short episodes of marine flooding and sediment deposition (Paterson et al., 2006, 2008).

Carbonate sediments formed in marine waters are at a state of disequilibrium when exposed to meteorically-derived waters. As such, it is not unusual for these periods of exposure to play a primary role in establishing the porosity and permeability architecture within a platform early in its diagenetic history. Meteoric diagenesis is characterized by a complex series of dissolution–precipitation reactions which are distributed within a spatial framework defined by hydrological zones (Figure 3.1). Sediments above the

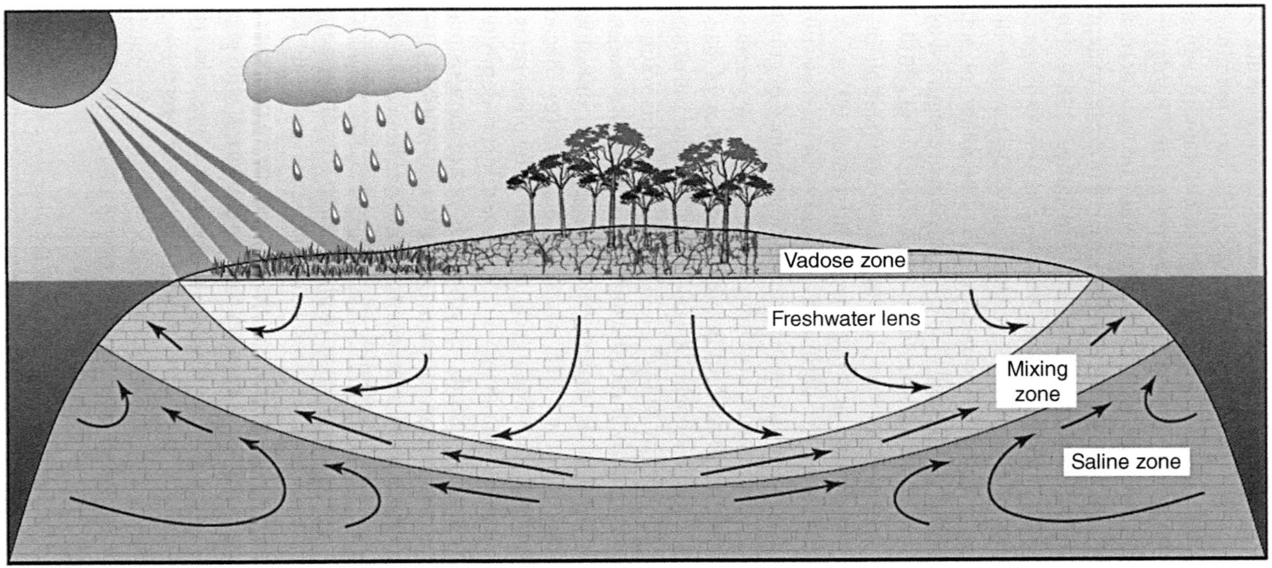

Figure 3.1 Schematic cross-section through a young carbonate island illustrating the distribution of hydrological zones established in response to meteoric recharge ('hydrozones'), within which develop distinct diagenetic products.

water table reside in the vadose or unsaturated zone where pores are filled by a mixture of water and air. The water table forms the upper boundary of a lenticular-shaped body of fresh water – the freshwater lens – that floats upon the underlying denser saline water. This saline zone is separated from the freshwater lens by a zone of intermediate salinity water termed the mixing zone, which provides a route for discharge of meteoric waters. It is widely accepted that each of these hydrological zones plays host to a distinct combination of diagenetic processes.

Fluid chemistries and fluxes within each of these hydrological zones provide a framework for understanding and predicting the distribution of diagenetic processes and products during periods of meteoric exposure (e.g. Whitaker et al., 1997). Because they are defined by hydrological processes, the extent and position of these zones are highly dynamic, with the potential to respond quickly to changes in boundary conditions (Whitaker and Smart, 2007a,b). The elevation of the water table is tied to sea-level, and the thickness of the lens is controlled by the effective recharge, the size and geometry of the exposed area, and the permeability of the rock hosting the lens. Discharge of meteoric water occurs via the mixing zones, increasing in magnitude towards the coast, and leading to compensatory inflow of seawater beneath the mixing zone.

This brief introduction illustrates how diagenetic systems are the product of a complex interplay between solute transport and reaction processes that occurs over a wide range of temporal and spatial scales. Reactive transport modeling (RTM) represents a powerful tool to understand the structural complexity of these systems, the coupling between hydrological and geochemical mechanisms, and the timescales and spatial distribution of diagenesis. Syn-depositional diagenesis presents a particular challenge, as a result of the interactions between episodes of sediment deposition and overprinting by repeated episodes of meteoric diagenesis. While our understanding of carbonate sedimentary processes and the distribution of their products has advanced significantly in recent years, aided through the use of forward sediment models, our capacity for quantitative prediction of diagenetic alteration has failed to advance at a similar rate. This is especially true for systems where the impact of chemical alteration is not clearly confined by depositional boundaries. However, even in systems where diagenesis has followed the original depositional framework, we often fail to predict accurately which facies will be most affected. The remaining parts of this chapter explore two approaches to addressing this problem, their contributions to understanding syn-sedimentary digenesis, their potential and their current limitations.

3.3 Understanding Syn-depositional Diagenesis through RTM

3.3.1 Marine Diagenesis

Whether carbonate sediments are formed in marine or lacustrine environments, diagenesis commences immediately upon deposition. At this earliest stage, cementation and micritization are generally recognized as the dominant rock-altering processes. The recent review of marine cementation by Christ et al. (2015) highlights a broad relationship between water depth and rates of marine cementation based upon several coincident factors. Specifically, near-surface waters, with their higher energy and temperatures, are likely to be more favourable to marine cementation. Thus in areas of shoaling, in particular at platform margins and on windward coasts, wave set-up and tidal action can pump warm (highly $CaCO_3$ supersaturated) seawater through the young sediments (Longuet-Higgins, 1983; Przyborska, 2014). Such higher-energy environments are often characterized by well-sorted and grainy sediments and/or reefs, and the high permeability enhances the flux of cementing fluid, promoting cementation throughout a thicker package of sediments (James et al., 1976; Marshall, 1985). Marine cements are currently dominated by aragonite and high-Mg calcite, and regional differences in seawater chemistry may account for the more rapid cementation rate reported from reefs in the Red Sea (Friedman et al., 1974) compared with the eastern Pacific (Cortés, 1997).

In contrast, in environments dominated by mud deposition, low sediment permeability reduces circulation and thus localizes cementation to form thin hardgrounds (Shinn, 1969). In environments where little to no sediment deposition occurs over long timescales, hardground formation is the primary process underway. Rates of marine cementation within near-surface sediments are often reported as a time to achieve lithification due to cement precipitation, and range from a year to many thousands of years (Christ et al., 2015). These timescale estimates, along with the dependence of marine cementation upon water temperature and hydrodynamic energy, suggests that this process is sensitive to fluctuations in climatic and oceanographic boundary conditions on an annual to millennial timescale.

RTMs have become an established tool for the simulation of the role of degradation of organic matter, directly or indirectly, in driving submarine diagenesis. This work focuses largely on the deep ocean, where solute transport is dominantly diffusive, and models incorporate a range of enzymatic reactions, organisms and oxidants (see review by Arndt et al., 2013). Studies are largely motivated by the importance of these reactions for global cycling of elements in the marine system, and carbonate dissolution and precipitation are often viewed simply as factors that need to be corrected for (Thullner et al., 2009). However, recent work has suggested that syn-depositional precipitation of marine carbonates may be a major player in the carbon cycle (Schrag et al., 2013), and these authigenic carbonates can provide important new insights into paleo-environmental conditions (e.g. Arndt et al., 2009).

RTMs have been applied to quantify diagenetic dynamics in marine sediments (Thullner et al., 2005; Jourabchi et al., 2005). The dominant diagenetic processes within the submarine environment reflect the redox environment, but are also affected by the sedimentary iron cycle (Stoessel, 1992; Soetaert et al., 2007). The sequential utilization of terminal electron acceptors (O_2, NO_3, Mn(VI), Fe(III), SO_4^{2-} and CH_4) results in a characteristic redox zonation, and associated shifts in carbonate saturation state, favouring either carbonate dissolution or precipitation. The former dominates in the oxic and nitrogenous zones, where degradation of organic matter and associated secondary redox reactions generally acidify porewaters (Pfeifer et al., 2002). Below the nitrogenous zone, the picture becomes more complex. A number of anaerobic biogeochemical reactions have been invoked as triggers of authigenic carbonate precipitation (Woods et al., 1999; Greene et al., 2012; Bergmann et al., 2013). Some anaerobic processes (e.g. Fe and Mn reduction) unambiguously increase porewater pH (Soetaert et al., 2007), promoting supersaturation. Other processes, like sulfate reduction, affect both pH and dissolved organic carbon (DIC) and have complex effects on carbonate saturation state dependent on both initial conditions and reaction rate (Meister, 2013; Soetaert et al., 2007). Crucially, none of these metabolic pathways function in isolated geochemical environments, and so RTM simulations are critical to tease apart the evolution of the carbonate saturation state in both the space and time domains.

Sea floor depth is a major factor affecting the processes controlling organic carbon oxidation and thus porewater chemistry (Thullner et al., 2009). Most RTM studies in the marine environment focus on the deeper benthic environment (>200m water depth). This is characterized by low rates of sediment and organic matter supply, relative homogeneity in types of organic matter, low rates of bioirrigation and low microbial diversity, and transport is dominantly by diffusion. RTM simulations in shallow marine environments need to incorporate the effects of temporal variations on timescales from semi-diurnal and diurnal to seasonal, as well as more complex sediment deposition/erosion dynamics, and advective exchange of fluids driven by tides and currents. In addition, organic matter loads tend to be much more heterogeneous, deriving from a range of different sources and including pre-processed terrestrial or fossil organic matter that may be more refractory (Arndt et al., 2013). Finally, in shallow systems, active bioirrigation promotes loss of reducing species to the water column and thus promotes reoxidation within the sediment (Thullner et al., 2009).

Simulations of redox-driven carbonate diagenesis are generally limited to a few tens of centimetres below the sea floor, driven by rapid decay of labile organic compounds, and fail to consider the effect of degradation of the most refractory compounds over much longer distances on geological timescales (Middelburg, 1989). In contrast, diagenesis driven by reflux of dense brines generated by evaporation of seawater in restricted settings has been simulated over distances of tens of kilometres and timescales up to millions of years (Xiao and Jones, this volume and references therein). However, there is clear scope for RTM simulations of reflux systems to consider organic mediation of reactions, given their potential impact on syn-sedimentary dolomitization.

3.3.2 Vadose Zone Diagenesis

When a fall in relative sea-level exposes the young carbonates to meteoric water, there is a radical shift in diagenetic environment. Vadose zone diagenesis occurs in a hydrologically and geochemically complex environment in which flow of diagenetic fluids occurs in pores where gas is often present, and relative permeability evolution can potentially play an important role in controlling diagenetic processes (Brooks and Whitaker, 1997). While net porosity generation is positively correlated with recharge (Whitaker and Smart, 2007b), the dynamics of CO_2 dissolution and degassing from recharge waters are responsible for modulating the spatial distribution of cementation and dissolution within the vadose zone and the top of the water table. Specifically, CO_2 dissolution within descending recharge waters increases their potential for $CaCO_3$ dissolution through the production of carbonic acid. Degassing of CO_2 from these waters can drive cementation through a reverse of this reaction.

In addition to CO_2 in rainfall sourced from the atmosphere, significant CO_2 is generated generated in the soil by microbial and root respiration processes. As a consequence, recharge waters that infiltrate through soil often have an inherently higher potential for diagenetic alteration than waters that flow over bare rock surfaces. Thick, low permeability and laterally continuous soils also act to limit upward loss of CO_2 to the atmosphere, increasing the net dissolution potential of recharge waters. However, where effective pathways for CO_2 degassing exist, recharge waters may re-equilibrate with atmospheric P_{CO2} and precipitation may result. In this case, net dissolution may be minor but local transport of $CaCO_3$ from regions of high to low P_{CO2} conditions may occur. As such, the distribution and extent of soil development on a carbonate island can play a significant role in determining the rates and patterns of diagenesis within the vadose zone.

The nature of fluid flow within the vadose zone also plays an important role in determining the character of diagenesis. In a vadose zone where fluids simply percolate downwards to the water table through matrix porosity, dissolution is concentrated within the upper vadose zone, where waters are at greatest disequilibrium. Solutes are then transported to the water table and freshwater lens. Through time, positive feedbacks between the flow rate and dissolution in the vadose zone lead to the development of connected vug, channel and fracture porosity. This enables a fraction of the recharge to bypass the vadose zone, thereby reducing rates of vadose zone diagenesis (Whitaker et al., 2006; Whitaker and Smart, 2007b; Smart et al., 2011). In thin vadose zone systems, where capillary rise above the water table extends to the surface, evaporation can drive upward transport of water and solutes into the bedrock or soil. Evaporation will drive precipitation of low-Mg calcite cements beneath and within the soil, forming a range of pedogenic calcretes and reducing porosity in the upper part of the vadose zone (Whitaker and Smart, 2007b).

Little attention has been paid to modelling carbonate diagenesis within the vadose zone above the water table, at least in part because of the additional complexities of a multi-phase system with temporal variations in fluid saturation. Xiao and Jones (2006) simulated exposure of aragonitic sediments under humid conditions and suggested that vadose dissolution may extend some 10 m below the exposure surface, whilst minimal alteration is reported under semi-arid conditions. Whitaker et al. (2011) presented a higher resolution study which also incorporated the effects of biotic CO_2 generation and transpiration, and temporal variation in rainfall at a range of timescales. These simulations suggested that rainfall has limited dissolution potential, and, in a semi-arid climate such as characterizes the northern Bahamas, can generate around 0.5% porosity over 10,000 years. Dissolution is focused in the upper 10–20 cm and declines rapidly with depth. However, shallow vadose dissolution is enhanced by CO_2 generated by soil microbial activity and root respiration at depths >20 cm. Deep-rooting vegetation, which characteristically develops in arid to semi-arid settings, can significantly increase the P_{CO2} of vadose air (Figure 3.2). However, total dissolution is primarily a function of total rainfall, due to both higher fluid flux and reduced CO_2 degassing with increasing porewater content.

Whitaker et al. (2011) also examined drives for precipitation of calcite in the vadose zone, in the absence of aragonite. A combination of CO_2 degassing and evaporation drives near-surface precipitation of calcite, resulting in cementation focused in the upper 10 cm of the vadose zone. A second drive for precipitation is solute exclusion during root water uptake, which can result in localized precipitation to depths >1 m below the exposure surface and may account for formation of rhizocretions. The vadose zone is highly dynamic, and responds to individual storm events, sequences of storms and seasonal climatic variations. Whilst the cumulative effect on total porosity may be limited, repeated cycles of dissolution and precipitation can result in significant changes to the pore structure within the vadose zone.

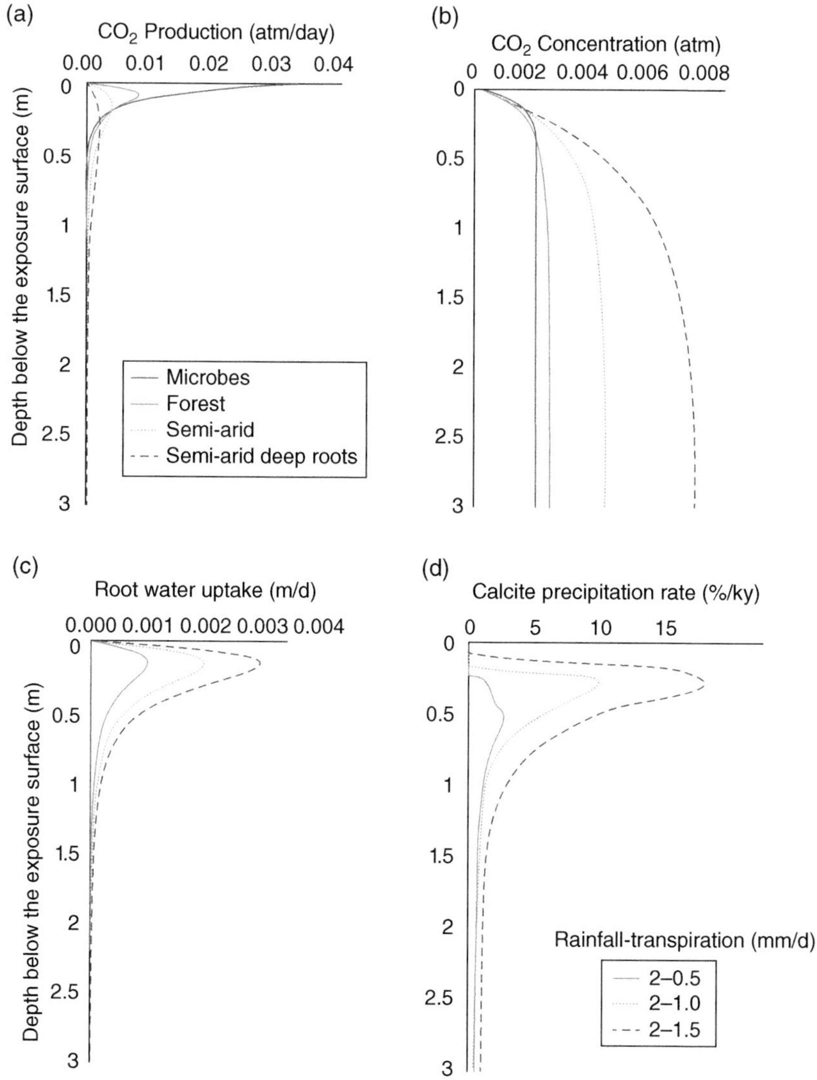

Figure 3.2 (a and b) RTM simulations of vadose zone diagenesis showing production and concentration of CO_2 resulting from bacterial and root respiration for three contrasting vegetation systems. (c and d) Root water uptake and resulting calcite precipitation rates for differing rates of transpiration. *Source:* Adapted from Whitaker et al. (2011).

The overall control of vadose zone thickness was studied by Xiao and Jones (2006), where a 50 m and 100 m vadose zone was simulated to represent exposure during ice-house lowstands, under a humid climate (represented by a 0.5 m/yr recharge flux on the top boundary of the domain). These simulations show that the dissolution potential of the waters recharging the vadose zone is largely depleted within the shallowest part of the vadose zone (Figure 3.3). As a result there is no drive for diagenesis within the deeper vadose zone or at the water table, and reactions within the underlying hydrozones remain largely unaffected.

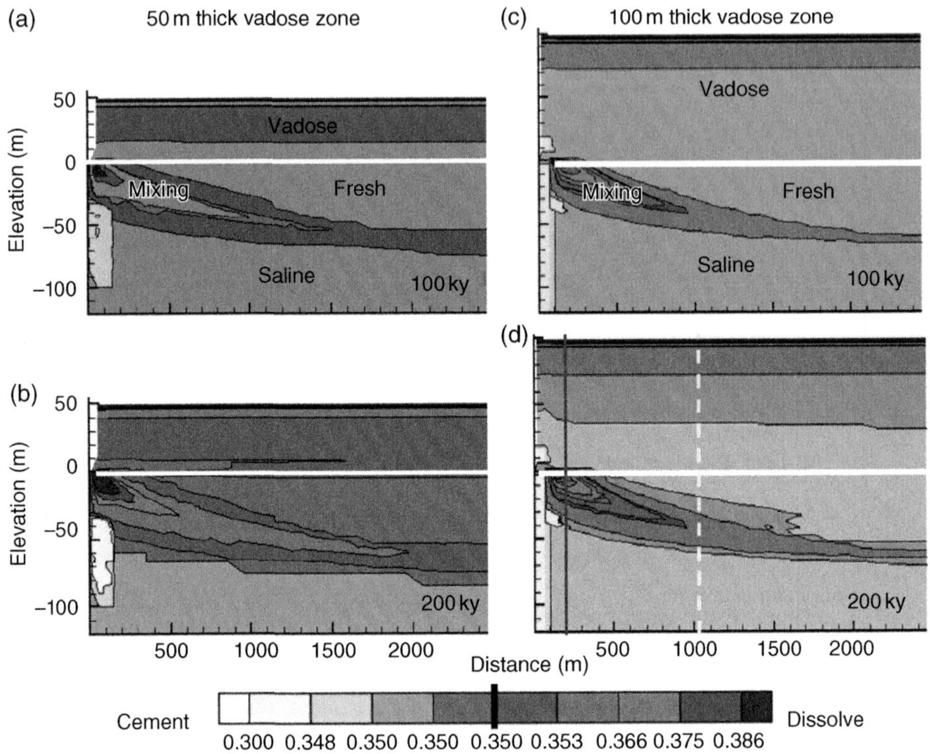

Figure 3.3 Porosity evolution of a carbonate platform at 100 and 200 ky under a humid climate for two sea-level lowstand scenarios, 50 m thick vadose zone (a, b) and 100 m thick vadose zone (c, d). The white horizontal line is the water table. Original fractional porosity is specified as 0.35 (35%). There is enhanced calcite dissolution and porosity increase below the exposure surface and also within the mixing zone near the coast, due to focused fluid flow, which is more obvious with a thinner vadose zone. *Source:* From Xiao and Jones (2006).

3.3.3 Freshwater Lens Diagenesis

Most recharge to the water table occurs following high-intensity storm events. High permeability conduits permit rapid transmission of calcite-undersaturated recharge waters to the water table, even with a thick vadose zone (Jocson et al., 2002; Whitaker et al., 2006). This effect is not shown in the simulations of Xiao and Jones (Figure 3.3), which assume flow through an equivalent porous media. Bypass flow also supplies terrestrial organic matter from the surface and soil to support CO_2 generation within the freshwater lens (Whitaker and Smart, 2007a). Rapid bypass flow can drive dissolution at considerable depths (>100 m) beneath the exposure surface (Whitaker et al., 2006). Studies of freshwater lens geochemistry in modern eogenetic carbonates universally suggest that aragonite stabilization and subsequent dissolution of $CaCO_3$ is widespread, and occurs at higher rates in more humid climates (Budd, 1988; Whitaker and Smart, 2007b). Mixing between recharge waters and those at the top of the freshwater lens has been invoked to account for observations of dissolution at the water table (Mylroie and Carew, 1990). However, simulations based on a range of waters from the modern

Bahamas suggest that this process has limited potential, and may even drive minor cementation depending on the composition of end-members (Whitaker and Smart, 2007b).

Notwithstanding the net increase in dissolved calcium in meteoric groundwaters in carbonate island aquifers, there are zones where cements are precipitated, at least during some time periods. At the water table, cementation can be driven by degassing of CO_2 generated within the freshwater lens, with precipitation forming a water-table calcrete (e.g. Perry et al., 1989). As discussed above for marine porewaters, high rates of organic matter oxidation by sulfate reduction can cause calcite supersaturation, generating cements that might be distinguished by a dull cathodoluminescence indicative of reducing conditions. Finally, in mixed-mineralogy settings, incongruent dissolution should be considered. Thus, for example, dolomites and evaporite minerals are more soluble than calcite in meteoric waters, and their dissolution will result in calcite precipitation (Wigley, 1973).

The recharge of calcite-undersaturated meteoric water to the water table provides the basis for simulations of calcite dissolution in the freshwater lens, which show dissolution focused at shallow depths below the water table. Sainz-Garcia et al. (2011) simulated rates of porosity generation of 2% in 10,000 years at the water table, with dissolutional potential consumed within 6 m. Higher vertical resolution simulations by Cooper and Whitaker (2011) suggest greater focus at the water table, with very rapid porosity generation (up to 17%/ky), declining rapidly with depth below the water table (see Figure 3.4). Cooper (2015) confirmed the importance of grid resolution, but also

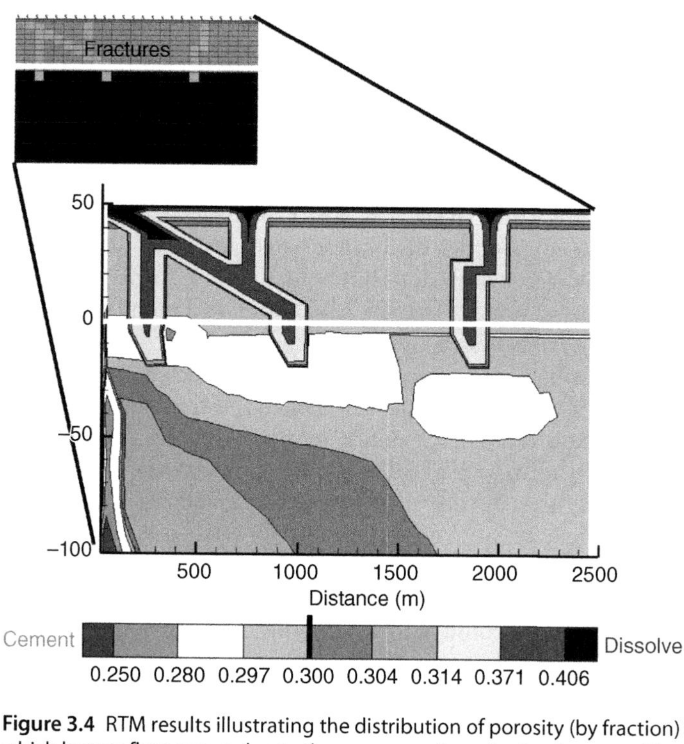

Figure 3.4 RTM results illustrating the distribution of porosity (by fraction) in a carbonate platform in which bypass flow occurs due to the presence of conductive fractures. The platform was specified with a starting porosity of 0.3 (30%) and an initial mineralogy of 100% aragonite. Bypass flow leads to porosity enhancement in the fractures and cementation around the locations where fractures meet the water table (white horizontal line) and thus deliver $CaCO_3$ saturated waters to the freshwater lens. *Source:* From Xiao and Jones (2006).

examined sensitivity to recharge rate and chemistry of recharge waters sampled in the Bahamas. This additional generation of dissolutional potential leads to dissolution rates that are three times greater than those generated by rainwater percolation alone.

The importance of constraining the geochemistry of waters delivered to the freshwater lens is highlighted by contrasting the results shown in Figure 3.3 with the work of Xiao and Jones (2006). The equilibration of recharge waters within the vadose zone results in little or no change in porosity within the lower vadose zone and freshwater lens. However, incorporating bypass-flow, though specifying discrete fractures through a 50 m thick vadose zone, allows for focused dissolution and aragonite stabilization (Figure 3.4). Under a humid climate this system generates porosity within fractures that reach the water table at a rate of some 0.5% in 10,000 years, whilst in the adjacent matrix flow parts of the aquifer, stabilization of aragonite to calcite reduces porosity by 0.4% in the same time interval. Simulations of a calcitic aquifer (Cooper and Whitaker, 2011; Sainz-Garcia et al., 2011) confirm that in the absence of aragonite the bulk of the freshwater lens is diagenetically inactive. However, this work does not consider the effect of degradation of organic matter, generated at or just below the land surface, which has the potential to generate additional acidity at depth with the freshwater lens (Whitaker and Smart, 2007a).

3.3.4 Mixing Zone Diagenesis

Dissolution and precipitation can occur as a result of mixing between waters of different chemical composition due to the non-linear dependence of $CaCO_3$ saturation on factors that include ionic strength and pH (Runnels, 1969). Mixing between meteoric water equilibrated with the limestone aquifer and seawater results in mixtures which are calcite undersaturated, providing the seawater fraction in the mixture is $<10\%$ (Plummer, 1975). This is expressed in modern mixing zones as pervasive dissolution of allochems and matrix, formation of a macroscopic 'swiss cheese' texture, and the development of laterally extensive cave networks (Back et al., 1979; Smart et al., 1988, 2006). Studies of modern mixing zones indicate that dissolution is enhanced both because meteoric waters mix with groundwater of seawater salinity that is close to calcite equilibrium (rather than supersaturated seawater), and because of the oxidation of organic matter suspended in the density gradients of the mixing zone (Smart et al., 1988; Whitaker and Smart, 1997). Oxidation of organic carbon exhausts the dissolved oxygen, but continued oxidation is possible by reduction of sulfate derived from mixing with underlying saline groundwater. Re-oxidation of sulfide in the overlying oxygenated waters will also drive dissolution by formation of sulfuric acid. Alternatively, if the fraction of seawater in the mix is high, other dissolved ions are present, and critically, if the P_{CO2} is close to atmospheric, mixing with fresh groundwaters can cause precipitation of carbonate cements (Frank and Lohmann, 1995; Csoma et al., 2006). Laboratory experiments to determine controls of calcite dissolution and precipitation in fresh/salt water mixing zones by Singurindy et al. (2004) confirm these controls, and generate a simple scale-independent transport theory that agrees well with observations from field studies at a number of sites.

RTM simulations of the fresh/salt water mixing zone indicate that dissolution increases with the P_{CO2} of the lens waters, and the fluid flux, which is controlled by the island width and recharge rate and increases significantly close to the coast (Sanford and Konikow, 1989; Cooper and Whitaker, 2011; Figure 3.5). Dissolution rates are greatest in

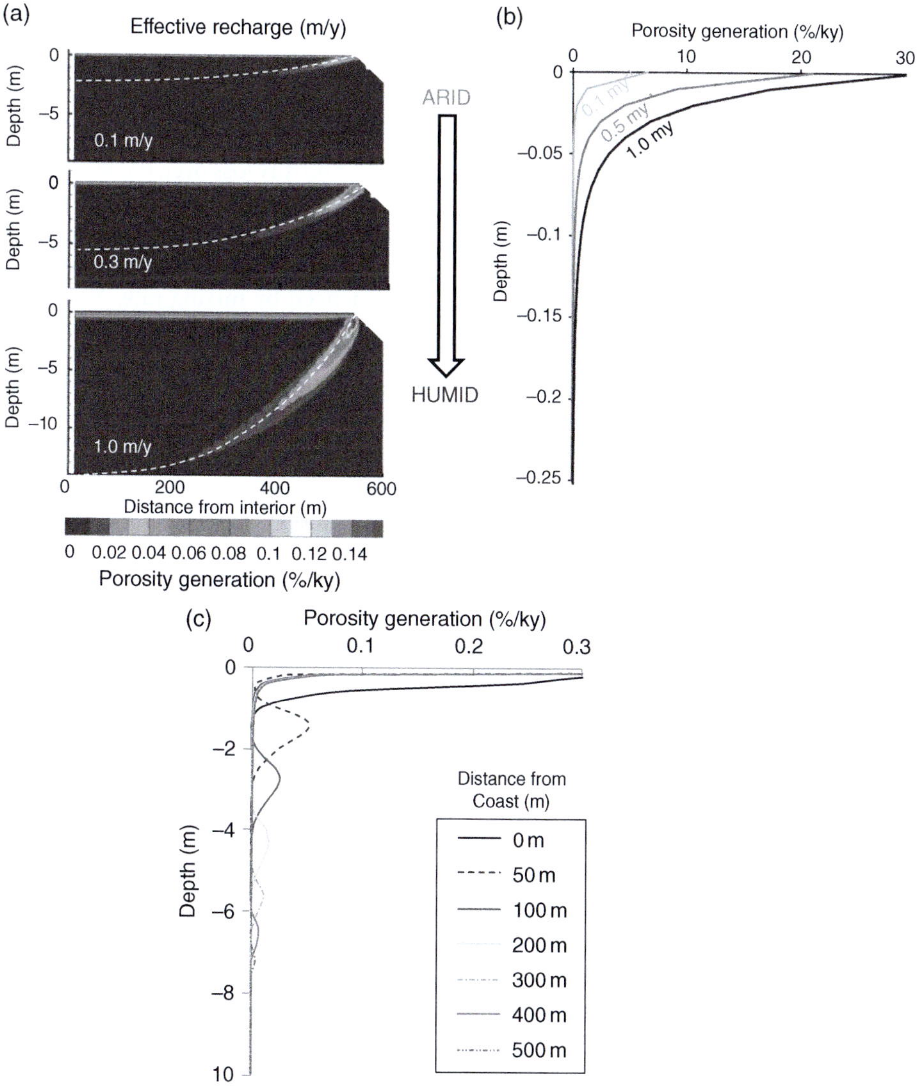

Figure 3.5 RTM simulations showing strong climatic control on meteoric dissolution for half of a small (1 km wide) strip island comprising calcite with 30% porosity and horizontal and vertical permeability of 10^{-10} and 10^{-11} m^2. Recharge chemistry is that of modern low-lying carbonate islands in the northern Bahamas, with a P_{CO2} of 1% at 25 °C (Whitaker and Smart, 2007b), mixing at the lens base with modern local seawater. (a) 2D simulations show an increase in the lens thickness in direct proportion to recharge. Porosity generation is focused at the water table and in the mixing zone adjacent to the coast, but in the interior the mixing zone is diagenetically inert. The dashed white line is the 50% isochlor, which marks the midpoint of the mixing zone. (b) Very high spatial resolution 1D simulations of the zone immediately beneath the water table show the exponential decay in rate of porosity generation with depth below the water table, with >99% of dissolution focused in the upper 10 cm of the freshwater lens, but an increase in dissolution in direct proportion to recharge rate. (c) Variation in mixing zone dissolution rate with depth (note change in vertical and horizontal scales) at increasing distance from the coast. *Source:* Adapted from Cooper and Whitaker (2011) and Cooper (2015).

the upper (fresher) part of the mixing zone where transport enhances mixing. However, simulations by Rezaei et al. (2005) suggest that close to the coast, maximum dissolution shifts to the lower (more saline) part of the mixing zone because of the high fluid flux, rather than where undersaturation is greatest. These results are largely confirmed by the work of Xiao and Jones (2006), but they are also extended by the observation of increased cementation near the platform margin of the order of 5% porosity loss over 100,000 years, due to the increased circulation of seawater beneath the mixing zone. This underlines the important role of RTM simulations for systems where the interplay between transport and reaction is non-trivial. Saturation index calculations are useful, but fail to show how much calcite is dissolved, as this is strongly influenced by mixing rate. No RTM simulations have attempted to incorporate the effect of organically mediated processes within the mixing zone, but a higher P_{CO2} saline end-member both increases dissolution rate and shifts the locus of maximum dissolution to the lower part of the mixing zone (Rezaei et al., 2005).

The fresh/salt water mixing zone has previously been invoked as a potential environment in which dolomite could form, as waters undersaturated with respect to calcite may remain dolomite-supersaturated, due to the high dolomite supersaturation of seawater. Badiozamani (1973) suggested that 'Dorag' dolomite could form from mixed waters (5–50% seawater) undersaturated with calcite yet supersaturated with dolomite. This model was invoked by Land (1973) and numerous subsequent workers to account for dolomites not associated with evidence of brine salinity reflux. However, Hardie (1987) demonstrated that the range of salinities that may be thermodynamically favourable to formation of non-stoichiometric dolomite is negligibly small. Kinetic considerations also cast doubt on mixing zone dolomites, based on significant differences between the rate of dolomite precipitation relative to that of calcite dissolution in the mixing zone (Plummer, 1975). Furthermore, modern mixing zones are characterized by either the absence of dolomite, or only very small amounts of dolomite (e.g. Plummer et al., 1976; Smart et al., 1988; Csoma et al., 2004). Not surprisingly, therefore, RTM simulations fail to show any dolomite forming within the mixing zone. Any minor amount of dolomite which may form within specific (generally higher salinity) zones within the mixing zone (e.g. Ward and Halley, 1985) may be related to organically mediated processes and occur in spite of, rather than because of, the process of groundwater mixing. Mixing zones may be related to dolomitization simply by providing the pump for circulation of saltwater in the near coastal zone (Whitaker and Smart, 1993), as illustrated by the RTM simulations of Xiao et al. (2013) (see also Xiao & Jones, Chapter 4 of this volume).

3.4 Challenges in Reactive Transport Modelling of Syn-depositional Diagenesis

Within the past 15 years, the application of RTM technology has provided quantitative constraints on the distribution and rate of syn-depositional diagenetic processes in carbonate systems, and potential responses to external drivers. Results of simulations have also been used to constrain sets of rules to represent the effects of diagenesis within forward sediment models of early diagenesis (see below). However, the number of workers applying these techniques remains small, and emphasis has been heavily upon understanding dolomitization, particularly by brine reflux (see Chapter 4 of this volume). RTM studies of meteoric systems are starting to provide quantitative insights into the role of climate and soil development in carbonate diagenesis, but this work is in its infancy. The important role of biological processes has been demonstrated by RTM simulations of the vadose zone, and is also known from biogeochemical studies of modern systems to extend below the water table. Simulations of marine porewater chemistry indicate a range of metabolic pathways that may be important for carbonate diagenesis, but the effects of these on the sediments remain to be fully explored. Significant scope exists for development of models of meteoric systems that evaluate these effects, incorporating the complexities of advective solute transport, which is assumed to be minor in marine systems. As our capacity to model individual hydrological zones develops, there is an opportunity for a more integrated approach, coupling processes at the bedrock surface and within the soil, with those in the vadose zone and in the underlying phreatic zone.

In addition to considering only a subset of reactions known to be influential, models developed to date tend to ignore the complexity of the pore structure and thus of solute transport in natural systems. Typically, only a single rock type is considered, or any spatial trends in initial physical or chemical properties are extremely simplistic. As a result, the simulated distributions of diagenetic products only resemble field observations in the most general terms, and fail to replicate features such as the heterogeneity in diagenetic alteration and sharp diagenetic fronts often seen in the field. This not only hinders the application of simulation results to further understanding the operation of diagenesis in more complex geological examples, but also undermines confidence of non-modellers in the utility of the modelling process.

Recent simulations of dolomitization have demonstrated the impact of incorporating greater geological realism in initial rock properties, at a range of spatial and temporal scales, in generating more complex and geologically realistic output (Gabellone et al., 2016). Where there are positive feedbacks between diagenesis, permeability and fluid flow, models with a simple initial distribution of porosity and permeability may be poorly suited for direct application to most simple natural systems. In addition, while many geologists may consider (at least qualitatively) the role of depositional facies in controlling heterogeneities in porosity and permeability, it is much rarer to consider the effects of the distribution of geochemical properties such as reactive surface area and mineralogy. It is still unclear to what degree inherited rock property distributions play a role in determining the locations of preferential flow paths and/or regions of enhanced diagenetic alteration. Simulations suggest that the initial rock parameters (reactivity or permeability) that are most predictive of final diagenetic rock properties likely reflect the Damkohler number (the ratio of reaction rate to the rate of solute advection; Phillips, 1991; Whitaker et al., 2012).

In general, RTM simulations model the change in porosity due to mineral transformations, dissolution and precipitation, but permeability is inferred, most commonly using the Carmen-Kozeny relationship. Simulations of dolomitization by Budd and Park (2017) have shown the effect of the positive feedbacks resulting from diagenetic porosity enhancement on replacement of calcite with dolomite. Heterogeneity in initial porosity and permeability fields, and consequent focusing of reactive fluids, result in the development of pronounced perturbations in the geometry of the dolomite front. In contrast, stabilization of aragonite to less dense low-Mg calcite might be

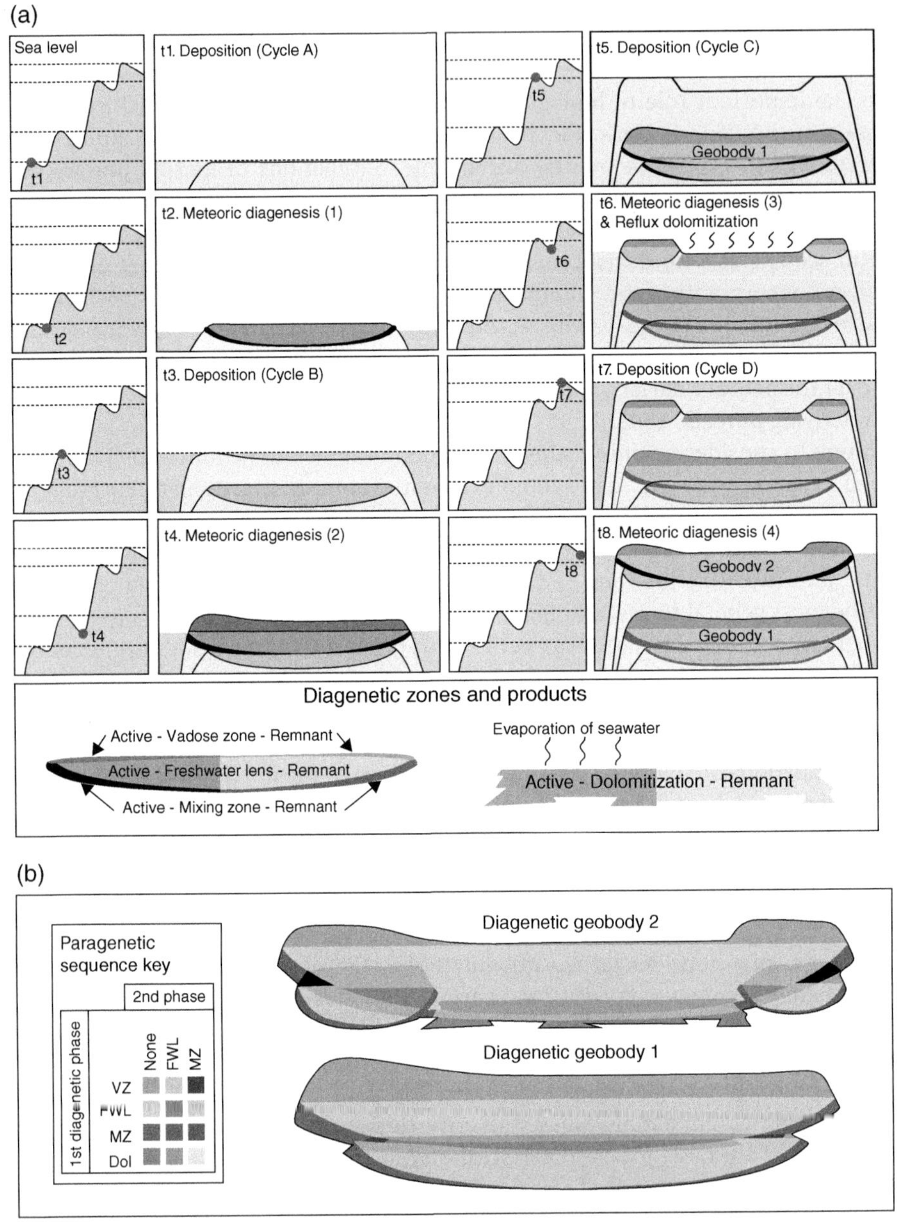

Figure 3.6 (a) A conceptual sequence stratigraphic scenario integrating syn-depositional diagenesis. **t1**: Cycle A is deposited as relative sea-level rises (transgressive sequence) and is then subjected to alteration by a meteoric diagenetic system as relative sea-level falls to **t2**. Cycle B is deposited during a second transgressive phase (ending at **t3**) but during the subsequent exposure event (**t4**) meteoric diagenesis cross-cuts the A/B cycle boundary. Cycle C is deposited during a significant rise in relative sea-level (**t5**), but the subsequent low stand (**t6**) is insufficient to expose the entire carbonate platform. Instead, meteoric alteration is restricted to two smaller hydrological systems beneath platform margin islands. These islands also restrict flow of ocean water to the platform interior, allowing evaporative concentration and dolomitization of platform interior sediments by brine reflux.

expected to become self-limiting, at least where precipitation occurs local to dissolution. However, such simplistic conclusions, based solely on the volume of the solids, ignore the effect of the change in pore geometry, for example, by dissolution of aragonitic grains and precipitation of pore-filling secondary calcite.

Arguably the most critical simplification of most RTM simulations of early carbonate diagenesis is that they fail to account for the influence of changing paleo-environmental conditions – perhaps the definitive aspect of syn-depositional diagenesis. The timescales involved in most diagenetic processes (thousands of years to hundreds of thousands of years) are extended relative to those of changes in boundary conditions known to play a role in controlling diagenesis. For example, shallow meteoric diagenesis is responsive to individual storms, wet/dry seasons and secular changes in climate from decadal timescales to those linked with changes in glacio-eustatic sea-level. We currently have little understanding of whether the effects of boundary condition variations on shorter timescales can be accurately represented over longer time periods using simple time-averaged boundary values.

Changing sea-level will have a dramatic effect on elevation of the water table, and thus the thickness of the vadose zone, as well as on the position and dimensions of all diagenetic hydrozones that develop below the water table. The dynamic and transient nature of these hydrological zones means that a sedimentary package may experience diagenesis within multiple zones over time, and may even be subject to repeated sequences of diagenetic alteration as these zones migrate vertically and laterally in response to changes in relative sea-level (Figure 3.6). The response of the hydrozones to changing sea-level also depends upon the platform geometry. Thus, for a steep-sided platform, the margin defines the coastline largely independent of the magnitude of sea-level fall. In platforms with a gently inclined surface, even a small magnitude drop in relative sea-level can result in a significant lateral shift in the position of the coastline and associated geometry and flux within the freshwater lens, and particularly the mixing zone (Read and Horbury, 1993).

Critical boundary conditions include not only the elevation of sea-level relative to the platform surface, but also the rate of meteoric recharge and the degree of soil development and vegetative cover, which determine the potential for meteoric dissolution (Whitaker and Smart, 1997). The meteoric hydrological system will also differ between a platform isolated from terrigenous input and a land-attached platform, where

Figure 3.6 (Cont'd) These processes do not penetrate deep enough within the succession to overprint previous diagenetic phases. These processes cease upon the re-flooding of the platform and deposition of Cycle D (**t7**), which occurs before the final meteoric alteration phase in **t8**, during which the meteoric system overprints the diagenetic phases that record the **t6** diagenetic system. These periods of deposition and exposure lead to the production of two internally complex but spatially related diagenetic geobodies. The internal architecture of these geobodies formed because of the events detailed in (a) can be seen in (b). The diagenetic events that produced Geobody 1 leads to seven possible paragenetic sequences being recorded within 11 different regions within the geobody. The spatial complexities of diagenesis at t6 lead to a more complex internal architecture within Geobody 2, which contains 12 paragenetic sequences arranged within 27 paragenetic regions. Such a conceptual exercise helps us understand the complexities that could arise from even a relatively simplistic scenario. In reality, the 3D nature of these diagenetic process, as well as their temporally and spatially transitory nature, will lead to greater levels of complexity. (See colour plate section for the colour representation of this figure.)

significant recharge may be sourced from the continental hinterland. The latter may include non-carbonate lithologies, which would influence both flow and fluid chemistry. As such, carbonate diagenetic products represent a record of fluid–rock equilibration that may be interpreted petrographically to constrain a sequence of fluid–rock encounters (Figure 3.6). However, the resulting complexity presents a significant challenge to the understanding of syn-depositional diagenesis.

A limited number of RTM simulations have recently explicitly considered carbonate diagenesis within an evolving sedimentary framework, focusing on dolomitization by seawater and seawater-derived brines (Garcia-Fresca et al., 2012; Palmer et al., 2014; Frazer et al., 2015; Gabellone et al., 2016). Whilst the availability of the requisite computational power to undertake such simulations is becoming less of a limitation, the ability to efficiently implement such developments currently provides a baffle to development. An alternative approach is to approximate the effect of diagenesis within the framework of forward stratigraphic models, and this is the focus of the remaining part of this chapter.

3.5 Coupled Forward Stratigraphic-Diagenetic Models

3.5.1 Stratigraphic Forward Models (SFMs)

Stratigraphic forward models (SFMs) simulate the spatial distribution of sediment accumulation as a function of sediment production, erosion, transport and depositional processes, and the evolution of accumulation over time to build sedimentary sequences. SFMs of carbonate systems vary in complexity both in terms of the granularity with which they consider different sediment factories, and the range and complexity of controls on sediment accumulation that are included. All recognize the primary importance of interactions between eustacy and subsidence that together determine accommodation, with the depth (light) dependence of production rate, and some also include the effects of wind and wave energy and/or climate on physical and biological systems. SFMs have been used by the sedimentological community to test conceptual sequence stratigraphic models by simulating the development of stratal relationships in response to the (often non-linear) interactions between this complex set of controls from first principles. These have generated insights ranging in scale from controls on development of high-frequency metre-scale depositional cycles, to predictions of seismic-scale sequence stratigraphic architecture. SFMs have also been used to replicate and understand the geological complexity of well-constrained outcrop examples, as well as to simulate subsurface data-sets and generate inputs for reservoir models and flow simulations.

3.5.2 Carbonate Diagenesis and Sequence Stratigraphy

The conceptual framework for interpreting carbonate depositional systems using a sequence stratigraphic approach was developed from observations of the geometries and attributes of siliciclastic systems (Sarg, 1988; Crevello et al., 1989). The changes in relative sea-level that control deposition also exert a fundamental control on early diagenetic processes (Goldhammer et al., 1990; Tucker, 1993; Read and Horbury, 1993, and references therein) via their effect on fluid composition and flux, as well as duration of subaerial exposure. These workers postulated that, for example, in a transgressive sequence tract (TST – sediments accumulated during periods of rising relative sea-level), marine calcite cements, increasing in abundance towards the maximum flooding surface, may be an important diagenetic feature. Sequences deposited during periods of high relative sea-level (HST) were thought to be characterized by formation of cements, vuggy and mouldic porosity and possibly dolomitization occurring during higher-frequency sea-level falls. Sequences deposited during falling and low stands (LST) should show a greater effect of diagenesis associated with subaerial exposure, with vadose dissolution and karstification overprinting meteoric cements, vuggy and mouldic porosity. Tucker (1993) stressed that within these simple temporal contrasts, lateral variation might be anticipated, with proximal areas showing more meteoric diagenesis than distal parts. Meteoric processes were thought to be more important under more humid conditions, with the indicative diagenetic feature of more arid conditions being dolomitization by brines. However, because of diagenetic overprinting, consideration of stacking patterns was also seen as important, with the suggestion that progradational, aggradational and retrogradational sequences are each characterized by distinct patterns of syn-sedimentary carbonate diagenesis on the larger scale.

The integration of a conceptual model for diagenesis within a sequence stratigraphic context provides a spatial and temporal framework and can significantly improve the interpretation of key geological observations. Such an approach has been applied to end-member humid and arid meteoric diagenetic systems by both Read and Horbury (1993) and Tucker (1993). Universally applying the learning from these simple end-member studies, however, is often not possible, as diagenetic processes do not act consistently through time. Temporal changes in the dominance of sediment-producing organisms, along with secular seawater chemistry variations, together determine sediment aragonite content, and thus the degree of potential syn-depositional diagenetic alteration. Changes in the magnitude and frequency of relative sea-level change can also dramatically change patterns of meteoric overprinting. As such, greater temporal resolution in geological characterization is required if conceptual models are to contribute to meaningful predictions of diagenesis.

Csoma et al. (2004, 2006) achieved this through the identification of 'diagenetic salinity cycles' within the paragenetic sequences preserved beneath unconformities formed during subaerial exposure. The diagenetic products record a progression from marine conditions, through the mixing zone to the meteoric environment and back through the mixing zone to marine conditions, suggesting a relative sea-level fall and rise. This provides a valuable framework for description and prediction of diagenetic responses to sea-level changes at the depositional cycle scale within the sequence stratigraphy. However, individual cycles may be more difficult to interpret where diagenetic overprinting occurs during repeated exposure cycles, characteristic of periods of higher amplitude eustacy. In addition, the nature of such cycles might be expected to vary systematically with depth below the water table, and systematically from proximal to distal locations, for example in response to variations in the diagenetic activity of the mixing zone (as shown in Figure 3.5). Thus, whilst such simple models are conceptually attractive, there is considerable complexity that may be critical to reservoir quality prediction, but it is difficult (perhaps impossible) to address without recourse to quantitative modelling.

3.5.3 Integrating Diagenesis into SFMs – 1D and 2D Modelling

Although conceptual models provide a clear indication of the potential of incorporating diagenetic processes into SFMs, development of such models has not been widely attempted. Matthews and Froelich (1987) simulated deposition of a 1D column of aragonitic sediment that filled accommodation generated during periods of rising relative sea-level. When sea-level fell, these sediments were subject to meteoric diagenesis within hydrological zones of fixed thickness that migrated vertically through the sediment column in response to changes in relative sea-level. The user specified the nature and rate of diagenesis, which was uniform within each hydrozone. This model showed that the relationship between diagenetic alteration and the subaerial exposure surface forming the top of the sediment column at the time of diagenesis was obscured by the complexity of the glacio-eustatic sea-level history. It would be impossible to unravel the diagenetic history of the simple stratigraphic sequence by traditional stratigraphic and sedimentological methods, and the authors concluded that forward modelling is the preferred approach to analysis of syn-depositional sequences.

Using a 1D model it is impossible to evaluate the impact of lateral variations in the thickness of the hydrozones, which would, for example, enhance mixing zone diagenesis close to the coast, whilst in the platform interior, diagenesis in the freshwater lens should be greater. This limitation was overcome by development of a 2D model (CARB2D$^+$), based on the SFM of Bosence and Waltham (1990), which allowed simulation of syn-depositional diagenesis within a time-varying framework of hydrologically defined diagenetic zones (Whitaker et al., 1997, 1999). CARB2D$^+$ includes both spatially varying sediment texture and spatial variations in the distribution and size of meteoric hydrozones controlling syn-sedimentary diagenesis. The water table varies with sea-level, whilst the geometry of the lens and mixing zone depend on the island size and geometry, meteoric recharge, and upscaled permeability. Whitaker et al. (1997) present simulations of the diagenetic evolution of an aggrading 6 km diameter carbonate platform subject to the same types and rates of diagenetic processes as modelled in 1D by Matthews and Froelich (1987). This includes mixing zone dolomitization, though given our understanding of the importance of carbonate dissolution in the mixing zone, this is perhaps more usefully viewed as mixing zone porosity.

The results are summarized in Figure 3.7. The glacio-eustatic sea-level curve from the past 500,000 years, combined with a simple linear rate of subsidence, resulted in short periods of subtidal sediment accumulation (total duration 65 ky) interspersed with extended periods of exposure (total duration 435 ky). Most of the depositional sequences that formed are separated by unconformities representing >60 ky exposure, during which meteoric diagenesis affected the accumulating platform sediments. Simulations show a clear stacked sequence of diagenetic zones, which were readily recognizable because of their lateral continuity, and distinct trends in diagenetic evolution from platform interior to margin, which could not be represented in a 1D model. Apparent spatial associations between unconformity surfaces and underlying diagenetic zones suggest causality, but these are misleading. Instead, each diagenetic geobody is a product of overprinting during a number of separate low stands. For a given sea-level curve, differences in subsidence rate (controlling accommodation) simply condense platform-top stratal sequences, but they exert a major control on their diagenetic evolution by increasing the extent of diagenetic overprinting. In this example, steep margins

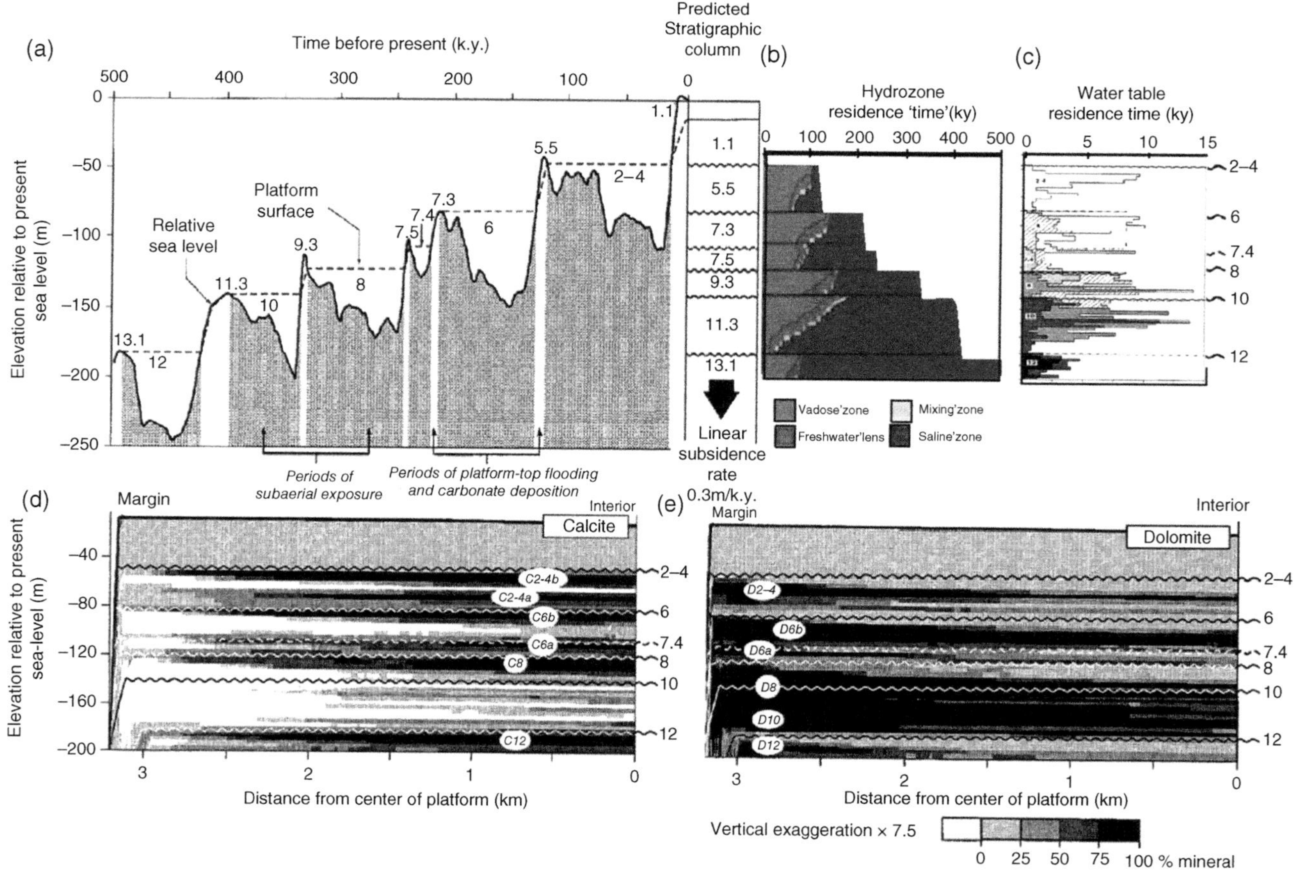

Figure 3.7 (a) Relative sea-level curve showing periods of exposure of the top of the aggrading platform (shading), with depositional units and subaerial unconformities numbered by oxygen isotope stages. (b) Residence time in diagenetic hydrozones and (c) at the water table, during periods of platform top exposure. Grey shading differentiates water table residence time during individual sea-level low stands. (d) Simulated distribution of calcite and (e) dolomite. Diagenetic geobodies are numbered for the depositional sequences defined by the overlying unconformity and to which they might be supposed to relate genetically, although as shown in (c), this association is coincidental and geobodies result from overprinting during several separate low stands. After Whitaker et al. (1997).

prevented significant progradation of the platform. This maintained a near-constant spatial distribution of hydrozone thickness over time relative to sea-level, and meant that diagenetic patterns resulted solely from vertical migration of hydrozones. This work does not aim to simulate a specific platform, but it is noted that stacked sequences of mixing zone caves are described in the margin of the eastern Yucatan by Smart et al. (2006), although given the more extensive platform and lower subsidence, the periods of time involved here are likely to be substantially greater.

Jones et al. (2004) assessed the utility of CARB2D$^+$ to predict early diagenesis in a Pennsylvanian reservoir example, the Northwest Extension of the Salt Creek Field in the Midland Basin of west Texas. The reservoir interval consists of cyclically stacked wackestones, packstones and oolitic-skeletal grainstones. Early diagenesis has significantly modified the reservoir quality by pore-filling meteoric cements and fabric-selective dissolution to generate abundant mouldic porosity. To investigate model application in a data-poor environment, a representative 2D section was defined using an initial surface, the external platform (decompacted) geometry, a regional-scale burial history, and literature constraints on Pennsylvanian sea-level and climate.

Most simulations using different sea-level amplitude and frequency cycles resulted in very different platform geometries from that observed. For the simulation that best fitted the seismic architecture (eustacy dominated by 400 ky cycles), cumulative residence time of sediments in different hydrozones was tracked for a semi-arid climate (Figure 3.8). Simulated freshwater lens and mixing zone residence times show distinct lateral trends, as expected from the thickness of these hydrozones, with distance from the coast, and distinct vertical trends in response to lower-order sea-level cycles. At the time of platform demise, the simulated (post early-diagenesis) porosity distribution is highly heterogeneous. The largest simulated increase in porosity (up to 30%) was associated with dissolution in the freshwater lens. Porosity occlusion (up to 35%) is dominated by cementation in the vadose zone and the freshwater lens, primarily due to the reprecipitation of calcite derived from surface dissolution, but also from the stabilization of aragonite to calcite. Early porosity distribution was generated using diagenetic rate data from RTMs and hydrozone residence times from CARB2D$^+$, and compacted (based on sediment texture) to reservoir depth. Comparison of predicted reservoir porosity with data from ooid grainstone intervals in seven cored wells shows a relatively good match, although with overestimation of both the lowest (<10%) and highest (30–40%) porosity classes. Using the proportion of different pore types and predicted reservoir porosity generated a good match to observed reservoir permeability distribution.

3.5.4 3D Forward Stratigraphic-Diagenetic Models (FSDMs)

CARB3D$^+$ was developed to enable simulation of more complex platform geometries, sedimentary architectures and diagenetic patterns in 3D (Figure 3.9). It differs from the earlier model in employing an explicit simulation of the evolving hydrodynamic environment resulting from interaction of waves and currents with the bathymetry, and resulting sediment transport. The new model also introduced an alternative mode of parameterization of diagenetic rates. Within each hydrozone, the rates of diagenesis are spatially variable and determined by the climate via its effects on the flux of reactive fluids and geochemically determined diagenetic potential. This potential is controlled

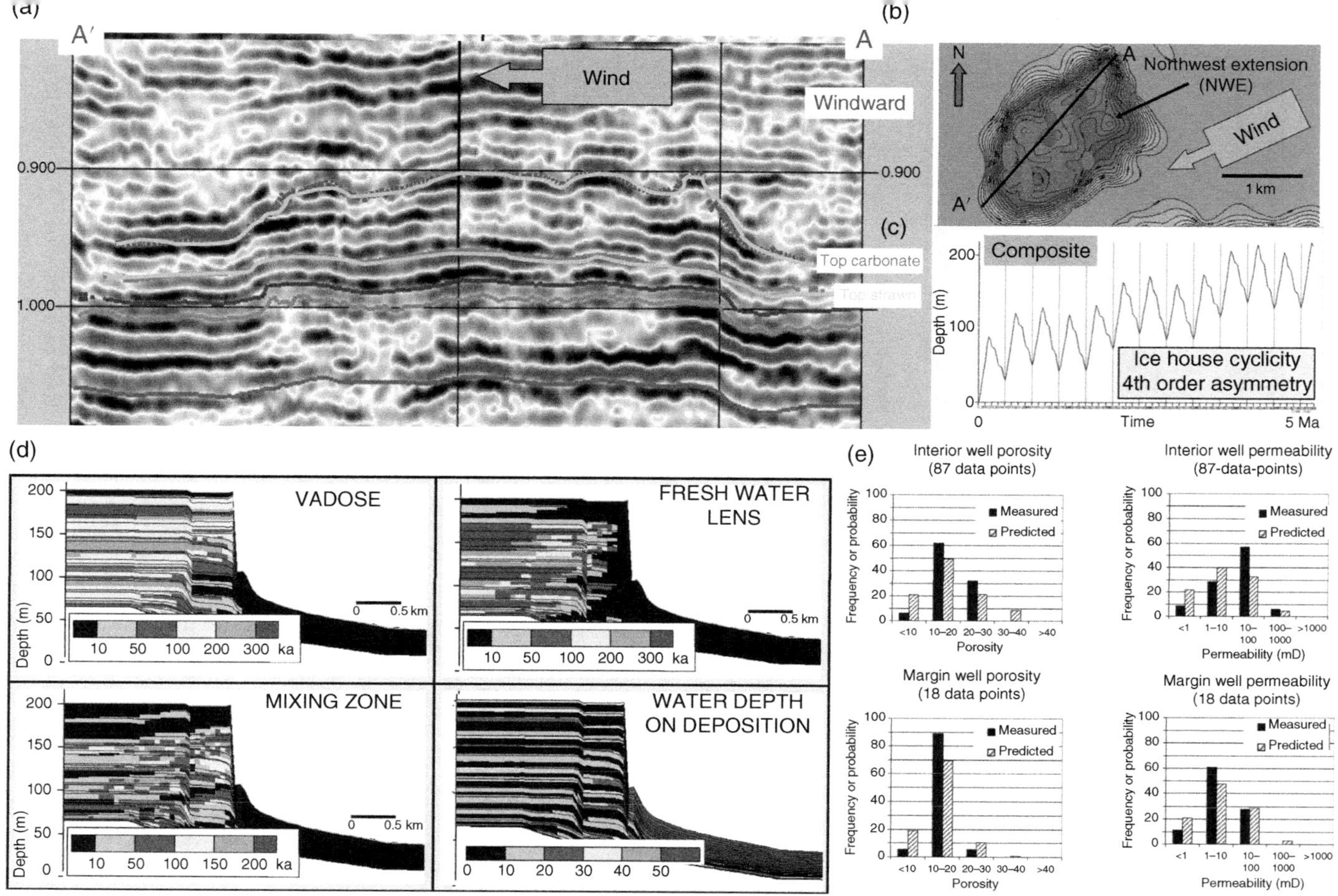

Figure 3.8 Architecture of NW Extension of Salt Creek defined from 3D seismic data shown for (a) 2D slice along and (b) plan view showing line of section A–A′. (c) Synthetic sea-level curve used in CARB2D$^+$ to model (d) residence time in meteoric diagenetic hydrozones, shown here for windward platform margin. (e) Predicted and observed porosity and permeability for ooid grainstones in seven cored interior and marginal wells. *Source:* After Jones et al. (2004), used with permission of AAPG.

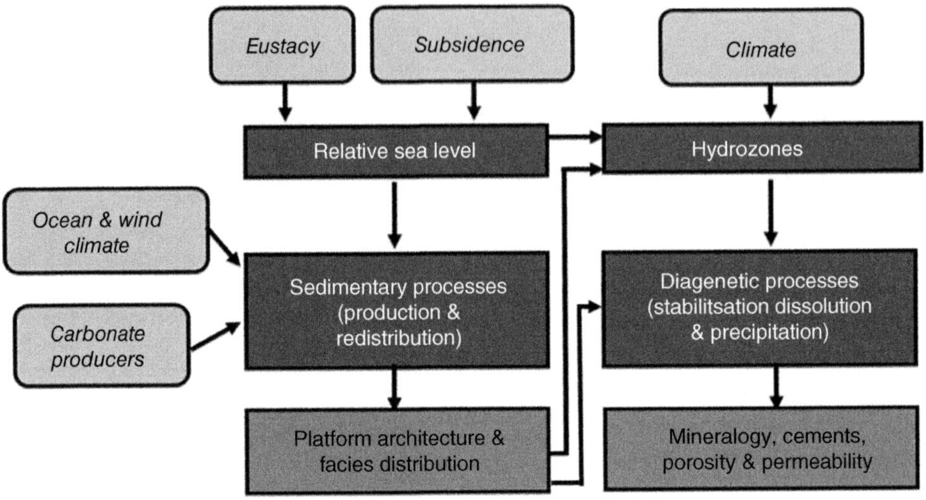

Figure 3.9 Outline structure of CARB3D$^+$ including extrinsic controls, model components and primary outputs.

by the degree of soil development and associated biological activity. Grain-size dependent reaction rates and feedback between the evolving porosity–permeability characteristics and distribution of the hydrozones are also included. The rules governing diagenesis are derived both from RTM studies (see references in Section 3.3 of this chapter) and from hydrochemical studies of modern carbonate islands (e.g. Whitaker and Smart, 2007a,b).

CARB3D$^+$ can also model the effects of early meteoric dissolution and precipitation of cements on porosity and permeability. The model tracks progressive changes in the volumes of cement and fabric-selective (matrix) and non-fabric-selective (secondary karstic) porosity. Primary permeability is determined directly from porosity and is dependent on grain size, following the approach of Lucia (2007). Dissolutional porosity is portioned between fabric-selective (matrix) and non-fabric-selective porosity, but cementation preferentially reduces fabric-selective porosity. Non-fabric-selective porosity is modelled as the progressive enlargement of capillary tubes (analogous to touching-vug porosity) with an assumed spacing, allowing the simple prediction of secondary permeability using the Kozeny-Carman model. Total porosity and permeability are the sum of both fabric-selective and non-fabric-selective components.

3.5.5 Application of CARB3D$^+$ to Understanding Carbonate Sedimentation and Syn-sedimentary Diagenesis

Over the last ten years, CARB3D$^+$ has been used in industry and academia for training, to investigate controls on platform development, and to understand specific reservoirs and outcrop analogues. Some applications have developed a single, large complex model of a specific reservoir, whilst others have involved rapid modelling of multiple model realizations. The resulting high vertical resolution geological models can be upscaled and input to standard reservoir simulators. More fundamentally, CARB3D$^+$ modelling has underlined the critical need to integrate studies of syn-depositional diagenesis with the stratigraphic evolution of carbonate buildups and identified gaps in our understanding of governing processes. Here we summarize a range of applications of the code by our group and collaborators in industry and academia, which are in the public realm (Table 3.1). These range from the Recent and the Cenozoic, for which we have more confidence that modern sedimentological and diagenetic processes can be reasonably applied, to the deeper past, where biota, soils, atmospheric and seawater composition have at times been quite different. Because of the important control on sediment texture of both flux of diagenetic fluids and sediment reactivity, we include studies of the distribution of sediment texture and exposure surfaces from SFM, as well as specific simulation of syn-sedimentary diagenesis.

3.5.5.1 Prediction of Sediment Distribution and Platform Architecture using CARB3D$^+$

Syn-depositional diagenesis is controlled both by extrinsic parameters (fluid flux and geochemistry) and intrinsic parameters, primarily the permeability and reactivity of the substrate. The latter are strongly facies controlled, with fine-grained materials having lower permeability, but also higher effective reactive surface area. The capacity of the model to generate geologically reasonable facies distributions was evaluated by Paterson et al. (unpublished data) by comparison of maps of the distribution of current speed and sediment texture generated by CARB3D$^+$ with observational data. Wind and open ocean current speeds and direction, and initial surface geometries based on modern bathymetry were specified for two platforms of contrasting size: south Cocos (Keeling) platform in the Indian Ocean and the rather larger Caicos platform in the Caribbean. Figure 3.10 shows good agreement between synthetic facies maps and those based on sediment sampling and analysis of Landsat imagery by Kaczmarek et al. (2010), which postdate the modelling work. Current speeds on the simulated south Cocos (Keeling) platform also show a good match with those measured for passages between the barrier islands and across the platform. However, to replicate the distribution of reefs requires simulation of more than one principal wind direction to reflect the monsoonal climate (not possible at the time of the original study).

Over time, the vertical stacking of these spatial patterns of sediment distribution gives rise to a complex 3D geometry of permeability and reactivity. The evolution of platforms during 'ice-house' periods of high frequency and high amplitude glacio-eustatic sea-level fluctuations is complex and gives rise to rhythmic patterns of carbonate sedimentation (Burchette and Wright, 1992; Read, 1985). These patterns are often used for lateral correlation between wells or outcrop sections, as well as to infer timescales for sediment accumulation. Paterson et al. (2006) used CARB3D$^+$ to systematically investigate the effects of key controls on accumulation and sedimentary architecture of isolated ice-house platforms. Longer period (4th order) sea-level cycles were shown to dominate over high-frequency (5th order) cycles. Subsidence rate exerts a fundamental control on stacking patterns and the number of 'missed beats' – when the platform top remains emergent at sea-level maxima – which create non-Milankovitch sequence boundary frequencies.

In greenhouse periods, the relatively low amplitude of sea-level fluctuations results in a quite different pattern of sediment accumulation for isolated carbonate platforms. Forkner et al. (2010) compared synthetic stacking patterns driven by eustatic oscillations that result from Milankovitch astronomical forcing, generated using CARB3D$^+$, with a

Table 3.1 Applications of CARB3D⁺.

Case study	Age	Focus	Data	Period	Platform dimensions	Figure and reference
Cocos (Keeling)	Modern	Control of hydrodynamics on facies distribution	Hydrodynamic data, sediment samples and satellite interpretation	1 ky	13 × 18 km, <10 m thick	Our Fig. 3.10a–f Paterson et al. (unpublished data)
Caicos	Modern	Control of hydrodynamics on facies distribution	Satellite interpretation and sediment samples	1 ky	100 × 70 km, <10 m thick	Our Fig. 3.10 g–i Paterson et al. (unpublished data)
Generic simulations compared with North Andros core	Pleistocene	Control of facies and sea-level on cementation below exposure surfaces	Porosity from Pleistocene core, modern hydrochemistry and RTM	0.5 My	3.5 km diameter, 40 m thick	Our Fig. 3.14 Smart & Whitaker (unpublished data)
La Molata, Cabo de Gata, south-east Spain	Late Miocene	Heterozoan to photozoan sedimentation; effects of eustacy on sediment accumulation over complex volcanic basement	Outcrop mapping, petrography geochemistry	3.4 My	>4 km diameter, 150 m thick	Our Fig. 3.12 Kaczmarek et al. (2009)
Natuna, Sarawak Basin, Indonesia	Mio-Pliocene	Control of eustacy and subsidence on sedimentation and hydrozone residence time	Seismic, core and well-logs from reservoir	4 My	4 × 8 km, 105 m thick	Our Fig. 3.13 Paterson et al. (2006, 2008)
North Madura, East Java Basin, Indonesia	Early Miocene	Controls on mound coalescence and platform demise	3D seismic from potential reservoirs	4 My	1–7 km diameter, 360 m thick	Our Fig. 3.11 Hughes et al. (2008)
Lombardy, Italy	Triassic	Controls on platform progradation and marine cementation, development of rigorous workflow.	Outcrop mapping, petrography geochemistry		8 × 10 km diameter, 500 m thick	Berra et al. (2016)
Latemar, Italy	Triassic	[1]Cyclicity in stacking patterns [2]Climatically controlled diagenesis and tracer experiments to evaluate pore connectivity	Outcrop mapping, petrography geochemistry	1 My	2.5 km diameter, 170 m thick	[1]Forkner et al. (2010) Our Fig. 3.15 [2]Whitaker et al. (2014)
Salt Creek, west Texas	Carboniferous	Porosity and permeability from meteoric diagenesis	Logs, 3D seismic and core	1 My	3.4 × 5 km 160 m thick	Our Fig. 3.8 Jones et al. (2004)
Tengiz, Caspian Sea	Carboniferous	Diagenetic response to switch from aggradation to progradation and implications for reservoir quality	Seismic, well log, core, petrography geochemistry, isotopes		17 × 20 km, 500 m thick	Our Fig. 3.16 Frazer et al. (2014b)

well-studied measured stratigraphy. Three sets of cyclic parameters were compared using Fischer plots created from 1D columns extracted from a 3D model of the well-studied Mid-Triassic Latemar Platform of northern Italy. They demonstrated how the use of Milankovitch insolation as a proxy for high-frequency eustacy can generate a succession with vertical stacking trends highly comparable with those observed in outcrop, and importantly provide a process-based approach for constraining this key input to SFMs. Although the suggestion that stacking patterns can be interpreted in terms of Milankovitch-driven ('allogenic') cyclicity in sea-level often underpins conceptual models of sequence stratigraphy, some have suggested that such patterns are illusory or related to auto-cyclic processes ('allogenic' cyclicity; Burgess, 2008). Cooper and Smart (2010) generated CARB3D$^+$ synthetic stratigraphies for isolated carbonate platforms using both stationary Milankovitch cyclic and random sea-level drivers to evaluate statistical approaches for distinguishing allogenic and autogenic cyclicity. This suggested that the exponential distribution test for Poisson-related distributions used by Burgess (2008) is not capable of reliably identifying stationary cyclic behaviour, whilst the layer thickness inventories test devised by Bailey and Smith (2005) generally detected cyclicity in the Milankovitch cyclic successions and, given a timescale calibration, the frequency of the main drivers.

Gulden and Simo (2012) used CARB3D$^+$ to test field-observation-derived hypotheses that describe functional relationships between carbonate system characteristics and platform geometry as defined by the progradation–aggradation ratio (P/A). The novel aspect of this study was the 'ensemble' approach employed, whereby the numerical engine of the code was decoupled from the graphical user interface to enable many model realizations. This approach was directed at understanding systems for which 'best' parameter values are unknown and verification data are limited. The P/A, which describes the temporal translation of the break in depositional profile that defines the platform margin, can be calculated from outcrop data or seismic profiles, and can be related to accommodation history and sediment supply. The characteristics of the ensemble of simulations show that aggrading and prograding platforms are the most common geometries, with a trade-off between a system's tendency to prograde and to aggrade. The initial bathymetry is more important than system characteristics in controlling P/A; deep, steeply sloping basins favour aggradation, while shallow, gently sloping basins favour progradation. Backstepping is less common, but most likely in deep, steeply sloped basins with weak, shallow reef production and in systems with increasing accommodation and many high-amplitude, high-frequency sea-level oscillations. Downstepping is rare, especially in deep basins with steep platform slopes. Strongly negative relative sea-level change is near-necessary for downstepping, together with small-amplitude, high-frequency sea-level oscillations and healthy margins. SFM using CARB3D$^+$ provides a link between seismically resolvable features (such as P/A), which characterize the platform architecture and facies proportions, and sediment characteristics on the platform that are key to reducing the uncertainty in reservoir quality prediction. However, because of their commercial potential, the nature of these relationships remains confidential.

A combination of seismic data and 3D SFM was also used by Hughes et al. (2008) to explore how interactions between tectonism, eustasy, carbonate accumulation rate, and environmental stresses impact the nucleation, growth, and demise of Early Miocene carbonate platforms from the East Java Basin (Figure 3.11). Combining seismic

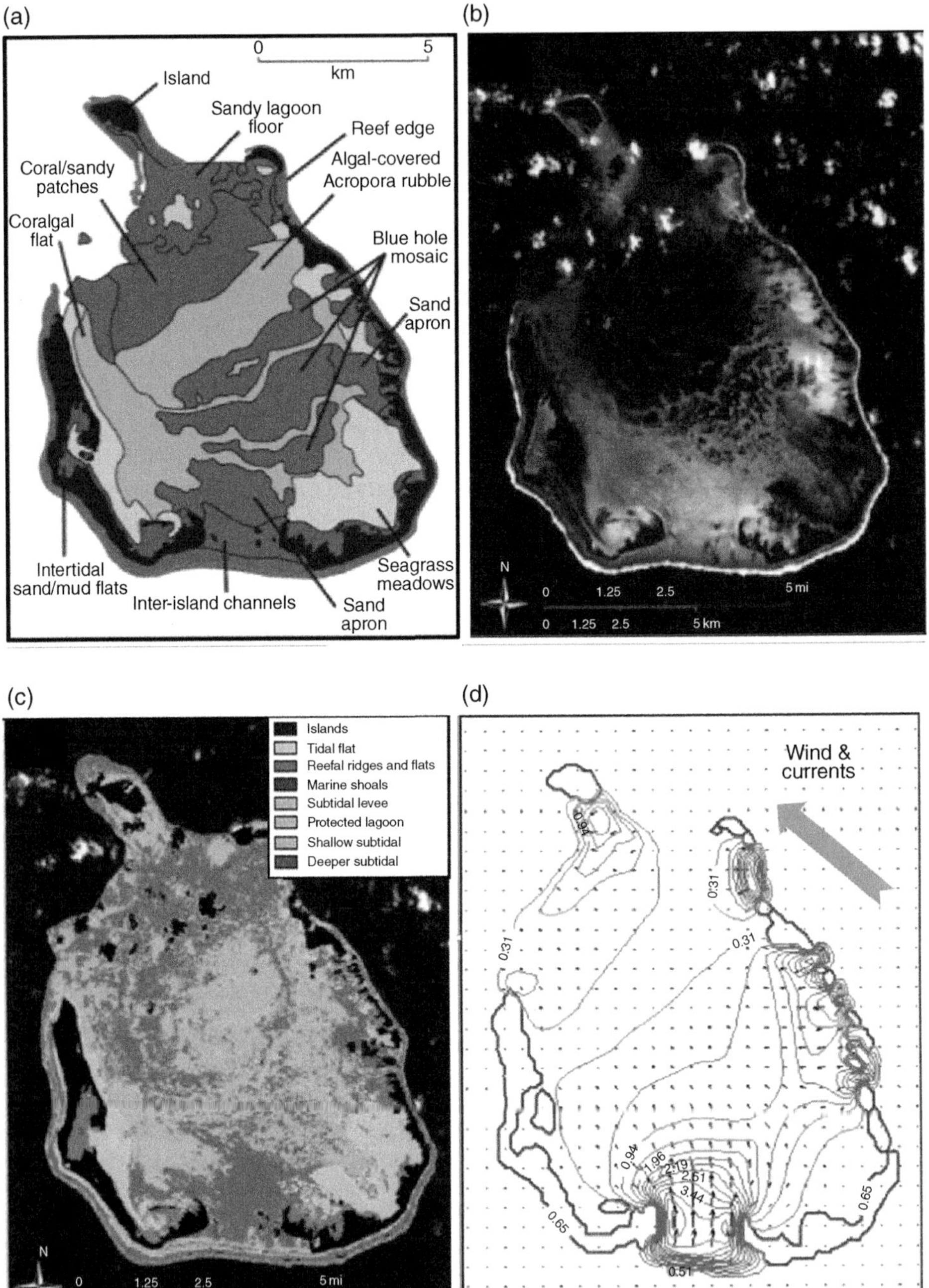

Figure 3.10 Evaluation of CARB3D$^+$ sediment modelling by comparison with the modern south Cocos (Keeling) platform (a–f) and the Caicos platform (g–i). Synthetic facies (f and i) are compared with direct observation of facies distribution based on sample characterization (a, after map of Smithers, 1994) and that inferred by an analysis of Landsat imagery (by Kaczmarek et al., 2010) (c and h). For Cocos (Keeling) we also show predicted current velocity (d), and the resulting redistribution of coarse and fine sediments (e). Paterson et al., unpublished data. (See colour plate section for the colour representation of this figure.)

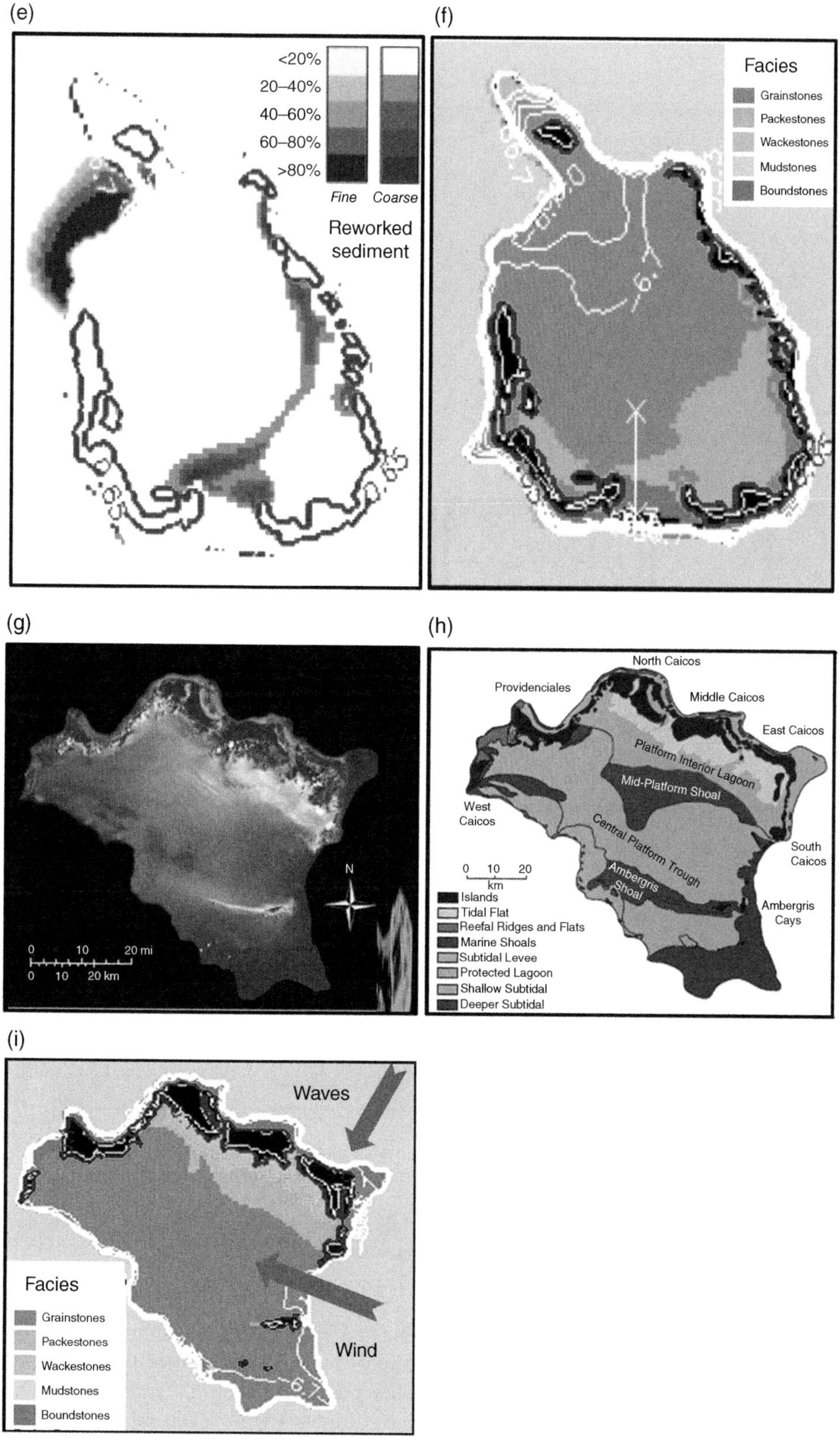

Figure 3.10 (Cont'd)

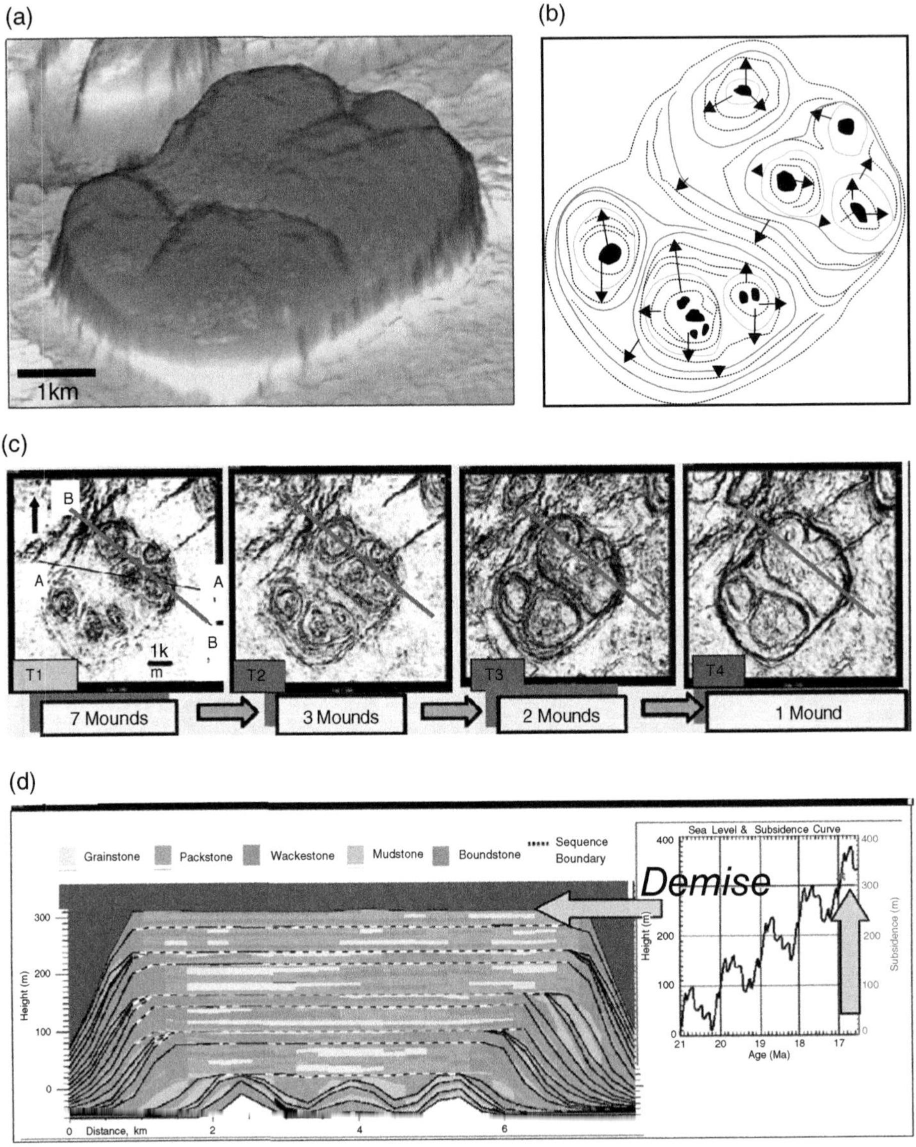

Figure 3.11 (a) PGS seismic data showing the amalgamated platform in North Madura with (b) the growth pattern interpreted from time slices. In (c), four selected slices illustrate stages in platform growth, with nucleation and amalgamation of separate mounds, then drowning caused by a reduction in production rate; (d) as simulated using CARB3D$^+$ and displayed as a 2D section marked as B-B in (c). *Source:* After Hughes et al. (2008), used with permission of AAPG. (See colour plate section for the colour representation of this figure.)

visualization with SFM sensitivity analysis enabled quantification of the controls on the platform evolution, and suggested that the growth and demise of the platforms was controlled more by carbonate productivity rates than tectonic subsidence. The carbonate platforms in this area appear to have initiated with seven mounds which, with high productivity and low accommodation, amalgamated to form a single large platform. Lower productivity causes isolated platforms or partially amalgamated platforms. After amalgamation, the platform appears to have drowned, and the SFM simulations suggest that this could have been driven by low productivity (less than 1 m/kyr), even at low subsidence rates (0.04 m/kyr). This indicates the key role of stressing the system, perhaps by additional nutrients causing a reduction in light levels and favouring a change in biota from coral-dominated to red-algal dominated. Such a change has been observed (Hallock et al., 1993) and associated with platform drowning of Tertiary build-ups in Turkey and Spain (Bassant et al., 2004).

The inverse of this scenario, a transition from heterotrophic to phototropic conditions, was simulated by Kaczmarek et al. (2009) for the Upper Miocene at La Molata, Cabo de Gata region of southeast Spain, constrained by detailed outcrop work, high-resolution constraints on stratal geometry, lithological heterogeneity and relative sea-level history (Franseen et al., 1993). Whilst CARB3D^{+} lacks the functionality to incorporate the input of volcaniclastics or megabreccia formation, the synthetic stratigraphy does honour most of the important aspects observed in outcrop, within the unconformity-bound depositional sequences (Figure 3.12). Geometries reflect the interactions between high-magnitude, high-frequency fluctuations in relative sea-level, and the substantial (~200 m) and complex paleotopography of the Neogene volcanic substrate, as well as changes in the character of the carbonate factory over time. Heterozoan-dominated sequences are most accurately modeled using lower maximum carbonate production rates over a broad range of water depths, and photozoan-dominated units are best modeled with production–depth curves that have higher maximum values but lower depositional depth ranges. Sequence stratigraphy, relative sea-level curves and magnetostratigraphy provide tight constraints on third- and fourth-order fluctuations which control sequence development. In addition, higher frequency fifth-order cyclicity is suggested by metre-scale cycles in both the lower heterozoan units, and the uppermost photozoan sequence.

3.5.5.2 FSDM – Simulation of Diagenetic Hydrozones

The importance of understanding the facies and texture of accumulating sediments is clear when simulations coupling sedimentology and early diagenesis are examined. Paterson et al. (2008) simulated the repetitive and cumulative overprint of meteoric diagenesis in icehouse platforms. Although the residence time within all meteoric hydrozones is characterized by a high degree of lateral continuity, there is more limited vertical continuity, with significant variations on a fourth-order sequence scale across the platform (Figure 3.13). The frequency and magnitude of sea-level variation, subsidence rate and climate are shown to be fundamental controls on residence time in diagenetic hydrozones, as well as on the distribution of sediment texture and platform architecture (Paterson et al., 2006). In addition to the sedimentological response to rates of dissolutional lowering of the land surface, climate also controls the freshwater-lens thickness and the depth and width of the mixing zone for a given distance from the coast.

There is a hierarchy of eustatically driven overprint cycles that accords with the principles of sequence stratigraphy, with higher-frequency cyclicity modifying the primary signal generated by lower-frequency cycles. Third-order cycles are identified as a critical control on the overall extent of meteoric diagenetic modification of the platform, whilst fourth-order cycles control the vertical distribution of diagenesis within the platform. Meteoric alteration is minimal for 3rd and 4th order transgressive system tracts (TST) and greatest for third- and fourth-order highstand system tracts (HST).

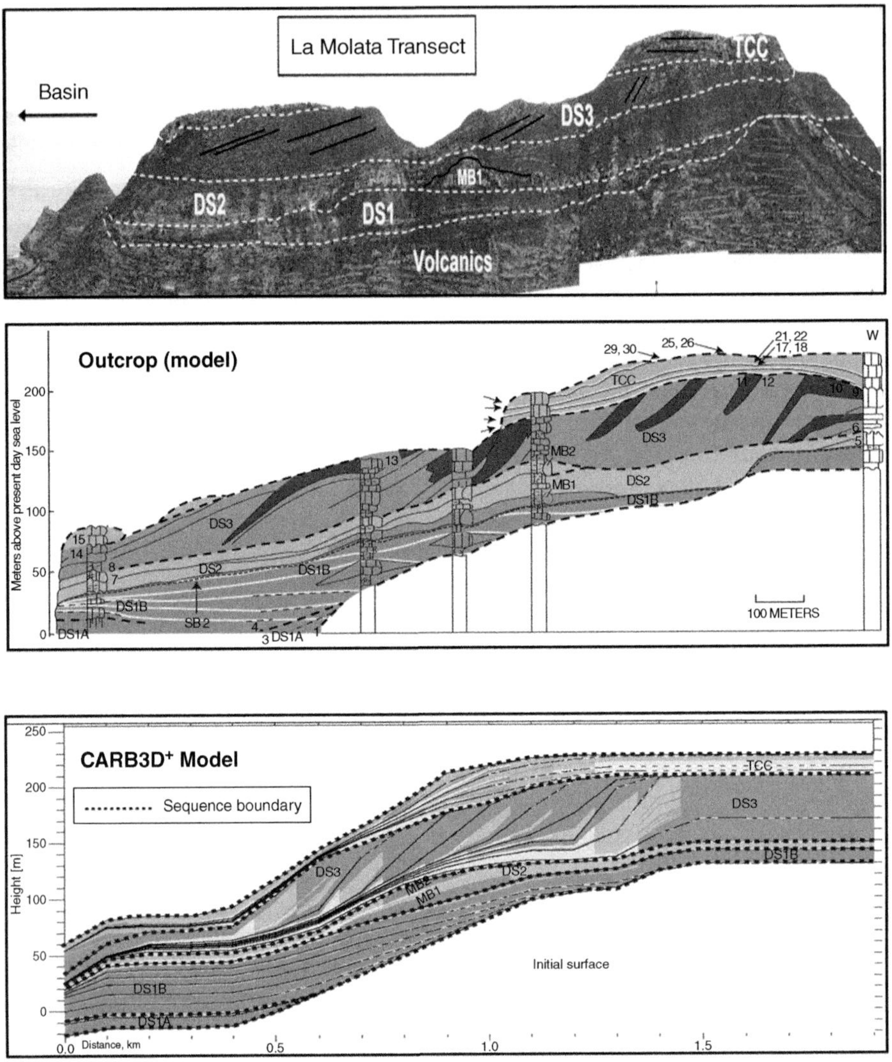

Figure 3.12 Outcrop photograph, static geological model built from outcrop data (Franseen et al., 1993), and forward sediment model for the La Molata Outcrop. *Source:* Data from Kaczmarek et al. (2009).

Fifth-order cycles are often missed beats, generating prolonged exposure and long vadose zone residence times, but have only a minor impact on freshwater lens and mixing zone residence times. Diagenetic overprinting is greater at lower subsidence rates; in high-subsidence regimes, sediment packages move below the zone of meteoric-water influence rapidly. Higher effective recharge can increase the thickness of the freshwater lens and mixing zones, which increases residence times significantly. Residence times in the vadose zone can be directly linked to a sequence-stratigraphic framework. However, for the freshwater lens and the mixing zone, this is only possible with good constraints of the elevation of sea-level minima.

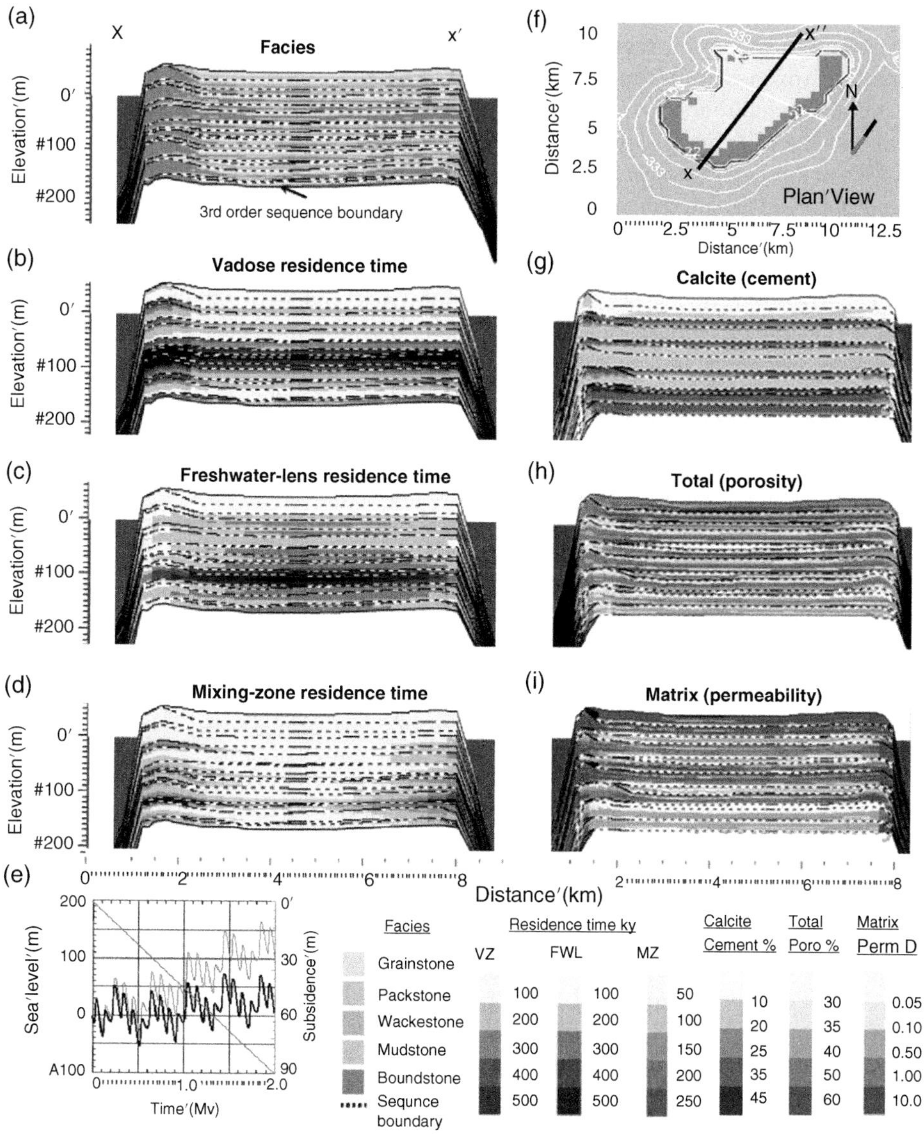

Figure 3.13 CARB3D+ simulations for an isolated platform based on Miocene Natuna, Sarawak Basin, Indonesia. 2D sections through a 3D model show distribution show facies (a), residence time in the vadose zone (b), freshwater lens (c), and mixing zone (d), given the sea-level curve (e) with eustatic (black curve), subsidence (red line) and 'relative' sea-level variations (red curve). Cross-sections (x–x′) parallel to the principal wind direction (red/blue line) shown in map (g). In addition, we show one set of potential distributions of calcite cement (h), porosity (i) and matrix permeability (j). *Source:* Based on Paterson et al. (2008), used with permission of SEPM. (See colour plate section for the colour representation of this figure.)

3.5.5.3 FSDM – Simulation of Diagenetic Processes

Although there is broad agreement on the nature of the products of diagenesis within different hydrozones (James and Choquette, 1990; Moore, 1989, and references therein), uncertainties remain about the distribution and the rate of diagenesis (e.g. compare Melim, 1996, and Whitaker et al., 2006). Paterson et al. (2008) thus focused on using CARB3D$^+$ cumulative residence times as a surrogate for the extent of meteoric diagenesis for an isolated platform. This version of the model required the user to specify processes and rates of syn-depositional diagenesis within each hydrozone, to explicitly simulate the evolution of early meteoric diagenesis through time. Thus, one possible distribution of calcite cementation, porosity and permeability is shown in Figure 3.13g–i. This shows well-developed lateral continuity, with differences from margin to interior and between windward (boundstone-dominated) and leeward margins, and significant vertical contrasts on a fourth-order sequence scale.

Subsequent model development included internal parameterization of diagenetic rates based on climate and soil type (see Section 5.3.2), parameters that can be inferred from sources such as global climate models and reconstructions of paleogeography, as well as characterization of paleo-exposure surfaces at outcrop or from core/downhole geophysics. The chemical potential and recharge rate at the surface provides the drive for diagenesis and the distribution of sediment texture and mineralogy in the underlying sediment column determines the vertical distribution of dissolution and precipitation The power of this approach is shown in Figure 3.14, which compares the change in

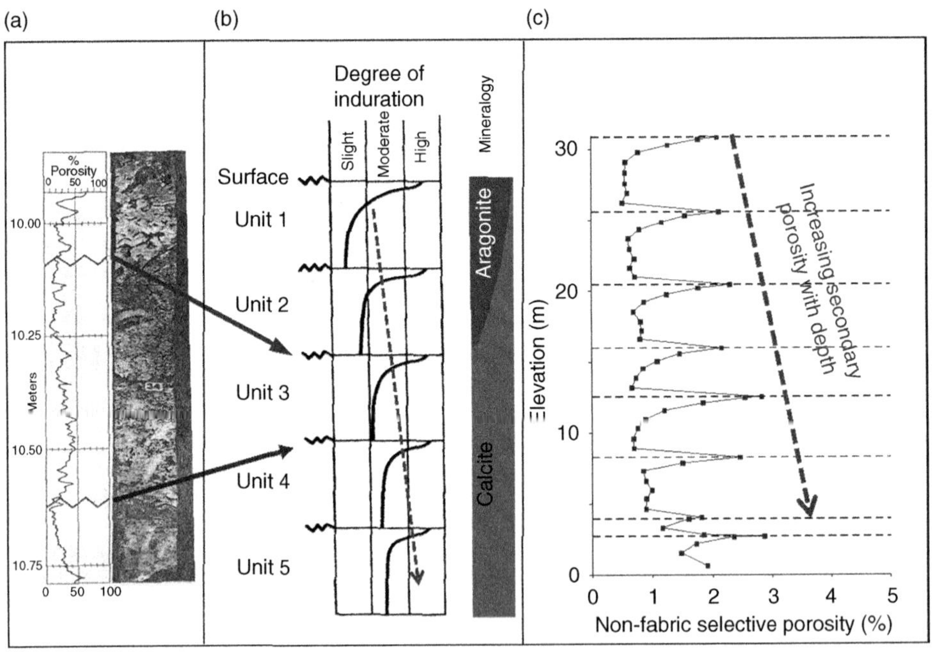

Figure 3.14 Comparison of (a) porosity data and core photo from a single depositional package, and (b) a stack of depositional packages separated by unconformities, from Pleistocene 5e deposits on North Andros Island, Bahamas (*Source:* adapted from Beach, 1995), with (c) synthetic non-fabric selective porosity data from a CARB3D$^+$ simulation, with unconformities shown as dashed lines (Smart and Whitaker, unpublished data).

non-fabric selective porosity due to increasing cement abundance simulated by CARB3D$^+$ and core data from the northern Bahamas (Smart and Whitaker, unpublished data, compared with data from Beach, 1995). Pleistocene and Holocene carbonates show peaks in cementation below individual paleo-exposure surfaces (Figure 3.14a), superimposed on a general increase in cementation with depth (Figure 3.14b). Field observations in modern systems (Whitaker and Smart, 2007a,b; Cooper et al., 2016) and RTM simulations (Whitaker et al., 2011; Cooper, 2015) suggest near-surface dissolution during recharge events/seasons driven by organic acids and high CO_2 production, combined with low CO_2 degassing. During subsequent dry periods, evapotranspiration and CO_2 degassing then drive calcite precipitation at and below the surface. These processes lead to preferential cementation of finer-grained sediments due to the combined effect of greater effective reactive surface area, higher retention of water against gravity drainage, and capillary rise from the top of the freshwater lens.

The model generates an exponential reduction in porosity occlusion by shallow vadose dissolution–cementation below the exposure surfaces. Furthermore, porosity loss in the muddy platform interior occurs at almost twice the rate simulated for the boundstone-dominated margins. Synthetic fifth-order sediment cycles are rather thicker (average 4.6 m) than those observed in core, due to the higher subsidence rate specified (4–10 times that on North Andros; McNeill, 2005). The synthetic sequences will be less altered than those observed in the cores, both because higher subsidence rates modelled will decrease the degree of overprinting, and because the 41 ky (obliquity-driven) cyclicity modelled is at a higher frequency that the 100 ky (eccentricity-driven) cycles that dominated over the last 0.9 My (Zachos et al., 2001). The net effect is laterally continuous zones of alteration associated with individual cycle tops, superimposed upon an increase in vadose zone diagenesis with depth (age) due to diagenetic overprinting.

Using CARB3D$^+$, Paterson et al. (2008) confirmed that high-amplitude eustatic sea-level fluctuations characteristically result in a high degree of overprinting which can obscure diagenetic signatures in ice-house platforms. Whitaker et al. (2014) evaluated conceptual models that suggest that greenhouse platforms are likely to show significantly less diagenetic overprinting because of the shorter duration of exposure to meteoric waters (Read, 1998). Simulations based on the Latemar platform, with progressive reduction of accommodation over time, demonstrate that overprinting may also occur in greenhouse carbonates at times of low accommodation generation (Figure 3.15). In the Cyclic unit, a combination of short duration of exposure and rapid subsidence below the influence of meteoric processes do limit the impact of meteoric diagenesis. Reducing accommodation in the Transition and Condensed units means that low-amplitude eustatic oscillations will produce multi-cycle superimposed diagenetic events overprinting the same volume of sediments, generating an apparently simple distribution of diagenesis which might readily be interpreted as resulting from a single exposure event.

These simulations also provide a comparison of the diagenesis that might be expected in a small isolated platform under contrasting climatic end-member diagenetic scenarios (Whitaker et al., 2014). Aridity effects include not only low recharge rate, but also a thin/discontinuous soil cover. In contrast, the humid climate simulation had a 20 times higher recharge and a thick, laterally continuous soil cover, which resulted in recharge waters with a high PCO2, and thus more rapid surface lowering and subsurface

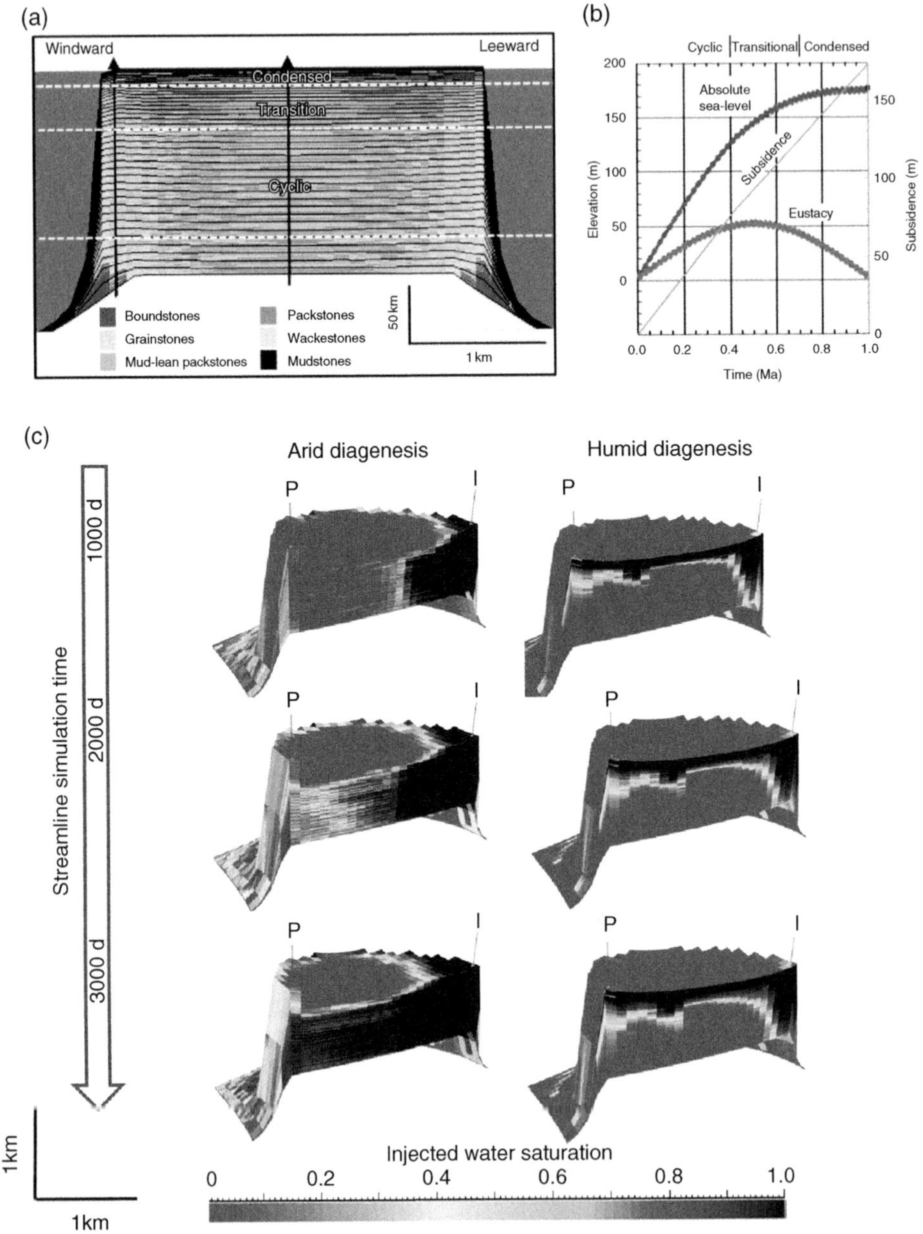

Figure 3.15 (a) 2D section oriented along principal wind direction through 3D simulation of the Latemar platform, showing platform architecture and facies distribution, arbitrary divisions into cyclic, transitional and condensed units (each 300 ky) reflecting an overall reduction in accommodation shown in (b) sea-level curve. (c) 3D distribution in half the platform (cut along line of 2D section) showing connate water saturation for arid and humid simulations at three time intervals during single-phase tracer experiment. Blue and red vertical lines mark the positions of injector (I) and producer (P) wells. *Source:* After Whitaker et al. (2014), used with the permission of The Geological Society. (See colour plate section for the colour representation of this figure.)

dissolution. Results show how the higher rate of dissolutional surface lowering during subaerial exposure under a more humid climate controls the graininess of the platform via its effect on energy levels across the platform top during the subsequent transgression. $CaCO_3$ dissolved at the surface is re-precipitated in the semi-arid simulation as vadose cements, whilst in the humid simulation cementation is more extensive and occurs also within the freshwater lens, reflecting more rapid percolation to the water table. However, associated subsurface meteoric dissolution generates significant secondary porosity under a more humid climate. This reflects both the additional drive for dissolution in the lens provided by subsurface oxidation of surface-derived organics in the lens, and enhanced mixing zone dissolution because of higher soil CO_2 and increased specific discharge. The impact of these effects is most marked during periods when slower subsidence causes diagenetic overprinting from repeated exposure events. Thus, whilst in our arid climate simulations meteoric diagenesis is only a second-order modifier of depositional porosity and permeability, humid diagenesis can completely dominate over control of reservoir quality by depositional facies.

The study of Whitaker et al. (2014) illustrates a further strength of SFM: the rapid generation of high vertical resolution, fully 3D representations of rock properties that are directly amenable to quantitative analysis. After accounting for the effects of compaction (to 2 km burial depth), synthetic stratigraphies from CARB3D$^+$ were used as a basis for streamline flow simulations to better understand the impact of depositional and diagenetic heterogeneity on flow behaviour. Simple tracer experiments were undertaken, using numerically efficient streamline simulations to track particles. This allowed quantitative comparison of sweep pattern, sweep efficiency and recovery factor to understand which elements of the permeability structure are important in channelling and baffling flow in complex reservoirs, and how this is impacted by differences in subsidence and climate for a greenhouse platform. Two sets of features appear to control post-burial fluid flow through the synthetic platforms: low permeability, laterally continuous horizons (such as separates the Transition zone from the overlying Condensed zone), and the development of karstic permeability. Simulations suggest that karstification under humid conditions could give rise to permeability several orders of magnitude higher than that observed in core, where they would likely be recorded as zones of low or no recovery. Such extreme permeability features are well known at shallow depth (Whitaker and Smart, 2000: Smart et al., 2006), and their persistence at depth has been observed from bit drops and loss of circulation (e.g. Walles, 1993). They can also be inferred from flow behaviour during production of reservoirs (e.g. Putney et al., 2008; Ronchi et al., 2009). The effect of karstification may be difficult to distinguish from that of fracturing, which can also be focused on platform margins.

The effects of dissolution and karstification during syn-depositional diagenesis are offset by the reduction of matrix porosity by compaction and precipitation of cement, both in meteoric hydrozones but also in the marine environment. Early marine cements are a particular feature of platform carbonates accumulating in seas with a high carbonate saturation, as proposed for Triassic seawaters in equatorial latitudes (e.g. Riding and Liang, 2005). Berra et al. (2016) used CARB3D$^+$ to simulate the development of a spectacularly well-exposed high-relief Triassic carbonate platform in the central Southern Alps, Italy, where early fibrous radiaxial isopachous marine cements occlude the fabric-controlled porosity in the mud-free slope facies and margin belt. Simulations confirm that meteoric diagenesis is negligible due to the arid climate and lack of soil, and the diagenetic history is dominated by a progressive reduction in the original porosity by marine cement precipitation and sediment compaction, in agreement with field observations. In the model the cementation rate is constant and is a function solely of duration of exposure to seawater. Efficiency of cementation is strictly related to the sediment type; cementation is uniform and extensive in the slope facies, but it is less homogeneous in the platform interior, where porosity is lower due to the abundance of more compressible fine-grained sediments.

CARB3D$^+$ has also been used to generate interpretative frameworks for understanding the potential impact of synsedimentary diagenesis within active carbonate oil fields. The Tengiz field in Kazakhstan is an isolated carbonate platform of Devonian and Carboniferous age, with a stratigraphic thickness of over 2 km (Weber et al., 2003; Collins et al., 2006). Its internal reservoir property distributions were thought to be strongly influenced by diagenetic processes (Kenter et al., 2010), and the extensive influence of syn-sedimentary meteoric diagenesis was not identified until recently (Dickson and Kenter, 2014). The application of CARB3D$^+$ to this platform has helped provide a broad prediction of the stratigraphic architecture, facies distribution and meteoric influence on this system within a spatial framework (Figure 3.16; Frazer et al., 2014a). Specifically, CARB3D$^+$ simulations, conditioned against well and log data, identified four spatial zones within which eustasy, platform subsidence and deposition/erosion combine to develop distinct internal patterns of meteoric alteration. The platform accumulated over an extended period, recording the transition from greenhouse to icehouse conditions. The associated patterns of meteoric alteration ranged from minor, para-sequence bound alteration, where all meteoric alteration within a sediment

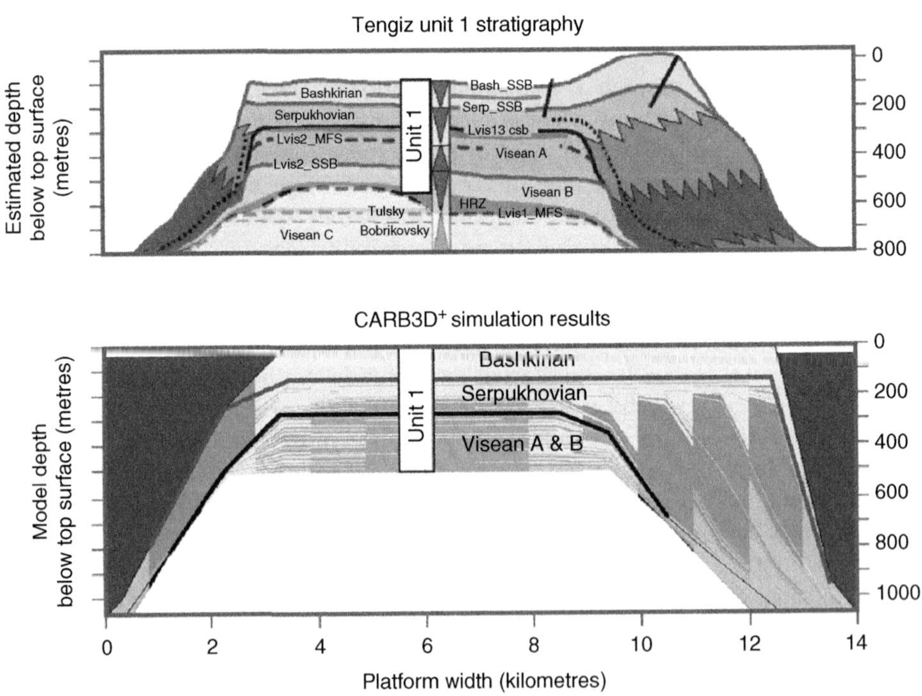

Figure 3.16 Stratigraphy of Unit 1 in Tengiz, after Collins et al. (2006), compared with synthetic stratigraphy generated using CARB3D$^+$. After Frazer et al. (2014b).

package could be directly tied to the exposure surface at its top, to boundary-transcendent, extensive meteoric alteration where repeated exposure to meteoric fluids associated with many overlying exposure surfaces led to complete recrystallization and extensive cementation. The simulations also suggest a contrast in the degree of diagenetic overprinting between areas of greater or lesser progradation in the Serpokhovian.

The range of examples presented above emphasize the strong control on deposition and diagenesis of platform carbonates exerted by the interplay of biological, climatic and environmental factors, as well as antecedent topography, tectonics and eustasy. Because of the considerable spatial and temporal variations in these controlling factors and the complex interactions between them, a wide range of depositional architectures is observed in the geological record, with an array of associated diagenetic overprints. Berra et al. (2016) systematically describe an iterative workflow that can be used to derive the parameters needed to produce a numerical model mirroring the development of a real platform, and illustrate this workflow by simulating development of the Triassic Esino Limestone succession of the central Southern Alps, Italy, using CARB3D$^+$. This study combined knowledge of the architectural, sedimentological and diagenetic features of the outcrop with data from modern and ancient analogues, using sensitivity analyses to tune parameter values within ranges defined from field, laboratory and uniformitarian constraints. In this and other examples described here, the development of the numerical model allowed an understanding of the effects of the interaction between productivity, environmental energy and creation of accommodation space on the architecture of the carbonate system and the depositional and diagenetic facies, which can be challenging to interpret from observation of the geological products alone. Equally important, they also provide a fully-3D digital numerical model that can be used for the parameterization of the petrophysical properties of selected complex geological bodies at outcrop or in the subsurface, with potential application to reducing uncertainty in exploitation of carbonate reservoirs. However, it is important to remember that models always retain properties that may limit their ability to mimic real systems, their results may be affected by the processes included, computational algorithms used, temporal and spatial scale of discretization, and uncertainties in parameterization and other input data.

3.6 Discussion and Conclusion

While diagenesis has long been a challenge for reservoir management in carbonate lithologies, many decades of characterization-focused studies have failed to provide a clear path to generating spatial predictions for the distributions of diagenetically controlled reservoir properties. This does not undermine the value of characterization studies as primary sources of data, but does clearly demonstrate the need for an integrated framework within which these complex and often confusing observational data-sets may be interpreted and extrapolated for reservoir prediction. Understanding the link between diagenetic products and their formative processes, along with the external influences on their operation, provides a methodology that intrinsically incorporates geological observations and concepts within a spatial framework defined by fundamental laws of chemistry and physics.

Young carbonate sediments are highly reactive and prone to syn-sedimentary diagenesis, which often has a significant impact on pore types and connectivity, and can also play a crucial role in establishing a template for later diagenetic fluid flow. Changes in relative sea-level, along with prevailing climatic conditions (arid vs. humid) are the primary controls on the duration, severity and spatial distribution of syn-sedimentary diagenesis at the platform scale. These overarching external conditions determine the size and position of the four primary hydrological zones: the vadose zone, the freshwater lens, the mixing zone, and the marine zone. Each presents different conditions for diagenetic alteration of the carbonate sediments. If relative sea-level fluctuates over a large vertical range at high frequency, rapid migration of these zones can lead to extensive 'overprinting' of meteoric diagenetic products across sequence boundaries, making the final products difficult (and often impossible) to separate into a clear parasequence. Factors such as dominant depositional facies, soil development, and spatial and temporal variability in recharge play important secondary roles in local conditions for diagenesis.

Moving from conceptual to numerical models of syn-sedimentary diagenesis requires careful geological characterization of diagenetic products, as well as quantitative formalization of existing knowledge in areas of hydrology and chemistry. This formalization process has revealed that our understanding of some key diagenetic processes is still incomplete. For example, there are knowledge gaps relating to the impact of biogeochemical processes on diagenesis. Furthermore, formalization frequently requires data on observable temporal and spatial timescales to be scaled appropriately to provide effective values suitable for the formulations inherent within numerical simulations. Some of these questions may be answered by direct observation of diagenesis under controlled laboratory conditions, whilst others require field characterization in modern carbonate systems to identify key aspects of, and interactions between, the hydrological, geochemical and biological systems driving diagenesis.

The quantitative rigour offered by computational reactive transport modelling allows us to critically evaluate alternative conceptual models for diagenesis and generate new predictive concepts. The ability to generate multiple chemically and physically constrained realizations of our diagenetic concepts allows us to explore underlying controls and determine the impact of our geological uncertainty on the results. However, much of this potential remains largely unrealized, partly because of the small community of specialist users with the expertise to apply the available technologies. Simple models can reveal new insights, but can be misleading if they fail to consider key processes, or to simulate behaviour at appropriate spatial and temporal scales (which might require much finer resolution than previously anticipated). There is also a critical need to incorporate feedbacks between diagenetic changes in mineralogy and porosity, and the resulting evolution of the pore geometry and hence both permeability and reactivity.

One of the key benefits offered by RTM simulations is the ability to predict diagenesis at timescales beyond those that can be examined in the laboratory. They allow us to explore the operation of processes in environments not represented currently, or under conditions that are challenging or impossible to observe directly. Once sufficient data are gathered on the operation of diagenetic processes through field and lab experiments, the physical and chemical rules that underpin RTM allow us to extrapolate (within uncertainty bounds) their operation to these more exotic conditions, and to geologically relevant timescales. This allows us access to an entire field of numerical experimentation driven by geological concepts. RTMs offer an improved capability to predict carbonate diagenesis by (i) helping to develop better conceptual models based on chemically and physically realistic scenarios; (ii) providing quantitative estimates of rates and distribution of diagenesis; and (iii) describing diagenetic trends and geobodies which can be used to populate reservoir models.

However, we have only recently started to realize the potential of this approach. While RTM has provided some key insights into the operation of diagenetic processes, the sophistication in representing geochemical and hydrological processes has often been offset by a reduction in geological realism, undermining their applicability. The most critical aspect of this lack of realism is the lack of stratigraphic development during diagenesis and thus, strictly speaking, no RTM of truly syn-depositional diagenesis has been developed to date. Sequence stratigraphy provides the simplest means of generating a spatial and temporal framework within which to understand the evolution of diagenesis in response to changes in relative sea-level and climate. However useful such conceptual models are, they remain fundamentally unsuitable for reservoir modelling due to their limited ability to generate specific predictions of the spatial distribution and extent of diagenetic and depositional processes.

The incorporation of algorithms that represent the syn-depositional diagenetic changes, resulting from hydrological and geochemical processes, within stratigraphic forward models offers a novel route to predict syn-sedimentary alteration of primary porosity and mineralogy. A major advantage over RTMs is that this allows explicit consideration of the controls on, and feedbacks between, sedimentological and diagenetic processes. The outcome is a prediction of the development of paragenetic sequences (as envisaged by Csoma et al.'s (2004) diagenetic salinity cycles) at locations distributed across the platform in full 4D (at multiple locations through time). We have described development of this approach, culminating in the model CARB3D$^+$, and application to a range of platform types, sea-level histories and climatic settings from Modern to Paleozoic times.

CARB3D$^+$ offers a means of distributing petrophysical data in the inter-well environment, and generating valuable 3D data-sets with 100% data coverage which enable us to extract dimensions, temporal and geometric relationships. The model enables assessment of sensitivity of diagenesis to controlling parameters and uncertainty, such as response to sea level and initial surface topography. CARB3D$^+$ also provides geologists and engineers with a shared mental picture of the subsurface geology and a common understanding of important concepts. Finally, by preserving insights from our understanding of process, CARB3D$^+$ can be used for building reservoir models which may better honour actual 3D distributions than do those that rely primarily on geostatistics.

There are inherent limitations to both RTM and FSDM, but also clear synergisms between the two approaches to modelling syn-sedimentary diagenesis. The explicit process-based simulations of diagenesis generated using RTMs have been used as one important source of data from which to develop rule-sets that govern diagenesis within the meteoric hydrozones of CARB3D$^+$. In the other direction, it is possible to use complex geometries and distribution of rock properties from CARB3D$^+$ as input to more geologically realistic RTMs. Each approach represents a different solution to the compromise between the ability, on the one hand, to explicitly simulate process (a strength of RTM), and on the other, to consider the effects of sedimentation and boundary conditions that vary in time and space (via coupled FSDM). The ultimate target must be to explicitly simulate diagenesis by RTM within the context of a forward stratigraphic model. Irrespective of the potential of such a tool, prediction of syn-sedimentary diagenesis for periods in the geological past (rather than in modern systems) will continue to be plagued by considerable uncertainty. This arises not least because the data to constrain model inputs remain highly uncertain, particular for reservoirs during the exploration phase. Irrespective of such limitations, process-based models remain the only real route through which 'soft' geological concepts can have a direct impact on the spatial distributions of properties that are deployed to reservoir applications.

Acknowledgements

We thank Yitian Xiao for editorial support, and for the long-standing collaboration in application of RTM to understanding carbonate diagenesis. We are grateful for the contribution of colleagues, including Peter Smart, Gareth Jones, Steve Kaczmarek, Fabrizio Berra and John Reijmar, and many more who are not named but also played a role in developing our ideas. The misconceptions and simplifications remain entirely of our own making. Development of CARB3D^{+} was carried out in collaboration with Peter Smart at the University of Bristol and David Waltham and colleagues at RHUL, and funded via the Bristol Carbonates Consortium by Chevron, Conoco-Phillips, Eni, ExxonMobil, Petrobras and Shell. We acknowledge the fruitful discussions and the integrity, persistence and good humour of colleagues who have worked with us in developing model applications for outcrop and reservoir examples. Miles Frazer thanks Chevron Energy Technology for their support and review of this work

References

Arndt, S., Hetzel, A., and Brumsack, H.-J. (2009). Evolution of organic matter degradation in Cretaceous black shales inferred from authigenic barite: A reaction-transport model. *Geochimica et Cosmochimica Acta* 73: 2000–2022.

Arndt, S., Jorgensen, B.B., LaRowe, D.E., et al. (2013). Quantifying the degradation of organic matter in marine sediments: A review and synthesis. *Earth Science Reviews* 123: 53–86.

Back, W., Hanshaw, B.B., Pyle, T.E., et al. (1979). Geochemical significance of groundwater discharge and carbonate solution to the formation of Caleta Xel Ha, Quintana Roo, Mexico. *Water Resources Research* 15: 1521–1535.

Badiozamani, K. (1973). The dorag dolomitization model, application to the middle Ordovician of Wisconsin. *Journal of Sedimentary Petrology* 43: 965–984.

Bailey, R.J. and Smith, D.G. (2005). Quantitative evidence for the fractal nature of the stratigraphic record: results and implications. *Proceedings of the Geological Association* 116: 129–138.

Bassant, P., Van Buchem, F., Strasser, A., and Lomando, A. (2004). A comparison of two Early Miocene carbonate margins: The Zhujiang Carbonate Platform (subsurface, South China Sea) and the Pirin Platform (outcrop, Southern Turkey). *American Association of Petroleum Geologists Memoir* 80: 153–170.

Beach, D.K. (1995). Controls and effects of subaerial exposure on cementation and development of secondary porosity in the subsurface of Great Bahama Bank. *American Association of Petroleum Geologists Memoir* 63: 1–33.

Bergmann, K.D., Grotzinger, J.P., and Fischer, W.W. (2013). Biological influences on seafloor carbonate precipitation. *Palaios* 28: 99–115.

Berra, F., Lanfranchi, A., Smart, P.L., et al. (2016). Forward modelling of carbonate platforms: Sedimentological and diagenetic constraints from an application to a flat-topped greenhouse platform (Triassic, Southern Alps, Italy). *Marine and Petroleum Geology* 78: 636–655.

Bosence, D.W.J. and Waltham, D.A. (1990). Computer modeling the internal architecture of carbonate platforms. *Geology* 18: 26–30.

Brooks, S.M. and Whitaker, F.F. (1997). Co-evolution of early diagenesis and vadose zone hydrology of Holocene carbonate sands. *Earth Surface Processes and Landforms* 22: 45–58.

Budd, D.A. (1988). Aragonite-to-calcite transformations during fresh-water diagenesis of carbonates: insights from pore-water chemistry. *Geological Society of America Bulletin* 100: 1260–1270.

Budd, D.A. and Park, A.J. (2017). Formation of bed-scale spatial patterns in dolomite abundance during early replacive dolomitization: I. Mechanisms and feedbacks revealed by reaction-transport modeling. *Sedimentology*. doi:10.1111/sed.12400

Burchette, T.P. and Wright, V.P. (1992). Carbonate ramp depositional systems. *Sedimentary Geology* 79: 3–57.

Burgess, P.M. (2008). The nature of shallow-water carbonate lithofacies thickness distributions. *Geology* 36: 235–238.

Choquette, P.W. and James, N.P. (1987). Diagenesis 12. Diagenesis in limestones – 3. The deep burial environment. *Geoscience Canada* 3–35.

Christ, N., Immenhauser, A., Amour, F., et al. (2012). Triassic Latemar cycle tops – Subaerial exposure of platform carbonates under tropical arid climate. *Sedimentary Geology* 265: 1–29.

Christ, N., Immenhauser, A., Wood, R.A., et al. (2015). Petrography and environmental controls on the formation of Phanerozoic marine carbonate hardgrounds. *Earth Science Reviews* 151: 176–226.

Collins, J.F., Kenter, J.A.M., Harris, P.M., et al. (2006). Facies and reservoir-quality variations in the late Visean to Bashkirian outer platform, rim, and flank of the Tengiz buildup, Precaspian Basin, Kazakhstan. *American Association of Petroleum Geologists Memoir* 88: 55–95.

Cooper, K.C. (2015). Biogeochemistry of an eogenetic karst island: implications for diagenesis. Unpublished PhD thesis, University of Bristol.

Cooper, K. and Smart, P. (2010). Is Milankovitch cyclicity recognisable in carbonate sequences? Numerical experiments using the forward model CARB3D$^+$. *American Association of Petroleum Geologists International Conference* (Abstract). Karst Waters Institution Special Publication 16.

Cooper, K.C. and Whitaker, F.F. (2011). Diagenetic patterns in a small carbonate island setting: a reactive transport study. *Karst Waters Institution Special Publication* 16: 17–19.

Cooper, K.C., Whitaker, F.F., Anesio, A., et al. (2016). Dissolved organics carbonate transformations and microbial community response to variations in recharge waters in a shallow carbonate aquifer. *Biogeochemistry* 128: 215–234.

Cortés, J. (1997). Biology and geology of eastern Pacific coral reefs. *Coral Reefs* 16 (suppl.): S39–S46.

Crevello, P., Sarg, J.F., Read, J.F., and Wilson, J.L. (ed.) (1989). Controls on carbonate platform to basin development. *Society of Economic Paleontologists and Mineralogists Special Publication* 44.

Csoma, A.E., Goldstein, R.H., Mindszenty, A., and Simone, L. (2004). Diagenetic salinity cycles and sea-level along a major unconformity. Monte Composauro, Italy. *Journal of Sedimentary Research* 74: 889–903.

Csoma, A.E., Goldstein, R.H., and Pomar, L. (2006). Pleistocene speleothems of Mallorca: implications for palaeoclimate and carbonate diagenesis in mixing zones. *Sedimentology* 53: 213–236.

Dickson, J.A.D. and Kenter, J.A.M. (2014). Diagenetic evolution of selected parasequences across a carbonate platform: Late Paleozoic, Tengiz Reservoir, Kazakhstan. *Journal of Sedimentary Research* 84: 664–693.

Forkner, R.M., Hinnov, L.A., and Smart, P. (2010). Use of insolation as a proxy for high-frequency eustasy in forward modeling of platform carbonate cyclostratigraphy – A promising approach. *Sedimentary Geology* 231: 1–13.

Frank, T.D. and Lohmann, K.C. (1995). Early cementation during marine-meteoric fluid mixing: Mississippian Lake Valley Formation, New Mexico. *Journal of Sedimentary Research* 65.

Franseen, E.K., Goldstein, R.H., and Whitesell, T.E. (1993). Sequence stratigraphy of Miocene carbonate complexes, Las Negras area, southeastern Spain: Implications for quantification of changes in relative sea level. *American Association of Petroleum Geologists Memoir* 57: 409–434.

Frazer, M., Whitaker, F., and Hollis, C. (2014a). Fluid expulsion from overpressured basins: Implications for Pb-Zn mineralisation and dolomitisation of the East Midlands platform, northern England. *Marine and Petroleum Geology* 55: 68–86.

Frazer, M.A., Whitaker, F.F., Katz D., et al. (2014b). Understanding the development of an early, diagenetically-controlled template for reservoir quality distribution using forward sediment/diagenetic models of the Tengiz Platform, Kazakhstan. *2014 AAPG Annual Convention and Exhibition*, Houston, Texas, April 6–9, 2014. AAPG Datapages/Search and Discovery Article #90189.

Frazer, M., Whitaker, F., and Hollis, C. (2015). Understanding the development of dolomite geobodies using reactive transport modelling: The importance of platform growth. *Proceedings of the 15th Bathurst Conference*, Edinburgh, UK (abstract).

Friedman, G.M., Amiel, A.J., and Schneidermann, N. (1974) Submarine cementation in reefs: example from the Red Sea. *Journal of Sedimentary Petrology* 44: 816–825.

Gabellone, T., Whitaker, F., Katz, D., et al. (2016). Controls on reflux dolomitisation of epeiric-scale ramps: insights from reactive transport simulations of the Mississippian Madison Formation (Montana and Wyoming). *Sedimentary Geology* 345: 85–102.

Garcia-Fresca, B., Lucia, F.J., Sharp, J., and Kerans, C. (2012). Outcrop-constrained hydrogeological simulations of brine reflux and early dolomitisation of the Permian San Andres Formation. *American Association of Petroleum Geologists Bulletin* 96: 1757–1781.

Goldhammer, R.K., Dunn, P.A., and Hardie, L.A. (1990). Depositional cycles, composite sea level changes, cycle stacking patterns, and the hierarchy of stratigraphic forcing – examples from platform carbonates of the Alpine Triassic. *Geological Society of America Bulletin* 102: 535–562.

Greene, S.E., Bottjer, D.J., Corsetti, F.A., et al. (2012). A subseafloor carbonate factory across the Triassic–Jurassic transition. *Geology* 40: 1043–1046.

Gulden, L. and Simo, A. (2012). Understanding progradation-aggradation ratios of isolated carbonate platforms: A forward sediment model approach. American Association of Petroleum Geologists Article 90142, Annual Convention, 22–25 April, Long Beach, California.

Hallock, P., Mullerkarger, F.E., and Halas, J.C. (1993). Coral-reef decline. *National Geographic Society Research and Exploration* 9: 358–378.

Hardie, L.A. (1987). Dolomitization: a critical view of some current views. *Journal of Sedimentary Petrology* 57: 166–183.

Hollis, C. (1998). Reconstructing fluid history: an integrated approach to timing fluid expulsion and migration on the Carboniferous Derbyshire Platform, England. *Geological Society of London Special Publication* 144: 153–159.

Hollis, C. and Walkden, G. (1996). The use of burial diagenetic calcite cements to determine the controls upon hydrocarbon emplacement and mineralization on a carbonate platform, Derbyshire, England. *Geological Society of London Special Publication* 107: 35–49.

Hughes, T.M., Simo, J.A.T., Ruf, A.S., and Whitaker, F. (2008). Forward sediment modeling of carbonate platform growth and demise, East Java Basin: Example North Madura. *Proceedings of the Indonesian Petrological Association 32nd Annual Convention*, IPA08-G-117.

James, N.P. and Choquette, P.W. (1990). Diagenesis 9. Limestones – the meteoric diagenetic environment. *Diagenesis* 11: 35–74. doi:10.1029/SC004p0045

James, N.P. and Jones, B. (2015). Origin of Carbonate Rocks. Chichester: Wiley.

James, N.P., Ginsburg, R.M., Marszalek, D.S., and Choquette, P.W. (1976). Facies and fabric specificity of early subsea cements in shallow Belize (British Honduras) reefs. *Journal of Sedimentary Petrology* 46: 523–544.

Jocson, J.M.U., Jenson, J.W., and Contractor, D.N. (2002). Recharge and aquifer response: Northern Guam Lens Aquifer, Guam, Mariana Islands. *Journal of Hydrology* 250: 231–254.

Jones, G.D., Whitaker, F.F., Barnaby, R., et al. (2004). Predictive modeling of sedimentology and early diagenesis in an icehouse isolated carbonate platform: Salt Creek Field, West Texas. American Association of Petroleum Geologists Hedberg Meeting on Structural Diagenesis, Austin, Texas.

Jourabchi, P., Van Cappellen, P., and Regnier, P. (2005). Quantitative interpretation of pH distributions in aquatic sediments: a reaction-transport modeling approach. *American Journal of Science* 305: 919–956.

Kaczmarek, S.E., Bova, J., Whitaker, F.F., et al. (2009). From conceptual models to outcrop-based forward models using CARB3D$^+$: Preliminary findings from La Molata, Cabo de Gata, S.E. Spain. *American Association of Petroleum Geologists Annual Convention*, Denver, Colorado.

Kaczmarek, S.E., Hicks, M.K., Fullmer, S.M., et al. (2010). Mapping facies distributions on modern carbonate platforms through integration of multispectral Landsat data, statistics-based unsupervised classifications, and surface sediment data. *American Association of Petroleum Geologists Bulletin* 94: 1581–1606.

Kenter, J.A.M., Tankersley, T., Skalinski, M.T., et al. (2010). *Tengiz Field (Republic of Kazakhstan) Unit 1 Platform Static Model: Using a Hybrid Depositional–Diagenetic Approach*. Society of Petroleum Engineers, SPE-139934.

Land, L.S. (1973). Contemporaneous dolomitization of middle Pleistocene reefs meteoric water, North Jamaica. *Bulletin of Marine Science* 23: 64–92.

Longuet-Higgins, M.S. (1983). Peristaltic pumping in water waves. *Journal of Fluid Mechanics* 137: 393–409.
Lucia, F.J. (2007). *Carbonate Reservoir Characterization.* New York: Springer-Verlag.
Machel, H.G. (2004). Concepts and models of dolomitization: a critical reappraisal. *Geological Society London Special Publication* 235: 7–63.
Machel, H.G. and Mountjoy, E.W. (1986). Chemistry and environments of dolomitization – A reappraisal. *Earth Science Reviews* 23: 175–222.
Mackenzie, F.T. (2005). *Sediments, Diagenesis, and Sedimentary Rocks: Treatise on Geochemistry*, 2nd ed., Vol. 7. Amsterdam: Elsevier.
Marshall, J.F. (1985). Cross-shelf and facies related variations in submarine cementation in the Central Great Barrier Reef. *Proceedings of the 5th Coral Reef Symposium* 3: 509–512.
Matthews, R.K. and Frohlich, C. (1987). Forward modeling of bank-margin carbonate diagenesis. *Geology* 15: 673–676.
McNeill, D.F. (2005). Accumulation rates from well-dated late Neogene carbonate platforms and margins. *Sedimentary Geology* 75: 73–87.
Meister, P. (2013). Two opposing effects of sulfate reduction on calcite and dolomite precipitation in marine, hypersaline and alkaline environments. *Geology* 41: 499–502.
Melim, L.A. (1996). Limitations on lowstand meteoric diagenesis in the Pliocene–Pleistocene of Florida and the Great Bahama Bank: Implications for eustatic sea-level models. *Geology* 24: 893–896.
Middelburg, J.J. (1989). A simple rate model for organic matter decomposition in marine sediments. *Geochimica et Cosmochimica Acta* 53: 1577–1581.
Moore, C.H. (1989). *Carbonate Diagenesis and Porosity.* Amsterdam: Elsevier.
Mylroie, J.E. and Carew, J.L. 1990. The flank margin model for dissolutional cave development in carbonate platforms. *Earth Surface Processes and Landforms* 15: 413–424.
Palmer, T., Whitaker, F., Gabellone, T., et al. (2014). Evaluating links between sequence stratigraphy and reflux dolomitisation of the Permian San Andres Formation. *AAPG Annual Convention and Exhibition*, Houston, Texas, 6–9 April, Article 90189.
Paterson, R.J., Whitaker, F.F., Jones, G.D., et al. (2006). Accommodation and sedimentary architecture of isolated icehouse carbonate platforms: insights from forward modeling with CARB3D$^+$. *Journal of Sedimentary Research* 76: 1162–1182.
Paterson, R.J., Whitaker, F.F., Smart, P.L., et al. (2008). Controls on early diagenetic overprinting in icehouse carbonates: Insights from modeling hydrological zone residence times using CARB3D$^+$. *Journal of Sedimentary Research* 78: 258–281.
Perry, E.C., Swift, J., Gamboa, J., et al. (1989). Geologic and environmental aspects of surface cementation, north coast, Yucatan, Mexico. *Geology* 17: 818–821.
Pfeifer, K., Hensen, C., Adler, M., et al. (2002). Modeling of subsurface calcite dissolution, including the respiration and reoxidation progresses of marine sediment in the region of equatorial upwelling off Gabon. *Geochimica et Cosmochimica Acta* 66: 4247–4259.
Phillips, O.M. (1991). Flow and Reactions in Permeable Rocks. Cambridge: Cambridge University Press.
Plummer, L.N. (1975). Mixing of sea water with calcium carbonate groundwater. In: *Quantitative Models in the Earth Sciences* (ed. E.H.T. Whitten), 219–236. Geological Society of America Memoir 142.
Plummer, L.N., Vacher, H.L., Mackenzie, T., et al. (1976). Hydrochemistry of Bermuda: A case history of ground-water diagenesis of biocalcarenites. *Geological Society of America Bulletin* 87: 1301–1316.

Przyborska, A. (2014). Attenuation of groundwater flow due to irregular waves in permeable sea bottom. In: *Insights on Environmental Changes* (ed. T. Zielinski et al.), 109–125. *GeoPlanet: Earth and Planetary Sciences*. doi:10.1007/978-3-319-03683-0_8

Putney, K.L., Collins, J.F., Harris, P.M., et al. (2008). Reservoir characterisation of the Korolev Field, North Caspian Basin, Kazakhstan. *American Association of Petroleum Geologists Search & Discovery* Article 20055.

Read, J.F. (1985). Carbonate facies models. *American Association of Petroleum Geologists Bulletin* 69: 1–21.

Read, J.F. (1998). Phanerozoic carbonate ramps from greenhouse, transitional and ice-house worlds: clues from field and modelling studies. *Geological Society of London Special Publication* 149: 107–135.

Read, J.F. and Horbury, A.D. (1993). Eustatic and tectonic controls on porosity evolution beneath sequence-bounding unconformities and parasequence disconformities on carbonate platforms. In: Diagenesis and Basin Development (ed. A.D. Horbury and A.G. Robinson), 155–197. American Association of Petroleum Geologists Studies in Geology 36.

Rezaei, M., Sanz, E., Raeisi, E., et al. (2005). Reactive transport modeling of calcite dissolution in the fresh-salt water mixing zone. *Journal of Hydrology* 311: 282–298.

Riding, R. and Liang, L. (2005). Seawater chemistry control of marine limestone accumulation over the past 550 million years. *Revista Española Micropaleontologia* 37: 1–11.

Roehl, P.O. and Choquette, P.W. (2012). *Carbonate Petroleum Reservoirs*. New York: Springer Science & Business Media.

Ronchi, P., Ortenzi, A., Borromeo, O., et al. (2009). Diagenetic processes and their impact on the petrophysical properties in Kashagan carbonate platform reservoir (Carboniferous Kazakhstan). *American Association of Petroleum Geologists Annual Conference*, Denver, June 2009.

Runnels, D.D. (1969). Diagenesis, chemical sediments, and the mixing of natural waters. *Journal of Sedimentary Petrology* 39: 1188–1201.

Sainz-Garcia, A.M., Molinero-Hughet, J.J., and Saaltink, W.M.W. (2011). Numerical modelling of coastal aquifer karst processes by means of coupled simulations of density-driven flow and reactive solute transport phenomena. *Carbonates and Evaporites* 26: 19–27.

Sandberg, P.A. (1983). An oscillating trend in Phanerozoic nonskeletal carbonate mineralogy. *Nature* 305: 19–22.

Sanford, W.E. and Konikow, L.F. (1989). Simulation of calcite dissolution and porosity changes in saltwater mixing zones in coastal aquifers. *Water Resources Research* 25: 655–667.

Sarg, J.F. (1988). Carbonate sequence stratigraphy. *Society of Economic Paleontologists and Mineralogists Special Publication* 42: 155–181.

Schrag, D.P., Higgins, J.A., Macdonald, F.A., and Johnston, D.T. (2013). Authigenic carbonate and the history of the global carbon cycle. *Science* 339 (80): 540–543.

Shekdar, R., Sahni, I., Benson, G.S., et al. (2014). Modelling and simulation of a Jurassic carbonate ramp outcrop, Amellago, High Atlas Mountains, Morocco. *Petroleum Geoscience* 20: 109–123.

Shinn, E.A. (1969). Submarine lithification of Holocene carbonate sediments in the Persian Gulf. *Sedimentology* 12: 109–144.

Singurindy, O., Berkowitz, B., and Lowell, R.P. (2004). Carbonate dissolution and precipitation in coastal environments: laboratory analysis and theoretical consideration. *Water Resources Research* 40: W4401.

Skalinski, M., Kenter, J.A., Jenkins, S., and Tankersley, T. (2013). Updated rock type definition and pore type classification of a carbonate buildup, Tengiz Field, Republic of Kazakhstan (Russian). *SPE Caspian Carbonates Technology Conference* 1–16.

Smart, P.L., Dawans, J.M., and Whitaker, F.F. (1988). Carbonate dissolution in a modern mixing zone, South Andros Island, Bahamas. *Nature* 335: 811–813.

Smart, P.L., Beddows, P.A., Coke, J., et al. (2006). Cave development on the Caribbean coast of the Yucatan Peninsula, Quintana Roo, Mexico. *Geological Society of America Special Paper* 404: 105–128.

Smart, P.L., Richards, D.A., Beck, W., and Whitaker, F.F. (2011). Hydrochemistry of drip waters in an intergranular flow carbonate aquifer: Turks and Caicos Islands. Carbonate Geochemistry, Carbonate Geochemistry: Reactions and Processes in Aquifers and Reservoirs, Madison, Wyoming. *Karst Waters Institute Special Publication* 16: 65–68.

Smithers, S.G. (1994). Sediment facies of the Cocos (Keeling) Islands. *Atoll Research Bulletin* 407: 1–34.

Soetaert, K., Hofmann, A.F., Middleburg, J.J., et al. (2007). The effect of biogeochemical processes on pH. *Marine Chemistry* 105: 30–51.

Stoessel, R.K. (1992). Effects of sulphate reduction on $CaCO_3$ dissolution and precipitation in mixing zone fluids. *Journal of Sedimentary Petrology* 62: 873–880.

Thullner, M., Van Cappellen, P., and Regnier, P. (2005). Modeling the impact of microbial activity on redox dynamics in porous media. *Geochimica et Cosmochimica Acta* 69: 5005–5019.

Thullner, M., Dale, A.W., and Regnier, P. (2009). Global-scale quantification of mineralization pathways in marine sediments: a reaction–transport modeling approach. *Geochemistry, Geophysics, Geosystems* 10: Q10012.

Tucker, M.E. (1993). Carbonate diagenesis and sequence stratigraphy. *Sedimentology Review*: 51–72.

Walles, F.E. (1993). Tectonic and diagenetically-induced seal failure within the south-western Great Bahamas Bank. *Marine Petroleum Geology* 10: 14–28.

Ward, W.C. and Halley, R.B. (1985). Dolomitization in a mixing zone of near-seawater composition, late Pleistocene Northeastern Yucatan Peninsula. *Journal of Sedimentary Petrology* 55: 407–420.

Weber, L.J., Francis, B.P., Harris, P.M.M., and Clark, M. (2003). Stratigraphy, lithofacies, and reservoir distribution, Tengiz Field, Kazakhstan. *American Association of Petroleum Geologists Memoir* 83: 351–394.

Whitaker, F.F. and Smart, P.L. (1993). Saline groundwater circulation in carbonate build-ups: an overview and case study from the Bahamas. In: *Diagenesis and Basin Development* (ed. A.D. Horbury and A.G. Robinson), 113–132. American Association of Petroleum Geologists Studies in Geology 36.

Whitaker, F.F. and Smart, P.L. (1997). Climatic control of hydraulic conductivity of Bahamian limestones. *Groundwater* 35: 859–868.

Whitaker, F.F. and Smart, P.L. (2000). Characterising scale-dependence of hydraulic conductivity in carbonates: evidence from the Bahamas. *Journal of Geochemical Exploration* 69: 133–137.

Whitaker, F.F. and Smart, P.L. (2007a). Geochemistry of meteoric diagenesis in carbonate islands of the northern Bahamas: 1. Evidence from field studies. *Hydrological Processes* 21: 949–966.

Whitaker, F.F. and Smart, P.L. (2007b). Geochemistry of meteoric diagenesis in carbonate islands of the northern Bahamas: 2. Geochemical modelling and budgeting of diagenesis. *Hydrological Processes* 21: 967–982.

Whitaker, F., Smart, P., Hague, Y., et al. (1997). Coupled two-dimensional diagenetic and sedimentological modeling of carbonate platform evolution. *Geology* 25: 175–178.

Whitaker, F., Smart, P., Hague, Y., et al. (1999). Structure and function of a coupled two-dimensional diagenetic and sedimentological model of carbonate platform evolution. *Society of Economic Paleontologists and Mineralogists Special Publication* 62: 339–356.

Whitaker, F.F., Paterson, R.J., and Johnston, V. (2006). Meteoric diagenesis during sea-level lowstands: evidence from modern hydrochemical studies on northern Guam. *Journal of Geochemical Exploration* 89: 420–423.

Whitaker, F., Cooper, K., Smart, P., et al. (2011). Predicting carbonate diagenesis: emerging insights from reactive transport modelling. *Proceedings of the 14th Bathurst Conference*, Bristol, UK (Abstract).

Whitaker, F.F., Al-Helal, A.B., Gabellone, T., and Griffiths, G. (2012). Integrated reactive transport modelling: challenges and opportunities for improved prediction of diagenetic impact on reservoir quality. *Proceedings of the TOUGH Symposium 2012*, LBNL, Berkeley, CA.

Whitaker, F.F., Felce, G.P., Benson, G.S., et al. (2014). Simulating flow through forward sediment model stratigraphies: insights into climatic control of reservoir quality in isolated carbonate platforms. *Petroleum Geoscience* 20: 27–40.

Wigley, T.M.L. (1973). The incongruent solution of dolomite. *Geochimica et Cosmochimica Acta* 37: 1397–1402.

Woods, A.D., Bottjer, D.J., Mutti, M., and Morrison, J. (1999). Lower Triassic large sea-floor carbonate cements: Their origin and a mechanism for the prolonged biotic recovery from the end-Permian mass extinction. *Geology* 27: 645–648.

Xiao, Y. and Jones, G.D. (2006). Reactive transport modeling of carbonate and siliciclastic diagenesis and reservoir quality prediction. *Abu Dhabi International Petroleum Exhibition and Conference* held in Abu Dhabi, U.A.E., SPE Paper 101669.

Xiao, Y., Whitaker, F.F., Al-Helal, A.B., et al. (2013). Fundamental approaches to dolomitization and carbonate diagenesis in different hydrogeological systems and the impact on reservoir quality distribution. *International Petroleum Technology Conference 2013*, IPTC paper 16579.

Zachos, J.C., Pagani, M., Sloan, L., et al. (2001). Trends, rhythms and aberrations in global climate 65 Ma to present. *Science* 292: 686–693.

4

Reactive Transport Modeling and Reservoir Quality Prediction

Yitian Xiao[1] *and Gareth D. Jones*[2]

ExxonMobil Upstream Research Company, Spring, TX, 77389, USA

4.1 Fundamental Challenges in Reservoir Quality Prediction

Reservoir characterization is the art and science of describing, analysing, interpreting, and assessing rocks and fluids for the purpose of understanding and predicting the spatial variation and temporal distribution of their physical properties and dynamic behaviours. A fundamental challenge in reservoir characterization is the accurate prediction of the impact of diagenesis on reservoir quality and heterogeneity at multiple spatial and temporal scales (Figure 4.1). For example, early carbonate diagenesis modifies primary porosity and permeability by processes that include the stabilization of metastable aragonite to calcite, calcite dissolution, calcite cementation and dolomitization. Fluid flow associated with geothermal convection can cause both early and late diagenesis. During petroleum production, enhanced oil recovery (EOR) operations such as CO_2 or steam injection and *in situ* combustion can cause chemical reactions between the injected fluids and the reservoir rocks. These reactions are sometimes called *artificial diagenesis* (Hutcheon and Abercrombie, 1990; Xiao and Jones, 2006; Chapter 5 of this volume) and may be either beneficial or detrimental to reservoir quality and enhanced oil recovery. Therefore, predictions of both natural and artificial diagenesis constrain the associated changes in porosity and permeability in the reservoir and help to optimize strategies for petroleum exploration, development, and production.

Traditional reservoir characterization approaches include lab-based analyses, core and outcrop description and measurements, well log interpretations, well tests, and seismic characterization. These methods of characterization all have different spatial scales and resolution (Figure 4.1). This information is then incorporated into a reservoir model to simulate fluid flow (Figure 4.1). Reactive transport modeling (RTM) is an emerging technology that can be used to simulate coupled multi-phase fluid flow and diagenetic

1 Currently with ExxonMobil Exploration Company, Spring, TX, 77389, USA
2 Currently with ExxonMobil Development Company, Spring, TX, 77389, USA

Reactive Transport Modeling: Applications in Subsurface Energy and Environmental Problems, First Edition. Edited by Yitian Xiao, Fiona Whitaker, Tianfu Xu and Carl Steefel.

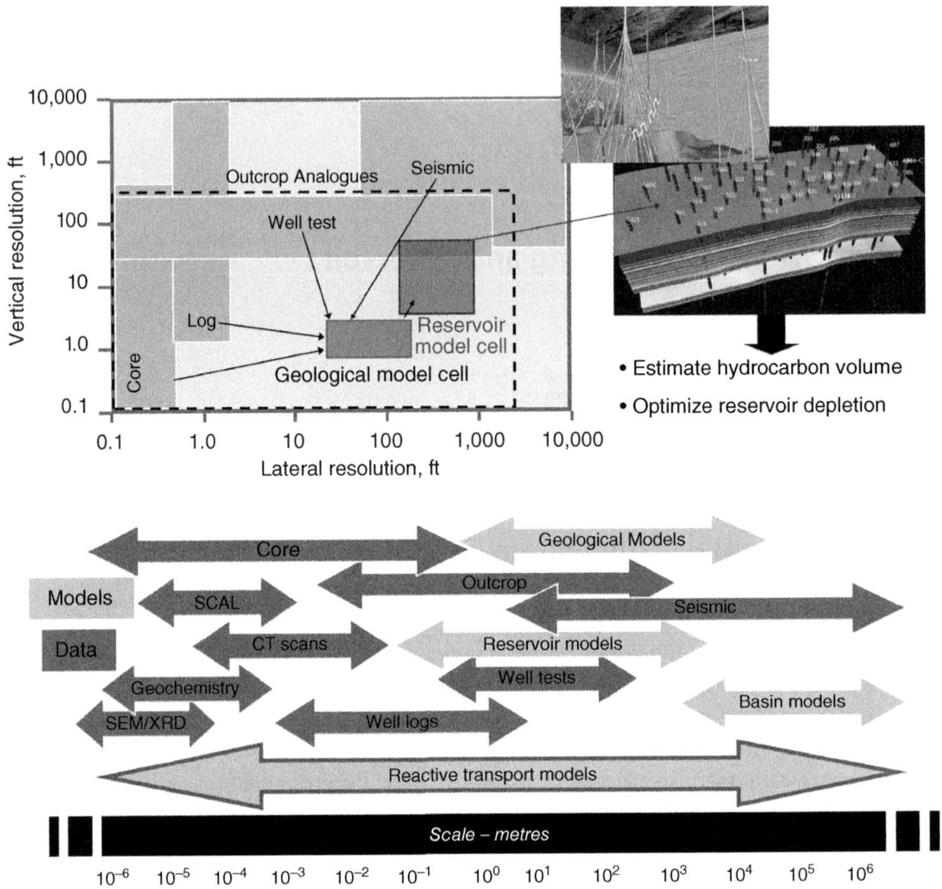

Figure 4.1 Schematic diagram showing an integrated approach to reservoir characterization by linking experimentation, observation, and modeling at multiple spatial scales.

reactions (Steefel et al., 2005; Xu et al., 2006). It has the unique capability to integrate experimental and observational data with other modeling approaches (e.g., basin and reservoir models) to predict the spatial and temporal variations in porosity and permeability and reservoir quality evolution at both geological and production timescales (Figure 4.1). This chapter synthesizes recent applications of reactive transport modeling to investigate different styles of diagenesis in both carbonate and siliciclastic rocks and selective outcrop and reservoir case studies (Table 4.1). These investigations include:

1) Dolomitization in different hydrological systems, such as in geothermal convection, freshwater–seawater mixing, brine reflux, and hydrothermal flow, including generic models, outcrop and reservoir based case studies.
2) Geothermal convection and burial diagenesis in a salt-buried isolated platform reservoir in Tengiz, Kazakhstan, and a pre-salt rift carbonate in the South Atlantic.
3) Fault-induced hydrothermal fluid flow and illitization based on the Rotligendes Formation, Germany.
4) Formation alteration associated with acid gas and steam injection and operation-induced diagenesis.

Style of diagenesis		Publications author/year	Flow mechanism	RTM approach	Key results	Reservoir quality impact/limitations
Dolomitization	Reflux	Jones and Xiao (2005)	Brine reflux in a generic carbonate platform	2D isothermal por/perm feedback	Brine reflux is an efficient dolomite mechanism. Anhydrite cements precipitate ahead of the dolomite front. Dolomite fingers extending several hundreds of metres ahead of the dolomite front developed with initial rock property heterogeneity.	Porosity increase from dolomite replacement and reduction from anhydrite cement down-dip (distal) of the dolomite front.
		Budd and Park, 2018	Circulation of hypersaline brines by unspecified flow mechanism	2D isothermal fully coupled with porosity / permeability feedback	Bed-scale simulations of dolomitization demonstrated the emergence of patterns, form a calcite grainstone random variations in porosity and permeability as a result of autogenic self-organization that is triggered by initial variations in reactive surface area.	Increases confidence that nested bed-scale patterns (short range and long range correlation trends) of dolomite evident in outcrops should be expected in reservoirs as result of self-organization. Reservoir simulations have previously shown that these bed-scale patterns can have an impact on reservoir performance.
		Xiao and Jones (2007), Xiao et al. (2013)	Brine reflux in a generic carbonate platform	3D non-isothermal por/perm feedback	Extensive 3D dolomite body develops from amalgamation of smaller dolomite bodies associated with multiple brine ponds.	Dolomite reservoir sweet spots are located close to the terminal dolomite front. Dolomite fingers are connected to the main dolomite body.
		Jones and Garcia-Fresca (2010); Garcia-Fresca and Jones (2011)	Brine reflux in an outcrop analogue: Cretaceous Glen Rose Formation	1D isothermal por/perm feedback	A dolomite body can be dislocated from a brine source due to chemical vs. physical controls on brine reflux. Reflux dolomite is likely to be stratigraphically discordant.	Need to consider chemical vs. physical controls along episodic reflux flow paths when predicting dolomite in a sequence stratigraphic framework.
		Garcia-Fresca et al. (2012)	Brine reflux in an outcrop analogue: Permian San Andres Formation	2D variable density flow with Mg mass balance calculations	Need to consider temporal evolution of brine reflux in a sequence stratigraphic framework. Impact of initial rock property heterogeneity on brine reflux flow patterns. Supports reflux dolomite interpretation for San Andres Fm outcrops.	Complex dolomite body distribution and connectivity. Did not model or consider effect of anhydrite cements related to dolomitization.

(*Continued*)

Table 4.1 (Continued)

Style of diagenesis	Publications author/year	Flow mechanism	RTM approach	Key results	Reservoir quality impact/limitations
	Lu and Cantrell (2016)	Brine reflux in a reservoir case study: Ghawar field, Arab-D reservoir zone	2D isothermal	Migration of brine ponds can result in discrete reflux dolomite geobodies. Geometry of dolomite front controlled by depositional heterogeneity. Simulated dolomite distribution compares favourably with field observations.	No anhydrite cements precipitate ahead of dolomite front with Jurassic seawater. Thick dolomite bodies can be spatially disconnected.
	Gabellone et al. (2016)	Reflux dolomitization in an outcrop analogue: Mississippian Madison ramp	2D non-isothermal	Dolomitization of epeiric scale ramps by brine reflux requires multiple brine sources that migrate up and down depositional dip in response to eustasy and antecedent topography.	Complex dolomite body distribution and connectivity along stratigraphic dip and strike. Up-dip to down-dip variations in reservoir quality more complex due to multiple brine sources, initial rock property heterogeneity and extensive overprinting.
Mixing zone	Sanford and Konikow (1989)	Freshwater–seawater mixing in a generic carbonate aquifer	2D isothermal model with brine and water	Mixing of seawater and calcite-saturated fresh groundwater can result in a solution that is undersaturated with respect to calcite, resulting in rapid carbonate dissolution along coastlines.	RTM results suggest that the mixing zone dissolution can generate a significant increase in porosity in tens of thousands of years.
	Xiao and Jones (2007)	Freshwater–seawater mixing in a generic carbonate platform	2D isothermal model with air, brine, and water	No replacement dolomite occurred in the mixing zone. Minor dolomitization at the coast beneath the mixing zone supports this variant of dolomitization by seawater. Dissolution of limestone in the mixing zone.	Rapid porosity generated from limestone dissolution in the mixing zone. Minimal impact of dolomitization by seawater beneath the mixing zone.
	Cooper et al. (2014)	Ascending freshwater-brine mixing in an outcrop analogue: La Molata, Spain	2D non-isothermal freshwater recharge and mesohaline fluid mixing	RTM results support rapid and extensive dolomitization when brines and freshwater mix, supporting a proposed variant of mixing-zone model based on outcrop observations.	Consider scenario of dolomitized reservoir where ascending flow from a freshwater aquifer interacts with residual brines in a land-attached carbonate platform.

	Geothermal	Wilson et al. (2001)	Geothermal convection in a generic carbonate platform	2D non-isothermal	Geothermal convection is a viable but relatively slow dolomitization mechanism. Anhydrite cements precipitate downstream of the dolomite front.	Dolomitization predicted to occur at depth in the interior of carbonate platforms (only 100 ky of geological time simulated).
		Whitaker and Xiao (2010)	Geothermal convection in a generic carbonate platform	2D non-isothermal model	Geothermal convection is a more efficient dolomite mechanism than previous RTM studies suggest. The rate of dolomitization is critically dependent on the abundance of dolomite. Heterogeneity in permeability and reactive surface area can produce complex patterns of dolomite.	Extensive reservoir-scale dolomite replacement is possible in ~10 My. Dolomite is predicted to be more abundant at the platform margin than in prior RTM models of geothermal convection.
		Al-Helal et al. (2012)	Geothermal convection and brine reflux in a generic carbonate platform	2D non-isothermal model	Reflux and geothermal convection can co-exist and operate in different regions of a carbonate platform. The rate of brine reflux dolomitization is significantly greater than geothermal convection.	Reflux can result in extensive platform interior dolomitization and anhydrite cementation with a close spatial relationship to the brine source. Geothermal convection results in minor dolomitization at the platform margin.
	Hydrothermal	Xiao and Jones (2007), Xiao et al. (2013)	Fault-controlled hydrothermal fluid flow in a generic carbonate platform	2D and 3D non-isothermal	Dolomitization from hot ascending brines is rapid. Continuous flow for tens of thousands of years is required to lead to large-scale dolomitization.	Extent and pattern of dolomite around fault reflects heterogeneity in initial permeability. Reservoir quality modification is variable; dolomitization enhances porosity whereas anhydrite cements occlude porosity.
		Gomez-Rivas et al. (2010, 2014)	Fault-controlled hydrothermal fluid flow in an outcrop analogue: Benicassim, Spain	2D non-isothermal	Overpressured dolomitizing fluids released during fault growth in a syn-rift tectonic period, may result in large lateral flow rates and dolomitization consistent with outcrop observations. Additional phase of dolomitization exploiting faults as flow conduits may occur in later sag phase of rifting.	Extent and pattern of dolomite around fault reflects heterogeneity in initial permeability. Reservoir quality modification is variable; dolomitization enhances porosity whereas anhydrite cements occlude porosity.
Other carbonate diagenesis	Early diagenesis	Xiao and Jones (2006)	Fluid flow in the vadose, freshwater lens, mixing, and saline zones in a generic carbonate platform	2D isothermal with por/perm feedback	Simulate flow and styles of diagenesis in the unsaturated vadose zone, freshwater lens, mixing zone and seawater.	Expect enhanced porosity associated with rapid mixing zone dissolution, particularly in a humid paleoclimate, up to 1 km from the coast. Fractures in the vadose zone and mixing zone focus flow become solution-enhanced to form a karst network.

(*Continued*)

Table 4.1 (Continued)

Style of diagenesis	Publications author/year	Flow mechanism	RTM approach	Key results	Reservoir quality impact/limitations
Geothermal convection and burial diagenesis	Jones and Xiao (2006)	Geothermal convection in a reservoir case study: Tengiz, Kazakhstan	2D basin model coupled with GeochemistWorkbench for calcite dissolution/precipitation	Geothermal convection can modify reservoir quality during deposition and after burial. Differences in facies between the platform and the rim and their stratigraphic partitioning controls intra-reservoir diagenetic patterns. Shale-filled salt withdrawal basins can locally drive convective flow in the underlying reservoir.	Enhanced porosity towards the centre of the Tengiz reservoir from ascending convective flow. Complex pattern of dissolution and cementation in syn-depositional fracture network development at platform margin.
	Jones and Xiao (2013)	Geothermal convection in generic pre-salt rift carbonates, South Atlantic	2D basin model coupled with GeochemistWorkbench for calcite dissolution/precipitation	Geothermal convective flow is feasible in a lacustrine carbonate reservoir based on the Pre-Salt reservoirs discovered in Brazil. Amount of net dissolution and cementation are low in the geological time available prior to burial. Permeable fault zones are an exception to this rule. Lake-derived pore fluids are rapidly flushed by seawater.	Diagenetic modification of reservoir quality by geothermal convection in pre-salt lacustrine carbonates will be associated with permeable fault zones. This will enhance vertical connectivity in the reservoir from dissolution. Fault zone cementation may create some local reservoir baffling, but not compartmentalization.
Clastic diagenesis	Xiao et al. (2007)	Fault-controlled flow and illitization in a reservoir case study, Rotliegendes	1D and 2D non-isothermal model	Hydrothermal fluid flows through the faults and fractures into the Rotliegendes formation, dissolves the K-feldspar and generates kaolinite, muscovite, illite, and quartz cement/overgrowth.	The diagenetic zonings (kaolinite, muscovite/illite, quartz, and K-feldspar) are closely associated with fluid flow along faults and fractures. Kaolinite and illite cementation significantly reduces reservoir permeability.

Siliciclastic and operation-induced diagenesis	**Operation induced Diagenesis**	Chapter 1 and of this volume, and references therein	CO_2 storage and geothermal exploration in conceptual and case studies	1D–3D non-isothermal models	Model for hydrogeochemical behavior of storage reservoirs at different spatial scales – from pore-scale to basin-scale, and temporal scales – from seconds to millennia, including geomechanical effects based on conceptual and real case studies (e.g. Slipner, Paris, Songliao).	Geochemical reactions induced by CO_2 and its feedback on physical properties need to be assessed for long-term CO_2 storage. Injecting acids could enhance connection between the production/injection wells and the fractured reservoir for geothermal operations.
		Xiao et al. (2009, 2010, 2011)	CO_2 and acid gas injection in different types of carbonate and clastic reservoirs	1D and 2D isothermal model	The behaviour of CO_2 and other acid gases during injection and storage is controlled by the gas mixture composition, reservoir mineralogy, timing, and injection design. Co-injection of H_2S yields similar behaviour compared with CO_2 injection in both siliciclastic and carbonate reservoirs.	Most acid gas injection scenarios result in a local porosity increase near the well. However, co-injection of SO_2 in a limestone-dominated carbonate reservoir can lead to significant anhydrite precipitation and porosity reduction in the near-well region.
		Meddaugh et al. (2012)	High pH steam injection in the Eocene carbonate reservoir, Wafra Field	2D multi-phase non-isothermal model	Steam injection can cause calcite and brucite precipitation and dolomite and anhydrite dissolution in the carbonate reservoir, affecting storage capacity, production and injectivity.	Injection of high-temperature, high-pH fluids may introduce fluid–rock interactions that affect fluid flow near the wellbore and with depth. Injection of high-quality steam (>95%, neutral pH) can minimize injectivity loss.
		Chapters 5 and 6 of this volume, and references therein	Multiple phase flow associated with enhanced oil recovery	1D and 2D non-isothermal model	Injection to and/or production from conventional and heavy oil reservoirs can induce drastic geochemical reactions and introduce risks for HC recovery.	Risk evaluation and management require thorough understanding and modeling of these geochemical systems based on the characterization of the reservoirs and injecting/producing fluids and gases.

Adopting a sensitivity analysis approach, RTMs are used to examine how key natural variables, such as climate, sea-level, temperature, salinity and fluid composition, and porosity/permeability heterogeneity impact upon different styles of diagenesis and reservoir porosity evolution. RTM results provide insights on key reservoir issues, such as the occurrence and distribution of dolomite geobodies, limestone to dolomite transition and anhydrite cementation, the distribution of high-permeability 'super-k' dolomite, and mineral transformation and impact on reservoir quality caused by fault-controlled fluid flow during both early and burial diagenesis. Other studies investigated how formation damage caused by acid gas injection and oil production can be predicted by reactive transport modeling. Linking fundamental geochemical processes to predict reservoir quality has improved our understanding of numerous conceptual diagenetic models, hypotheses, and paradigms. The novel approach demonstrates that reactive transport models, when sufficiently integrated with traditional methods and calibrated with field and production data, have the potential to improve our ability to predict carbonate and siliciclastic reservoir quality from first principles.

RTM is still considered an emerging technology and faces several challenges before it can be routinely used by industry to help predict and risk reservoir quality. At the end of the chapter we identify some key barriers to the penetration and commercial application of RTM in the energy industry: (1) demonstrated predictive capability; (2) business impact and efficiency; (3) access to specialist users; and (4) tool compatibility. We also discuss potential RTM applications and relative applicability and impact at different hydrocarbon exploration and exploitation business stages, and offer our perspective of the present and future role of RTM for reservoir quality prediction in the energy industry.

4.2 Reactive Transport Modeling Approach

Reactive transport modeling is a rapidly growing discipline that has drawn much attention in the geochemical, hydrological, and engineering communities in the past 20 to 30 years (Wood and Hewett, 1982; Phillips, 1991; Steefel and Lasaga, 1994; Litchner, 1996; Steefel et al., 2005; Xu and Pruess, 2001; Xu et al., 2006). The approach relies on knowledge of geology, geochemistry, hydrology, physics, and, in particular, the rapid development of computing power to model coupled fluid–rock interactions. Some of its key applications include nuclear waste disposal, CO_2 sequestration, biogeochemical processes and enhanced oil recovery, which are covered extensively in this book. Its applications also include reservoir engineering, carbonate and siliciclastic diagenesis, and environmental remediation. The basic concept of reactive transport modeling for reservoir quality prediction can be illustrated with a simple model for dolomitization of limestone (Figure 4.2). The diagram shows that dolomitization requires the supply and transport of Mg^{2+} into the hydrological environment, the chemical reaction replacing Ca^{2+} with Mg^{2+} in calcite, transport of Ca^{2+} out of the system, and the associated modifications to the pore system (porosity and permeability).

A reactive transport model relies on a mathematical formulation to describe the coupled transport and geochemical processes. The general governing equation can be written as:

$$\frac{\partial}{\partial t}(\phi C_i) = \frac{\partial}{\partial x}\left(\phi D \frac{\partial C_i}{\partial x}\right) - \phi v \frac{\partial C_i}{\partial x} + \phi \sum_k \left(\frac{\partial C_i}{\partial t}\right)_k \tag{4.1}$$

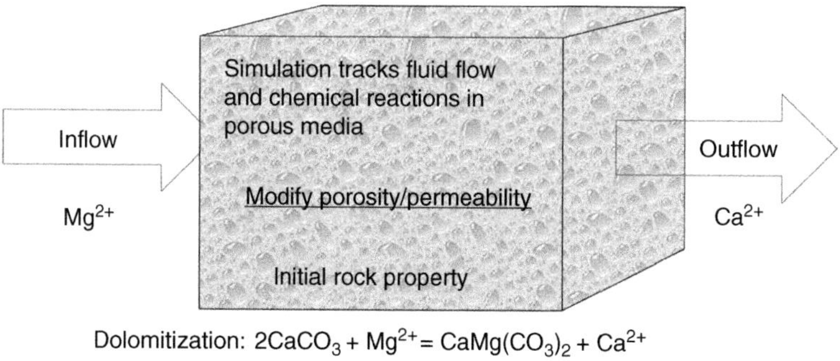

Figure 4.2 Schematic diagram showing the reactive transport process of dolomitization. *Source:* From Xiao and Jones (2006), used with permission of AAPG.

where C_i is the concentration of a specific species in the pore fluid, D is the combined diffusion and dispersion coefficient term, v is the linear fluid flow rate, and ϕ is the porosity. The first two terms on the right-hand side describe the transport process (diffusion, dispersion, and advection), while the last term describes the effect of chemical reactions including detailed treatment of the kinetics and thermodynamics of the system.

Due to complex boundary conditions and complicated coupling between the transport and reaction terms, it is impossible to provide analytical solutions to Equation (4.1) for even the simplest geochemical system. Therefore numerical solutions are used. Fortunately, due to the exponential increase in computational power available, geologically realistic reactive transport models have been developed that are begining to provide exploitable insights to reservoir quality prediction at sufficient spatial and temporal scales to impact petroleum exploration, development, and production strategies. In our investigations, several well known and extensively tested numerical programs, such as Geochemist's Workbench (Bethke, 1996), Basin2 (Bethke et al., 1999) and ToughReact (Xu et al., 2006), were used for the RTM simulations. The chapter highlights results from recent efforts in applying this technology to investigate both carbonate and siliciclastic diagenesis and the impact on reservoir quality, as summarized in Table 4.1.

4.3 Modeling Dolomitization in Different Hydrogeological Systems

4.3.1 Dolomitization and Impact on Carbonate Reservoir Quality: From Reservoir to Outcrop Observations

Significant volumes of world hydrocarbon resources occur in dolomite reservoirs. For example, major Middle East reservoirs like the Arab-D in the Ghawar field are partially dolomitized. The spatial distribution of dolomite presence and in particular the diagenetic transition between limestone and dolomite is often a critical control of reservoir quality in many carbonate reservoirs (Powers, 1962; Cantrell et al., 2004). The Jurassic Ghawar field in Saudi Arabia is the most notable and prolific Middle East example where the distribution of dolomite is often associated with high rates of oil production (Figure 4.3). Depending on the original depositional texture, style of dolomitization, and

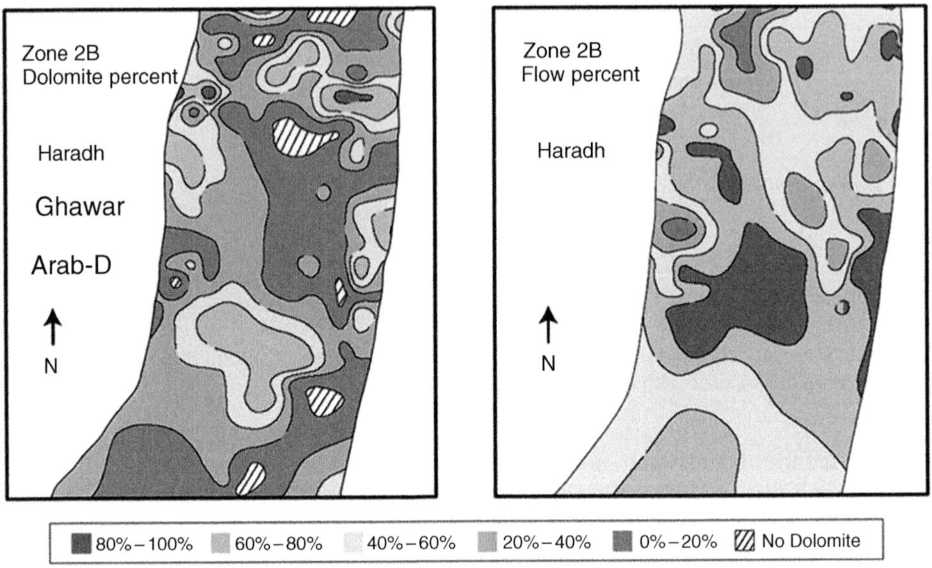

Figure 4.3 Correlations of dolomite distribution and flow percentages in the Arab-D Formation of the Ghawar Field, Saudi Arabia. *Source:* From Cantrell et al. (2004), used with permission from GeoArabia.

presence of anhydrite cement, dolomite can behave as a reservoir or as a baffle/barrier to flow (Powers, 1962; Tucker, 1993; Cantrell et al., 2004). High permeability, or so-called 'Super K' dolomites with multi-Darcy permeability sporadically occur in thin zones (sometimes <5 ft) and are vital contributors to individual well and overall reservoir performance (Cantrell et al., 2004). Laterally continuous, low permeability 'tight' dolomites can also have significant impact on subsurface flow and thus reservoir management strategies. The transition between limestone and dolomite is readily identified in both wireline logs and cores and is dominantly observed as a sharp interface described as a 'dolomite front' (*sensu* Wilson et al., 1990). Unfortunately, dolomite fronts in carbonate reservoirs are often below seismic detection, particularly those associated with 'thin' beds of dolomite of the order of a few to several tens of feet thick. Predicting the spatial distribution or connectivity of dolomite and the associated petrophysical properties between wells and away from well control is thus a fundamental challenge in carbonate reservoir characterization. In contrast to depositional facies, there are no defined rules or guidelines for correlating dolomite bodies in the subsurface. Existing correlations of dolomite tend to be lithostratigraphical and contain considerable uncertainty in the degree of interpreted reservoir geobody and connectivity (Figures 4.3 and 4.4).

One way to understand the distribution and connectivity of sedimentary bodies is to study outcrop analogues (Van Wagoner et al., 1988; Posamentier et al., 1992). Many outcrop studies have focused on dolomitized carbonates. However, there are few predictive dolomitization models, and among them very few have specifically focused on transitions between limestone and dolomite and/or described the connectivity of dolomite bodies. Outcrop-based studies by Wilson et al. (1990) and Yao and Demicco (1997) that mapped dolomite fronts associated with geothermal circulation and ascending hydrothermal fluids, respectively, are notable exceptions. These analogues are, however, not

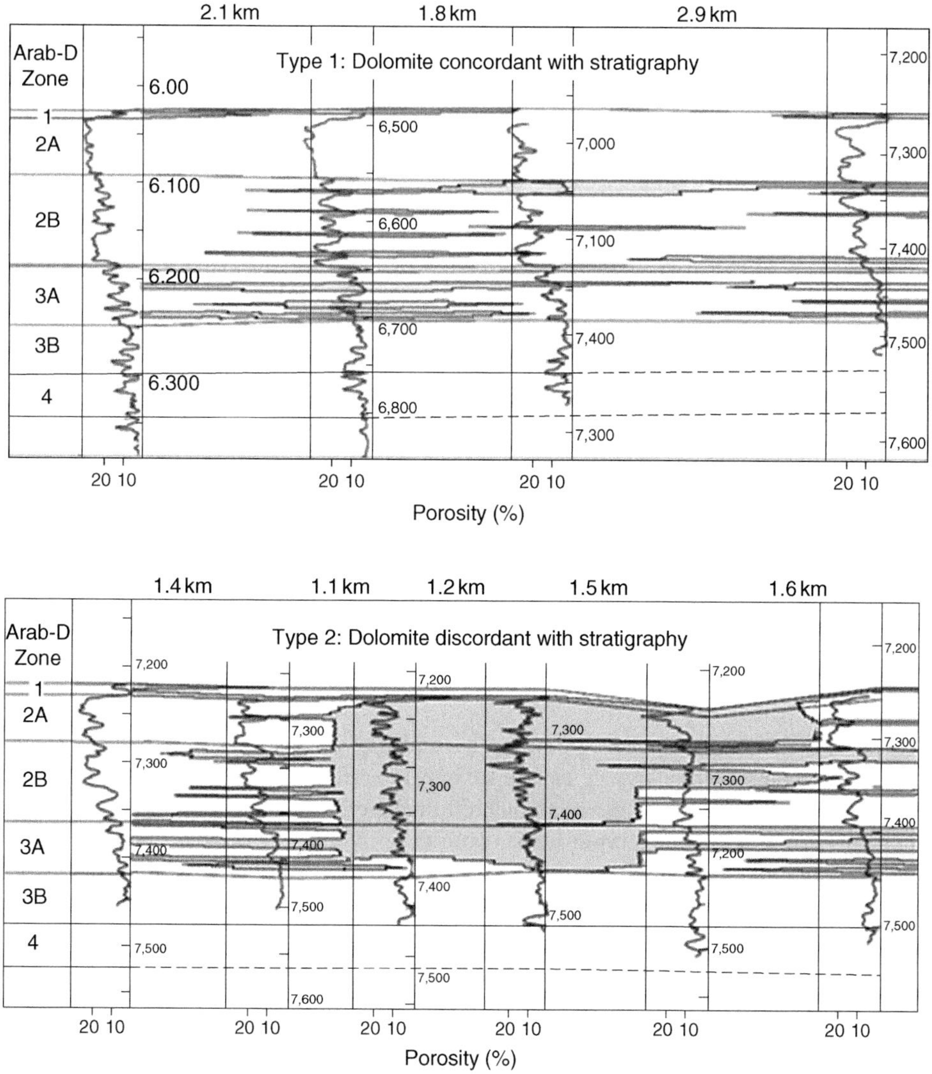

Figure 4.4 Correlations of limestone to dolomite (shaded grey) transitions from wireline well logs in the Ghawar Field, Saudi Arabia. Note 1–2 km well spacing and vertical scale in feet. *Source:* From Cantrell et al. (2004), used with permission from GeoArabia.

appropriate for reservoirs where the majority of dolomite is interpreted to have formed by the reflux of brines from evaporated seawater, such as Ghawar field in the Middle East (Cantrell et al., 2004). Dolomite fronts mapped in Plio-Pleistocene outcrops on Bonaire by Lucia and Major (1994) are interpreted to have formed by brine reflux (Figure 4.5). The Bonaire study provides insights on the propagation of a dolomite front and the distribution of petrophysical properties in a reflux system, but is of limited use in helping to develop dolomite correlation strategies in extensive carbonate ramp reservoirs like the

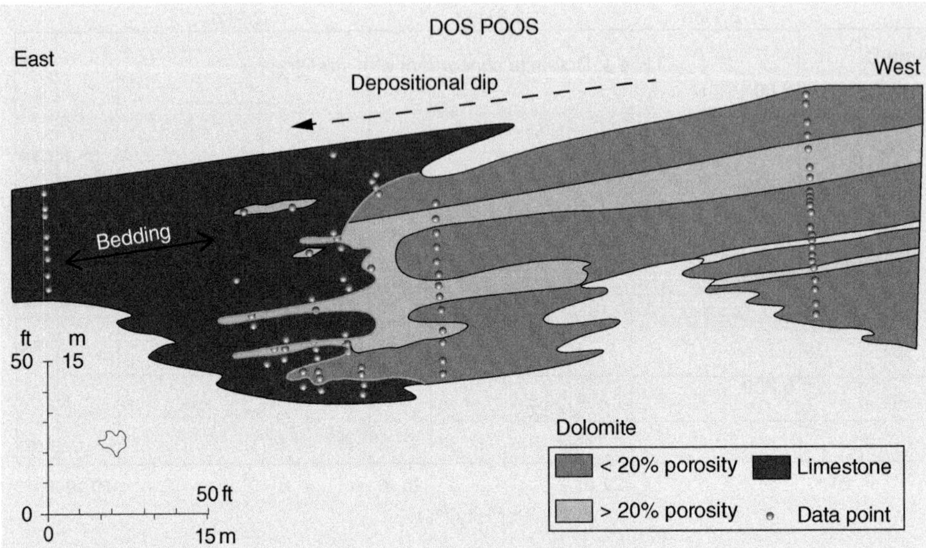

Figure 4.5 Limestone to dolomite transition mapped in a Plio-Pleistocene outcrop on Bonaire, Lesser Antilles. Note dolomite 'fingering' in the direction of depositional dip. *Source:* From Lucia and Major (1994), used with permission from International Association of Sedimentologists Special Publication.

Ghawar field. Other smaller-scale outcrop examples of limestone/dolomite transitions interpreted to have formed as a consequence of brine reflux are shown in Figure 4.6.

Moreover, the presence of other associated diagenetic products, such as anhydrite, adds more complexity to reservoir quality prediction. Therefore, a process-based modeling approach that can link the depositional and burial environment and the hydrogeological systems to limestone to dolomite transformation and distribution and anhydrite occurrence could provide significant insight for carbonate reservoir characterization.

4.3.2 Conceptual Hydrological Models of Dolomitization

It is well established that large-scale dolomitization requires both the active circulation of Mg-rich fluids (commonly derived from seawater) and favourable geochemical conditions to overcome kinetic barriers to dolomitization (Land et al., 1975; Zenger et al., 1980; Land, 1985; Arvidson and Mackenzie, 1999). Simms (1984) was one of the first studies to recognize that most models for the formation of replacement dolomites are essentially hydrological models (Figure 4.7). Flow of dolomitizing fluids may result from gradients in density generated by contrasts in salinity/brine reflux (Jones and Xiao, 2005) or temperature/geothermal convection (Sanford et al., 1998; Wilson et al., 2001; Whitaker and Xiao, 2010; Al-Helal et al., 2012), or pressure gradients generated by elevation head, tectonic or sedimentary compaction (Whitaker et al., 2004), or fault-induced hydrothermal fluid flow (Machel, 2004; Davies and Smith, 2006; Gomez-Rivas et al., 2010). It was hypothesized that different flow systems might generate distinctive distributions of dolomite (Wilson et al., 2001; Whitaker and Smart, 1993; Whitaker et al., 2004). However, flow of diagenetic fluids is critically dependent upon spatial

Figure 4.6 Outcrops of limestone/dolomite transitions. (a) Devonian, Cripple Creek, Canada; (b) Ordovician, Franklin Mountains, USA; (c) and (d) Ordovician, Franklin Mountains, USA. In all images, dolomite is in light shade and limestone is grey. The arrows show the interpreted direction of brine reflux paleo-fluid flow. Images a, b, and d courtesy of Jerry Bellian, and D from Tillotson (2003), used with permission from AAPG.

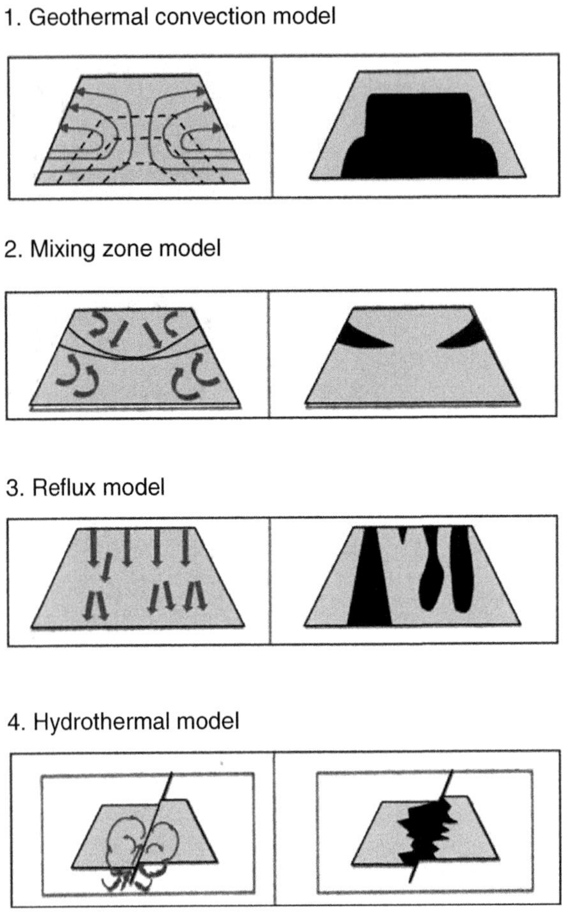

Figure 4.7 Conceptual hydrological models of dolomitization illustrating the flow and potential dolomitization patterns. Arrows represent fluid flow patterns and shaded areas represent potential dolomite distribution.

variation in permeability, generating complex dolomite geobodies at a range of scales. In addition, whilst simplistic models of individual flow systems are conceptually valuable, hybrid circulation systems commonly develop where more than one flow system operates simultaneously (Jones et al., 2004; Al-Helal et al., 2012). Finally, natural flow systems will evolve over time in response to changes in rock properties and/or geological boundary conditions, such as relative changes in sea-level, or the feedback to the fluid flow from the altered rock properties (Jones and Xiao, 2005; Xiao et al., 2013). Meaningful prediction of fluid flow systems capable of dolomitization thus depends upon exploring such complexities.

The next section summarizes results from reactive transport models employed to investigate dolomitization mechanisms in different hydrogeological systems: (1) geothermal convection; (2) brine reflux; (3) mixing zone and sub-mixing zone; and (4) fault-controlled hydrothermal fluid flow. The RTM results provide new insights on the transformation and distribution of dolomite, limestone, and anhydrite, and implications for reservoir connectivity and subsurface correlation and interpretation strategies.

4.3.3 Geothermal Convection Models

Geothermal ('Kohout') convection of seawater cools the margins of carbonate platforms, warming the interior at shallow depths (Kohout et al., 1977), and has been invoked to explain dolomitization during early burial. Geothermal convection can occur as 'free' or 'forced' convection. Free convection is characterized by counter-rotating fluid flow cells which develop in anisotropic carbonate platforms (Sanford et al., 1998). However, a substantial thickness (3–500 m) free of laterally continuous mudstone/shale beds is required for free convection to occur. Furthermore, unless these free convection cells are partially hydraulically open ('mixed convection'), the available magnesium would be rapidly consumed and dolomitization would cease. In contrast, forced convection of cold seawater through carbonate platforms during early burial (also termed 'Kohout convection') draws on a nearly limitless source of magnesium. Forced convection affects large volumes of carbonates for extended time periods and is a mechanism essentially independent of relative sea-level (Jones et al., 2000; Wilson et al., 2001; Whitaker et al., 2004).

Early short-term (0.1 My) RTM simulations (Wilson et al., 2001) suggested that this process can drive replacement dolomitization and anhydritization, mostly at >50 °C. However, complete dolomitization would require more than 70 My based on a simple linear extrapolation of the short-term simulation results (Figure 4.8). The RTM simulations over a more extended time-frame (up to 30 My) using the same hydrogeological and geochemical parameters indicate significant non-linearities in the system (Whitaker and Xiao, 2010), meaning that the rate of dolomitization would accelerate as more precursor limestone is dolomitized, which is consistent with the results of high-temperature experiments. Geothermal convection may completely dolomitize parts of the platform in around 15 My (Figure 4.8). As dolomitization proceeds, the process becomes predominantly flux-controlled, with the development of a wedge-shaped dolomite body that gradually thins from the platform margin to the interior and is formed at considerably shallower depths and cooler temperatures (20–30 °C) than suggested by simple extrapolation of short term simulations (Wilson et al., 2001).

Dolomitization is relatively insensitive to changes in model boundary conditions such as relative sea-level and platform geometry, but occurs significantly slower in circular compared with more elongate platforms (Wilson et al., 2001). Sediment permeability and reactive surface area, often inversely related, are key controls on dolomitization. The restricted fluid flux limits geothermal dolomitization to the platform margin of low permeability muddy platforms, despite having a high reactive surface area (Whitaker and Xiao, 2010). Dolomitization in more permeable grainy platforms is limited by the reactive surface area, and occurs only in the platform centre due to widespread cooling around the margin. Vertical layering of sediment types results in a complex vertical diagenetic stratigraphy, favouring the formation of more reactive beds at shallow depths where permeability is not the limiting factor (Tillotson, 2003), but shifting to more permeable beds at depth. Bank-marginal fracturing limits dolomitization of the platform interior, regardless of whether the fractures are baffles or conduits for flow (Figure 4.9; Whitaker and Xiao, 2010).

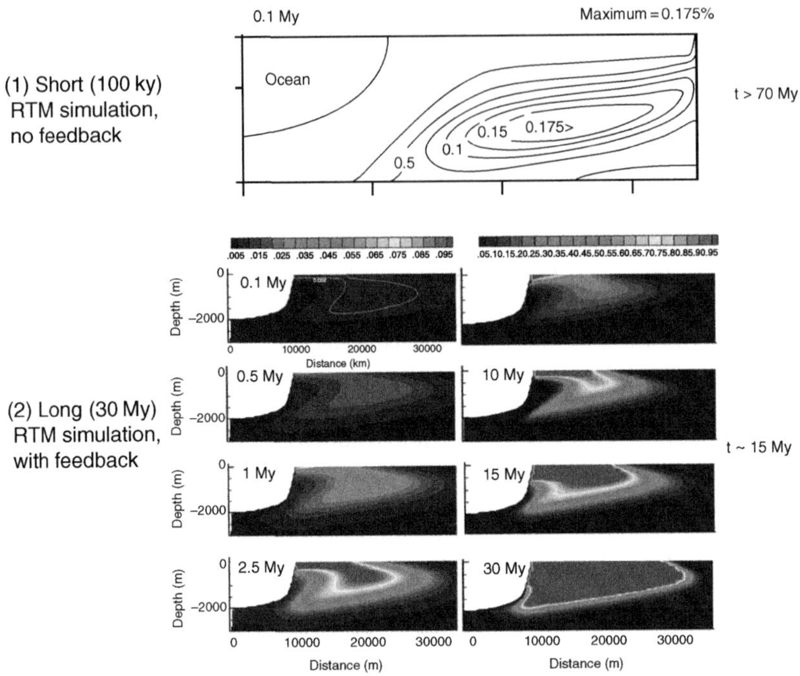

Figure 4.8 Distribution of dolomite as a fraction of rock volume predicted from RTM. (1) Early short-term simulation (100 ky) results from Wilson et al. (2001) with contours showing dolomite %; (2) long-term baseline simulation at time steps from 0.1 to 30 My. Note × 10 adjustment in scale bar from 0.1–2.5 My displays compared with those for 5–30 My time-steps. From Whitaker and Xiao (2010).

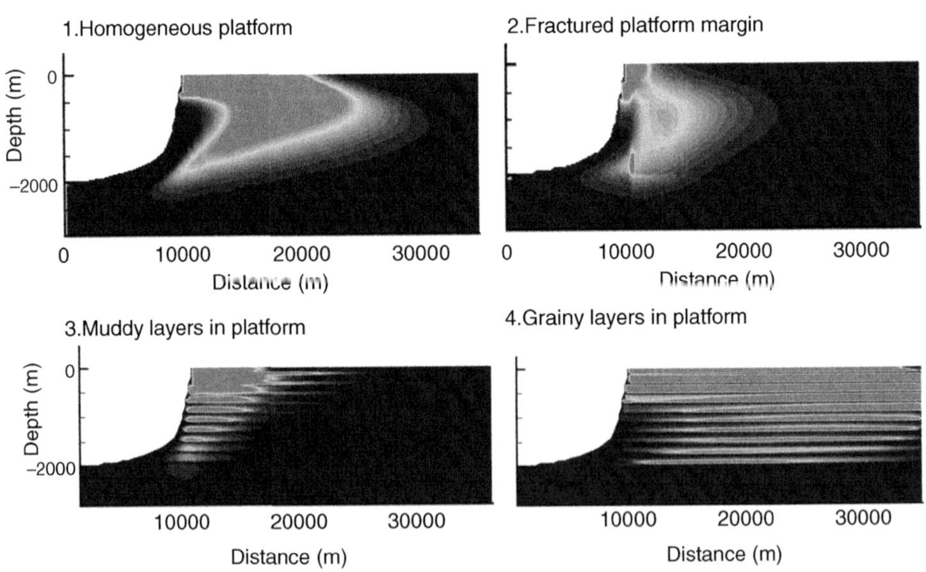

Figure 4.9 Comparison of dolomite (light shades) distribution as a fraction of rock volume from: (1) a homogeneous platform; (2) a fractured platform margin; (3) muddy layers in a platform; (4) grainy layers in a platform. From Whitaker and Xiao (2010). (See colour plate section for the colour representation of this figure.)

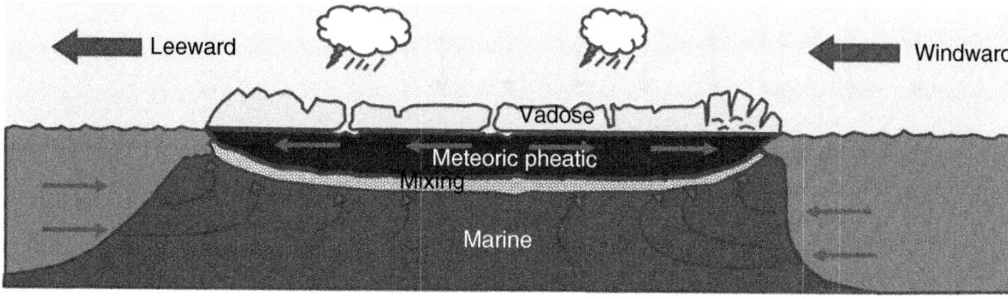

Figure 4.10 Conceptual model of an emergent isolated carbonate platform showing the distribution of four different hydrological zones: vadose, meteoric phreatic, mixing, and marine. From Moore (2001).

4.3.4 Mixing Zone Models

The 'mixing zone' model of dolomitization is perhaps the most controversial (Hanshaw et al., 1971; Hardie, 1987; Humphrey and Quinn, 1989; Land, 1991; Oswald, 1992; Wicks and Herman, 1996; Lu and Meyers, 1998; Machel, 2004; Melim et al., 2004). The mixing of seawater and freshwater theoretically results in a fluid that is supersaturated with respect to dolomite and undersaturated with respect to calcite (Badiozamani, 1973); however, the thermodynamic calculation of fluid mixing did not take into account the transport control, which could quickly change the saturation state along the fluid flow path (Hardie, 1987; Bethke, 1996). A series of qualitative predictive concepts were developed that related different early carbonate diagenetic processes and products to sequence stratigraphy, platform type, and climate (Figure 4.10; see summary in Moore, 2001, pp. 245–291). These conceptual models illustrate how the spatial distributions of diagenetic styles are related to different hydrological zones in a carbonate platform (Figure 4.10); however, these simplistic models are limited with respect to making quantitative diagenetic predictions.

4.3.4.1 Traditional Mixing Zone Model

RTMs were developed to investigate early diagenesis in a generic isolated carbonate platform (Rezaei et al., 2005; Xiao and Jones, 2006; Xiao et al., 2013). The 2D RTM results, including the simulated fluid flow and distribution of hydrological zones (vadose zone, freshwater lens, mixing zone, and saline zone) in the left half of an isolated carbonate platform, are shown in Figure 4.11. In the salinity distribution map (Figure 4.11a), the horizontal bold line represents sea-level, and the arrowed lines are streamlines. The vadose zone is about 4 metres thick across the platform. The thickness of the freshwater lens varies, and is deepest at the platform centre (model boundary on right side). The mixing zone is thinner in the middle of the platform and thicker at the platform margin. The overall hydrological zone distribution is in excellent agreement with other simulation results with the same boundary conditions (Sanford and Konikow, 1989; Lee, 2003), as well as with the Ghyben-Herzberg model (Vacher, 1988; Vacher et al., 1990; Vacher and Quinn, 1997).

The simulation results suggest that, due to the lack of Mg^{2+} supply from limited transport plus slow reaction rates, rain and seawater mixing zone dolomitization is not a viable mechanism for large-scale dolomitization in isolated platforms (Figure 4.12). As pointed

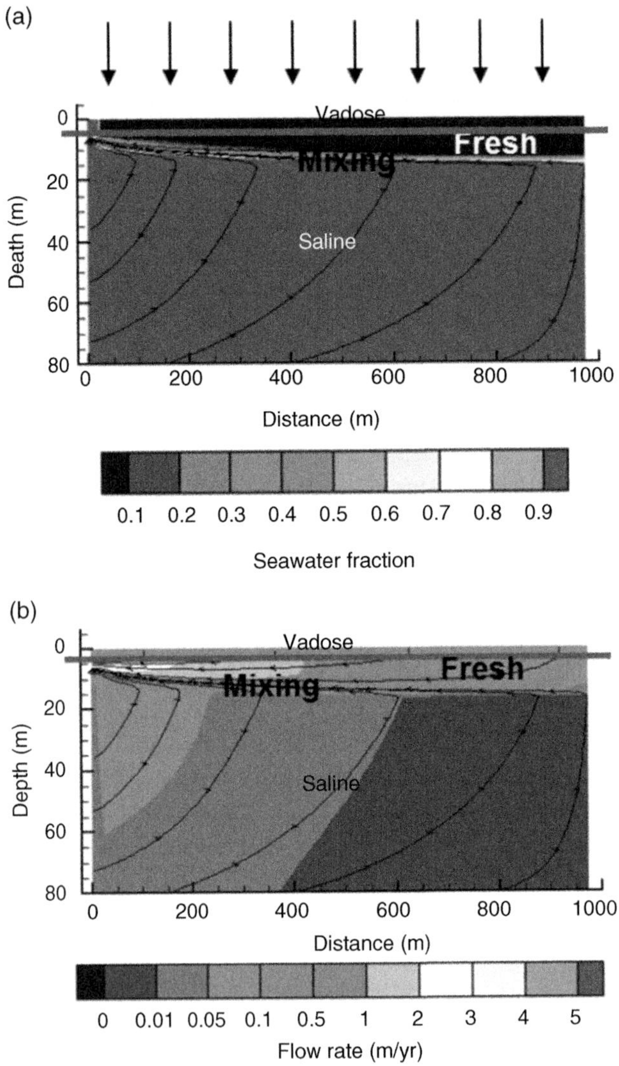

Figure 4.11 Simulation results of fluid flow and hydrological zone distribution in an isolated carbonate platform. (a) Freshwater–seawater distribution; (b) Flow rate in m/yr. Arrows represent streamlines. All distance units are in metres and flow rate units in m/yr. From Xiao et al. (2013).

out by Hardie (1987), due to non-linear mixing behavior of freshwater–seawater according to the thermodynamics, mixing zone fluids in the isolated carbonate platform are often undersaturated with respect to calcite and dolomite and can lead to significant dissolution (Sanford and Konikow, 1989; Figure 4.12a). Circulation of seawater in the sub-mixing zone (Figure 4.12), on the other hand, might result in more dolomitization (Whitaker and Smart, 1993). This is supported by RTM results, with up to 3% dolomite forming at the platform margin in 100 ky (Figure 4.12b). This is a good illustration of how RTM can be used to evaluate proposed conceptual models like the mixing zone and identify/support alternative dolomitization models in the same hydrological system.

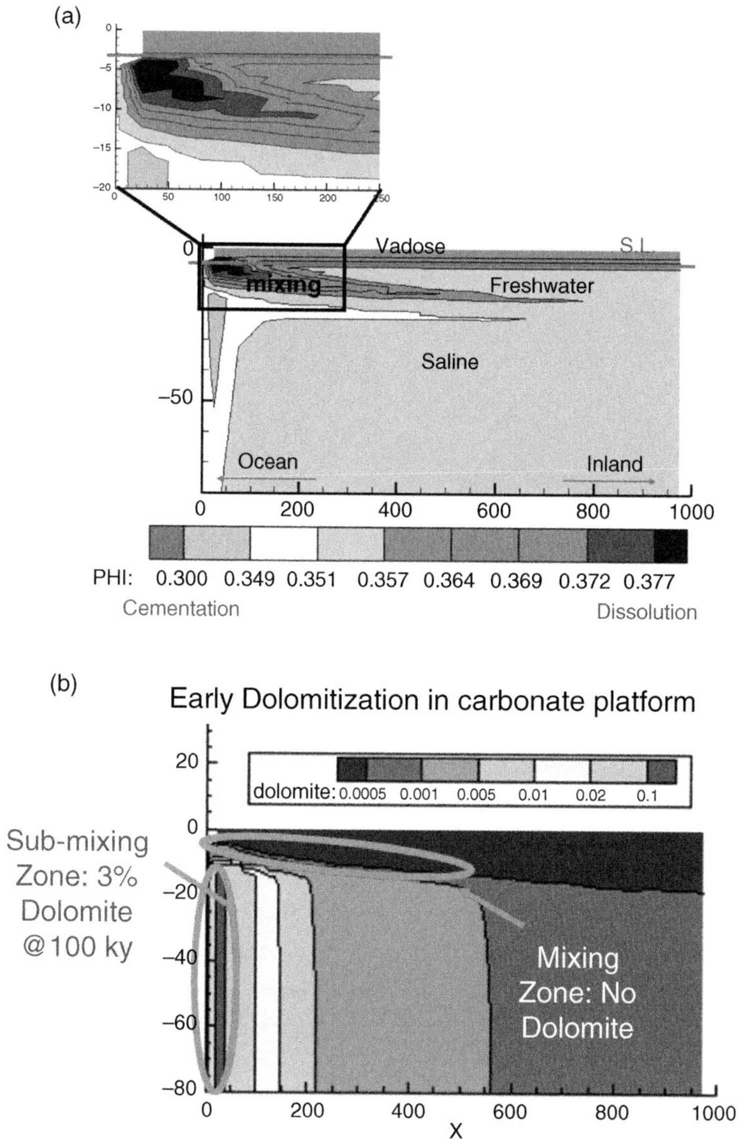

Figure 4.12 Simulation results showing (a) potential dissolution and (b) lack of early dolomitization (Xiao and Jones, 2006) in the traditional mixing zone in an isolated carbonate platform after 100 ky (Xiao et al., 2013).

4.3.4.2 Ascending Freshwater–Mesohaline Brine Mixing Model: La Molata Miocene Outcrop Case Study

A recent RTM study tested a hypothesis for dolomitization from mixing of ascending fresh groundwater with mesohaline porewaters based on patterns of observed extensive dolomitization at the Miocene La Molata outcrop in Spain (Li et al., 2013). Figure 4.13a shows a cross-section of the La Molata platform with contours of dolomite percentage

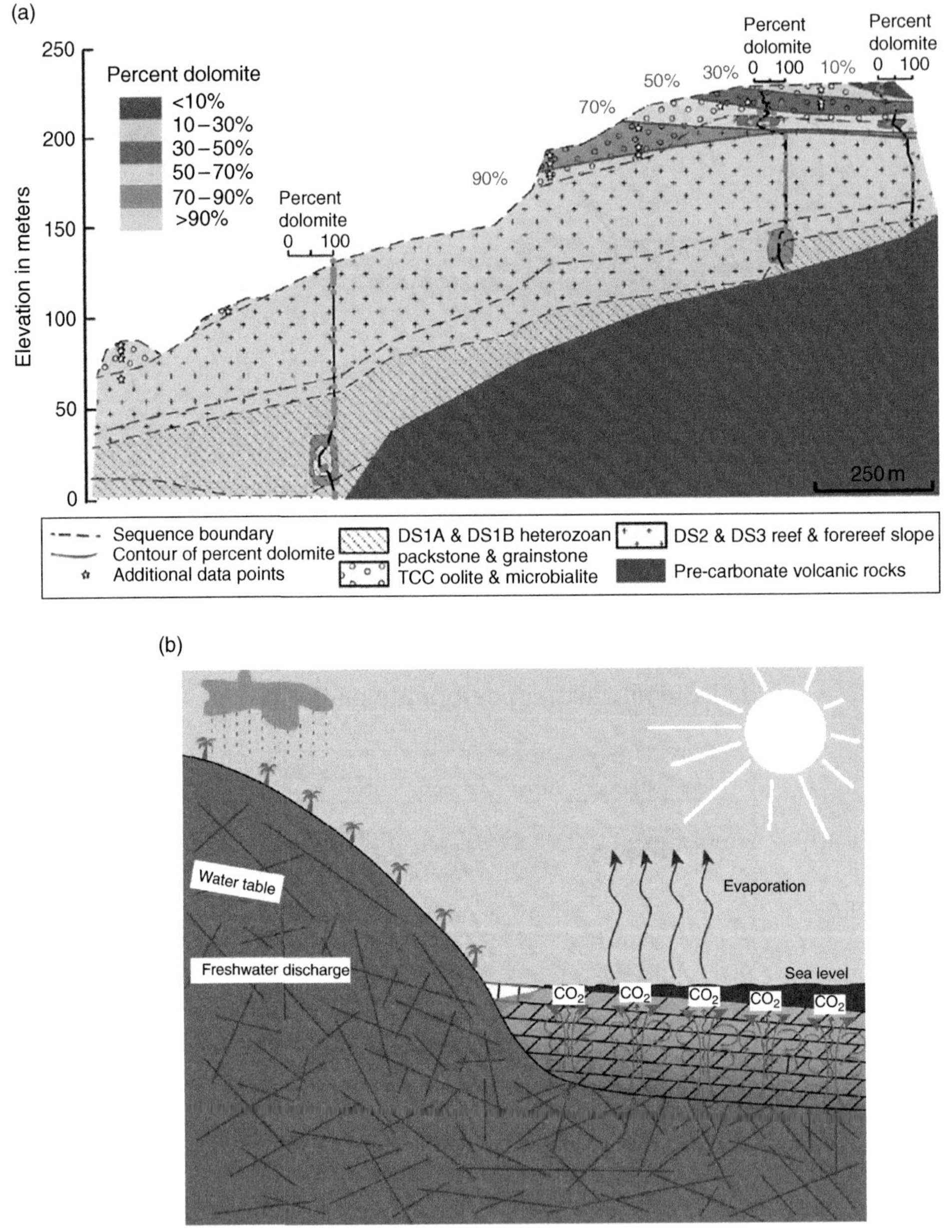

Figure 4.13 Proposed ascending freshwater–mesohaline water mixing mechanisms of observed extensive dolomitization at La Molata, Miocene. (a) Cross-section of La Molata platform with contours of dolomite percentage superimposed to illustrate the dolomite distribution. (b) Freshwater flows upwards through fractured volcanic aquifer to discharge into a carbonate platform from below. *Source:* From Li et al. (2013), used with permission of AAPG.

superimposed to illustrate dolomite distribution. Dolomite abundance increases basinwards and with stratigraphic age (Li et al., 2013). Li et al. (2013) argued that fractures in the underlying volcanics provided pathways for fluid flow driven by a large hydraulic head from the hinterland, plus the density buoyancy required to force freshwater to flow upward and discharge into a carbonate platform from below (Figure 4.13b). This mechanism would create a zone of active fluid mixing far exceeding that observed in traditional fluid mixing zones associated with coastal aquifers (Figures 4.11–4.13).

RTMs were developed to investigate the concept that ascending freshwater–mesohaline water mixing has the potential for significant flow, transport of Mg^{2+}, and generation of extensive dolomitization in this setting (Cooper et al., 2014). Geochemical mixing calculations show that, for a wide range of freshwater and seawater end-member compositions, non-linear mixing behaviour generates the so-called 'Dorag zone' where it is thermodynamically favourable to dissolve calcite and precipitate dolomite (Hardie, 1987). However, RTM did not predict pervasive dolomite in such a freshwater–seawater mixing system within 1 million years, as mixing freshwater/rainwater tends to dilute the Mg^{2+} concentration, reducing the dolomitization potential of the seawater end-member (Xiao et al., 2013; Cooper et al., 2014). Substituting seawater for mesohaline brines in the RTM increased the concentration of Mg^{2+} available for dolomitization. Freshwater discharge from the basal volcanic aquifer generates freshwater–seawater convective mixing flow, and the elevated saline water temperature (40 °C) helps overcome the kinetic barriers (Cooper et al., 2014; Figure 4.14). The modeling results suggest that this mechanism can lead to significant dolomitization in the carbonate platform within 100 ky and supports the modified brine mixing zone dolomitization hypothesis proposed by Li et al. (2013).

4.3.5 Reflux Dolomitization Models

Although there are numerous dolomitization models and hypotheses, in a comprehensive global survey, Sun (1995) concluded that the majority of dolomite reservoir examples are associated with evaporites or evaporitic conditions and are interpreted as being reflux in origin. Furthermore, Sun (1995) proposed that reflux of seawater evaporated to less than gypsum saturation (mesohaline reflux) may account for a number of dolomite reservoirs classified as 'non evaporitic'.

Reflux-driven flow results from differences in fluid density generated by lateral spatial variations in salinity. Isolation or restriction of platform-top seawater can generate brines that flow downwards under the influence of gravity as potential dolomitizing fluids (Figure 4.15). The reflux model was originally proposed by Adams and Rhodes (1960) to explain dolomitization of Permian carbonates in West Texas, which have a close stratigraphic association with platform interior evaporites.

4.3.5.1 2D Simulations of Brine Reflux Dolomitization

Numerous hydrogeological models have been used to investigate density-driven fluid flow in various geological settings (Simms, 1984; Kaufman, 1994; Jones et al., 2003; Whitaker et al., 2004). These models provided significant insight about reflux fluid flow and the significance for extensive reflux dolomitization based on Mg mass balance calculations. More recently, fully coupled 2D RTM models were applied to investigate reflux dolomitization (Jones and Xiao, 2005). To capture the heterogeneous porosity and permeability of carbonate sediments in the shallow subsurface (0–500 m), due to

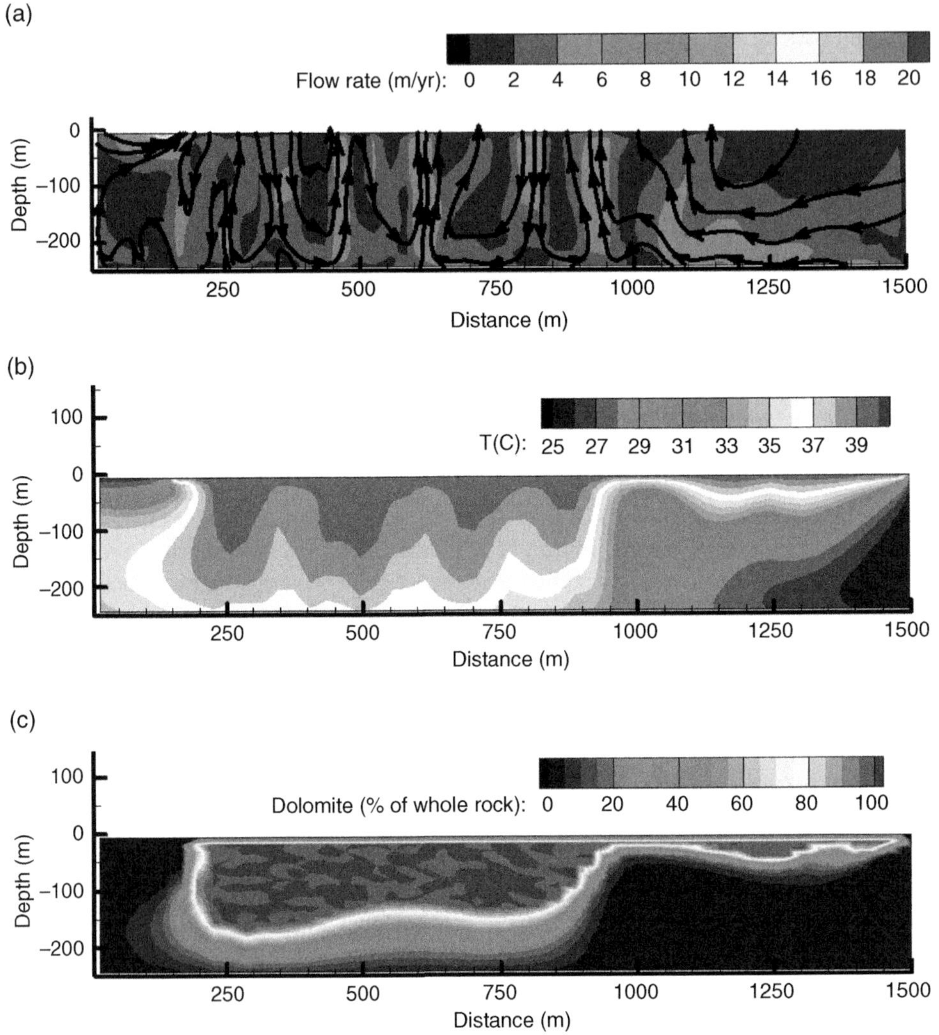

Figure 4.14 RTM results illustrating ascending freshwater–mesohaline water mixing mechanisms of observed extensive dolomitization at La Molata, Miocene. (a) Fluid flow rate; (b) temperature distribution; and (c) dolomite distribution. Arrows represent streamlines of fluid flow. Dolomite is in volume % of rock. From Cooper et al. (2014). (See colour plate section for the colour representation of this figure.)

either variations in sediment texture or early diagenesis (Enos and Sawatsky, 1981; Melim et al., 2001, 2002), models incorporated heterogeneous porosity and permeability distributions to investigate the effects of rock property heterogeneity on reflux dolomitization (Figure 4.15).

The simulation results demonstrated that the initial porosity and permeability heterogeneity in the precursor sediment has a significant effect on the pattern of dolomitization (Figure 4.16). The applied heterogeneity results in the formation of pronounced perturbations in the geometry of the dolomite front, which extended several kilometres

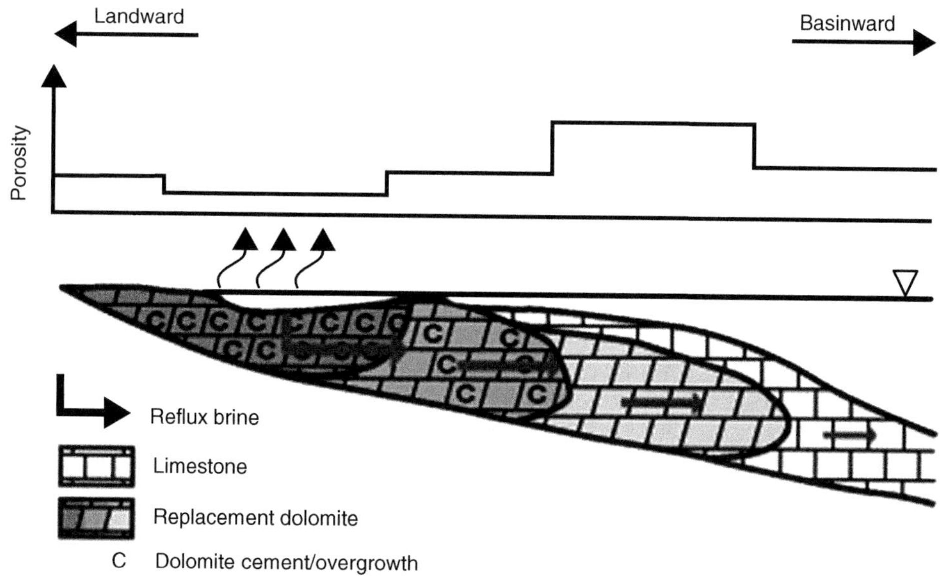

Figure 4.15 Conceptual model of dolomitization (replacement) and the precipitation of dolomite cements (overdolomitization) in a reflux system. Porosity increases basinwards due to extensive dolomite cementation proximal to the brine source. From Jones and Xiao (2005).

basinward from the main dolomite body after 2 My of active reflux (Figure 4.16a–e). These perturbations, described as dolomite 'fingers' by Wilson et al. (2001), became more pronounced with time, suggesting a positive feedback between the initial exploitation of preferential flow paths and subsequent changes in rock properties by dolomitization that enhances flow (Figure 4.16a–e), as well as negative feedback on flow associated with anhydrite cementation (Figure 4.16 f–j).

The simulation results also revealed the complex association of anhydrite cementation and dolomite distribution. Anhydrite cement formed due to the interaction of SO_4^{2-} from the brine with excess Ca^{2+} released from the dolomitization. As dolomitization proceeded to completion 'upstream', no Ca^{2+} was released, which led to undersaturation with respect to anhydrite in the brine. This is the primary reason for the remobilization of the initially formed anhydrite cement (Figure 4.16 f–j). Focusing of reflux flow resulted in enhanced anhydrite dissolution in the dolomite fingers, whereas anhydrite had a greater preservation potential adjacent to the fingers (Figure 4.16 j). Dolomitization enhances porosity by up to 8% and anhydrite cements occlude porosity by up to 14% (Figure 4.16 k–o). The dolomite fingers are porosity 'sweet zones' surrounded by non- to partially dolomitized sediments that were locally cemented with anhydrite and have a lower porosity.

The 2D reactive transport simulations demonstrate the complex and dynamic nature of dolomitization, associated anhydrite cementation and porosity evolution in a simplified reflux system. Incorporation of initial porosity and permeability heterogeneity resulted in complex dolomite geometry with pronounced dolomite 'fingers' extending ahead of the main dolomite body. The simulated dolomite distribution changes from thick and extensive to thin and separated in the direction of the reflux flow, consistent with outcrop observations (Figures 4.5 and 4.6). The distribution of limestone to

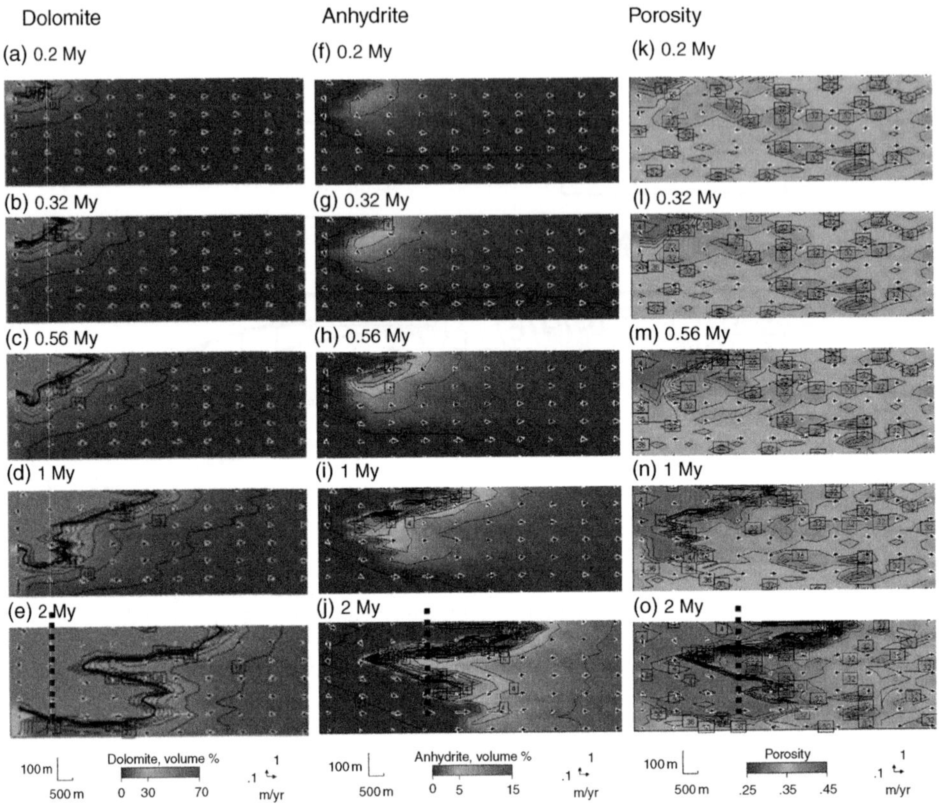

Figure 4.16 Effect of initial porosity and permeability heterogeneity on dolomite, anhydrite and porosity. (a–e) Dolomite (contours are percentage dolomite with a contour interval of 5); (f–j) anhydrite (contours are percentage anhydrite with a contour interval of 2); (k–o) porosity (contours are porosity [fraction] with a contour interval of 0.02). From Jones and Xiao (2005). (See colour plate section for the colour representation of this figure.)

dolomitization transition and potential location of reservoir 'sweet' spots can have important implications for reservoir connectivity. For example, the extensive main dolomite body in the platform interior (Figure 4.16E) suggests good lateral and vertical reservoir connectivity, whereas in the limestone to dolomite transition zone, the layered dolomite 'fingers' and anhydrite cement indicates complex reservoir connectivity, with good reservoir intervals (uncemented dolomite) sandwiched by tight zones with extensive anhydrite cementation (Figure 4.16 J, O). Bed-scale RTM of dolomitization by brine circulation have also demonstrated the emergence of distinct dolomite patterns as a result of self-organization from initial variations in porosity and permeability in the precursor limestone (Budd and Park, 2018).

Reactive transport models support previous conclusions from stratigraphic relationships, geochemical studies, hydrological measurements and magnesium mass balance calculations that hypersaline reflux is capable of extensive (tens of km in lateral extent and hundreds of metres in depth) pervasive dolomitization, on time-scales of hundreds of thousands of years. Results also suggest that the concept of mesohaline reflux dolomitization originally proposed by Simms (1984) is viable. Our simulations successfully replicated the processes and general patterns observed in many dolomite reservoirs interpreted to be of reflux origin. For example, our simulations predict that the dolomitization would thin basinward, and also the precipitation of dolomite cements proximal to the brine source post-dolomitization (overdolomitization): both phenomena observed in reservoir studies by Saller and Henderson (1998). Anhydrite precipitation and dissolution in our simulations provides a process-based framework for interpreting anhydrite cements or their absence. Based on our study, we have developed a new conceptual model of reflux dolomitization that incorporates the spatial and temporal distribution of two additional rock types to the Saller and Henderson (1998, 2001) model: (1) limestone with anhydrite cements and (2) dolomite with anhydrite cements (Figure 4.17).

4.3.5.2 3D Simulations of Brine Reflux Dolomitization

2D reactive transport models of dolomitization are limited by their ability to incorporate spatial variations in natural variables that control brine reflux, and thus extrapolation to the distribution of dolomite in 3D, including geological models developed for simulating production of hydrocarbons and reservoir management that require 3D distribution of petrophysical properties. Fully coupled 3D reactive transport models of brine reflux were developed for the first time to investigate the effect of permeability and spatial distributions in the magnitude and distribution of platform-top brines (Xiao and Jones, 2007; Xiao et al., 2013). Salinity variations observed in modern environments, for example Lake Macleod (Logan, 1987) in addition to those interpreted from the thickness and type of evaporite deposits in the rock record, were used to condition model boundary conditions and initial rock and fluid properties. In the model a flow domain of 5 km × 5 km × 100 m with a grid node spacing of 200 m horizontally and 2 m vertically were specified. Reflux was simulated from three platform-top brine ponds approximately 500 m in diameter with different maximum salinities of 210, 175, and 140 ppt (Figure 4.18) The basal boundary was specified as a no-flow boundary (refluxing brines never reached this depth in the geological time simulated) and the sides of our flow domain cube were specified as hydrostatic and thus open to flow. All other parameters and boundary conditions were specified as those used in the earlier 2D reactive transport models of brine reflux that incorporated a heterogeneous distribution of permeability described above in Jones and Xiao (2005).

The RTM results showing the 3D evolution of brine reflux up to 5000 years are depicted in Figure 4.18. Individual reflux brine plumes sourced by the three different ponds, evident after 500 years of flow, eventually coalesce to form a single brine plume (Figure 4.18). At 5000 years the mesosaline to hypersaline brine plume extends over the 25 km^2 area modeled to a maximum depth of approximately 50 m, but with considerable variation in vertical extent related to brine pond proximity and local permeability heterogeneity. Modeled maximum rates of brine reflux were of the order of a few metres per year, with spatial variations controlled by the proximity to the different brine ponds and permeability. The resultant flow field is the net product of both

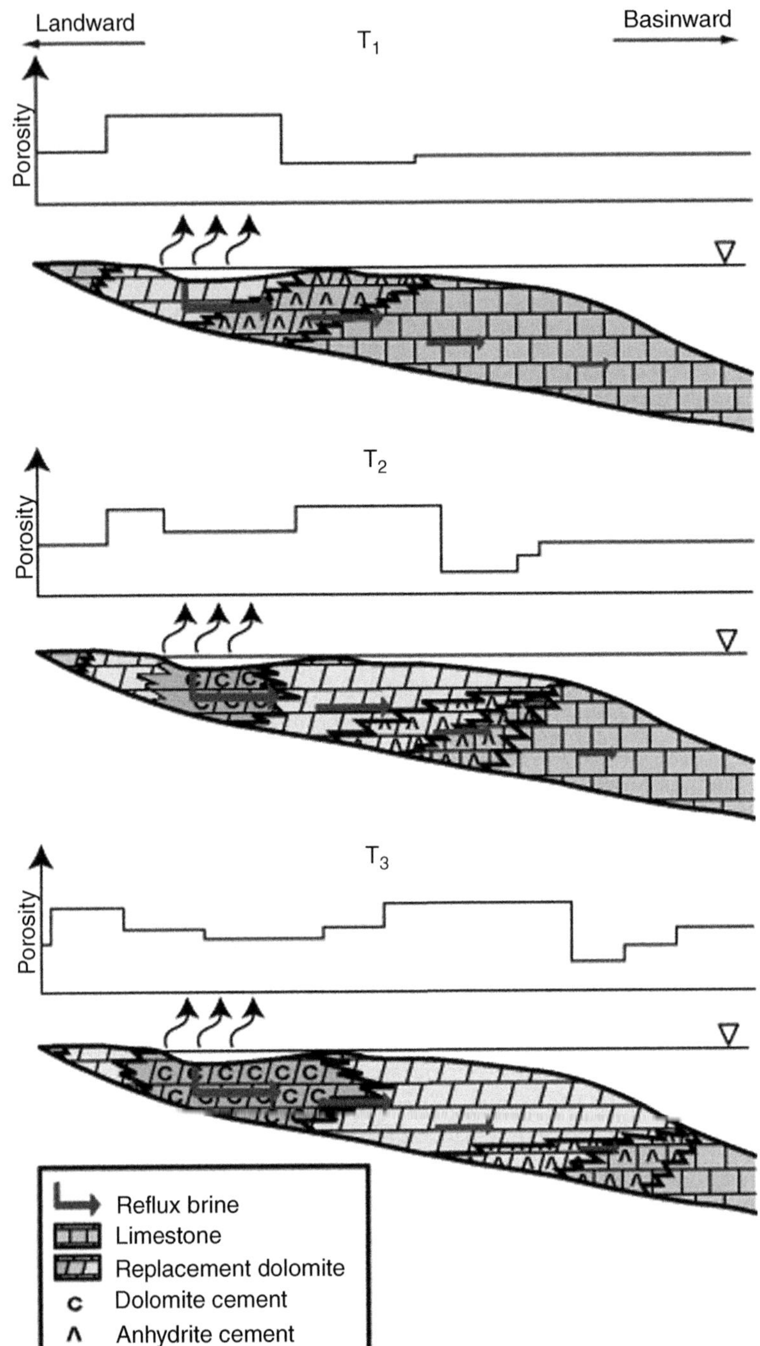

Figure 4.17 New conceptual model of reflux dolomitization with brines close to gypsum saturation that incorporates the spatial and temporal distribution of two additional rock types to the Saller and Henderson (1998) model, and trends in associated porosity evolution in a dip-section model. T_2 and T_3 are snapshots of the system at times later than T_1. From Jones and Xiao (2005).

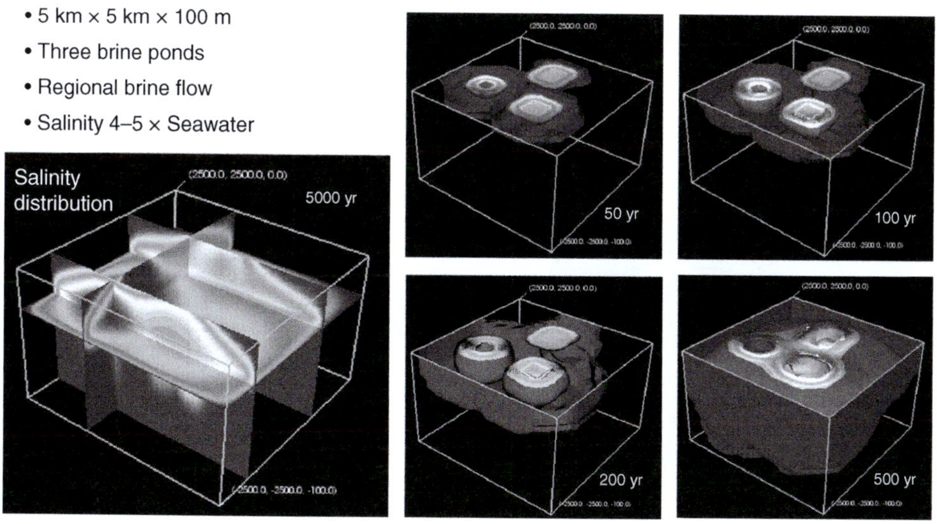

Figure 4.18 3D simulation showing the distribution of reflux brines at 50, 100, 200, 500, and 5000 yr. Seawater in black, hypersaline brines in light shades, and mixed fluids in between. From Xiao and Jones (2007).

reinforcing and opposing flows from the three separate ponds. In contrast to unidirectional flow in our 2D results, 3D simulations demonstrate that refluxing brines, at least locally in the vicinity of brine ponds, have the potential to flow up and down depositional dip and along depositional strike. After approximately 5000 yrs the brine plume appeared to reach a quasi-steady state relative to the effective hydraulic head in the brine ponds that drives flow. Beyond 5000 years, advection is limited, and the slower process of diffusion controls brine plume spreading. It is unlikely that steady-state reflux would be obtained in nature due to variations in climate, sea-level and geomorphology that act at shorter timescales to change the distribution of the platform-top salinity that drives flow.

The 3D simulations generated more complex patterns of brine reflux and distributions of dolomite and anhydrite than observed in our 2D study, and have significant implications for understanding connectivity in carbonate reservoirs (Figures 4.19–4.21).

3D evolution of dolomite - 'main body' and 'fingers'

The spatial evolution of dolomite from the RTM over a period of 100 ky, for the reflux brine plume depicted in Figure 4.19, demonstrates how initially small isolated dolomite bodies located beneath and adjacent to the brine ponds grow and amalgamate to form a single dolomite body that extends over the entire 25 km^2 (Figure 4.19). Results show that dolomite bodies grow at different rates. In a similar manner to our 2D simulations, the rate of dolomite body propagation is a function of the Mg^{2+} concentration of the brine and the flow rate. In a reflux system this is controlled by the salinity of the brine ponds, distance from the pond and the magnitude of permeability (assuming a uniform temperature and reactive surface area). Simulated dolomite body growth occurs by the

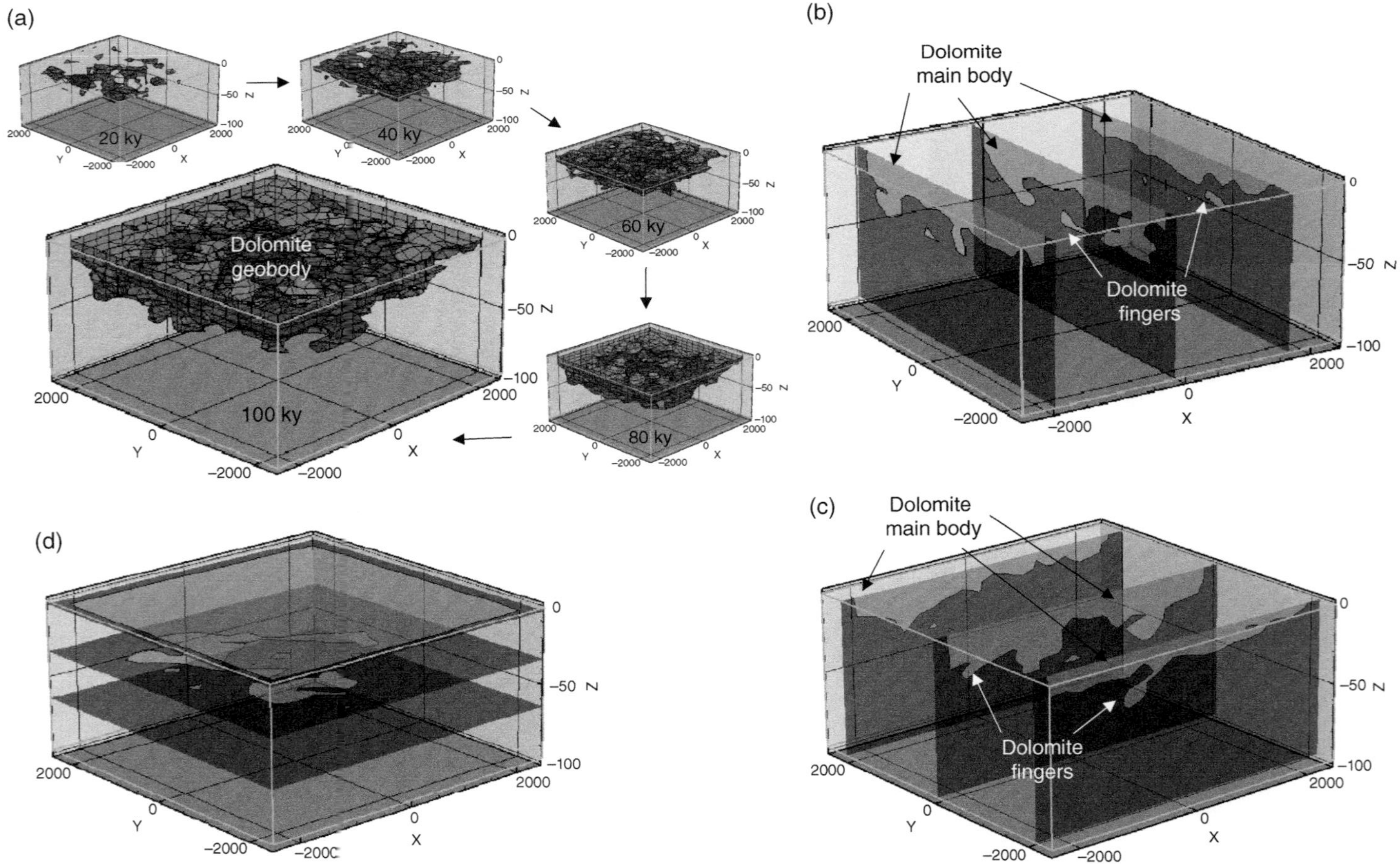

Figure 4.19 RTM 3D distribution of dolomite. (a) Dolomite evolution from 20 to 100 ky. (b)–(c) 2D slices showing the dolomite main body and fingers in x- and y-directions. (d) 2D slices showing dolomite in z-direction. From Xiao and Jones (2007).

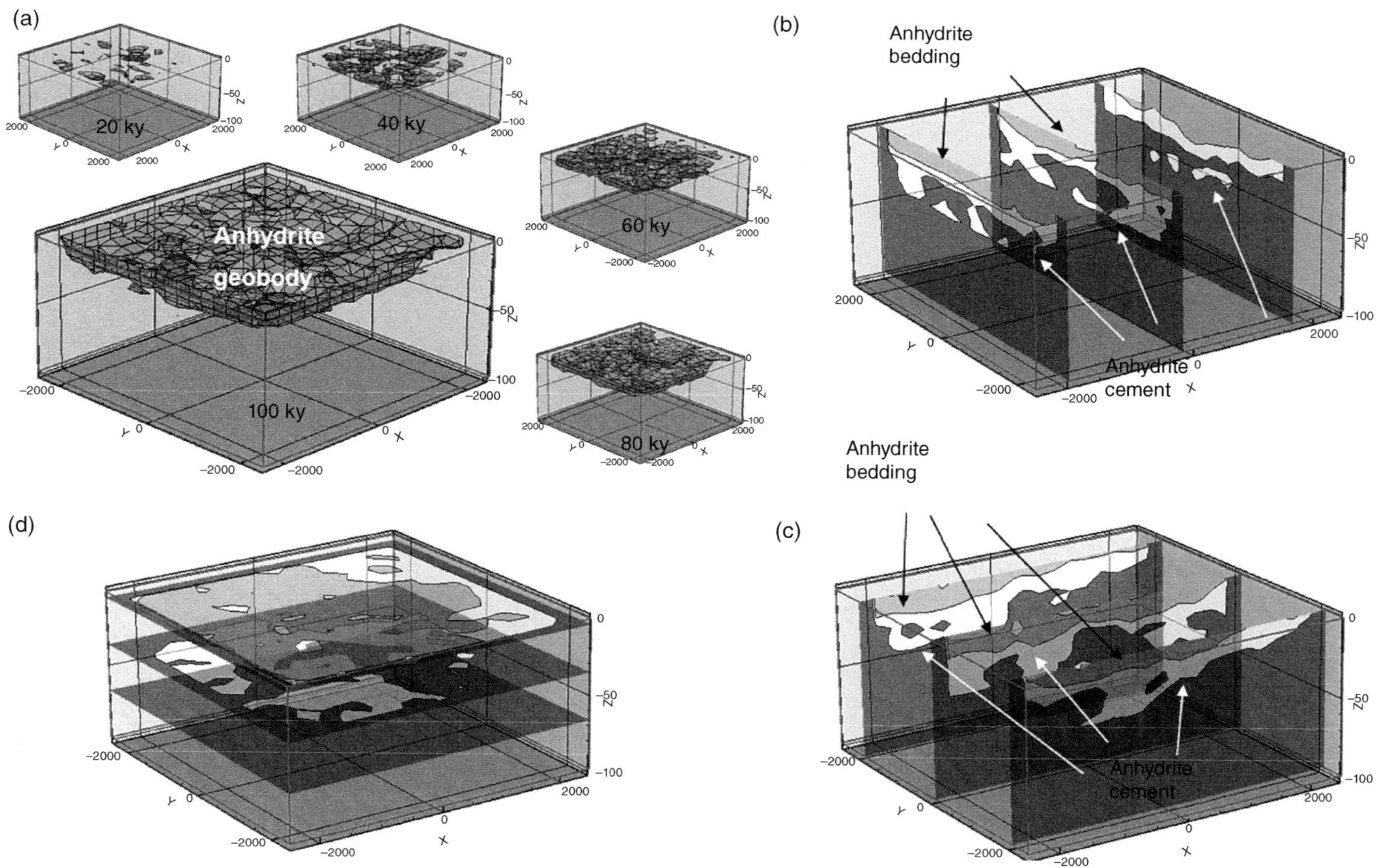

Figure 4.20 RTM 3D distribution of anhydrite (bedded and cement). (a) Anhydrite evolution from 20 to 100 ky. (b)–(c) 2D slices showing the bedded anhydrite (top layer) and anhydrite cement (middle layer) in *x*- and *y*-directions. (d) 2D slices showing anhydrite in *z*-direction. From Xiao and Jones (2007).

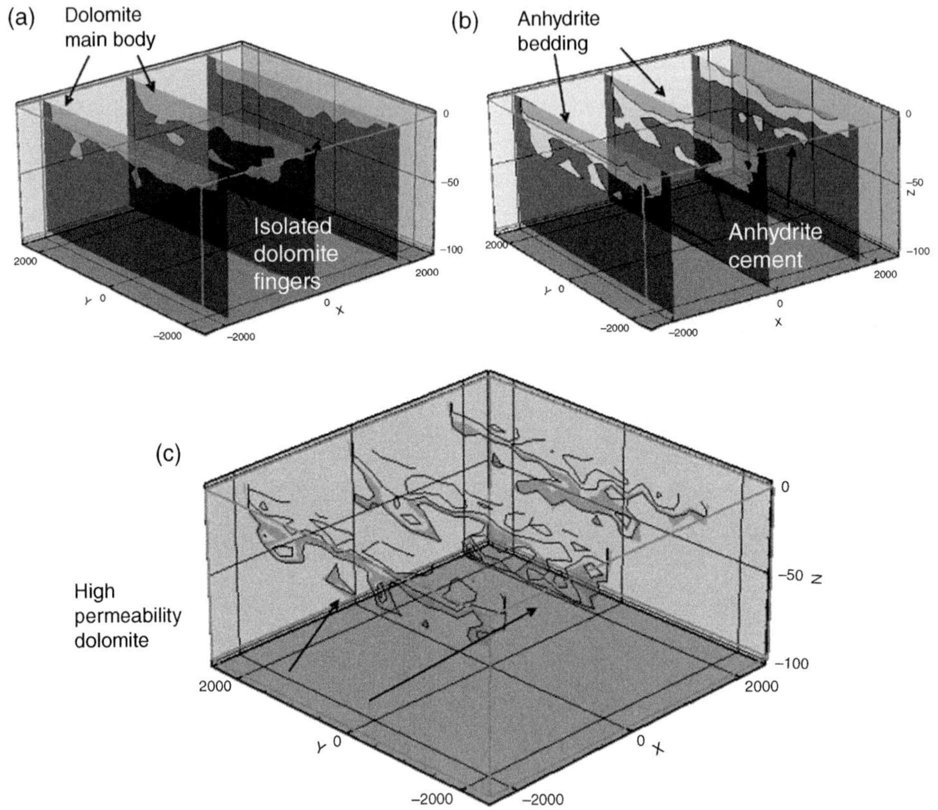

Figure 4.21 RTM 3D distribution of (a) dolomite; (b) anhydrite; and (c) high-permeability dolomite (dolomite with <5% anhydrite and >15% porosity). All properties are shown in 2D slices in *x*-direction. Porosity intervals are shown in contours and 'high-K' dolomite is shaded in red. From Xiao and Jones (2007). (See colour plate section for the colour representation of this figure.)

propagation and eventual amalgamation of dolomite 'fingers' (Wilson et al., 2001). The reaction front from limestone to dolomite is sharp. Visualization of the 3D dolomite body after 100 ky of reflux shows the significant variation in the dimensions and shape of dolomite 'fingers' that extend beyond the 'main' dolomite body, here described as dolomite that is laterally continuous over the areal extent of the flow domain. Dolomite fingers thin in the direction of flow and tend to become more pronounced with time, depending on the specified feedback of both dolomitization and anhydrite cementation on permeability and the initial permeability of limestone the dolomitization front intersects with along a reflux flow path. Examination of model results using 2D slices shows dolomite bodies that are apparently isolated from the main body (Figure 4.19).

However, these bodies are connected to the main body in 3D and were generated by the transport of Mg^{2+} along a flow path out of the 2D plane visualized (Figure 4.19). Thus reflux associated with local brine ponds has the potential to generate complex distributions of dolomite with 'finger' orientations that reflect multidirectional fluid flow. This has three important implications for subsurface correlation and geological modeling of reflux dolomites: (1) understanding the processes of reflux and paleofluid flow can provide a framework for correlation scenarios and guidelines for geological modelers; (2) reflux flow can be locally multidirectional with respect to depositional dip and strike; and (3) dolomite 'fingers' tend to connect to the 'main body' in a single reflux episode and are rarely isolated 'pods'. The occurrence of isolated dolomite pods is restricted to the early termination of reflux, as illustrated by the distribution of dolomite at times less than 60 ky (Figure 4.19), multiple episodes of brine reflux separated in time and space, and dolomitization by other mechanisms (Jones et al., 2003; Whitaker et al., 2004; Garcia-Fresca et al., 2012; Gabellone et al., 2016).

3D evolution of anhydrite distribution (bedded and cements)

Dolomites generated in a reflux system have a multi-faceted relationship with evaporites that are depositional and/or occur as cements (Jones and Xiao, 2005; Xiao and Jones, 2007; Xiao et al., 2013). The 3D simulations of brine reflux dolomitization show two types of anhydrite: (1) anhydrite cements that precipitate ahead of a migrating dolomite front due to the release of calcium, as explained in our 2D study; and (2) the direct precipitation of 'bedded' anhydrite from evaporated seawater (Figure 4.20b,c). Bedded anhydrite is distributed as a laterally continuous body (Figure 4.20b,c) with rare isolated 'holes' that may act as potential fluid conduits (Figure 4.20d). Anhydrite cement has a more complex distribution that is related to the distribution of dolomite. Greater preservation of anhydrite cement is evident in the 3D models of dolomitization (compare Figures 4.19 and 4.20). This is due to the shorter duration of brine reflux simulated and the precipitation of bedded anhydrites that reduce permeability, retard flow and slow the rate of anhydrite dissolution. Future simulations will explore the relationship between dolomitization and anhydrite and the implications for identifying dolomite reservoir sweet spots where anhydrite cements are rare/absent.

High permeability dolomite distribution

Understanding the distribution and connectivity of dolomite bodies in the subsurface is the first step towards building more reliable reservoir models. The magnitude and distribution of permeability in a dolomite body is an equal, if not more important, consideration. Contrasts in dolomite permeability can be extreme and occur over short distances. The problem is exemplified by the occurrence of so-called 'Super-K' high permeability (>100 mD to several D) dolomites that are often thin (a few feet), such as those in the Ghawar field, which account for a significant portion of total production (Cantrell et al., 2004). Reactive transport models provide a process-based method of investigating the origin and spatial distribution of dolomite permeability (Jones and Xiao, 2005; Xiao and Jones, 2007; Xiao et al., 2013). The 3D models were interrogated to illustrate reservoir sweet spot identification by applying the following rules: dolomite with >15% porosity and <5% anhydrite (Figure 4.21). Dolomite 'sweet spots' (shown in red) are located close to the dolomite front, have a variable thickness and are generally well connected in this case (Figure 4.21c). These sweet spots occur in both the dolomite 'main body' and the 'fingers' (Figure 4.19a,c). This general pattern is similar, but not exclusive to several dolomite reservoirs, although our model is a simplified hypothetical example (authors' personal observation). Permeability reduction via the

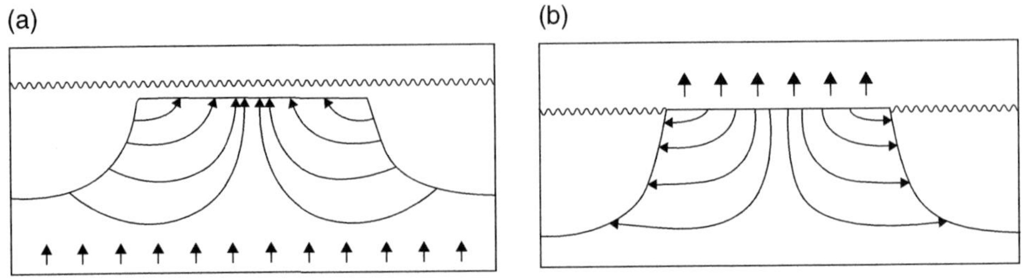

Figure 4.22 Conceptual models of geothermal (Kohout) convection driven by basal flux of geothermal heat (a) and reflux of brines driven by evaporation of platform top seawater (b), for a generic isolated carbonate platform.

overdolomitization process was not observed in our 3D model, perhaps due to the relatively short simulation time.

Geothermal convection and brine reflux driven dolomitization

Both geothermal convection and brine reflux drive the circulation of seawater-derived fluids through carbonate platforms during early burial (Figure 4.22), but the dynamic interactions between heat and solute transport and resulting diagenesis are at present poorly understood (Jones et al., 2004; Whitaker et al., 2004). In this section, we discuss the results of high-resolution RTM simulations to investigate the interaction between geothermal convection and brine reflux for dolomitization at the platform scale (Al-Helal et al., 2012).

The RTM revealed that a subsurface brine plume rapidly develops with the onset of reflux of 85‰ brines in the platform interior. This reverses the flow of cold ocean waters into the platform and restricts geothermal convection to a small zone <350 m in depth within 2.5 km of the margin (Figure 4.23). Brines may penetrate to >2 km depth within 1 My, but fluid flux is more rapid (>1 m/yr at shallow depths due to the initial permeability–depth relationship, specified anisotropy, and positive diagenetic enhancement of porosity and permeability) (Al-Helal et al., 2012).

Whilst chloride provides a conservative tracer for brine flux, water–rock interactions driven by this flux are reflected in changes in calcium, magnesium, and sulfate through time (Al-Helal et al., 2012). Reflux dolomitization leads to a rapid decline in Mg^{2+} at relatively shallower depths. This is mirrored by an increase in Ca^{2+}, indicating dolomitization occurs through the replacement of calcite, and no significant primary dolomite precipitation occurs. Dolomitization is focused beneath the brine pool, where a tabular dolomite body progressively develops and reaches a thickness of 200 m within 1 My (Figure 4.24). There is a broad front of dolomitization, with a steady decline in dolomite abundance with increasing depth and distance towards the margin. In contrast, the minor geothermal convection cell at the platform margin forms a small, partially dolomitized body with no more than 60% dolomite. At depths >100 m within the zone partially dolomitized by reflux, anhydrite cements form up to 20% of the total rock fraction after 1 My (Figure 4.24). Precipitation of $CaSO_4$, assumed to be anhydrite (anhydritization), occurs at 40–50 °C, where Ca^{2+} released by dolomitization reacts with SO_4^{2-} in the brines. Thus, while reflux dolomitization enhances reservoir quality at shallow

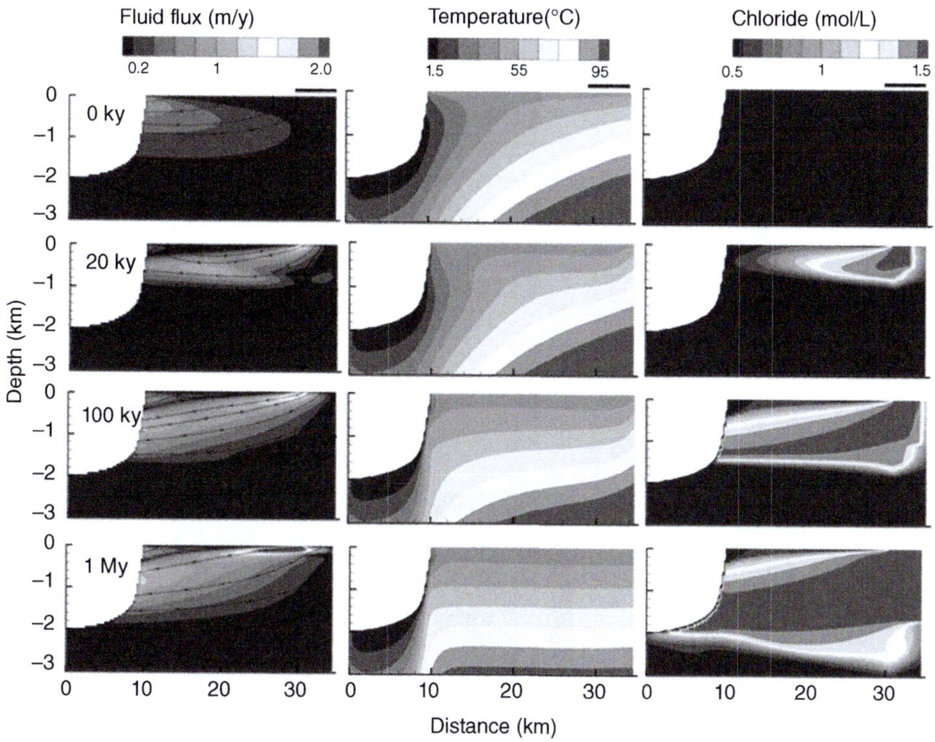

Figure 4.23 Temporal evolution of fluid flux (illustrative streamlines), temperature and chloride concentration of fluids for the baseline simulation, from initial steady-state geothermal convection (0 ky), over 10 ky, 100 ky and 1 My reflux from a 5 km wide platform interior brine pool (black bar). From Al-Helal et al. (2012).

depths, anhydrite reduces porosity (by up to 7%) and permeability below the reflux-driven dolomitization layer. This has the effect of focusing flow laterally and extending the dolomite body towards the platform margin (Al-Helal et al., 2012).

4.3.5.3 Brine Reflux Dolomitization Case Studies

The RTMs of reflux dolomitization presented were based on conceptual generic carbonate platforms at multiple spatial scales (from full field reservoir to sector scale). The natural progression of these foundational studies is to evalaute RTM capability to investigate dolomite patterns in a specific outcrop analogue or reservoir. Recent RTM applications to outcrop analogues have included the Permian San Andres Formation (Garcia-Fresca et al., 2012), and the Mississippian Madison Formation (Gabellone et al., 2016). More recently, RTM of reflux dolomitization has been applied to the Arab-D, Ghawar Field, Saudi Arabia (Lu and Cantrell, 2016).

Dolomitization of the Permian San Andres Formation

Garcia-Fresca et al. (2012) developed a hydrogeological model to test the effectiveness of brine reflux as the mechanism behind early dolomitization of the Permian San Andres Formation. Brine reflux is constrained by sequence stratigraphy and a

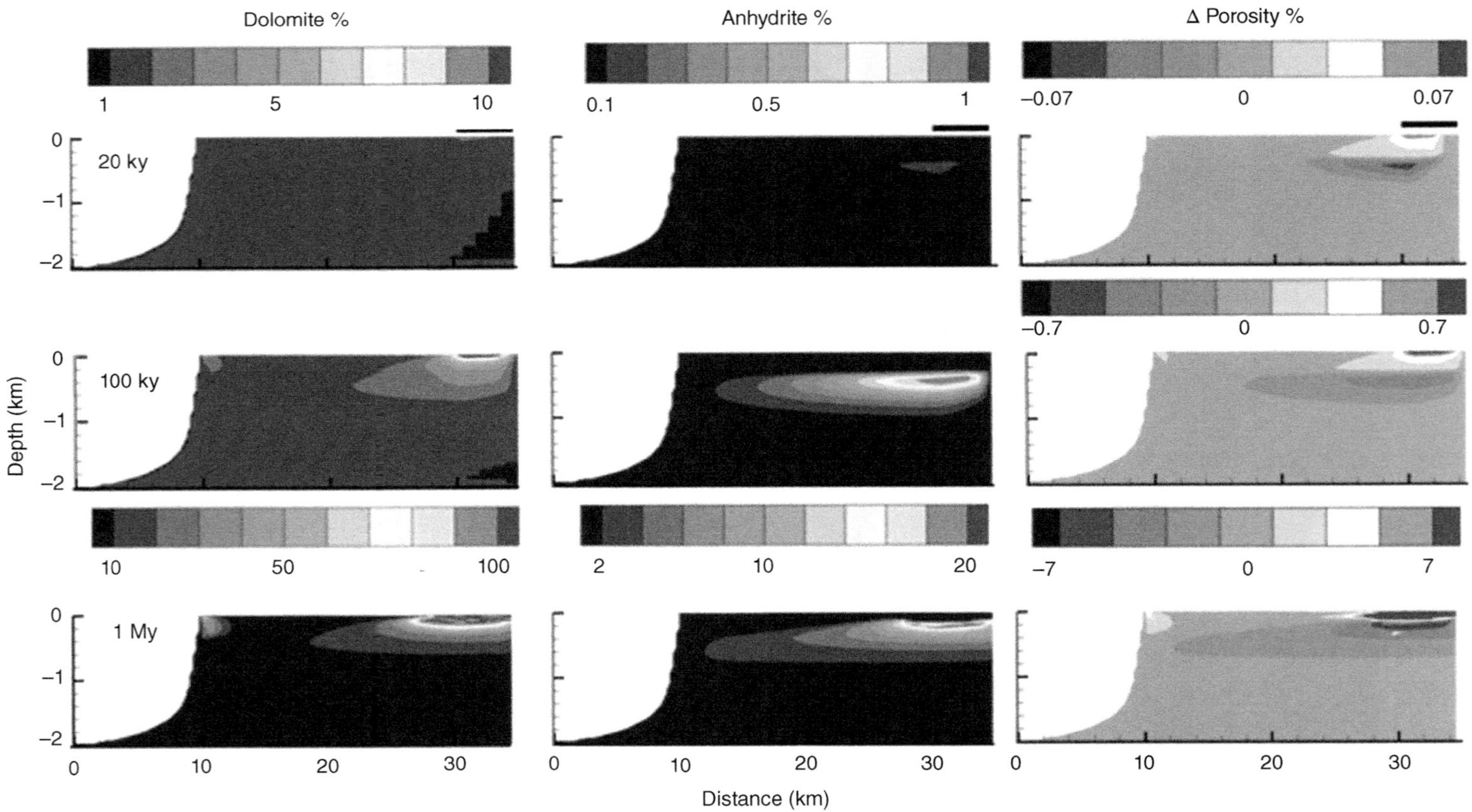

Figure 4.24 Temporal evolution of the dolomite and anhydrite abundance (mineral %) and change from initial porosity (volume %) after 20 ky, 100 ky and 1 My. From Al-Helal et al. (2012).

heterogeneous distribution of petrophysical properties based on outcrop data. Brine reflux was simulated during the deposition and erosion of the San Andres platform, and fluid flow and solute transport were calculated in response to relative sea-level fluctuations. Compaction-related porosity loss, concomitant porosity and permeability feedback were tracked. Reactive transport was not used to simulate the process of replacement dolomitization. Instead, the quantity of dolomite formed and its distribution were estimated using magnesium mass balance calculations for a range of Mg exchange efficiencies.

The modeling results (Figure 4.25) indicate that: (1) brine reflux is an effective mechanism of magnesium delivery to dolomitize large carbonate successions; (2) relative sea-level-controlled transient boundary conditions result in intricate flow and salinity patterns that can generate irregular dolomite bodies with complex spatial distributions; (3) pervasive dolomitization can result from several short-lived reflux events by the amalgamation of brine plumes sourced in different locations and times; and (4) the model successfully recreates the general dolomite and limestone patterns observed. The model confirms the relevance of 'latent reflux', where brines continue to dolomitize subsurface strata even after the platform-top brine source becomes inactive (Garcia-Fresca, 2008; Garcia-Fresca et al., 2012).

Reflux of an epeiric-scale ramp: the Mississippian Madison Formation

The Garcia-Fresca et al. (2012) study investigated the stratigraphic complexity and its control on reflux fluid flow and dolomite formation, but the chemical reactions were not explicitly coupled. Gabellone et al. (2016) expanded the investigation by exploring the chemical interactions between complex stratigraphical and geological controls (spatial distribution of brine composition and temperature) and intrinsic controls (permeability and reactivity) using local- and regional-scale reactive transport models of sequential episodes of brine reflux that resulted in the partial dolomitization of the Mississippian Madison ramp (Figure 4.26). The results show preferential early dolomitization of fine-grained, more reactive beds, and that pervasive dolomitization is focused beneath the brine pools where flow is perpendicular to bedding (Gabellone et al., 2016). Down-dip of the brine pool, bedding-parallel flow is focused in relatively permeable coarse-grained beds, providing reactants for the preferential alteration of intervening, more reactive fine-grained beds. In contrast, their results suggest that thicker sequences of fine-grained beds experience a slower rate of dolomitization, limited by the supply rate of magnesium. Regional-scale models, with brines of increasing salinity towards the ramp interior, produced a similar pattern of dolomitization to that interpreted by Gabellone et al. (2016) (Figure 4.26). However, their simulations, in which reflux is driven by lateral density contrasts, generate flow rates that are several orders of magnitude too low for pervasive dolomitization. RTM results suggest that a single platform interior brine source can generate dolomite locally, but is not sufficient to dolomitize large eperic-scale platforms, which simulations suggest would require multiple brine sources across the platform-top (Gabellone et al., 2016).

Reflux dolomitization in the Arab-D reservoir, Ghawar field

It has long been recognized that the Arab-D reservoir in the Ghawar field has a variable distribution of dolomite, which is a critical control on reservoir performance. In Ghawar, dolomite occurs with either a stratigraphically concordant or discordant distribution

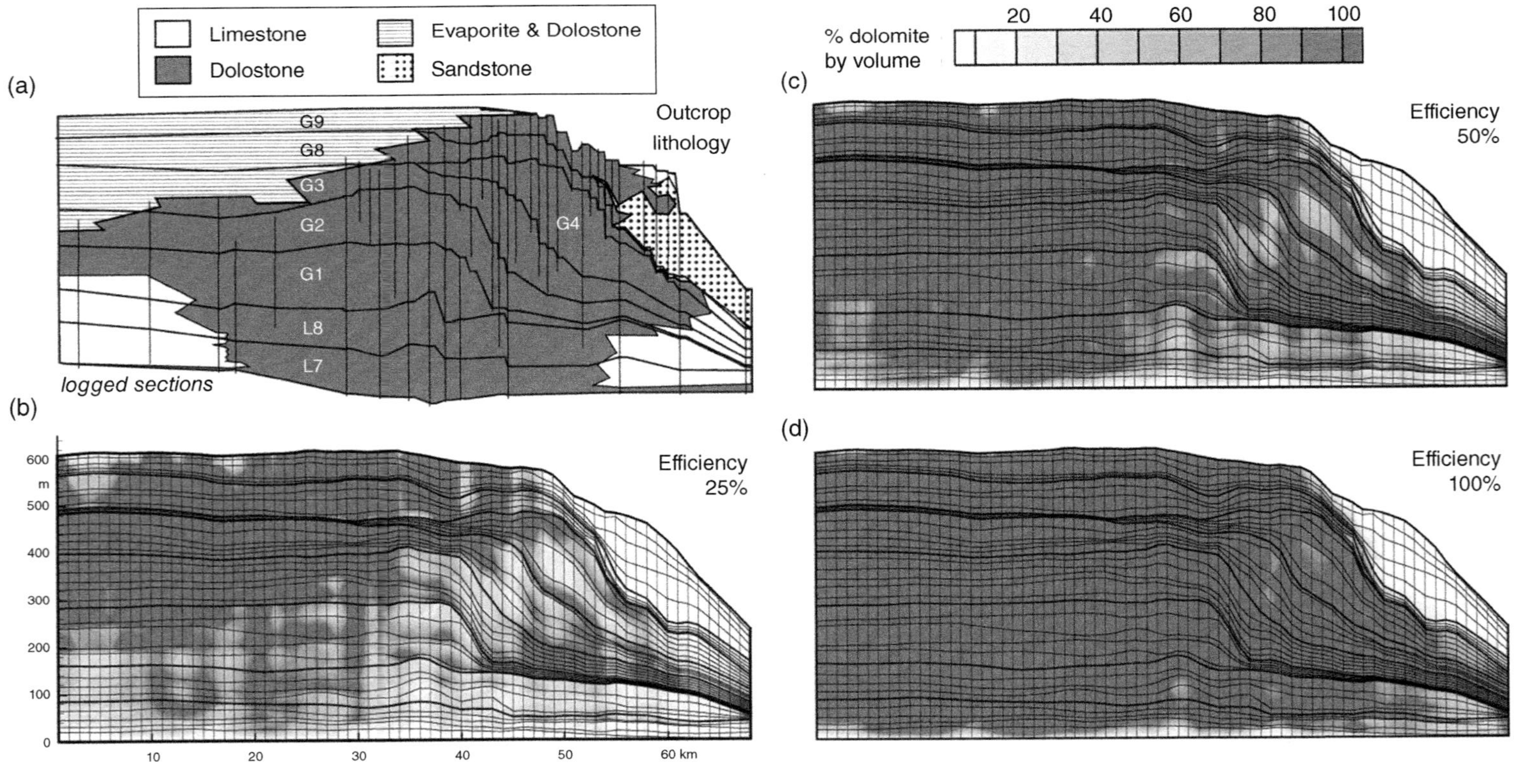

Figure 4.25 Model results for different magnesium–calcium replacement efficiencies compared with outcrop lithologies. Depth unit in metres and distance unit in km. (a) Outcrop lithology. (b) 25% replacement efficiency. (c) 50% replacement efficiency. (d) 100% replacement efficiency. The variability in results for different efficiencies reflects the sensitivity of reflux dolomitization to the kinetics of the magnesium-calcium replacement reaction. Dolostone bodies across the platform result from different combinations of favourable conditions, such as proximity to the brine source, zones of high permeability, permeability contrasts, and latent reflux (personal communication, *Source:* courtesy of Garcia-Fresca).

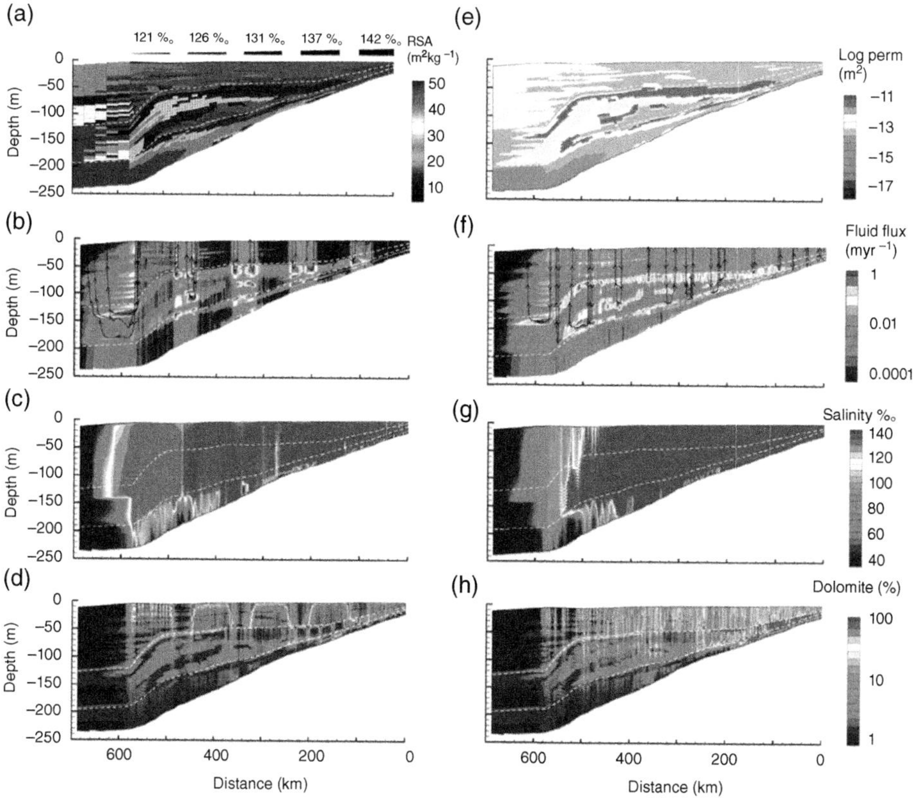

Figure 4.26 Initial rock properties (a and e) and results of 2D regional-scale models of density-driven reflux sourced from five brine pools (black bars at the top; b–d; Simulation 5), and with a user-specified brine salinity distribution (f–h; Simulation 6). Dashed lines correspond to sequence boundaries. The total simulated duration of reflux for the three sequences is 3 My. Dolomite abundance is in percentage of the total rock volume, using a log scale. From Gabellone et al. (2016). Used with permission by Journal of Sedmentary Research.

(Cantrell et al., 2004). Stratigraphically concordant dolomite was originally interpreted to have been formed early from brine reflux. Stratigraphically discordant dolomite was interpreted to have been generated primarily from ascending hydrothermal fluids (Cantrell et al., 2004).

Lu and Cantrell (2016) applied RTM to study the stratigraphically discordant dolomites of the Ghawar and concluded that: (1) these dolomites initially formed via brine reflux, but were overprinted by late-stage hydrothermal dolomitization; and (2) reflux is responsible for mapped parallel, linear trends of dolomite. Their reflux model hypothesizes that an evaporative lagoon (which is the source of dolomitizing fluids) formed during the falling stage systems tract of a depositional sequence, and that with continuing sea-level fall, this lagoon progressively migrated towards the deeper parts of an intrashelf basin adjacent to the Ghawar field, leaving behind linear dolomite bodies along a series of abandoned coastlines (Lu and Cantrell, 2016). The 2D RTM predicted pattern of dolomite bodies generally agrees with both the observed vertical distribution of non-stratigraphic dolomite, as well as the mapped lateral distribution of the dolomite trends (Figure 4.27). In addition, the major ion composition of Late Jurassic seawater

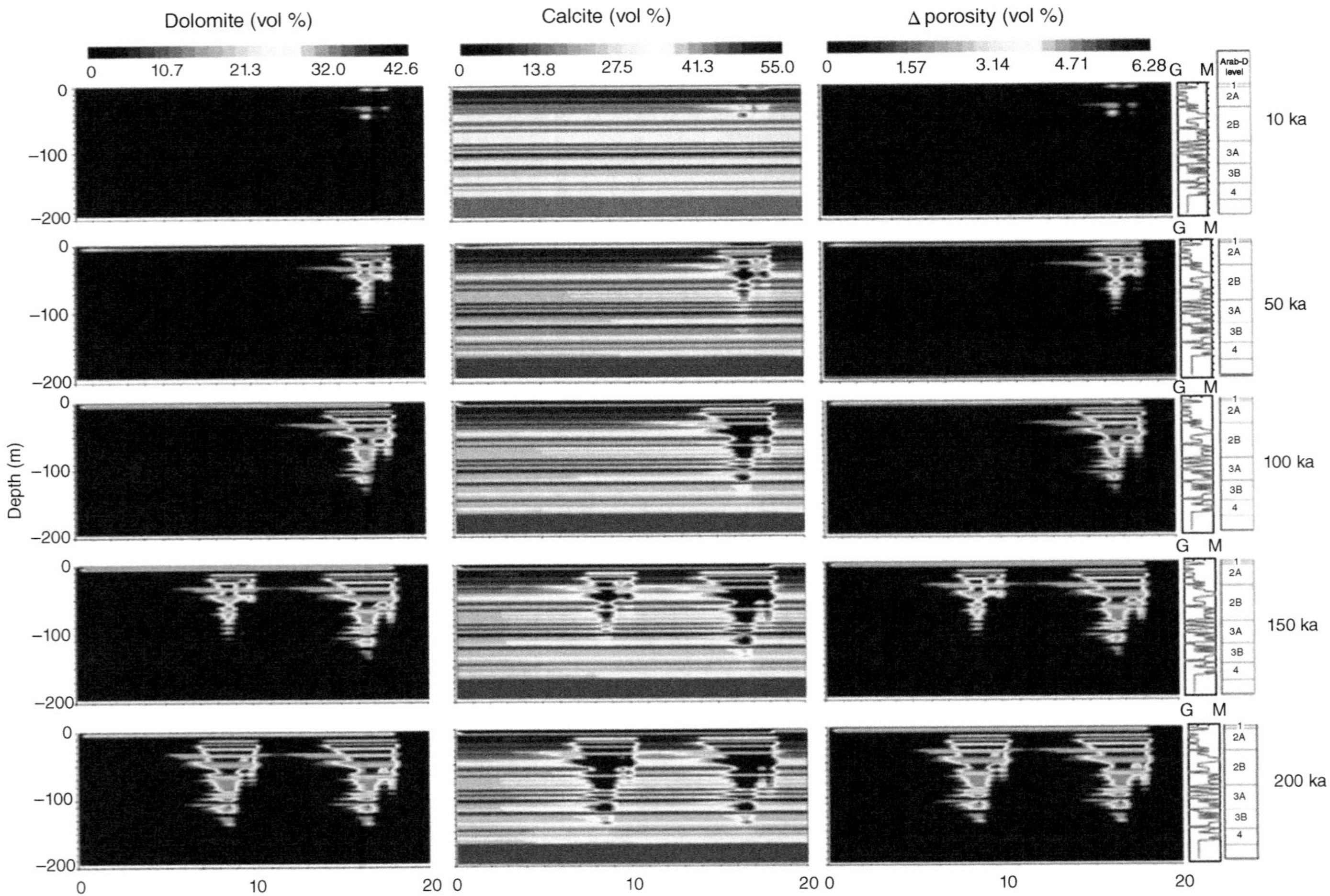

Figure 4.27 Temporal and spatial evolution of dolomite and calcite abundance (volume %) and Δ porosity (volume %) for base case simulation at 10 ka, 50 ka, 100 ka, 150 ka and 200 ka. The Arab-D reservoir zonation scheme and simplified lithological textures are shown on the right for reference. G, grainstone; M, mudstone. The vertical exaggeration is 409. *Source:* From Lu and Cantrell (2016), used with permission from *Sedimentology*.

was calculated based on fluid inclusion data. In contrast to previous RTM studies that used evaporated brines from modern seawater, no anhydrite cements associated with dolomitization precipitated when specifying Jurassic seawater.

4.3.6 Fault-Controlled Hydrothermal Models

Fault zones and adjancent strata are often dolomitized. Such dolomites are commonly described as 'hydrothermal' or high-temperature dolomite (HTD) irrespective of whether they meet the strict definition of hydrothermal (Machel, 2004; Davies and Smith, 2006). From a reservoir perspective, two key questions are: (1) How far does the dolomite and associated diagenetic modification extend away from the fault? and (2) How do stratigraphic heterogeneity (including precursor dolomite) and fault zone properties control the occurrence and distribution of HTD? Cantrell et al. (2004) observed that hydrothermal dolomite in the Ghawar field follows NE–SW and NW–SE trending normal faults that were active after the deposition of Arab-D carbonates, which allowed hydrothermal fluids to rise to the Arab-D and form baroque dolomites (Figure 4.28). These local pathways for vertical fluid movements would most likely lie along the NE–SW dolomite trends, and would typically be quite narrow in scale and difficult to detect with the current well spacing and core control from vertical wells.

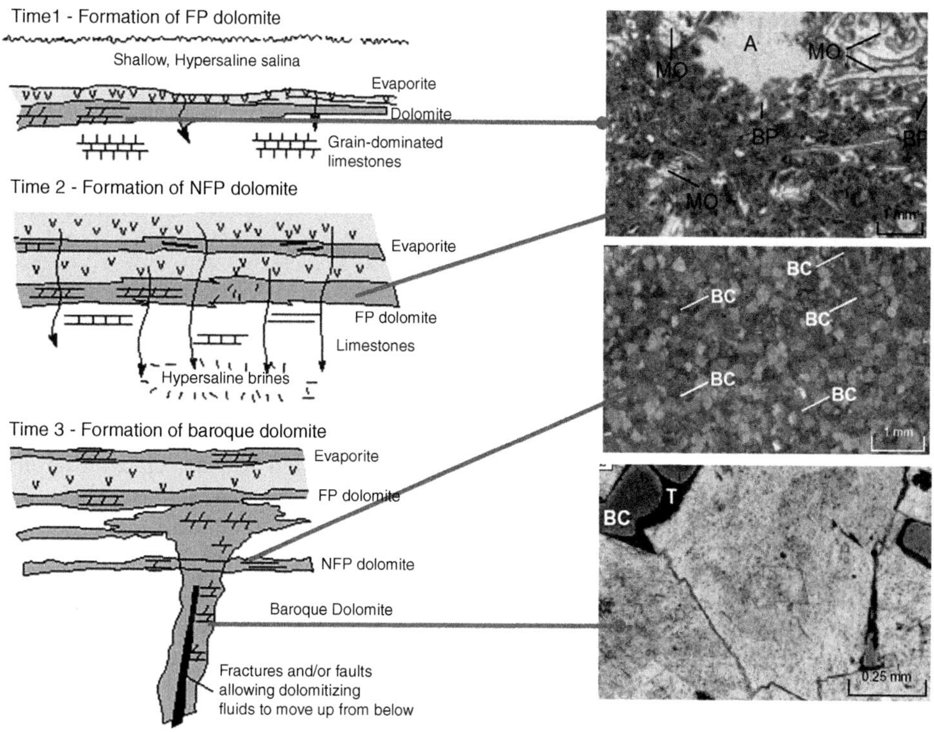

Figure 4.28 Model of dolomitization for the Arab-D reservoir in the Ghawar field, showing reflux and hydrothermal fluids responsible for fabric-preserving and non-fabric-preserving dolomite fomation. *Source:* From Cantrell et al. (2004), used with permission from GeoArabia.

Attempts to directly identify and characterize these postulated faults and/or fracture zones in seismic data and correlate them with the observed high dolomite trends have unfortunately been unsuccessful to date, because most observed faults at the Arab-D level tend to have a very short vertical throw and are only marginally visible in seismic profiles (Cantrell et al., 2004).

4.3.6.1 2D and 3D Conceptual HTD Models

2D and 3D RTM models were developed to investigate the effects of fault-associated Mg^{2+}-bearing hydrothermal fluids on carbonate reservoirs. Figure 4.29 shows the distribution of brine, dolomite, anhydrite, temperature, and calcite after 100 ky. The results suggest that with continuous hydrothermal flow over tens of thousands of years, burial diagenesis can significantly alter the reservoir properties and lead to large-scale HTD formation and anhydrite precipitation (Xiao et al., 2013). The RTM simulation used concentrated modern seawater as a proximy for the hydrothermal brine that led to significant anhydrite precipitation. A more recent study (Lu and Cantrell, 2016) used Jurassic seawater to represent the hydrothermal brines, and anhydrite precipitation was absent, probably due to the difference in brine composition and temperature.

The results from the 2D and 3D RTM investigation suggest that, depending on the initial rock composition and timing of the HTD event, the reaction can lead to no anhydrite cement formation in a lower dolomite interval (formed earlier), anhydrite formation in the middle limestone interval, and anhydrite dissolution in the upper anhydritic dolostone interval (Figure 4.30). This is because the HTD fluid was initially in equilibrium with the lower dolomite rocks and therefore no anhydrite cementation occurs; as the replacement dolomite takes place in the middle limestone interval, the sulfate ions will interact with the excess Ca^{2+} to form anhydrite cement. As the HTD fluid enters the top anhydritic dolostone interval, it is depleted in sulfate and Mg^{2+} ions and therefore could lead to anhydrite dissolution, a potential mechanism to enhance the permeability of dolomite that was plugged with early anhydrite cement, potentially making or widening 'holes' in bedded anhydrite around faults.

4.3.6.2 Fault-controlled Dolomitization at the Benicassim Outcrop in Maestrat Basin, Spain

The generic study presented above assumed continuous fault zone flow (Xiao et al., 2013). Another RTM investigation introduces more geological realism, with pulses of fluid movement related to fault movement and then later open flow driven by different mechanisms, and was applied to an outcrop study (Figure 4.31). Results suggest that overpressured dolomitizing fluids released during active fracturing in the form of fluid pulses, as expected in a syn-rift tectonic period, may result in large lateral flow rates along the most permeable layers (Figure 4.32a,b). Figure 4.32c shows that fluids released from geopressured zones lose their temperature very rapidly because thermal conduction is more dominant than advection. This implies that although solution fluxes are high through the permeable layers, the solutions do not keep their high temperatures (120–130 °C) when they are released upwards. The results of this study suggest that these high fluid velocities are not maintained for periods longer than 2–5 years, according to the configuration of the Benicassim half-graben.

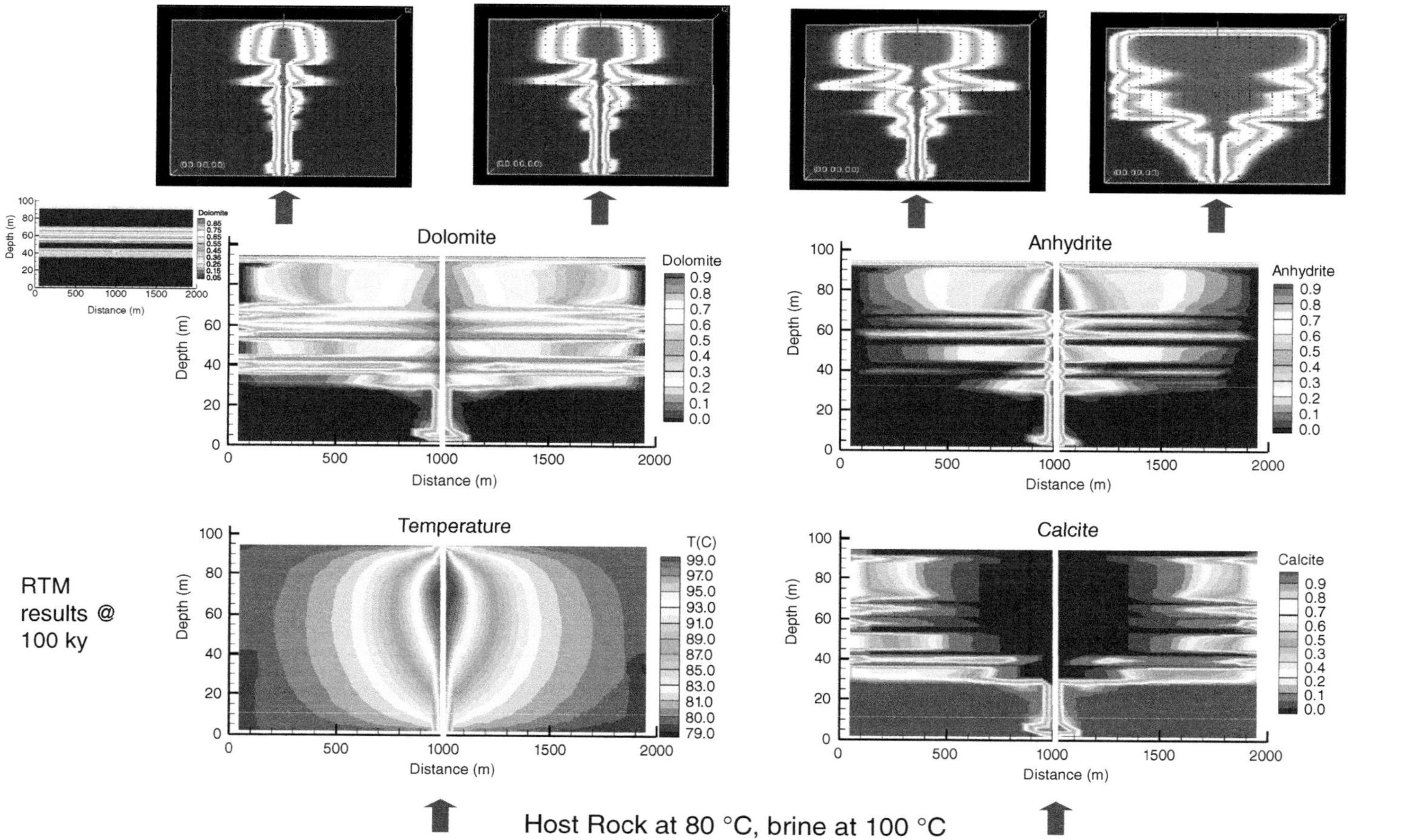

Figure 4.29 2D RTM results showing the distribution of brine, dolomite, anhydrite, temperature, and calcite after 100 ky of fault-associated hydrothermal diagenesis. The hydrothermal brine flow and distribution reflects the initial permeability contrast from stratigraphic layering (interbedded limestone/anhydrite). Initial formation temperature is 80 °C and the hydrothermal brine is 100 °C. From Xiao et al. (2013).

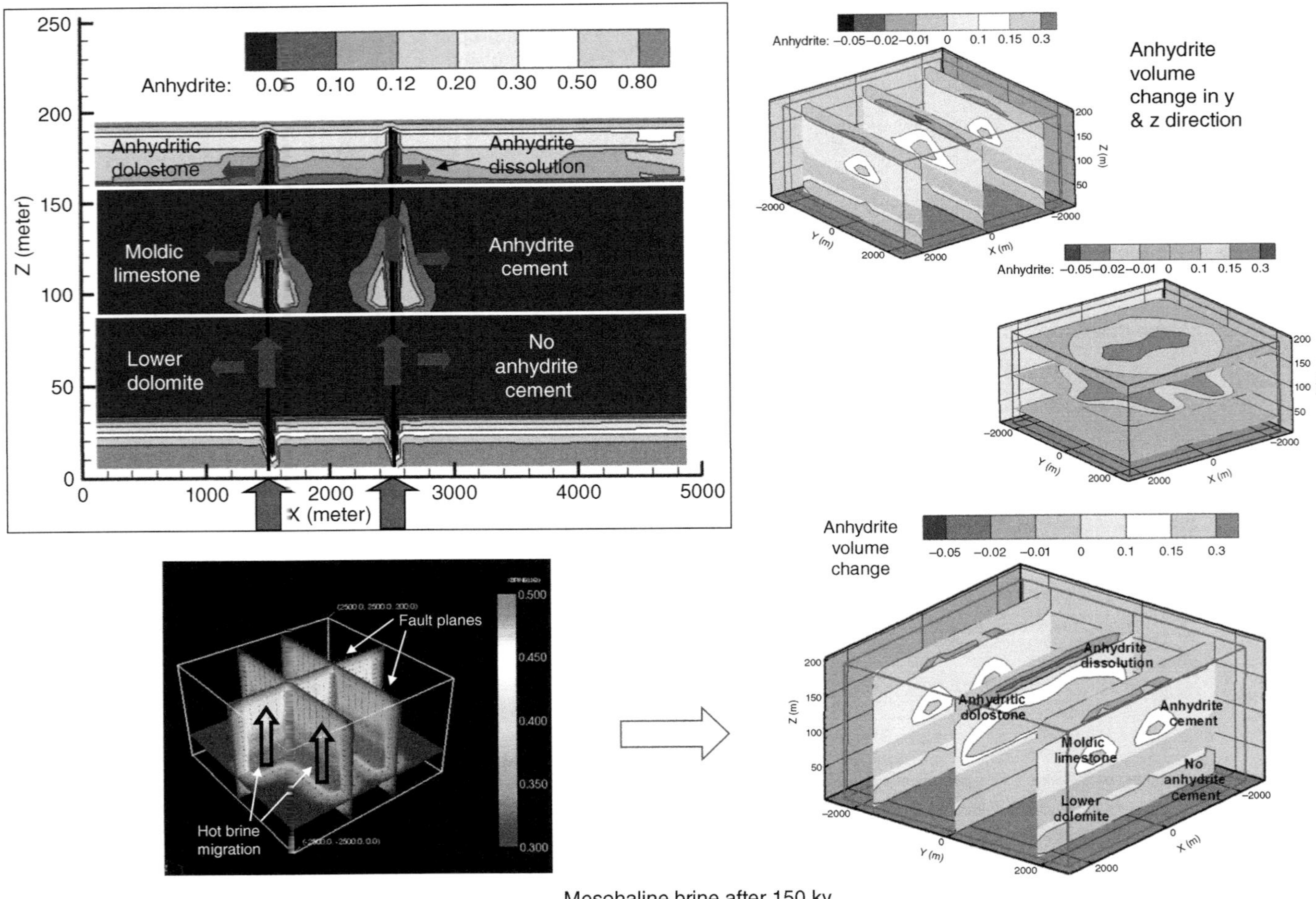

Figure 4.30 3D RTM results showing that HTD fluids entering a carbonate reservoir can lead to no anhydrite cement in the lower dolomite interval, anhydrite cement in the middle limestone interval, and anhydrite dissolution in the top anhydritic dolostone interval. From Xiao et al. (2013). (See colour plate section for the colour representation of this figure.)

Figure 4.31 Panoramic view of the field study area near Benicassim (NE Spain). Lighter layers are limestone and darker layers are sub-stratiform dolomite, with dotted lines marking two faults in the field of view. The lateral field of view is ~6.5 km. *Source:* From Gomez-Rivas et al. (2010), used with permission from EAGE.

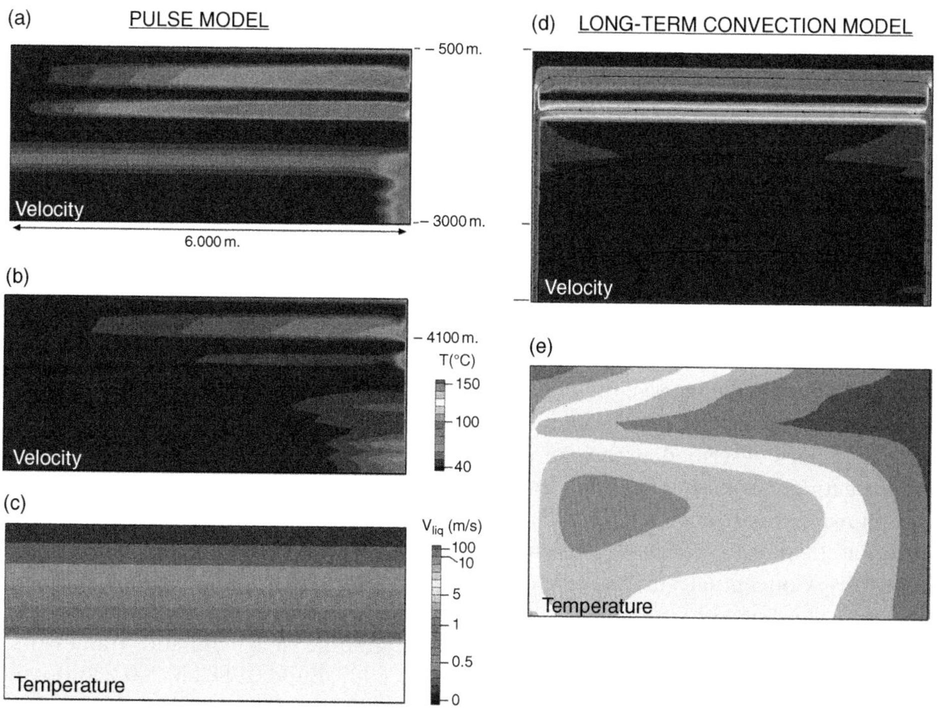

Figure 4.32 Simulation results: (a) fluid velocity vector module after 1 year; (b) fluid velocity vector module after 10 years; (c) temperature after 1 year; (d) fluid velocity vector module after 20,000 years in an open system; (d) temperature after 20,000 years in an open system. (a) to (c) show results of pulse simulations, while (d) and (e) display outputs from long-term convection simulations. Black arrowed lines in (d) represent streamtraces. From Gomez-Rivas et al. (2010), used with permission from EAGE.

In contrast, a longer-term fluid circulation system in the post-rift environment can provide enough warm solution to the reactive shallow layers whenever a driving pressure and/or temperature gradient is maintained over the time period in an open system. In this scenario, dolomitizing solutions flow laterally along high-permeability layers, heating up the host and surrounding rocks over long time periods (Figure 4.32e,f). This system must be open in order to have enough Mg for significant dolomitization to occur. A difference in horizontal permeability of two orders of magnitude is required to account for fluid flow in specific layers (Gomez-Rivas et al., 2010; Corbella et al., 2006, 2014).

4.3.7 Summary of Dolomite RTM Results

Reactive transport models (RTM) have been applied to investigate key controls on reflux dolomitization and reservoir quality in four different hydrogeological systems: (1) brine reflux; (2) mixing zone and submixing zone; (3) geothermal circulation; and (4) fault-controlled hydrothermal circulation. The models predict the spatial and temporal distribution of key geological parameters, such as limestone, dolomite, anhydrite and porosity, and how these parameters were affected by depositional and diagenetic variables (initial lithology, porosity and permeability, diagenetic fluid compositions, temperature and pressure, etc.). The 2D and 3D modeling results were consistent with regional and reservoir trends in dolomite/anhydrite distribution published in the literature, compared and calibrated with outcrop analysis (Garcia-Fresca et al., 2012; Gabellone et al., 2016; Gomez-Rivas et al., 2010), and supported by real reservoir-based investigation and observations (Lu and Cantrell, 2016), and provided a better understanding of key diagenetic processes controlling reservoir quality in dolomitized carbonate reservoirs. The modeling results, together with outcrop and well-based observations, generated predictive diagenetic rules or concepts that can be used to guide porosity and permeability distribution in 3D reservoir connectivity models.

4.4 Early Diagenesis in Isolated Carbonate Platforms

During carbonate diagenesis, the potential for porosity modification is greatest during periods of exposure and the development of the four distinct hydrological and geochemical zones described as the vadose zone, the meteoric phreatic zone (freshwater lens), the mixing zone and the saline zone (Figure 4.33). These diagenetic environments, also known as hydrological zones (Smart et al., 1988; Moore, 2001; Whitaker et al., 2004; Paterson et al., 2006), provide a template for predicting different styles of early diagenesis. Conceptual qualitative models of early diagenesis in carbonate platforms have been developed by relating the position of hydrological zones, each with associated styles of diagenesis, in a sequence-stratigraphic and climatic framework (Reeckmann and Gill, 1981; Mathews and Froelich, 1987; Budd, 1988a,b, 2002; Rudolph and Lehmann, 1989; Loucks, 1999; McClain et al., 1992; Matsuda et al., 1995; Moore, 2001). These conceptual models provide a foundation for reservoir quality prediction, but are limited because: (1) the modification of reservoir quality by early diagenesis is often the cumulative effect of multiple cyclic exposure events, and (2) the conceptual models suggest spatial variations in different styles of diagenetic processes, but not their rates and thus changes in rock properties. Furthermore, recent observations from well-constrained studies in Neogene carbonates of South Florida and the Great Bahama Bank have questioned some of the established paradigms upon which the conceptual diagenetic models of Moore (2001) are based (Melim et al., 2002).

Reactive transport models were developed to investigate the 2D spatial distribution of rates of early diagenesis, including the stabilization of aragonite to calcite, calcite dissolution, calcite cementation and dolomitization in four hydrological zones (vadose, meteoric phreatic, mixing and marine phreatic) in an isolated carbonate platform (Figure 4.33). Specifically, the model investigated the sensitivity of early diagenetic rates to mineralogy, climate (dissolved carbon dioxide (P_{CO2}) and the rate of meteoric

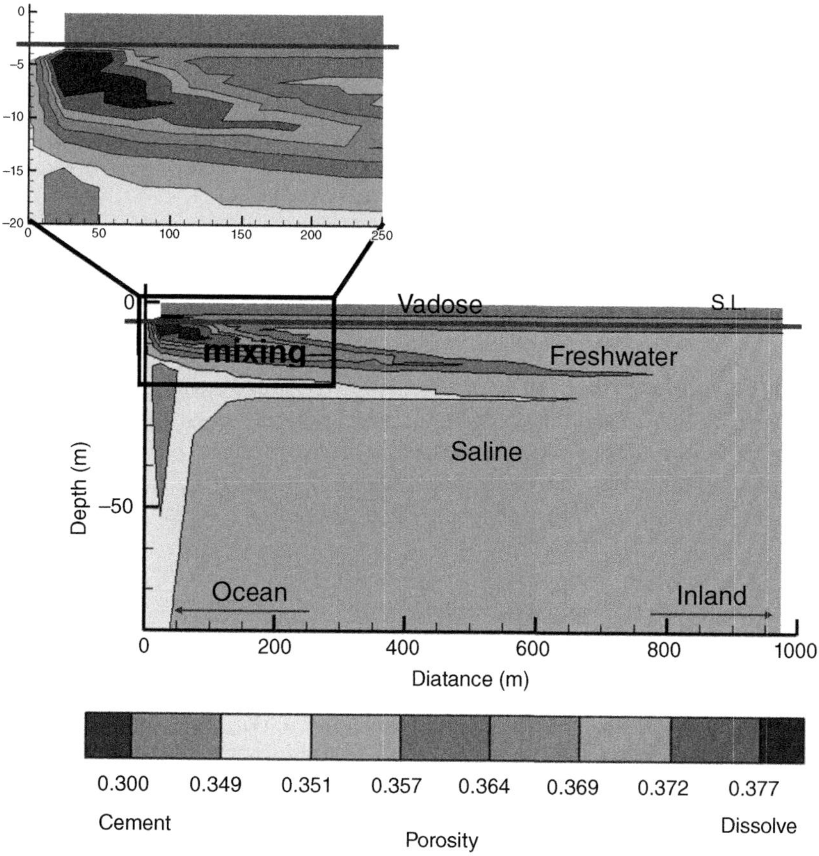

Figure 4.33 Porosity distribution in a carbonate platform after 100 ky of early diagenesis in a semi-arid climate. Enhanced calcite dissolution and porosity increases occur in the mixing zone along the coast due to focused fluid flow. Initial porosity is uniformly specified as 0.35. From Xiao and Jones (2006).

recharge), the position of relative sea-level, exposed platform width and the presence of platform margin fractures and/or karst.

More recently, Whitaker and Frazer (Chapter 3 in this volume) have developed process-based models that take into account the interactions between episodes of sediment deposition and overprinting by repeated episodes of meteoric diagenesis to predict the syn-depositional diagenesis. Interested readers can refer to their chapter for detailed discussion.

4.5 Geothermal Convection and Burial Diagenesis

As described above, geothermal circulation of seawater in carbonate platforms has grown in popularity as a viable dolomitization mechanism based on supporting RTM results (Machel, 2004; Whitaker and Xiao, 2010). Dissolution and cementation of limestone can also occur as seawater changes saturation state in response to variations in temperature

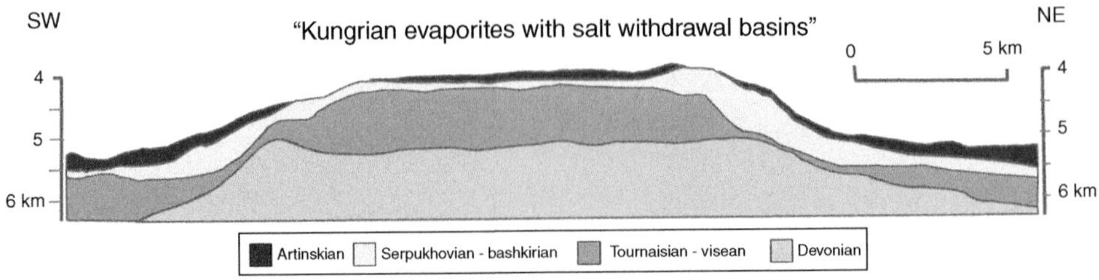

Figure 4.34 Seismic scale stratigraphic cross-section of the Tengiz reservoir used to design RTM models of geothermal convection. From Jones and Xiao (2006).

along a flow path. The potential for geothermal circulation to modify reservoir quality was investigated in two carbonate reservoir examples (Table 4.1): (1) the Tengiz field (Jones and Xiao, 2006) and (2) the South Atlantic subsalt carbonate fields like Lula in Brazil (Jones and Xiao, 2013). These two reservoirs have distinctly different depositional geometries (platform vs. ramp), but both have salt seals with shale-filled withdrawal basins.

4.5.1 Geothermal Convection and Reservoir Quality in Tengiz Field, Kazakhstan

Tengiz is a 'super-giant' oil field located in Kazakhstan in an isolated carbonate platform reservoir (Lisovsky et al., 1992; Weber et al., 2003; Figure 4.34). The Jones and Xiao (2006) 2D RTM study of Tengiz was based on a seismic stratigraphic cross-section and general descriptions of depositional facies from Weber et al. (2003). Additional information on the burial history, such as heat flux and thermal properties, was available from a regional basin model designed to investigate hydrocarbon charge. In this sense, despite the field being on production for a number of years, the reactive transport model inputs and design criteria were more typical of an early or 'greenfield' reservoir development business stage (Jones and Xiao, 2006).

Compared with previous more generic simulations of geothermal circulation in carbonate platforms, for example Wilson et al. (2001), the Tengiz RTM model has a geologically realistic sequence-stratigraphic-derived variable reservoir geometry (backstepping, aggradation and progradation) that was used to model the distribution of facies and porosity and permeability. This aspect of model design proved important because the platform region of Tengiz, dominated by grainstones and packstones, has distinctly different rock properties than the lower-porosity fractured microbial boundstones of the rim (Collins et al., 2005, 2006; Kenter et al., 2006, 2010).

Simulation results demonstrated that geothermal circulation in Tengiz would have been long-lived, occurring both during carbonate deposition and after burial (Jones and Xiao, 2006). Ascending flow is focused in the more permeable upper section of the reservoir (known as Unit 1) sourced from reservoir-scale convective cells (Figure 4.35). The pattern of convective flow is notably more complex in the platform margin boundstone facies, with local counter-rotating flow cells resulting from fracture-enhanced vertical permeability (Figure 4.35). Dissolution occurs where fluids rise and cool due to the retrograde solubility of calcite. The simulated magnitude of porosity enhancement ranges from $<1\%$ to $>10\%$ depending on stratigraphic position in the reservoir and the hydrostratigraphy, with

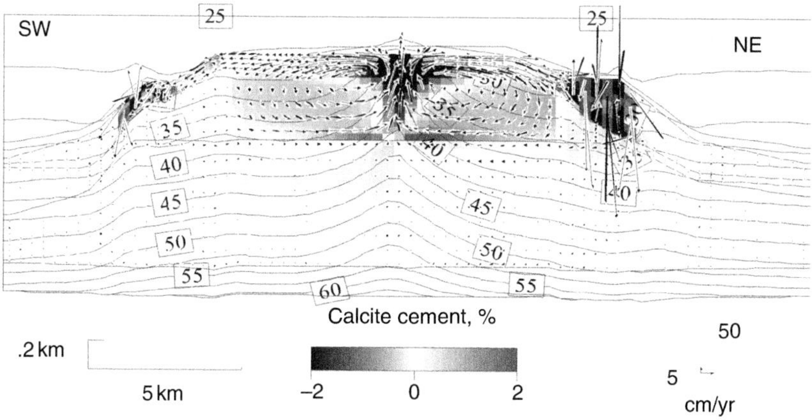

Figure 4.35 RTM of geothermal convection with calcite dissolution and cementation in the Tengiz reservoir after 10 My of burial. Contours are isotherms (°C). From Jones and Xiao (2006).

permeability specified as a function of facies and burial history (Jones and Xiao, 2006). The fastest rates of calcite dissolution by geothermal convection occur before reservoir burial towards the centre of the reservoir. In contrast, cementation is predicted to occur where fluids descend and warm. In general the descending flow is more dispersed, resulting in minor but more extensive porosity degradation by cementation (Figure 4.35).

The RTMs of Tengiz with different overburden scenarios predicted that the presence of shale-filled salt withdrawal basins above the reservoir would disrupt the pattern of convective flow and the distribution of diagenesis (Figure 4.36). The contrast in thermal conductivity between salt and shale generates a local thermal anomaly that acts to focus ascending convective flow and shift the zone of dissolution away from the centre of the carbonate platform reservoir (compare Figures 4.35 and 4.36).

Based on RTM results from the different viable geological scenarios simulated, it is possible to identify five distinct predictive diagenetic concepts for the Tengiz reservoir (Figure 4.37). Subsequent reservoir characterization studies and additional field data have been used to evaluate some of these predictive reservoir quality concepts (Jones et al., 2007; Kenter et al., 2010; Dickson and Kenter, 2014; Collins et al., 2013). For example, Tengiz exhibits enhanced porosity towards the platform centre that is not related to a change in depositional facies and postdates meteoric diagenesis (Jones et al., 2007). Geothermal circulation is the leading predictive geological concept to explain this reservoir quality pattern observed in Tengiz and other important reservoirs in Kazakhstan, for example, Kashagan (Ronchi et al., 2010). In contrast, enhanced porosity is not observed in the Tengiz reservoir beneath salt withdrawal basins (Jones et al., 2007). The rates of dissolution driven by convection after burial are too low compared with those prior to burial, and/or are masked by differences in burial compaction due to variations in overburden loading (Jones and Xiao, 2006; Jones et al., 2007).

4.5.2 Geothermal Convection in South Atlantic Pre-Salt Rift Carbonates

Prolific hydrocarbon discoveries in the pre-salt section of the Santos and Campos Basins, offshore Brazil and the conjugate African margin, have created a business

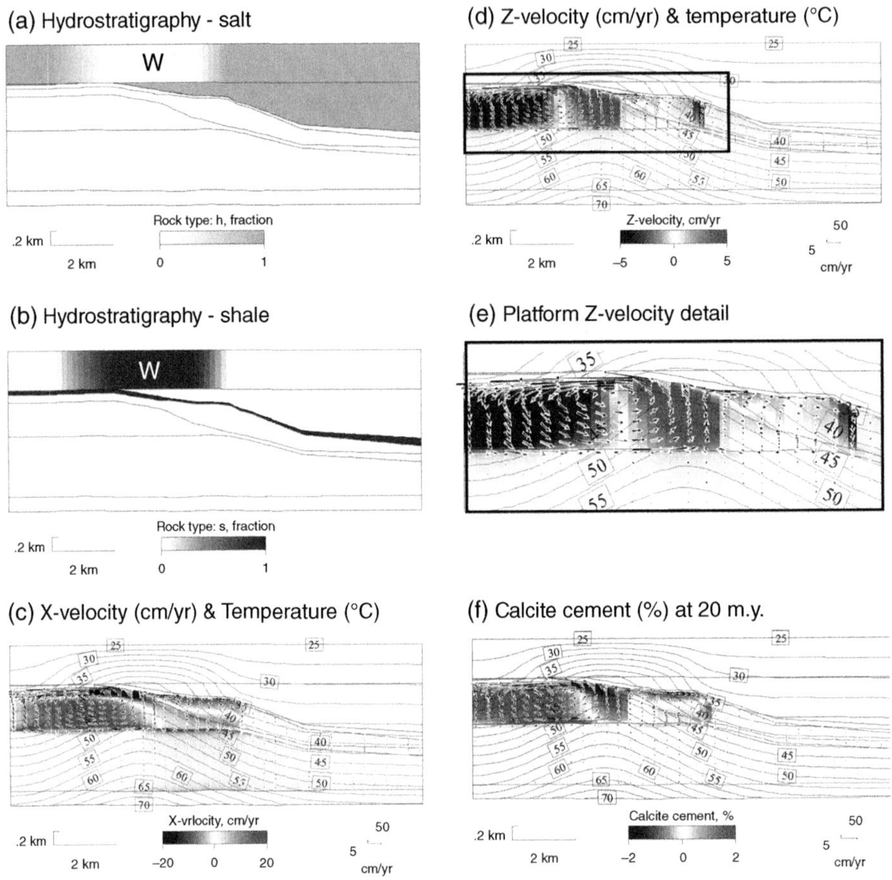

Figure 4.36 Geothermal circulation after platform burial and development of a 5.1 km shale-filled salt withdrawal basin that is 500 m deep. (a) Halite, 'W', withdrawal basin; (b) shale; (c) horizontal fluid velocity, velocity vectors and temperature (contour interval of 2.5 °C); (d) vertical fluid velocity, fluid velocity vectors, and temperature; (e) enlargement of vertical fluid velocity distribution in platform; (f) calcite cementation/dissolution at 20 My. Note: to enhance visualization of diagenetic patterns, the colour bar scale does not reflect the full range of the data, as discussed in the text. From Jones and Xiao (2006). (See colour plate section for the colour representation of this figure.)

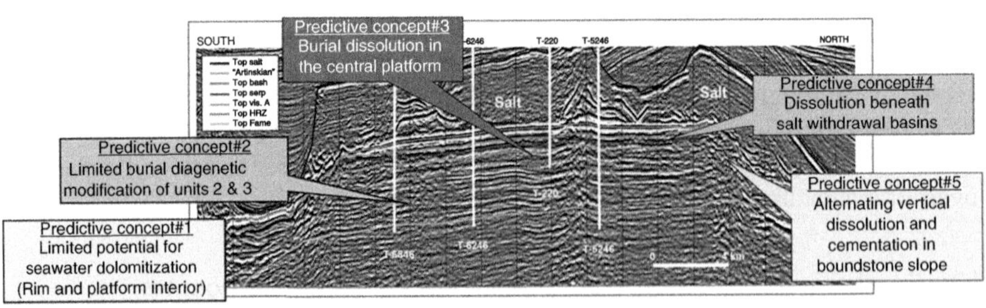

Figure 4.37 Five predictive concepts for diagenetic modification of reservoir quality in the Tengiz field developed from reactive transport models of geothermal convection. From Jones et al. (2007).

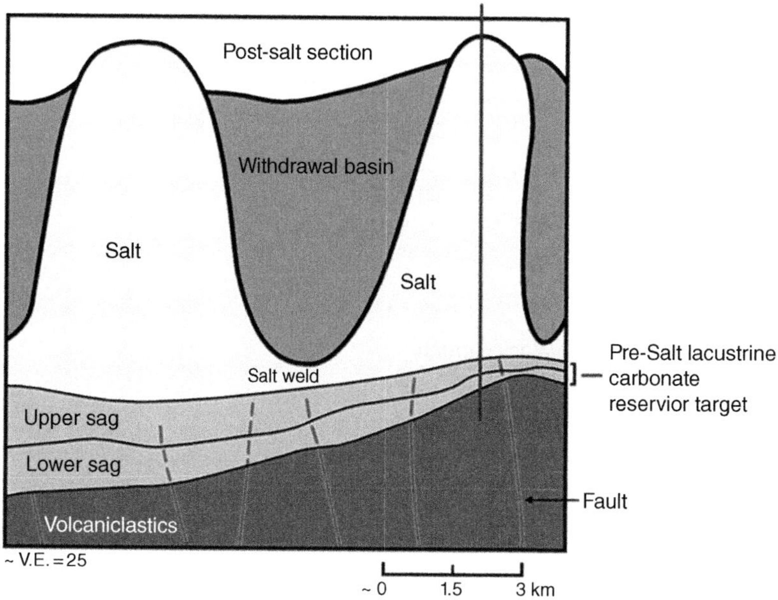

Figure 4.38 Conceptual half-graben tilted fault block based on the Lula field in Brazil, used to design reactive transport models to investigate the diagenetic potential for geothermal convection to modify reservoir quality in subsalt lacustrine carbonate reservoirs (upper and lower sag intervals). From Jones and Xiao (2013).

imperative to predict reservoir presence and quality in lacustrine carbonates (Karner and Gamboa, 2007; Mello et al., 2011; Wright, 2012; Jones and Xiao, 2013; Saller et al., 2016). RTM can play a useful role in developing digital analogues and predictive concepts because reservoir analogues for this exploration play are rare.

A generic conceptual model based on a published seismic cross-section of the Lula field in Brazil was used to explore the diagenetic potential of geothermal convection using 2D RTM models (Figure 4.38). Rock properties were estimated from modern carbonates and lacustrine facies guided by sparse reservoir data. Waters sampled from Lake Tanganyika, an East African Rift area with an active carbonate factory, were specified prior to burial, with a later switch to seawater to reflect marine transgression of the South Atlantic rift basins.

RTM results demonstrate that the thermal contrast between cool sub-thermocline and shallow, warm lake water is sufficient to generate convective flow in the underlying carbonates and volcaniclastics (Figure 4.39). Flow is focused in the more permeable sublittoral carbonates, which results in a dip-orientated stratigraphic diagenetic gradient anchored by the descending and rising limbs of the convective cells (Figure 4.40). Resulting diagenetic patterns are distinct, but rates of the order of 0.01 vol.%/My are too low to materially impact reservoir quality in the residence time available prior to marine transgression and deposition of the seal. This conclusion proved robust to alternative permeability and heat flux scenarios tested (Jones and Xiao, 2013). Introducing faults per the conceptual model proved the exception. Faults

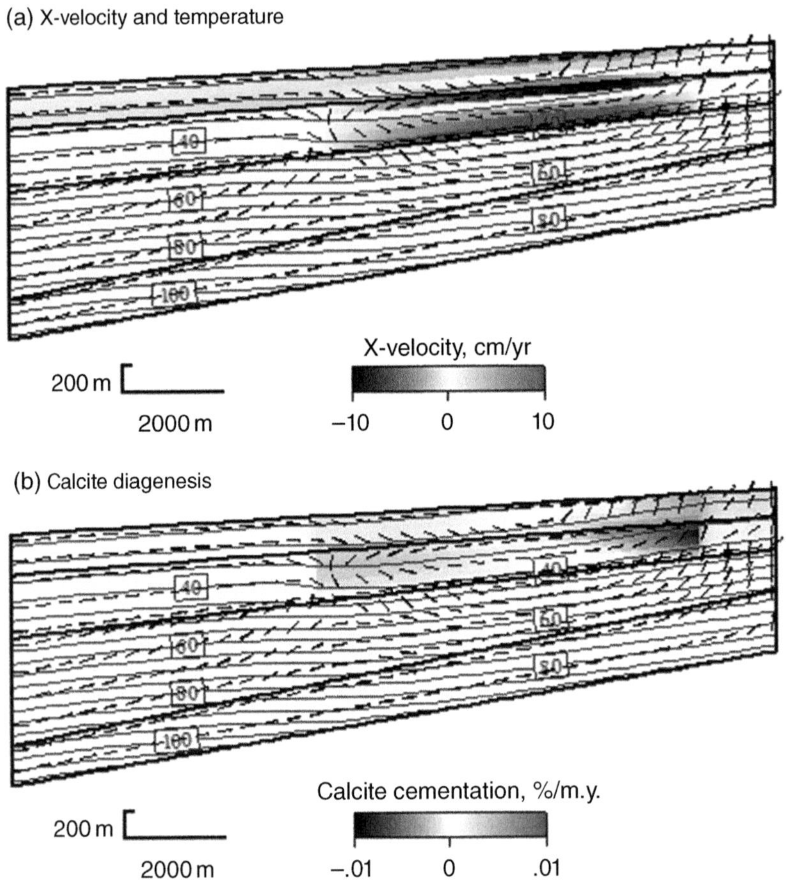

Figure 4.39 RTM of geothermal convection of lake water in lacustrine carbonates (Upper and Lower Sag reservoirs) prior to burial by salt. Contours are temperature, and arrows show direction of flow. Post-lacustrine depositional environments are for the upper and lower sag carbonates. From Jones and Xiao (2013).

and their associated damage zones, with enhanced vertical permeability, locally disrupt the pattern of convective flow (Figure 4.40). Where faults intersect more permeable facies, the convection-driven inflection of isotherms results in adjacent zones of cementation and dissolution (Figure 4.40). Rates of dissolution and cementation of the order of 1–10 vol.%/My are sufficient to create local baffles and high-permeability conduits in the pre-salt reservoirs. Post-burial RTM demonstrated that rates of convection-driven diagenesis were generally too low to materially impact reservoir quality (Jones and Xiao, 2013). Furthermore, simulations that incorporated solute transport demonstrated that reflux of seawater and brines associated with the marine transgression would rapidly flush evolved lake-derived pore fluids in the reservoir, necessitating a more sophisticated RTM model to explore fluid mixing and diagenesis (see Figure 4.14).

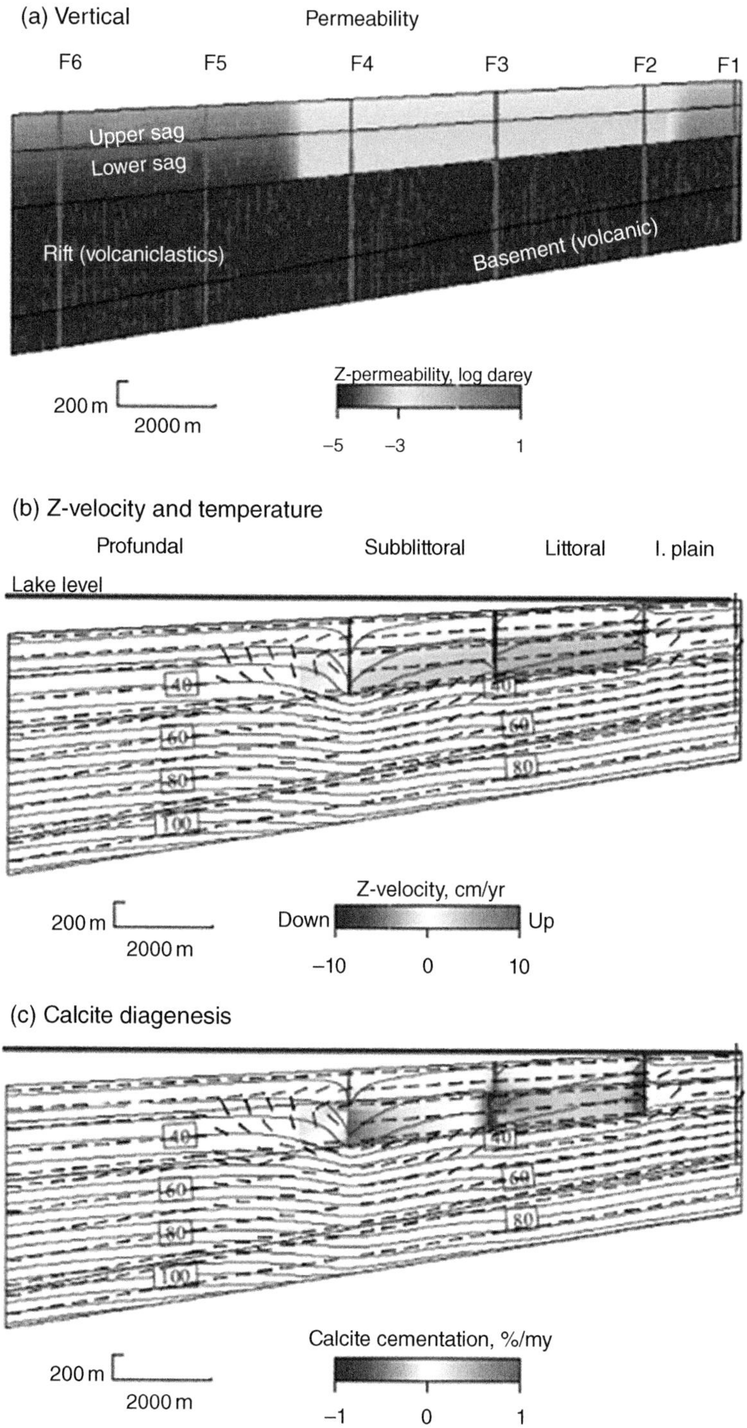

Figure 4.40 RTM of geothermal convection of lake water in lacustrine carbonates (upper and lower sag reservoirs) prior to burial by salt. Effect of introducing six permeable vertical faults and associated damage zones on flow and diagenesis. From Jones and Xiao (2013).

4.6 Burial Diagenesis: Fault-Controlled Illitization

4.6.1 Illitization and Permeability Reduction in Rotliegendes Play, Germany

The Rotliegendes Play in northern Germany is one of the major gas-producing fields in Europe. The complexity of reservoir properties in the Rotliegendes play has been a critical factor for gas production. Production and rock data indicate that areas with moderate porosity can have extremely low permeability, largely due to clay mineral cements that plug the pore throats. Other areas with permeability sufficiently high for production can have relatively low bulk porosities. Thus, porosity distribution differs significantly from permeability distribution. Reservoir quality is considered as a combination of primary depositional attributes and secondary diagenetic alterations.

One of the proposed conceptual diagenetic models is so-called 'fault-controlled illitization' (Gaupp et al., 1993, 2005). The conceptual model suggests that penetration of the Rotliegendes sequence by acidic fluids from the Carboniferous coal seams produced a zoned clay mineral halo around tectonic and stratigraphic Rotliegendes and Carboniferous stratigraphic contacts. An inner zone, hundreds of metres wide, is dominated by extensive dickite growth and feldspar dissolution. An outer zone displays a dense meshwork of illite rich in Al. The intensity of illite growth decreases with increasing distance from the coals (Figure 4.41).

4.6.2 1D and 2D Reactive Transport Models

Xiao et al. (2007) developed 1D and 2D reactive transport models to investigate the mechanisms of illite formation and to predict the spatial distribution of K-feldspar,

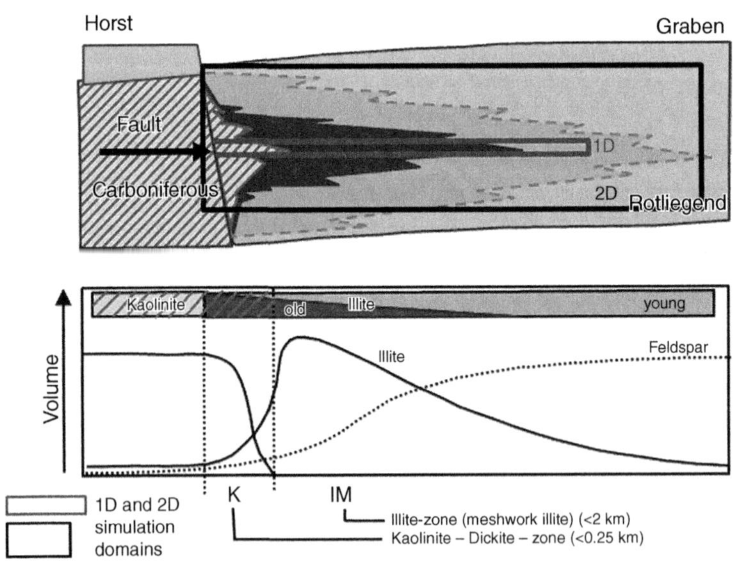

Figure 4.41 Schematic diagram showing fault-controlled illitization and mineral zoning in the 1D and 2D simulation domains.

quartz, illite, and kaolinite in the Rotliegendes, and the implications for reservoir quality. We investigated the effects of key natural variables (initial mineralogy, fluid chemistry, K, Al and SiO_2 sources, temperature, CO_2 and fluid migration, and faults/fractures) on illite formation and distribution. The key reaction that we tested was:

$$3KAlSi_3O_8 + 2H^+ \rightarrow KAl_3Si_3O_{10}(OH)_2 + 6SiO_2 + 2K^+$$

The model input for the initial system (pre-illitization) consists of 50% quartz, 15% K-feldspar, and 35% porosity with a 2 Darcy permeability. The fault-induced flow is specified to have a rate between 0.1–0.5 m/yr. The acidic fluid from the Carboniferous coal has a pH from 5.0 to 6.5. The illitization temperature ranges from 115 °C to 150 °C.

The simulated 1D distribution of mineral alteration after one million years is shown in Figure 4.42. The results indicate that the slightly acid fluid from the Carboniferous formation dissolves the K-feldspar and leads to the generation of kaolinite near the fault, and muscovite (a thermodynamic analogue for illite) away from the fault. This process also leads to quartz overgrowth cementation in the kaolinite/illite zone, and together they are partly responsible for the diminished reservoir quality.

Xiao and Jones (2006) also conducted a 2D reactive transport simulation, and the distribution of minerals after two million years of illitization is shown in Figure 4.43 (map view). In the 2D model we incorporated the major fault at the left side of the domain and some faults/fractures extending from the boundary fault into the Rotliegendes formation. The 2D results show that the hydrothermal fluids ascend from the boundary fault at the left, preferentially flow through the fractures into the formation, dissolve K-feldspar, and generate kaolinite, muscovite, and quartz cements. The K-feldspar away from the faults is unaltered, and consequently any portion of the fault

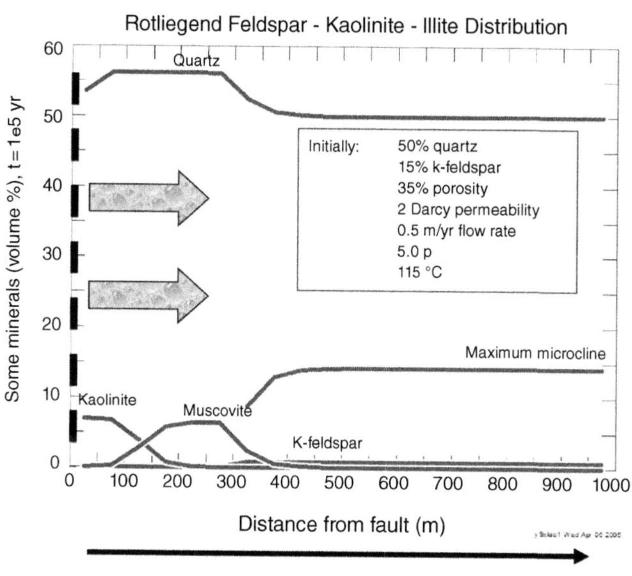

Figure 4.42 1D reactive transport modeling results showing the mineral zoning in Rotliegendes after 1 My of illitization. From Xiao and Jones (2006).

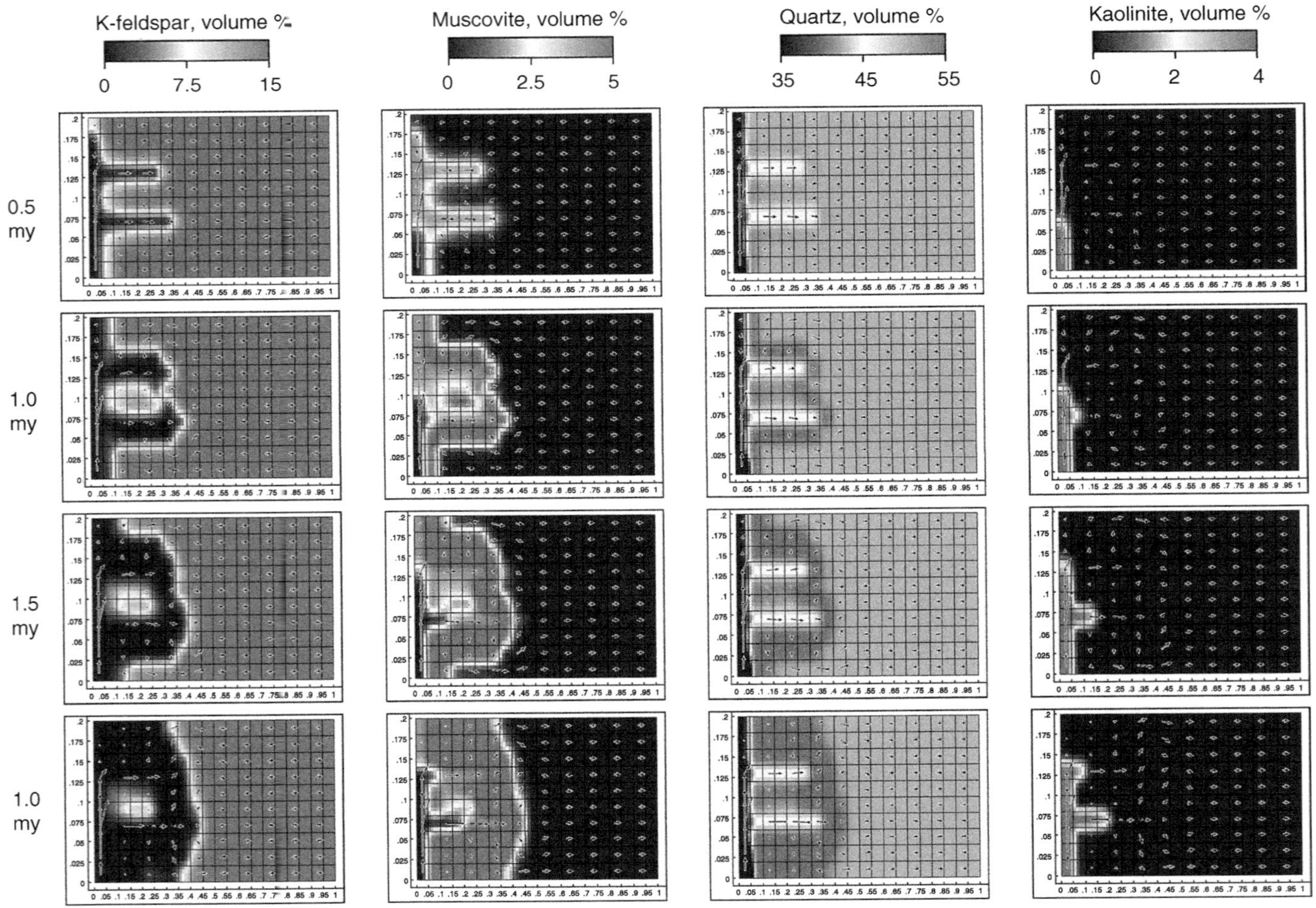

Figure 4.43 2D reactive transport modeling results showing the mineral zoning associated with fault-controlled burial diagenesis representing Rotliegendes after 2 My of illitization. From Xiao and Jones (2006). (See colour plate section for the colour representation of this figure.)

that is sealed prior to fluid flow will shelter portions of the reservoir and reduce permeability degradation from fluid flow and associated kaolinite, illite, and quartz cements.

Both the 1D and 2D simulation results are consistent with the proposed 'fault-controlled illitization' model and field observations (Gaupp et al., 1993, 2005). This demonstrates that moderate amounts (~10%) of K-feldspar can provide sufficient K, Al, and SiO_2 for illite, kaolinite, and quartz cementation (no need for K input from external sources). Sufficient fluid flow (>1 cm/yr) and temperature (>100 °C) are needed to drive K-feldspar dissolution and illite formation. The diagenetic zonation (kaolinite, muscovite, quartz, and K-feldspar) is predicted to be closely associated with fluid flow along faults and fractures. More acidic fluids (due to dissolution of CO_2 and organic acid) would cause faster illitization and more kaolinite formation near the fault zone. Unopened (sealed) faults serve as flow barriers and may preserve porosity/permeability.

4.7 Diagenesis and Reservoir Alteration Associated with Oil and Gas Operations

4.7.1 CO_2 and Acid Gas Injection (AGI) in Siliciclastic and Carbonate Reservoirs

A commonly proposed option for reducing atmospheric CO_2 emissions is to inject CO_2 into petroleum reservoirs for enhanced oil/gas recovery (EOR/EGR), or to store CO_2 in deep saline aquifers. Sequestering bulk gases from combustion containing CO_2, H_2S and/or SO_2 requires less energy than separation from flue gases or a coal gasification process, and therefore might be the preferred disposal option (Xu et al., 2007; Knauss et al., 2005; Chapters 1 and 2, this volume; and references therein). Large-scale injection of CO_2 and other gases into subsurface reservoirs may induce a complex interplay of multiphase flow, dissolution, precipitation, diffusion, convection, and other chemical reactions (Johnson et al., 2001; Pruess et al., 2003; Audigane et al., 2007, 2008; Xiao et al., 2008, 2009, 2010; Liu et al., 2011, 2013). Injected CO_2 can be retained in the reservoir via hydrodynamic, residual and solubility trapping. Over the long term, however, mineral trapping of CO_2 is potentially more attractive because it could immobilize CO_2 for long timescales and increase storage security. Depending on the spatial distribution and temporal evolution of the CO_2 injection and associated mineral dissolution and precipitation (artificial diagenesis), the gas–fluid–rock interactions could have a significant impact on injection performance and storage capacity and security. Alteration of the predominant host rock minerals is usually very slow and therefore is not experimentally accessible under deep reservoir or aquifer conditions. Reactive transport modeling is a promising approach that can be used to investigate the spatial and temporal evolution of injected $CO_2/H_2S/SO_2$ and associated formation alteration (Johnson et al., 2001; Pruess et al., 2003; Knauss et al., 2005; Audigane et al., 2007, 2008; Xu et al., 2007; Xiao et al., 2008, 2009, 2010; Liu et al., 2011, 2013).

Readers interested in detailed discussions of RTM and CCS should refer to chapters 1 (Xu et al.) and 2 (Audigane et al.) in this book, and extensive references therein. The following section summarizes recent results from RTM of acid gas injection (AGI) in representative siliciclastic and carbonate reservoirs, with a focus on reservoir quality alteration associated with AGI operations. In particular, Xiao et al. (2008, 2009) investigated AGI in feldspar-rich and quartz-dominated sandstone reservoirs, and in limestone-dominated and dolomite-dominated carbonate reservoirs. The investigations studied key technical issues on (1) the short- and long-term behaviour of gas–fluid–rock interactions in geological formations; (2) the role of reservoir mineralogy on injection performance and storage security; (3) the effect of gas mixtures (e.g., H_2S and SO_2) on CO_2 storage; and (4) the physical and chemical processes associated with CO_2 and AGI in the storage reservoir.

4.7.2 Reactive Transport Model Setup

The 1D radial reactive transport models represent CO_2 injection into a siliciclastic and carbonate reservoir at 2 km depth and 70 °C. CO_2 and other gases were injected in the reservoir at a rate of 31,500 tons per year (1 kg/s) over a period of 100 years. The reactive transport models simulate the system from 0 to 10,000 years. There are three scenarios of mixed gas injected: CO_2 only; $CO_2 + H_2S$; and $CO_2 + SO_2$ in which CO_2 is injected as a gas phase while both H_2S and SO_2 (~5% each) are injected as aqueous solutes. The reservoirs are specified to have an initial porosity of 0.30 and initial permeability of 100 mD. The siliciclastic and carbonate reservoirs were defined by hypothetical mineral assemblages, representing: (1) an oligoclase feldspar-rich sandstone reservoir; (2) a quartz-dominated sandstone reservoir; (3) a limestone-rich reservoir; and (4) a dolomite rich carbonate reservoir, respectively (Table 4.2). Other primary and secondary minerals are also listed in Table 4.2.

4.7.3 Simulation Results: Injection in Siliciclastic Reservoirs

4.7.3.1 Feldspar-Rich Sandstone Reservoir

The simulation results from the CO_2-only case are similar to those for $CO_2 + H_2S$. The results from the $CO_2 + SO_2$ case are, however, significantly different (Figure 4.44). CO_2 and $CO_2 + H_2S$ injection leads to lower pH (~4) near the well bore, resulting from the dissolution of CO_2 and H_2S into the formation water. Once injection stops, the pH is quickly buffered by the reservoir minerals and goes back to a near-neutral value (~6). Co-injection of SO_2 with CO_2 leads to very low pH (~0), presumably due to the dissolution of SO_2 and the formation of a sulfuric acid zone close to the injection well. Corrosion and well abandonment are potential issues. After injection, the pH is buffered but still remains considerably lower near the well bore.

The gas injection leads to an increase in porosity close to the well due to net mineral dissolution, and a decrease at distance due to mineral trapping in all three cases (Figure 4.45). However, the porosity gains associated with CO_2 and $CO_2 + H_2S$ injection are much smaller (0.30 to 0.32) compared with the $CO_2 + SO_2$ injection (0.30 to 0.50) at 100 years. The same trend is observed for mineral trapping (Xiao et al., 2009).

4.7.3.2 Quartz-Dominated Sandstone Reservoir

The pH evolution in a quartz-dominated sandstone reservoir is similar to that in a feldspathic reservoir, with a slight difference in pH buffering capacity due to the lack of more reactive feldspar minerals (Figure 4.46).

Similarly, AGI leads to an increase in porosity close to the well due to net mineral dissolution, and a decrease at distance due to mineral trapping in all three cases

Table 4.2 Initial mineral compositions of the siliciclastic and carbonate reservoirs and secondary minerals considered in the simulation. From Xiao et al. (2009).

		Volume percentage			
Mineral	Chemical formula	Feldsparic reservoir	Sandstone reservoir	Limestone reservoir	Dolomite reservoir
***Primary*:**					
Quartz	SiO_2	40.6	60.3	1.0	0.06
Kaolinite	$Al_2Si_2O_5(OH)_4$	1.41	0.7	1.5	0.2
Calcite	$CaCO_3$	1.35	1.05	63.0	3.50
Illite	$K_{0.6}Mg_{0.25}Al_{1.8}(Al_{0.5}Si_{3.5}O_{10})(OH)_2$	0.7	0.6	0.6	0.1
Oligoclase	$Ca_{0.2}Na_{0.8}Al_{1.2}Si_{2.8}O_8$	13.86	1.96	0.5	0.1
K-feldspar	$KAlSi_3O_8$	5.74	2.24	1.2	0.1
Na-smectite	$Na_{0.29}Mg_{0.26}Al_{1.77}Si_{3.97}O_{10}(OH)_2$	4.8	2.03	0.6	0.1
Chlorite	$Mg_{2.5}Fe_{2.5}Al_2Si_3O_{10}(OH)_8$	1.19	0.7	1.6	0.39
Hematite	Fe_2O_3	0.35	0.3	0.0	0.03
Dolomite	$CaMg(CO_3)_2$	0.0	0	0.0	65.8
Porosity		30	30	30	30
***Secondary*:**					
Anhydrite	$CaSO_4$				
Magnesite	$MgCO_3$				
Low-albite	$NaAlSi_3O_8$				
Siderite	$FeCO_3$				
Ankerite	$CaMg_{0.3}Fe_{0.7}(CO_3)_2$				
Dawsonite	$NaAlCO_3(OH)_2$				
Ca-smectite	$Na_{0.145}Mg_{0.26}Al_{1.77}Si_{3.97}O_{10}(OH)_2$				
Alunite	$KAl_3(OH)_6(SO_4)_2$				
Pyrite	FeS_2				
Opal-A	SiO_2				

(Figure 4.47). However, the porosity gains associated with CO_2 and $CO_2 + H_2S$ injection are much smaller (0.30 to 0.33) compared with the $CO_2 + SO_2$ injection (0.30 to 0.40) at 100 years. Due to the lower concentration of initial reactive feldspar minerals, there is less mineral trapping and corresponding porosity reduction.

4.7.4 Simulation Results: Injection in Carbonate Reservoirs

4.7.4.1 Limestone Reservoir

The simulation results from the CO_2-only case are similar to those for $CO_2 + H_2S$. The results from the $CO_2 + SO_2$ case are, however, significantly different (Figure 4.48). CO_2 and $CO_2 + H_2S$ injection leads to lower pH ($4.5 \sim 5$, less so comparing to siliciclastic

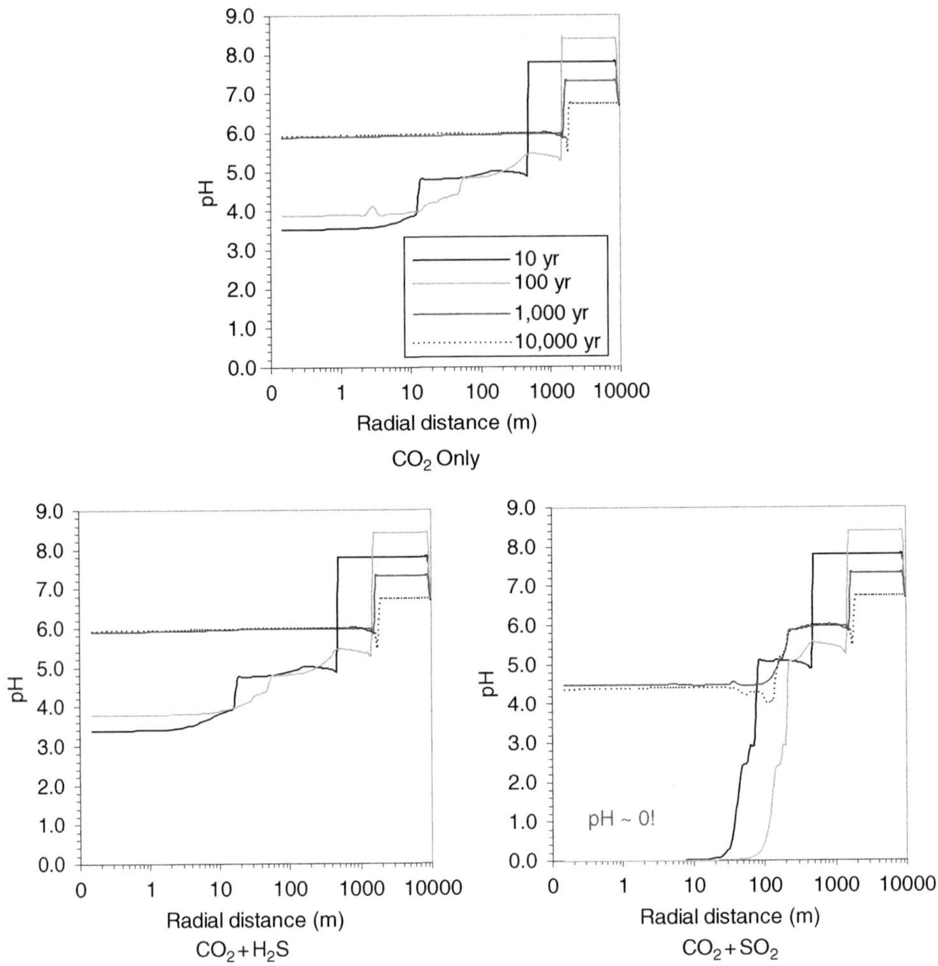

Figure 4.44 Simulated pH distribution as a function of radial distance at 10, 100, 1000, and 10,000 years for the three mixed gas injection scenarios in a feldspar-rich sandstone reservoir. From Xiao et al. (2008).

reservoir) near the well bore, resulting from the dissolution of CO_2 and H_2S in the formation water and faster limestone buffering reactions. Once the injection stops, the pH increases slightly (~5), but does not rebound to a more neutral pH due to limited buffering capacity. Co-injection of SO_2 with CO_2 leads to very low pH (~0), presumably due to the dissolution of SO_2 and the formation of a sulfuric acid zone close to the injection well. Corrosion and well abandonment can be potential issues. After the injection, the pH is buffered but still remains considerably lower near the well bore.

Simulation results indicate that there are significant porosity increases (0.30 to 0.40 at 100 years) associated with CO_2 and $CO_2 + H_2S$ injection, due to calcite dissolution near the well bore (Figure 4.49). However, there is a significant porosity decrease (0.30 to 0.20) associated with $CO_2 + SO_2$ injection because of anhydrite precipitation. The latter is due to the interaction of sulfate ions with the excess Ca^{2+} released from limestone dissolution.

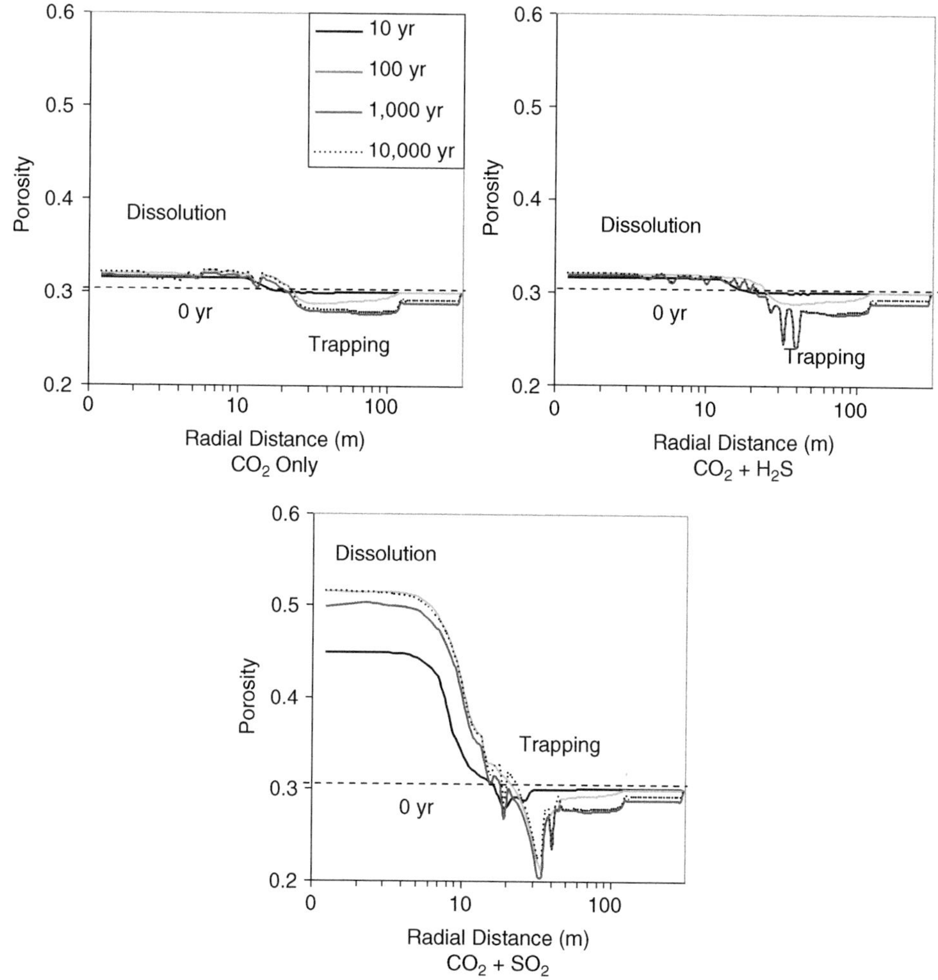

Figure 4.45 Simulated porosity evolution as a function of radial distance at 10, 100, 1000, and 10,000 years for the three mixed gas injection scenarios in a feldspathic reservoir. From Xiao et al. (2008).

There is little mineral trapping in all three mixed gas injection scenarios, suggesting limited CO_2 mineral trapping capacity in a limestone-dominated carbonate reservoir.

4.7.4.2 Dolomite Reservoir

During AGI, the pH buffering capability of a dolomite reservoir appears to be smaller and lagging due to slower dolomite reaction rates, hence leading to lower pH (Figure 4.50) compared with a limestone reservoir. Future work will investigate the behaviour of AGI in a mixed limestone–dolomite reservoir. We will also evaluate the effect due to the presence of anhydrite, either as bedded anhydrite or anhydrite cement.

Simulation results indicate that there is less porosity increase (0.30 to 0.34 at 100 years) associated with CO_2 injection, largely due to slower dolomite dissolution rates (Figure 4.51). $CO_2 + H_2S$ injection leads to a similar amount of dolomite dissolution, but also some calcite precipitation, and hence a lower porosity increase (0.30 to 0.32). There

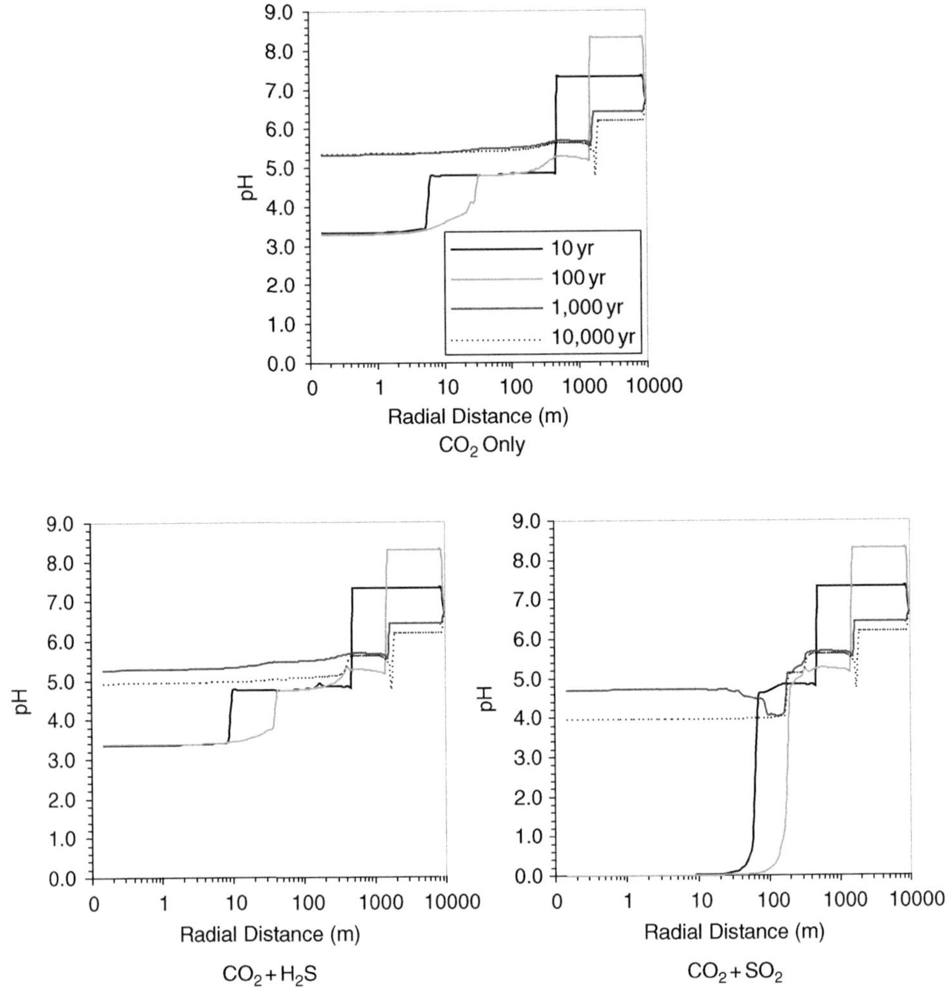

Figure 4.46 Simulated pH distribution as a function of radial distance at 10, 100, 1000, and 10,000 years for the three mixed gas injection scenarios in a quartz-dominated sandstone reservoir. From Xiao et al. (2009).

is significant porosity increase (0.30 to 0.40 at 100 years) associated with $CO_2 + SO_2$ injection due to progressive dolomite dissolution. However, there is no excess Ca^{2+} released from dolomite dissolution, hence no anhydrite precipitation occurs in this case. Therefore dolomite reservoirs may offer a better opportunity to store mixed $CO_2 + SO_2$ gas. There is little mineral trapping in all three mixed gas injection scenarios, suggesting limited CO_2 mineral trapping capacity in a dolomite-dominated carbonate reservoir.

4.7.5 Summary of CO_2 and Acid Gas Injection and Reservoir Alteration

- The behaviour of CO_2 and other acid gases during injection and storage is controlled by the gas mixture composition, reservoir mineralogy, timing, and injection design.

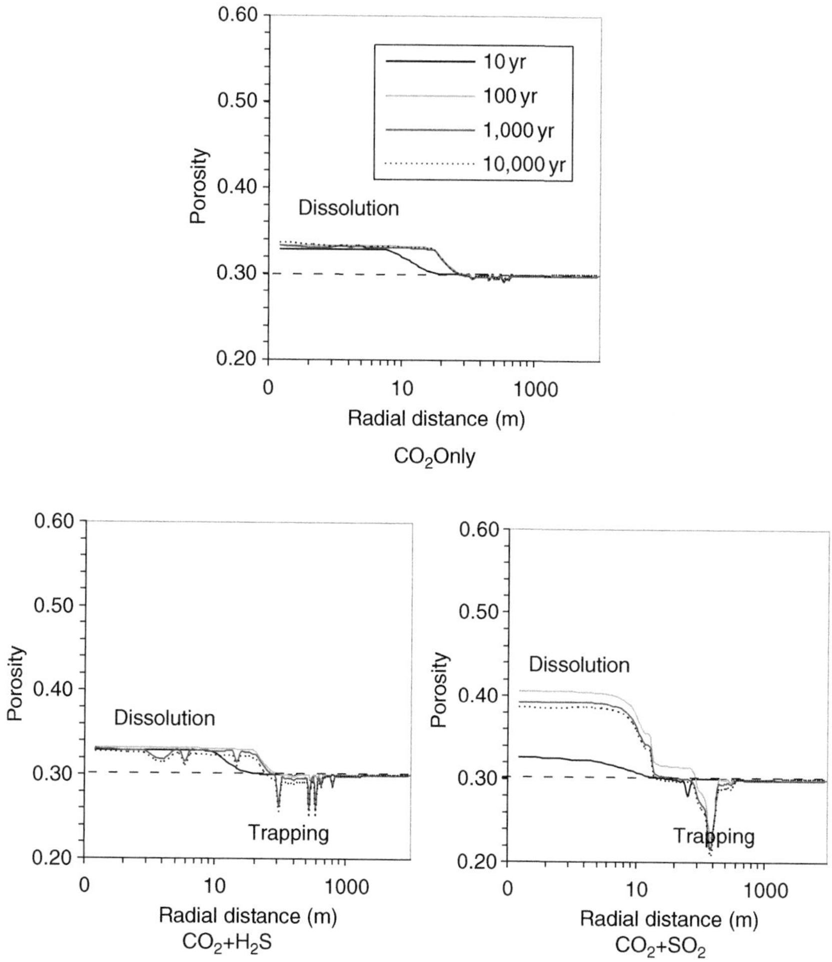

Figure 4.47 Simulated porosity evolution as a function of radial distance at 10, 100, 1000, and 10,000 years for the three mixed gas injection scenarios in a quartz reservoir. From Xiao et al. (2009).

Co-injection of H_2S yields similar behaviour compared with CO_2 injection in both siliciclastic and carbonate reservoirs.

- Co-injection of SO_2 can potentially result in the formation of sulfuric acid and extremely low pH close to the well. Corrosion of pipes and well abandonment are potential risks.
- Most AGI scenarios result in porosity increase near the well and a decrease at distance. However, co-injection of SO_2 in a limestone-dominated carbonate reservoir can potentially lead to significant anhydrite precipitation and porosity reduction in the near-well region.
- The results from reactive transport models provide valuable insights for describing, analysing, interpreting, and assessing the physical properties and dynamic behaviours of injected CO_2, and facilitating the screening and evaluation of CO_2 storage strategy.

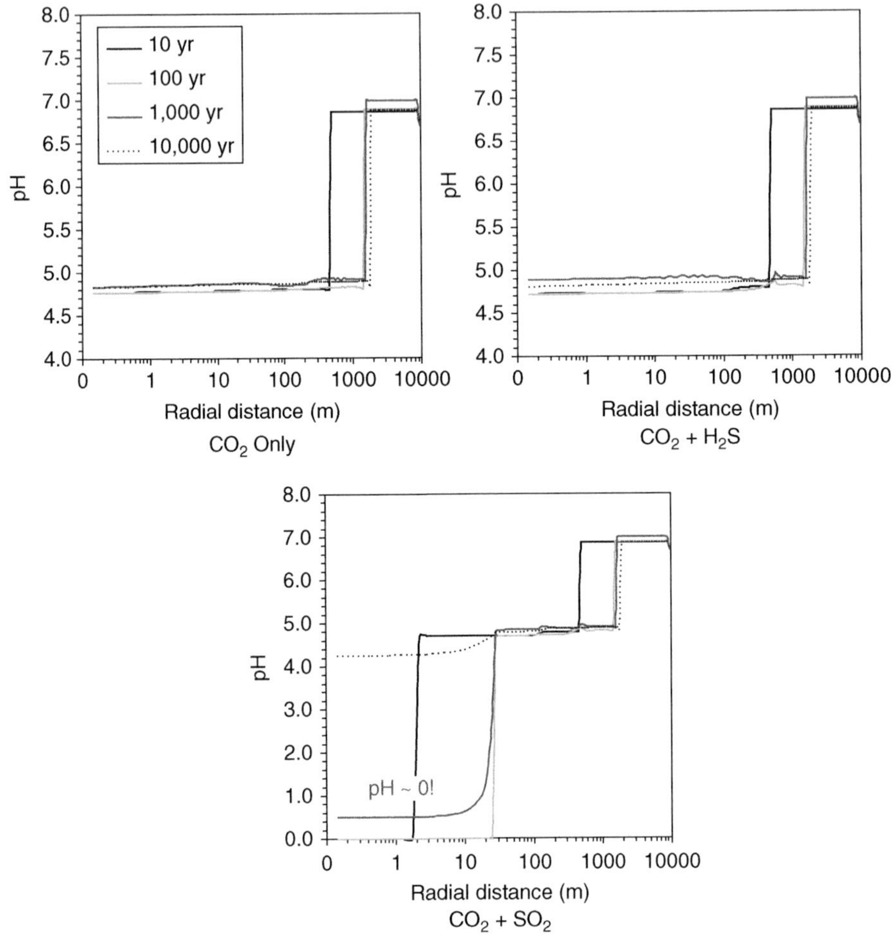

Figure 4.48 Simulated pH distribution as a function of radial distance at 10, 100, 1000, and 10,000 years for the three mixed gas injection scenarios in a limestone reservoir. From Xiao et al. (2008).

4.7.6 Reservoir Alteration from Steam and Acid Injection

Steam, acid injection, and *in situ* combustion are the common methods for enhanced oil recovery. The extreme temperature of steam injection often causes chemical reactions between the injected fluids and the reservoir rocks, which is called artificial diagenesis (Hutcheon and Abercrombie, 1990). Previous discussions focused on AGI. This section focuses on steam and acid injection. For example, injection of steam/hot water into a clastic reservoir such as in Cold Lake formation tends to dissolve quartz, calcite, and kaolinite, and produce smectite (Ca-beidellite: Hutcheon and Abercrombie, 1990; Bethke, 1996). There have been many observations that smectite precipitation plugs the critical pore throats and significantly reduces the reservoir permeability (Bethke, 1996; Xiao and Jones, 2006).

Artificial diagenesis can, on the other hand, be utilized to our advantage to enhance reservoir quality and oil recovery. One of the common strategies is to apply acid

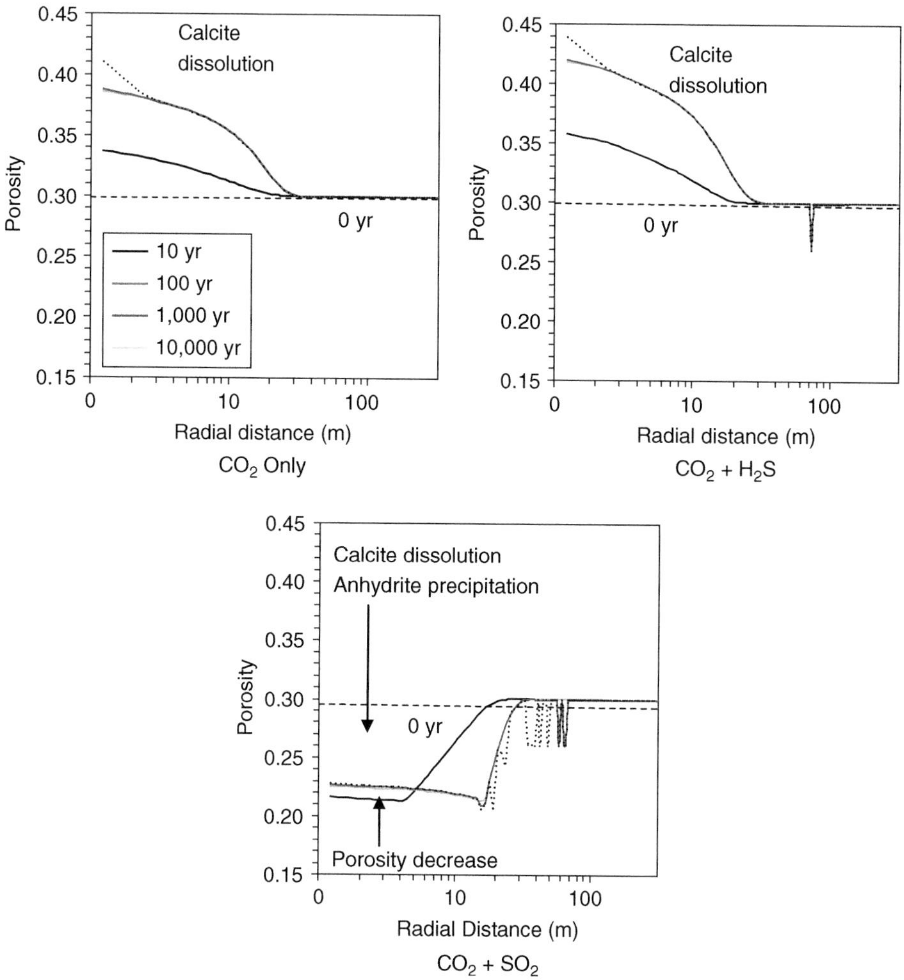

Figure 4.49 Simulated porosity evolution as a function of radial distance at 10, 100, 1000, and 10,000 years for the three mixed gas injection scenarios in a limestone reservoir. From Xiao et al. (2008).

stimulation in carbonate reservoirs to increase the porosity and permeability near the well bore (Kalfayan, 2001). Our reactive transport model shows the porosity evolution of a carbonate reservoir (with an initial porosity of 0.2) subject to acid stimulation and the so-call 'wormhole' development at metre and centimetre scales (Figure 4.52). The results show significant porosity increases (from 0.2 to 0.6) near the well bore due to carbonate dissolution from the acid.

The reactive transport simulations suggest that the injected fluid (e.g. hot water and acid) may react with minerals and fluids in the reservoir and cause desirable or undesirable effects (artificial diagenesis). Other reservoir disposal/injection/EOR operations that may cause artificial diagenesis include geothermal exploration or steam injection (Chapter 1, this volume), (waste) water injection for disposal or pressure maintenance purpose (Al-Abduwani et al., 2005; Van Berk et al., 2015), and polymer injection

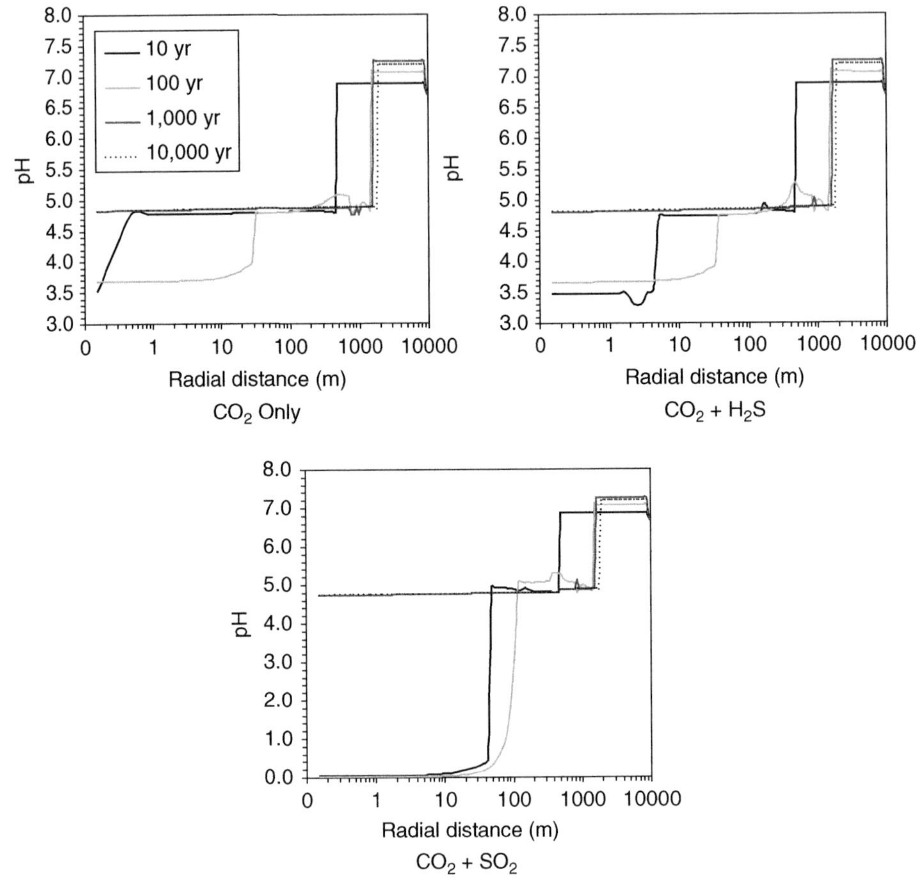

Figure 4.50 Simulated pH distribution as a function of radial distance at 10, 100, 1000, and 10,000 years for the three mixed gas injection scenarios in a dolomite reservoir. From Xiao et al. (2009).

(Donaldson et al., 1985). Reactive transport simulations should be used in connection with reservoir simulations to predict the distribution and degree of artificial diagenesis as a function of time and injection rate (steady or cyclic) to provide additional information to optimize the recovery strategy.

4.7.6.1 Case Study: RTM of Steam Flood in Eocene Carbonate Reservoir, Wafra Field

Meddaugh et al. (2012) applied RTM to simulate high-pH steam injection into the First Eocene reservoir in the Wafra Field, Saudia Arabia and Kuwait. The purpose was to understand possible changes in reservoir quality such as mineralogy, porosity, and potential scale formation, so that injection performance can be maintained and recovery can be enhanced. RTM results show mineral distribution in the vicinity of a well bore for a variety of injected fluid compositions (Figure 4.53). Near the zone of injection, dissolution and precipitation reactions dominate. Calcite and brucite are the primary preipitates, while dolomite and anhydrite are dissolved. Both dissolution and

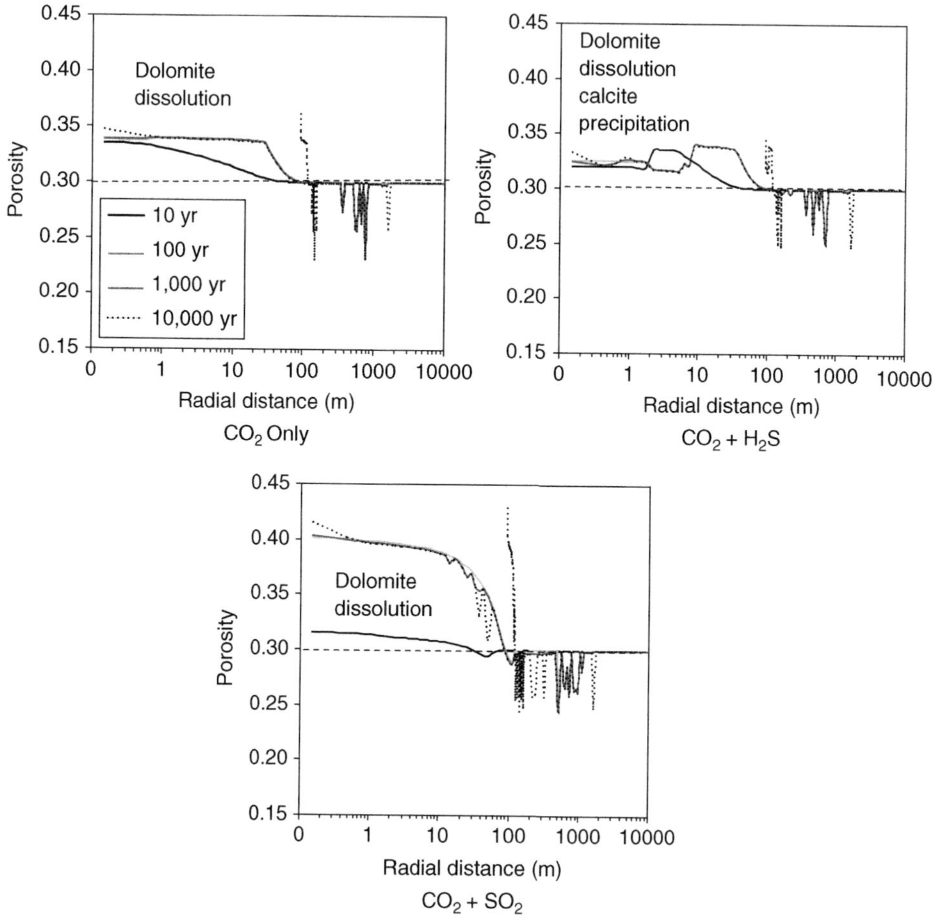

Figure 4.51 Simulated porosity evolution as a function of radial distance at 10, 100, 1000, and 10,000 years for the three mixed gas injection scenarios in a dolomite reservoir. From Xiao et al. (2009).

precipitation appear to occur within a few metres of the injection face, reflecting relatively fast reaction rates and a sharp reaction front. Adjusting injection steam quality can largely reduce performance degrading near well bore reactions, and mitigate reservoir alteration away from the well, to optimize injection performance.

4.8 The Present and Future Role of Reactive Transport Models for Reservoir Quality Prediction

The past decade and a half has seen rapid growth in the use of RTMs to investigate different styles of diagenesis (Table 4.1). Based on a business imperative for quantitative, reliable reservoir quality predictions, RTMs have permeated from universities and government research laboratories to the upstream technology divisions of several major energy companies. RTM numerical experiments and reservoir characterization studies

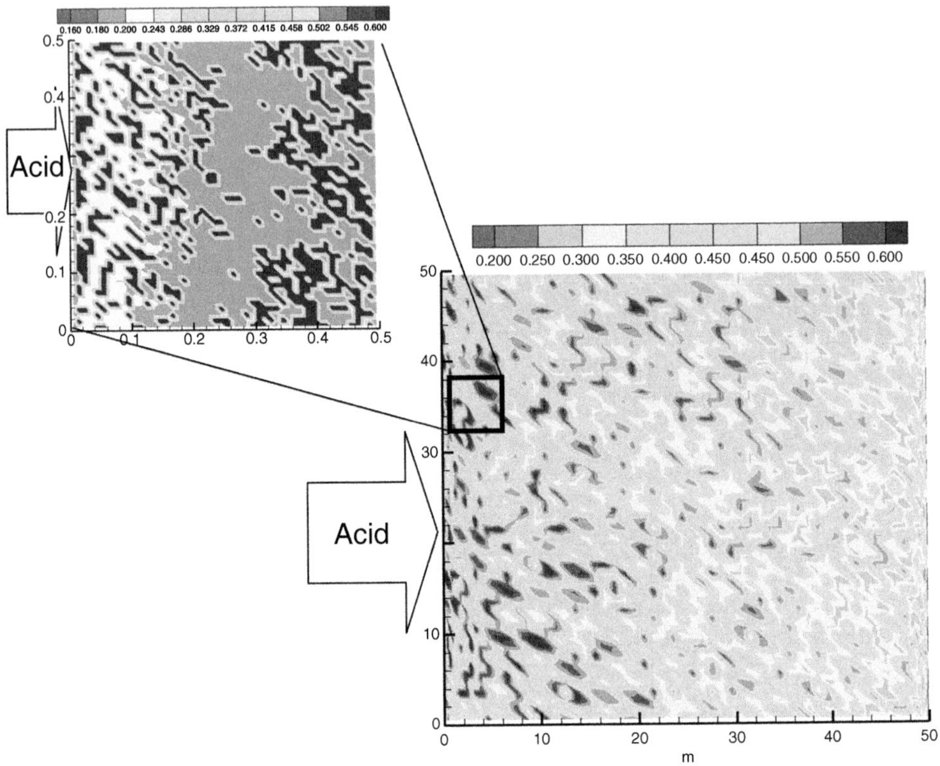

Figure 4.52 Reactive transport model of acid stimulation and porosity (wormhole) development in a carbonate reservoir. From Xiao and Jones (2006).

that integrate RTM results have become increasingly common at energy industry technical meetings and in scientific publications. Without a doubt, RTMs have furthered our knowledge of diagenetic processes, products and their intrinsic and extrinsic controls. With the benefit of RTM experience and knowledge, we are now able to evaluate quantitatively a wide range of conceptual diagenetic models, originally depicted as sketches, in a way that was not practically possible only a decade ago. This has resulted in the refinement and rejection of existing diagenetic paradigms, the development of new predictive concepts, and a fresh perspective on process-based predictions of reservoir quality.

Despite this recent progress, RTM is still considered an emerging technology and, in the authors' experience, is not routinely being used by industry to help predict and risk reservoir quality. Four key barriers to the penetration and commercial application of RTM in the energy industry are: (1) demonstrated predictive capability; (2) business impact and efficiency; (3) access to specialist users; and (4) tool compatibility.

Demonstrated predictive capability is the single biggest hurdle for future RTM business application. Published studies have primarily focused on using RTMs to investigate diagenesis in generic models, whereas application to outcrop and subsurface reservoir case studies are comparatively rare. Furthermore, the accuracy of published RTM reservoir quality predictions are generally described in terms of broad qualitative

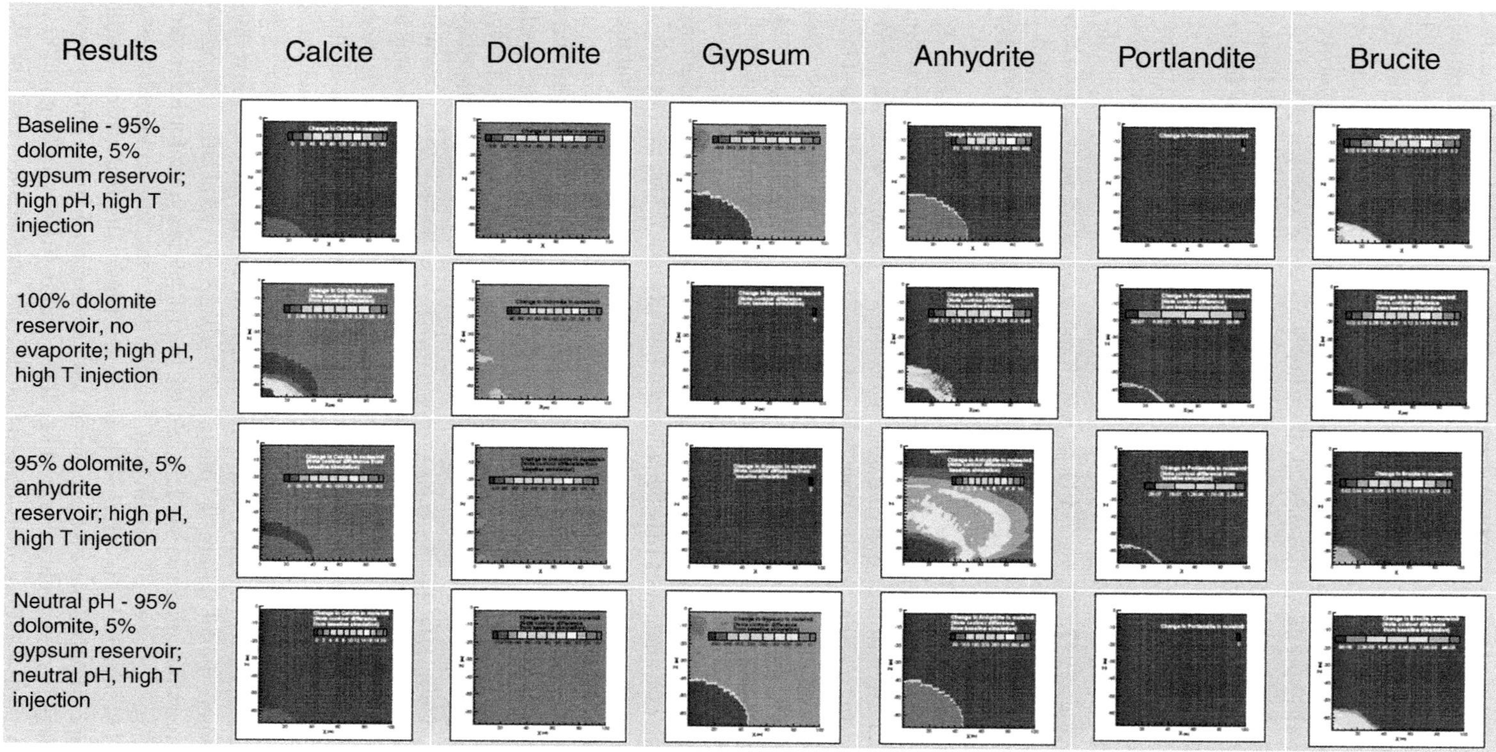

Figure 4.53 Summary of 2D RTM showing the impact of rock composition and injected steam pH on reservoir alteration after 6–12 months. Well bore is on the lower left and the scale of the model is 100 m. From Meddaugh et al. (2012). Used with permission from SPE.

outcrop- to reservoir-scale patterns of diagenetic products, and typically for only a subset of processes (Table 4.1). Quantitative comparison of 2D RTM results of dolomite abundance to that observed in reservoir zones for a region of the Ghawar field is a timely progression towards more transparent benchmarking of predictions (Lu and Cantrell, 2016). However, this is still a significant step removed from demonstrating predictive capability beyond prior knowledge. In the energy industry, rigorous comparison of pre-drill prediction to post-drill reality from static and dynamic data is the norm by which reservoir characterization business value is assessed. RTM has yet to build an inventory of demonstrated success from which to justify and grow future business application.

The majority of RTM studies have focused on carbonate reservoir quality issues over the past 15 years, as evidenced in this chapter. Application to siliciclastic reservoir diagenesis and reservoir quality appeared to have drawn less attention (e.g., Gallo et al., 1998; Bolton et al., 1999; Thyne, 2001). In fact, RTM models for siliciclastic reservoirs could face more challenges: more mineral phases require consideration, and uncertainty in extrapolating thermodynamic and kinetic data from the laboratory to geological conditions.

Business impact and efficiency describes where in the lifecycle of an asset RTM is likely to have the most influence, and any notable challenges associated with technology deployment. In exploration, reservoir quality is one of a number of hydrocarbon play elements used for risking. Application of RTM to predict diagenesis, itself only one component of reservoir quality risking, would be extremely uncertain, especially in a frontier basin, and rarely capable of confidently informing exploration risk. Using RTM to investigate the extent (local vs. field-wide) of unexpected porosity occluding diagenesis would be one exception that could impact the decision to drill additional exploration wells in the play and their location. Exploration discoveries with unusual styles of diagenesis with rare analogues may benefit from investigative RTM, for example, lacustrine carbonates of Brazil deposited in highly alkaline waters with the precipitation and dissolution of stevensite (Tosca and Wright, 2015).

In the asset lifecycle, new-field development, so called greenfields, present the greatest opportunity for RTM to influence multi-billion dollar investment decisions. Subsurface uncertainty in greenfields is high, because data are relatively sparse, typically consisting of a 3D seismic survey and a few appraisal wells. When the diagenetic modification of reservoir quality is a key uncertainty, for example in many carbonate reservoirs, RTM has the potential to help generate distinct alternative reservoir quality scenarios for production forecasting and development planning. There is relentless pressure to optimize capital efficiency and reduce 'cycle time' between field discovery and first production. This presents a challenge for RTM deployment, which would introduce an additional step to the current reservoir characterization workflow and requires access to specialist staff. Furthermore, RTM often provides the most informative results when model design is integrated with geochemical and petrographical analysis of core, which often arrives late relative to project deadlines for reservoir characterization prior to reservoir simulation. Finally, it is tough to build business confidence in RTM realizations that depart from those derived from traditional characterization methods biased by reservoir and outcrop analogues. This is because of the acknowledged uncertainly in several key RTM inputs (Jones and Xiao, 2005, 2006), the entrenched perception that diagenesis is not predictable, and the lack of documented RTM predictive success. The situation is logical, but unfortunate, because modeling to illustrate viable subsurface uncertainty with RTM may identify material business risk.

Development planning for greenfields often includes investigating production mechanisms that involve injection of fluids into the reservoir or aquifers. Depending on the mineralogy of the reservoir and the injected fluid composition, operationally induced diagenesis can modify injectivity (formation damage or well bore scaling) and/or fluid breakthrough to production wells. Examples include reinjection of sour gas in the Tengiz field (Darmentaev et al., 2010) and steam injection in the Wafra field (Meddaugh et al., 2012). In contrast to predicting reservoir quality that evolved over geological timescales, application of RTM to investigate diagenesis induced by field operations is a more tractable problem with much less uncertainty. The composition, properties and rates of injected fluids are known, as are the rock properties around the borehole. In this sense the application of RTM to investigate operational diagenesis is similar to current reservoir flow simulations conducted at the borehole scale.

In the early- to mid-production phase of an asset, the quantity of data (static and dynamic) combined with traditional characterization methods and a simulation model that matches production history is generally sufficient for effective reservoir management. It is unlikely that RTM is capable of reducing uncertainty in production decisions related to the distribution of reservoir quality. Infill drilling tactics are typically tested by-the-bit supported by forecasts from reservoir models and sometimes 4D seismic data. Evaluating the expansion of production to undrilled regions of a field has more in common with greenfield scenarios, but with the added benefit of characterization concepts being refined by additional static and particularly dynamic data. Application of RTM to support future phases of field development is best justified when the area being evaluated is expected to include different styles of diagenesis.

Later in field life, so-called 'brownfields' may consider investing in improved oil recovery (IOR) or enhanced oil recovery (EOR) schemes with injected fluids to improve hydrocarbon recovery. Uncertainty in performance and investment costs for IOR/EOR schemes are both high, creating an ideal situation for RTM to explore operationally induced diagenesis as described above for greenfields. Furthermore, field pilot projects are often used to evaluate the effectiveness of an IOR scheme prior to full field investment, and these provide rich high-density data-sets to evaluate RTM predictions of operationally induced diagenesis (Meddaugh et al., 2012).

RTM is a highly specialist skill-set that is currently rare in the energy industry, where static geological models of reservoirs are constructed by geoscientists and then simulated for dynamic performance by engineers. Effective RTM model design also requires broad geological knowledge of the reservoir, including its burial history (structure, stratigraphy and fluid distribution), and the ability to specify rock and fluid properties that are significantly different from those measured in the reservoir today, based on appropriate analogues. Presently, oil industry technical consulting companies are not offering RTM services, so the current skill gap will take time to bridge if demand for this technology grows.

Tool compatibility is an issue because today RTM softwares are primarily research codes that were not originally designed for integration with energy industry software, which itself has migrated towards single software platforms that span geological interpretation to reservoir simulation. Experience demonstrates that frequent porting between software is inefficient and, when compounded with a population of few specialist users, can significantly impede project progress. Furthermore, the present suite of RTM software is primarily designed for simulations of the present earth state. This is useful for investigating operational diagnesis, but presents a challenge for simulating reactive flow over geological timescales where changes in the earth state (stratigraphic framework, rock and fluid properties) occur due to deposition, erosion and tectonics. Stonger coupling of functionality found in basin models with that of RTM models is needed to address this issue.

What is the future role of RTM technology to predict reservoir quality? In one future, and probably the most likely based on projection of the status quo, RTM will remain largely a research tool predominantly used in academia with rare commercial application in the energy industry. The majority of this research will continue to focus on carbonates because of their susceptibility to diagenesis and reservoir quality uncertainty. Knowledge from RTM research, mostly based on generic examples and outcrop analogues, will slowly trickle down to indirectly influence interpretations of reservoir quality. Industry geoscientists will generally be wary of how and when to leverage digital analogues of diagenesis developed with RTM. Outcrop and reservoir analogues will continue to sway expert opinion in the interpretation and risking of reservoir quality. In contrast, reservoir engineers will be more open to RTM investigations of injected fluids, but it may be harder for them to access this technology, which to date has tended to reside with geoscientists.

In an alternative future, RTM disrupts the status quo and becomes a routine technology in the reservoir characterization workflow. Growing industry confidence in value-adding RTM predictive capability will be the tipping point. Demonstration will require disciplined benchmarking of RTM to predict diagenesis in laboratory experiments and in outcrop and reservoir analogues, at a scale and precision that is relevant for business decisions. Academia will be the principal driving force in this effort, but success will require strong collaboration with industry partners. Interest from industry will be skewed towards the super majors and large national companies in which carbonate reservoirs feature prominently in their portfolios. The current business landscape is persuasive, with numerous multibillion barrel carbonate greenfields awaiting development (for example, pre-Salt fields in Brazil) and the consideration of IOR/EOR schemes to boost recovery in several super giant Middle East reservoirs. However, the rapid growth of unconventional resources, the continued exploration success in deepwater siliciclastic reservoirs, combined with fluctuating oil and gas prices, all present a challenge to greater investment in RTM research. Any early RTM predictive success may trigger the emergence of industry-savvy start-up companies providing dedicated consulting with compatible user-friendly software. This is a well-trodden path that other technologies have successfully followed to disrupt traditional subsurface interpretation work flows.

In summary, RTM will continue to positively influence reservoir quality prediction. The reach and impact of this technology is described in two alternative futures. Which of these will transpire will depend on demonstrating RTM predictive capability that adds value to subsurface investment decisions.

Acknowledgements

We thank our ExxonMobil colleagues for their contributions to this chapter, including Lee Esch, Sean Guidry, Joel Collins, Mike Nicholas, Jeremy Jameson, Jon Kauffman, Sherry Stafford, Mauro Lo Cascio, Dave Leary, Mike Hayes, Shawn Fullmer, Bob Alway,

Scott Parker, Will Maze, Gary Teletzke, John Wilkinson, Mike Weaver, Jim Markello, and Jim Anderson. We also thank our academic colleagues for the fruitful collaborations in this work, including Fiona Whitaker, Al-Hedel Anwar, Katherine Cooper from University of Bristol, Faye Liu, Peng Lu and Chen Zhu from Indiana University, Enrique Gomez-Rivas from Universitat Autònoma de Barcelona, and Robert Goldstein, Zhaoqi Li, and Evan Franseen from University of Kansas. We also would like to thank Drs Tianfu Xu and Karsten Pruess from Lawrence Berkeley National Laboratory for their generous support in using the ToughReact program, and Fitra Rochendi from ExxonMobil for redrawing some of the figures. Careful reviews by numerous reviewers for our previously published works, as well as recent reviews by Lee Esch, Fiona Whitaker, and Tianfu Xu significantly improved the manuscript. ExxonMobil's support and permission to publish this paper is greatly appreciated.

References

Adams, J.E. and Rhodes, M.L. (1960). Dolomitization by seepage reflux. *American Association of Petroleum Geologists Bulletin* 44: 1912–1920.

Al-Abduwani, F.A.H., Shirzadi, A., van den Broek, W.M.G.T., and Currie, P.K. (2005) Formation damage vs. solid particles deposition profile during laboratory-simulated produced-water reinjection. *SPE Journal* 10 (2): 138–151.

Al-Helal, A., Whitaker, F., and Xiao, Y. (2012). Modeling dynamic interactions between geothermal and reflux circulation: The importance of temperature in controlling dolomitization and anhydrite cementation. *Journal of Sedimentary Research* 82: 196–215.

Arvidson, R. and Mackenzie, F. (1999). The dolomite problem: Controls of precipitation kinetics by temperature and saturation state. *American Journal of Science* 299: 257–288.

Audigane, P., Gaus, I., Czernichowski-Lauriol, I., et al. (2007). Two dimensional reactive transport modeling of CO_2 injection in a saline aquifer at the Sleipner site, North Sea. *American Journal of Science* 307: 974–1008.

Audigane, P., Lions, J., Gaus, I., et al. (2008). Geochemical modeling of CO_2 injection into a methane gas reservoir at the K12-B field, North Sea. In: Carbon Dioxide Sequestration in Geological Media – State of the Science (ed. M. Grobe, J.C. Pashin, and R.L. Dodge), 1–20. AAPG Studies 59.

Badiozamani, K. (1973). The dorag dolomitization model, application to the middle Ordovician of Wisconsin. *Journal of Sedimentary Petrology* 43: 965–984.

Bethke, C.M. (1996). Geochemical Reaction Modeling. Oxford: Oxford University Press.

Bethke, C.M., Lee, M.-K., and Park, J. (1999). *Basin modeling with Basin2 (release 4).* Hydrogeology Program, Urbana Champaign, University of Illinois.

Bolton, E.W., Lasaga, A.C., and Rye, D.M. (1999). Long-term flow/chemistry feedback in a porous medium with heterogeneous permeability: kinetic controls of dissolution and precipitation. *American Journal of Science* 299: 1–68.

Budd, D.A. (1988a). Aragonite-to-calcite transformation during fresh-water diagenesis of carbonates: insights from pore-water chemistry. *Geological Society of America Bulletin* 100: 1260–1270.

Budd, D.A. (1988b). Petrographic products of freshwater diagenesis in Holocene ooid sands, Schooner Cays, Bahamas. *Carbonates and Evaporites* 3: 143–163.

Budd, D.A. (2002). The relative roles of compaction and early cementation in the destruction of permeability in carbonate grainstones: a case study from the Paleogene of west-central Florida, USA. *Journal of Sedimentary Research* 72: 116–128.

Budd, D. A., and Park, A. J. (2018). Formation of bed-scale spatial patterns in dolomite abundance during early dolomitization: Part I. Mechanisms and feedbacks revealed by reaction–transport modelling. *Sedimentology* 65: 209–234.

Cantrell, D., Swart, P., and Hagerty, R. (2004). Genesis and characterization of dolomite Arab-D reservoir, Ghawar Field, Saudi Arabia. *GeoArabia* 9: 11–36.

Collins, J.F., Putney, K.L., Harris, P.M., et al. (2005). A comparison of stratigraphic architecture, lithofacies, and reservoir quality in adjacent large and small isolated carbonate platforms: Tengiz and Korolev Fields, North Caspian Basin, Kazakhstan. *American Association of Petroleum Geologists Annual Meeting*, v. 89, Program Abstracts.

Collins, J.F., Kenter, J.A.M., Harris, P.M., et al. (2006). Facies and reservoir quality variations in the Late Visean to Bashkirian outer platform, rim, and flank of the Tengiz buildup, Pricaspian Basin, Kazakhstan. In: *Giant Hydrocarbon Reservoirs of the World: From Rocks to Reservoir Characterization and Modelling* (ed. P.M. Harris and L.J. Weber), 55–95. American Association of Petroleum Geologists Memoir 88.

Collins, J., Narr, W., Harris, P.M., et al. (2013). Lithofacies, depositional environments, burial diagenesis, and dynamic field behavior in a carboniferous slope reservoir, Tengiz Field (Republic of Kazakhstan), and comparison with outcrop analogs. In: Deposits, Architecture, and Controls of Carbonate Margin, Slope, and Basinal Settings (ed. K. Verwer, T.E. Playton, and P.M. Harris). SEPM Special Publication 105.

Cooper, K., Xiao, Y., Whitaker, F., et al. (2014). Reactive transport modeling of mixed fluid dolomitisation: Traditional coastal mixing zones and ascending mixing of fresh-mesohaline water. AAPG, April 4–9, Houston, USA.

Corbella, M., Ayora, C., Cardellach, E., and Soler, A. (2006). Reactive transport modeling and hydrothermal karst genesis: The example of the Rocabruna barite deposit (Eastern Pyrenees). *Chemical Geology* 233 (30): 113–125.

Corbella, M., Gomez-Rivas, E., Martin-Martin, J.D., et al. (2014). Insights to controls on dolomitization by means of reactive transport models applied to the Benicassim case study (Maestrat Basin, eastern Spain). *Petroleum Geoscience* 20 (1): 41–54.

Darmentaev, S., Yessaliyeva, A., Azhigaliyeva, A.M., et al. (2010). *Tengiz Sour Gas Injection Project*, SPE 139851-MS.

Davies, G.R, and Smith, L.B.J. (2006). Structurally controlled hydrothermal dolomite reservoir facies: An overview. *AAPG Bulletin* 90: 1641–1690.

Dickson, J.A.D. and Kenter, J.A.M. (2014). Diagenetic evolution of selected parasequences across a carbonate platform: late Paleozoic, Tengiz reservoir, Kazakhstan. *Journal of Sedimentary Research* 84: 664–693.

Donaldson, E.C., Chilingarian, E.V., and Yen, T.F. (1985). Enhanced Oil Recovery I – Fundamentals and Analyses. Amsterdam: Elsevier.

Enos, P. and Sawatsky, L.H. (1981). Pore networks in Holocene carbonate sediments. *Journal of Sedimentary Petrology* 51: 961–985.

Gabellone, T., Whitaker, F.F., Katz, D., et al. (2016). Controls on reflux dolomitisation of epeiric-scale ramps: insights from reactive transport simulations of the Mississippian Madison Formation (Montana and Wyoming). *Sedimentary Geology* 345: 85–102.

Gallo, Y.L., Bildstein, O., and Brosse, E. (1998). Coupled reaction-flow modeling of diagenetic changes in reservoir permeability, porosity, and mineral compositions. *Journal of Hydrology* 209: 366–388.

Garcia-Fresca, B. (2008). Paleohydrology and reflux seepage dolomitization in the Permian San Andres Fm (Guadalupe Mountains and Algerita Escarpment). PhD thesis, University of Texas, Austin, TX.

Garcia-Fresca, B. and Jones, G.D. (2011). The apparent stratigraphic concordance of reflux dolomite: New insights from synsedimentary reactive transport models. In: Carbonate Geochemistry: Reaction and Processes in Aquifers and Reservoirs (ed. S.A. Engel, P. J. Moore, and H. DuChene), 27–30. Karst Waters Institute Special Publication 16.

Garcia-Fresca, B., Lucia, F.J., Sharp, J.M., and Kerans, C. (2012). Outcrop-constrained hydrogeological simulations of brine reflux and early dolomitization of the Permian San Andres Formation. *AAPG Bulletin* 96 (9): 1757–1781.

Gaupp, R., Matter, A., Platt, J., et al. (1993). Diagenesis and fluid evolution of deeply buried Permian (Rotliegende) Gas Reservoirs, Northwest Germany. *AAPG Bulletin* 77 (7): 1111–1128.

Gaupp, R., Baunack, C., Pudlo, D., et al. (2005). Paleo oil- and gasfields in the Rotliegend of the North German Basin: effects upon hydrocarbon reservoir quality. *DGMK-Forschungsbericht* 593–8: 1–242.

Gomez-Rivas, E., Stafford, S.L., Lee, A.G.K., et al. (2010). Flow patterns of dolomitizing solutions in a buried carbonate ramp: the Benicassim case study (Maestrat Basin, NE Spain). *72nd European Association of Geoscientists and Engineers Conference and Exhibition, EarthDoc*, Barcelona, Spain, pp. K025.

Gomez-Rivas, E., Corbella, M., Martin-Martin, J.D., et al. (2014). Reactivity of dolomitizing fluids and Mg source evaluation of fault-controlled dolomitization at the Benicàssim outcrop analogue (Maestrat Basin, E. Spain). *Marine and Petroleum Geology* 55: 26–42.

Hanshaw, B.B., Back, W., and Deike, R.G. (1971). A geochemical hypothesis for dolomitization by ground water. *Economic Geology* 66: 710–724.

Hardie, L.A. (1987). Dolomitization: a critical view of some current views. *Journal of Sedimentary Petrology* 57: 166–183.

Humphrey, J.D. and Quinn, T.M. (1989). Coastal mixing zone dolomite, forward modeling and massive dolomitization of platform-margin carbonates. *Journal of Sedimentary Petrology* 59: 438–454.

Hutcheon, I. and Abercrombie, H.J. (1990). Fluid-rock interactions in thermal recovery of bitumen, Tucker Lake Pilot, Cold Lake, Alberta. In: Prediction of Reservoir Quality through Chemical Modeling, 161–170. AAPG Special Volumes.

Johnson, J.W., Nitao, J.J., Steefel, C.I., and Knaus, K.G. (2001). Reactive transport modeling of geologic CO2 sequestration in saline aquifers: The influence of intra-aquifer shales and the relative effectiveness of structural, solubility, and mineral trapping during prograde and retrograde sequestration. *Proceedings, First National Conference on Carbon Sequestration*, Washington, DC.

Jones, G.D. and Xiao, Y. (2005). Dolomitization, anhydrite cementation and porosity evolution in a reflux system. *American Association of Petroleum Geologists Bulletin* 89: 577–601.

Jones, G.D. and Xiao, Y. (2006). Geothermal convection in the Tengiz carbonate platform, Kazakhstan: Reactive transport models of diagenesis and reservoir quality. *American Association of Petroleum Geologists Bulletin* 90: 1251–1272.

Jones, G.D., Whitaker, F.F., Smart, P.L., and Sanford, W.E. (2000). Numerical modeling of geothermal and reflux circulation in Enewetak Atoll: Implications for dolomitization. *Journal of Geochemical Exploration* 69–70: 71–75.

Jones, G.D., Smart, P.L., Whitaker, F.F., et al. (2003). Numerical modeling of reflux dolomitization in the Grosmont Platform complex (Upper Devonian), Western Canada Sedimentary Basin. *AAPG Bulletin* 87: 1273–1298.

Jones, G.D., Whitaker, F.F., Smart, P.L., and Sanford, W.E. (2004). Numerical analysis of seawater circulation in carbonate platforms: II. The dynamic interaction between geothermal and brine reflux. *American Journal of Science* 304: 250–284.

Jones, G.D., Collins, J.F., Xiao, Y., et al. (2007). Geothermal convection at Tengiz: Reactive transport models of predictive diagenesis and evidence from the rocks. *AAPG 2007 Annual Meeting*, Long Beach California (abstract).

Jones, G.J. and Garcia-Fresca, B. (2010). Preservation of limestone in dolomitized carbonate evaporite reservoirs. 2010 AAPG International Convention and Exhibition, September 12–15, Calgary, Alberta, Canada.

Jones, G.J. and Xiao, Y. (2013). Rates of burial diagenesis driven by geothermal convection in pre-salt lacustrine carbonate reservoirs. *AAPG Bulletin* 97 (8): 1249–1271.

Kalfayan, L. (2001). *Production Enhancement with Acid Stimulation*. Houston: PennWell.

Karner, G.D. and Gamboa, L.A.P. (2007). Timing and origin of the South Atlantic pre-salt sag basins and their capping evaporites. In: Evaporites Through Space and Time (ed. B.C. Schreiber, S. Lugli and M. Babel), 15–35. Geological Society, London, Special Publication 285.

Kaufman, J.K. (1994). Numerical models of fluid flow in carbonate platforms: implications for dolomitization. *Journal of Sedimentary Research* A64: 128–139.

Kenter, J.A.M., Harris, P.M., Weber, L.J., et al. (2006). Late Visean to Bashkirian platform cyclicity in the central Tengiz buildup, Pricaspian Basin, Kazakhstan. In: *Depositional Evolution and Reservoir Development, Giant Hydrocarbon Reservoirs of the World: from Rocks to Reservoir Characterization and Modeling* (ed. P.M. Harris and L.J. Weber), 7–54. American Association of Petroleum Geologists, Memoir 88.

Kenter, J.A.M., Tankersley, T., Skalinski, M.T., et al. (2010). Tengiz Field (Republic of Kazakhstan) Unit 1 platform static model: Using a hybrid depositional–diagenetic approach. *Society of Petroleum Engineers* 139935: 1–7.

Knauss, K.G., Johnson, J.W., and Steefel, C.I. (2005). Evaluation of the impact of CO_2, co-contaminant gas, aqueous fluid, and reservoir rock interactions on the geologic sequestration of CO_2. *Chemical Geology* 217: 339–350.

Kohout, F.A., Henry, H.R., and Banks, J.E. (1977). Hydrology related to geothermal conditions of the Floridan Plateau. In: The Geothermal Nature of the Floridan Plateau (ed. K.L. Smith and G.M. Griffin), 1–34. Florida Department of Natural Resources Bureau Geology Special Publication 21.

Land, L. (1985). The origin of massive dolomite. *Journal of Geoscience Education* 33: 112–125.

Land, L. (1991). Dolomitization of the Hope Gate Formation (north Jamaica) by seawater: reassessment of mixing-zone dolomite. In: *Stable Isotope Geochemistry: A Tribute to Samuel Epstein* (ed. H.P. Taylor, J.R. O'Neil and I.R. Kaplan), 121–130. *Geochemical Society (London) Special Publication 3.*

Land, L., Salem, M., and Morrow, D. (1975). Paleohydrology of ancient dolomites: geochemical evidence. *American Association of Petroleum Geologists Bulletin* 59: 1602–1625.

Lee, A. (2003). 3-D numerical modeling of freshwater lens on atoll islands. *Proceedings, Tough Symposium 2003*, Lawrence Berkeley National Laboratory, Berkeley, CA, pp. 1–7.

Li, Z., Goldstein, R.H., and Franseen, E. (2013). Freshwater mesohaline mixing: A new scenario for dolomitization. *Journal of Sedimentary Research* 83: 277–283.

Lichtner, P.C. (1996). Continuum formulation of multicomponent-multiphase reactive transport. In: Reactive Transport in Porous Media (ed. P.C. Lichtner, C.I. Steefel, and E.H. Oelkers), 1–79. Reviews in Mineralogy, Mineral Society of America, 34.

Lisovsky, N.N., Gogonenkov, G.N., and Petzoukha, Y.A. (1992). The Tengiz oil field in the Pre-Caspian Basin of Kazakhstan (Former USSR) Supergiant of the 1980s. In: *Giant Oil and Gas Fields of the Decade 1978–1988* (ed. M.T. Halbouty), 101–122. AAPG Special Publication.

Liu, F., Ellett, K., Xiao, Y., and Rupp, J. (2013). Assessing the feasibility of CO_2 storage in the New Albany Shale gas play with potential enhanced gas recovery using reservoir simulation. *International Journal of Greenhouse Gas Control* 17: 111–126.

Liu, Y., Lu, P., Zhu, C., and Xiao, Y. (2011). Coupled reactive transport modeling of CO_2 injection in Mt. Simon Sandstone Formation, Midwest USA. *International Journal of Greenhouse Gas Control* 5: 295–307.

Logan, B.W. (1987). *The MacLeod Evaporite Basin, Western Australia*. AAPG Memoir 44. Tulsa, OK, American Association of Petroleum Geologists.

Loucks, R.G. (1999) Paleocave carbonate reservoirs: origins, burial-depth modifications, spatial complexity, and reservoir implications. *American Association of Petroleum Geologists Bulletin* 83: 1795–1834.

Lu, F.H. and Meyers, W.J. (1998). Massive dolomitization of a late Miocene carbonate platform: a case of mixed evaporative brines with meteoric water, Nijar, Spain. *Sedimentology* 45: 263–277.

Lu, P. and Cantrell, D. (2016). Reactive transport modelling of reflux dolomitization in the Arab-D Reservoir, Ghawar Field, Saudi Arabia. *Sedimentology* 63 (4): 865–892.

Lucia, F.J. and Major, R.P. (1994). Porosity evolution through hypersaline reflux dolomitization. In: *Dolomites* (ed. B. Purser, M. Tucker, and D. Zenger), 345–360. International Association of Sedimentologists Special Publication 21.

Machel, H.G. (2004). Concepts and models of dolomitization: a critical reappraisal. In: *The Geometry and Petrogenesis of Dolomite Hydrocarbon Reservoirs* (ed. C.J.R. Braithwaite, G. Rizzi, and G. Darke), 7–64. Geological Society Special Publication No 235.

Mathews, R.K. and Froelich, C. (1987). Forward modeling of bank margin carbonate diagenesis. *Geology* 15: 673–676.

Matsuda, H., Tsuji, Y., Honda, N., and Saotome, J. (1995). Early diagenesis of Pleistocene carbonates from a hydrochemical point of view, Irabu Island, Ryuku Islands: porosity changes related to early diagenesis. *American Association of Petroleum Geologists Memoir* 63: 35–54.

McClain, M.E., Swart, P.K., and Vacher, H.L. (1992). The hydrogeochemistry of early meteoric diagenesis in a Holocene deposit of biogenic carbonates. *Journal of Sedimentary Petrology* 62: 1008–1022.

Meddaugh, W.S., Osterloh, W.T., Gupta, I., et al. (2012). The Wafra Field First Eocene carbonate reservoir steamflood pilots: geology, heterogeneity, steam/rock interaction, and reservoir response. *SPE Annual Technical Conference and Exhibition*, San Antonio, Texas, 8–10 October 2012. SPE 158324-MS.

Melim, L.A., Anselmetti, F.S., and Eberli, G.P. (2001). The influence of pore type on permeability of Neogene carbonates. In: Subsurface Geology of a Prograding Platform Margin: Results of the Bahamas Drilling Project (ed. R.N. Ginsburg), 217–238. SEPM Special Publication 70.
Melim, L.A., Westphal, H., Swart, P.K., et al. (2002). Questioning carbonate diagenetic paradigms: evidence from the Neogene of the Bahamas. *Marine Geology* 185: 27–53.
Melim, L.A., Swart, P.K., and Eberli, G.P. (2004). Mixing-zone diagenesis in the subsurface of Florida and the Bahamas. *Journal of Sedimentary Research* 74 (6): 160–169.
Mello, M.R., Bender, A.A., Azambuja Filho, N.C., and de Mio, E. (2011). Giant sub-salt hydrocarbon province of the Greater Campos Basin, Brazil. *Offshore Technology Conference Brazil Paper #22818.*
Moore, C.H. (2001). Carbonate reservoirs: porosity evolution and diagenesis in a sequence stratigraphic framework. *Developments in Sedimentology* 55.
Oswald, E.J. (1992). Dolomitization of an upper Miocene reef complex, Mallorca, Spain. PhD dissertation, State University of New York, Stony Brook, NY.
Paterson, R.J., Whitaker, F.F., Jones, G.D., et al. (2006). Accommodation and sedimentary architecture of isolated icehouse carbonate platforms: insights from forward modeling with CARB3D+. *Journal of Sedimentary Research* 76 (10): 1162–1182.
Phillips, O.M. (1991). *Flow and Reactions in Permeable Rocks.* Cambridge: Cambridge University Press.
Posamentier, H.W., Allen, G.P., James, D.P., and Tesson, M. (1992). Forced regressions in a sequence stratigraphic framework: Concepts, examples, and exploration significance. *American Association of Petroleum Geologists Bulletin* 76: 1687–1709.
Powers, R.W. (1962). Arabian Upper Jurassic carbonate reservoir rocks. In: Classification of Carbonate Rocks (ed. W.E. Ham), 122–192. American Association of Petroleum Geologists Memoir 1.
Pruess, K., Xu, T., Apps, J., and García, J. (2003). Numerical modeling of aquifer disposal of CO2. *SPE (Society of Petroleum Engineers) Journal* March: 49–59.
Rezaei, M., Sanz, E., Raeisi, E., et al. (2005). Reactive transport modeling of calcite dissolution in the salt water mixing zone. *Journal of Hydrology* 311 (1–4): 282–294.
Reeckmann, S.A. and Gill, E.D. (1981). Rates of vadose diagenesis in Quaternary dune and shallow marine calcarenites, Warrnambool, Victoria, Australia. *Sedimentary Geology* 30: 157–172.
Ronchi, P., Ortenzi, A., Borromeo, O., et al. (2010). Depositional setting and diagenetic processes and their impact on the Visean–Bashkirian Kashagan carbonate platform (Pricaspian Basin, Kazakhstan). *American Association of Petroleum Geologists Bulletin* 94: 1313–1348.
Rudolph, K.W. and Lehmann, P.J. (1989). Platform evolution and sequence stratigraphy of the Natuna Platform, South China Sea. In: *Controls on Carbonate Platforms and Basin Development* (ed. P.D. Crevello, J.L. Wilson, J.F. Sarg, and J.F. Read), 353–361. SEPM Special Publication 44.
Saller, A.H. and Henderson, N. (1998). Distribution of porosity and permeability in platform dolomites: insight from the Permian of West Texas. *American Association of Petroleum Geologists Bulletin* 82 (8): 1528–1550.
Saller, A.H. and Henderson, N. (2001). Distribution of porosity and permeability in platform dolomites: Insight from the Permian of west Texas: Reply. *American Association of Petroleum Geologists Bulletin* 85: 530–532.

Saller, A., Rushton, S., Buambua, L., et al. (2016). Presalt stratigraphy and depositional systems in the Kwanza Basin, offshore Angola. *AAPG Bulletin* 100 (7): 1135–1164.
Sanford, W.E., Whitaker, F., Smart. P.L., and Jones, G.D. (1998). Numerical analysis of seawater circulation in carbonate platforms: I. Geothermal circulation. *American Journal of Science* 298: 801–828.
Sanford, W. and Konikow, L. (1989). Porosity development in coastal carbonate aquifers. *Geology* 17: 249–252.
Simms, M. (1984). Dolomitization by groundwater-flow system in carbonate platforms. *Transactions of the Gulf Coast Association of Geological Societies* 24: 411–420.
Smart, P.L., Dawans, J.M., and Whitaker, F. (1988). Carbonate dissolution in a modern mixing zone. *Nature* 335 (6193): 811–813.
Steefel, C.I. and Lasaga, A.C. (1994). A coupled model for transport of multiple chemical species and kinetic precipitation/dissolution reactions with applications to reactive flow in single phase hydrothermal system. *American Journal of Science* 294: 529–592.
Steefel, C.I., DePaolo, D.J., and Lichtner, P.C. (2005). Reactive transport modeling: An essential tool and a new approach for the Earth sciences. *Earth and Planetary Science Letters* 240: 539–558.
Sun, S.Q. (1995). Dolomite reservoirs: porosity evolution and reservoir characteristics. *American Association of Petroleum Geologists Bulletin* 79: 249–257.
Thyne, G, 2001. A model for diagenetic mass transfer between adjunct sandstone and shale. *Marine and Petroleum Geology*, 18, p. 743–755
Tillotson, B. (2003). Bed and facies scale selectivity during late-stage dolomitization: Lower Ordovician El Paso Group, Franklin Mountains, West Texas. AAPG Annual Convention, Salt Lake City, Utah.
Tosca, N.J. and Wright, V.P. (2015). Diagenetic pathways linked to labile Mg-clays in lacustrine carbonate reservoirs: a model for the origin of secondary porosity in the Cretaceous pre-salt Barra Velha Formation, offshore Brazil. In: Reservoir Quality of Clastic and Carbonate Rocks: Analysis, Modelling and Prediction (ed. P. Armitage et al.). Geological Society, London, Special Publication.
Tucker, M.E. (1993). Carbonate diagenesis and sequence stratigraphy. *Sedimentology Review* 1: 51–72.
Vacher, H.L. (1988). Dupuit-Ghyben-Herzberg analysis of strip-island lenses. *Geological Society of America Bulletin* 100: 580–591.
Vacher, H.L. and Quinn, T.M. (1997). Geology and Hydrogeology of Carbonate Islands. Amsterdam: Elsevier.
Vacher, H.L., Bengtsson, T.O., and Plummer, L.N. (1990). Hydrology of meteoric diagenesis: Residence time of meteoric ground water in island fresh-water lenses with application to aragonite-calcite stabilization rate in Bermuda. *Geological Society of America Bulletin* 102: 223–232.
Van Berk, W., Fu, Y., and Schulz, H-M. (2015). Temporal and spatial development of scaling in reservoir aquifers triggered by seawater injection: Three-dimensional reactive mass transport modeling of water–rock–gas interactions. *Journal of Petroleum Science and Engineering* 135: 206–217.
Van Wagoner, J.C., Posamentier, H.W., Mitchum, R.M., et al. (1988). An overview of the fundamentals of sequence stratigraphy and key definitions. In: *Sea-Level Changes: An Integrated Approach* (ed. C.K. Wilgus et al.), 39–45. Society of Economic Paleontologists and Mineralogists Special Publication 42.

Weber, J.L., Francis, B.P., Harris, P.M., and Clark, M. (2003). Stratigraphy, lithofacies and reservoir distribution, Tengiz field Kazakhstan. In: Permo-Carboniferous Carbonate Platform and Reefs (ed. W.M. Ahr, P.M. Harris, W.A. Morgan, and I.D. Sommerville), 351–394. SEPM Special Publication 78 and AAPG Memoir 83.

Whitaker, F.F. and Smart, P.L. (1993). Geochemistry of meteoric waters and porosity generation in carbonate islands of the Bahamas. *Geofluids II, 2nd International Conference on Fluid Evolution, Migration and Interaction in Sedimentary Basins and Orogenic Belts*, University of Belfast.

Whitaker, F. and Xiao, Y. (2010). Reactive transport modeling of early burial dolomitization of carbonate platforms by geothermal convection? *American Association of Petroleum Geologists Bulletin* 94 (6): 889–917.

Whitaker, F.F., Smart, P.L., and Johnston, V. (2004) A review of rates of surface lowering and early carbonate diagenesis. *Appendix to Carb3D+ User's Manual*. University of Bristol, UK.

Wicks, C.M. and Herman, J.S. (1996). Regional hydrogeochemistry of a modern coastal mixing zone. *Water Resource Research* 32: 401–407.

Wilson, A.M., Sanford, W.E., Whitaker, F.F., and Smart, P.L. (2001). Spatial patterns of diagenesis during geothermal circulation in carbonate platforms. *American Journal of Science* 301: 727–752.

Wilson, E.N., Hardie, L.A., and Phillips, O.M. (1990). Dolomitization front geometry, fluid flow patterns, and the origin of dolomite: The Triassic Latemar buildup, northern Italy. *American Journal of Science* 290: 741–796.

Wood, J.R. and Hewett, T.A. (1982). Fluid convection and mass transfer in porous sandstones – A theoretical model. *Geochimica et Cosmochimica Acta* 47: 1707–1713.

Wright, V.P. (2012). Lacustrine carbonates in rift settings: the interaction of volcanic and microbial processes on carbonate deposition. In: *Advances in Carbonate Exploration and Reservoir Analysis* (ed. J. Garland, J.E. Neilson, S.E. Laubach, and K.J. Whidden), 39–47. Geological Society Special Publication 370.

Xiao, Y. and Jones, D.J. (2006). Reactive transport modeling of carbonate and siliciclastic diagenesis and reservoir quality prediction. SPE paper #101669, *Conference Proceedings of the Abu Dhabi International Petroleum Exhibition and Conference* held in Abu Dhabi, UAE, 5–8 November 2006.

Xiao, Y. and Jones, D.J. (2007). Reactive transport modeling of limestone-dolomite transitions: Implications for reservoir connectivity. SPE paper #11199, *Conference Proceedings of the International Petroleum Technology Conference* held in Dubai, UAE, 4–6 December 2007.

Xiao, Y., Esch, W.L., Welton, J.E., et al. (2007). Predicting fault controlled diagenesis and porosity/permeability evolution in Rotliegendes Reservoir, Germany: Insights from reactive transport models. *2007 AAPG Convention*, Long Beach, California.

Xiao, Y., Xu, T., and Pruess, K. (2008). The effects of gas-fluid-rock interactions on CO_2 injection and storage: Insights from reactive transport modeling. *Proceedings of the 9th International Conference on Greenhouse Gas Technologies*, 16–20 November 2008, Washington, DC. *Energy Procedia* 1 (1): 1783–1790.

Xiao, Y., Maze, W., Teletzke, G., et al. (2009). Coupled reactive transport models of acid gas injection (AGI) in siliciclastic and carbonate reservoirs: Understanding fundamental controls on injection performance and storage security. *2009 International Petroleum Technology Conference* (IPTC), Doha, Qatar. SPE 13279.

Xiao, Y., Macleod, G., Advocate, D.M., et al. (2010). Controls on natural CO2 occurrence in geological formations and the implications on CO2 storage capacity and site selection. *Proceedings of the 10th International Conference on Greenhouse Gas Control Technologies*, September 19–23, Amsterdam, the Netherlands. *Energy Procedia* 4: 4688–4695.

Xiao, Y., van den Bosch, R., Liu, F., and Steffen, K. (2011). Performance evaluation in data rich Fayetteville shale gas plays – integrating physics-based reservoir simulations with data driven approaches for uncertainty reduction. *Conference Proceedings of the International Petroleum Technology Conference* held in Bangkok, Thailand, 7–9 February, SPE 14940: 1–19.

Xiao, Y., Jones, G.D., Whitaker, F.F., et al. (2013). Fundamental approaches to dolomitization and carbonate diagenesis in different hydrogeological systems and the impact on reservoir quality distribution. *The Sixth International Petroleum Technology Conference (IPTC)*, March 26–28, Beijing, China. SPE paper 16579.

Xu, T. and Pruess, K. (2001). Modeling multiphase fluid flow and reactive geochemical transport in variably saturated fractured rocks: 1. Methodology. *American Journal of Science* 301: 16–33.

Xu, T., Sonnenthal, E., Spycher, N., and Pruess, K. (2006). ToughReact's User's Guide: A Program for Non-Isothermal Multiphase Reactive Transport Geochemical Transport in Variely Saturated Geologic Media. Laurence Berkeley National Laboratory, Berkeley, CA.

Xu, T., Apps, J.A., Pruess, K., and Yamamoto, H. (2007). Numerical modeling of injection and mineral trapping of CO_2 with H_2S and SO_2 in a sandstone formation. *Chemical Geology* 242 (3–4): 319–346.

Yao, Q. and Demicco, R.V. (1997). Dolomitization of the Cambrian–Ordovician carbonate platform, southern Canadian Rocky Mountains: dolomite front geometry, fluid inclusion geochemistry, isotopic signature, and hydrogeological modeling studies. *American Journal of Science* 297: 892–938.

Zenger, D.H., Dunham, J.B., and Ethington, R.L. (ed.) (1980). *Concepts and Models of Dolomitization*. SEPM Special Publication 28.

5

Modeling High-Temperature, High-Pressure, High-Salinity and Highly Reducing Geochemical Systems in Oil and Gas Production

Guoxiang Zhang[1], Jeroen Snippe[2], Esra Inan-Villegas[3] and Paul Taylor[3]

[1] *Shell Exploration and Production Inc., Houston, TX, USA*
[2] *Shell Global Solutions International B.V., Amsterdam, The Netherlands*
[3] *Shell Global Solutions (UK), London, UK*

Geochemical processes taking place in hydrocarbon reservoirs of high temperature, high pressure, high salinity and highly reducing conditions during production, as well as the strategy and the approaches of modeling such geochemical systems, are discussed in this chapter. Three reactive transport modeling examples are presented. The first one focuses on well/reservoir scaling and clay swell during water injection; the second illustrates how microbial and/or thermochemical sulfur reduction is enhanced to generate H_2S during water injection; and the last is a modeling evaluation of wellbore cement corrosion.

5.1 Introduction

Since 'easy oil' is running out, hydrocarbon (HC) exploration and production are focused on technically challenging HC systems. One important challenge is to produce, effectively and economically, from deeper geological settings (e.g. up to ~8000 metres). Deep reservoirs are naturally at high temperature (High-T, up to ~200 °C), high pressure (High-P, up to ~1000 atmospheres) and high salinity conditions (High-S, formation brine salinity can reach up to halite saturation or higher). In addition, most hydrocarbon reservoirs are also under geochemically highly reducing (High-R) conditions. Therefore, the geochemical system of deep HC reservoirs can be characterized with four High (4-High) conditions: High-T, High-P, High-S, and High-R. Moreover, even 'easy oil' or shallow HC reservoirs could be relatively high in one or more of these parameters, compared with shallow geological systems (e.g. groundwater systems). These characteristics are quite unique and distinguishing from most other geochemical systems accessible to human beings. Therefore, understanding and modeling the HC reservoirs of 4-High conditions or 'less-than-4' high conditions is the objective of this chapter.

Reactive Transport Modeling: Applications in Subsurface Energy and Environmental Problems, First Edition. Edited by Yitian Xiao, Fiona Whitaker, Tianfu Xu and Carl Steefel.

From a geochemical perspective (i.e. the water–rock–HC interactions), 4-high reservoirs are normally either at thermodynamic equilibrium or at a stabilized diagenetic evolution in which geochemical reactions must be underway at very slow kinetic rates and engage a stabilized kinetic network. The geochemical system appears usually as a real or pseudo-steady-state condition as a result of the thermodynamic equilibrium or the stabilized diagenetic evolution. Production of HCs from 4-High reservoirs may have to rely on injection of water, gases (e.g. CO_2) or other chemicals as secondary or tertiary methods of HC recovery. Injection of those fluids introduces foreign chemicals into the reservoirs, which have gone through stabilized geochemical evolution for geological timescales. Therefore, the injection behaves as a strong driver capable of triggering geochemical reactions in the reservoir. Some of these reactions can be drastic. The geochemical reactions include re-speciation of reservoir brine, neutralization of acidic compounds and alkaline compounds, redox, dissolution/precipitation of minerals, cation exchange, as well as adsorption/desorption. As the result of these reactions, the composition of the reservoir brine changes, volumes of minerals or mineral composition of the reservoir rock changes, and consequently, pores can be clogged. Scaling occurs if precipitation takes place within the wellbore or the near-wellbore region, and wellbore integrity can be affected if significant dissolution takes place at the wellbore (dissolution of tubing and casing material as well as cement leads to corrosion under the in-situ condition), or near the wellbore region. Strong dissolution near wellbore regions can loosen up the wellbore and lead to loss of the well. Injection may also cause an enhanced sulfate reduction under bacterial or thermochemical driving forces, leading to reservoir souring problems. Clay may swell if fresh water or brackish water is introduced into a saline reservoir, and clay swelling is known to cause formation damage by reducing the porosity and permeability, but may also pressurize the reservoir, leading potentially to positive effects on production.

Understanding and evaluation of the geochemical reactions during HC production from 4-High reservoirs are necessary to control and manage the relevant risks. Geochemical reactive transport modeling is a useful and unique tool for understanding and evaluating such processes.

5.2 Drivers of the Geochemical Reactions in 4-High Reservoirs During Oil and Gas Production

5.2.1 High Temperature

Geochemical reactions are either endothermic or exothermic. Endothermicity or exothermicity determines the temperature dependency of a reaction in terms of the status of thermodynamic equilibrium and the reaction rate. It also appears as the temperature-dependent solubility of minerals and gases. Changes in temperature must drive the reaction going either forwards or backwards from one thermodynamically stable status (i.e. thermodynamic equilibrium) to reach another. Normally, the temperature of 4-High reservoirs under discussion is higher than 50 °C, and the injection water is normally at ambient temperature (~25 °C), and often lower than 20 °C if seawater is used. Mixing of the waters at different temperatures will lead to redistribution of the aqueous species and, in turn, change the solubility or the saturation status of the relevant

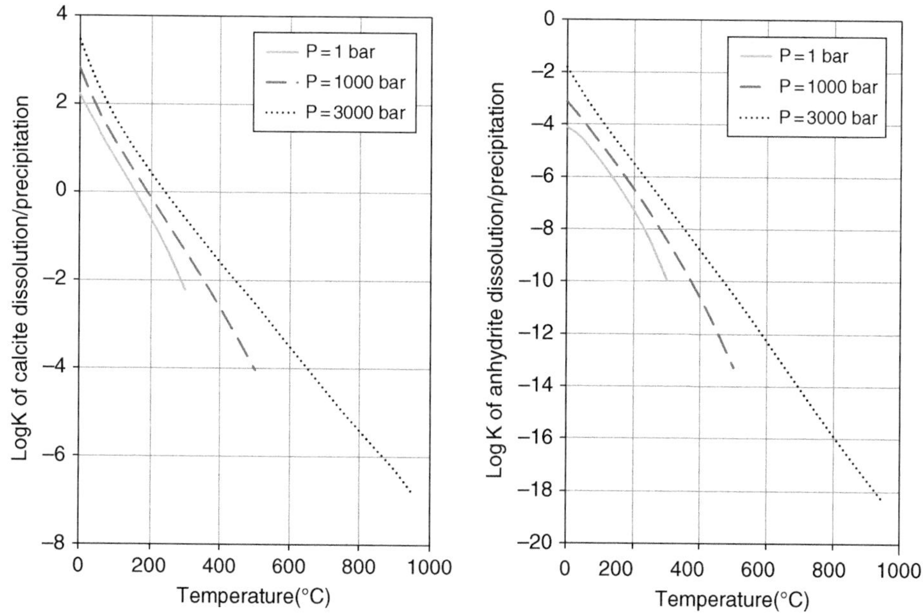

Figure 5.1 Thermodynamic equilibrium constant (K) of dissolution/precipitation of calcite ($CaCO_3 \rightleftharpoons Ca^{2+} + CO_3^{2-}$) and anhydrite ($CaSO_4 \rightleftharpoons Ca^{2+} + SO_4^{2-}$) as a function of temperature and pressure. The trends of LogK's are evaluated with SUPCRT92 (Johnson et al., 1992) based on the thermodynamic properties available in the paper. *Source:* Data from Johnson 1992.

minerals, most sensitively carbonates and sulfates (calcite and anhydrite), leading to precipitation and/or dissolution. We know that solubility of some minerals (e.g. anhydrite) changes significantly when temperature varies from ambient to high and vice versa. The thermodynamic equilibrium constant of anhydrite dissolution/precipitation is plotted as a function of temperature in Figure 5.1, showing that the temperature dependency of this mineral is very pronounced. Significant dissolution or precipitation can be expected, should injection or production cool down the reservoir. Solubility of calcite becomes much lower at higher temperatures, thus injection of calcite-saturated water (this is the most likely case) into a 4-High reservoir can induce strong precipitation of calcite, causing scaling problems. Similar effects also exist for other minerals and must be evaluated when producing from these reservoirs.

5.2.2 High Pressure

Effects of pressure on the thermodynamic equilibrium of dissolution/precipitation of calcite and anhydrite are also shown in Figure 5.1. The mass action of a reaction involving reactants and/or products of gaseous phase is determined with the partial pressure (under ideal condition) or fugacity (otherwise) of the reactants/products in gaseous phase. It can easily be understood and described as that the concentration of gases (appears as partial pressure under ideal conditions) increases as the gases are compressed. This phenomenon can be approximately described using Henry's law (or like formulations). However, under high-pressure conditions, better evaluation requires use of gas fugacity, rather than partial pressure, to calculate the mass action term, and therefore sophisticated PVT models considering interactions of gas molecules are employed in recent numerical simulators to evaluate the gas fugacity. Obviously, pressure is a key variable in the PVT models. Effects of pressure on reactions without gaseous phases involved are less pronounced but not negligible within the pressure range of 4-High reservoirs (i.e. can change from atmospheric for injection water up to 1000 bars or higher in the reservoir). Precipitation of halite due to depressurization from halite-saturated brine is a good example. Another example is the calcite precipitation at producer where pressure is lowered and CO_2, as well as other gases, escapes from the brine, elevating the pH. This phenomenon is not only a result of CO_2 degassing at lower pressure, but also a result of calcite solubility decrease at lower pressure. A lot of minerals possess similar behaviours (i.e. the solubility decreases at lower pressure). Precipitation caused by the pressure change is considered a common cause of producer scaling.

5.2.3 Salinity, pH and Alkalinity

Water used for injection is normally pH neutral (e.g. pond water) or slightly alkaline (e.g. seawater), at pH7.5–8.5. pH of the water in most reservoirs (e.g. most Gulf of Mexico reservoirs and Middle East reservoirs) is normally slightly acidic. The acidic condition is a result of (1) souring from kerogen cracking or diagenesis, (2) formation water being concentrated during sedimentation, burial and/or diagenetic evolution to high salinity, and/or (3) formation water being buffered by reservoir formation minerals. The pH of these reservoirs ranges normally between 5.0 and 6.0. Mixing of the injection water at pH neutral or slightly alkaline and the slightly acidic formation waters in the formation must lead to rebalancing of the pH-sensitive species, with some drastic reactions including dissolution and precipitation of relevant minerals.

Formation water in deep reservoirs (e.g. most Gulf of Mexico reservoirs and Middle East reservoirs) is normally very saline. The salinity is normally higher than 100,000 ppm, and can reach halite saturation or halite oversaturation (i.e. total dissolved ions up to 300,000 ppm and higher). Normally, concentrations of divalent cations (e.g. Ca^{2+} and Mg^{2+}) are high. Mixing of injection water at a salinity of seawater or lower with formation water of salinity 100,000 ppm up to 300,000 ppm or higher can promote a series of reactions, including precipitation of carbonate and sulfate minerals as well as cation exchanges.

Carbonates (CO_3^{2-} and HCO_3^-) in injection water are normally at CO_2 saturation under atmospheric pressure and the CO_2 content in the atmosphere. Carbonates in the reservoir formation water can be depleted due to the high concentration of Ca^{2+} and the high temperature (at which solubility of carbonate minerals is much lower), or enriched if the HC is oxidized/biodegraded, releasing CO_2. At high temperature, the solubility of carbonate minerals such as calcite becomes very low, and the precipitation of carbonate minerals removes most of the carbonates from the water. Under such conditions, calcite self-scaling can be possible during depressurization at producer while CO_2 degasses, leading to pH increase and rebalancing of the carbonate species.

5.2.4 Contrast in Redox Potential

Most HC reservoirs are geochemically very reducing because the HC is the strongest reducing agent (electron donor). The appearance of redox-sensitive minerals in the

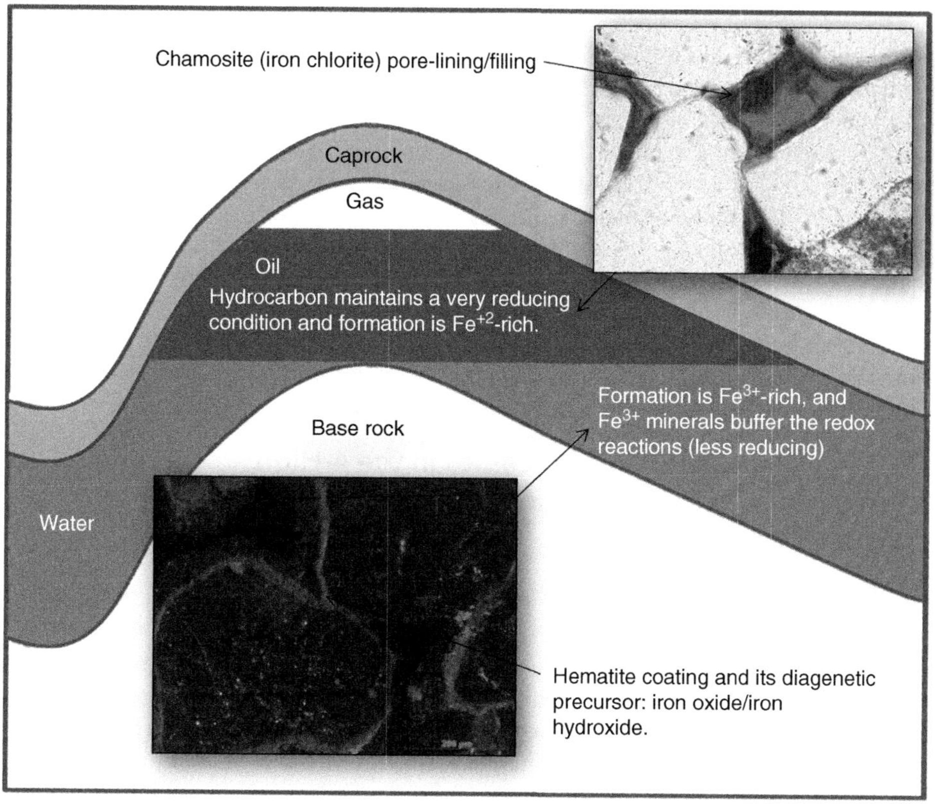

Figure 5.2 Illustration of dominant redox profile for a typical Gulf of Mexico reservoir.

reservoir formation, such as hematite (bearing iron-iii, also noted as Fe^{3+}, Fe(III) or iron (III)), siderite (bearing iron-ii, also noted as Fe^{2+}, Fe(II) or iron (II)) and pyrite (bearing iron-ii), can usually be considered as a sign of such extreme reducing conditions. Iron-ii minerals, such as iron-chlorite, siderite and/or pyrite, are normally products of reducing their diagenetic precursor in oxidized form, such as hematite. This conversion happens at least at the surface of the minerals (i.e. the HC–mineral interface), and the result is the alteration of minerals in the oxidized form into a reduced form at the mineral surface.

As an example, hematite in a Gulf of Mexico (GoM) deep-water reservoir (M-formation) was found altered, thoroughly, into iron-chlorite within the oil leg (interval filled up with HC), while the hematite within the water leg (an interval just below the oil leg and filled up with water) remains unchanged (Figure 5.2).

Difference in the redox potential between injection water and formation water (brine) can usually be reflected by the difference in concentrations of redox-sensitive species, such as iron. Concentration of iron, mainly ferrous iron (Fe^{2+}), can be used as an indication of the redox status of the geochemical system. In oxidizing water, such as seawater as a common injection water, iron appears as Fe^{3+}, which has very low solubility, and thus the concentration of total iron can be as low as 10^{-8} molal (mol/kg water) or lower. In a reducing geochemical system, iron appears as Fe^{2+} and the concentration can reach 10^{-3} molal or higher.

Concentrations of another set of redox-sensitive species, sulfur species, are also different under different redox conditions. In seawater, dissolved sulfur appears mainly as sulfate (up to 10^{-2} molal or higher), and concentration of the reduced sulfur species is negligible (10^{-8} molal or lower). In the HC reservoir formation water, dissolved sulfur can also appear as sulfate, but with concentrations much lower than the concentration in seawater, but the concentrations of reduced sulfur species (e.g. dissolved H_2S or HS^-) can be higher (10^{-4} molal or higher).

Mixing of the waters with significant contrast in redox potential can induce a rebalance of electrons between species of different redox status, leading to dissolution and precipitation of redox-sensitive minerals, such as hematite, magnetite, or pyrite, and to activation or acceleration in the kinetic rate of slow reactions, such as thermochemical sulfate reduction (TSR), leading potentially to souring and other risks.

5.3 Typical Geochemical Processes in the 4-High Reservoir During HC Production and the Impacts on Production

Geochemical reactions induced and promoted by injecting foreign fluids (fresh or brackish water) during HC production from 4-High reservoirs can cause well/reservoir scaling, wellbore corrosion enhancement, reservoir caprock integrity degradation, reservoir souring, and clay swelling, leading to permeability reduction and reservoir pressurization. The impact from these reactions on oil and gas production can be huge and potentially lead to loss of wells or even abandonment of the field. Without thorough understanding and appropriate evaluation, risks of well-loss due to scaling, failure of wellbore integrity, reservoir souring, clay swelling and/or failure of caprock integrity could not be effectively managed. Typical scenarios of these geochemical reactions taking place are elaborated as follows.

5.3.1 Scaling of Wells and Near Wellbore Formation Rocks by Carbonate Precipitation

As calcite-saturated water is injected into a hot formation and heated up, calcite precipitates in the injector wellbore and the near-wellbore zone of the reservoir. Precipitation of calcite can be enhanced as an increase in calcium concentration occurs when injecting water meets the formation water in which the calcium could have been enriched during the deposition, burial and compaction of the formation sediments. Mixing of the injection water and the formation water at downhole can thus immediately lead to precipitation of calcite within the wellbore or in the near-wellbore zone. Although calcite solubility can slightly increase at higher pressure and the pH can be lowered when mixing with the acidic formation water, calcite can still precipitate because the temperature dependency of calcite solubility is much stronger than that of pressure dependency. Calcite solubility at downhole temperature (e.g. up to 200 °C) can decrease significantly, up to 1000 times than that at ambient temperature, while calcite solubility increases less than two-fold when pressure increases from 1 bar to 2000 bars. Thus the temperature dependency overcomes, usually, the effect of the pressure dependency. The effects of a decrease in pH can also be easily compensated/overcome by the effects of the increase in calcium concentration. Therefore, at downhole conditions, the overall effect will be a decrease in calcite solubility and an increase in calcite saturation due to the high calcium concentration that normally exists in 4-High reservoirs. The scaling potential of calcite and other carbonate minerals can easily be underestimated in the production technology community, because the evaluation is traditionally performed based on the calculated saturation indices of the minerals using batch geochemical simulators (basically a speciation simulator), and this parameter for calcite or other carbonate minerals is usually not higher than that of another typical scaling mineral, barite. However, the amount of calcite precipitation can be huge – many times more than that of barite – especially under fluid-dynamic condition (i.e. carbonate is replenished from the injection water and calcium is usually available at a high concentration in the reservoir formation brine).

5.3.2 Well Scaling by Precipitation of Sulfate Minerals

In addition to calcite, barite ($BaSO_4$) is considered as another common scaling mineral because: (1) the solubility of this mineral is very low, and precipitation is very rapid; and (2) Ba^{2+} and SO_4^{2-} are rarely balanced because SO_4^{2-} is usually dominant in most geochemical system over Ba^{2+}. Mixing of Ba^{2+}- and SO_4^{2-}-unbalanced waters (e.g. concentrations of SO_4^{2-} and Ba^{2+} are not molal-equivalent to each other) will always lead to $BaSO_4$ precipitation. However, the concentration of barium in the formation water of most reservoirs is low relative to the concentration of major divalent cations such as calcium and magnesium. Therefore, the total precipitation of barite can be limited compared with the precipitation of calcite. Nevertheless, a small but localized precipitation volume can be sufficient to damage a well (e.g. complete plugging of tubing, or loss of operability of a safety-critical device in the well). Due to the low and almost pH-independent solubility of barite, the mineral is almost impossible to remove once it has formed, so that remediation options are limited to very expensive operations such as milling out the scale from the well, rather than relatively inexpensive solutions such as pumping down an acidic fluid (usually effective for calcite scale).

As barite is commonly used to increase the density of the drilling mud in order to maintain wellbore mechanical stability during drilling, water samples from downhole are usually contaminated by the barite mud. Therefore, measured barium concentration of most water samples is unrealistically high, and the barite scaling potential is normally overestimated within the production technologists (PT) community based on the calculated saturation index using scaling software such as OLI/ScaleChem (http://www.olisystems.com).

5.3.3 Scaling Due to Precipitation of Other Minerals

Solubility of quartz is much higher at lower temperature. Where cool water is injected, this mineral will dissolve in the low-temperature zone (i.e. the near-wellbore zone) and release silica into the formation water. Solubility of feldspar minerals (e.g. K-feldspar and albite) is lower at low temperature and they can precipitate at the cooling zone. Silica released from dissolution of quartz can, in turn, be utilized in precipitation of feldspars. This scenario is common if injection is performed in a clastic reservoir. At the same time, clay minerals (e.g. smectite and illite) can also precipitate due to the increase in silica concentration. Precipitation of clay minerals will take place inside the reservoir more than inside the injector wellbore, due to the reaction kinetic rates of these minerals being much slower than those of the carbonate, sulfate and felsic minerals. Precipitation of feldspar and clay minerals can also consume some cations such as Ca^{2+}, Na^{+}, K^{+} and Mg^{2+}, but this consumption of cations would not significantly influence the water composition because the precipitation rate is slow and the rate of concentration change is usually smaller than the changes caused by flow and transport. The amount of precipitation can be orders of magnitude lower than the amount of calcite precipitation. Therefore, the effects of dissolution and precipitation of these minerals on porosity change can be limited.

5.3.4 Scaling Due to Combined Precipitation of Multiple Minerals, Solid Solution and/or Fines Migration

Multiple minerals can become oversaturated under the same condition, such as when injection water meets formation water in the reservoir. Therefore multiple minerals can potentially precipitate at the same time. However, the kinetic rates of these reactions can be very different: the kinetic rate constant of one mineral (e.g. calcite) can be many orders of magnitudes higher than for another (e.g. smectite). Interference can take place when precipitation of two minerals competes for the same dissolved species. In this case, the one with faster kinetics will dominate the precipitation and consumes the dissolved species in competing precipitation with other (slower) minerals. The one with slower kinetics will quickly reduce the degree of oversaturation (i.e. by lowering the saturation index), and may eventually be equilibrated or unsaturated, and will stop precipitation or start dissolution.

Barite and calcite can precipitate in the same geochemical system. Precipitations of both minerals are rapid (faster than most other minerals) and thus can take place within the injector wellbore during the injection of cool, fresh or seawater/brackish water into a hot, saline formation water. Barite forms very quickly, while calcite also forms quickly and in much larger amounts. The result will be co-precipitation of barite and calcite, or some solid solution involving other metals, such as Sr and Cd, forming $(Ba,Sr)CO_3$, $Ba(SO_4,CrO_4)$, $(Ba,Sr)SO_4$, and $(Cd,Ca)CO_3$ (Prieto et al., 1997), or $(Ba, Ca)(CO_3, SO_4)$ solid solutions (Tesoriero and Pankow, 1996).

Barite can form as tiny suspended particles (mainly in nano-scale) at the early nucleation stage and stays in solution (rather than precipitates) within a certain range of saturation. As calcite forms, usually in much larger amounts than barite, calcite nuclei are more concentrated than those of barite, and eventually solid calcite is formed and dropped out first from the solution with some barite nuclei within it, or forming a solid solution. This co-precipitation of barite and calcite can be seen very commonly. In this case, the scale is usually identified as barite scale because: (1) the scale evaluation performed before the injection usually shows that barite saturation index is high and much higher than calcite, which is true; and (2) the XRF (X-ray-fluorescence) results show barium clearly, which is also true. However, calcite content is usually much higher than barite in calcite co-precipitations, or Ca^{2+} is usually the dominant cation in the solid solution.

Another scenario is that the calcite or barite glues the suspended nano- to micrometre-scale (NMS) clay particles to form aggregates of clay in the solution or scales.

Clay minerals stay for a long time in solution as a colloidal phase, which appears as NMS particles. Obviously, the solution must be oversaturated with respect to some of the clay minerals. Within the wellbore or the near-wellbore region, clay minerals can also derive from the drilling mud, in which the clay could have been remained in the colloidal condition for a long time. In addition, larger clay particles (aggregates) can enter the producer wellbore. Precipitation kinetic rates for clay minerals are much lower than for calcite and barite. Without involvement of calcite and/or barite, many of these clay components (NMS clay particles and aggregates) will stay in the solution forever without formation of scales; however, they can be cemented by calcite or barite as these minerals are oversaturated. In this case, clay minerals contribute a larger portion of volume, and calcite and/or barite acts as glue to cement the NMS clay particles and aggregates. Looking at this scale, people may recognize this scale as clay minerals (cemented NMS particles and or aggregates) at first sight, but what forms the solid scale is not only clay, but more essentially, calcite and/or barite.

5.3.5 Souring by Thermochemical Sulfate Reduction (TSR) during HC Production

TSR is a thermochemically driven redox reaction that reduces S^{6+} species (SO_4^{2-}) by strong reducing reagents such as hydrocarbons and organic acids to S^{2-} species (H_2S gas, dissolved H_2S or hydrogen sulfide ions, HS^-). In TSR, sulfate is an electron accepter (oxidizing reagent, to be reduced during TSR) and hydrocarbons (or any other reducing reagent) are electron donors (reducing reagent, to be oxidized during TSR). Sulfate is reduced as the S–O bond is broken under high temperature. The S–O bond can also be broken at lower temperatures by microbes, which use the energy from methane and/or other organic matters for their metabolism; this process is called bacterial sulfate reduction (BSR), to be discussed later. TSR can be exergonic or endergonic, depending mainly on the presence or absence of specific organic reactants. From large numbers of observations and laboratory experiments (Cross et al., 2004; Thom and Anderson, 2008), both TSR and BSR can be considered irreversible under the conditions of a HC reservoir within geological timescales. TSR has been found to obey seemingly first-order kinetics (Orr, 1977; Cross et al., 2004; Thom and Anderson, 2008; Machel, 1987, 1998, 2001; Machel et al., 1995; Manzano et al., 1997).

For TSR, it appears to be common in geological settings with temperatures over 100 °C, but in some settings, temperatures above 160 °C appear to be necessary (Machel, 2001) for the reaction to be ongoing at meaningful rates within human timescales. TSR does not have a sharply defined and generally valid minimum temperature because the onset and rate of TSR is governed by several factors that vary from place to place, such as the composition of the available organic reactants, kinetic inhibitors and/or catalysts, recharge rate of sulfate, anhydrite dissolution rates, wettability, as well as migration and diffusion rates of the major reactants toward one another (Machel, 2001). This reaction requires a high temperature to be activated because the S–O bond in SO_4^{2-} can be broken only by high energy (thermal or microbial). The most important factor to promote TSR is temperature.

H_2S, often called 'sour gas', in the reservoir can be removed while precipitating into minerals, such as reduced sulfur (H_2S, HS^- or other forms of reduced sulfur) precipitating with ferrous iron (Fe^{2+}) into pyrite. This is called the sour gas scavenging process.

A HC reservoir can be fully converted to a sour gas reservoir (CO_2 and H_2S), with H_2S content up to 30% or higher, should the scavenging capacity of the reservoir be low or depleted. Normally, scavenging capacity is made up of cations in their reduced form, such as ferrous iron. These cations are common in clay minerals, but they can be absent if the reservoir is clay-free. Carbonate reservoirs contain fewer clay minerals than clastic reservoirs, and thus the scavenging capacity is usually lower than for clastic reservoirs (unless significant iron-bearing carbonates are present, such as siderite, ankerite or iron-rich dolomite). This is the reason why almost all sour gas reservoirs are carbonate reservoirs: it is simply the result of being short of reduced cations (e.g. Fe^{2+}) to precipitate with sulfide to be removed from the reservoir fluid (H_2S gas or dissolved H_2S and HS^-).

TSR has received much attention as more sour reservoirs have been discovered in recent decades. Some reservoirs have also become soured during production. However, research on TSR in petroleum systems only started few decades ago, and the research has focused mainly on the diagenetic process for understanding the reservoir H_2S level or to predict the H_2S level of a reservoir under discovery. Reservoir souring during production due to ongoing TSR has not yet been addressed in the literature because TSR is normally believed to have waned as the reservoir cools down during production. In this chapter, we will demonstrate how TSR can be promoted/enhanced during injection of cool water into a hot reservoir. Modeling studies of this cool water-induced TSR have been performed and validated with lab experiments.

Solubility of a typical sulfate-bearing mineral, anhydrite, is highly temperature-dependent. At ambient temperature, the solubility can be hundreds to thousands times higher than at higher temperature (160–200 °C) (Figure 5.1). Because of the strong temperature dependency of anhydrite solubility, and a rapid kinetic rate of anhydrite dissolution (Palandri and Kharaka, 2004), sulfate concentration would rise hundreds to thousands times in the reservoir formation water as the reservoir is cooled down from a temperature above 160 °C to ambient (~25 °C). A sharp increase in sulfate concentration at the cooled-down zone (near the injection well) will establish a sharp concentration gradient, in turn, leading to a strong diffusion flux of sulfate from the high-concentration and low-temperature zone to the low-concentration and high-temperature zone. Therefore, the concentration of sulfate at the high-temperature zone is elevated. That enrichment in sulfate promotes TSR, increasing the rate from a background level to a level hundreds to thousands times higher at some locations in the reservoir. This elevated rate can push the H_2S level of the reservoir from the background level (determined by the background TSR rate and the scavenging capacity) to a significantly higher level, which then becomes a risk. This TSR and souring processes is called waterflood-induced TSR and reservoir souring.

Kinetic rates of TSR have been measured in lab experiments by several authors (Cross et al., 2004; Thom and Anderson, 2008). These measurements aimed to understand the sour reservoir as a result of diagenetic processes lasting for geological timescales. However, so far there are no data reported in the literature that can validate the lab-measured TSR kinetic rates or rate constants derived from field observations, that is, measured H_2S concentrations or an abundance of scavenging products such as pyrite or other sulfide mineral phases.

Based on most of the observations from lab experiments, within the temperature range of 100–320 °C and with the sulfate concentrations in normal reservoirs

(10^{-4}–10^{-2} molal) and an unlimited supply of organic matter, which is true in HC reservoirs, TSR appears as first-order kinetics:

$$R_{\text{sulfate}} = KC_{\text{sulfate}} \tag{5.1}$$

where R_{sulfate} is the rate of sulfate being reduced, C_{sulfate} is the concentration of sulfate, and K is the rate constant as a function of temperature and can be obtained from lab experiments.

This can be considered as a reduced form of Monod kinetics (Monod, 1949), which is usually used to describe kinetic reactions involved in microbial activity. The rate of this type of reaction is usually limited by the availability of one or more reactants in the solution. TSR is a similar type of kinetic reaction in which the rate is constrained by the concentration of dissolved electron acceptors and/or donors. The rate then can be evaluated with Monod kinetics as shown in the following equation:

$$R_{\text{sulfate}} = \mu_{\max}\left(\frac{C_{\text{sulfate}}}{C_{\text{sulfate}} + S_{\text{sulfate}}}\right)\left(\frac{C_{\text{organic}}}{C_{\text{organic}} + S_{\text{organic}}}\right) \tag{5.2}$$

where $\mu_{\max}$ is the maximum rate of sulfate to be reduced at the given temperature while all constraining factors are in the most favourable conditions; S is the half-saturation constant, subscript $_{\text{organic}}$ denotes organic matter which is the electron donor (reducing reagent) of the reaction, and C stands for concentration.

This equation will reduce to a pseudo first-order if organic matter is plentiful, where the third term of the above equation, $C_{\text{organic}}/(C_{\text{organic}} + S_{\text{organic}})$ stays constant:

$$R_{\text{sulfate}} = KC_{\text{sulfate}} \tag{5.3}$$

where $K = (\mu_{\max}/(C_{\text{sulfate}} + S_{\text{sulfate}}))(C_{\text{organic}}/(C_{\text{organic}} + S_{\text{organic}}))$ is usually considered as a constant when variation of sulfate concentration is limited. Therefore, K is usually considered as a constant although, theoretically, it is certainly not, leading to the TSR kinetics found and reported as first-order kinetics in almost all the literature. This explains the gaps obtained between the TSR rate constants from different lab experiments using different sulfate concentrations (e.g. Orr, 1977; Cross et al., 2004; Thom and Anderson, 2008, Machel, 2001).

Rigorously, TSR is *not* a first-order kinetic reaction. All the suggested linear rate laws and the rate constant are approximate and should be used carefully, keeping in mind the sulfate concentration used in the experiment. We would suggest using the Monod kinetics rate law to evaluate the TSR kinetics, having defined and provided the corresponding parameterization. The redefined rate constant will be independent of sulfate concentration.

5.3.6 Souring by Bacterial Sulfate Reduction (BSR) During HC Production

As discussed above, sulfate can also be reduced by bacteria and archaea within a temperature range from ambient levels up to ~80 °C. Multiple bacteria and archaea have been found to be able to obtain energy for their metabolism from oxidizing organic matter, while using sulfate as an electron acceptor to produce CO_2 and H_2S. Sulfates can also be reduced by many other bacteria and used as cell material to grow sulfur-containing cells, and this process is known as assimilatory sulfate reduction from which no sulfide is released. The amount of sulfate consumed in the assimilatory sulfate reduction is also very small compared with the amount of sulfate reduced by sulfur reducers, such as *Desulfovibrio vulgaris*, among others, which obtain energy (rather than assimilate as cell material) for growth from reducing sulfate and expel the resulting sulfide as waste.

Based on most lab and field observations, sulfate-reducing bacteria and archaea (here we call them sulfate reducers) are active at temperature ranges from ambient to about 50 °C. Above 50 °C the microbial activity declines, and ceases to metabolize when temperature is above 80 °C or below water freezing point. Microbial activity is a function of temperature within the temperature range 0–80 °C. The temperature of most HC reservoirs is significantly higher than ambient level. Injection of water at ambient temperature can cool down the reservoir, resulting in certain regions to be at a favourable temperature range for bacterial metabolism, and thus BSR can be active.

Desulfovibrio vulgaris is the most common and best studied species of this type of bacteria. Growth of these bacteria can be quantitatively evaluated using Monod kinetics:

$$R_{Desulfovibrio\ vulgaris} = \mu_{\max} C_{Desulfovibrio\ vulgaris} \left(\frac{C_{\text{sulfate}}}{C_{\text{sulfate}} + S_{\text{sulfate}}} \right) \left(\frac{C_{\text{organic}}}{C_{\text{organic}} + S_{\text{organic}}} \right) \tag{5.4}$$

where $R_{Desulfovibrio\ vulgaris}$ is the growth rate of *Desulfovibrio vulgaris* biomass; $\mu_{\max}$, similarly, is the maximum growth rate, while all constraining factors are in the most favourable status.

Accordingly, the rate of sulfate to be reduced is calculated as a linear function of the biomass growth:

$$R_{\text{sulfate}} = R_{Desulfovibrio\ vulgaris} Y \tag{5.5}$$

where R_{sulfate} is the rate of sulfate reduction and Y is the yield coefficient, which is defined as the sulfate to be reduced by growth of unit biomass.

5.3.7 Scavenging – An Overview of the Sulfur Mass Balance in the HC Reservoir During TSR or BSR

Where temperature is favourable, the BSR (or TSR) would be ongoing at a rate constrained, primarily, by the availability of sulfate as electron acceptor and HC as electron donor. In HC reservoirs, the HC can be considered unlimited, thus the most important constraint is the availability of sulfate. Sulfate is a common component in any water, and can be higher in seawater or some surface waters than in other waters. In the reservoir, sulfate can be quickly consumed by BSR and fall to a low level at which BSR could only be sustained at a low level. In most clastic reservoirs, as the sulfate is consumed by BSR it could be quickly replenished by diffusive transport from outside of the boundary or dissolution of sulfate-bearing minerals, such as anhydrite. A balance between consumption and replenishment of sulfate can be reached and sustained if concentration of sulfate tends to stabilize in geological timescales, or sulfate is consumed should no replenishment exist. For the scenario that the sulfate concentration stabilizes, we have following mass balance:

$$R_{\text{sulfate}}^{\text{BSR/TSR}} = R_{\text{sulfate}}^{\text{dsln-prcpn}} + R_{\text{sulfate}}^{\text{trspt}} \tag{5.6}$$

where $R_{\text{sulfate}}^{\text{BSR/TSR}}$ is the rate of sulfate consumed by BSR, $R_{\text{sulfate}}^{\text{dsln-prcpn}}$ is the rate of sulfate generated from dissolution of sulfate-bearing minerals (minus the rate of precipitation into other sulfate-bearing minerals), and $R_{\text{sulfate}}^{\text{trspt}}$ is the rate of sulfate refilled by flow and/or transport mechanisms (advection, diffusion and hydraulic dispersion).

The above equation is a typical reactive transport equation. For a closed system, or a hydrostatic system, sulfate can be refilled only by diffusion if there are no sulfate-bearing minerals existing in the formation rock. In a fully closed system (no diffusion across the boundary), the sulfate mass balance equation can be established between the rates of BSR/TSR and dissolution of sulfate mineral(s) only:

$$R_{\text{sulfate}}^{\text{BSR/TSR}} = R_{\text{sulfate}}^{\text{dsln}} \tag{5.7}$$

This equation illustrates a typical BSR/TSR dissolution scenario in which sulfate is consumed by BSR and replenished by dissolution of anhydrite. On the other hand, the sulfide generated from BSR/TSR will be accumulated in the system, and most could be dissolved in the brine, causing an increase in sulfide concentration leading to precipitation of sulfide minerals in the case of existing cations (e.g. Fe^{2+}) in the system. The overall effect is a transformation of sulfate mineral into sulfide mineral at a rate depending mainly on the temperature:

$$\text{anhydrite} \rightarrow SO_4^{\ 2-} \rightarrow H_2S \rightarrow \text{pyrite}\left(\text{or pyrrhotite}\right) \tag{5.8}$$

In the above equation, as the product of BSR/TSR, H_2S can be removed from the mobile phase (dissolved or gaseous) by precipitation into minerals, mostly pyrite or pyrrhotite. The whole process can be viewed as a diagenetic process converting S^{6+} minerals into S^{2-} or S^{-} minerals. The rate of this diagenetic conversion can be significant (i.e. recognizable within the lifetime of a field, should sulfate be replenished easily) or very slow (i.e. negligible should sulfate replenishment be very limited). During this diagenetic conversion, the concentration of sulfate is determined by (1) the ease of replenishment by dissolution of minerals as well as transport, and (2) the temperature of the geochemical system, which determines the rate of BSR or TSR.

As the product H_2S generated from BSR/TSR can either be accumulated in the system, increasing concentration in the reservoir formation water/brine, building up H_2S fugacity and escaping from the brine/water solution as a free gas if the concentration is higher than the solubility in the brine/water at the temperature and pressure (i.e. the bubble point), or partitioned into a pre-existing reservoir oil or gas phase (if present). As a result of the accumulation, precipitation of S^{2-}/S^{-} minerals can be promoted as H_2S is accumulated in the reservoir brine/water and the concentration increases, where cations, especially ferrous iron, are available in the water/brine. Precipitation of iron sulfide minerals can be sustained if iron can be replenished by dissolution of ferrous-bearing minerals such as iron-chlorite, siderite and others, or by transport from the outside through flow and transport mechanisms (diffusion and dispersion). Thus the H_2S level in the reservoir (concentration in the reservoir brine/water or partial pressure in the reservoir gas or concentration in oil) is the result of balancing the generation (source) of H_2S in the reservoir and the removal (sink) from the dissolved phase within the reservoir. Accounting also for the effects of transport processes, the mass balance equation of H_2S under the BSR (as well as TSR) context in the reservoir can be written as:

$$R_{H_2S}^{BSR/TSR} = R_{H_2S}^{prcpn} + R_{H_2S}^{trspt} \tag{5.9}$$

where $R_{H_2S}^{BSR/TSR}$ is the H_2S generated from BSR/TSR per unit time, $R_{H_2S}^{prcpn}$ is the H_2S removed from the mobile phase by precipitating iron sulfide minerals per unit time, and $R_{H_2S}^{trspt}$ is the H_2S transported into or out of the system by flow and/or diffusion per unit time.

This process of H_2S generation is called souring, and the removal or immobilization by precipitating into sulfide minerals is called scavenging. As long as a balance between souring and scavenging is reached, the H_2S level in the reservoir stays unchanged. At this condition, all the relevant chemical reactions are still ongoing normally at a rate defined by the flow and transport, the abundance and effective surface area of the sulfate-bearing minerals (e.g. anhydrite and pyrite), and the sustainable concentration of iron in the brine/water. Of course, other geochemical conditions may also have effects.

This situation can be reached rapidly or slowly, depending on how quickly the key processes can be stabilized. At this condition, the reservoir geochemical system might be (incorrectly) interpreted as in thermodynamic equilibrium, but it is indeed never a thermodynamic equilibrium condition because BSR/TSR are irreversible reactions. The system can be approximately considered at thermodynamic equilibrium condition only if all other reactions (other than the reactions directly related to BSR or TSR) are at thermodynamic equilibrium, no flow and transport processes are involved, and the rate of BSR or TSR is very low, due to the limited availability of sulfate (i.e. the concentration of sulfate is low). In this case, the geochemical system can be considered as a pseudo-equilibrium condition, and approaches based on the thermodynamic equilibrium assumption can be applied in modeling practice.

As discussed, sulfide generated from BSR/TSR can be scavenged, precipitating into sulfide minerals should cations such as Fe^{2+} exist in the system. This precipitation will not be sustainable if Fe^{2+} or other necessary cations could not be replenished. In the 4-High HC system, the concentration of the most favourable cation, Fe^{2+}, in brine is normally high (up to 10^{-4} molal) if iron-bearing minerals (including ferric iron, Fe^{3+}) are present. Typical Fe^{2+} minerals are siderite, ankerite, and chamosite. Many clay minerals normally bear Fe^{2+} in the interlayer. Therefore, another balance between consumption of Fe^{2+} by precipitation into iron sulfide minerals and the generation of Fe^{2+} from dissolution of Fe^{2+}-bearing minerals (e.g. chamosite) can be achieved:

$$R_{Fe^{2+}}^{dsln} = R_{Fe^{2+}}^{prcpn} + R_{Fe^{2+}}^{trspt} \tag{5.10}$$

The concentration of sulfate, sulfide and iron (Fe^{2+}) in a 4-High reservoir will eventually be determined by balancing the rates of sulfate consumption by BSR/TSR and generation by dissolution of sulfate minerals, the rates of sulfide generation by BSR/TSR and the rate of sulfide consumption by precipitating into iron sulfide minerals, and the rates of Fe^{2+} consumption by precipitating into iron sulfide minerals and Fe^{2+} generation from dissolution of Fe^{2+} minerals such as siderite, ankerite, and chamosite.

5.3.8 Clay Swelling Due to Cation Exchange During Injection of Water

Clay minerals are hydrous aluminium phyllosilicates, with variable amounts of metal ions (cations) which either replace Al in the structure, leading to negative charges for the aluminium-silicate crystal layer surface (e.g. Mg as the most common cation), or act as 'glue' (most common cations such as Na, K, Ca, Mg ...) to attract two or three negatively charged aluminium-silicate crystal layers to form double-layer or triple-layer crystal structures. Due to the polarized behaviour, a certain amount of water molecules is absorbed into the gap between the aluminium phyllosilicate layers (the interlayer). The amount and the charge status of the cation, and the amount of water in the interlayer, determine the width of the interlayer (from 7.9 Å to wider), and in turn, determine the bulk density where the clay stays in an expandable (unconfined) condition, or determines the pressure (stress) if the clay is confined within a constant volume. In mechanically unconfined conditions, the interlayer gap would become wider while more water is absorbed from pore-spaces into the interlayer, leading to an increase in chemical potential or activity of water within the interlayer. Under confined conditions, the confining pressure will increase as water molecules enter the interlayer, leading to excess stress. Other light and polarizable molecules such as CO_2 and NH_3 may also be absorbed into the interlayer similarly to water molecules. The amount of water and cation within the interlayer is determined by balancing the total chemical potential of the ions and the molecules in the interlayer and of those in the porewater. The process of transferring cations between porewater and the interlayer to balance the chemical potential is called cation exchange (Appelo and Postma, 2005). Models for handling cation exchange are well developed and available in many literatures (Appelo and Postma, 2005).

The change in volume or confining pressure of clay as water composition changes is called clay swelling/shrinkage. It is a common phenomenon for some clay minerals, such as montmorillonite (including smectite as a member of the group), which can normally be found in young/shallow formations (typically Mesozoic or younger). Montmorillonite group minerals are mostly altered into illite in older formations. Interbedded smectite and illite are found commonly in Mesozoic formations, indicating an ongoing illitization process as a long-term diagenetic trend. Illitization of the montmorillonite group can be complete in older (Mesozoic or older) formations, and thus the swelling potential of that formation has waned. The montmorillonite group minerals appearing in younger formations (Mesozoic or younger) are normally considered to be swelling clays (or expandable clays).

Cation exchange takes place where the chemical composition of the clay porewater changes. Normally the formation porewater is desalinized as the injection water invades the pore space mixing with the indigenous porewater, leading to cation exchange, reducing the ionic strength (salinity), increasing the water activity in the interlayer, and, possibly, causing clay swelling if the formation contains such type of clay. Processes in the opposite direction (i.e. clay shrinking) can take place if the salinity of the injection water is higher than that of the formation water, but it is very rare.

The effect of clay swelling on HC production is usually recognized as formation damage which is common as clay swells near the wellbore location, reducing the porosity and permeability of the formation and thus impairing injectivity. Clay swelling at the far-wellbore region inside the formation can pressurize the formation and therefore increase the pressure gradient of the HC fluid. Reports are found that injection of 'low-salinity water' in some swelling clay-bearing formations can produce exceptionally high productivity, and this benefit is normally attributed to the changes of fluid wettability, which could help the displacement of HC from pore-space (Bernard, 1967). However, the pressurization by clay swelling might have also contributed to the productivity improvement. Further research on minimizing the formation damage, and maximizing the pressurization is necessary for productivity improvement.

5.3.9 Wellbore Cement Corrosion by Acid Attack from Formation Water/Brine

Portland cement, as the most common type of cement, is used exclusively in well completion. Chemically, cement clinker is composed of a series of oxides, including CaO, SiO_2, Al_2O_3, Fe_2O_3 and SO_3 as the constituent components. These oxides appear in complex forms as $Ca_3Al_2O_6$ (tricalcium aluminate or C_3A in short, 10%), $Ca_4Al_2Fe_2O_{10}$ (tetracalcium aluminoferrite or C_4AF, 8%), Ca_2SiO_5 (belite or dicalcium silicate or C_2S, 20%), Ca_3SiO_4 (alite or tricalcium silicate or C_3S, 55%), Na_2O and K_2O (sodium oxide, potassium oxide, <2%) and $CaSO_4 \cdot 2H_2O$ (gypsum or CSH_2, 5%). The cement clinker is ground to a powder to be used. Most of these components are highly soluble.

Hydration is a process by which the anhydrous cement complexes (above phases) dissolve when cement particles first come into contact with water, and then precipitate into hydroxide phases called hydration products (illustrated with equations below) when the water is oversaturated. The process continues dissolving the anhydrous cement phases (complexes), and precipitating hydration products, as elaborated first by a famous chemist, Henry Le Chatelier (1850–1936), in 1882 as Le Chatelier's principle (Desch, 1938).

All complexes (C_3S, C_2S, C_3A, C_4AF) start dissolving simultaneously as water is available to the anhydrous cement particles, but at different kinetic rates, and produce different hydrates. The cement particles of the complexes become smaller and smaller during dissolution and the hydration stops when 95% of the particles become smaller than 45 micrometres (μm). An average size of the particles is 15 μm when hydration products are formed and accumulate at the surface of the anhydrous cement particles, preventing the water from contacting the anhydrous cement complexes. Also, water can no longer be mobile because capillary suction becomes very high when the particle size reduces to an average diameter of 15 μm. The water remaining is then held hostage. This water is very alkaline, with a pH determined mainly by the following reactions, and at the high end of the pH spectrum (11–14):

$$2Ca_3SiO_5 + 7H_2O \rightarrow 3CaO \cdot 2SiO_2 \cdot 4H_2O + 3Ca(OH)_2 \tag{5.11}$$

$$Ca(OH)_2 = Ca^{2+} + 2OH^- \tag{5.12}$$

A huge amount of $Ca(OH)_2$ (portlandite) remains as the major product of the hydration processes. Other hydration products are calcium silicate hydrate phases in different forms, ettringite ($Ca_6Al_2(SO_4)_3(OH)_{12} \cdot 26H_2O$) and monosulfoaluminate ($Ca_4Al_2O_6SO_4$), or in its hydration form ($Ca_4Al_2O_6SO_4$-$14H_2O$). As the main components in the hydrated

Table 5.1 Geochemical composition and properties of minerals in hydrated cement.

Phases	Chemical composition	Mol. weight (g)	Mol. volume (cm^3)	Mass ratio in H-class cement
Portlandite	$Ca(OH)_2$	74.092	33.06	0.671
CSH0.8	$SiO_2Ca_{0.8}(OH)_{1.6}$	119.357	37.89	0.164
CSH1.1	$SiO_2Ca_{1.1}(OH)_{2.2}$	141.858	45.03	
CSH1.8	$SiO_2Ca_{1.8}(OH)_{3.6}$	193.449	61.41	
Ettringite	$Ca_6Al_2(SO_4)_3(OH)_{12}\cdot 26H_2O$	1255.083	705.10	0.117
Hydrotalcite	$Mg_6Al_2CO_3(OH)_{16}\cdot 4(H_2O)$	543.961	265.96	0.038

cement and most important phases that give the strength to the concrete, calcium silicate hydrate phases are abbreviated as CSH or C-S-H in the literature. The stoichiometry of CSH in cement paste is variable, and the state of chemically and physically bound water in its structure is not transparent. Therefore CSH is differentiated with the stoichiometry and suffixed with the stoichiometric coefficient of Ca in the formula (such as CSH0.8, CSH1.1, CSH1.8, etc.) for geochemical modeling purposes. Typical composition and thermodynamic properties of the constituent minerals of hydrated cement were collected from the literature (Stronach and Glasser, 1997; De Windt et al., 2004; Kurtis, 2015) and provided in Table 5.1 for the reader's convenience.

Within hydrated cement (i.e. concrete) all phases are thermodynamically stable at the high pH condition (pH ~ 11–14). The high pH is determined and maintained by portlandite ($Ca(OH)_2$) through reaction $Ca(OH)_2 = Ca^{2+} + 2OH^-$. Portlandite is, indeed, a pH buffer of the cement chemical system. The extreme alkaline condition can be sustained when the cement chemical system is closed, otherwise intake of external, less alkaline fluids (water or moisture) can neutralize the system by dissolving portlandite as well as other phases that can stay thermodynamically stable only in extremely alkaline conditions. This condition is self-satisfied in the concrete because the hydration products seal the pore spaces and the hydrated cement is almost impermeable. Porosity is a few percent and permeability is in the nano-Darcy range. Therefore, external water and/or moist air can only invade the cement by diffusion, which is known to be very slow, so that the effect of the water/moist air intake can be seen only to a few centimetres depth in the concrete within a human lifespan. The diffusion of water and/or moist air in the hardened cement is largely retarded due to hydration of the remaining unhydrated phases, which utilizes the water in the diffusive flux, leading to a retardation of the diffusion front. Fineness of the hydrated cement establishes a high capillary barrier to thoroughly prevent water/moist air penetration into the hydrated cement. Therefore, the hydrated cement can present high enough mechanical strength and thermodynamic stability under various subsurface (temperature and pressure) conditions.

Diffusion is a phenomenon that can transfer mass through almost any media, whether permeable or impermeable. Diffusion of water/moist air into cement is inevitable, although can be very slow and limited. The results of diffusion of water/moist air into cement are (1) acidity is brought into the cement, leading to a neutralization of the cement chemical system, and (2) carbonate (CO_2 dissolved in water) and/or CO_2 (in air) is brought into the cement, leading to a precipitation of calcite. This process is called carbonation. The precipitated calcite is volumetrically larger than that of the portlandite, due to the fact that the molar volume of calcite is greater than the molar volume of the portlandite. One of the results of carbonation is a further reduction of the porosity and permeability of the cement. Carbonation is then another guard against water penetration, in addition to hydration, preventing the invasion of external fluids (water/moist air).

Crevices, micro-fractures or even open fractures can be created if the cement is under varying thermal and/or mechanical conditions or for other reasons. The crevices and micro-fractures can then be used by fluid as a flow path invading the cement. Involvement of external fluids can either further dissolve portlandite and other phases, leading, potentially, to a widening of the crevices and micro-fractures, or precipitate calcite (carbonation), sealing the open pathways, depending on the water flux and how rapidly the precipitation can happen. The flow path can be sealed by the precipitation of calcite if the crevice/micro-fracture is small enough and the precipitation of calcite is fast enough. This scenario can be true only if the flow path can be clogged by the precipitation, leading to a capillarity barrier. However, there are cases in which the developing sealing material within the flow path cannot compensate for the dissolution, and the flow path becomes wider. Fluid-conductive crevices/micro-fractures are usually widened due to the dissolutions, and the released ions (mainly Ca^{2+}) from the dissolution are lost into outflows (Ranaivomanana et al., 2013; Ylmén and Jäglid, 2013). Flow can be enhanced in the widened flow path, and therefore the dissolution and flow activity form a positive feedback cycle. Cement can quickly collapse (within days to months) if such a positive feedback cycle is formed. The flow path can be continuously widened until failure occurs (totally loss of mechanical strength). Such a condition is the risk that we need to avoid.

Major phases of the hydrated cement can dissolve when contacting external water as:

$$Ca(OH)_2\ (\text{portlandite}) + 2H^+ = 2H_2O + Ca^{2+} \tag{5.13}$$

$$SiO_2Ca_{1.1}(OH)_{2.2}(\text{CSH1.1}) + 2.2H^+ = 2.2H_2O + 1.1Ca^{2+} + SiO_2(aq) \tag{5.14}$$

$$\begin{aligned} Ca_6Al_2(SO_4)_3(OH)_{12}\cdot 26H_2O\ (\text{ettringite}) + 4H^+ &= 34H_2O + 6Ca^{2+} \\ &+ 3SO_4^{2-} + 2AlO_2^- \end{aligned} \tag{5.15}$$

$$\begin{aligned} Mg_6Al_2CO_3(OH)_{16}\cdot 4H_2O\ (\text{hydrotalcite}) + 13H^+ &= 12H_2O + 6Mg^{2+} \\ &+ HCO_3^- + 2AlO_2^- \end{aligned} \tag{5.16}$$

These are the major reactions causing cement corrosion. All the above phases can be thermodynamically stable only under alkaline conditions ($pH > 10$), and tend to dissolve as the alkalinity is neutralized. Therefore, one important factor that we must consider for risk evaluation of the cement corrosion is the pH of the external fluids. The external fluid can be non-corrosive if the pH is as high as that of the cement pore fluid (>10) because portlandite (67% of the total cement constituent mass) and all other solid phases will not dissolve in fluid of the same pH as the pore fluid. However, the dissolution of portlandite and most other phases in an invading external water of lower pH is faster than in that of higher pH. The lower the pH of the external water, the faster the dissolution progress. Therefore, to preserve the cement in a highly alkaline condition is key to guarding against corrosion. It is extremely important to protect the cement from access by low pH water. Model evaluation of the effects of invasion by external water/moist air through diffusion only or flow are not common. Some modeling exercises on the potential of cement corrosion will be presented in the later sections of this chapter.

5.4 Modeling Approaches and Numerical Simulators

5.4.1 Gaps of the Simulators in the Oil and Gas Production Technology Community

5.4.1.1 Scale Simulators

For evaluation of geochemistry-related risks during HC production, modeling approaches and tools (e.g. numerical simulators) have been developed in the oil and gas industry. Typical geochemical model evaluation has been performed to assess the potential risk of wellbore scaling for more than two decades. Multiple numerical simulators have been developed, commercialized and utilized by the industry for many years. Most of them are software for evaluation of scaling tendencies of wellbore and/or near-wellbore regions, and souring potentials within reservoirs. OLI/ScaleChem is one of this type of simulator and is widely used in the production chemistry community (http://www.olisystems.com/oli-studio-analyzers-scalechem). Scaling tendencies are represented by two quantities in this type of evaluation procedure: the saturation index of the potential scaling minerals, and the potential amount of the precipitations (called excess solutes). Simulators of this type are normally batch geochemical simulators, which calculate the saturation index of the minerals to be potentially precipitated and the amount of precipitate (without kinetics, i.e. assuming local thermal equilibrium conditions). The scaling potential is then evaluated based on the calculated saturation index and the excess solutes. It is believed that the larger the saturation index, the higher the potential for precipitation, and the greater the excess solutes, the more severe the scaling will be. This approach seems logical and appropriate, but could be biased due to the fact that the model evaluation is based on a batch mode. The actual amount of the precipitation (of the scaling minerals) cannot be accurately calculated using batch simulators, although the functionality for quantitatively evaluating the mass (or volume) of the precipitation is usually available in the scaling simulators because the amount of minerals precipitated inside and near the wellbore is not only determined by the local thermodynamics, but is also controlled by the water flow passing through that region and, in most cases, the amount of precipitation is rather dictated by the water flux rate. Therefore, coupling flow and transport processes is key to better simulation of the precipitation of scaling minerals.

5.4.1.2 Souring Simulators

Souring is normally evaluated in the production technology community by using souring simulators. Similar to the scaling simulators, souring simulators are mainly batch simulators or batch simulators with limited flow and transport functionality, or batch simulators with consideration of flow and transport components from reservoir simulator outputs. A typical souring simulator, SourSimRL (ABRACO – Associação Brasileira de Corrosão, available in North America at www.esss-usa.com/), is designed mainly for evaluating the souring process carried out by bacterial sulfate reduction (BSR). This simulator is widely used in oil majors for evaluation of risks induced by BSR processes (Evans et al., 2013). The simulator calculates the BSR based on sulfate concentrations of the water in place and the concentration of electron donors (e.g. HC, fatty acids and other organic matter). Effects of flow and transport in the reservoir can be considered using the calculated temperature and pressure from a reservoir flow model. The dissolved sulfate in place is the only sulfate in the model available to be reduced, and no replenishment is considered. Possible replenishment of sulfate by dissolution of the sulfate minerals, such as anhydrite, is ignored. This means that the simulator only accounts for a portion of sulfate in the whole system for BSR and TSR. This can be fine only when the BST/TSR is not sulphate limited. The simulator also handles the H_2S scavenging process represented by precipitation of pyrite or other iron sulfide phases. The iron (in reduced form, Fe^{2+}) required by the precipitation of these minerals is considered in the simulator as either available only in the formation water (the dissolved iron), or replenished from dissolving one nominal mineral or a representing mineral, normally siderite, and actual iron minerals are to be considered, explicitly, in the recent development. Obviously, these assumptions are hardly appropriate for evaluation of BSR or TSR, as we know that the sulfate available for BSR or TSR is normally not the only dissolved sulfate in place. The sulfate for BSR and TSR can also be available from flow and transport replenishment and dissolution of minerals. Further, iron as the cation required for precipitating iron sulfide mineral phases is not only available from the dissolved iron in place, but can also be replenished by flow and transport and from dissolution of iron minerals, mostly siderite, iron-chlorite, iron dolomite, and iron-bearing clays. In many cases, in reality, sulfate can be replenished by flow and transport and dissolution of sulfate minerals. By ignoring these important sources of sulfate, the reliability of the model predictions using the present souring simulators of this type is limited. Indeed, the souring processes by BSR and/or TSR must be evaluated considering the effects of flow, transport and a complete mass balance of dissolved and/or solid sources of sulfate. For the scavenging side, replenishment of Fe^{2+} by flow and transport as well as dissolution of other Fe^{2+} minerals should be considered in an appropriate manner. Geochemical reactive transport modeling is just the modeling technology that can handle the souring process. An RTM simulator able to handle geochemical systems of high temperature, high pressure, high salinity and highly reducing conditions is capable of better modeling the souring process.

5.4.2 Clay Swelling Evaluation Approaches

In the present production technology community, clay swelling is assessed by evaluating the salinity and the composition of the cations in the clay porewater. The total amount of cations and the ratio of monovalent to divalent cations are used as indices to estimate the swelling potential of clay (Scheuerman and Bergersen, 1990). This is an empirical method (grey-box model) with mechanisms hidden underneath. There is no practically useful tool reported in the oil and gas industry to simulate the swelling or contraction directly. An algorithm to calculate the clay swelling pressure due to changes in porewater chemical composition has been developed, discussed and elaborated by Wayllace (2008) and installed in TOUGHREACT (Zheng et al., 2014). However, with the implementation of this functionality, TOUGHREACT becomes a very useful simulator to handle clay swelling due to chemical composition change of the porewater, and even more useful if a geomechanics solver is also coupled to handle deformation by clay swelling. Further development of TOUGHREACT is ongoing at Lawrence Berkeley National Laboratory, and information about the TOUGHREACT simulator is available at http://esd1.lbl.gov/research/projects/tough/.

5.4.3 Reactive Transport Modeling Simulators Applicable to Petroleum Geochemical Systems

A geochemical reactive transport modeling simulator can be used to model high temperature, high-pressure, high-salinity systems at highly reducing conditions only if the simulator is capable of handling:

- non-isothermal multiphase fluid flow and solute transport in real petroleum systems which are characterized normally with complexity in geometry, geological formation structure (stratigraphic variation, heterogeneity, baffles and faults) and texture of rocks (grain/crystal size and orientation), and mineral compositions (also including surface and interlayer) in a comprehensive and effective manner;
- non-ideality of fluids, that is, property (PVT) changes and phase partitioning under 4-High conditions in an effective and accurate way;
- non-ideality of the geochemical components in water solution under 4-High conditions (e.g. activity coefficients of dissolved geochemical species in aqueous solution) using appropriate approaches;
- geochemical reaction networks in an effective manner (efficiently solving geochemical reactions);
- integration of the solution of flow and transport and solution of geochemical reactions with high numerical stability and computational efficiency.

Of course, high temperatures (significantly higher than ambient) and change of temperature during production (including waterflood, CO_2 EOR and thermal/chemical EOR) must be taken into account in modeling practice. Properties of most constituent components (e.g. fluids, chemical components) and processes (e.g. chemical reaction rate and equilibrium constants) are significantly temperature-dependent, and therefore it is of no value to model the HC reservoir without considering temperature change and the variables as a function of the temperature. Indeed, most available RTM simulators are capable of handling heat transfer and temperature change.

Over the typical ranges of temperature and pressure, fluid (HC and water) properties, especially, density, viscosity and mutual solubility of HC and water, change significantly. Change in the properties has a significant effect on fluid mobility, as well as reactivity, and thus must be evaluated in the RTM. As an RTM simulator, the capability to handle the property change is essential. Without these capabilities, an RTM simulator, such as those developed for applications of environmental research and remediation, water resource research and other shallow systems under ambient temperature and pressure conditions, would not be able to be applied in a petroleum geochemical system. An appropriate 'PVT model' is usually a necessary installation in reservoir fluid flow simulators, and therefore also necessary for RTM simulators for application in oil and gas production. For quantitative evaluation purpose, the effects of temperature and/or pressure are taken care of by using temperature- and/or pressure-dependent thermodynamic equilibrium constants. These are usually made available as a function of temperature and/or pressure for specific reactions (for details, please see below).

Geochemical reaction networks can be mathematically denoted by a nonlinear algebraic equation system. The Newton-Raphson Iterative Method (NRI) is usually used to solve the nonlinear algebraic equation system. NRI requires calculation of a Jacobian matrix, which entirely comprises the derivative terms of the mass of each component with respect to the concentrations of each species. The derivative terms can be calculated analytically (i.e. the calculation of each term is based on the actual mathematical formulation of the derivative terms) or numerically (i.e. the derivative terms are approximated using finite difference). The analytical calculation of the derivative terms is the most intensive work of NRI and consumes more than 60% of the total computational power needed by NRI, while numerical calculation of the derivative terms may need more than 90% of the computing power. For low concentration systems, the Debye-Hückel (Debye and Hückel, 1923) model is usually used for calculation of the ionic activity coefficients, and in this case, the numerical calculation of the derivative terms is slower than the analytical calculations, but the difference is not too big and the performance of the simulator can still be acceptable. For high concentration systems, an ion-interaction model is usually required in RTM simulators, and the numerical calculation of the derivative terms can be hundreds of times slower than the analytical calculation, especially for large geochemical systems: the more geochemical components considered in the model, the bigger the difference in efficiency for calculating the derivative terms between the numerical calculation and the analytical calculation. When handling a complex geochemical system (characterized by a large number of geochemical components, i.e. minerals, ionic solutes, surfaces, interlayers and complexes), a geochemical solver using numerical calculation for the derivative terms is ridiculously slower than geochemical solvers using analytical calculation. For oil and gas reservoirs, a high-concentration complex geochemical system is the main objective of the RTM; therefore, to calculate the derivative terms effectively is key for the solution of the non-linear algebraic equation system. For the high ionic strength (salinity) conditions, simulators with fully analytical calculation of the derivative terms, such as TOUGHREACT-PITZER (Zhang et al., 2006a; Xu et al., 2006) using the Pitzer ion-interaction model (Pitzer, 1973, 1991; Pitzer and Mayorga, 1973; Pitzer and Kim, 1974; Harvie and Weare, 1980; Harvie et al., 1984), can usually perform much more efficiently than most of the other simulators with numerical calculations of the derivative terms of the NRI.

Another constraint of the capability and/or the numerical efficiency is the approach of integrating the solution of the geochemical system and the flow of fluids and transport of the solutes. As discussed in previous chapters (Chapter 1), there are three major approaches to integrating the solution of flow and transport and the solution of a geochemical reaction network:

- operator-split approach (OSA) or no-iteration approach (NIA)
- sequential iteration approach (SIA)
- direct substitution approach (DSA)

Comparison of these three approaches can be found either in this book (Chapter 1) or elsewhere (Xu, 1996; Xu et al., 1999; Samper et al., 2009).

The OSA (NIA) is referred to as the full explicate coupling approach, and for application in 4-High geochemical systems, is more likely to suffer from poor numerical stability or poor efficiency because of the high non-linearity of such systems. OSA (or NIA) is easy for programming and economic in computer memory requirements. This is the reason why some RTM simulators are still produced using this approach. For simple flow, transport and geochemical systems, simulators deploying OSA could still be useful, but are certainly inefficient for 4-High geochemical systems.

On the other hand, DSA, as a fully implicit coupling approach, can handle non-linearity with good numerical stability and efficiency for small geochemical systems (small numbers of geochemical components), but becomes inefficient when dealing with large geochemical systems (large number of geochemical component) due to the huge memory requirement, which is computationally expensive.

The best approach for the 4-High geochemical system should be one that can handle the non-linearity of multiphase flow and the high non-linearity of the high salinity conditions with reasonable numerical stability and the highest numerical efficiency, using reasonable computational power and memory requirements. SIA uses almost the same size of memory as the NIA and performs with much better numerical stability, thanks to the iteration algorithm which leads to a certain level of implicity. A better sequential iteration approach was developed by Tebes-Stevens under supervision of Valocchi (Tebes-Stevens et al., 1998), which takes account of the dominant derivative terms of the reaction rate with respect to the component concentrations, adding some implicity into the solution procedure of the coupled system. This is the so-called SIA-1 approach, which has been implemented in BIO-CORE (Zhang and Samper, 2001) and systematically tested by Zhang (2001). SIA-1 is considered much better than the regular SIA in terms of numerical stability as well as efficiency when solving for geochemical systems with kinetics (Zhang, 2001). SIA-1 can also be deployed in a local equilibrium system without kinetic constraint in the geochemical reactions. Unfortunately, SIA-1 has not yet been adopted by major RTM simulators other than BIO-CORE and FEREACT (Tebes-Stevens et al., 1998).

Regular SIA is used in most RTM simulators, and has performed effectively for most applications. SIA-1 is better when dealing with system of high non-linearity, and thus can be better for RTM simulators handling 4-High geochemical systems. However, this hypothesis remains to be further tested and confirmed.

Mathematical equations/formulations, algorithms and solution technologies have been developed for handling the above-mentioned objectives, and most of them have been integrated into a few numerical simulators. We are going to discuss the method of handling High-T, High-P, High-S and High-R conditions.

5.4.4 Handling High Temperature

As discussed above, the effects of temperature and pressure on geochemical reactions are considered using temperature- and/or pressure-dependent thermodynamic equilibrium constants. The status of a given geochemical system at a given temperature and pressure tends to reach a steady-state. A thermodynamic equilibrium is defined as when all reactions in the system are reversible and the results of the reaction are balanced out by their reverse reaction at the same rate. At the equilibrium, Gibbs free energy (to be noted as free energy hereafter) of the system is minimized and the total chemical potential of the system becomes zero. In a closed aqueous system (without mass and energy exchange with the surroundings), thermodynamic equilibrium can be easily sustained. The mass action law prescribes a thermodynamic equilibrium at which the minimum of the free energy of the whole system is sustained at a given temperature and pressure, and a balance between the reactants and products of a reaction is reached in terms of chemical potential (e.g. the summation of chemical potential of all reactants and products in the system reduces to zero.) At any given temperature and pressure, the ratio of active amount of the products over the active amount of the reactant that can sustain a zero summation of chemical potential between two sides of a reaction is defined as the thermodynamic equilibrium constant:

$$K = \frac{\Pi a_i^{\lambda_i}}{\Pi a_j^{\lambda_j}} = \frac{\Pi \gamma_i^{\lambda_i} m_i^{\lambda_i}}{\Pi \gamma_j^{\lambda_j} m_j^{\lambda_j}} \quad (5.17)$$

where K is the thermodynamic equilibrium constant, a is the active amount, i.e., the activity of the product i, or of the reactant j, which are evaluated based on the mass of the reactant or products (number of moles per specific weight/volume/total moles of solvent or solution), and m_i and m_j multiply the non-ideality, a quantity to define how much of the molar free energy of the reactants or products can be effectively accounted for in the given system under the given temperature and pressure. This quantity, γ_i or γ_j, is called the activity coefficient of the reactant i or product j. The activity coefficient is a complex function of the solution composition, which is determined by the interaction effects between ionic or molecular particles in the chemical system (e.g. solution, gas mixture or solid solution). λ_i and λ_j are the stoichiometric coefficients of the reactant i or product j.

Based on the above definition, we know that the thermodynamic equilibrium constant only depends on the species in their reference state, and therefore is a thermodynamic property that only depends on pressure and temperature (and not on species concentrations). It takes a unit of concentration with a power of x, for example $[m]^x$ or $[m]^x[bar]^x$ if a gaseous species is involved. An equilibrium constant can be dimensionless only if the total numbers of moles are the same between reactants and products, and the product of activity coefficients of all the reactants and products is assumed to be 1 (in this case the power, x, takes a value of 0.) Practically, the equilibrium constant is defined exclusively for the specific reaction for which it was defined, with specific stoichiometry, unit of concentration and gas partial pressure (usually molal for concentration and bars for partial pressure), as well as the activity coefficient calculated with specific ionic activity models. The unit of the thermodynamic equilibrium constant is and should remain consistent with the reaction equation. We may not need to care about the unit of LogK in a dedicated thermodynamic database as long as the LogKs were given for the specific reactions with specific stoichiometry, in specific units of concentration and/or partial pressure, as well as ionic activity model. A thermodynamic equilibrium constant is usually given as a function of temperature in popular thermodynamic databases, either as a fitting equation or at the temperature grid. The database from the Lawrence Livermore National Laboratory, associated with the EQ3/6 software package for batch geochemical systems and normally generated using software SUPCRT92 (Johnson et al., 1992), is considered the most comprehensive and is used by not only EQ3/6 but also TOUGHREACT with a reformatted version. Most of the other typical geochemistry and/or RTM simulators also share data, fully or partially, with this database. The most common fitting equation of thermodynamic equilibrium constant (normally written in logarithm scale, LogK) is:

$$\text{LogK} = a_1 + a_2 \ln T + a_3 T + \frac{a_4}{T} + \frac{a_5}{T^2} \quad (5.18)$$

where K is the thermodynamic equilibrium constant and T is temperature in kelvin; a_1 to a_5 are fitting parameters evaluated using regression methods based on experimental data. Similar equations are used to define other temperature-dependent parameters.

5.4.5 Handling High Pressure

Change of pressure within the pressure range of shallow systems leads to very small and ignorable effects on the thermodynamic equilibrium constant. The effects become visible or significant as pressure changes within a much bigger range, such as the range of a HC reservoir. However, effects of pressure change are significantly less than the effects of temperature change within HC reservoirs. For most RTM practice, omission of the pressure effects on the thermodynamic equilibrium constant is acceptable, depending on the pressure range of the actual problem and the purpose of the modeling practice. Pressure dependency of the thermodynamic equilibrium constant is handled in TOUGHREACT version 3.0 and a Shell in-house simulator, MoReS-PHREEQC. Numerical stability of RTM while handling pressure dependency has not been well evaluated. Please refer to TOUGHREACT version 3.0 manual (Xu et al., 2014) and/or Wei (2012) to obtain further information.

When working with RTM simulators without the capability to handle pressure dependency, ad hoc adjustment can be considered when the temperature and pressure range is known. A possible ad hoc method would be to find the temperature–pressure distribution over the time and space domain of the geochemical system, then find the most influential temperature–pressure combination within the domain, determine a monotonic correlation equation to relate the temperature and pressure of the most influential temperature–pressure combination, and, adjust the equilibrium constant to reflect the pressure effect for the given temperature. This approach uses an altered temperature dependency function to reflect the effects of pressure, that is, combining the temperature dependency and pressure dependency should the temperature and the pressure correlate, monotonically. Injection of cool water into a hot reservoir can always induce a low temperature–high pressure (at the injector) and high temperature-low pressure (producer) T–P profile in which the temperature and the pressure are monotonically correlated. Therefore, to consider the pressure effect, calibrate the temperature-dependent thermodynamic equilibrium constant (LogK) to the relevant pressure for all the T–P combinations and the pressure effect is accounted for, parasitically, in the temperature dependence function.

5.4.6 Handling High Salinity

Salinity is defined as the content of salts dissolved in a specific amount of water. In most of the conditions another term, ionic strength, is used for a more accurate description of the dissolved salt by only accounting for the ionic contents to demonstrate the thermodynamic activity of the dissolved salts, excluding the portion of the salts in complex form. Ionic strength is defined as:

$$I = \frac{1}{2}\sum_{i=1}^{n} m_i z_i^2 \qquad (5.19)$$

where I is the ionic strength, n is the number of ionic species in the system, and m and z are the molality and charge of ion i, respectively. Obviously, ionic strength positively correlates with salinity. In fresh water or brackish water, dissolved salts exist in either ionic (cation and anion), or complex (tightly associated cation and anion) forms. Amount of the complexes can be clearly determined as a function of temperature (may also be slightly dependent on pressure). Distribution of free ions or ions in complexes can be calculated based on the complexation reaction, and the free ions are available for further reactions with a minor non-ideality which can be evaluated using traditional ionic activity models such as the Debye-Hückel model (Debye and Hückel, 1923). In contrast, in concentrated waters or brines, most ions are engaging in interaction, rather than forming complexes, the ionic activity is largely determined by the interaction, and the non-ideality becomes very significant. It was found that the non-ideality is caused by the interactions of the ions and molecules due to the charge carried by the ions or the polarity of the molecules. It is also found that the higher the concentration, the larger the non-ideality.

Normally, when ionic strength is lower than 0.1 molal, the non-ideality (i.e. the ionic activity) can be evaluated using the classic Debye-Hückel equation (Debye and Hückel, 1923), as systematically parameterized for a useful range of temperatures and pressures by Helgeson and Kirkham (1974). Variations are also possible in the form of b-dot or the Davies equation (Davies, 1962), which can be accurate up to approximately 1 molal. These equations determine the ionic activity by taking the effects of solution ionic strength and the charge into a coefficient, namely the ionic activity coefficient, γ, as $\ln(\gamma_i) = Az_i^2\sqrt{I}$ where A is a temperature-dependent number and z is the charge of ion i.

For concentrated solutions, specific ion interaction must be considered when solution ionic strength is higher than 1 molal. Pitzer's model (Pitzer, 1973, 1991; Pitzer and Mayorga, 1973; Pitzer and Kim, 1974; Harvie and Weare, 1980; Harvie et al., 1984) takes account of the effects for specific ion pairs, triplets of cation–anion, cation–cation, anion–anion, cation–cation–anion, cation–anion–anion, as well as effects from neutral species. A series of temperature-dependent coefficients are used to evaluate the effects of specific ion pairs/triplets. This model is considered the most effective model to deal with the high ionic strength problems. For detailed equations, please see the work of Pitzer (Pitzer and Kim, 1974; Pitzer, 1991) and Harvie et al. (1984). The Pitzer model had been installed in some geochemical batch simulators such as EQ3/6 (Wolery, 1992, and the updated versions) and PHREEQC (Parkhurst and Appelo, 1999, 2013) and RTM simulators such as TOUGHREACT–Pitzer (Zhang et al., 2006a) and to be available in newer versions of TOUGHREACT (Xu et al., 2014). PHREEQC has been coupled with flow and transport simulators such as a Shell in-house reservoir simulator, MoReS, leading to MoReS-PHREEQC as a robust RTM simulator for modeling geochemical processes taking place within a HC reservoir during HC production (Farajzadeh et al., 2012; Wei, 2012). Other RTM simulators of the same type also exist but are only available internally in the hosting enterprise and have seldom been reported. For details of the Pitzer ion-interaction model and the host simulators, please refer to respective manuals of the simulators.

The Pitzer model is much more complicated than the Debye-Hückel equation and its variants (called Debye-Hückel type equations in general). Using Pitzer's model in RTMs, significantly more computational power is needed than for the RTM using Debye-Hückel type equations. Simplifications of Pitzer's model for a single ion-pair (NaCl) dominant brine was developed by Helgeson et al. (1981). Their so-called HKF model can be viewed as a simplified Pitzer model in which only the interaction between Cl and Na is considered. Therefore, this model is effective for NaCl-dominated brines concentrated up to halite saturation.

5.4.7 Handling Highly Reducing Conditions

As we know, the HC reservoir stays in a highly reducing condition due to the reducing potential of the HC species. Oxidation can take place in the reservoir if oxidizing minerals exist in the formation (in the case that HC migrates into a host formation that bears oxidizing minerals) or any oxidizing reagent is brought into the reservoir. The result of the oxidation process is to generate some organic acid (e.g. acetic acid) as the product. Nevertheless, the oxidation is limited because the amount of HC is huge compared with the oxidizing reagent, or the rate of the oxidation is limited. Therefore, most HC reservoirs stay in a very reducing condition.

Under such conditions, common chemical components all appear in their reducing form in the reservoir formation brine. For example, carbon appears, in addition to HC, as organic acids, concentration of carbonate species remains low, and carbonate minerals such as calcite may disappear. Sulfur can still exist as sulfate (S^{6+}) but the concentration is low due to low solubility of anhydrite at high temperature and possible consumption by bacterial or thermochemical reduction into reduced form (S^{2-} or S^{-}). Concentration of sulfide species can be high if scavenging capacity is low (clastic formation with low clay content) or very high if scavenging capacity is absent (e.g. some pure carbonate reservoirs); However, in HC reservoirs, the sulfur species are usually out of thermodynamic equilibrium. At the same condition, concentration of ferrous iron (Fe^{2+}) appears in water at a high level while ferric iron (Fe^{3+}) disappears.

In highly reducing HC reservoirs, the geochemical system is dominated by components (elements) in their reducing status (species) and the reactions taking place among them. Nevertheless, in a regular thermodynamic database, geochemical reactions are defined using elements of regular (i.e. mild redox) conditions, rather than using their reducing forms, and thus are not convenient for use in a reduced-species-dominant system. For example, sulfur species in a regular database is represented using sulfate (SO_4^{2-}), as it is the dominant species in the whole family of sulfur species, and reactions involving sulfur species are all defined using sulfate. This is fine if the system is not in a reducing condition, and there is no change in the redox condition (i.e. no redox reactions are taking place), but a sulfide species, or sulfur species in another chemical valence, is needed if redox is going on. Correspondingly, relevant reactions also need to be redefined using reducing species. As an example, pyrite precipitation is usually defined in most of thermodynamic databases as:

$$\text{Pyrite}\left(FeS_2\right) + 3.5O_{2(aq)} + H_2O = Fe^{2+} + 2H^+ + 2SO_4^{2-} \tag{5.20}$$

This reaction takes place when pyrite is exposed to oxidized water, leading to dissolution. In reducing conditions, especially where oxygen is extremely depleted, pyrite would precipitate rather than dissolve. The above reaction would proceed backwards and sulfate is reduced first and then pyrite precipitates. Apparently, this reaction describes a thermodynamic equilibrium, but the reverse direction (i.e. sulfate reduction) does not happen under the same condition, because sulfate reduction is involved as the first step, which can take place only when catalysed by microbial activity or high temperature (thermochemical sulfate reduction, TSR). Sulfate reduction is kinetically constrained and irreversible under the same condition. The overall reaction network can also be written in a different form such as the following, if kinetics is omitted:

$$\text{Pyrite}\left(FeS_2\right)+H_2O=0.25H^+ +0.25SO_4^{2-} +Fe^{2+} +1.75HS^- \tag{5.21}$$

However, the above equations describe systems in a thermodynamic equilibrium condition, skipping the kinetic constraints that actually dictate the progress of the reactions. Multiple irreversible kinetic steps were needed to approach this condition as determined by the equations. Therefore, sulfate reduction should be treated as irreversible and handled in a kinetic network of reactions. In a HC reservoir of extremely reducing conditions, pyrite precipitates where there is replenishable Fe^{2+} and (S^{2-} or S^-) sulfide. Precipitation of pyrite (rate and amount) is dictated by the rate of sulfide replenishment, while Fe^{2+} can usually be recharged faster than sulfide. Under such reducing conditions, pyrite or pyrrhotite group minerals precipitate from sulfide, as illustrated by the following equation, rather than $SO_4{}^{2-}$ as described in Equation (5.21).

$$2HS^- +Fe^{2+} =FeS_2 +H_2 \quad \text{or} \quad 2HS^- +Fe^{2+} =FeS_2 +2e^- +2H^+ \tag{5.22}$$

$$HS^- +Fe^{2+} =FeS+H^+ \tag{5.23}$$

In summary, the equation written in the oxidation form is not a natural choice to model pyrite precipitation because: (1) the oxygen concentration (or fugacity) is extremely low (or essentially zero), so dealing with this reaction is numerically challenging; (2) the reverse reaction can only take place if $SO_4{}^{2-}$ can be effectively reduced (into H_2S), and sulfate reduction must rely on microbial activity or high temperatures and is considered irreversible; (3) precipitation of pyrite or pyrrhotite phases under reducing conditions from sulfide does not require transfer of electrons, or may only need to transfer an electron from S^{2-} (to produce S^-) to H^+ to form H_2 gas. Nonetheless, precipitation of pyrite as an S^- mineral may take place on the sulfate-reducing pathway, thus there is no need to transfer an electron from S^{2-} to H^+. Reducing S^{6+} to S^- or S^{2-} takes place in the TSR/BSR which is handled in RTM with a dedicated TSR/BSR kinetics. The value of getting into the reducing pathway is obvious: we can better understand and handle the precipitation of solid sulfide phases, which is important for H_2S risk evaluation because the scavenging is made up of the precipitation. Unfortunately, there is no detailed research reported in modeling the reducing pathway of TSR/BSR processes.

Redox processes of sulfur species and minerals are the most important reactions during waterflood, although a similar situation may apply to other redox-sensitive elements. To model these reactions properly, actual reactions in the reducing form that are suitable for sulfate reduction and pyrite precipitation should be installed in the thermodynamic database.

5.4.8 Numerical Simulators Available for Modeling 4-High Reservoirs

5.4.8.1 TOUGHREACT and TOUGHREACT-PITZER

TOUGHREACT (Xu et al., 2006, 2011, 2014) is a general-purpose multiphase flow and geochemical reactive transport numerical simulator that was originally developed by integrating geochemical transport into the framework of TOUGH2, a multiphase flow simulator (Pruess et al., 1999). TOUGHREACT also simulates diffusive and advective transport of gases and solutes, and geochemical reactions including aqueous speciation, mineral dissolution and precipitation under equilibrium conditions and kinetic constraints, cation exchange, surface complexations, and kinetics in the dissolved phases such as biodegradation and slow redox reactions. These processes can be simulated in complex flow and transport systems, such as in variably saturated environments and multiple porosity/permeability media (e.g. rock matrix–fracture systems). TOUGHREACT has been used to simulate reactive geochemical transport processes at various scales and over a wide range of geochemical conditions (Xu et al., 2006, 2009, and references therein; Xiao and Jones, 2006, 2007; Zhang et al., 2009, 2011; Whitaker and Xiao, 2010).

In general, TOUGHREACT can handle dilute to moderately concentrated solutions (<1 molal), concentrated NaCl-dominant solutions up to halite saturation (typically <6 molal), moderately concentrated $CaCl_2$-dominant solutions (typically <1 molal), concentrated $MgSO_4$-dominant solutions (typically <2 molal) and moderately concentrated $NaSO_4$-dominant solutions (typically <0.5 molal) by using an extended Debye-Hückel ionic activity model (i.e. Helgeson-Kirkham-Flowers, or HKF, model; Helgeson et al., 1981). The HKF ionic activity model has been extensively used/validated (Xu et al., 2011) in dilute to moderately concentrated solutions of salt mixtures up to halite saturation and 0.5–1.0 molal for other salts as the maximum.

There are presently no apparently defined HC phases considered in TOUGHREACT as fluids. HC can be considered as a generalized fluid phase: Non-Aqueous Phase Liquids (NAPLs) or gas mixture in RTM using two dedicated fluid modules, TMVOC and TMGS. The HC phase is generally omitted when dealing with reactions mainly in the water phase. This is acceptable for studies focusing on the souring and scaling effects due to water injection, and the HC is considered as not directly involved in the aqueous reactions that lead to reservoir souring and scaling. Exceptions occur when the exchange of chemical compounds between water and HC cannot be ignored, for instance, when acidity is released from HC as it is degraded (oxidized). The acidity affects calcite dissolution and produces secondary effects on other Ca-related minerals, including anhydrite, and as a source of partitioning of H_2S, which will slow down its transport to the producers. In some conditions (e.g. other dominant factor of calcite precipitation is weaker than the impacts of the acidity released from the HC, and/or TSR rate constant is much smaller than the present level) this effect can be significant.

TOUGHREACT-PITZER (Zhang et al., 2006a,b) is an extended version of TOUGHREACT v.1.2. Pitzer's ionic activity model is implemented in this version to handle highly concentrated brine. With this version, most brine in the HC reservoir can be modeled without limitation in ionic strength. This functionality, in addition to the comprehensive capability of TOUGHREACT v.1.2, leads to TOUGHREACT-PITZER being one of the simulators usable for the RTM of geochemical systems in 4-High conditions. One advantage of TOUGHREACT-PITZER – high numerical efficiency and stability compared with other RTM simulators using a PITZER ionic activity model – arises from the analytical calculation of the derivative terms in the Jacobean matrix in solving the nonlinear algebraic system.

5.4.8.2 PHREEQC-based Simulators

PHREEQC (Parkhurst and Appelo, 1999, 2013) is a batch geochemical simulator that has lately been extended with one-dimensional flow and transport simulation functionality. From a geochemical point of view, it is a robust simulator that can handle high temperatures, high pressures, high salinity and highly reducing conditions. This simulator has been used as a stand-alone geochemical solver to be combined with a multiphase flow and transport simulator to work as an RTM simulator. COMSOL-PHREEQC Interface (Nardi et al., 2014), PHAST (Parkhurst et al., 2010), HYDRUS-PHREEQC (Jacques et al., 2008) and MoReS-PHREEQC (Farajzadeh et al., 2012; Wei, 2012) are examples of this type of simulator. Among them, MoReS is the Shell in-house HC reservoir simulator, which employs comprehensive equations of state (PVT models, e.g. Peng–Robinson equation of state) for multiple fluid phase and solute transport in the HC reservoir at high temperature, high pressure conditions for complex geological structures. Including the capability of PHREEQC for 4-High conditions (with a dedicated thermodynamic database), MoReS-PHREEQC has become one of the few simulators that can effectively handle the RTM of a 4-High system. The numerical efficiency and stability, while dealing with high concentrations (salinity), is not ideally optimized due to the algorithm used in calculating the derivative terms in the geochemical solver of PHREEQC, which is slow, as well as the coupling scheme which is numerically loose, except for tight coupling features for HC–water component partitioning (Snippe and Wei, 2014). Simulators such as PHAST and HYDRUS-PHREEQC are considered to behave similarly to MoReS-PHREEQC, but may not be suitable for reservoirs at 4-High conditions. Other RTM simulators suitable for 4-High reservoirs may exist, possibly in the oil and gas industry, but no reports of them are available.

5.5 Applications of RTM in Evaluating Risks Related to Geochemical Processes in 4-High Reservoirs

In this section, three previously unpublished 4-High RTMs using TOUGHREACT-PITZER are presented. Models using MoReS-PHREEQC have been published elsewhere, and include applications in water flooding, enhanced oil recovery (EOR), CO_2 sequestration and sour gas injection (Al Mayahi et al., 2012; Farajzadeh et al., 2012; Wei, 2012; Snippe and Tucker, 2014; Snippe and Wei, 2014; Kampman et al., 2016; Tambach et al., 2017; Snippe et al., 2017).

5.5.1 RTM Evaluation of Well and Reservoir Scaling and Clay Swelling During Waterflood

Water injection is performed in an offshore sandstone reservoir (referred to as 'TBC' in this book). Well/reservoir scaling and clay swelling had been identified as the major risks. RTM was performed to predict the water composition change, dissolution and precipitation of minerals caused by the water injection, as input for the risk assessment. Predicted mineral precipitation was directly used to evaluate the scaling potential, and the predicted water composition change was used as input to evaluate clay swelling potential using an empirical approach (Scheuerman and Bergersen, 1990).

5.5.1.1 Geological, Hydrogeological and Geochemical Setting

The TBC reservoir lies at 1650 m water depth, and approximately 2925–2960 m subsea (mss), at 58 °C, consisting predominantly of medium to coarse-grained sands in massive, amalgamated bodies. A radial 2D domain covering the whole reservoir formation interval and spanning a radial distance of 1000 m (estimated distance to define a constant pressure and constant water composition boundary under the injection rate) is considered for setting up the reactive transport model. Twenty-five model layers, at 1.575 m thick/layer on average, are defined to give a proper vertical resolution of the RTM to reflect the vertical heterogeneity in mineralogy and hydraulic properties, such as permeability, porosity and capillarity. Lateral variation in these properties is small within the model domain, and therefore ignored. Permeability averages over all model layers at $1.18 \times 10^{-12}\,m^2$, with a minimum of $1.6 \times 10^{-14}\,m^2$ and a maximum $4.2 \times 10^{-12}\,m^2$. Porosity varies between 0.22 and 0.32, with an average of 0.286. Water saturation ranges between 0.36 and 0.45 in the shale, and between 0.09 and 0.15 in the sandstone layers. Along with the porosity, the amount of the water within the pore space is determined by the product of porosity and water saturation, called water-occupied porosity or water porosity for short. The water porosity varies between 0.024 and 0.103, with an average of 0.028. Water (effective) permeability is calculated using the cubic law based on the water-occupied porosity.

Mineralogical composition, based on XRD and SEM data, is dominated by a felsic component. Quartz and feldspar content was observed at above 80 wt.% in most samples. Content of total clay minerals varies from 6–11%. Three clay-rich (clay content 27–57%) layers were found at the top, middle and bottom of the formation. Minor minerals include carbonates such as calcite (2–12%) and siderite (0–3%), sulfates such as gypsum (2%, detected only in two samples), and pyrite (from trace amounts up to 6%). Clinoptilolite (a natural zeolite comprising a microporous arrangement of silica and alumina tetrahedral, 3%) is observed at the topmost layers. The actual specific reactive surface area can be determined using different technologies, such as gas sorption; however, none of them is believed reliable due to various reasons such as heterogeneity. There is no direct measurement available for mineral reactive surface area, and it was estimated based on SEM image. Calibration of the mineral surface area is necessary because literature data and the estimates are not consistent, and therefore neither are sufficiently reliable. The calibration of the surface area was performed together with the calibration of cation exchange selectivity coefficient by perusing a good match between model-simulated steady-state formation water composition and the measured background formation water composition. The calibrated specific reactive surface area and other kinetic parameters are listed in Table 5.2.

Several water samples were collected; S-78A and S-81A are considered most representative of TBC reservoir formation water. Salinity ranges between 169,181 and 229,560 mg total solids (TDS)/litre brine, and pH ranges between 6.13 and 6.9. The difference in cation concentrations relates to the difference in content and type of clay minerals of the interval of perforations. Measured concentrations of trace metals, except for Sr and Ba, are usually lower than the detection limit, and they have very limited effects on defining the overall chemical character. Sr and Ba are chemically similar to other divalent cations such as Ca^{2+} and Mg^{2+} and also engage in exchange with clay interlayer cations and precipitation of carbonate and sulfate minerals. The sulfate concentration of sample S-71A is higher than others, and correspondingly, concentration of Ba from this sample is significantly lower than the others, reflecting the control of barite solubility. However, barite may come from the drilling fluid. Thus the reliability of the measured concentrations of sulfate and Ba is questionable.

Table 5.2 Mineral specific reactive surface area, kinetic rate constants and other kinetic parameters.

Mineral	Specific reactive surface area (cm^2/g mineral)	Dissolution/ precipitation rate constant (mol/s/m^2)	Activation energy (kJ/mol)	Parameters η/θ (Equation B.5 in Xu et al., 2006)	Dissolution rate constant (mol/s/m^2) in acid mechanism and its activation energy (kJ/mol), species and parameter n_{ij} in Xu et al. (2006)
Barite	*	*	*	*	*
Calcite	*	*	*	*	*
Gypsum	*	*	*	*	*
Dolomite (disordered)	*	*	*	*	*
Siderite	*	*	*	*	*
Magnesite	*	*	*	*	*
Quartz	0.255	1.2589×10^{-14}	38.0**/87.5***	1.0/1.0	No addition mechanism
K-feldspar	0.185	3.8905×10^{-13}	38.0**/87.5***	1.0/1.0	8.7096×10^{-11}/51.7/H^+/0.5
Albite	0.185	2.7542×10^{-13}	69.8	1.0/1.0	6.9183×10^{-11}/65.0/H^+/0.457
Clinoptilolite	55	1.6596×10^{-13}	35.00	1.0/1.0	1.0471×10^{-11}/23.6/H^+/0.34
Smectite	55.23	1.6596×10^{-13}	35.00	1.0/1.0	1.0471×10^{-11}/23.6/H^+/0.34
Illite	10.22	1.6596×10^{-14}	35.00	1.0/1.0	1.0471×10^{-19}/23.6/H^+/0.34
Muscovite	32.25	1.0184×10^{-13}	22.00	1.0/1.0	No addition mechanism
Kaolinite	8.03	6.6069×10^{-9}	65.90	1.0/1.0	4.8978×10^{-8}/65.90/H^+/0.777
Chlorite	18.1	3.01995×10^{-13}	88.0	1.0/1.0	7.7624×10^{-12}/88.0/H^+/0.50
Pyrite	1.992	2.8184×10^{-8}	56.9	1.0/1.0	No addition mechanism

Notes:
* Mineral assumed to be constrained by local equilibrium.
** Activation energy for dissolution.
*** Activation energy for precipitation.

The present chemical composition of the formation water is a result of water–rock interaction during geological time, and at least since the latest significant charge event which is considered to be multiple million years ago. Interactions between porewater and minerals lead to dissolution of felsic (primary) minerals at one place and/or time, and precipitation of other minerals, such as calcite, sulfate and ferric and ferrous minerals, at another place and/or time due to changes in temperature and pressure during burial and subsidence. Cation exchange between dissolved cations in the porewater and clay interlayer cations also influences the water composition, especially where significant amounts of clay minerals are present in the water–rock system. At present the formation water must be, at least locally, in quasi-equilibrium with respect to the formation minerals, while some slow kinetic reactions are still ongoing, making up the slow diagenetic trend. This local equilibrium status determines the water chemical composition. Geochemical simulations can verify the understanding of the geochemical condition, that is, the current water composition and water–rock integration. Speciation models and batch reaction models are useful tools for this purpose. At first we conducted speciation simulations as well as geochemical steady-state simulations to build a sound initial condition for the subsequent simulation with injection of water.

5.5.1.2 RTM Setup using TOUGHREACT-PITZER and Model Calibration

The simulation was performed using TOUGHREACT-PITZER with EOS3 fluid module, without explicit consideration of the oil as a fluid phase. A pragmatic approach was used to account for the displacement of oil by injecting water. The oil phase, compared with the water, is assumed immobile in the initial condition in the model, and the displacement of the oil is mimicked by assuming it 'dissolves' by making up the porosity, causing 'water porosity' increase during the injection of water. The universal kinetic rate law was used to calculate the 'displacement' of oil, and the parameters of the rate laws were calibrated based on the expected displacement progress.

Four types of water were used for the injection at a rate of 5 kg/s/well: untreated seawater, treated seawater I, treated seawater II, and a fresh water (low salinity). The composition of these waters and the formation brine is given in Table 5.3.

Thermodynamic equilibrium constants are read directly from the TOUGHREACT database, which was developed based on the thermodynamic data from the EQ3/6 package. Kinetic rates and other relevant parameters are from Palandri and Kharaka (2004). Ion exchange selectivity coefficients are salinity-dependent. Values from the literature are mostly for fresh water and not useable for our case, in which ionic strength reaches 2.6 molal and calcium concentration is above 0.1 molal. Thus, we calibrated this parameter together with adjustment of the mineral reactive surface area within the uncertainty range. Initial values are from Birkholzer et al. (2008) and calibrated by reproducing the formation water composition of S-78A by the modeled concentrations from the steady-state simulation using a vertical one-dimensional geochemical model covering all the model layers. Values of the calibrated selectivity coefficient and typical values from the literature as the initial values for our calibration are listed in Table 5.4. However, the selectivity coefficients need further validation with experiments to account for scenarios with different water salinities than the present case. For example, modeling injection of fresh water may require different selectivity coefficient values.

Comparison of the modeled steady-state water composition and the pH with the measured water composition, showing that the steady-state water composition matches

Table 5.3 Chemical composition of injection and formation waters.

Components	Unit	Untreated seawater	Treated seawater I	Treated seawater II	Fresh water	Formation water
Ca^{2+}	mg/kg	500	316	121.58	200	4750
Mg^{2+}	mg/kg	1300	400	88.10	5	990
Na^{+}	mg/kg	12,018	11,277	11,530.32	561	53,350
K^{+}	mg/kg	400	367	379.56	18	450
Cl^{-}	mg/kg	21,406	19,408	18,488.21	1245	95,546
HCO_3^{-}	mg/kg	142	92	101.94	5	Tiny***
SO_4^{2-}	mg/kg	2800	25	37.72	1*	7.5
Temperature	°C	18–30	18–30	18–30	18–30	58
pH	pH unit	8.45	7.74**	8.21	7–8 (7.8**)	6.5
TDS	mg/kg	38,632	31,883	30,765.95	2038	161,250

Notes:
* Analyses suggests 'almost' no sulfate; 1 ppm is assumed in the RTM.
** Calculated pH based on the compositions.
*** Bicarbonate concentration is constrained by calcite solubility and is low, as calcium concentration is very high.

Table 5.4 Selectivity coefficients from the literature and calibrated values from the present work.

Parameters	Calibrated	Literature
Master cation	H+	Na+
H^{+}	1.00	1.00
Na^{+}	0.24	1.00
Ca^{2+}	0.76	0.4
Mg^{2+}	0.88	0.45
K^{+}	0.27	0.2
Ba^{2+}	0.46	0.35
Sr^{2+}	0.45	n/a

well with the measured composition in terms of concentration of most components, leaves the concentrations of the redox-sensitive components (e.g. iron) and trace components (e.g. aluminium) discrepant from each other, possibly due to (1) the fact that the sample might have been oxidized (e.g. Fe^{2+} to Fe^{3+}) during sampling, handling, transport and analysis; and (2) the concentration of the component is lower than the detection limit (such as Al). The good match indicates that the RTM catches the deterministic reactions of the water–rock system within the reservoir formation, forming a good initial condition for the RTM prediction (to predict changes of water composition and mineral dissolution/precipitation) during water injection.

Table 5.5 Saturation indices of relevant minerals with respect to formation water and four injection waters.

Mineral	S-78A	S-78A*	Untreated seawater	Treated seawater I	Treated seawater II	Fresh water
Barite	0.99	0	0	−2.02	−1.76	−2.68
Calcite	1.38	0	0.94	−0.03	0.09	−0.88
Dolomite (disordered)	2.45	−3.20	2.12	−0.11	−0.12	−3.51
Siderite	−6.72	−8.09	−0.21	−0.93	−0.42	−1.55
Anhydrite	−2.19	−3.18	−0.73	−2.80	−2.96	−3.67
Gypsum	−2.41	−3.40	−0.57	−2.64	−2.80	−3.45
K-feldspar	−2.84	0	−10.09	−9.58	−9.83	−10.58
Pyrite	−51.89	0	−62.33	−64.34	−64.63	−69.67
Quartz	−0.12	−0.08	−3.65	−3.47	−3.56	−3.42
Albite (low)	−2.66	0.18	−11.14	−10.62	−10.88	−11.61
Smectite (Na)	−4.10	1.17	−13.93	−12.65	−13.63	−13.12
Smectite (K)	−4.58	0.68	−14.13	−12.84	−13.82	−13.31
Smectite (Ca)	−4.31	0.95	−14.20	−12.92	−13.96	−13.02
Smectite (Mg)	−4.23	1.04	−14.01	−12.78	−13.86	−13.12
Montmorillonite (Na)	−4.03	0.88	−13.44	−12.37	−13.27	−12.99
Montmorillonite (K)	−4.58	0.33	−13.66	−12.60	−13.49	−13.21
Montmorillonite (Ca)	−4.27	0.64	−13.73	−12.68	−13.65	−12.87
Montmorillonite (Mg)	−4.17	0.74	−13.53	−12.53	−13.53	−12.99
Illite	−4.33	2.43	−10.40	−9.04	−10.07	−9.87
Muscovite	−9.29	−0.39	−12.13	−10.19	−11.39	−11.15
Kaolinite	−6.32	−0.19	−9.17	−7.39	−8.51	−7.24

Note:
* Mineral constraints are applied: bicarbonate concentration is determined by calcite saturation, sulfate concentration is determined by barite saturation, aluminium concentration is determined by K-feldspar saturation, and redox status is determined by pyrite saturation.

Given the calculated saturation indices of minerals (the third column, S-78A*, in Table 5.5 to which appropriate mineral constraints are applied), felsic minerals (quartz and K-feldspar), carbonate minerals (calcite, dolomite) and sulfate minerals (pyrite and barite) are in (or close to) an equilibrium with respect to the formation water, while most of clay minerals are oversaturated except for kaolinite and muscovite, which are undersaturated but still close to an equilibrium with respect to the water. This result makes sense considering the actual geochemical conditions of the water–rock system (i.e. water with the chemical composition contacting the reservoir formation minerals at that pressure and temperature). Primary minerals, mostly felsic, reached equilibrium with respect to the water because of the long reaction time (multiple millions of years) and moderately fast reaction kinetics. Secondary minerals such as carbonates and sulfates are in or close to equilibrium because of their rapid reaction kinetics.

Clay minerals (mainly illite and smectite) have slow kinetic rates and keep precipitating and removing the products from dissolution of felsic minerals. Meanwhile, clay minerals convert from one to another: alteration of smectite (close to saturation) and kaolinite (slightly unsaturated) into illite (highly oversaturated), alteration of mica (muscovite, undersaturated) into kaolinite (slightly unsaturated) and alteration of mica (undersaturated) into illite (highly oversaturated) take place. These findings (from the speciation model) are consistent with the SEM observations. However, K-feldspar (in equilibrium) can still precipitate (overgrowths) where conditions are appropriate (Section 5.5.1.3).

A good match implies good representation of the geochemical reactions in the system, as well as a good calibration of the relevant parameters used in the simulation. Figure 5.3 (right) shows model-simulated pH as a function of time and depth approaching stability within 2000 years. At 2000 years, the average pH over the entire column is very close to the measured pH of S-78A, implying a good representation of the whole geochemical system by the model. A good match is also obtained for most other components (Figure 5.3, left plots).

5.5.1.3 Model-Predicted Scaling Risk

As discussed earlier, injection of seawater into 4-High reservoirs always leads to a risk of calcite scaling. Calcite precipitates, largely, as the calcite-saturated/oversaturated seawater is injected and heated up downhole as it mixes with the high-calcium formation water (4750 ppm, see Table 5.3). RTM results as shown in Figure 5.4 indicate that calcite precipitates at an amount up to 50 mol/m^3 within the near-wellbore region (5 kg calcite/m^3 of rock). Note that this RTM is a reservoir-scale model focused on the reactive transport processes within the reservoir, rather than a high-resolution wellbore model (there is no wellbore explicitly discretized in the model), so this amount of calcite precipitation is huge. In reality, the calcite precipitation would be more concentrated inside the wellbore, so that concentration within the wellbore must be higher than the volume-averaged figure of 5 kg/m^3 that this model predicted. Obviously, this is a severe scaling issue. The porosity change is illustrated in Figure 5.5.

Model results also indicate that injection of treated seawater I and treated seawater II can largely reduce the scaling potential of calcite, as shown in Figures 5.6 and 5.7. The predicted amount of calcite is at least ten times lower than that for injecting untreated seawater. Injection of fresh water leads to a dilution of the reservoir formation brine and dissolution of calcite and other minerals at the wellbore and the near-wellbore region (Figure 5.8). Scaling is no longer a potential risk.

5.5.1.4 Model-Predicted Clay Swelling Risk

As shown in Figure 5.9, model-predicted water compositions in all gridblocks in the model domain, and from the beginning of injection through the end of the injection, are all plotted on the Scheuerman diagram for one smectite end-member (the potential swelling clay component in the TBC reservoir). When injecting untreated seawater, the model predicted that the water compositions (diamond data-points in Figure 5.9, top left plot) of all the times and locations all fall within the 'no impairment' region, and that means injection of the untreated seawater will not result in any significant clay swelling risk. For injection of fresh water, the model predicted water composition falls within the 'severe impairment' zone, indicating a definite clay swelling risk as shown in Figure 5.9 (bottom right). The swelling potential largely increases when injecting fresh water

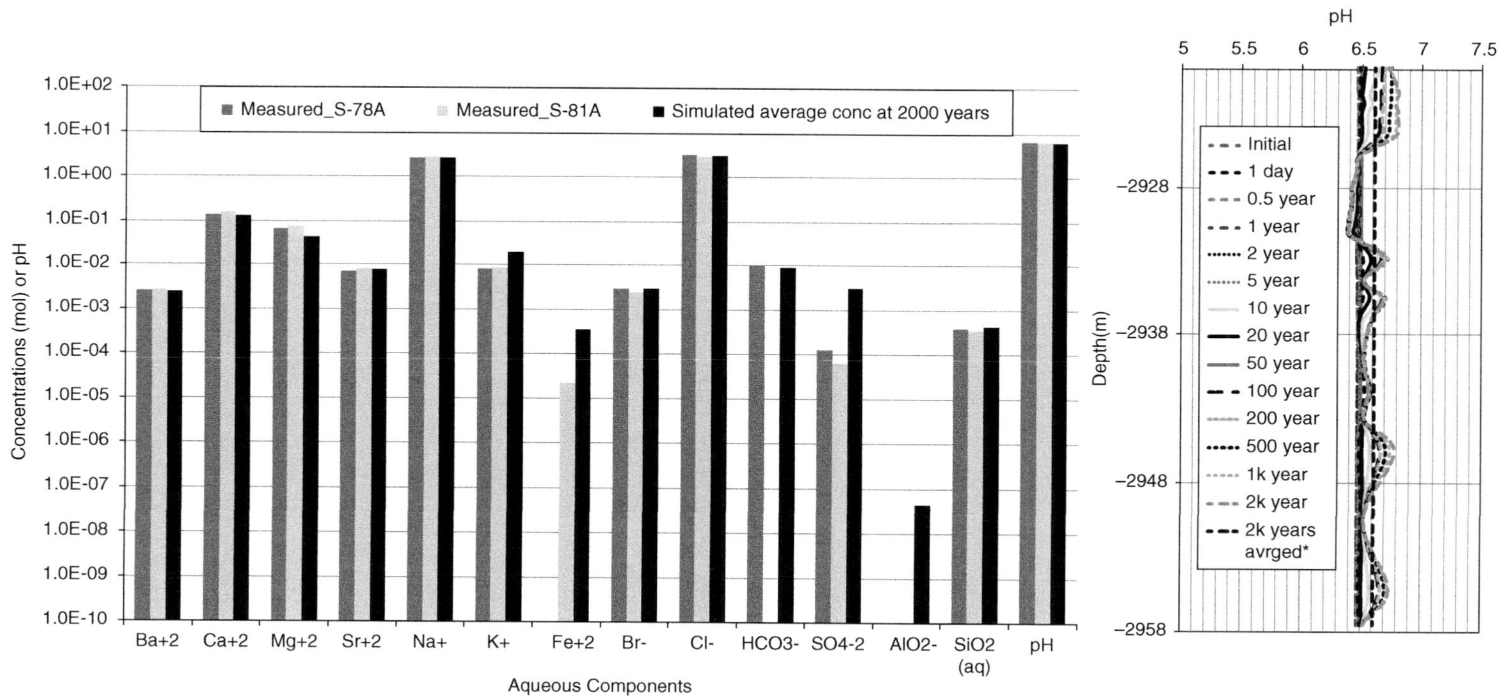

Figure 5.3 Comparison of the modeled average concentrations over the formation and the measured background concentrations from two representative samples (left), and modeled evaluation of pH over time (right). Results indicate that measured concentrations of most cations, carbonate (only one sample) and Si, and pH are well matched. The modeled average pH over the whole interval (2000 year averaged*) is 6.6 and the measured pH is 6.5: an insignificant difference. Measured Fe and Al are too low (1 mg/kg) or not available, thus simulated concentration is considered as a reconciliation of these species (note that aluminium concentration is constrained by K-feldspar solubility, carbonate concentration is constrained by calcite solubility, sulfate concentration by barite and redox by pyrite). The modeled concentration of sulfate deviates from the measured concentration. This is probably due to water samples being contaminated by drilling fluids containing barium sulfate. Cl^- and Br^- are conservative, and thus not affected by the reactions in the simulation.

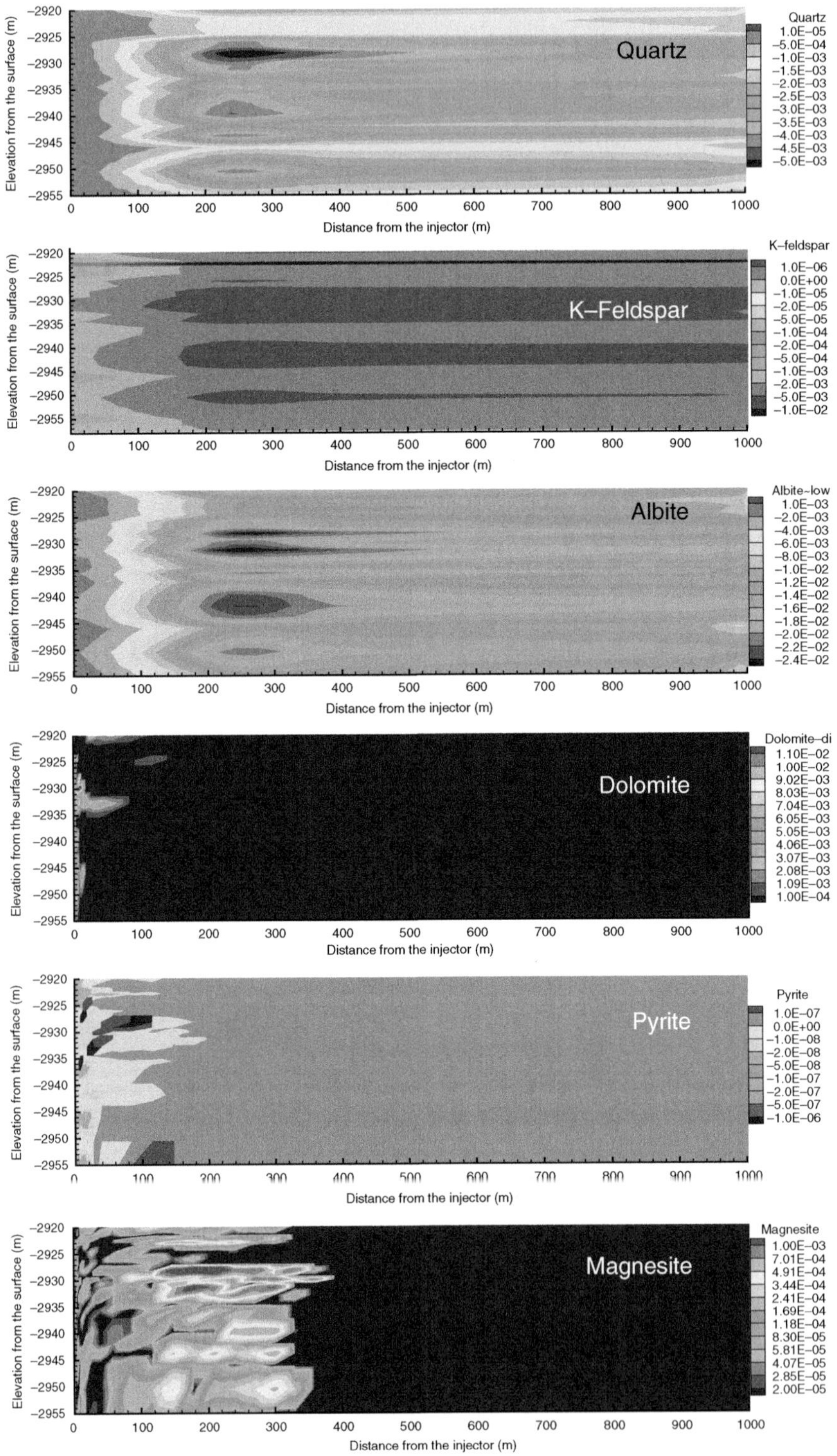

Figure 5.4 Model-predicted dissolution/precipitation of minerals (mol mineral/m^3 of rock) at the 20th year of the injection of untreated seawater. The colour scales represent the distribution of concentrations, from high (red) to low (blue). Note that precipitation or dissolution of most of these minerals is considered minor in terms of the amount of change except for calcite. (See colour plate section for the colour representation of this figure.)

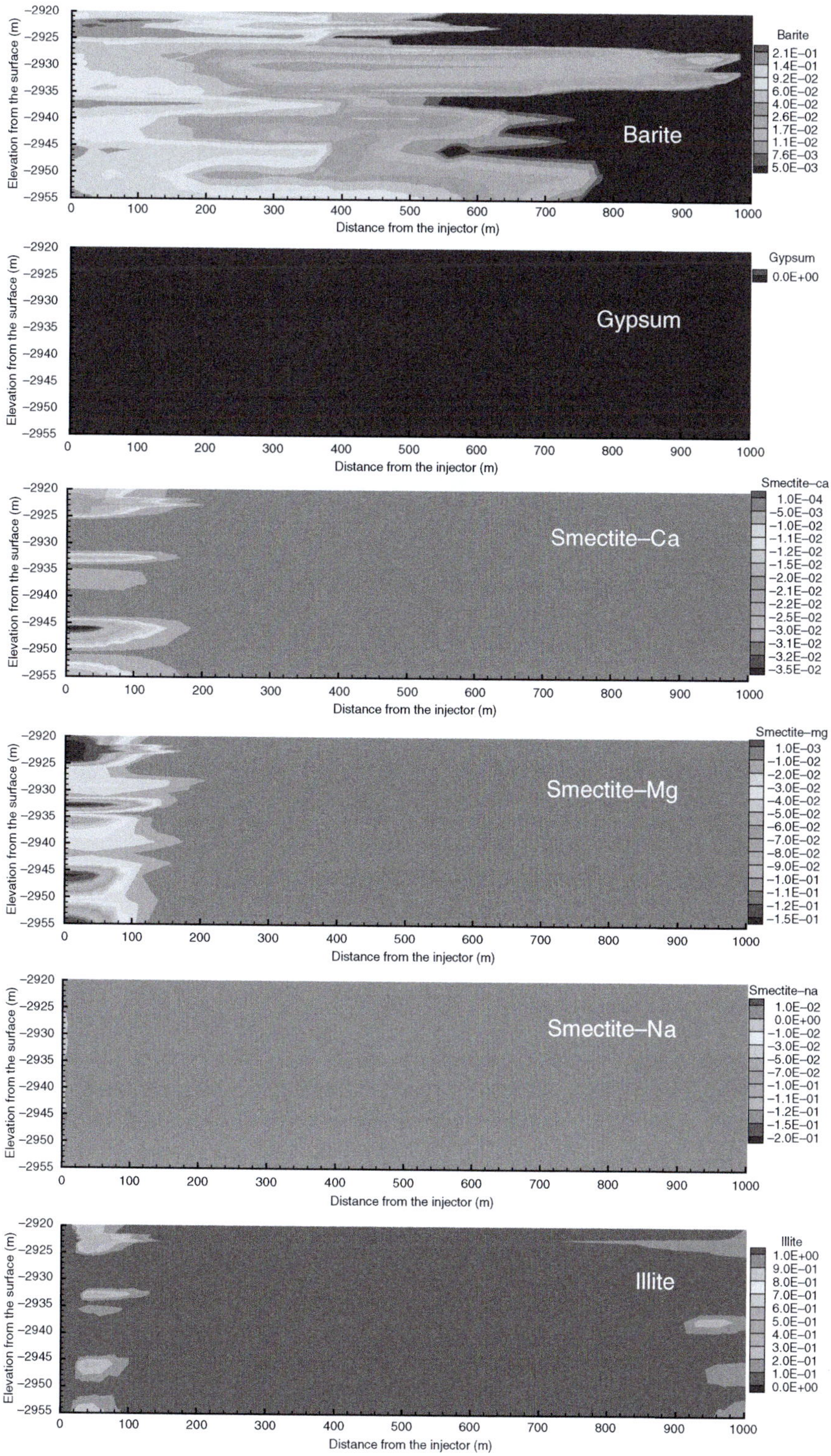

Figure 5.4 (Cont'd)

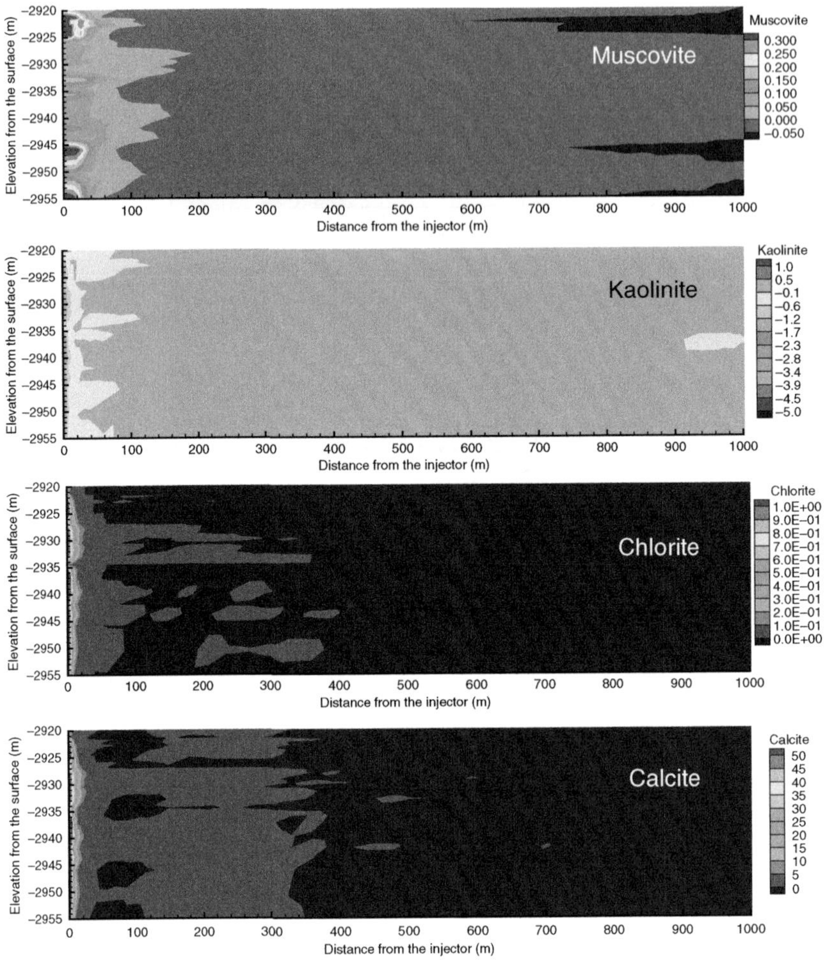

Figure 5.4 (Cont'd)

mixing with the formation water in a transition zone in which the background water–rock equilibrium has been destroyed and a new equilibrium has not yet been established. This transition zone/time will be succeeded by a stable phase dominated by the injection water. Clay swelling potential of injection of treated seawater I and treated seawater II also remains low, as illustrated in Figure 5.9 (top right and bottom left), although the treatment removes, largely, Ca^{2+} and Mg^{2+} from the seawater.

5.5.1.5 Summary and Limitations

RTM results show that injection of untreated seawater will not cause clay swelling, but large amounts of calcite will precipitate, forming scale in the wellbore and near-wellbore region; Injection of treated seawater I has very low or no potential for clay swelling and a low potential for scaling. Slow precipitation of clay minerals, illite, muscovite and chlorite, can lead to an average porosity change at about −0.3 pu at or near the wellbore

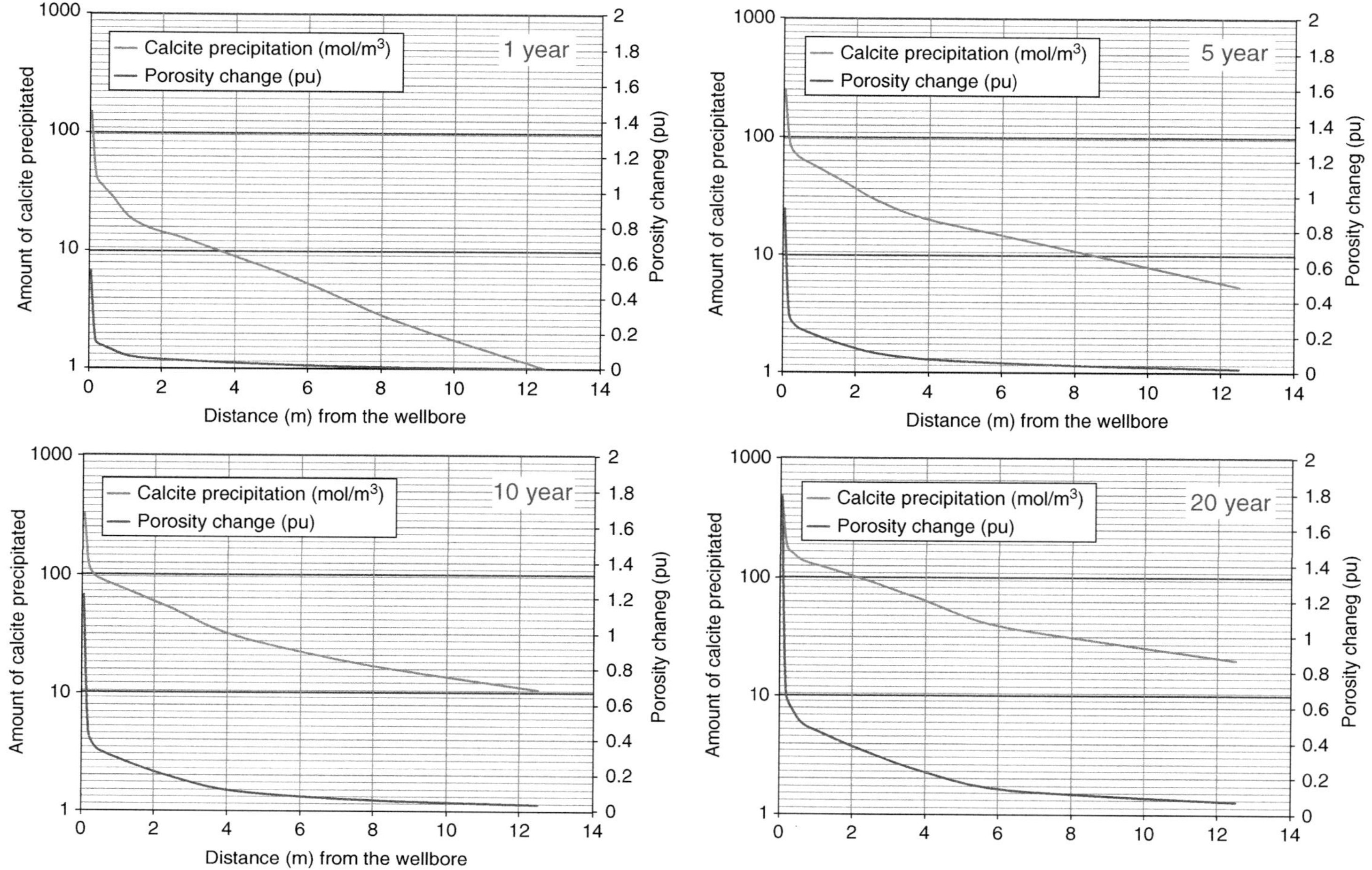

Figure 5.5 Model-predicted amount (mol/m^3 rock) of calcite precipitation and the resulting porosity reduction (porosity unit, 1 pu = 1% volume fraction) as a function of distance (m) from the injector at 1st, 5th, 10th and 20th years of untreated seawater injection.

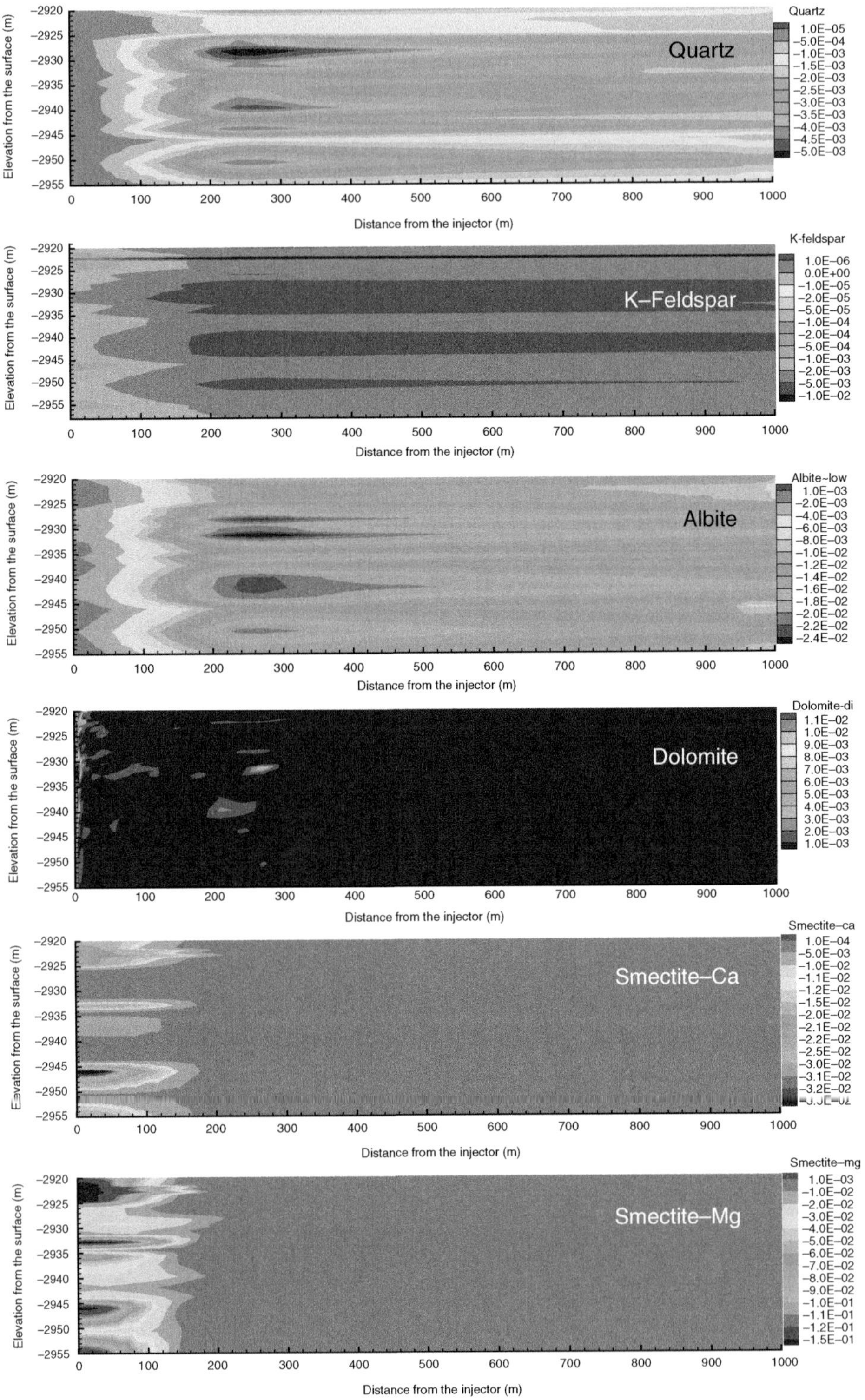

Figure 5.6 Model-predicted dissolution/precipitation of minerals (mol mineral/m^3 of rock) at the end of injection of treated seawater I. Note that precipitation or dissolution of these minerals is minor: less than 0.01 pu porosity change.

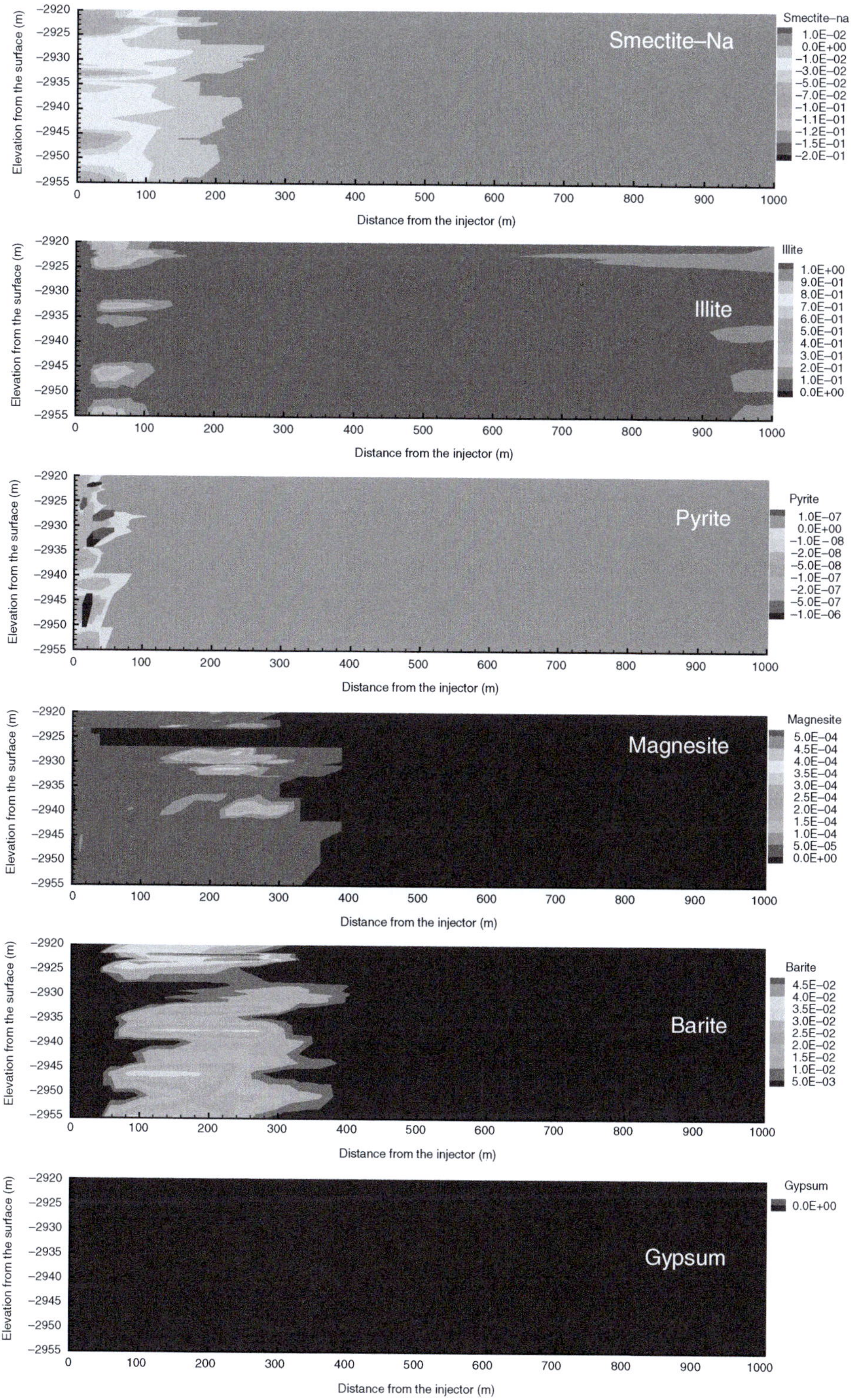

Figure 5.6 (Cont'd)

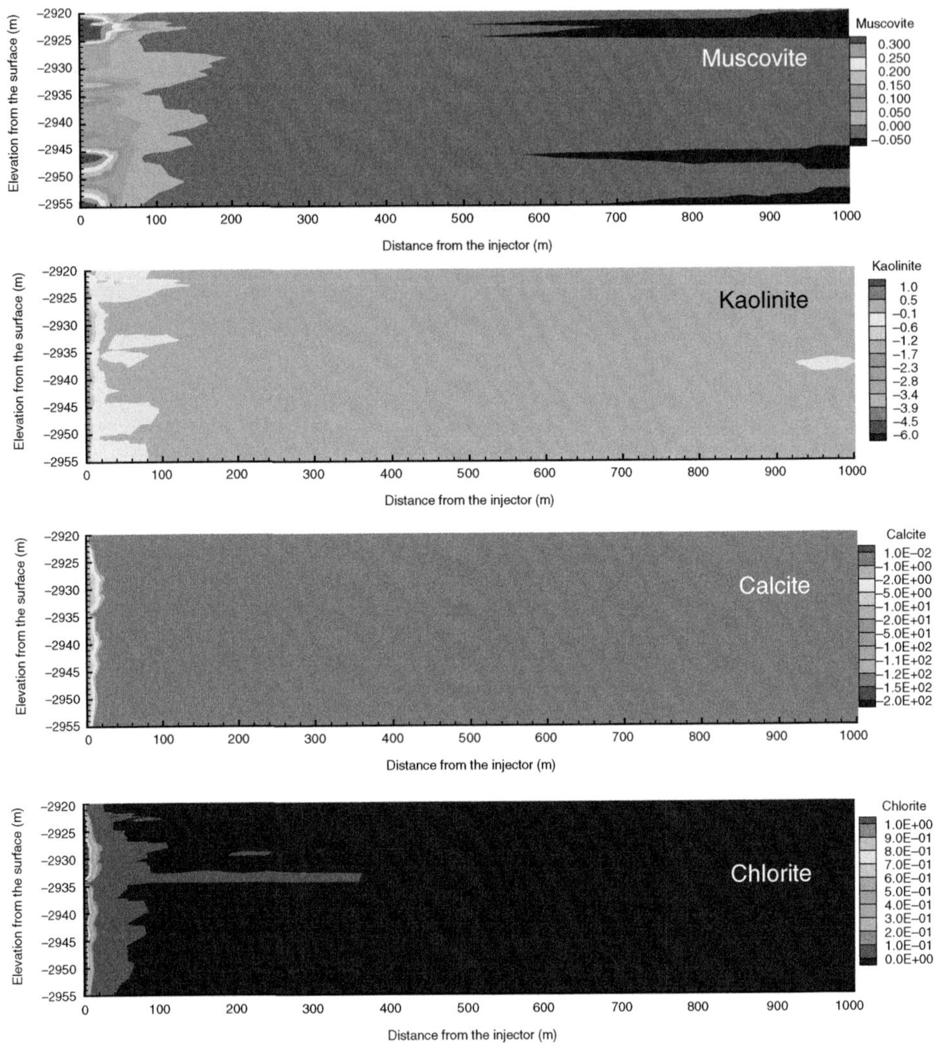

Figure 5.6 (Cont'd)

region, but precipitation is slow and the porosity reduction can be compensated for by calcite dissolution (+0.7 pu). Potential wellbore instability issues may occur. Injection of treated seawater II can induce low clay swelling potential at or near the wellbore and a low scaling potential. Slow precipitation of clay minerals, illite, muscovite and chlorite is predicted, leading to average porosity change of about −0.3 pu at or near the wellbore region, but the precipitation is slow and the porosity reduction can be compensated for by calcite dissolution (+0.35 pu). There is a potential risk for wellbore instability. Injection of fresh water (low salinity water) will lead to significant potential of clay swelling at the mixing zone of the formation and a low scaling potential. However, clay swelling within the formation away from the wellbore may pressurize the formation leading, potentially, to an improvement in the well productivity. Injection of this water

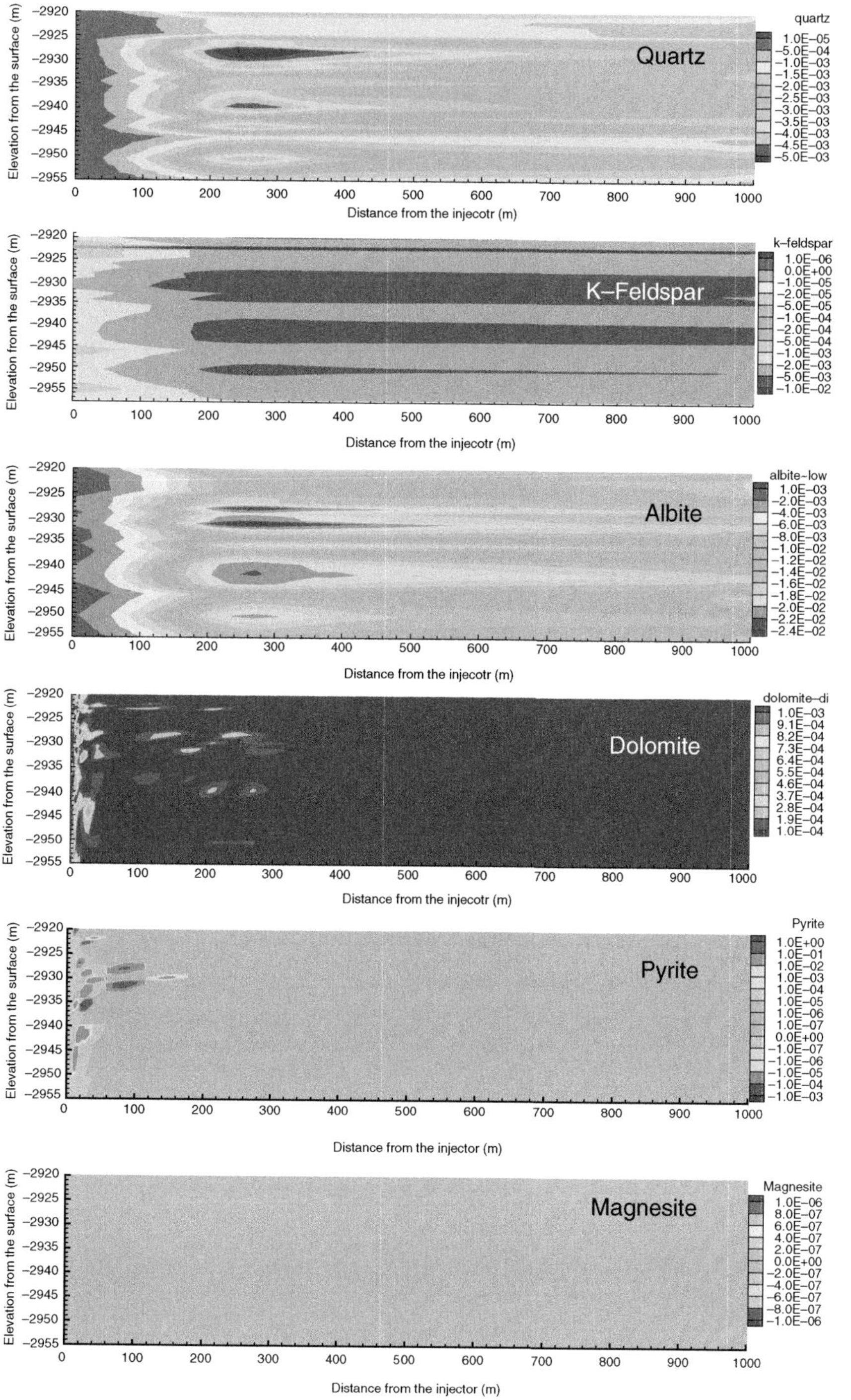

Figure 5.7 Model-predicted dissolution/precipitation of minerals (mol mineral/m^3 of rock) at the end of 20 years of injection of treated seawater II. Note that precipitation or dissolution of these minerals is minor in terms of mineral abundance or porosity.

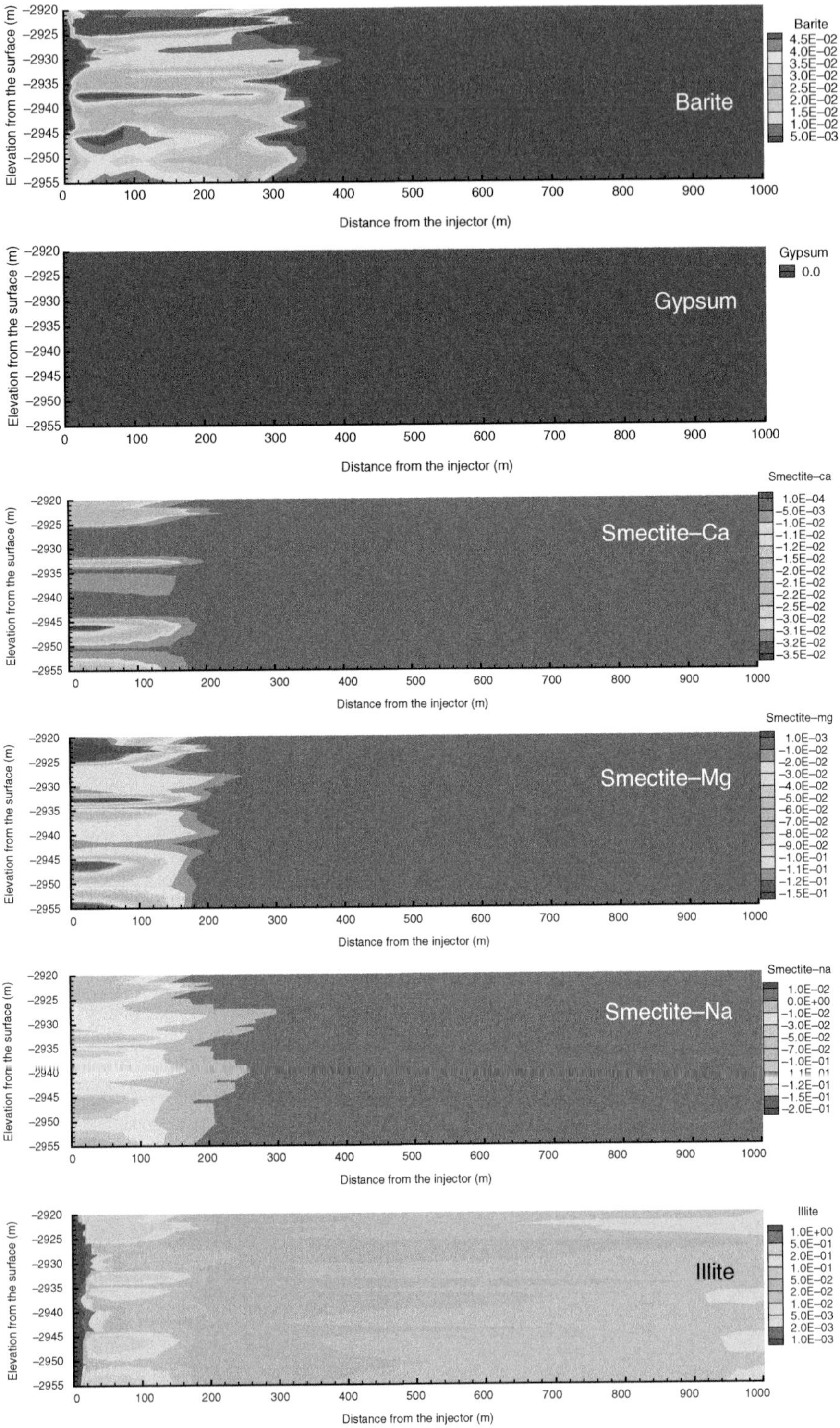

Figure 5.7 (Cont'd)

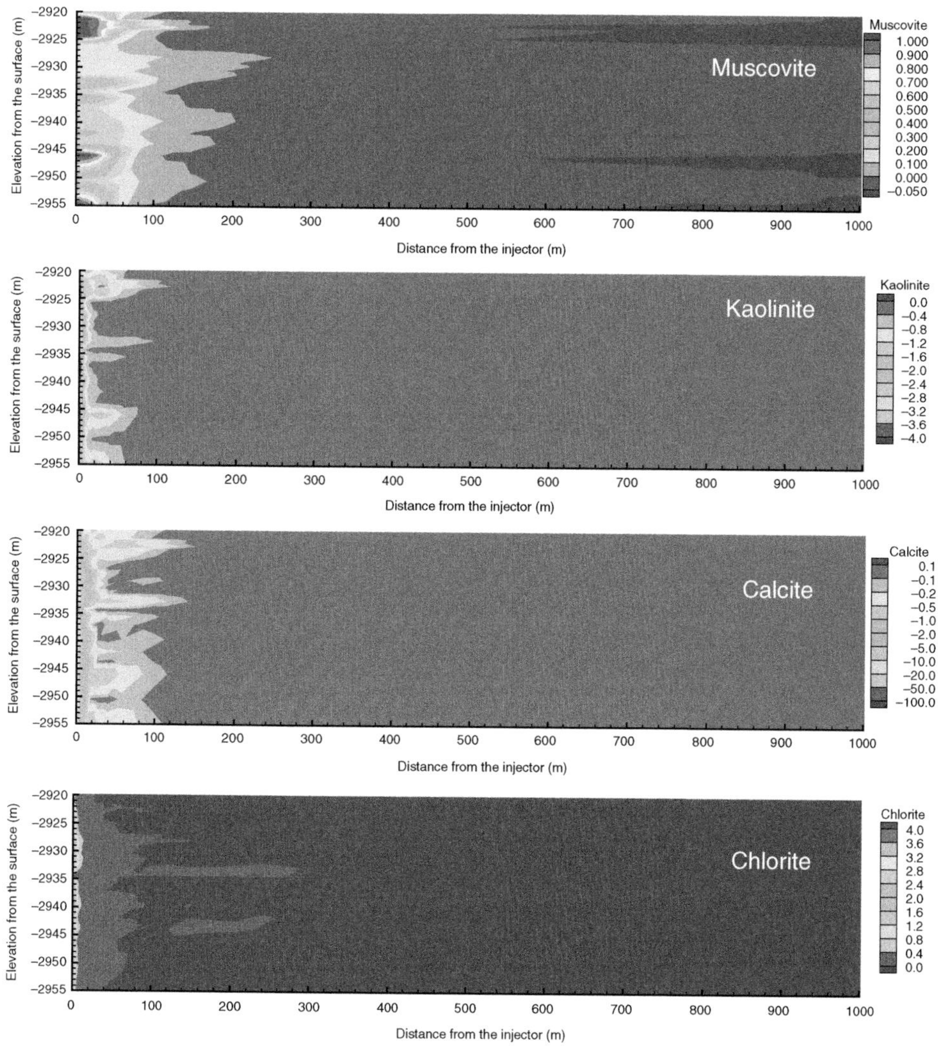

Figure 5.7 (Cont'd)

can induce slow precipitation of clay minerals, illite, muscovite and chlorite, leading to average porosity change of about −0.1 pu at or near the wellbore region, but calcite dissolution at the wellbore can enhance the porosity (+0.4 pu) and wellbore instability needs to be checked.

Thermodynamic data used in the present RTM are built mainly for ambient pressure (atmospheric pressure). The thermodynamic equilibrium constants and other parameters are mostly pressure dependent. Uncertainties of RTM results by using the current pressure-independent data, such as thermodynamic equilibrium constants of carbonate minerals and CO_2 gas, have been evaluated (Zhang et al., 2011) and believed acceptable. Although the present RTM was carefully calibrated by matching the present formation water composition, caveats still remain for most of

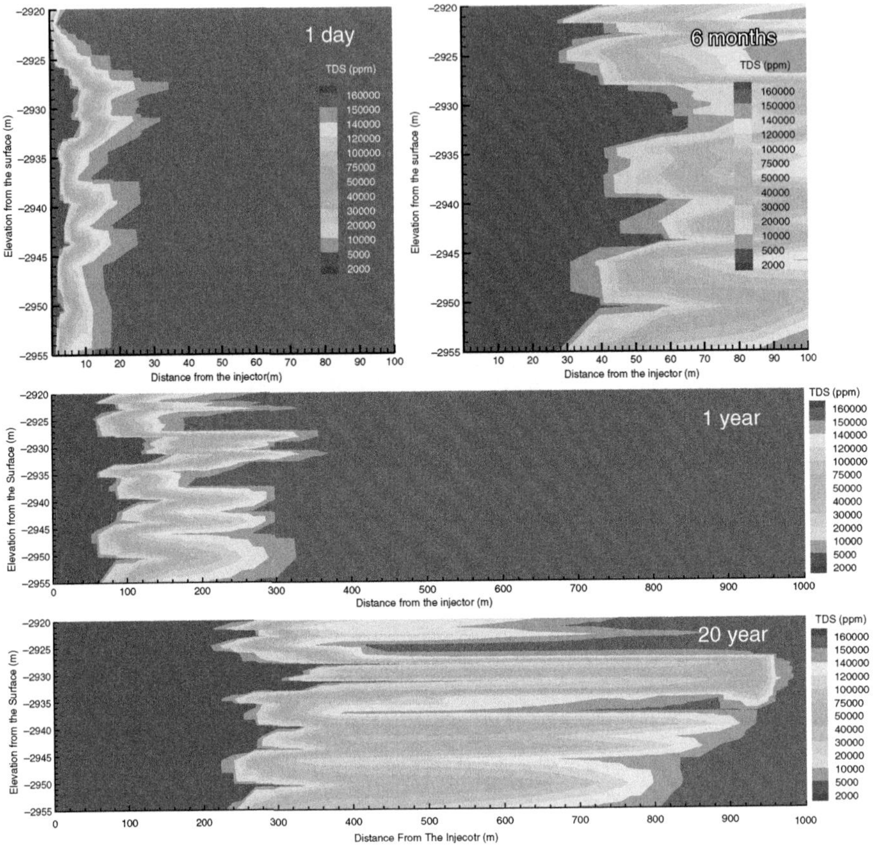

Figure 5.8 Model-predicted total dissolved solids (TDS in ppm) at 1 day, 6 months,1 year and 20 years during injection of fresh water, with geochemical reactions between the injected water and the pristine reservoir porewater, mineral dissolution/precipitation and cation exchange all considered. Note that the x-axis extent of the two top figures is 0–100 m and of the middle and bottom figures is 1000 m.

the other parameters. The present version of TOUGHREACT cannot explicitly consider oil as a fluid phase, although the oil mobilization is simulated using a pragmatic approach. Uncertainties from manipulating this fluid phase still need to be investigated, although effects of this uncertainty are considered to be limited because: (1) the key objective of the present work is the geochemical reactions and the consequences; (2) the geochemical reactions are strongly dependent on temperature, geochemical compositions of the formation water and injection water, and mineral compositions of reservoir rock, so effects of the pressure are considered secondary; and (3) temperature and the geochemical system are properly captured in the RTM. Some of the model parameters used are from the literature and calibrated based on present-day water composition; however, the entire model is not yet fully validated. Therefore, confidence will be built up by using the results for validation based on dedicated experimental data.

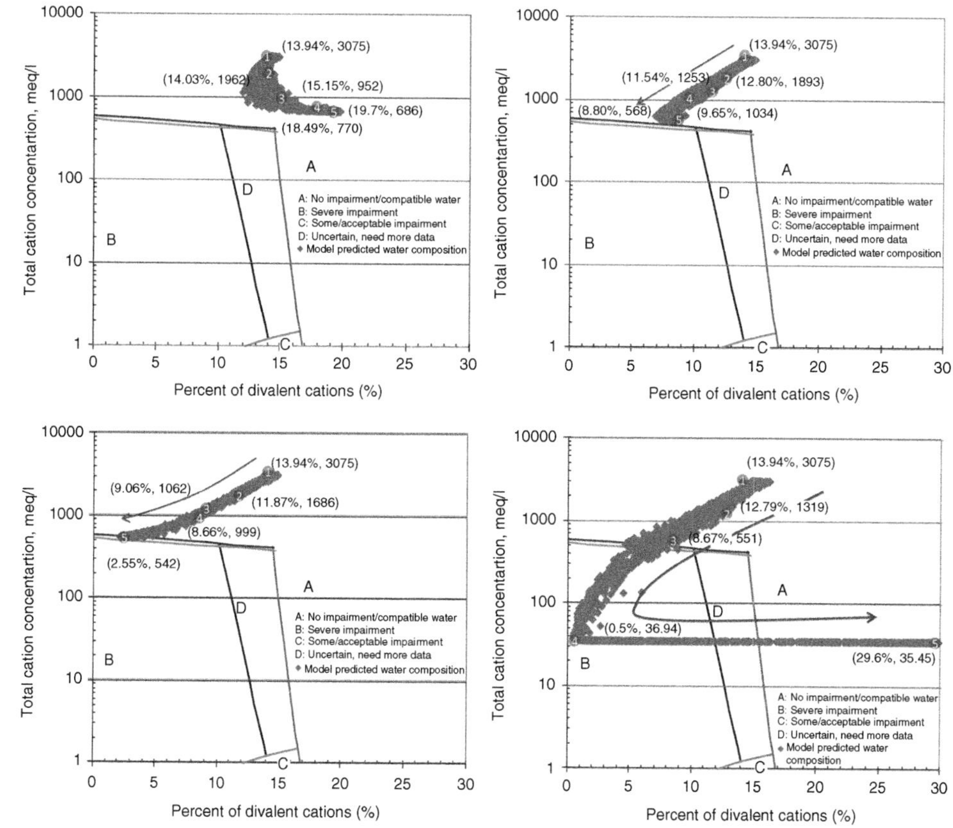

Figure 5.9 Evaluation of smectite (montmorillonite) swelling potential based on model-predicted reservoir water composition over a 20-year injection duration and over the whole domain (◆) on a Scheuerman diagram (a Cartesian space of total cation concentration and the percentage of divalent cations over total cations), using untreated seawater (top left), treated seawaters I (top right) and II (bottom left), and fresh water (bottom right). ①: background formation water chemical composition; ②: transient water chemical composition 5.7 m away from the injector and at the first day of injection; ③: transient water at 57 m away from the injector and at 6 months of injection; ④: transient water at 84 m away from the injector and at 1 year of injection; ⑤: injected untreated seawater in contact with formation rock after fully displacing the pristine formation water.

5.5.2 Modeling Reservoir Scaling and Souring by TSR During Waterflood

An offshore reservoir (referred to as CPX) comprising the M-formation at a depth of more than 8000 m below sea-level and temperature of about 180 °C is under plan for development with waterflood. The rock is not heavily fractured and the main HC volume is made up of rock matrix. Injection water will be a 'sulfate-removal' treated seawater at about 18 °C. Based on the temperature, pressure, formation mineralogy, formation water composition and injection water composition, potential risks relevant to the waterflood have been foreseen, including scaling and souring. Clay minerals in this formation are mostly no-swelling clay members, and thus the swelling risk is considered minor. In order to evaluate these potential risks, RTM has been performed. The first RTM effort is a single-well sector RTM covering a horizontal section of the reservoir

(1300 m radius) to screen the potential risk. Results indicate that, due to the large contrast in temperature and water chemical composition, precipitation of calcite as well as some clay minerals can lead to wellbore and near-wellbore region scaling. In addition, both microbial sulfate reduction (BSR) and thermochemical sulfate reduction (TSR) can take place within the reservoir during the waterflood. BSR will take place where the temperature is favourable (cooled down to 20–50 °C by the injection water), but will seize where the temperature is higher than 80 °C. The H_2S concentration within a zone close to the injector but where BSR is active will be also dictated (diluted) by the injection water in which the H_2S concentration is about 10^{-9} molal. Therefore, the injection water dilutes the formation water, lowering the H_2S concentration from 10^{-4} molal down towards 10^{-9} molal near the injector. A dedicated evaluation of the effects of the BSR was performed before the present RTM study, and concluded that the risk of souring by BSR is very limited. However, TSR is identified to take place actively within a zone of temperature at 150–180 °C, and H_2S from TSR can be accumulated ahead of the injection water and arrive at the producer before breakthrough. Therefore, TSR was identified as a non-ignorable risk. Further evaluation was conducted; laboratory experiments to test the potential TSR mechanism and measure the kinetics were performed and the results were used to validate the RTM prediction. A second stage of RTM was performed to calibrate TSR kinetic parameters based on the observations from the laboratory TSR experiments. A field-scale RTM covering the same section of the reservoir was perused to predict the H_2S concentration at a producer within the reservoir formation and during the waterflood using the calibrated kinetic parameters. In short, the RTM was developed in three phases: *Screening model*, phase 1; *Validation model*, phase 2 and *Prediction model*, phase 3. TOUGHREACT was used in this modeling practice.

5.5.2.1 Geochemical Setting

The M-formation at CPX is composed of cross-laminated, predominantly finegrained sandstones with bimodal grain distribution. These sandstones are composed of subrounded to rounded framework grains that are mainly well to moderately sorted. Differential compaction, ductile clay and mica grains are observed. Based on thin-section point counting, scanning electron microscopy (SEM), and X-ray diffraction (XRD) analyses, typical mineral composition was observed and is listed in Table 5.6.

5.5.2.2 Formation Brine Composition

Multiple water samples from the wells in the greater CPX area have been collected (Table 5.7), and discrepancies between the measured concentrations of different samples are within 10% of each other (note that the typical measurement error in TDS is 1%). This would not be considered significant from a geochemical point of view. Uncertainties exist in the measured brine composition because:

- Some brine samples are collected by centrifuging core, and thus the water composition may evolve (e.g. become oxidized) during core storage under preservation (pressure and temperature are different from the *in situ* condition);
- No brine sample has been collected from the water leg, where the brine is expected to be more saline. An indication exists that the salinity increased by almost 7600 mg/l (~3%) over a depth of just 79 ft, and halite-saturated brine is expected at the bottom of the water leg.

Table 5.6 Typical mineral composition of the M-formation.

Mineral	Content (vol.%)
Calcite	0–1
Barite	0–0.5
Anhydrite	0.6
Dolomite	0–1
Quartz	60–75
Plagioclase	10
K-feldspar	5
Mixed-layer illite/smectite	1
Mica/illite	1
Chlorite group including iron-chlorite	0.5–6
Hematite	1–6
Pyrite	0–1

Table 5.7 Typical chemical compositions of brines from the CPX reservoir at pressure 18,000 psi and temperature 163 °C.

Solutes	Test method	Concentration (molal)
Ba^{2+}	ICP	8.60×10^{-5}
Ca^{2+}	ICP	7.41×10^{-1}
Fe^{2+}	ICP	2.12×10^{-4}
Total Fe	ICP	3.85×10^{-3}
Li^{+}	ICP	6.81×10^{-3}
Mg^{2+}	ICP	2.0×10^{-2}
Mn^{2+}	ICP	1.43×10^{-3}
K^{+}	ICP	1.16×10^{-1}
Si^{4+}	ICP	1.49×10^{-3}
Na^{+}	ICP	2.57
Sr^{2+}	ICP	1.45×10^{-2}
HCO_3^{-}	Titration	1.94×10^{-3}
Br^{-}	Titration/IC	1.26×10^{-2}
CO_3^{2-}	Titration	0.0
Cl^{-}	Titration/IC	4.16
SO_4^{2-}	IC	2.13×10^{-3}
S^{2-}	IC	0.00
pH	In pH unit	5.578

5.5.2.3 Geochemical Reactions Induced by Waterflood

Having gone through multiple million years of geochemical evolution, the formation water and the rock minerals in the CPX reservoir are in a steady state, meaning that the water composition may have stayed (almost) unchanged for a quite long time, although slow kinetic reactions might still be ongoing, making up the diagenetic trend. The kinetic reactions must be very slow because there should be no reactions very far from equilibrium. Some key diagenetic reactions are the dissolution of anhydrite and chlorite – TSR – scavenging the reaction network (Figure 5.10). Anhydrite is slowly dissolving, releasing sulfate to refill the system as TSR consumes it. On the other hand, pyrite or iron sulfide (e.g. pyrrhotite) precipitates as the concentration of sulfide (S^- and S^{2-}) rises due to TSR. Iron-chlorite in the system dissolves to release ferrous iron. These three reactions sooner or later will find a balance to stabilize the concentrations of sulfide, iron and sulfate in the system. The result of this reaction network is to transform anhydrite and chlorite, potentially, into carbonate, pyrite and pyrrhotite as well as quartz, kaolinite and feldspars.

Considering the potential risk in sulfate scaling as well as souring, sulfate removal treatment for seawater as an injection water was designed. Sulfate concentration was expected to reduce to 40 ppm from about 2000 ppm, the original level in the seawater. Concentration of other ionic components remained unchanged except for calcium and magnesium, which were lowered to maintain a charge balance. The temperature is estimated at 18 °C before injection into the reservoir. Injection of this water into a reservoir at 180 °C will drastically cool down the reservoir and induce a series of reactions: precipitation of carbonate, and dissolution of feldspars, some clay minerals as well as sulfate, especially anhydrite. Dissolution of anhydrite will release sulfate and sharply increase the concentration of sulfate and calcium in the formation brine at the cooled-down zone of the reservoir. A cooling zone is developed from the near injector towards far field in the radial direction. Besides advective transport, diffusive and dispersive transport will take place because the sulfate concentration increases from about

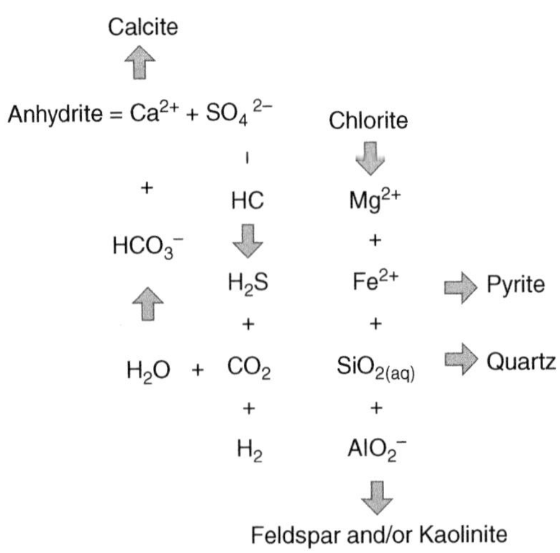

Figure 5.10 Geochemical reaction network in the CPX reservoir during waterflood.

180 ppm to more than 1000 ppm caused by dissolution of anhydrite. As soon as the sulfate reaches a hotter zone by advective/diffusive/dispersive transport, the TSR is enhanced/triggered. The enhanced TSR elevates the sulfide concentration and in turn accelerates the scavenging that requires a faster (than background level) dissolution of chlorite and faster (than background) precipitation of pyrite and/or pyrrhotite. Under the TSR – the dissolution of anhydrite and chlorite – the scavenging reaction network will sooner or later reach a new balance with an elevated level of sulfide (H_2S) concentration in the system.

To find the elevated level of H_2S (transient or stabilized) in the formation water becomes the objective of the RTM. To set up such an RTM, the key thermodynamic equilibrium constants as well as key kinetic parameters are evaluated as follows.

5.5.2.4 Temperature-Dependent and Pressure-Dependent Thermodynamic Data

TOUGHREACT uses a dedicated thermodynamic database that contains data for most geochemical constituents and reactions (Xu et al., 2011). The thermodynamic equilibrium constants of most of the geochemical reactions to be taking place during the waterflood are available in a popular thermodynamic database for the temperature range 0–300 °C, and normally for pressure at 1 bar (<100 °C) or vapour saturation pressure (≥100 °C). As mentioned above, this database was developed at the Lawrence Livermore National Laboratory (LLNL) by Wolery (1992) and Johnson et al. (1992) in direct association with a batch geochemical simulator, EQ3/6. During the waterflood, the CPX reservoir will go through a varying temperatures (180 °C → 18 °C) and varying pressures (800–1200 bar), and these parameters vary noticeably over these ranges. Thermodynamic equilibrium constants for key reactions within this wide temperature and pressure ranges can be evaluated with SUPCRT (Johnson et al., 1992), and compiled to be used for the RTM simulation. Temperature-dependent and pressure-dependent thermodynamic equilibrium constants (log units) for anhydrite dissolution and precipitation are shown as an example.

As discussed in Section 5.2.4, the pressure dependence can be handled by modifying the temperature-dependent fitting curve, should a correlation between temperature and pressure in the domain can be monotonically determined, and the temperature- and pressure-dependent thermodynamic equilibrium constant can be found as shown in Figure 5.11. However, the actual fitting parameter should be determined using the regression algorithm and is not elaborated here.

5.5.2.5 Handling Solid Reduced Sulfur (Pyrite or Pyrrhotite) Under Reduced Conditions

As discussed in Section 5.2.4, pyrite is formed directly from S^- and pyrrhotite from S^{2-} species. Considering that the pyrite is more stable than pyrrhotite phases (may not necessarily be true in the extreme reducing condition), pyrite is used in the model to numerically represent all the possible solid sulfide phases to be formed when H_2S is generated from TSR. The reaction is shown by the following equation:

$$2HS^- + Fe^{2+} = FeS_2 + H_2 \quad (5.24)$$

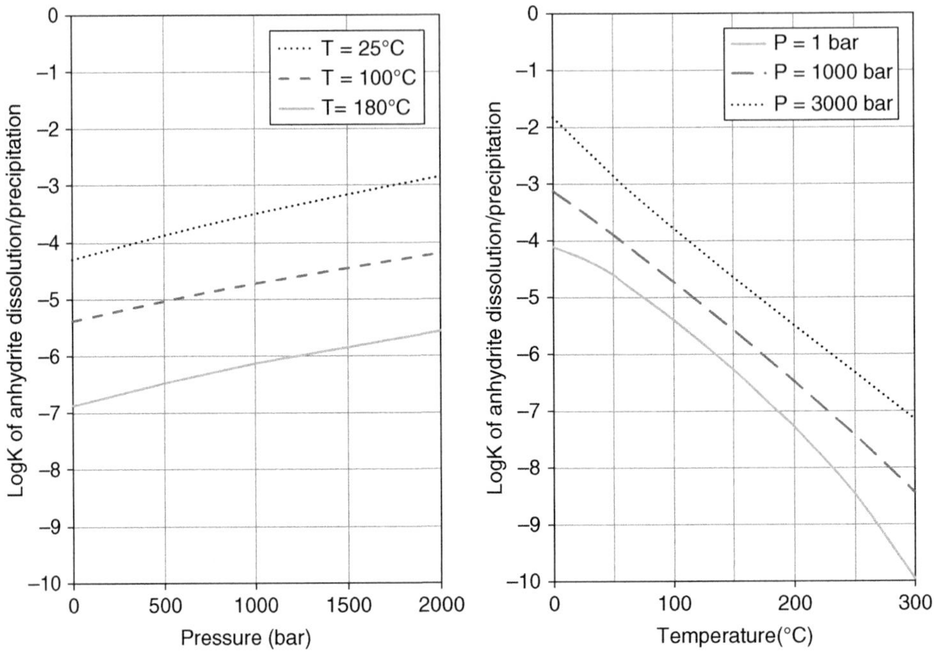

Figure 5.11 Temperature-dependent and pressure-dependent thermodynamic equilibrium constants of anhydrite dissolution and precipitation.

Generation of H_2 gas from TSR was observed from our TSR experiment, verifying the above reaction, and H_2 may stay in the formation brine in an appropriate form as formation pressure (18,000 psi) is much higher than the experimental condition (4000 psi). It is not easy to handle this reaction because there is H_2 gas fugacity involved in the calculation, and there is no a handy model to calculate the fugacity of H_2 gas under such conditions. To avoid handling H_2 gas fugacity calculation, which is not available in TOUGHREACT (like in most other RTM simulators), the reaction can be rewritten as:

$$FeS_2 + 2e^- + 2H^+ = 2HS^- + Fe^{2+} \tag{5.25}$$

Handling pH and Eh is much easier than handling H_2 gas in RTM simulations with most available simulators. In our present RTM this reaction is handled even avoiding the Eh calculation by considering the redox condition stabilizes at reducing status. A new parameter was defined to handle the reaction under stabilized redox conditions:

$$Ke = Ka_e^{\,2} \tag{5.26}$$

where K is the equilibrium constant of the reaction, $K = \dfrac{a_{HS^-}^2 a_{Fe^{2+}}}{a_{H^+}^2 a_{e^-}^2}$, a_e is the activity of electron, therefore *Ke* is the equilibrium constant of the following partial reaction:

$$FeS_2 = 2HS^- + Fe^{2+} - 2H^+ \tag{5.27}$$

$$Ke = \frac{a^2_{HS^-}\, a_{Fe^{2+}}}{a^2_{H^+}} \tag{5.28}$$

Directly handling this partial reaction is feasible using TOUGHREACT if the reaction is properly formulated in the database. Ke is then calibrated using the experimental results (measured precipitation of sulfide phases) as well as the background H_2S level (representing a stabilized scavenging condition) in the reservoir. $Ke = 10^{-9.99}$ at 180 °C and 18,000 psi, and $10^{-21.575}$ at 300 °C and 4000 psi. Note that this manipulation is only for numerical ease in the RTM, avoiding handling the calculation of H_2 gas fugacity and activity of electron or Eh; however, it can only be done when the Eh of the system is quasi-stabilized.

5.5.2.6 TOUGHREACT RTM Phase 1: Screening Phase (Risk Screening)

The H_2S level in the CPX reservoir is a result of balancing H_2S generation and scavenging. The key kinetic properties dictating the balance are: TSR rate constant and rate constant of anhydrite dissolution, chlorite dissolution and pyrite (as well as other iron sulfide phases) precipitation, effective surface area of anhydrite and iron-chlorite. These parameters determine (1) how rapid the TSR can be, (2) how rapidly the H_2S can be removed from the liquid phase. Thus, these parameters are carefully evaluated first based on literature data for the screening phase RTM, Phase 1, and then calibrated using experimental data (calibration model, Phase 2), and eventually used in the prediction model, Phase 3.

Using the TSR kinetic rate constant from Cross et al. (2004), sulfate half-life time (HLT) at a range of 35–500 years with 100 years as the best representative, and other kinetic parameters from Palandri and Kharaka (2004), a TOUGHREACT RTM was developed based on the measured/validated reservoir stratigraphic data and hydraulic data, with preliminary geochemical data, established data (e.g. thermodynamic database) and literature data (e.g. TSR kinetics data) for screening purposes. Risks were identified from the TOUGHREACT Phase 1 RTM, including well/reservoir scaling and souring.

Scaling

Scaling risks had been identified from TOUGHREACT simulations: calcite scaling at the injector and anhydrite scaling at the producer. This result is similar to what had been identified in the production chemistry/production technology community prior to the TOUGHREACT RTM. However, the quantitative TOUGHREACT model prediction is a good addition because the model predicts the actual amount of the scaling minerals precipitating at a given location and time (Figure 5.12) based on well capture of the flow, transport, as well as geochemical reactions within the reservoir under the *in situ* condition.

Calcite also precipitates within a zone ahead of the injection water front where temperature is elevated and carbonate and calcium are transported from the cooled near-injector zone (Figure 5.13). By the same mechanism, anhydrite precipitates at the zone behind and around the cooling front. A producer located at this region will first receive significant amounts of calcite and later anhydrite, causing a scaling risk. Appropriate-scale squeeze plans are needed.

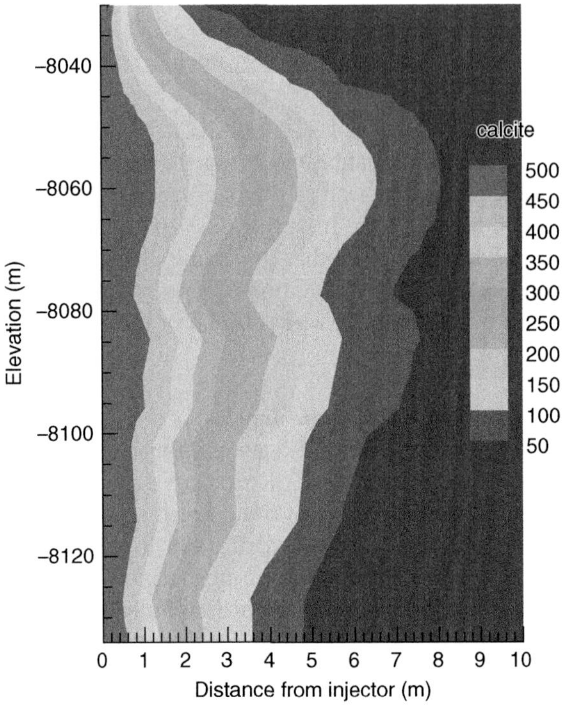

Figure 5.12 TOUGHREACT RTM (Phase 1, screening) predicted the precipitation of large amounts of calcite at the injector and the near-injector region (the contours shows the mol mineral/m^3 rock for calcite at 40 years).

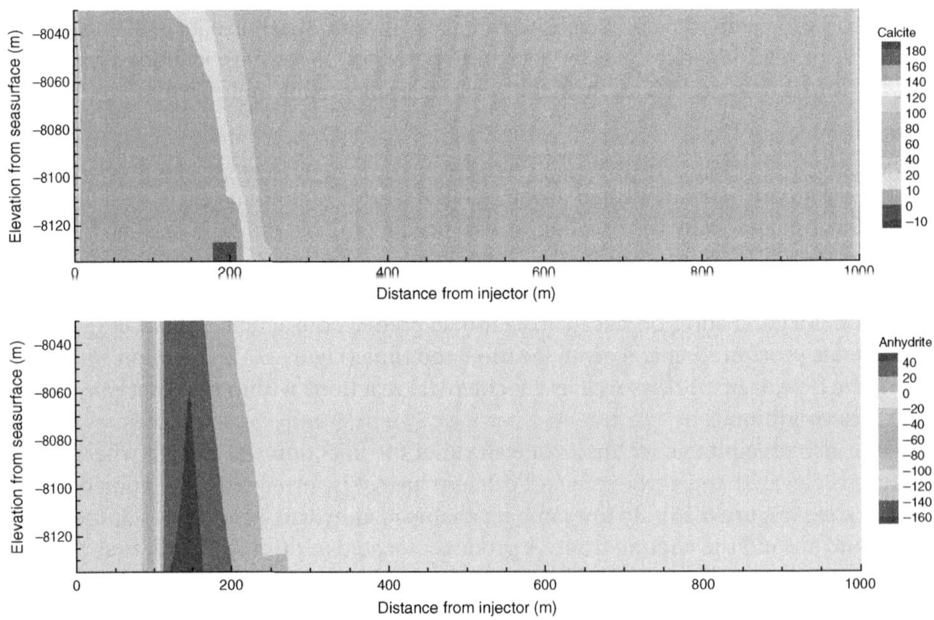

Figure 5.13 The model predicted precipitation of calcite (top figure) and anhydrite (bottom figure) at 20 years of production under waterflood.

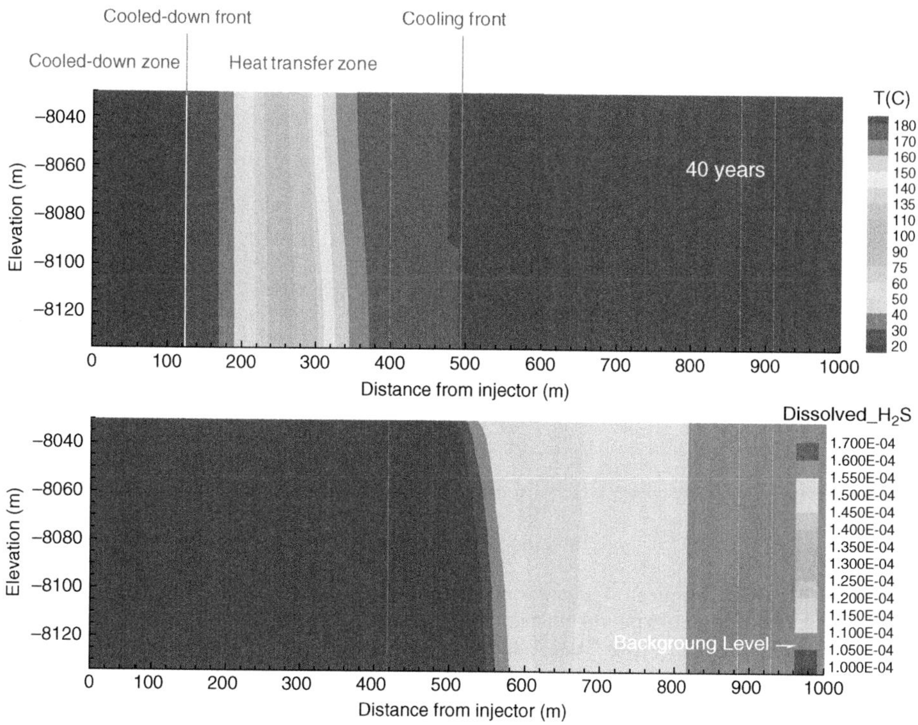

Figure 5.14 TOUGHREACT RTM (Phase 1, screening) predicted the H_2S (mol/kg of water) in the formation water during waterflood (the contour maps shows the concentration at 40 years). (See colour plate section for the colour representation of this figure.)

Souring

A souring risk was identified considering that thermochemical sulfate reduction (TSR) can be promoted at the reservoir in-situ temperature of 180 °C. The aqueous sulfate can be replenished from dissolution of anhydrite as the reservoir temperature is lowered by the injection water. TOUGHREACT RTM predicted that the H_2S concentration in the formation water could increase by 12% as a base case and 78% in the worst-case scenario (Figure 5.14). This result warns us to pay more attention to the potential TSR, and a subsequent experimental validation working phase is triggered.

5.5.2.7 TOUGHREACT Validation Model, Phase 2: Anhydrite Leachability Experiment to Validate the Kinetic Parameters of Anhydrite Dissolution

Key uncertainties in the TOUGHREACT Phase 1 RTM were identified: the kinetic rates of anhydrite dissolution and thermochemical sulfate reduction (TSR) reactions. Therefore, two lab experiments focused, respectively, on anhydrite dissolution and TSR were conducted. The anhydrite leaching experiment was designed as a coreflood experiment with a flow rate comparable to the flow rate in the reservoir during waterflood. Based on the experimental design, TOUGHREACT RTM was set up to simulate the experiment prior to the actual experimental investigation. The pre-experimental model prediction was performed using a range of values for the effective surface area of anhydrite, and different levels of average anhydrite content in the M-formation. The SO_4^{2-} concentration of the

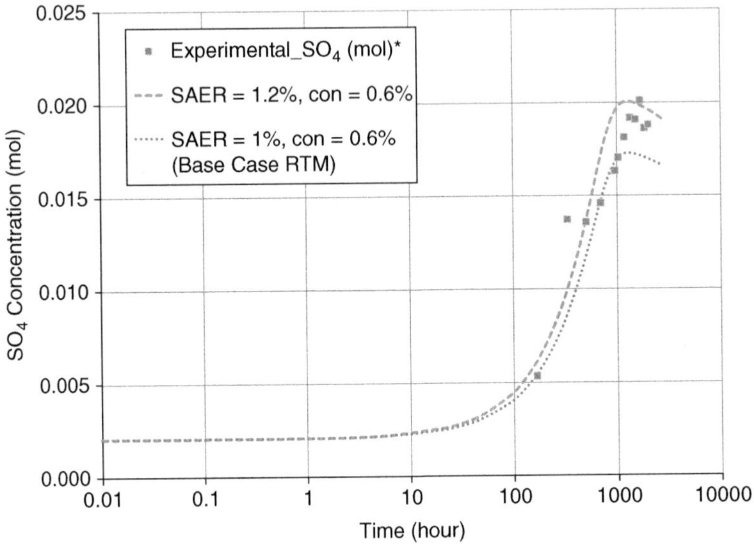

Figure 5.15 TOUGHREACT simulated sulfate concentration and the measured sulfate concentration from the anhydrite dissolution experiment (leaching experiment), showing that using the kinetic parameters in the TOUGHREACT RTM Phase 1 (screening) model can well reproduce the experimental results, indicating that the kinetic parameters used in TOUGHREACT were valid. Note that SAER=the ratio of effective surface area over the total surface area; con= content of anhydrite in the formation.

effluent has been predicted using the kinetic parameters used in the screening RTM. Figure 5.15 shows the pre-experimental RTM prediction which was later compared to the experimental observation, indicating that the model predicted sulfate concentration due mainly to anhydrite dissolution was nicely reproduced by the experiment.

As discussed above, anhydrite dissolution can be enhanced when the reservoir is cooled down during water injection, and releases sulfate. Therefore, kinetic rate parameters of anhydrite dissolution are key to this process. The apparent rate of anhydrite dissolution is determined by following factors/parameters.

Rate constant and activation energy of the dissolution/precipitation reactions are well-established parameters and available in the literature and databases (Palandri and Kharaka, 2004). The rate constant and the activation energy of the anhydrite dissolution/precipitation reaction are well established and were considered validated, since a good match between the predicted sulfate concentration and the observed sulfate concentration from the experiment is reached.

The reactive surface area of anhydrite was measured from SEM images; however, a large portion of the anhydrite surface area is coated by HC or isolated from pore fluids by other mineral grains (Figure 5.16). The proportion of the anhydrite surface area exposed to the pore fluid is estimated based on SEM images at about 1%. This ratio is considered as a new parameter in the model and will be considered validated if a good match is obtained between the modeled and the experimental sulfate concentration. The uncertainty in the measured surface area from SEM is largely reduced when the ratio of effective surface area over the total surface area (SAER) is calibrated. The calibrated SAER is about 1% of the total surface area. The product of these two parameters define a new quantity, 'effective surface area,' which is directly incorporated into the kinetic law for

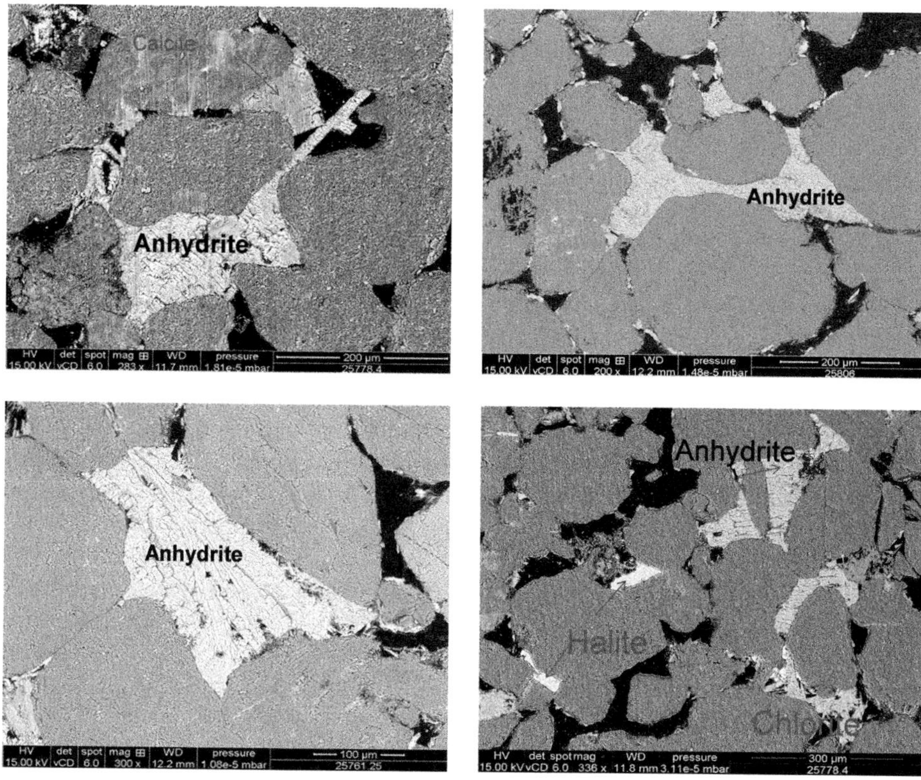

Figure 5.16 Scanning electronic microscope image of the CPX M-formation rock showing that only a very small portion of the anhydrite surface is available for dissolution.

calculating the dissolution rate in the RTM simulation. An RTM prediction using SER = 1.2% can produce a better match to the experimental results (Figure 5.15), which is very close to the initial estimate, 1%, determined based on the measurement.

Another uncertainty remains with the content of the anhydrite in the rock. The average value over the whole formation is 0.6% vol. Thus, 0.6% is used throughout the Phase 2 RTM and was considered validated as long as the pre-experimental model prediction was matched by the experimental results prediction for the case of anhydrite content = 0.6% and effective ratio of the surface area (SAER) = 1.2% (Figure 5.15).

This result fully validated the kinetic parameters used in the previous RTM for anhydrite dissolution under such a geochemical condition. Thus, all kinetic parameters for determining the anhydrite dissolution are valid, and will be kept unchanged in the next phase of the RTM.

5.5.2.8 TOUGHREACT Validation Model, Phase 2: Evaluation Uncertainties in the TSR Rate Constant, Anhydrite Leachability, and Iron-Chlorite Leachability

Another set of key parameters is that involved in the kinetic rate calculation of the TSR reaction, such as the TSR rate constant as well as the activation energy, which determine, directly, how fast the TSR reaction could be.

These parameters ware adopted from the literature (Cross et al., 2004) in Phase 1 of the TOUGHREACT model. Confidence in applying these parameters without any validation in the CPX reservoirs is low. Therefore, another lab experiment testing the TSR using reservoir rock and reservoir fluid was conducted. The experiment was performed at an elevated temperature, 300 °C, rather than the reservoir temperature, 180 °C, in order to accelerate the TSR to be measurable within a reasonable time. The TSR experiment was also simulated using TOUGHREACT RTM with actual experimental setup before performing the experiment. Experimental results show that with the TSR kinetic parameters used in Phase 1, the model underestimated the H_2S concentration, meaning the TSR rate constant from Cross et al. (2004) is too low for our case.

Cross et al. (2004) presented laboratory TSR experiments and measured TSR rates at about 300 °C. The experimental conditions in this study were previously considered the most representative of the CPX reservoir amongst others (e.g. Thom. and Anderson, 2008) in terms of pH, sulfate concentration and pressure. Thom and Anderson (2008) presented another set of experimental data. The rate constants from their experiments are higher than the data of Cross et al. (2004) and most others. Uncertainties exist in these experiments, especially when their results are extrapolated from 300 °C and higher to a lower temperature (180 °C). Cross et al. (2004) shows a rate constant as sulfate half-life time (HLT) extrapolated from their results at 180 °C ranges between 35 years and 500 years, while Thom and Anderson (2008) shows a HLT much shorter than that of Cross et al. (2004) – about 0.1–1 year at 180 °C – illustrating that the TSR rate constant of Thom and Anderson (2008) at 180 °C is hundreds of times higher than that of Cross et al. (2004). The Cross et al. (2004) data represent the low end of the uncertainty range, while Thom and Anderson (2008) data could represent the high end. The range of the TSR rate constant at 180 °C spans about three orders of magnitude. This is a big range which cannot be directly used in the RTM without a careful evaluation based on our actual reservoir.

The TOUGHREACT RTM result calibrated to the measurement of our TSR experiment suggested a TSR rate constant about five times higher than what was used in the previous RTM (the data from Cross et al., 2004), and 20-200 times lower than the data from Thom and Anderson (2008). Correspondingly, the scavenging capacity from iron-chlorite dissolution needs to be much higher than we had assumed in the previous RTM, in which iron-chlorite was not considered as having so high a potential to dissolve when encountering the injection water as observed in the experiment. The kinetic parameters for the dissolution of chlorite and precipitation of pyrite were also calibrated by matching the total solid sulfide change, 0.06% wt., in the system.

The measured TSR rate from our lab experiment is about 20 years as a sulfate HLT at 180 °C (Figure 5.17). This value is in between the value of Cross et al. (2004), 35–500 years HLT, and Thom and Anderson (2008), 0.1–1 year HLT. Modeled H_2S concentrations are compared with the measurement for the best calibrated value, HLT = 20 years. The H_2S level is a result of TSR as well as scavenging by precipitation of pyrite (representing all iron sulfide phases) and dissolution of chlorite, providing ferrous iron for iron pyrite precipitation. An uncertainty range of the TSR rate is determined by the model sensitivity analysis (Figure 5.17), which was performed aiming to reproduce the range of the measured H_2S concentration. The determined uncertainty range

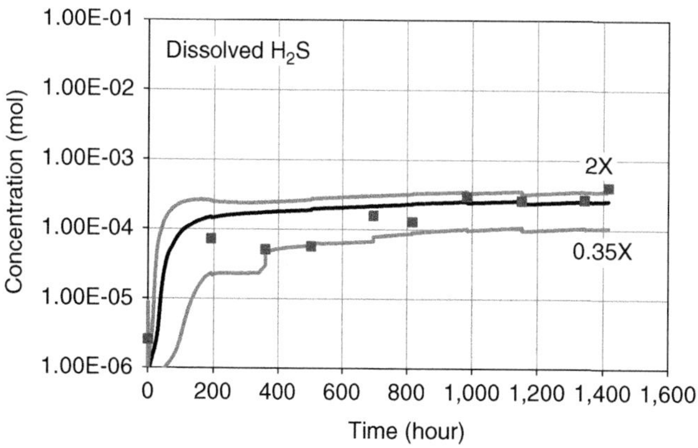

Figure 5.17 TOUGHREACT RTM simulation of the laboratory TSR experiment. The observed H_2S concentration is reproduced by the RTM with the calibrated TSR kinetic parameters; the RTM result captures (the simulated value is higher than) 50% of the measured H_2S concentrations. Sensitivity analyses indicated that with a rate constant double the calibrated value of the parameter, the RTM results capture 90% of all the measurement, defining a scenario of an accumulative probability at 90% (P90). Using a rate constant 35% of the calibrated TSR rate constant, the RTM results capture 10% of the measured H_2S concentration, defining a scenario of an accumulative probability at 10% (P10).

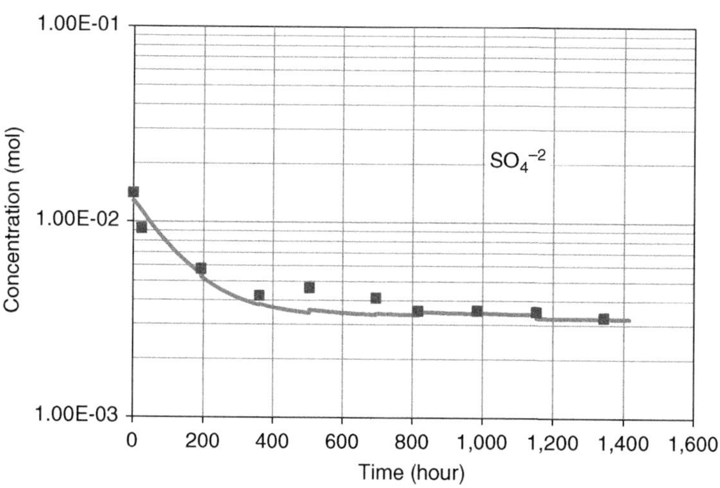

Figure 5.18 TOUGHREACT RTM simulation of the in-house laboratory TSR experiment. The observed sulfate concentration is reproduced by the RTM with the calibrated TSR kinetics parameters.

of the TSR rate spans from 35% of the best calibrated value (HLT = 20 years) to 200%. The low end (i.e. 35% of the best calibrated value) represents 10% of the accumulative probability (of the measurements) and the high-end value represents 90% of the accumulative probability. The model results also capture well the sulfate declining processes caused by (1) the TSR and (2) the reprecipitation of anhydrite as the water is heated up (Figure 5.18).

5.5.2.9 TOUGHREACT RTM Phase 3: Prediction

As the lab experiments are successfully modeled with TOUGHREACT, the key kinetic parameters, such as anhydrite dissolution rate constant, reactive surface area, and TSR rate constant, are all validated. Confidence in TOUGHREACT RTM is increased. Therefore, the validated parameters are used in the TOUGHREACT RTM to predict the H_2S concentration in the water. Results show that the H_2S concentration in the reservoir formation water can increase by 9% due to TSR when using the P90 TSR rate constant (Figure 5.19). TOUGHREACT RTM also predicted a worst-case scenario in which the H_2S concentration increases 78% from the background level. This increment is then translated into an increment of H_2S in produced gas.

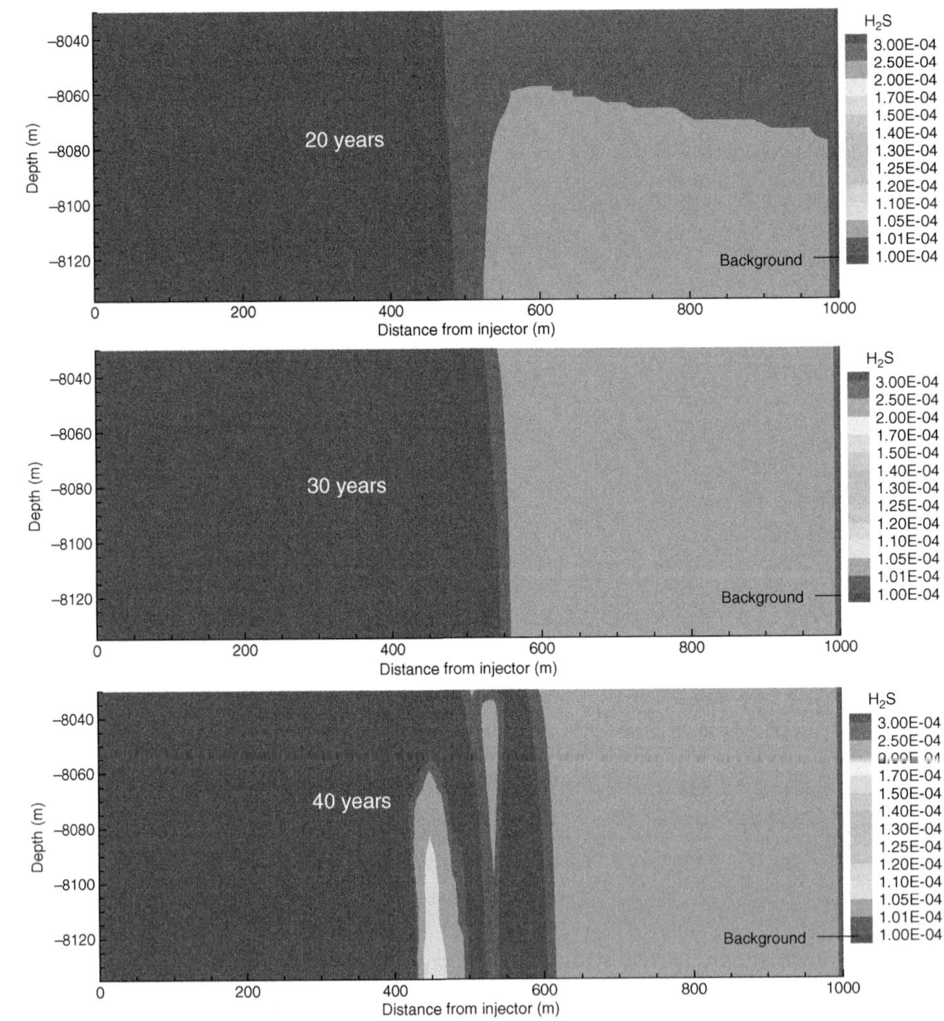

Figure 5.19 TOUGHREACT (using the TSR rate constant of P90 as base case) predicted H_2S concentration in the water during waterflood, showing that the H_2S concentration in the water will increase about 9% at 40 years from the background level as a result of TSR.

5.5.3 RTM Evaluation of Wellbore Cement Corrosion of a Legacy Well in CO_2 and CO_2/Acid Gas Storage

Well leakage can usually be attributed to wellbore cement job failure, and less often to tubing or casing metal material (steel) failure. However, cement, both hydrated and carbonated, is stable only in an alkaline environment (pH > 11). Maintaining the highly alkaline condition relies on (1) the pH buffer effects of the hydrated calcium silicate phases (CSH) and other mineral phases in the hydrated cement or carbonate phase, and (2) the ultra-low permeability of the cement, preventing it from intrusion of external water of generally neutral or acidic pH. Cement can be quickly corroded if the highly alkaline environment (pH > 11) is destroyed mainly by intrusion of external pH-neutral or acid waters. However, intrusion of acidity of any kind into the cement cannot be absolutely prevented because diffusion of protons (H^+) into cement takes place constantly. Intrusion of water by advection through flow paths (any mechanical defect in the cement, e.g. micro-fractures or crevices) can take place if such flow paths exist. The strength of the advective intrusion depends on the pressure gradient along the flow path, typically the path across the annular cement structure. The pressure gradient across the annular cement could be small or negligible if the well is fully sealed, and large if the well is opened (connected to the atmosphere). On the other hand, the chemical composition, especially the pH, of the water around the cement is another key factor for the dissolution of the cement phases. The dissolution process is controlled by the diffusive or advective intrusion of water of a given chemical composition. A geochemical reactive transport simulation is capable of capturing the process and quantitatively evaluating the cement corrosion.

Chemical composition of a regular Portland cement is given in Section 5.3.9, and the dissolution reactions are also discussed in that section. Modeling damage of concrete structures has focused on mechanical processes (Ožbolt et al., 2013), and very few efforts have been made with modeling geochemical processes. De Windt et al. (2004) analysed, evaluated and modeled the geochemical interaction of cement with clay minerals; Huet et al. (2010) presented a reactive transport model of Portland cement in CO_2-saturated water. These studies are good examples available in the literature. The RTM for wellbore corrosion presented in this Section with synthetic scenarios considers the reactions of the cement phases, minerals within the surrounding sandstone, and the formation brine at meso-high salinity, which is a popular scenario in oil and gas reservoirs.

One generic scenario of a legacy well being approached by a CO_2 plume (Figure 5.20) is considered in the following modeling practice. During CO_2 or other acid gas injection, a legacy well located within a certain distance from the injector can be approached by a pressure propagation front first (pressure front), and the pressurized wellbore can potentially induce hydraulic flux from outside into the inside of the wellbore, penetrating the annular cement. If the well is not sealed properly, the pressure gradient over the annular cement can be significant. In this case, if the wellbore cement is poorly mechanically integrated with the surrounding rock, advection flux of the formation water can form through the annular cement, causing significant dissolution of either hydrated cement phases (e.g. CSH) or carbonated phases (e.g. calcite).

After pressurization, the CO_2 diffusion front arrives at the legacy well and the formation water is acidified. The dissolution of the cement phases will be enhanced due to the lower pH. The CO_2 plume will eventually reach the wellbore, leading to additional dissolution. RTMs are performed to simulate the effects of the dissolution of these cement phases under

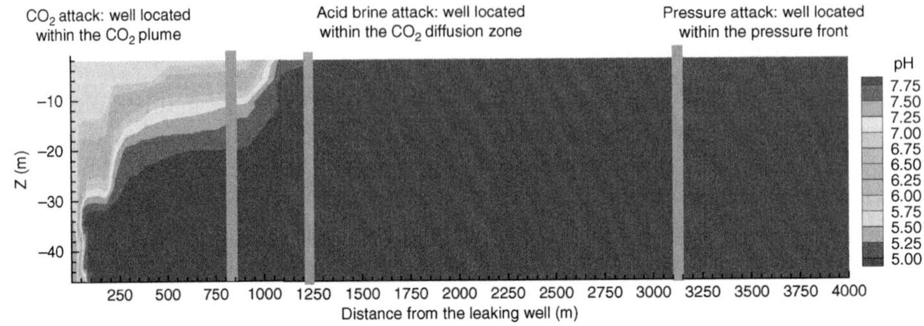

Figure 5.20 In a synthetic CO_2/acid gas injection scenario, a legacy well is located at a certain distance from the injection, and sequentially approached by a pressure front, CO_2 diffusion front (acid brine), and advection front (CO_2 plume) during CO_2 injection. Note that the injector is located at the left edge of the figure. (See colour plate section for the colour representation of this figure.)

pressurization (pressure attack), acidified water intrusion (acid water attack) and CO_2 intrusion. The formation temperature is assumed to be 60 °C and the pressure is 200 bar.

5.5.3.1 Mineralogical Composition and Water Composition of the Wellbore Intervals

A clastic reservoir with a variety of mineral compositions of andesitic origin is considered in this study. Modeling one end-member of the possible mineral composition package, a clean sandstone formation with formation water at 75,000 ppm halite salinity is considered. The formation contains quartz (85%), albite (4%), K-feldspar (6%), illite (2%) smectite (2%) and kaolinite (1%). A water composition is obtained by equilibrating a water at 75,000 ppm halite salinity with the reservoir mineral package, and the resulting water composition is given in Table 5.8. The mineral composition of the cement phases is given in Table 5.1.

5.5.3.2 Model Setup

A 2D radial model domain was determined to represent a legacy well and the near-wellbore formation rock (Figure 5.21), which it is assumed has stayed in the gas reservoir for 50 years, then was approached by a pressure front, a CO_2 diffusion front and, ultimately, a CO_2 advection front, i.e. the CO_2 plume. The radius (R) of the domain is 5.6 m. Wellbore annular cement/concrete is located within a radial interval of $R = 0.13$ m to $R = 0.21$ m. Tubing/casing is located within $R < 0.13$ m and assumed impermeable. Formation rock is located outside the annular where $R > 0.21$ m.

Permeability of cement/concrete ranges between 1 nD to 10 nD (Roy et al., 1993). Additives (e.g. fly ash) to the application can significantly reduce the permeability. Porosity of cement/concrete ranges between 10–15% (Roy et al., 1993). In the present study, the permeability is assumed to be 1 nD and the porosity is assumed to be 10%, i.e. 10 porosity units (pu) as the base case.

Note that this modeling exercise only takes into account the reservoir section and focuses on the chemical reactions of the cement corrosion process, rather than well leakage processes. To simulate the geochemical reactions in the cement, formation water and formation mineral systems are the objective of this model practice. Also, no

Table 5.8 Water composition generated by an equilibrating water of 75,000 ppm halite with a clean sandstone, for RTM of wellbore cement corrosion (concentration is given by molal).

Components	Concentration
pH	6.94
Ca^{2+}	0.0326
Mg^{2+}	0.015
Na^{+}	0.903
K^{+}	0.00183
Cl^{-}	1.09
HCO_3^{-}	9.55×10^{-3}
SO_4^{2-}	1.21×10^{-4}
AlO_2^{-}	1.22×10^{-8}
$SiO_{2(aq)}$	3.75×10^{-4}

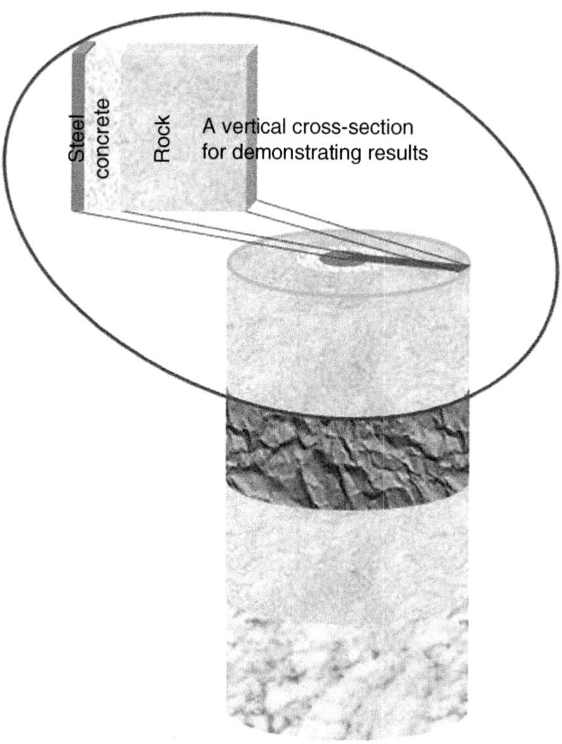

Figure 5.21 A radial domain of a legacy well for modeling wellbore cement corrosion, with a vertical cross-section for demonstrating results.

caprock section is included. This means that the model results are not directly applicable to evaluations of leakage (through the caprock section) into shallower formations. For that purpose, the caprock mineralogy and cement–caprock mineral–water interaction need to be included in the model. The case with CO_2 exposure in particular could behave differently from the current case, in which the surrounding reservoir rock is permeable so that CO_2 can quickly travel to and from the rock–cement interface, while for the cement–caprock interface this is not the case; the situation would be more transport-dictated and may not be similar to the modeled scenarios.

5.5.3.3 Modeled Wellbore Cement Corrosion Processes

Using the RTM setup with above-mentioned mineral compositions and water composition, potential wellbore cement corrosion processes are simulated for four different scenarios: (1) the undisturbed condition, i.e. a wellbore buried in the formation without any disturbance from any injection; (2) pressure attack; (3) acid water/brine attack; and (4) CO_2/acid gas attack.

Corrosion in undisturbed conditions (corrosion of wellbore without CO_2 injection)

In this scenario, the pH of the formation porewater (6.94) is very close to neutral. Where the wellbore cement with the very alkaline cement porewater (pH 11.2) contacts the formation, drastic reactions must take place at the cement–formation interface, forming a cement-rock bound by calcium and silicate phases sealing the wellbore. A pH gradient is thus built up from the cement–rock interface to far away in the formation where the pH is still neutral (Figure 5.22). A pH transition zone (declining

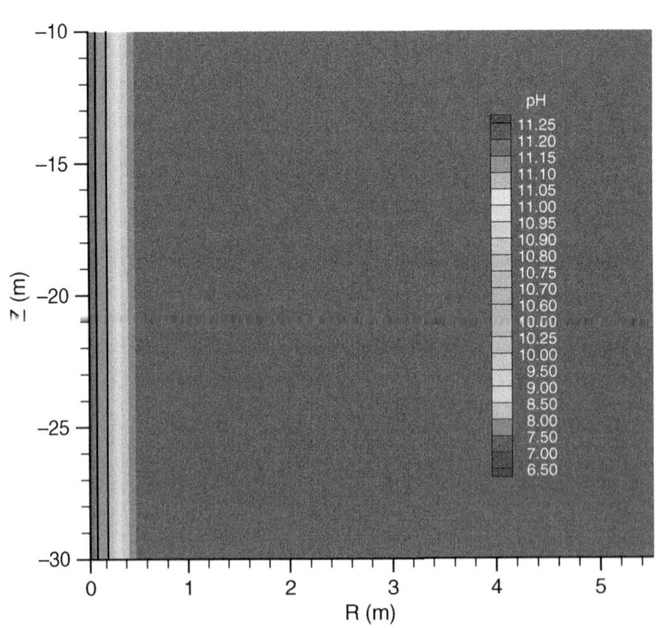

Figure 5.22 Modeled pH profile around a legacy well of 50 years old in a clean sand formation with brine at a salinity of 75,000 ppm and pH at 6.94.

zone) will form in the formation porewater, within which the pH gradually drops from 11.2 at the cement–rock interface down to 6.94 at a certain distance in the formation. This transition zone presents a standard diffusion profile (Figure 5.22) extending from $R = 0.21$ m to $R = 1.2$ m at 10 years, to $R > 5.6$ m at 20 years. Sooner or later, acidity transported by diffusion from the formation towards the cement–rock interface can destroy the cement-rock bound and corrosion of the wellbore cement starts. However, as discussed above, these reactions are limited by the availability of formation porewater in contacting the cement, which is controlled by diffusion processes because the cement holds an extremely low permeability (nano-Darcy's), and thus advective flux is very limited.

Clearly, the strong alkaline mineral, portlandite, will dissolve when the formation porewater of pH 6.94 seeps into the cement, or the 'acidity' of the formation porewater (of pH 6.94) diffuses into the cement. Model results indicate that during the first ten years, at the interface, the amount of the dissolution is <1000 mol/m^3, and a corresponding porosity change is <3.3 pu, and thus hardly visible from the figure. The amount of the dissolution increases up to 1000–2000 mol/m^3 (porosity change 3.3–7 pu) at 20 years, which becomes easily visible from the figure. Dissolution becomes obvious at 50 years (right column of plots in Figure 5.23) and reaches 4000–5000 mol/m^3 of rock at the cement–formation rock interface (porosity change 13–17 pu).

Along with the dissolution of portlandite, from which large amounts of calcium is released, calcite will precipitate if any carbonate is available *in situ* (Figure 5.24). Carbonate is very low in the cement matrix and thus can only be fed from the outside, normally by diffusion. Calcite can also precipitate outside the cement as portlandite dissolves, releasing calcium and alkalinity into the surrounding rock where carbonate can be easily available. This phenomenon is very common and can be seen at most cement structures buried in soil.

In response to the dissolution of portlandite and all other minerals, porosity of the wellbore cement will increase from 10 pu up to 25–30 pu at the cement–rock interface at 50 years of service without of CO_2 injection (right-hand column of plots in Figure 5.23). As long as the porosity increases at the cement–rock interface, the cement–rock bond is destroyed, and further/faster corrosion will be facilitated.

Alkalinity released from the dissolution of portlandite and ettringite diffuses into the formation, leading to an elevation of pH (Figure 5.22) under such a condition. Quartz and smectite will dissolve at the near wellbore zone (i.e. the high pH zone), and albite will precipitate at the same location. Further away from the wellbore where pH recovers to the baseline value around neutral, quartz and smectite will reprecipitate and albite will redissolve (Figure 5.25)

These results clearly verify the following observations from the common (pH neutral) oil/gas reservoirs (Carey et al., 2007, 2010; Kutchko et al., 2007; Friedmann et al., 2009): (1) most wells can survive for two to three decades without major leakage; and (2) most wells of 30 years or more do leak to some extent.

In order to understand the effects of time, the simulation was further run until 500 years. Results indicate that dissolution of portlandite continues and the dissolved amount of the mineral at the interface reaches 10,000 mol/m^3 and higher, while the porosity increases up to 35 pu. pH in the cement drops by 0.2 due to the diffusion of formation porewater (Figure 5.26). Leakage can, definitely, take place at the cement–rock interface.

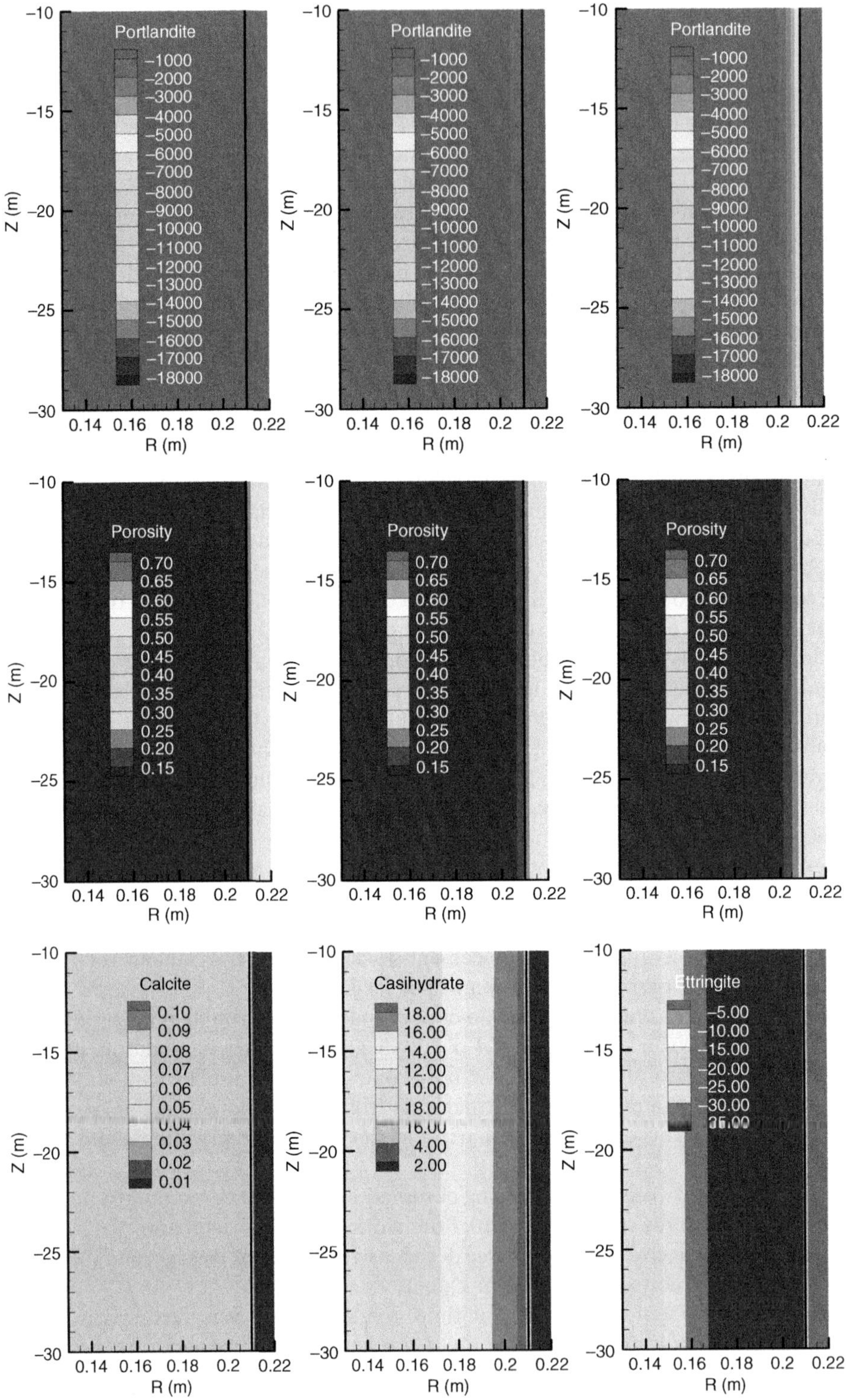

Figure 5.23 Amount (mol/m^{3})) of portlandite dissolution (negative number on the scale) in wellbore cement (top row) and the induced porosity change (middle row) at 10 years (left), 20 years (middle) and 50 years (right), respectively, due to the precipitation of calcite, CSH (marked as 'casihydrate' on the figure) and dissolution of ettringite (bottom row).

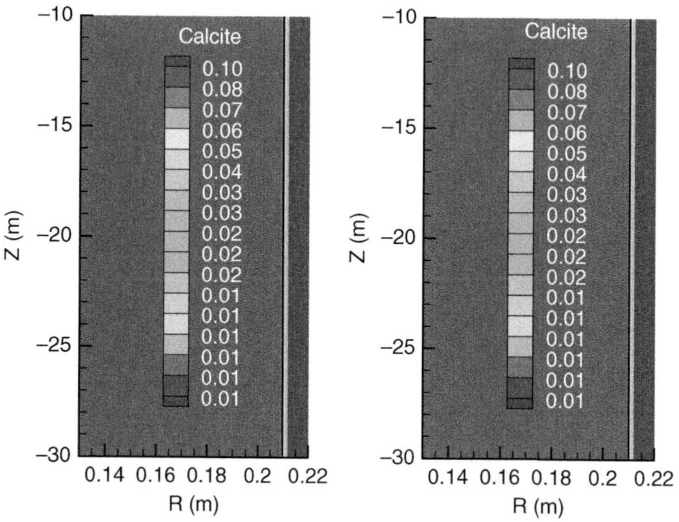

Figure 5.24 Calcite precipitated within the cement matrix as a result of carbonation due to diffusion of water bearing carbonate into the cement in undisturbed conditions. The amount of calcite precipitated is limited compared with the large dissolution amount of the cement phases (e.g. portlandite and CSH) under disturbed conditions (later subsections). The left plot shows results at 50 years and right figures show results at 500 years, indicating that the carbonation processes take place at the early stage and are significantly slowed down.

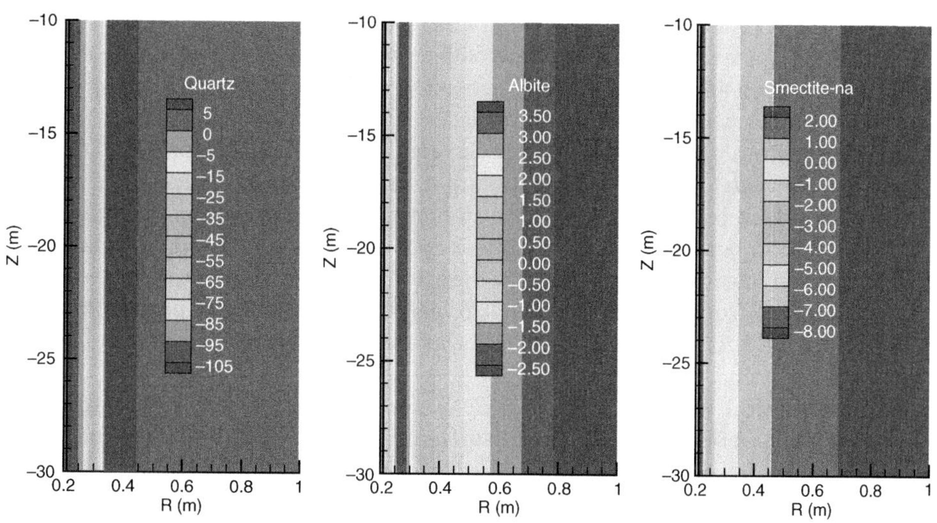

Figure 5.25 Modeled dissolution of quartz (left) and smectite (right) at the near-wellbore zone ($R = 0.21$ m to $R = 0.33$ m) due to leached alkalinity from dissolution of portlandite in the cement, and reprecipitation of these minerals further away from the wellbore ($R > 0.33$ m) when PH recovers back to the baseline value, 6.94. Albite (middle) is simulated precipitating at the near-wellbore zone and dissolving further away from it. All units are in mol/m^3.

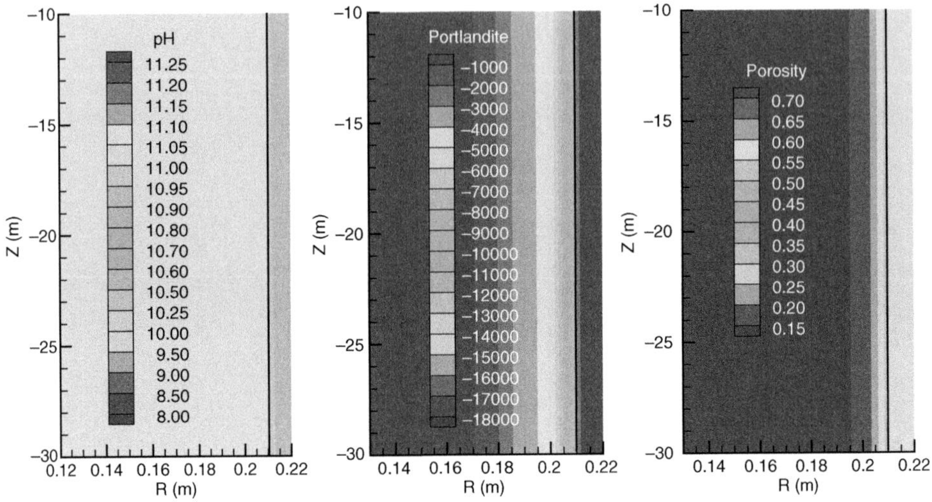

Figure 5.26 Modeled pH of the wellbore cement (left), dissolution (negative number on the scale) of portlandite (middle) and changes in porosity (right) at 500 years, in contact with a clean sand formation and porewater at 75,000 ppm salinity and pH 6.94.

In summary:

- Wellbore cement corrosion is ongoing at most oil/gas reservoirs under common conditions, due to dissolution of portlandite and ettringite, because these minerals are strongly alkaline. They can be chemically stable only in strongly alkaline environments (pH > 11) and would dissolve otherwise.
- Carbonation can take place at the surface of the cement structure or within the cement matrix as carbonate diffuses into the cement, leading to further tightening of the cement. However, intrusion of strongly acidic water through mechanical defects (meaning higher permeability) will dissolve the cement phases and cause more rapid failure.
- The corrosion rate is controlled by the rate of fluid exchange between cement and formation, which is determined by diffusion at an early stage (when the cement is fresh) as the low permeability of the cement does not allow significant fluid flux.
- The corrosion rate is also determined by the acidity (pH) of the surrounding environment. Leachate from the cement can also lead to dissolution and precipitation of other minerals, such as quartz, albite and clay minerals, altering the porosity of the near-wellbore zone.
- Good mechanical integrity (meaning no water-accessible crevices/fractures and extremely tight permeability at a few nano-Darcys) can increase the lifetime of a well and effectively prevent potential failure by corrosion during the field lifetime (tens to hundreds of years).

Results of attack by pressure front (increased water flux penetrating the cement)

When CO_2 is injected into a depleted gas reservoir, a legacy well in the reservoir can first be approached by the pressure front (elevation of pressure). Then, a CO_2 plume may or may not reach the well, depending on the distance of the well from the injector.

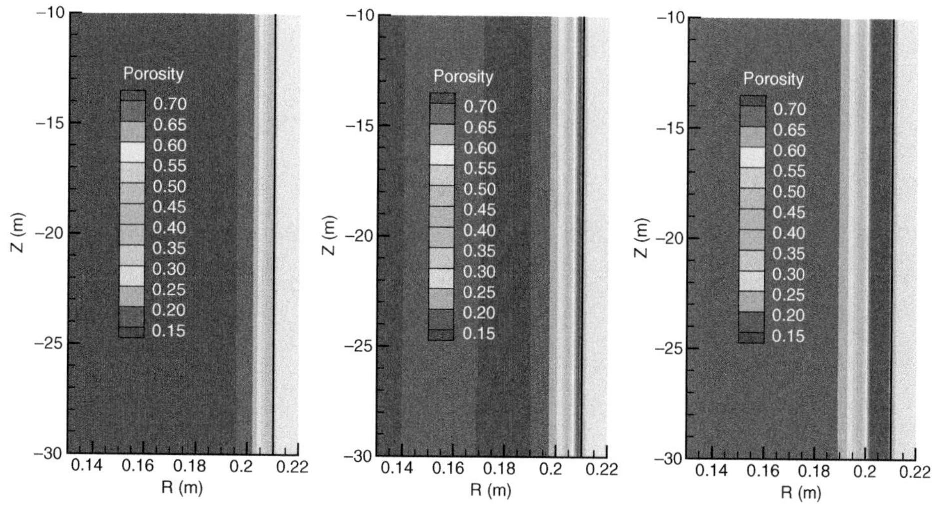

Figure 5.27 Modeled porosity change caused by the dissolution of portlandite and ettringite under attack by a pressure front (at a constant pressure elevation of 10 bars) for 50 years (left), 150 years (middle) and 450 years (right), respectively.

In order to understand the effects under pressure attack, a simulation was performed to follow the 50-year pre-injection, with a 10 bars pressure increase at the wellbore (i.e. pressure at the outer surface of the annular cement is elevated by 10 bars while pressure at the inner surface of the cement annular stayed unchanged – an open well may satisfy this condition) for an additional 450 years. A pressure increase of 10 bars for 450 years seams unrealistic because the pressure wave can quickly dissipate during injection, but a long-term simulation considering the elevated pressure at the outer annular surface is performed to help us understand the effects of pressurization, that is, the pressure front attack and the process taking place under the pressure attack.

Results show that the corrosion rates increase due to pressure attack. Comparing the left plot of Figure 5.27, modeling porosity after 50 years (without injection) + 50 years pressure attack during injection, to the right plot of middle raw in Figure 5.23, the modeled porosity after 50 year pre-injection only, one can see that for the second 50 years, porosity increase fasters, up to 30–50% (0.3–0.5 pu) at the cement–rock interface. Correspondingly, during the first 50 years (i.e. pre-injection), the porosity increases from 10% to 25% (0.1–0.25 pu). Comparing the right plot of Figure 5.27, 50-year pre-injection + 450-year pressure front attack, to the right plot of Figure 5.26, one can see that the modeled porosity change with a pressure front attack is much larger than that without any pressure front attack. Therefore, we can conclude that pressure elevation at the legacy well can enhance the corrosion of the wellbore cement because the higher pressure can promote a greater adjective flux porewater into the cement, supplying more acidity to dissolve the cement minerals.

Results of attack by diffusion front (dissolved CO_2)

In order to check the effects of dissolved CO_2 on the wellbore corrosion, a simulation was performed with formation brine saturated with CO_2 at a pressure 3 bars higher

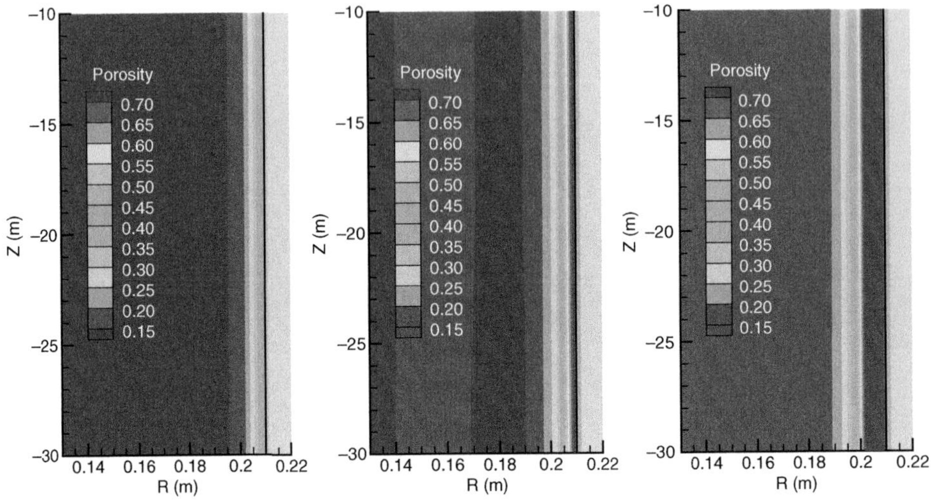

Figure 5.28 Modeled porosity change under attack by dissolved CO_2 (at CO_2 saturation) for 50 years (left), 150 years (middle) and 450 years (right).

than the background (hydrostatic) pressure. This scenario represents realistic circumstances taking place at a legacy well located at an intermediate distance (few hundred metres to a few km away from the injector). Results are shown in Figure 5.28. Porosity change in this case presents a similar enhancement for corrosion as with the scenario of a pressure attack (Figure 5.27). These results indicate that similar effects on wellbore cement corrosion can occur by increasing the flux into the cement or by adding acidity. Leakage can definitely develop after a certain period of either of these attacks.

Results of attack by advection front of CO_2 (supercritical and dissolved CO_2)

One simulation was performed for a scenario in which CO_2 directly attacked the legacy well with pressure elevated by 10 bars. This scenario assumes that an injector is located very close (tens to hundreds of metres) to a legacy well (although an injector would never intentionally be put so close to a legacy well). The purpose of this case study is to demonstrate the effects of the scenario. Results show that cement corrosion is strongly accelerated. Wellbore cement can reach total failure within a few months (Figure 5.29).

This result is consistent with the wellbore cement damage observed in experiments described in Duguid and Scherer (2010) and Sauki and Irawan (2010). Similar damage to the wellbore cement has also observed by Carey et al. (2007) from a 55-year-old oil well in SACROC field, Texas. This well has been exposed to CO_2 for 30 years during the EOR. Casing and cement samples are collected from the shale caprock–cement interface and the cement–casing interface of the legacy well, and found that: (1) casing did not show significant corrosion; (2) cement samples from the cement–caprock interface were orange-coloured, a sign of significant corrosion; (3) cement samples from the

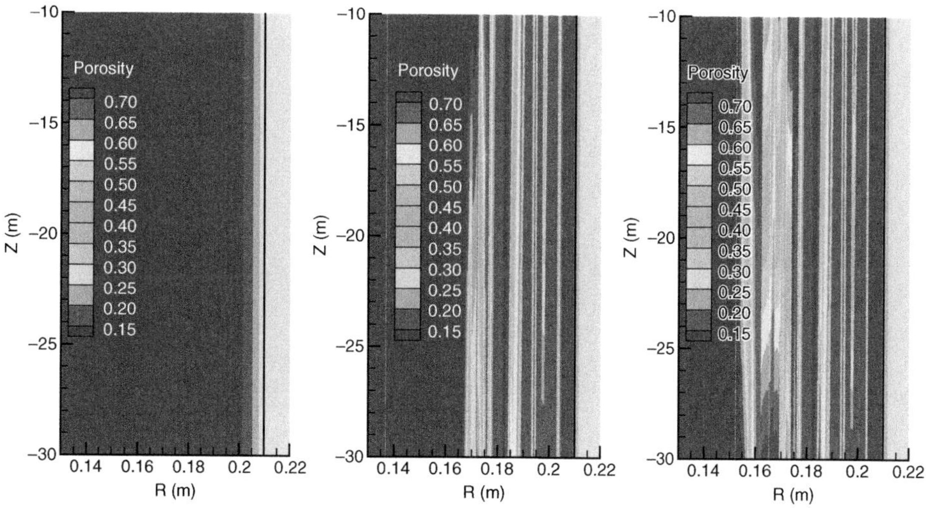

Figure 5.29 Modeled porosity change of wellbore cement under CO_2 attack for 1 month (left), 6 months (middle) and 1 year (right).

cement–casing interface indicated carbonation; (4) the permeability of the corroded cement is measured at 0.1–2 mD. The expectation from these observations is that the annular cement–rock interface at the perforated intervals may have been leaking due to wellbore cement corrosion under CO_2 attack.

5.5.3.4 Sensitivity Studies

As discussed in the previous subsections, rate of cement corrosion is determined by (1) the rate of fluid exchange between cement and the surrounding, and (2) pH of the formation. However, there are other factors that can also essentially affect the wellbore cement corrosion, such as wellbore integrity. Sensitivity studies presented in this subsection address the effects of higher cement permeability and poor well integrity (open borehole).

Without fractures, cement permeability falls within a range of 1 nD to 1 mD (Roy et al., 1993). We know that the permeability of cement is low enough to protect the wellbore, in most scenarios, from invasion by the formation porewater through advection. This can be further verified by following sensitivity analyses: simulations of wellbore cement corrosion without CO_2 injection for 50 years using cement permeabilities at 1 nD, 10 nD and 1 mD, respectively. Results show that the rate of corrosion is not sensitive to the permeability within the range of 1 nD to 1 mD (Figure 5.30) when the boundary at the cement-tubing side is assumed impermeable. This result can be easily understood because, if the well is fully abandoned, meaning that the bond between tubing and cement is in force and the cement as a permeable continuum is closed at the tubing side, a non-zero hydraulic gradient cannot be maintained across the wellbore cement. Pressure elevation at the cement–rock interface can be quickly transferred to the cement–tubing interface just few inches away, and the hydraulic gradient across the

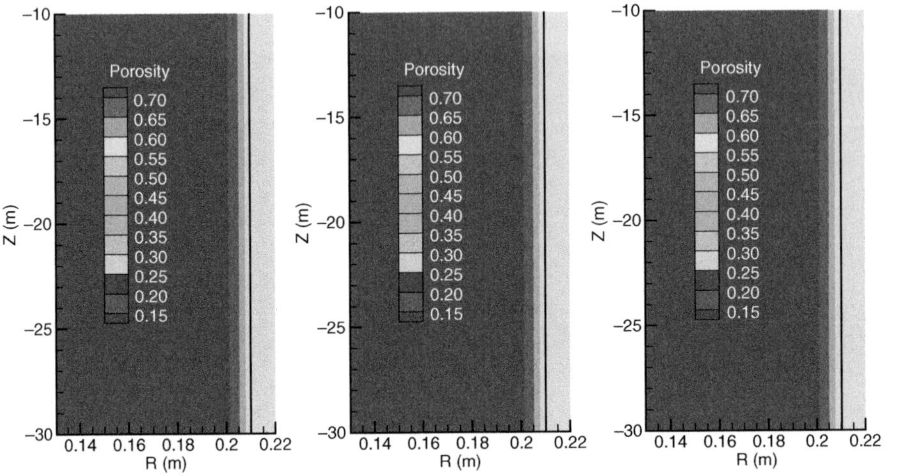

Figure 5.30 Modeled porosity change of wellbore cement due to corrosion in contact with a clean sand formation with brine salinity at 75,000 ppm for 50 years, using cement permeability at 1 nD (left), 10 nD (middle) and 1 mD (right), respectively.

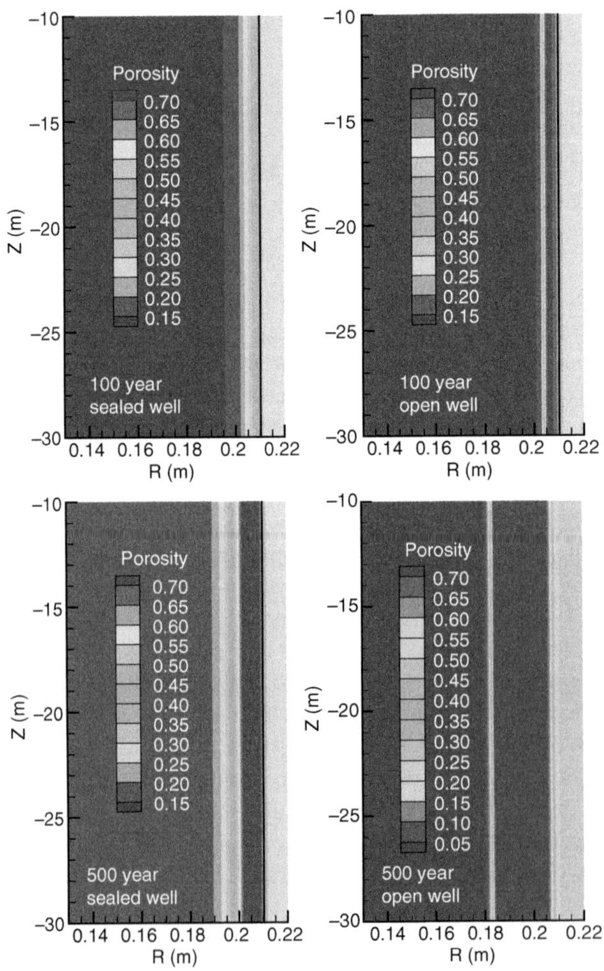

Figure 5.31 Modeled wellbore cement porosity change of an abandoned (sealed) well (top left and bottom left plots) and an open well (cement–rock interface is not fully sealed, top right and bottom right plots) due to corrosion in contact with formation brine under elevated pressure at 10 bars for 100 years (two top plots) and 500 years (two bottom plots), respectively, using the same permeability, 1 nD.

cement is always maintained close to zero. Thus, there is no significant advective flux across the cement–rock interface. The mass flux (transferring acidity) across this interface is always dominated by diffusion when the well is fully sealed.

However, when the wellbore is opened, the cement–rock interface is hydrodynamically connected to the atmosphere or any other boundary controls of lower pressure, and the pressure at the interface is controlled by the hydrostatic gradient, and so a constant hydraulic gradient can be maintained across the cement, and a significant advective flux can seep into the cement from the formation when injected CO_2 approaches the legacy well. Thus, corrosion can be significantly accelerated (Figure 5.31).

5.6 Summary

A hydrocarbon reservoir can normally be considered as a geochemical system at high temperature, high pressure, high salinity and highly reducing conditions. Having gone through multiple millions of years of geological evolution, a thermodynamic equilibrium or pseudo-equilibrium is reached between the reservoir formation water, HC and rock. The chemical composition of reservoir water stays constant or very slowly evolving, reflecting the diagenetic trend of the reservoir chemical system. However, the rate of this evolution is ignorable compared with the drastic reactions induced by injecting the reservoir with water, CO_2, or other fluid for enhanced oil recovery. Risks of scaling (of both wellbore and near-wellbore regions of the reservoir), souring (by BSR and/or TSR), clay swelling and wellbore cement corrosion can become key issues that need to be carefully evaluated. These processes can be simulated using geochemical reactive transport modeling technology; however, the capability of efficiently handling the high temperature, high pressure, high salinity and highly reducing conditions is necessary. Few RTM simulators are considered capable of handling a HC reservoir under 4-High conditions. Examples presented for modeling scaling, souring, clay swelling and wellbore cement corrosion demonstrated that TOUGHREACT or TOUGHREACT-PITZER, as one of the RTM simulators, can be used to simulate effectively the geochemical processes for evaluation of the relevant risks. MoReS-PHREEQC and other similar simulators are also capable of handling this type of geochemical system. Special attention paid to the temperature-dependent thermodynamic equilibrium constants, no-longer-negligible pressure dependency of the thermodynamic equilibrium constant, reformatting the database to define properly the key reactions in the highly reducing condition, and careful calibration of key kinetic parameters may become necessary when modeling such geochemical systems. Results show rich information that is very useful for evaluating the relevant risks.

Acknowledgements

The authors would like to thank Tim Tambach, Niko Kampman and Shawn Gao for a close working relationship and continuous communication on RTM topics. Thanks are also given to Erik Tegelaar, Chen Bao and Judith Sims, for their constructive Shell internal review.

References

Al Mayahi, N., Snippe, J., Rucci, F.D., and Linthorst, S.J. (2012). Water injection subsurface challenges and reactive transport modelling. *SPE EOR Conference at Oil and Gas West Asia*, 16–18 April, Muscat, Oman. SPE-154457-MS.

Appelo, C.A.J. and Postma, D. (2005). *Geochemistry, Groundwater and Pollution*, 2nd Ed. Leiden: A.A. Balkema.

Bernard, G.G. (1967). Effect of floodwater salinity on recovery of oil from cores containing clays. *SPE California Regional Meeting*, 26–27 October, Los Angeles, California, SPE-1725-MS. doi:10.2118/1725-MS

Birkholzer, J., Apps, J., Zheng, L., et al. (2008). Water quality effects caused by CO_2 intrusion into shallow groundwater. *Research Project on CO_2 Geological Storage and Groundwater Resources*, Lawrence Berkeley National Laboratory, Technical Report LBNL-1251E, Berkeley, CA.

Carey, J.W.M., Wigand, S.J., Chipera, G., et al. (2007). Analysis and performance of oil well cement with 30 years of CO_2 exposure from the SACROC Unit, West Texas, USA. *International Journal of Greenhouse Gas Control* 1: 75–85.

Carey, J.W.R., Svec, R., Grigg, J., et al. (2010). Experimental investigation of wellbore integrity and CO_2–brine flow along the casing–cement microannulus. *International Journal of Greenhouse Gas Control* 4: 272–282.

Cross, M.M., Manning, D.A.C., Bottrell, S.H., and Worden, H. (2004). Thermochemical sulphate reduction (TSR): experimental determination of reaction kinetics and implications of the observed reaction rates for petroleum reservoirs. *Organic Geochemistry* 35: 393–404.

Davies, C.W. (1962). *Ion Association*. London: Butterworths.

Debye, P. and Hückel, E. (1923). Zur Theorie der Elektrolyte. I. Gefrierpunktserniedrigung und verwandte Erscheinungen [The theory of electrolytes. I. Lowering of freezing point and related phenomena]. *Physikalische Zeitschrift* 24: 185–206.

Desch, C.H. (1938). Henry Louis Le Chatelier, 1850–1936. *Obituary Notices of Fellows of the Royal Society* 2 (6): 250–226. doi:10.1098/rsbm.1938.0005

De Windt, L., Pellegrini, D., and van der Lee, J. (2004). Coupled modeling of cement/claystone interactions and radionuclide migration. *Journal of Contaminant Hydrology* 68: 165–182.

Duguid, A. and Scherer, G.W. (2010). Degradation of oilwell cement due to exposure to carbonated brine. *International Journal of Greenhouse Gas Control* 4: 546–560. doi:10.1016/j.ijggc. 2009.11.001

Evans, P.J., Ambrus, C.R., Maliska Jr., E.S.S.S., and Kabche, J.P. (2013). Simulation of reservoir souring considering multiple carbon sources and environmental aspects. *4th International Symposium on Applied Microbiology and Molecular Biology in Oil Systems*, Rio de Janeiro, Brazil.

Farajzadeh, R., Matsuura, T., van Batenburg, D.W., and Dijk, H. (2012). Detailed modelling of the alkali/surfactant/polymer (ASP) process by coupling a multipurpose reservoir simulator to the chemistry package PHREEQC. SPE-143671-PA. *SPE Reservoir Evaluation and Engineering* 15(04).

Friedmann, S.J. (2009). Emerging technical challenge of 5-million ton/year CO_2 injection. *SPE Technical Proceedings, SPE-126942-MS, SPE International Conference on CO_2 Capture, Storage and Utilization*, San Diego.

Harvie, C.E. and Weare, J.H. (1980). The prediction of mineral solubilities in natural waters: The Na-K-Mg-Ca-Cl-SO_4-H_2O system from zero to high concentration at 25 °C. *Geochimica et Cosmochimica Acta* 44 (7): 981–997.

Harvie, C.E., Moller, N., and Weare, J.H. (1984). The prediction of mineral solubilities in natural waters: The Na-K-Mg-Ca-H-Cl-SO_4-OH-HCO_3-CO_3-H_2O-system to high ionic strengths at 25 °C. *Geochimica et Cosmochimica Acta* 44 (7): 723–751.

Helgeson, H.C. and Kirkham, D.H. (1974). Theoretical prediction of the thermodynamic behavior of aqueous electrolytes at high pressures and temperatures: II. Debye-Hückel parameters for activity coefficients and relative partial molal properties. *American Journal of Science* 274: 1199–1261.

Helgeson, H.C., Kirkham, D.H., and Flowers, D.C. (1981). Theoretical prediction of the thermodynamic behavior of aqueous electrolytes at high pressures and temperatures: IV. Calculation of activity coefficients, osmotic coefficients, and apparent molal and standard and relative partial molal properties to 600 °C and 5 kb. *American Journal of Science* 281: 1249–1516.

Huet, B., Prevost, J., and Scherer, G.W. (2010). Quantitative reactive transport modeling of Portland cement in CO_2-saturated water. *International Journal of Greenhouse Gas Control* 4: 561–574.

Jacques, D., Šimůnek, J., Mallants, D., and van Genuchten, M.Th. (2008). Modeling coupled hydrological and chemical processes: Long-term uranium transport following mineral phosphorus fertilization. *Vadose Zone Journal Special Issue* 7 (2): 698–711. doi:10.2136/VZJ2007.0084

Johnson, J.W., Oelkers, E.H., and Helgeson, H.C. (1992). SUPCRT92: A software package for calculating the standard molal thermodynamic properties of minerals, gases, aqueous species, and reactions from 1 to 5000 bar and 0 to 1000 °C. *Computer and Geosciences* 18 (7): 899–947.

Kampman, N., Busch, A., Bertier, P., et al. (2016). Observational evidence confirms modelling of the long-term integrity of CO2-reservoir caprocks. *Nature Communications* 7. doi:10.1038/ncomms12268

Kurtis, K. (2015). *Portland Cement Hydration*. Atlanta, Georgia: School of Civil Engineering, Georgia Institute of Technology.

Kutchko, B., Strazisar, B., Dzombak, D., et al. (2007). Degradation of well cement by CO_2 under geologic sequestration conditions. *Environmental Science and Technology* 41: 4787–4792.

Machel, H.G. (1987). Some aspects of diagenetic sulphate-hydrocarbon redox-reactions. In: *Diagenesis of Sedimentary Sequences* (ed. J.D. Marshall), 15–28. Geological Society Special Publication 36.

Machel, H.G. (1998). Gas souring by thermochemical sulfate reduction at 140 °C: discussion. *American Association of Petroleum Geologists Bulletin* 82: 1870–1873.

Machel, H.G. (2001). Bacterial and thermochemical sulfate reduction in diagenetic settings – old and new insights. *Sedimentary Geology* 140: 143–175.

Machel, H.G., Krouse, H.R., Riciputi, L.R., and Cole, D.R. (1995). Devonian Nisku sour gas play, Canada: a unique natural laboratory for study of thermochemical sulfate reduction. In: *Geochemical Transformations of Sedimentary Sulfur* (ed. M.A. Vairavamurthy and M.A.A. Schoonen), 439–454. ACS Symposium Series 612.

Manzano, B.K., Fowler, M.G., and Machel, H.G. (1997). The influence of thermochemical sulfate reduction on hydrocarbon composition in Nisku reservoirs, Brazeau River area, Alberta, Canada. *Organic Geochemistry* 27: 507–521.

Monod, J. (1949). The growth of bacterial cultures. *Annual Review of Microbiology* 3: 371.

Nardi, A.A., Idiart, P., Trinchero, L., et al. (2014). Interface COMSOL-PHREEQC (iCP), an efficient numerical framework for the solution of coupled multiphysics and geochemistry. *Computers and Geosciences* 69: 10–21.

Orr, W.L. (1977). Geologic and geochemical controls on the distribution of hydrogen sulfide in natural gas. In: *Advances in Organic Geochemistry* (ed. R. Campos and J. Goni), 571–597. Madrid: Enadisma.

Ožbolt, J., Oršanić, F., Kušter, M., and Balabanic, G. (2013). Modeling damage of concrete caused by corrosion of steel reinforcement. *VIII International Conference on Fracture Mechanics of Concrete and Concrete Structures*, FraMCoS-8, Toledo, Spain.

Palandri, J. and Kharaka, Y. (2004). A compilation of rate parameters of water-mineral interaction kinetics for application to geochemical modeling. US Geological Survey Open File Report 2004-1068. Menlo Park, CA: USGS.

Parkhurst, D.L. and Appelo, C.A.J. (1999). *User's Guide to PHREEQC (Version 2) – A Computer Program for Speciation, Batch-Reaction, One-dimensional Transport, and Inverse Geochemical Calculations.* US Department of the Interior, Geological Survey, Denver, CO, Water-Resources Investigations Report 99–4259.

Parkhurst, D.L. and Appelo, C.A.J. (2013). *Description of Input and Examples for PHREEQC Version 3 – A Computer Program for Speciation, Batch-Reaction, One-Dimensional Transport, and Inverse Geochemical Calculations.* US Department of the Interior, Geological Survey, Denver, CO, Techniques and Methods 6–A43.

Parkhurst, D.L., Kipp, K.L., and Charlton, S.R. (2010). PHAST Version 2 – A program for simulating groundwater flow, solute transport, and multicomponent geochemical reactions. In: *Book 6, Modeling Techniques* (Chapter 35 of Section A Groundwater) (ed. US Department of the Interior, Geological Survey). Denver, CO: USGS.

Pitzer, K.S. (1973). Thermodynamics of electrolytes, *I. Theoretical basis and general equations. Journal of Physical Chemistry* 77: 268–277.

Pitzer, K.S. (1991). Ion interaction approach: Theory and data correlation. In: *Activity Coefficients in Electrolytes Solutions*, 2nd ed. (ed. K.S. Pitzer), 75–155. Boca Raton, FL: CRC Press.

Pitzer, K.S. and Mayorga, G. (1973). Thermodynamics of electrolytes. II. Activity and osmotic coefficients for strong electrolytes with one or both ions univalent. *Journal of Physical Chemistry* 77: 2300–2307.

Pitzer, K.S. and Kim, J.J. (1974). Thermodynamics of electrolytes, IV. Activity and osmotic coefficients for mixed electrolytes. *Journal of the American Chemical Society* 96: 5701–5707.

Prieto, M., Fernandez-Gonzalez, A., Putnis, A., and Fernandez-Diaz, L. (1997). Nucleation, growth, and zoning phenomena in crystallizing (Ba,Sr)CO_3, Ba(SO_4,CRO_4), (Ba,Sr)SO_4, and (Cd,Ca)CO_3 solid solutions from aqueous solutions. *Geochimica et Cosmochimica Acta* 61 (16): 3383–3397.

Pruess, K., Oldenburgg, C., and Moridis, G. (1999). *TOUGH2 User's Guide Version 2.* Berkeley, CA: Ernest Orlando Lawrence Berkeley National Laboratory.

Ranaivomanana, H., Verdier, J., Sellier, A., and Bourbon, X. (2013). Sealing process induced by carbonation of localized cracks in cementitious materials. *Cement and Concrete Composites* 37 (1): 37–46.

Roy, D.M., Brown, P.W., Ski, D., et al. (1993). *Concrete Microstructure Porosity and Permeability*. Materials Research Laboratory, The Pennsylvania State University, SHRP-C-628.

Samper, J., Xu, T., and Yang, C. (2009). A sequential partly iterative approach for multicomponent reactive transport with CORE2D. *Computing and Geosciences* 13: 301. doi:10.1007/s10596-008-9119-5

Sauki, A.B. and Irawan, S. (2010). Effects of pressure and temperature on well cement degradation by supercritical CO_2. *International Journal of Engineering and Technology* 10 (4).

Scheuerman, R.F. and Bergersen, B.M. (1990). *Injection-Water Salinity, Formation Pretreatment, and Well-Operations Fluid-Selection Guidelines*. SPE-18461. doi:10.2118/18461-PA

Snippe, J. and Tucker, O. (2014). CO_2 fate comparison for depleted gas field and dipping saline aquifer. *Energy Procedia* 63: 5586–5601.

Snippe, J. and Wei, L. (2014). An efficient and internally consistent reactive transport modelling scheme for sour gas injection simulations. *Abu Dhabi International Petroleum Exhibition and Conference*, Abu Dhabi, UAE.

Snippe, J., Gdanski, R., and, Ott, H. (2017). Multiphase modelling of wormhole formation in carbonates from core scale to well scale. *Energy Procedia* 114: 2972–2984. doi:10.1016/j.egypro.2017.03.1426

Stronach, S.A. and Glasser, F.P. (1997). Modeling the impact of abundant geo-chemical components on phase stability and solubility of the $CaO–SiO_2–H_2O$ systems at 25 °C: Na^+, K^+, SO_4^{2-}, Cl^- and CO_3^{2-}. *Advances in Cement Research* 9: 167–181.

Tambach, T., Lonnee, J., and Snippe, J. (2017). Forecast of reactive CO_2 injection into a carbonate formation, Middle East. *Energy Procedia* 114: 2847–2855. doi:10.1016/j.egypro.2017.03.1405

Tebes-Stevens, C., Valocchi, A.J., VanBriesen, J.M., and Rittmann, B.E. (1998). Multicomponent transport with coupled geochemical and microbiological reactions: Model description and example simulations. *Journal of Hydrology* 209: 8–26.

Tesoriero, A.J. and Pankow, J.F. (1996). Solid solution partitioning of Sr^{2+}, Ba^{2+}, and Cd^{2+} to calcite. *Geochimica et Cosmochimica Acta* 60 (6): 1053–1063.

Thom, J. and Anderson, G.M. (2008). The role of thermochemical sulfate reduction in the origin of Mississippi Valley-type Deposits. I. Experimental results. *Geofluids* 8: 16–26. doi:10.1111/j.1468-8123.2007.00201

Wayllace, A. (2008). Volume change and swelling pressure of ansive clay in the crystalline swelling regime. PhD dissertation, University of Missouri.

Wei, L. (2012). Sequential coupling of geochemical reactions with reservoir simulations for waterflood and EOR studies. *SPE Journal* 17 (2): 469–484.

Whitaker, F. and Xiao, Y. (2010). Reactive transport modeling of early burial dolomitization of carbonate platforms by geothermal convection. *American Association of Petroleum Geologists Bulletin* 94 (6): 889–917. doi:10.1306/12090909075

Wolery, T.J. (1992). *EQ3/6, A Software Package for Geochemical Modeling of Aqueous Systems: Package Overview and Installation Guide. Lawrence Livermore National Laboratory*, UCRL-MA-110662 PT I.

Xiao, Y. and Jones, G.D. (2006). Reactive transport modeling of carbonate and siliciclastic diagenesis and reservoir quality prediction. *International Petroleum Technology Conference* held in Abu Dhabi, UAE, 5–8 November 2006. SPE 101669.

Xiao, Y. and Jones, G.D. (2007). Reactive transport models of limestone-dolomite transitions: implications for reservoir connectivity. *International Petroleum Technology Conference* held in Dubai, UAE, 4–6 December 2007. IPTC11199.

Xu, T. (1996). *Modeling non-isothermal multicomponent reactive solute transport through variably saturated porous media*. PhD dissertation, University of La Coruña, Spain.

Xu, T., Samper, J., Ayora, C., et al. (1999). Modeling of non-isothermal multicomponent reactive transport in field scale porous media flow systems. *Journal of Hydrology* 214: 144–164.

Xu, T., Sonnenthal, E.L., Spycher, N., and Pruess, K. (2006). TOUGHREACT: a simulation program for non-isothermal multiphase reactive geochemical transport in variably saturated geologic media. *Computers and Geosciences* 32, 145–165.

Xu, T., Sonnenthal, E., Spycher, N., et al. (2009). TOUGHREACT Version 2. *TOUGH Symposium 2009*, Lawrence Berkeley National Laboratory, Berkeley, CA.

Xu, T., Spycher, N., Sonnenthal, E., et al. (2011). TOUGHREACT Version 2.0: A simulator for subsurface reactive transport under non-isothermal multiphase flow conditions. *Computers and Geosciences* 37: 763–774.

Xu, T., Sonnenthal, E., Spycher, N., and Zheng, L. (2014). *TOUGHREACT V.3.0-OMP Reference Manual: A Parallel Simulation Program for Non-Isothermal Multiphase Geochemical Reactive Transport*. Lawrence Berkeley National Laboratory. Available at http://esd1.lbl.gov/FILES/research/projects/tough/documentation/TOUGHREACT_V3-OMP_RefManual.pdf

Ylmén, R. and Jäglid, U. (2013). Carbonation of Portland cement studied by diffuse reflection Fourier transform infrared spectroscopy. *International Journal of Concrete Structure and Material* 7: 119–125. doi:10.1007/s40069-013-0039-y

Zhang, G. (2001). *Nonisothermal hydrobiogeochemical models in porous media*. PhD dissertation, University of La Coruña, Spain.

Zhang, G. and Samper, J. (2001). Biohydrogeochemistry: a new face of groundwater chemistry. In: *Las Caras del Agua Subterrañea*, vol. I (ed. A. Medina, J. Carrera, and L. Vives), 63–69. Ministerio de Ciencia y Tecnologia [Department of Science & Technology], Spain.

Zhang, G., Spycher, N., Xu, T., et al. (2006a) *Reactive geochemical transport modeling of concentrated aqueous solutions with TOUGHREACT – Supplement to TOUGHREACT Users' Guide for Pitzer ion-interaction model*. Berkeley, CA: Lawrence Berkeley National Laboratory LBNL-57873.

Zhang, G., Spycher, N., Sonnenthal, E., and Steefel, C. (2006b). Implementation of a Pitzer activity model for modeling concentrated solutions. *TOUGH Symposium 2006*, Lawrence Berkeley National Laboratory, Berkeley, CA, May 15–17, 2006.

Zhang, G., Spycher, N., Sonnenthal, E., and Steefel, C. (2009). *Geochemical Transactions* 10: 11. doi:10.1186/1467-4866-10-11

Zhang, G., Taberner, C., Cartwright, L., and Xu, T. (2011). Injection of supercritical CO_2 into deep saline carbonate formations: predictions from geochemical modeling. *SPE Journal* 16 (4): 959–967. SPE-121272-PA. doi:10.2118/121272-PA
Zheng, L., Rutqvist, J., Liu, H.H., et al. (2014). Model evaluation of geochemically induced swelling/shrinkage in argillaceous formations for nuclear waste disposal. *Applied Clay Science* 97–98 (0): 24–32.

6

Multiphase Fluid Flow and Reaction in Heterogeneous Porous Media for Enhanced Heavy Oil Production

Xinfeng Jia[1,2], *Xiaohu Dong*[1,2], *Jinze Xu*[1] *and Zhangxin Chen*[1]

[1] *University of Calgary, Canada*
[2] *China University of Petroleum (Beijing), China*

Viscosity reduction through heat or mass transfer is one of the most important mechanisms of heavy oil *in situ* recovery processes. This chapter focuses on the transport phenomena in pure steam, hybrid solvent–steam, and gas–solvent–steam co-injection processes, and the effects of reservoir heterogeneity on recovery performance.

6.1 Introduction

6.1.1 Heavy Oil Reserve Distribution

Heavy oil and bitumen are characterized by their high viscosities (i.e. higher than 100 mPa·s for heavy oil and 10,000 mPa·s for bitumen) and low American Petroleum Institute (API) gravities (i.e. lower than 20.0 API for heavy oil and 10.0 API for bitumen) (Speight, 2009). In the world, the total crude oil resources are approximately 9 – 11 trillion barrels, among which more than two-thirds are unconventional heavy oil and bitumen. Out of the total 8 trillion barrels of heavy oil and bitumen resources, Canada and Venezuela each possesses 2–3 trillion barrels. In Canada, most of Western Canadian heavy oil and bitumen deposits are located in the three major regions in northern Alberta: Athabasca, Cold Lake, and Peace River. With the depletion of conventional oil resources and increasing demand for energy, effective and economical recovery of heavy oil deposits has practical importance.

6.1.2 Current Exploitation Methods

Heavy oil resources have been developed through two kinds of recovery methods: (1) open-pit mining methods are used to recover bitumen deposits that are less than 100 m deep; (2) in-situ recovery methods are applicable to heavy oil resources at a

Reactive Transport Modeling: Applications in Subsurface Energy and Environmental Problems,
First Edition. Edited by Yitian Xiao, Fiona Whitaker, Tianfu Xu and Carl Steefel.

greater depth. In addition, in-situ recovery methods are more cost-effective and environmentally friendly since they disturb less of the surface environment and avert the need for sand disposal (Das and Butler, 1994).

In general, in-situ recovery processes are classified into two categories: thermal and non-thermal recovery processes. The key to heavy oil recovery is to reduce oil viscosity and hence improve oil mobility. Thermal recovery processes, using heat to reduce oil viscosity in-situ, have been most successful, and currently account for the production of over half a million barrels of oil per day in the US, Venezuela, and Canada (Farouq-Ali, 2003). Thermal recovery techniques are effective as oil viscosity is very sensitive to temperature. Increasing the temperature of oil in a reservoir leads to the reduction of oil viscosity and the improvement of oil mobility. As seen in Figure 6.1, heating heavy oil can reduce its viscosity by several orders of magnitude.

At present, the most commonly implemented thermal recovery techniques are cyclic steam stimulation (CSS) (Vittoratos et al., 1990), steam flooding, steam-assisted gravity drainage (SAGD) (Butler et al., 1981), hot water flooding, and in-situ combustion (Moore et al., 1995). CSS is also called the steam 'huff and puff' process, and each cycle has three phases, as shown in Figure 6.1:

1) *Injection*: Steam is injected into a formation at a high pressure to heat the formation and decrease oil viscosity.
2) *Soaking*: After injecting steam for a certain period, the well is shut in for a while to allow the steam to soak into the formation and further decrease the oil viscosity.
3) *Production*: After soaking, oil is produced in the same well for a certain period.

CSS was first used in Venezuela in 1959. Since then, this method has been applied in many oil fields across the world, such as the San Joaquin Valley in the United States (Cline and Basham, 2002), Cold Lake in Canada (Donnelly, 2000), Lake Maracaibo in Venezuela (Escobar et al., 1997), and Liaohe in China (Peng et al., 2010). With decades of development, the average recovery factor (RF) of CSS has been increased from 15% of the original oil-in-place (OOIP) in the 1980s to approximately 40% in recent years (Alvarez and Han, 2013). CSS is attractive because it provides a quick payout at a relatively high success rate as a result of the cumulative field-development experience. However, it is still uncompetitive in terms of the ultimate recovery factor, in comparison with other thermal recovery processes such as SAGD (60–70% of OOIP) (Alvarez et al., 2014).

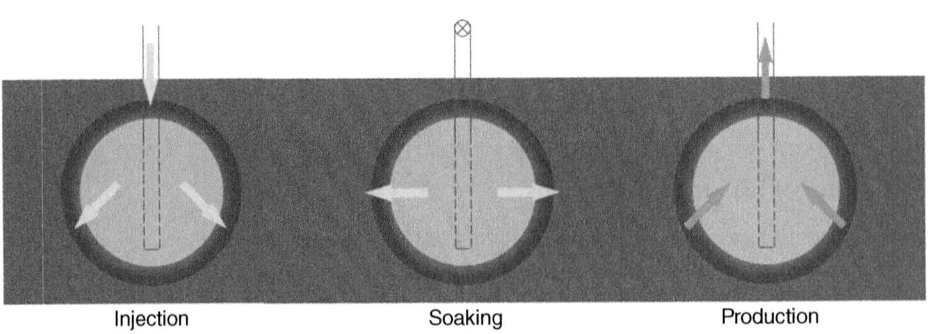

Figure 6.1 Schematics of a CSS process.

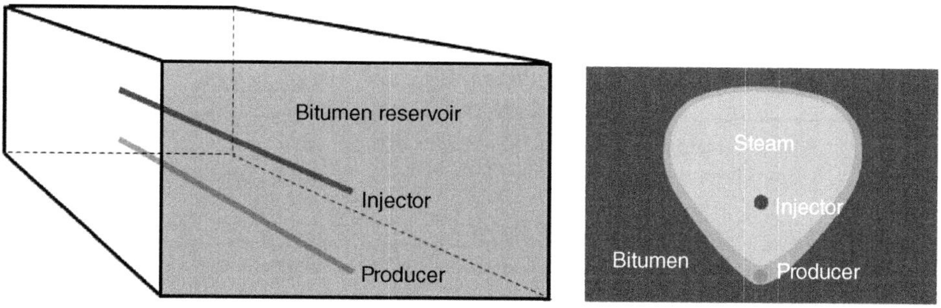

Figure 6.2 Schematics of a SAGD process for bitumen development.

In a typical SAGD process (Butler et al., 1981) (see Figure 6.2), two parallel horizontal wells, an upper injector and a lower producer, are drilled at the bottom of a formation. The injector is usually five metres above the producer. Steam is injected continuously into the formation and expands vertically and laterally to form a steam chamber. At the edge of the chamber, the heated bitumen and condensate drain down to the lower producer to be pumped out. As oil and condensate drain down, steam grows upwards and sideway. Since its proposal in the early 1980s, SAGD has gone through several phases in the Alberta Oil Sands Technology and Research Authority (AOSTRA) Underground Test Facility with success. To date, SAGD has been commercially applied in most new thermal projects in the Athabasca Oil Sand Deposits, because it can be operated at a relatively low pressure in reservoirs with low solution gas content. Efficiently utilizing steam and increasing oil production are two critical issues in SAGD. Thus, an oil production rate and a cumulative steam–oil ratio (SOR) are two important economic criteria for the evaluation of a SAGD process.

The SAGD process is sensitive to the heterogeneity of an oil sand formation. A shale layer limits the growth of a steam chamber and increases SOR (Shin and Choe, 2009). Gas cap and top water greatly increase steam injection once a steam chamber penetrates into such a top zone, which leads to a high SOR (Nasr et al., 2003b). A lean zone behaves as a thief zone during SAGD production, which increases SOR and decreases oil production (Xu et al., 2014). Published studies on the role of lean zones in the SAGD process remain limited. A steam chamber is a focus of SAGD studies because a good SAGD production process requires uniform and stable development of the steam chamber. Butler (1987) hypothesized that steam goes up and exchanges heat with cold oil at the border of the chamber, and the heated oil goes down along the lower edge of the chamber. Ito and Ipek (2005) considered that the steam chamber grows toward the top and lateral sides simultaneously. Butler (1991) likewise believed that the chamber possesses a higher rate upon initial upward expansion and it reaches the top of a reservoir. The sideward expansion then has a dominant function in chamber growth.

6.1.3 Potential in the Post-Steam Injection Era

6.1.3.1 Hybrid Steam–Solvent Processes

A number of new processes have been proposed to improve the performance of SAGD in the past two decades (Butler, 1997; Polikar et al., 2000; Nasr et al., 2003a; Leaute et al., 2016). In particular, a hybrid steam–solvent process has attracted tremendous

Table 6.1 Summary of the solvent – steam co-injection field implementations.

Technique	Theory	Scheme	Field pilot test
SAP	Butane – steam co-injection	Addition of ~15 wt.% of C_4 at $P = 2500 - 4000$ kPa	2002, Senlac 2004, Christina Lake, $\Delta q\uparrow = 30\%$, $\Delta SOR\downarrow = 35\%$
LASER	Pentane – steam co-injection at late cycles of CSS	Addition of 6 vol.% of C_{5+} into steam at CSS Cycle #7	2002, Cold Lake, $\Delta q\uparrow = 35\%$, $\Delta SOR\downarrow = 32\%$ 2011, Imperial Oil-Cold Lake, 240 well large pilot
ES-SAGD	Hexane – steam co-injection	Addition of 5–10 vol.% of $C_7 - C_{12}$ into steam at $P = 1400$ kPa	2006, Nexen-Long Lake pilot, $\Delta q\uparrow = 6\%$, $\Delta SOR\downarrow = 7\%$

research attention in recent years (Nasr et al., 2003a; Leaute et al., 2016). In such a process, a small amount of vaporized but condensable hydrocarbon solvent is added to steam. The injected solvent condenses along with steam at the bitumen–vapour interface and mixes with bitumen to further reduce the oil viscosity and enhance the drainage rate. The factual mechanisms of hybrid solvent–steam processes are rather complex, among which two major ones are described as: steam condensation to transfer latent heat to a formation; and solvent condensation to dissolve liquid solvent to bitumen. Besides, light solvents (i.e. propane, pentane) may cause asphaltene precipitation and in-situ upgrading.

Some of the hybrid solvent–steam techniques, such as a liquid addition to steam for enhancing recovery (LASER) have been successfully tested at both laboratory and field scales (Leaute, 2002; Leaute et al., 2016). Compared with conventional SAGD, hybrid solvent – steam processes usually demonstrate higher oil production rates and lower steam – oil ratios. Table 6.1 lists the theory, scheme, and field performance of three representative steam – solvent hybrid processes.

6.1.3.2 Steam – Solvent – Gas Co-injection Processes

Different from the conventional saturated steam, a steam – solvent – gas co-injection process, here referred to as multiple thermal fluids, is developed from a multiple thermal fluids generator (Tang et al., 2011). This generator utilizes the combustion and jetting mechanisms of a rocket engine. Considering the unique generation method of multiple thermal fluids, this heat-carrier has higher temperature and enthalpy compared with the conventional steam. For the development process of a heavy oil reservoir, it is a high-potential enhanced oil recovery (EOR) process. Furthermore, as a new heat-carrier, multiple thermal fluids are also different from the conventional gas mixture of steam and non-condensable gases. Firstly, multiple thermal fluids are produced from the combustion process in the generator, and it is a high-temperature and high-pressure gas mixture. The non-condensable gas fraction in the multi-thermal fluids is the gas mixture of N_2, CO_2, CH_4, CO, and so on. That is different from the conventional case. Secondly, in field tests, multiple thermal fluids are always injected into a reservoir directly after generation. This is different from the separate injection method in the conventional case. Thirdly, the steam fraction in multiple thermal fluids is always the superheated steam. It has higher enthalpy than the conventional saturated steam (Leaute and Carey, 2007; Berg, 2011; Delamaide and Parra Moreno, 2015). This new heat-carrier will be a high-potential EOR fluid in heavy oil reservoirs and oil sands.

6.1.4 Transport Equations

As mentioned above, heavy oil can be diluted either through steam heating or solvent dissolution. Both the heat transfer between heavy oil – steam and mass transport between heavy oil – solvent can be described by a convection – diffusion equation. The energy equation for fluid flow through a porous medium can be derived by using the first law of thermodynamics. For simplicity, the radioactive effects, viscous dissipation, and the work due to pressure changes are neglected (Bejan and Kraus, 2003).

In most cases, it is acceptable to assume that there is local thermal equilibrium where $T_s = T_f$, and T_s and T_f are the temperatures of the solid and fluid phases, respectively. A further assumption is that there is a parallel conduction heat transfer taking place in the solid and fluid phases, so there is no net heat transfer from one phase to the other. Taking an average over an elemental volume in a porous medium, we have the following energy equations for the solid and fluid phases, respectively (Nield and Bejan, 1999):

$$(1-\phi)\rho_s c_s \frac{\partial T_s}{\partial t} = (1-\phi)\nabla\cdot(K_s \nabla T_s) + (1-\phi)Q_s \tag{6.1}$$

$$\phi \rho_f c_f \frac{\partial T_f}{\partial t} + \rho_f c_f V\cdot\nabla T_f = \phi\nabla\cdot(K_f \nabla T_f) + \phi Q_f \tag{6.2}$$

Here the subscripts s and f refer to the solid and fluid phases, respectively; ρ is density, kg/m^3; c is the specific heat capacity, J/kg/K; ϕ is the porosity of a reservoir, fraction; Q is the rate of heat generation per unit volume, W/m^3; V is Darcy's velocity, m/s; K is thermal conductivity, W/m/K; and T is temperature, K.

The conductive heat flux through the solid is $K_s\nabla T_s$, and thus $\nabla(K_s\nabla T_s)$ is the net rate of heat conduction into a unit volume of the solid. In Equation (6.1), this appears multiplied by the factor $(1-\phi)$, which is the ratio of the cross-sectional area occupied by the solid to the total cross-sectional area of the medium. The other two terms in Equation (6.1) also contain the factor $(1-\phi)$, because this is also the ratio of the volume occupied by the solid to the total volume of the element. In Equation (6.2) there appears a convective term due to the seepage velocity. It is recognized that $V\cdot\nabla T_f$ is the rate of change of temperature in the elemental volume due to the convection of a fluid into it. Therefore, the second term in Equation (6.2) must be the rate of change of thermal energy, per unit volume of the fluid, due to convection.

Assuming local thermal equilibrium ($T_s = T_f$) and adding Equations (6.1) and (6.2) yields:

$$\rho c \frac{\partial T}{\partial t} + \rho_f c_f V\cdot\nabla T = \nabla\cdot(K\nabla T) + Q \tag{6.3}$$

where ρc is the overall volumetric heat capacity of the reservoir/fluid system, J/m^3/K; K is the overall thermal conductivity of the reservoir/fluid system, W/m/K; V is Darcy's velocity of the injection fluid, m/s; and Q is the overall rate of heat generation per unit volume of the porous medium, W/m^3. All these parameters are defined as:

$$\rho c = (1-\phi)\rho_s c_s + \phi \rho_f c_f \tag{6.4}$$

$$K = (1-\phi)K_s + \phi K_f \tag{6.5}$$

$$V = -\frac{k}{\mu}\nabla P \tag{6.6}$$

$$Q = (1-\phi)Q_s + \phi Q_f \tag{6.7}$$

Suppose that the porous medium is homogeneous and has a spatially constant value of thermal conductivity. Equation (6.1) may then be further simplified as:

$$\frac{\partial T}{\partial t} = \frac{K}{\rho c}\nabla^2 T - \frac{\rho_f c_f}{\rho c} V \cdot \nabla T + \frac{Q}{\rho c} \tag{6.8}$$

where two important parameters are introduced: the thermal diffusivity of the reservoir/fluid system and the thermal convection velocity of the fluid:

$$D = \frac{K}{\rho c} \tag{6.9}$$

$$u = \frac{\rho_f c_f}{\rho c} V \tag{6.10}$$

Then the energy balance equation becomes

$$\frac{\partial T}{\partial t} = D\nabla^2 T - u \cdot \nabla T + \frac{Q}{\rho c} \tag{6.11}$$

Neglecting any internal heat generation (i.e. $Q = 0$), the energy balance equation is a typical convection – diffusion equation:

$$\frac{\partial T}{\partial t} = D\nabla^2 T - u \cdot \nabla T \tag{6.12}$$

6.2 Thermal Recovery Processes

6.2.1 Modeling Assumptions

In a SAGD process, many different mechanisms are involved and they affect the drainage rate to varying degrees. It is difficult to account for all of these complex mechanisms involved in the process. Therefore, only the major mechanisms are considered and others are neglected to simplify the modeling procedure. The following assumptions are made in a transition zone:

1) The reservoir is homogeneous in porosity and permeability.
2) Temperature and pressure in a steam chamber are assumed constant.

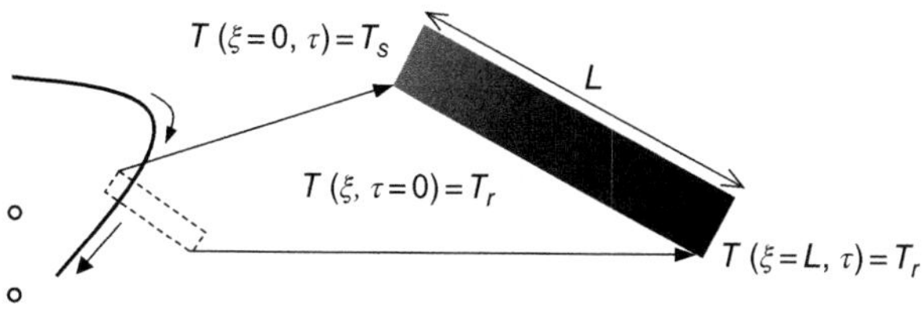

Figure 6.3 Schematics of one-dimensional heat transfer model in SAGD.

3) Line-sources are assumed for both steam and solvent injection.
4) The steam condensate does not penetrate into an oil zone and it drains parallel along chamber interface.
5) Heat losses into overburden and underburden layers are neglected.
6) Heat and mass transfer occur in the direction normal to the chamber boundary, and heat and mass transfer along the chamber boundary is negligible.
7) Heat is transported from steam to bitumen only through conduction and convection, and the thermal conductivity is a function of temperature.
8) The drainage of oil is in the liquid phase and it is caused purely by gravity.

On the basis of these assumptions, the heavy oil dilution process ahead of a steam chamber boundary is modeled as a one-dimensional transient heat transfer problem, as shown schematically in Figure 6.3.

6.2.2 Heat Transfer in SAGD

A one-dimensional heat conduction equation is used to describe temperature distribution in heavy oil normal to a steam chamber boundary. The left boundary is at constant temperature and the right boundary is closed. Initially the heavy oil temperature remains constant at the original reservoir temperature, T_r, and the heat transfer process in a transition zone can be described as:

$$\begin{cases} \frac{\partial T}{\partial t} = D\nabla^2 T - u\cdot\nabla T \\ T\left(x=s(U),t\right)=T_s \\ T\left(x=L,t\right)=T_r \\ T\left(x,t=0\right)=T_r \end{cases} \tag{6.13}$$

where U is the boundary moving velocity normal to the solvent chamber boundary, m/s. It should be noted that $x=s(U)$ represents the position of the left boundary of a transition zone segment that advances towards the untouched heavy oil zone due to the depletion of solvent-diluted heavy oil. This makes the mathematical model a moving boundary problem or a Stephen problem (Crank, 1984). To solve such a problem, there are generally three categories of methods: front tracking, front fixing, and fixed domain methods (Joshi et al., 2006). Here we use a front tracking method, in which a new coordinate is defined to immobilize the front:

$$\xi = x - \int_0^t U(t)\,dt \tag{6.14}$$

Apply the chain rule to obtain the first- and second-order derivatives of T with respect to ξ:

$$\frac{\partial}{\partial x}=\frac{\partial}{\partial \xi}\frac{\partial \xi}{\partial x}=\frac{\partial}{\partial \xi} \tag{6.15}$$

$$\frac{\partial^2}{\partial x^2}=\frac{\partial}{\partial x}\left(\frac{\partial}{\partial x}\right)=\frac{\partial}{\partial \xi}\left(\frac{\partial}{\partial \xi}\right)\frac{\partial \xi}{\partial x}=\frac{\partial^2}{\partial \xi^2} \tag{6.16}$$

Take the derivative of T with respect to t:

$$\frac{\partial}{\partial t}=\frac{\partial}{\partial \xi}\frac{\partial \xi}{\partial t}+\frac{\partial}{\partial t}\frac{d\tau}{dt}=\frac{\partial}{\partial \xi}(-U)+\frac{\partial}{\partial t}=-U\frac{\partial}{\partial \xi}+\frac{\partial}{\partial t} \tag{6.17}$$

Apply the above transformation to the governing equation, an IC (initial condition), and BCs (boundary conditions), a transformed heat conduction model in the new coordinates is:

$$\begin{cases} D\dfrac{\partial^2 T}{\partial \xi^2}+\sigma\dfrac{\partial T}{\partial \xi}=\dfrac{\partial T}{\partial \tau} \\ T(\xi=0,\tau)=T_s \\ T(\xi=L,\tau)=T_r \\ T(\xi,\tau=0)=T_r \end{cases} \tag{6.18}$$

where $\sigma=U-u$. Define dimensionless variables as follows:

$$T_D=\frac{T-T_r}{T_s-T_r},\ \xi_D=\frac{\sigma}{D}\xi,\ \tau_D=\frac{\sigma^2}{D}\tau \tag{6.19}$$

Then the model in dimensionless form is:

$$\begin{cases} \dfrac{\partial^2 T_D}{\partial \xi_D{}^2}+\dfrac{\partial T_D}{\partial \xi_D}=\dfrac{\partial T_D}{\partial \tau_D} \\ T_D(\xi_D=0,\tau_D)=1 \\ T_D(\xi_D=L^*,\tau_D)=0 \\ T_D(\xi_D,\tau_D=0)=0 \end{cases} \tag{6.20}$$

where:

$$L^*=\frac{\sigma}{D}L$$

The solution is given by Carslaw and Jaeger (1959):

$$T_D=\frac{1}{2}\left[erfc\left(\frac{\xi_D+\tau_D}{2\sqrt{\tau_D}}\right)+e^{-\xi_D}erfc\left(\frac{\xi_D-\tau_D}{2\sqrt{\tau_D}}\right)\right] \tag{6.21}$$

6.2.2.1 Gravity Drainage in a Transition Zone

With known temperature distributions, the oil viscosity can be computed. Then the gravity drainage velocity across a transition zone can be determined by using Darcy's law:

$$v = -\frac{K}{\mu}\nabla p \tag{6.22}$$

The total quantity of diluted oil that drains along the interface can be calculated by integrating Equation (6.3) across the transition zone:

$$q = \int_{\xi_{\min}}^{\xi_{\max}} v\,d\xi \tag{6.23}$$

where $\zeta_{\max}$ and $\zeta_{\min}$ are locations of the boundaries of the transition zone, respectively.

6.2.2.2 Boundary Movement

Suppose that oil flows into and drains out of a transition zone segment at rates of q_{in} and q_{out}, respectively. Within a short period of time, δt, the left boundary of the transition zone moves by $\delta\xi$ due to the depletion of solvent-diluted heavy oil from the segment. Therefore, a mass balance equation can be formulated:

$$\nabla q \delta t = l\phi \Delta S_o \delta\xi \tag{6.24}$$

If the time snippet is small enough, it is reasonable to regard the segment boundary length of a transition zone as constant and the residual oil saturation in the depleted area as uniform. Thereby the boundary moving velocity can be approximated as a constant value over that time interval:

$$U \approx \frac{\delta\xi}{\delta\tau} = \frac{\nabla q}{l\phi\Delta S_o} \tag{6.25}$$

6.2.2.3 Boundary Position

Given a constant boundary moving velocity U during a short time period Δt, the movement of the transition zone can be easily calculated by:

$$\Delta\xi_{mv} = U\Delta t \tag{6.26}$$

The respective advancement of a transition zone segment in the horizontal and vertical directions can be determined by using a trigonometric correlation:

$$\Delta x = \frac{\Delta\xi_{mv}}{\sin\theta}, \Delta y = \frac{\Delta\xi_{mv}}{\cos\theta} \tag{6.27}$$

The cumulative heavy oil production can be achieved by integrating the oil saturation difference over the steam chamber in an (x, y) coordinate system. The instantaneous oil production can be calculated by taking the derivative of Q_o with respect to t:

$$Q_o = \int_0^W \phi\Delta S_o (H - y)dx \tag{6.28}$$

Table 6.2 Properties of reservoir, fluid, thermal and solvent for a thermal recovery process.

Property	Value
Reservoir	
Height, H (m)	20
Width, W (m)	50
Length, L (m)	500
Porosity, ϕ (%)	25
Permeability, k (Darcy)	1
Relative oil permeability, k_r	0.9
Reservoir temperature, T_r (°C)	15
Bitumen	
Bitumen viscosity, μ_o (cP)	960,000
Bitumen density, ρ_o (kg/m^3)	1000
Bitumen molecular weight, M_o (g/mol)	570
Initial oil saturation, S_{oi} (%)	80
Thermal	
Density, ρ (kg/m^3)	4532.7
Heat capacity, C (J/kg/°C)	858.5
Steam temperature, T_s (°C)	250
Steam pressure, P (MPa)	2
Steam quality	0.7
Solvent	
Solvent type	Hexane
Viscosity, μ_s (cP)	0.294
Density, ρ_s (kg/m^3)	654.8

Table 6.2 lists the properties of a reservoir, fluid, thermal and solvent for a thermal recovery process. Figure 6.4 shows the evolution of a steam chamber boundary throughout a SAGD process. The profiles to the top left represent the steam chamber spreading phase, and those to the bottom right show its falling phase. Figure 6.5 shows the cumulative oil production and oil production rate in a SAGD process. The oil production rate during the steam chamber spreading phase remains nearly constant, which is in line with the conclusions generated from analytical studies. It declines continuously during the falling phase, which agrees well with experimental observations.

Along the steam chamber boundary, the transition zone properties change with position (Fisher et al., 2002). Figure 6.6 shows the temperature distributions at the top, middle, and bottom of the transition zone. It can be seen that the temperature profile at the top grows rapidly during the spreading phase (Figure 6.6a). At the middle part of the steam chamber boundary, the temperature grows slowly during the spreading period (Figure 6.6b). The transition zone at the middle part seems stabilized throughout the

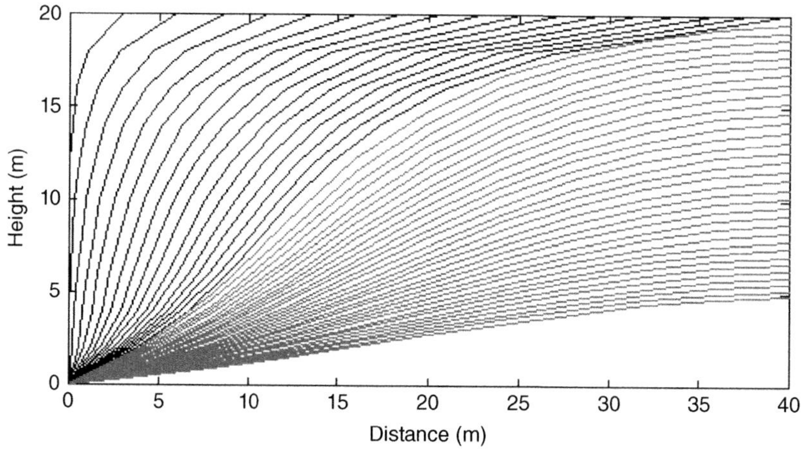

Figure 6.4 Steam chamber boundary evolution throughout a SAGD process.

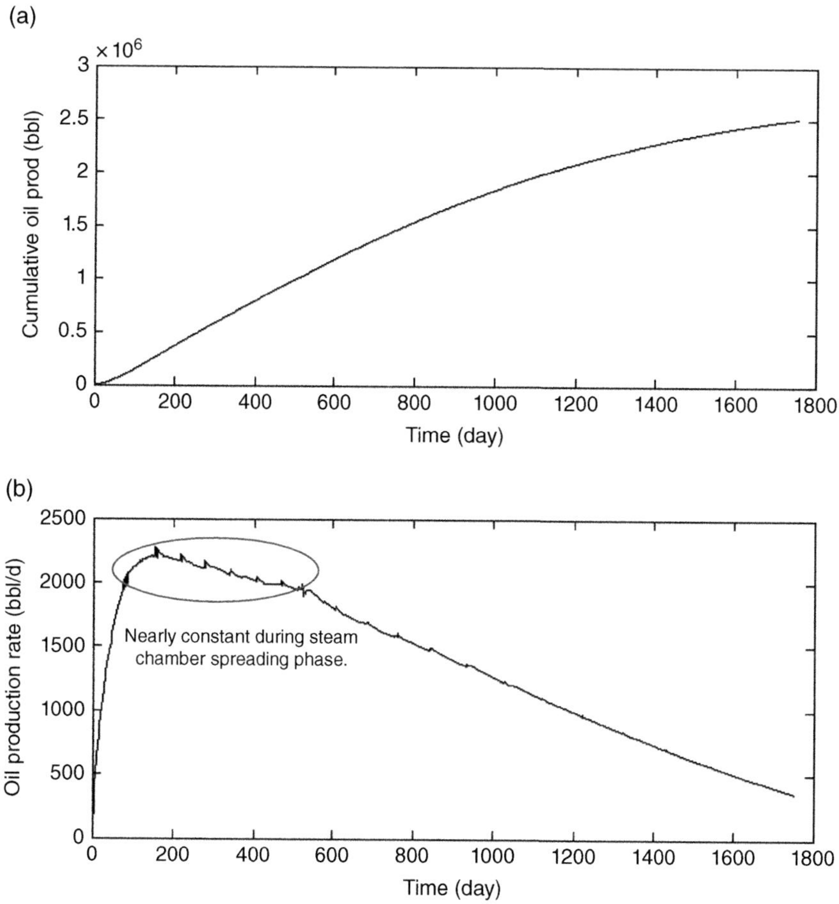

Figure 6.5 (a) Cumulative oil production and (b) oil production rate of SAGD.

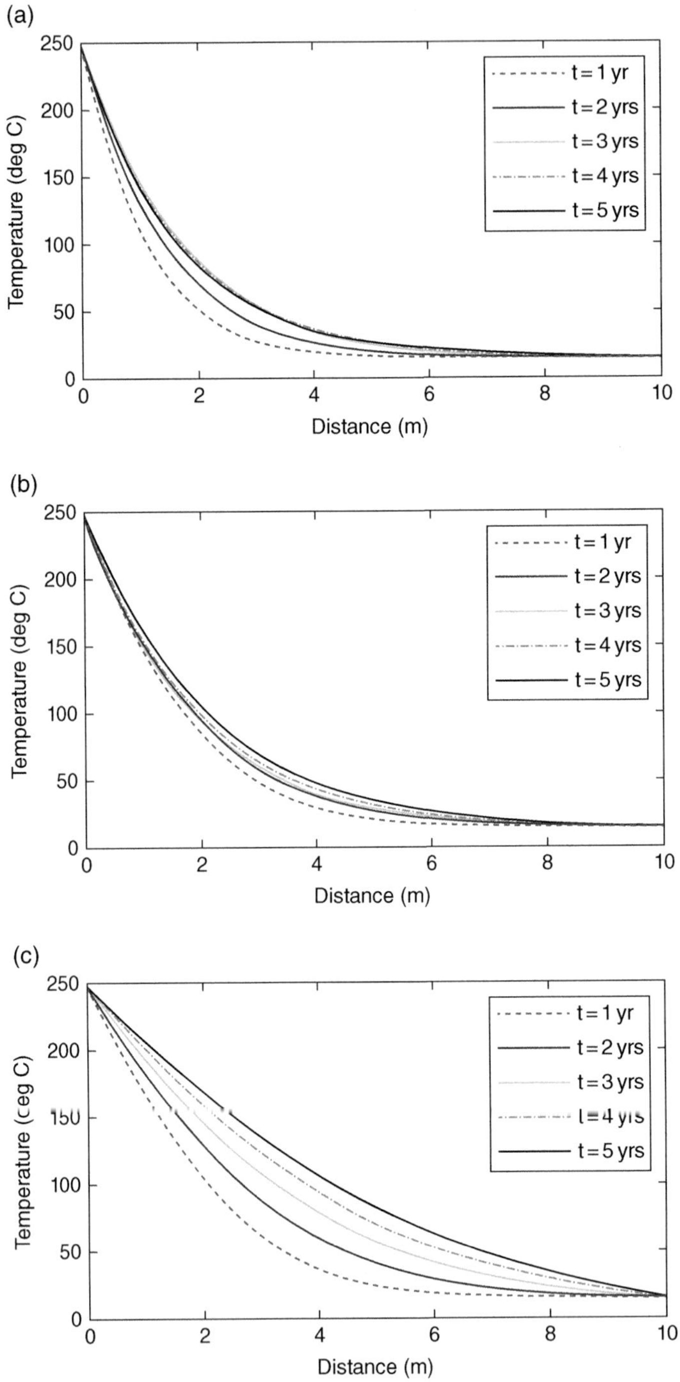

Figure 6.6 Temperature profile at (a) top, (b) middle, and (c) bottom part of steam chamber boundary.

spreading and falling phases, which validates the assumption of steady-state conduction and a constant boundary moving velocity in the previous analytical modeling. Figure 6.6c shows that the temperature profile at the lower part of the model grows quickly and steadily throughout the SAGD process, which is related to a small slip angle and a closed boundary at the bottom.

6.2.3 Heat Transfer in CSS

As mentioned before, throughout a CSS process, the operating pressure is cyclically drawn down and built up. This may cause a significant pressure gradient across the transition zone and lead to a heat convection effect in addition to conduction. Therefore, the dynamic pressure gradient/convection velocity should be taken into account for an accurate description of the heat transfer process. More specifically, the velocity term in Equation (6.12) should be considered as a variable rather than a constant value. The convection velocity across the transition zone, V_c, can be calculated by using Darcy's Law:

$$V_c = \frac{k}{\mu_c}\frac{\partial p}{\partial \xi} \tag{6.29}$$

Apparently, the computation of V_c requires knowledge of the pressure distribution. Figure 6.7 schematically shows that the temperature and pressure domains need to be considered for an accurate description of heat transfer in CSS.

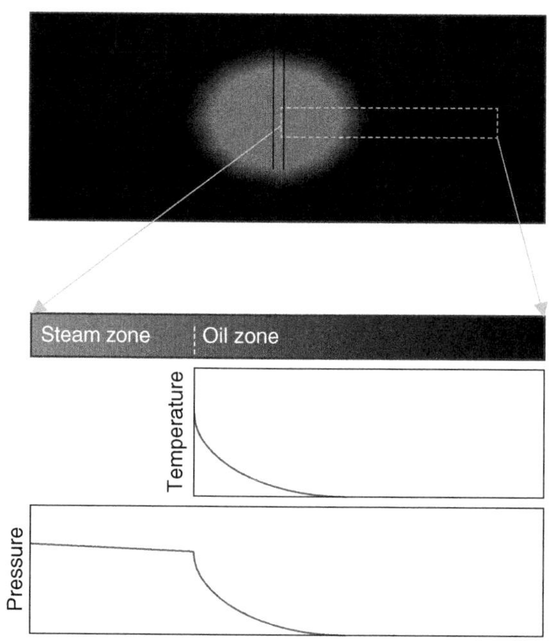

Figure 6.7 Coupled modeling of temperature and pressure of CSS.

The pressure transfer process in the transition zone can be expressed by using a diffusion equation. Suppose that the pressure at the left boundary or wellbore is maintained at constant and at the right boundary remains as the original reservoir pressure. Then the pressure diffusion model can be developed as:

$$\begin{cases} \dfrac{\partial}{\partial x}\left(\dfrac{k}{\mu}\dfrac{\partial p}{\partial x}\right) = \phi C_t \dfrac{\partial p}{\partial t} \\ p\left(x = s(U), t\right) = P_{inj} \\ p\left(x = L, t\right) = P_{ini} \\ p\left(x, t = 0\right) = P_{ini} \end{cases} \tag{6.30}$$

Similarly to the heat transfer model, the left boundary of a steam chamber also changes with time at the same speed, U, making the diffusion process a moving boundary issue. To solve the moving boundary problem, Equation (6.14) is also applied to transform the pressure model to the new coordinate system:

$$\begin{cases} \mu \dfrac{\partial^2 p}{\partial \xi^2} + \left(\dfrac{\phi C_t U}{k}\mu^2 - \dfrac{\partial \mu}{\partial \xi}\right)\dfrac{\partial p}{\partial \xi} = \dfrac{\phi C_t}{k}\mu^2 \dfrac{\partial p}{\partial \tau} \\ p(\xi, 0) = P_{ini} \\ p(\xi, 0) = P_{ini} \\ p(\xi = L, t) = P_{ini} \end{cases} \tag{6.31}$$

Define such dimensionless variables:

$$P_D = \frac{P_{ini} - p}{P_{ini}}, \ \xi_D = \frac{\xi}{L}, \ \mu_D = \frac{\mu}{\mu_{inlet}}, \ t_D = \alpha t, \ \delta^* = \frac{k}{\phi \mu_{inlet} C_t L^2} \tag{6.32}$$

Then the model is non-dimensionalized as:

$$\begin{cases} \mu_D \dfrac{\partial^2 p_D}{\partial \xi_D^2} + \left(\delta \mu_D^2 - \dfrac{\partial \mu_D}{\partial \xi_D}\right)\dfrac{\partial p_D}{\partial \xi_D} = \mu_D^2 \dfrac{\partial p_D}{\partial t_D} \\ p_D(\xi_D, t_D = 0) = 0 \\ p_D(\xi_D = 0, t_D) = \eta \\ p_D(\xi_D = 1, t_D) = 0 \end{cases} \tag{6.33}$$

where

$$\delta = \frac{U}{\delta^* L} = \frac{\phi \mu_{inlet} C_t L U}{k}, \ \eta = 1 - \frac{P_{inj}}{P_{ini}}$$

Due to the complex relationship between pressure, viscosity, and temperature, the governing equation of a pressure diffusion model is a second-order nonlinear partial

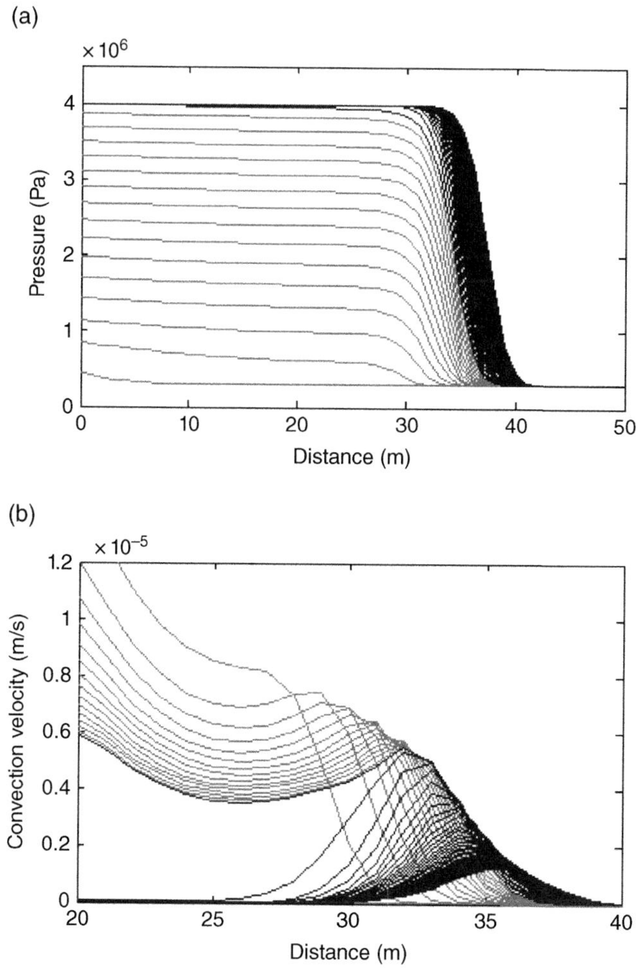

Figure 6.8 (a) Pressure buildup and (b) convection velocity ahead of the steam chamber of a CSS process.

differential equation, whose analytical solution is not readily available. Therefore, the pressure diffusion model is solved by using a numerical scheme, and the Newton – Raphson iterative method is used to compute the pressure distributions.

Figure 6.8 shows the pressure buildup process during the injection period of a CSS process, and the corresponding convection velocity and temperature ahead of a steam chamber. Profiles to the left represent a pressure buildup process, and lines to the right are pressure profiles after the bottom-hole pressure reaches a constant value. It is found that the convection velocity stays a considerable level ahead of the steam chamber boundary during the pressure buildup process. Once the pressure reaches a preset value, the convection velocity quickly declines to nearly zero.

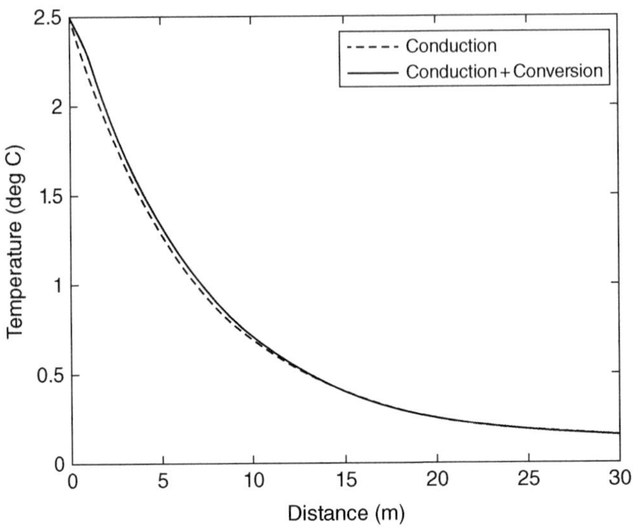

Figure 6.9 Comparison of heat conduction, and heat conduction plus convection.

6.2.4 Conductive and Convective Heat Transfer

It is commonly believed that conduction dominates heat transfer at the edge of a steam chamber. Convective heat transfer is not considered in the classical SAGD mathematical models, such as the one derived by Butler et al. (1981). While some researchers (Reis, 1992; Akin, 2005; Liang, 2005; Nukhaev et al., 2006; Azad and Chalatumyk, 2010) considered the conduction from steam to a cold reservoir to be the only heat transfer component, others questioned the assumption that thermal conduction dominates heat transfer at the edge of a SAGD chamber (Edmunds, 1991; Ito and Suzuki, 1996; Farouq-Ali, 1997; Ito et al., 1998; Sharma and Gates, 2010, 2011). A better understanding of the heat transfer process is critical for a successful SAGD project design. Figure 6.9 shows a temperature profile for conduction, and conduction plus convection. It can be seen that convection can provide a greater contribution to heat transfer than conduction at the edge of a steam chamber, but drops rapidly over a short distance normal to the interface. This clearly shows that convective heat transfer may exist only in the vicinity of a steam chamber boundary.

6.2.5 Multiple Phase Flow

The gravity drainage velocity across the transition zone can be determined by using Darcy's law:

$$v = -\frac{Kk_{ro}}{\mu}(\rho - \rho_s) g \sin\theta \tag{6.34}$$

where g is the gravitational acceleration, m/s^2, and θ is the slip angle, degree. A linear relative permeability is expected for the multiphase flow in the transition zone:

$$k_{ro} = \left(1 - S_g^*\right)^4, S_g^* = \frac{S_g}{1 - S_{wc}} \tag{6.35}$$

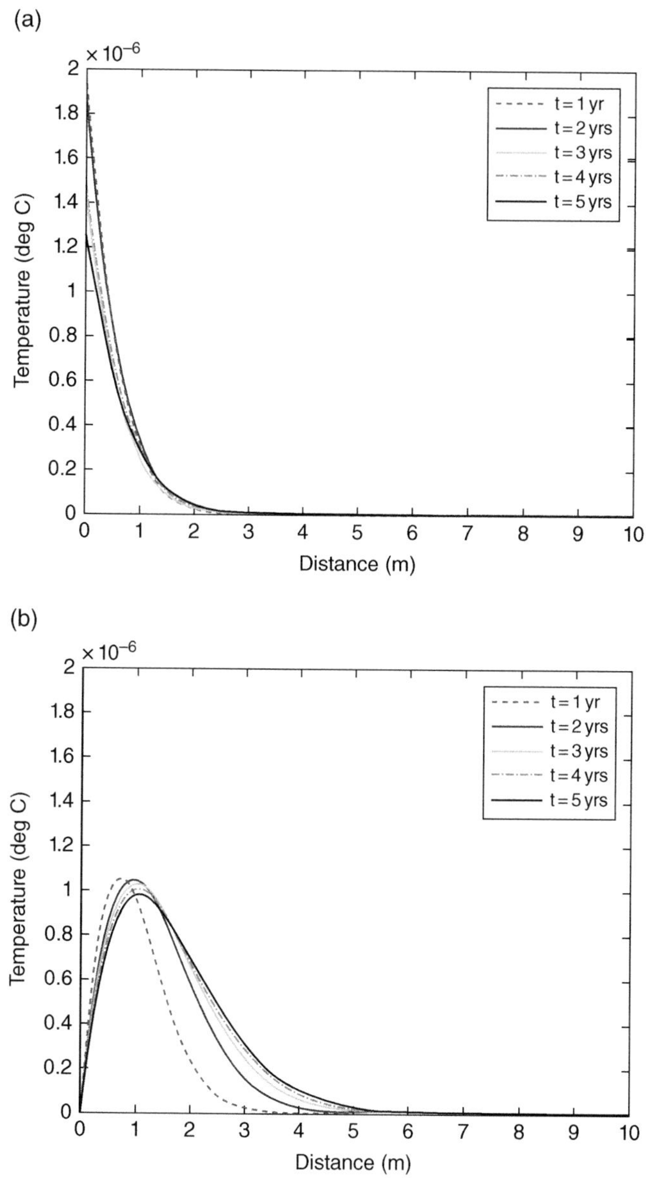

Figure 6.10 Drainage velocity profiles for (a) single and (b) multiple phase flow.

where S_g^* is the effective gas saturation and the capillary pressure is neglected, given its small magnitude in a high permeability environment (Yazdani and Maini, 2005). Also, the effective gas saturation is assumed to be linear with solvent concentration across the transition zone (Hayduk et al., 1973).

Figure 6.10 shows the drainage velocity profiles calculated by using single and multiple phase relative permeability, respectively. It is found that the former is thinner and higher but the latter is wider and smaller. This is because the higher gravity drainage velocity leads to a stronger flow rate at the interface, and consequently a faster boundary moving velocity normal to the interface, and thus a thinner transition zone.

6.3 Hybrid Thermal-Solvent Process

6.3.1 Mass Transfer

The transient heavy oil–solvent mixing process within a transition zone can be described by using Fick's Second law. Initially, the heavy oil is free of any solvent. The oil at the left boundary is fully saturated with solvent and the right boundary is impermeable. Thus the heavy oil–solvent mass transfer model can be developed as:

$$\begin{cases} \dfrac{\partial c}{\partial t} = D\dfrac{\partial^2 c}{\partial x^2} \\ c\left(x,\ t=0\right) = 0 \\ c\left(x = s(U),\ t\right) = c_{\max} \\ \dfrac{\partial c}{\partial x}\left(x = L,\ t\right) = 0 \end{cases} \tag{6.36}$$

where D is defined as an effective diffusion coefficient that incorporates all the mechanisms contributing to heavy oil–solvent mixing, such as molecular diffusion, convective dispersion, improved interfacial contact, and enhanced surface renewal by capillary imbibition. Similarly to the heat transfer model, the left boundary of the pressure model also changes with time. To solve the moving boundary problem, Equation (6.14) is also applied to transform the pressure model into the new coordinate system:

$$\begin{cases} \dfrac{\partial c}{\partial t} = D\dfrac{\partial^2 c}{\partial \xi^2} + U\dfrac{\partial c}{\partial \xi} \\ c\left(\xi = 0,\ t\right) = c_{\max} \\ \dfrac{\partial c}{\partial \xi}\left(\xi = L,\ t\right) = 0 \\ c\left(\xi,\ t=0\right) = 0 \end{cases} \tag{6.37}$$

The analytical solution to this model was derived from the general solution given by Carslaw and Jaeger (1959) and has the same form as that in Equation (6.21).

As mentioned above, the viscosity of heavy oil can be diluted mainly by two methods: solvent dissolution or heating by steam. Figure 6.11 compares the viscosity profiles by the two methods after one day's heating and dissolution, respectively. Obviously, the oil is diluted much faster through steam heating than solvent dissolution. In terms of production performance, the oil production rate of SAGD is two orders higher than its solvent analogue, vapour extraction.

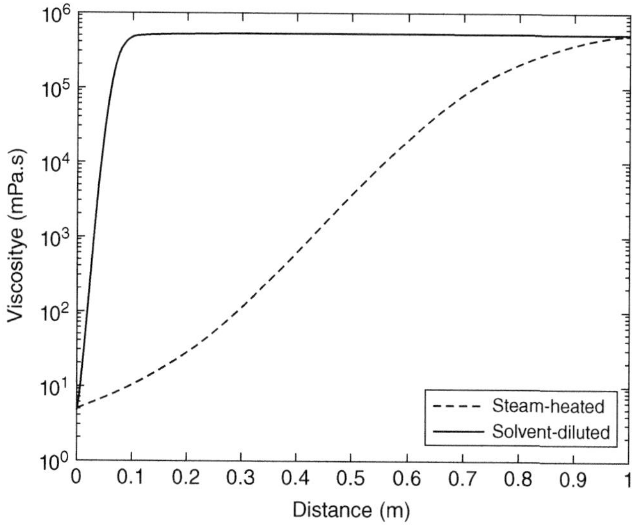

Figure 6.11 Viscosity profiles due to steam heating and solvent dissolution.

6.3.2 Coupled Heat and Mass Transfer

Hybrid steam–solvent processes have attracted tremendous research attention to improve the performance of SAGD in recent years. For example, in expanding-solvent SAGD (ES-SAGD), hexane is co-injected with steam into a steam chamber. One of the alleged mechanisms of such a process is the secondary viscosity reduction through solvent dissolution into heavy oil, in addition to bitumen dilution by heating.

Therefore, viscosity is a key coupling point of the heat transfer and mass transfer processes in hybrid steam–solvent processes. First, the heavy oil–solvent effective diffusivity can be more reasonably regarded as a variable of the mixture viscosity. Hayduk et al. (1973) and Hayduk and Minhas (1982) concluded that the diffusion coefficient of a solvent – oil system (ethane in n-hexane, heptanes, octane, dodecane, and hexadecane, or carbon dioxide in hexadecane) is dependent on the mixture viscosity, which was commonly expressed in a power law equation:

$$D(\mu)=\alpha\mu^{-\beta} \tag{6.38}$$

where α and β are constants depending on the crude oil and solvent properties as well as the operating conditions, and μ is the viscosity of the oil – solvent mixture, mPa·s. β is less than unity. Then the viscosity of a heavy oil–solvent system is commonly modeled in terms of solvent concentration by using the Lederer–Shu correlation (Shu, 1984):

$$\mu=\mu_S^{f_s}\mu_O^{f_o} \tag{6.39}$$

where $f_s=\dfrac{c}{\lambda(1-c)+c}$, $f_o=1-f_s$

$$\lambda=\frac{17.04(\gamma_o-\gamma_s)^{0.5237}\gamma_o^{3.2745}\gamma_s^{1.6316}}{\ln(\mu_o/\mu_s)}$$

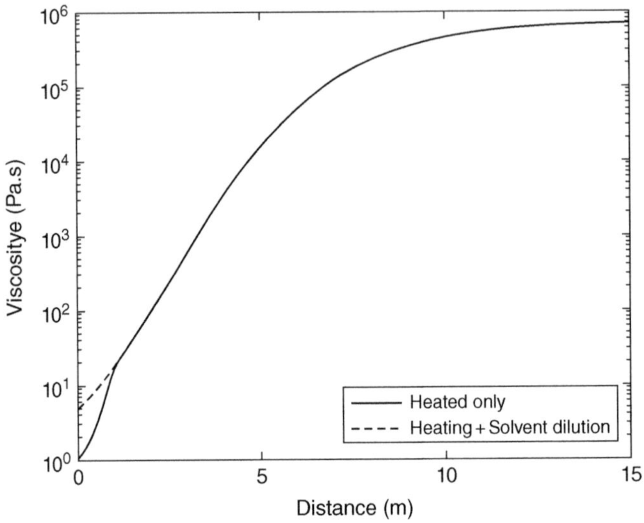

Figure 6.12 Viscosity profiles due to steam heating, and steam heating plus solvent dissolution.

where μ_o, μ_s, and μ are the viscosities of the crude oil, liquid solvent, and mixture of the two, respectively, mPa·s; f_o and f_s are the weighted volume fractions, vol.%; λ is a weight factor; and γ_o and γ_s are specific gravities of the crude heavy oil and liquid solvent, respectively. Figure 6.12 compares the respective viscosity profiles after pure steam heating and solvent dissolution following steam heating. It can be seen that the oil viscosity mainly takes effect at a short distance away from the interface, and it further dilutes the oil and greatly accelerates the oil drainage at the interface.

6.3.3 SAGD vs. ES-SAGD

Figure 6.13 compares the temperature profiles of SAGD and ES-SAGD at three points. Obviously, the ES-SAGD temperature profiles are thinner than those of SAGD. This is because the solvent dissolution accelerates the oil depletion rate at the interface, which can be evidenced by their respective boundary moving velocities, as shown in Figure 6.14. The boundary moving velocity of ES-SAGD is significantly higher than that of SAGD, especially in the early stage, and slightly slower than that of SAGD in the very late stage.

Figure 6.15 compares the steam chamber profiles of ES-SAGD and SAGD at $t = 2.5$, 5.0 and 7.5 yrs. It shows that the steam chamber of ES-SAGD grows much faster than that of SAGD due to its quicker boundary moving speed. As a result, the oil production rate of ES-SAGD is much higher than that of SAGD. Figure 6.16 shows that if solvent is added in the early and middle stages, the production rate can be nearly doubled.

6.4 Thermal–Solvent–Gas Co-injection Process

6.4.1 PVT Behaviour

For the mutiple thermal fluids (MTFs) injection process in heavy oil reservoirs, the reaction between thermal fluids and heavy crude oil will dominate the flowing process of MTFs in a reservoir (Liu et al., 2012; Dong et al., 2014; Delamaide and Parra Moreno,

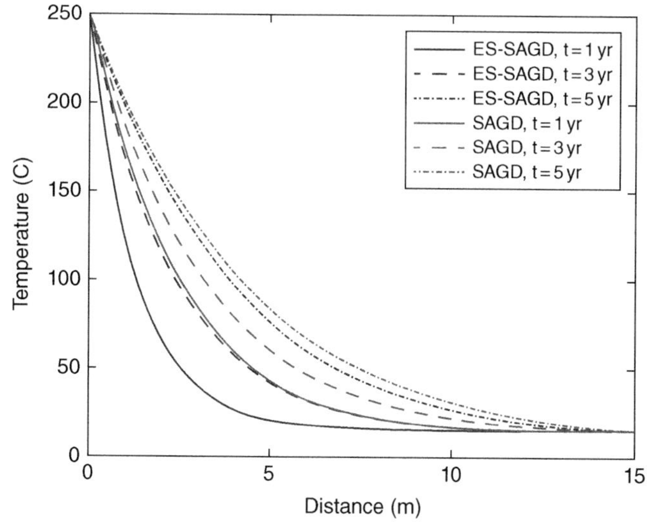

Figure 6.13 Comparisons of temperature distributions of SAGD and ES-SAGD at three different moments.

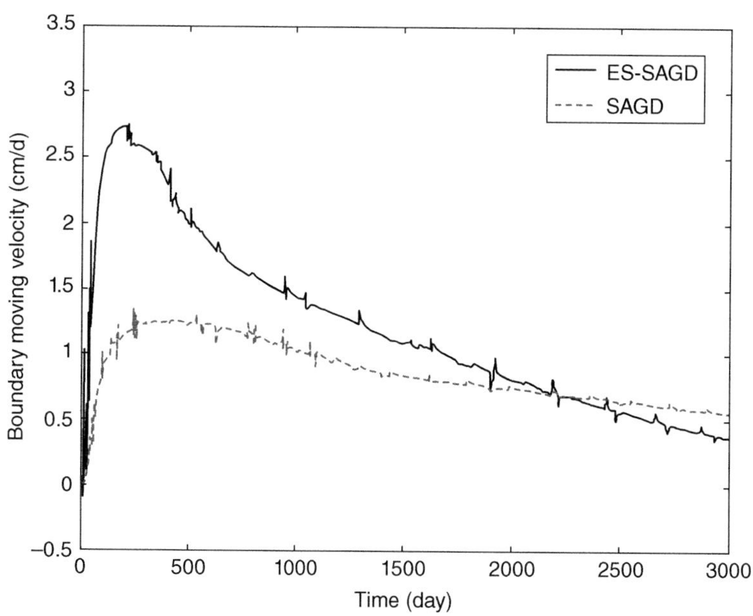

Figure 6.14 Comparisons of boundary moving velocity of ES-SAGD and SAGD.

2015). In addition, the PVT behaviour of heavy crude oil in a porous medium is improved. That is mainly caused by the dissolution of the non-condensable gas (NCG) fraction in heavy oil. The CO_2 fraction in MTFs has high solubility in heavy crude oil. Its dissolution can further reduce the oil viscosity, increase the oil volume, enhance the reservoir expansibility and improve the oil mobility. Different from the CO_2 fraction, N_2 has a low solubility, but it is helpful to recover the reservoir energy. Figure 6.17 shows

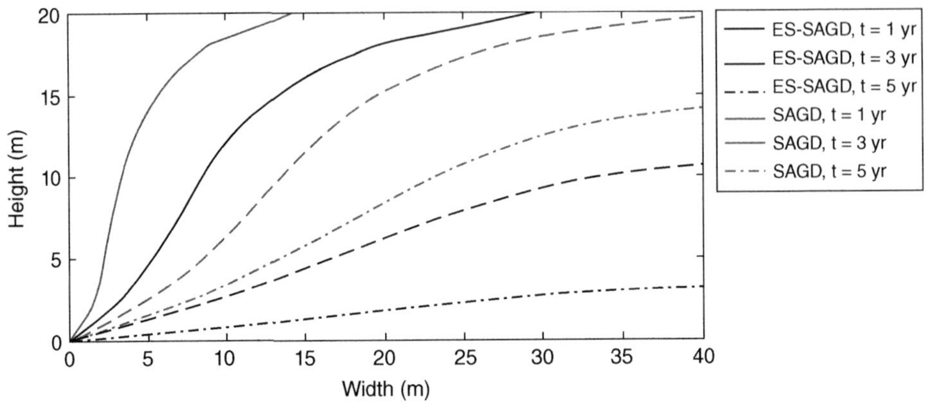

Figure 6.15 Comparison of steam chamber profiles of ES-SAGD and SAGD.

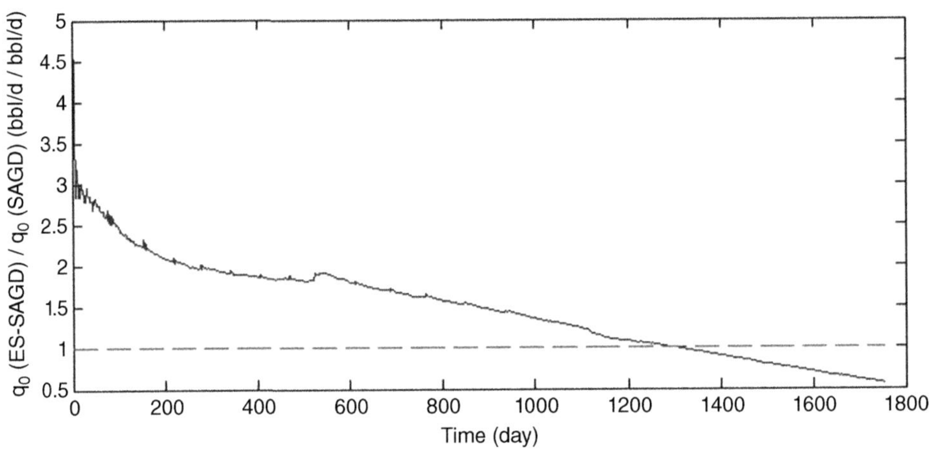

Figure 6.16 Ratio of oil production rate of ES-SAGD to that of SAGD.

the experimental data for a heavy oil – NCG mixture in a PVT cell. As shown, with a pressure increase, both the solubility of NCG in crude oil and the swelling factor of an oil–gas system are improved. In comparison with the oil without NCG, the presence of NCG can tremendously increase the oil mobility and swelling capacity of heavy crude oil. From the results, it is also observed that with an increase in temperature, both the solubility of NCG and the swelling factor of the oil–gas system are reduced. That is mainly caused by the effect of temperature on the gas molecular movement. High temperature is not useful for the gas dissolution in crude oil. Also, from this figure, it is shown that the NCG addition to crude oil also reduces the oil viscosity. Under the same temperature, on account of the effect of pressure on the NCG solubility in heavy oil, the viscosity-reducing ratio of heavy crude oil increases gradually with pressure.

The dissolution of NCG in heavy crude oil is very useful for the effective development of heavy oil reservoirs. CO_2 is a highly soluble gas in crude oil and tremendously improves the PVT characteristics of crude oil. For the MTFs injection project in an

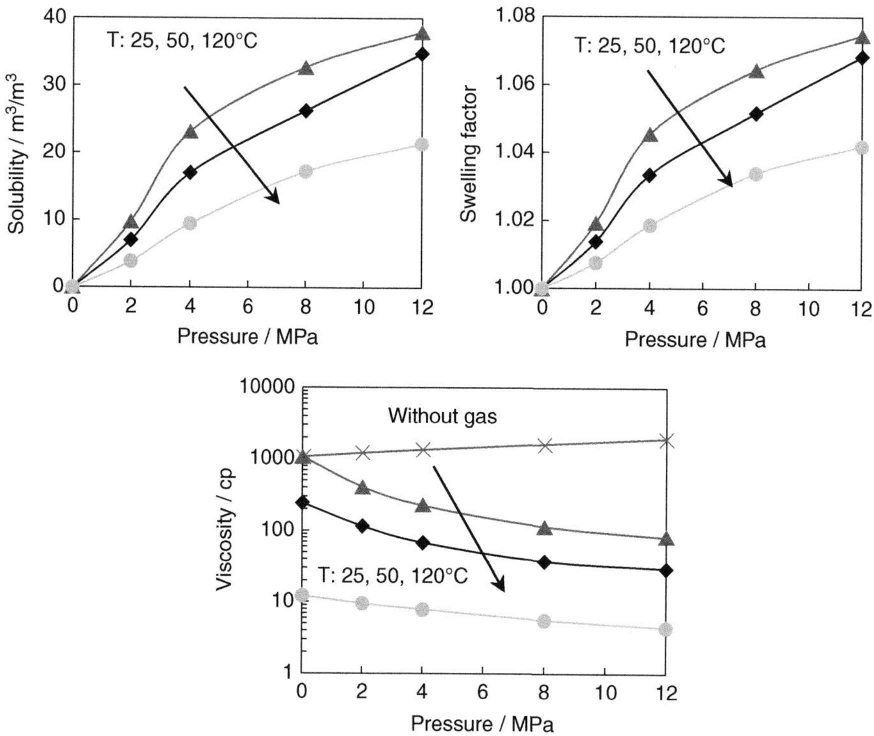

Figure 6.17 PVT performance of heavy oil–NCG (CO_2+N_2) mixture (API gravity of this heavy oil is 15.51°).

actual field, in order to effectively improve the PVT behaviour of heavy oil, rich air (with high O_2 concentration) should be used to produce MTFs with higher CO_2 concentrations.

6.4.2 MTFs Stimulation Process

MTFs are a new heat-carrier proposed in recent years for the EOR process of heavy oil and oil sand reservoirs. They are produced based on the combustion and jetting mechanisms of a rocket engine. The main components in MTFs include steam, nitrogen gas (N_2) and carbon dioxide (CO_2). Because of the high solubility of CO_2 in heavy crude oil, it can act as a solvent. Compared with conventional saturated steam, these new thermal fluids have higher temperature and enthalpy. Since 2009, many MTFs-based recovery projects have been carried out in heavy oilfields in China, and an obvious oil production increment is achieved.

A MTFs stimulation process is a new EOR process similar to the cyclic steam stimulation (CSS) process in heavy oil reservoirs and oil sands. The only difference from a CSS process is that the saturated steam used in the CSS process is replaced by MTFs. For the MTFs stimulation process, during operation the low-dissolved N_2 will carry the heat to a region far from the well bottom. Thus this process will have a larger effective drainage area than the CSS process. Simultaneously, the existence of CO_2 can also tremendously improve the displacement efficiency. Furthermore, compared with a CSS process, the

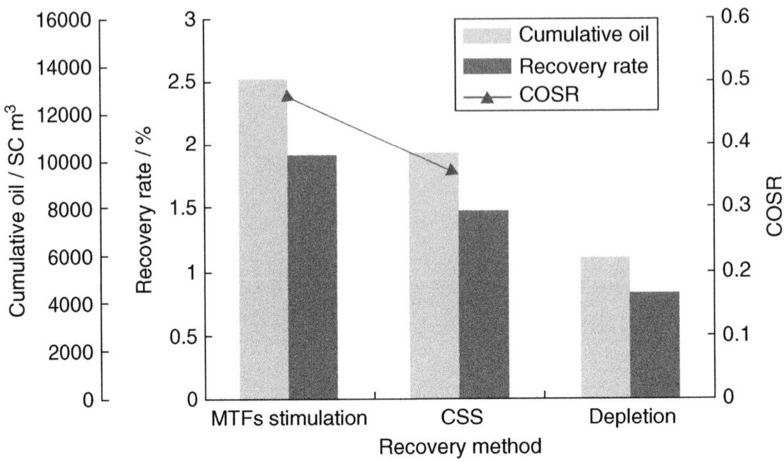

Figure 6.18 The performance of different recovery methods in heavy oil reservoirs.

amount of steam consumption is reduced in a MTFs stimulation process, and thus the SOR is also lower. Figure 6.18 compares the performance of the CSS, MTFs stimulation and depletion recovery processes in a typical heavy oil block from Bohai oilfield, China. The oil viscosity at reservoir conditions (56 °C) in this model is about 578 mPa·s. As shown, the MTFs stimulation process has the highest cumulative oil and recovery rate among the three methods. Compared with the CSS process, MTFs stimulation has a higher cumulative oil – steam ratio (cOSR). Because of the influence of NCG on heavy oil properties, this new stimulation method tremendously reduces the amount of steam consumption. Moreover, on account of the existence of insoluble N_2, this new method will have a higher recovery rate than CSS. It is very attractive for offshore heavy oil production. MTFs stimulation will be a high-potential EOR method in heavy oil or oil sands reservoirs.

6.4.3 MTFs-Assisted Gravity Drainage Process

A MTFs-assisted gravity drainage (MTFs-AGD) process is a new thermal gravity drainage process for a thick heavy oil reservoir. It fully combines the advantages of a SAGD process and a GAGD (gas-assisted gravity drainage) process, and thus tremendously improves the performance of a thermal recovery process in heavy oil reservoirs. This technique can act as an additional recovery stage of heavy oil reservoirs after the SAGD process. Compared with the conventional SAGD process, this new process has a higher recovery rate, and it is highly efficient. Their differences are numerically investigated through the performance of the commercial reservoir simulator, CMG STARS®. The results are shown in Figure 6.19. As shown, at the initial stage of a drainage process, there is no significant difference between the two processes. Then, as the thermal fluid injects, the chamber is expanded. Because of the effect of NCG, the vertical expanding rate of the steam chamber in the SAGD process is much higher than that in the MTFs-AGD process. For the MTFs-AGD process, after steam and NCG are injected into the reservoir, NCG will rapidly rise to the top of the reservoir. Thus it hinders the heat

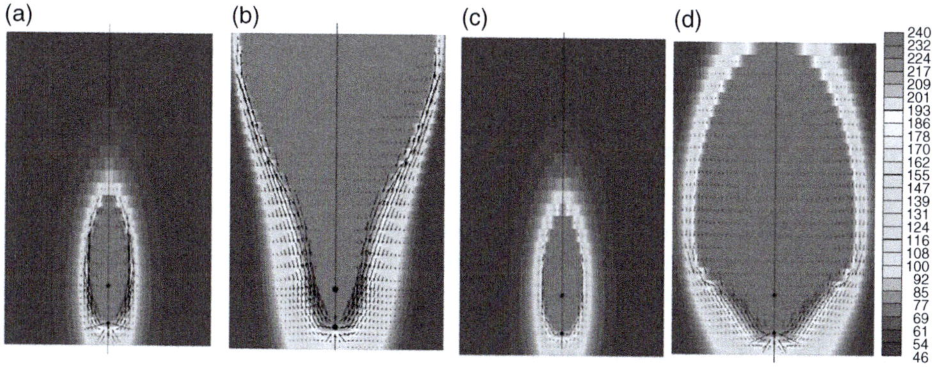

Figure 6.19 The temperature distributions of the SAGD and MTFs-AGD processes. (a) Initial stage of SAGD process; (b) steam chamber shape of SAGD process; (c) initial stage of MTFs-AGD process; (d) steam chamber shape of MTFs-AGD process. (See colour plate section for the colour representation of this figure.)

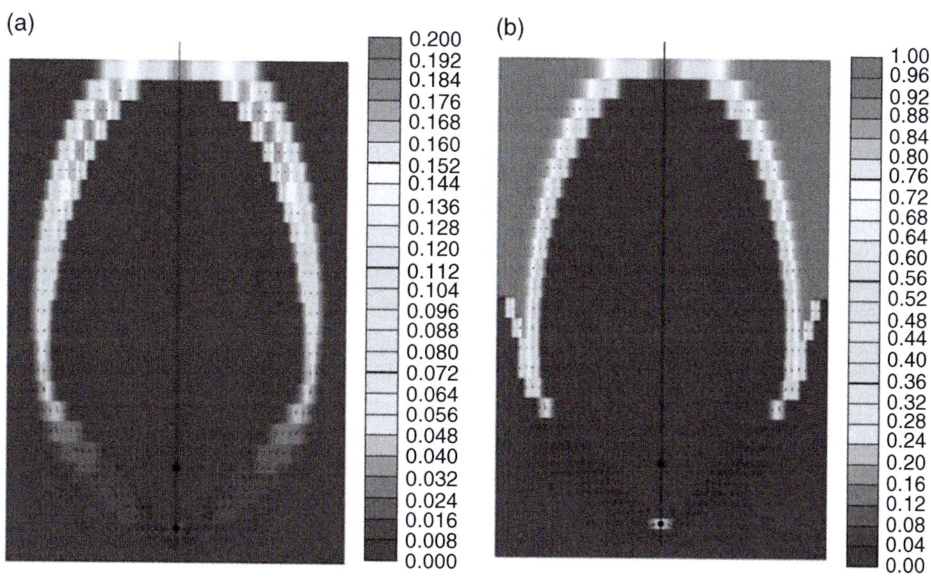

Figure 6.20 The NCG concentration during the MTFs-AGD process: (a) CO_2 and (b) N_2. (See colour plate section for the colour representation of this figure.)

transfer process between the steam chamber and caprock. Then the horizontal expansion of the steam chamber is improved. As shown, the steam chamber after the MTFs-AGD process was shaped like a 'liquid drop' instead of the conventional 'inverted triangle' shape in the SAGD process. Figure 6.20 shows the NCG concentration after the MTFs-AGD process. Due to the high dissolving capacity of CO_2 in heavy oil, the physical properties of heavy oil are improved. Therefore, the CO_2 fraction in MTFs is mainly gathered at the front of the steam chamber. But the injection of N_2 plays the role of heat insulation. It has low solubility in heavy oil, and thus N_2 will gather at the top of a heavy oil reservoir. Furthermore, its injection tremendously improves the heat efficiency of steam injected.

6.4.4 Recovery Mechanisms

The new thermal fluids combine the virtues of steam and non-condensable gases. The steam fraction will play an important role in the viscosity-reducing mechanism of heavy oil. The NCG fraction aims to improve the mobility of heavy oil and recover the reservoir energy. After the injection of MTFs, there is not just the heating effect of steam, but also the chemical and gas-operated effects.

CO_2 is a highly soluble gas in heavy crude oil. Its dissolution in heavy crude oil can further reduce the oil viscosity, increase the oil-phase volume, enhance the reservoir expansibility, and improve the oil mobility. The dissolution of CO_2 in heavy crude oil will result in the formation of foamy oil. It is an important mechanism to enhance heavy oil recovery. The dissolution of CO_2 in formation water can react with carbonate minerals in the formation to increase the reservoir permeability. N_2 has low solubility in heavy crude oil. It is helpful to recover the reservoir energy and can expand the effective heating zone. In addition, it has the role of auxiliary cleanup. For the MTFs-assisted gravity drainage process, because of the low dissolution, the N_2 fraction in MTFs can act as a heat insulation layer (man-made gas cap) at the top of a reservoir. It improves the heat efficiency of steam injected.

Furthermore, the MTFs-assisted gravity drainage process is also significantly different from the steam-alternative-gas push (SAGP) process. Firstly, on the fluid composition, the mole fraction of NCG in the SAGP process is usually less than 2 mol.%, and N_2 is the commonly used gas. But for MTFs it is produced from a combustion process, and the NCG component in MTFs is a gas mixture of N_2, CO_2, CH_4, and CO. Secondly, on the operation method, for a SAGP field project, steam and NCG are commonly injected separately (tube for steam and annulus for gas), but for MTFs they are injected directly after generation. Thirdly, the operation of the SAGP process has little influence on the steam chamber pressure, but for the MTFs-assisted gravity drainage process, because of the high injection rate of NCG, with the exception of a heat insulation mechanism, it still has the mechanisms of energy recovery, gas dissolution, foamy oil and auxiliary cleanup, and so this new technique has a higher recovery rate.

6.5 Uncertainty Analysis for Reservoir Heterogeneity

6.5.1 Bottom Water

The existence of bottom water will reduce the oil recovery of a SAGD process. During the recovery process of SAGD, the reservoir pressure usually presents a smoothly changing characteristic. But for heavy oil reservoirs with a higher oil viscosity, the highly viscous heavy oil between a well-pair and a bottom aquifer will act as a barrier to prevent the occurrence of water coning.

For the thermal recovery process of heavy oil reservoirs, the existence of bottom water will tremendously affect the normal development of a reservoir. Especially for some thin heavy oil reservoirs with bottom water, the recovery of CSS with horizontal

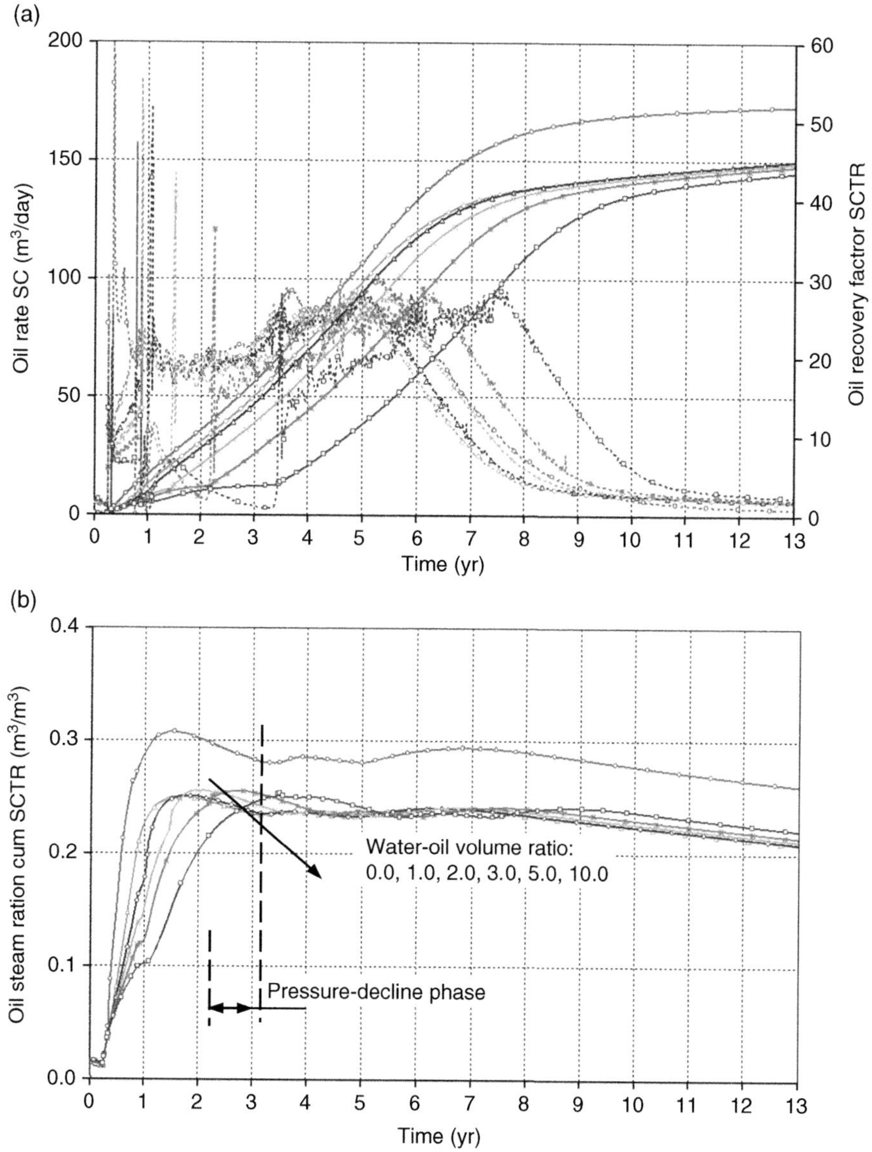

Figure 6.21 The effect of bottom water on the performance of the SAGD process: (a) oil rate and oil recovery factor and (b) cumulative oil – steam ratio.

wells is usually below 10%. Simultaneously, with an increase in the steam injection cycle, the oil–water interface is rising gradually. As a result, a significant water channel will occur in a horizontal well. Figure 6.21 shows the effects of bottom water on the performance of the SAGD process. In comparison with the no-water case, the existence of bottom water will reduce the ultimate recovery of the SAGD process by about 10% of the OOIP. With an increase in the water–oil volume ratio, the daily oil rate and recovery rate are reduced. Simultaneously, the starting time of the drainage phase is also delayed.

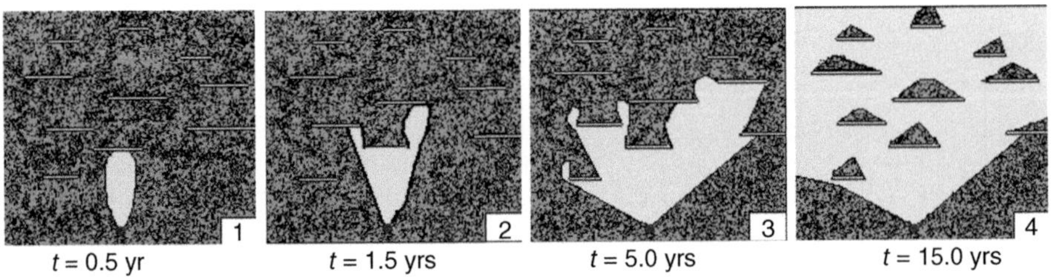

Figure 6.22 The effect of a small shale barrier on the steam chamber.

Moreover, compared with the model without an aquifer, bottom water reduces cOSR by about 0.05 – 0.07. The startup pressure-decline rate of the SAGD process is also reduced, and the startup time is prolonged. As shown, the startup time of the SAGD process with a 3.0 water–oil volume ratio is prolonged by about 0.47 years compared with the no-water model. The thicker the bottom water, the longer the startup time of the SAGD process.

6.5.2 Shale Barrier

A shale barrier is another important issue to affect the SAGD process (Liu, 1998; Doan et al., 2003). The existence of a shale barrier will hinder the formation and expansion of a steam chamber. In thick oil reservoirs, usually there are many scattered low permeable layers (shale sandwiches). These layers tremendously influence the normal formation and expansion of a steam chamber. But if the sandwich layers are small and distributed widely in a reservoir, their influences on the SAGD process will be weakened. The existence of scattered shale sandwiches can even benefit the heat conduction process of steam injection, as shown in Figure 6.22.

In order to describe the features of a shale barrier accurately, the shale parameters, including position, range and permeability, should be taken into account. For the position of a shale barrier, both vertical and horizontal positions should be included. For the distribution range, the simulation process should consider the width and length of the shale barrier. Using the commercial simulator CMG STARS®, the effect of a shale barrier on SAGD performance is investigated. It is observed that the existence of a shale barrier tremendously reduces the oil recovery and oil rate. In comparison with the shale width, the influence of the shale length on SAGD performance is more significant. Therefore, if a SAGD project is performed in a reservoir with a shale barrier, the position of a well pair should be perpendicular to the principal axis of the shale barrier. Simultaneously, in order to improve the SAGD performance in a heavy oil reservoir, the vertical distance between a well pair and a shale barrier should also be increased.

6.5.3 Lean Zone

In Canada, both Nexen's Long Lake and Suncor's Firebag SAGD projects reported the existence of intersecting lean zones that behave as thief zones. Thoroughly understanding the effect of lean zones and improving SAGD operations with such heterogeneity are critically important in reducing the disadvantages of lean zones during SAGD production.

Nasr et al. (2003b) used a high-temperature/high-pressure experimental facility to inject steam and observe the oil rate and chamber growth in the presence of top water or a gas cap. Their data revealed the existence of overlying thief zones. Heat loss is greater if a zone is saturated with water rather than gas, which is attributed to the difference in thermal conductivity. Doan et al. (2003) observed that water sand hampers the SAGD process. The existence of a top water layer worsens the scenario compared with a bottom water layer. An increased thickness of the top water layer likewise causes problems. Fairbridge et al. (2012) employed a nodal model and a channel model to simulate the behaviour of intra-formational zones with high water saturation during SAGD production. They concluded that the intermediate water zones impair the efficiency of an oil production process. Xu et al. (2014) performed a detailed numerical and analytical study on the effects of lean zones on SAGD performance in Athabasca, Canada. They indicated that extensive convective heat leads to the low efficiency of lean zones, and the effects of the location, thickness, size, and interval distance of lean zones on SAGD are significant.

With the presented reservoir settings, pressure, temperature, the water velocity and heat flux tend to increase in the early period and become steady after five years (Figure 6.23). With an increase in the volume of mobile water or water saturation, the pressure, temperature, water velocity, and heat flux can be large. The case with high water saturation consumes substantial steam mainly because of the larger convective heat flux. Such a case also produces minimal oil because of the high water velocity and low oil saturation. Thus, SOR for this case is large.

The effect of lean zones has a close relationship with their location, distribution, thickness, interval distance, and size. For a single-layer lean zone, the case with a lean zone above an injector has the highest SOR. For the single-layer lean zone with the same location, large size, low interval distance, and considerable thickness, we have obtained a high SOR. For multiple-layer lean zones, we have used a triangular periodical function to describe the lean zones. We study the parameters of connate water saturation, a period, and thickness. With minimal connate water saturation, a large period,

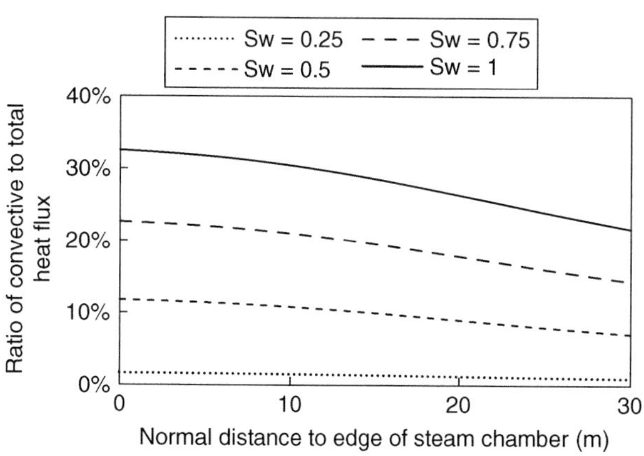

Figure 6.23 Ratio of convective to total heat flux.

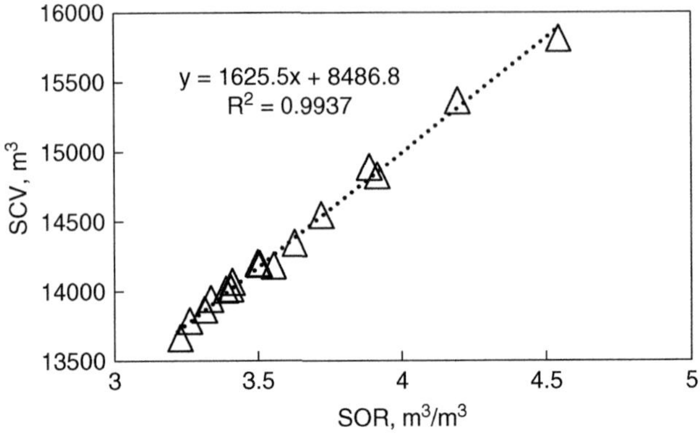

Figure 6.24 Correlation between SCV and SOR.

and large thickness, we can obtain high SOR. Given the strong linear correlation between steam chamber volume (SCV) and SOR (Figure 6.24), we can predict SCV on the basis of SOR.

6.6 Conclusions

This chapter discusses the dynamic heat and mass transfer in several heavy oil recovery processes, such as SAGD, hybrid steam–solvent, and steam–solvent–gas co-injection. The respective contribution of conduction and convection to the viscosity reduction of heavy oil is analysed. Also, the effects of heterogeneous reservoir properties are investigated. The following conclusions can be summarized:

1) Heat conduction dominates the heat transfer process in SAGD, whereas convective heat transfer may take place only near a steam chamber boundary.
2) In a hybrid steam–solvent process, solvent dissolution into the heated heavy oil can further reduce the oil viscosity, especially at the nearby area to a chamber interface. As a result, fluid drainage and chamber evolution can be accelerated so it increases the oil production rate. The oil production rate of ES-SAGD is higher than that of SAGD throughout the process. The enhancement is most significant in the early and middle stages of the process.
3) Compared with a thermal process, a pure solvent recovery process is extremely slow. The oil dilution through solvent dissolution is orders of magnitude slower than that through steam heating.
4) Multiple thermal fluids are a new heat-carrier proposed recently, and their main components include steam, N_2 and CO_2. They combine the virtues of steam and non-condensable gases. The dissolution of NCG in heavy oil can improve the mobility of heavy oil and recover the reservoir energy. For the MTFs stimulation process, after the injection of MTFs, there is not only the heating effect of steam, but also the chemical effect and gas-operated effect. A MTFs-AGD process is a new thermal gravity drainage process for thick heavy oil reservoirs. It combines the advantages of

the SAGD process and GAGD process, and it can act as an additional recovery stage of heavy oil reservoirs after the SAGD process.

5) The SAGD process is sensitive to the heterogeneity of an oil sand formation. A shale layer limits the growth of a steam chamber and increases SOR. Gas cap and top water greatly increase steam injection once a steam chamber penetrates into such a top zone, which leads to high SOR. A lean zone behaves as a thief zone during SAGD production, which increases SOR and decreases oil production. Published studies on the role of lean zones in the SAGD process remain limited.

6.7 Recommendations

6.7.1 Effects of Non-Condensable Gases on Heat and Mass Transfer

The addition of non-condensable gases (i.e. CO_2, N_2, air) into a steam chamber has widely been considered as a SAGD wind-down strategy. However, due to the lack of relative experimental data, the respective effect of non-condensable gases on the heat and mass transfer in a SAGD chamber boundary is still rather ambiguous. More research work, including experimental and modeling efforts, needs to be conducted for a better design of SAGD wind-down.

6.7.2 Effects of Reservoir Heterogeneity on Heat and Mass Transfer

This chapter investigates several scenarios of reservoir heterogeneity through numerical simulation. However, in reality a reservoir can be much more complex. More efforts are needed to address the effects of reservoir heterogeneity on heat and mass transfer, for example, the formation dip angle, presence of solution gas inside bitumen, and vertical and horizontal permeability heterogeneity.

Acknowledgements

The authors would like to acknowledge financial support from NSERC/AIEES/ Foundation CMG and AITF Chairs Funds, and an Imperial Oil Research Award.

References

Akin, S. (2005). Mathematical modeling of steam-assisted gravity drainage. *SPE Reservoir Evaluation and Engineering* 8 (5): 372–376.

Alvarez, J. and Han, S. (2013). Current overview of cyclic steam injection process. *Journal of Petroleum Science Research* 2 (3): 116–127.

Alvarez, J., Moreno, R., and Sawatzky, R.P. (2014). Can SAGD be exported? Potential challenges. *SPE Heavy and Extra Heavy Oil Conference: Latin America*, Society of Petroleum Engineers.

Azad, A. and Chalatumyk, R.J. (2010). SAGD Laboratory experimental and numerical simulation studies: Review of current status and future issues. *Journal of Petroleum Science and Engineering* 68 (3–4): 135–150.

Bejan, A. and Kraus, A.D. (2003). Heat Transfer Handbook. Hoboken, NJ: Wiley.
Berg, E.A. (2011). Challenges and opportunities, heavy oil development offshore UK. *SPE EUROPEC/EAGE Annual Conference and Exhibition*, Vienna, Austria, May 23–26.
Butler, R.M. (1987). Rise of interfering steam chambers. *Journal of Canadian Petroleum Technology* 26 (3): 70–75.
Butler, R.M. (1991). *Thermal Recovery of Oil and Bitumen*. Prentice Hall.
Butler, R.M. (1997). The steam and gas push (SAGP). *Petroleum Society's 48th Annual Technical Meeting*, Calgary, June 8–11.
Butler, R.M., McNab, G.S., and Lo, H.Y. (1981). Theoretical studies on the gravity drainage of heavy oil during in-situ steam heating. *Canadian Journal of Chemical Engineering* 59 (4): 455–460.
Carslaw, H. and Jaeger, J. (1959). *Conduction of Heat in Solids*. Oxford: Oxford University Press.
Cline, V.J. and Basham, M. (2002). Improving project performance in a heavy oil horizontal well project in the San Joaquin valley, California. *SPE International Thermal Operations and Heavy Oil Symposium and International Horizontal Well Technology Conference*, Society of Petroleum Engineers.
Crank, J. (1984). Free and Moving Boundary Problems. Oxford: Clarendon Press.
Das, S.K. and Butler, R.M. (1994). Effect of asphaltene deposition on the Vapex process: A preliminary investigation using a Hele – Shaw Cell. *Journal of Canadian Petroleum Technology* 33 (6): 39–45.
Delamaide, E. and Parra Moreno, W. (2015). Enhanced oil recovery of heavy oil in reservoirs with bottom aquifer. *SPE Western Regional Meeting*, Garden Grove, California, 27–30 April, SPE 174050.
Doan, L.T., Baird, H., Doan, Q.T., and Ali, S.M. (2003). Performance of the SAGD process in the presence of a water sand – A preliminary investigation. *Journal of Canadian Petroleum Technology* 42 (1): 25–31.
Dong, X., Liu, H., Pang, Z., et al. (2014). Flow and heat transfer characteristics of multi-thermal fluid in a dual-string horizontal well. *Numerical Heat Transfer, Part A: Applications* 66 (2): 185–204.
Donnelly, J.K. (2000). The best process for Cold Lake: CSS vs. SAGD. *Journal of Canadian Petroleum Technology* 39 (8): 5–7.
Edmunds, N. (1991). On the difficult birth of SAGD. *Journal of Canadian Petroleum Technology* 38 (1): 14–17.
Escobar, M.A., Valera, C.A., and Perez, R.E. (1997). A large heavy oil reservoir in Lake Maracaibo basin: cyclic steam injection experiences. *International Thermal Operations and Heavy Oil Symposium*, Society of Petroleum Engineers.
Fairbridge, J.K., Cey, E., and Gates, I.D. (2012). Impact of intraformational water zones on SAGD performance. *Journal of Petroleum Science and Engineering* 82: 187–197.
Farouq-Ali, S.M. (1997). Is there life after SAGD? *Journal of Canadian Petroleum Technology* 36 (6): 20–24.
Farouq-Ali, S.M. (2003). Heavy oil – ever more mobile. *Journal of Petroleum Science and Engineering* 37 (1–2): 5–9.
Fisher, D., Singhal, A., Goldman, J., et al. (2002). Insight from MRI and micro-model studies of transport of solvent into heavy oil during Vapex. *International Thermal Operations and Heavy Oil Symposium and International Horizontal Well Technology Conference*, Calgary, Canada.

Hayduk, W. and Minhas, B.S. (1982). Correlations for prediction of molecular diffusivities in liquids. *Canadian Journal of Chemical Engineering* 60 (2): 295–299.
Hayduk, W., Castaneda, R., Bromfield, H., and Perras, R.R. (1973). Diffusivities of propane in normal paraffin, chlorobenzene, and butanol solvents. *AIChE Journal* 19 (4): 859–861.
Ito, U. and Suzuki, S. (1996). Numerical simulation of the SAGD process in the Hangingstone Oil Sands Reservoir. *Journal of Canadian Petroleum Technology* 38 (9): 27–35.
Ito, U., Suzuki, S., and Yamada, H. (1998). Effect of reservoir parameter on oil rates and steam oil ratios in SAGD projects. *7th UNITAR International Conference on Heavy Crude and Tar Sands*, Beijing, China.
Ito, Y. and Ipek, G. (2005). Steam fingering phenomenon during SAGD process. *SPE International Thermal Operations and Heavy Oil Symposium*, Society of Petroleum Engineers.
Joshi, M., Pani, A., and Sabnis, S. (2006). Industrial Mathematics. New Delhi: Narosa Publishing House.
Leaute, R.P. (2002). Liquid addition to steam for enhancing recovery of bitumen with CSS: Evolution of technology from research concept to a field pilot at Cold Lake. *SPE/Petroleum Society of CIM/CHOA 79011, ITOHOS/CHWT Symposium*, Calgary, Alberta, November 4–7.
Leaute, R.P. and Carey, B.S. (2007). Liquid addition to steam for enhancing recovery (LASER) of bitumen with CSS: Results from the first pilot cycle. *Journal of Canadian Petroleum Technology* 46 (9): 22–30.
Leaute, R.P., Corry, K.E., and Pustany, B.P. (2016). Liquid addition to steam for enhancing recovery of cyclic steam stimulation or LASER-CSS. Canadian Patent No. 2,342,955.
Liang, L. (2005). An analytical model for cyclic steaming of horizontal wells. MS thesis, Stanford University, Stanford, CA.
Liu, D., Zhao, C., Su, Y., et al. (2012). New research and application of high efficient development technology for offshore heavy oil in China. *Offshore Technology Conference*, Houston, Texas, 30 April–3 May, OTC 23015.
Liu, W.Z. (1998). *The Development Modes of Heavy Oil Reservoirs by Thermal Recovery*. Beijing: Petroleum Industry Press.
Liu, X., Zhang, F., Huang, K., et al. (2011). Discussion about the thermal recovery of NB35-2 offshore heavy oilfield. *Reservoir Evaluation and Development* 1 (1–2): 61–63.
Moore, R.G., Laureshen, C.J., Belgrave, J.D.M., et al. (1995) In-situ combustion in Canadian heavy oil reservoirs. *Fuel* 74 (8): 1169–1175.
Nasr, T.N., Beaulieu, G., Golbeck, H., and Heck, G. (2003a). Novel expanding solvent-SAGD process "ES-SAGD". *Journal of Canadian Petroleum Technology* 42 (1): 13–16.
Nasr, T.N., Law, D.H.S., Beaulieu, G., et al. (2003b). SAGD application in gas cap and top water oil reservoirs. *Journal of Canadian Petroleum Technology* 42 (1): 32–38.
Nield, D.A. and Bejan, A. (1999). *Convection in Porous Media*, 2nd Ed. New York: Springer-Verlag.
Nukhaev, M., Pimenov, V., and Shandrygin, A. (2006). A new analytical model for the SAGD production phase. *SPE ATCE*, San Antonio, Texas, September 24–27.
Peng, S., Yuan, A., Jiang, H., and Zhang, S. (2010). Cyclic preheating test in SAGD horizontal wells of Liaohe Oilfield. *International Oil and Gas Conference and Exhibition in China*, Society of Petroleum Engineers.

Polikar, M., Cyr, T.J., and Coates, R.M. (2000). Fast-SAGD: half the wells and 30% less steam. *2000 SPE/Petroleum Society of CIM International Conference on Horizontal Well Technology*, Calgary, Alberta, Canada, November 6–8.
Reis, J.C. (1992). A steam-assisted gravity drainage model for Tar Sands: Linear geometry. *Journal of Canadian Petroleum Technology* 31 (10): 14–20.
Sharma, J. and Gates, I.D. (2010). Multiphase flow at the edge of a steam chamber. *Canadian Journal of Chemical Engineering* 88 (3): 312–321.
Sharma, J. and Gates, I.D. (2011). Convection at the edge of a steam-assisted gravity drainage steam chamber. *SPE Journal* 16 (3): 503–512.
Shin, H. and Choe, J. (2009). Shale barrier effects on the SAGD performance. *SPE/EAGE Reservoir Characterization and Simulation Conference*, Abu Dhabi, UAE, October 19–21.
Shu, W.R. (1984). A viscosity correlation for mixtures of heavy oil, bitumen, and petroleum fractions. *SPE Journal* 24 (3): 227–282.
Speight, J.G. (2009). *Enhanced Recovery Methods for Heavy Oil and Tar Sands*. Houston: Gulf Publishing Company.
Tang, X., Ma, Y., and Sun, Y. (2011). Research and field test of complex thermal fluid huff and puff technology for offshore viscous oil recovery. *China Offshore Oil and Gas* 23: 185–188.
Vittoratos, E., Scott, G.R., and Beattie, C.I. (1990). Cold Lake cyclic steam stimulation: a multiwell process. *SPE Reservoir Engineering* 5 (1): 19–24.
Xu, J. (2015). Effects of lean zones on SAGD performance. Master's dissertation, University of Calgary.
Xu, J., Chen, Z.J., Cao, J., and Li, R. (2014). Numerical study of the effects of lean zones on SAGD performance in periodically heterogeneous media. *SPE Heavy Oil Conference Canada*, Society of Petroleum Engineers.
Yazdani, A. (2007). Physical and numerical modeling of permeability and drainage height effects in VAPEX. PhD dissertation, University of Calgary, Canada.
Yazdani, A. and Maini, B.B. (2005). Effect of drainage height and grain size on production rates in the Vapex process: experimental study. *SPE Reservoir Evaluation Engineering* 8 (3): 205–213.

7

Modeling the Potential Impacts of CO_2 Sequestration on Shallow Groundwater: The Fate of Trace Metals and Organics and the Effect of Co-injected H_2S

Liange Zheng and Nicolas Spycher

Lawrence Berkeley National Laboratory, Berkeley, CA, 94720, USA

In this chapter, the results from reactive transport models are presented to illustrate the potential impact of CO_2 and brine leakage on shallow groundwater, in terms of the fate of trace metals and organics, and the effect of co-injected H_2S. We present two sets of simulations. The first set is based on a hypothetical leakage of CO_2 and brine into the High Plains Aquifer, focusing on the concentration changes in trace metals and organic compounds. The second set is developed assuming a generic sandstone aquifer intruded by a CO_2+H_2S mixture, to examine the interaction of CO_2+H_2S with metal-bearing sediments in the aquifer.

7.1 Introduction

CO_2 capture and deep geological storage is being considered as a potential mitigation measure against rising anthropogenic emissions of greenhouse gases. One potential risk from CO_2 geological storage is that CO_2 and brines could leak from a deep storage formation into overlying shallow, potable groundwater aquifers. Investigators considering the feasibility of geological storage must therefore look at the impact of potential CO_2 and brine leakage from deep subsurface reservoirs into overlying groundwater aquifers, to ensure the safety of drinking-water resources at these locations.

The primary concern regarding potential leakage into shallow aquifers is the mobilization of trace metals, which has been the focus of many studies, as reviewed by Lemieux (2011) and Harvey et al. (2012), and references therein. The dissolution of CO_2 into an aquifer would increase the concentration of dissolved carbonic acid in groundwater and thus increase its acidity, which could mobilize trace elements through mineral dissolution, desorption reactions, and/or exchange reactions involving H^+ and other mobilized constituents (e.g. Wang and Jaffe, 2004; Carroll et al., 2009; Zheng et al., 2009, 2012, 2013; Apps et al., 2010; Wilkin and Digiulio, 2010; Viswanathan et al., 2012; Trautz et al., 2013; Kirsch et al., 2014; Wunsch et al., 2014). The transport of deep metal-bearing

Reactive Transport Modeling: Applications in Subsurface Energy and Environmental Problems, First Edition. Edited by Yitian Xiao, Fiona Whitaker, Tianfu Xu and Carl Steefel.

brines along leakage pathways, with or without CO_2, in response to an increased formation pressure from injection operations could also directly mobilize salts and metals to shallow aquifers. In addition, because supercritical CO_2 is an excellent solvent for organic compounds (Anitescu and Tavlarides, 2006; Kharaka et al., 2009), concerns have been raised about the potential mobilization of organic constituents from depth and subsequent transport to shallow drinking water bodies via leakage pathways (Zheng et al., 2013; Zhong et al., 2014).

Last but not least, the co-injection of H_2S in geological sequestration operations is being considered because the burning of fossil fuels can emit significant amounts of H_2S and SO_2 (which disproportionates into H_2S and H_2SO_4), in addition to CO_2. Because the separation of these gas impurities from power plant effluents is quite costly, and their release to the atmosphere is environmentally harmful, it may be advantageous to consider co-injecting these gases with CO_2. Therefore, the potential impact of $CO_2 + H_2S$ leakage from deep saline reservoirs to overlying shallow aquifers is also a question that the scientific community must address to assess the potential risks to the public from CO_2 storage operations.

The potential mobilization of trace elements in response to CO_2 intrusion into potable groundwater has been investigated by laboratory experiments (Smyth et al., 2009; Lu et al., 2010; Little and Jackson, 2010; Wei et al., 2011; Viswanathan et al., 2012; Humez et al., 2013; Varadharajan et al., 2013; Wunsch et al., 2014; Lawter et al., 2016), natural analogues (e.g. Aiuppa et al., 2005; Flaathen et al., 2009; Keating et al., 2010; Arnórsson et al., 2010), and field experiments (Kharaka et al., 2010; Peter et al., 2012; Yang et al., 2013; Cahill and Jakobsen, 2013; Trautz et al., 2013). Numerical models have also been used to perform generic evaluations of the potential impact of CO_2 leakage on the water quality of shallow aquifers (e.g. Wang and Jaffe, 2004; Carroll et al., 2009; Zheng et al., 2009; Apps et al., 2010; Wilkin and Digiulio, 2010), to identify potential issues such as the leaching out of organic compounds at depth by supercritical CO_2 and their release in shallow aquifers (Zheng et al., 2013), and to interpret laboratory experiments (e.g. Viswanathan et al., 2012; Kirsch et al., 2014; Wunsch et al., 2014; Zheng et al., 2016) as well as field tests (Zheng et al., 2012; Trautz et al., 2013). These interpretations enable understanding of the underlying transport and chemical processes that control the mobilization of major and trace elements upon CO_2 dissolution in groundwater.

To the best of our knowledge, few studies have been conducted with respect to the potential impact on groundwater from organic constituents and co-injectants such as H_2S. For example, while the possible extraction of organic compounds by supercritical CO_2 inside deep reservoirs is a well-known phenomenon (e.g. Kharaka and Hanor, 2007), the potential for groundwater contamination from mobilized organic constituents transported to shallow aquifers has evidently not been studied. Kharaka et al. (2009) advocated further investigation of this issue when elevated dissolved organic carbon (DOC) concentrations were observed in the injection formation at the Frio test site in Texas. Zheng et al. (2013) and Zhong et al. (2014) investigated the transport of organic compounds along with leaking CO_2 via leakage pathways, but not the fate of organics in shallow aquifers. The effect of co-injected H_2S on mineral trapping of CO_2 and mineralogical alteration of reservoir host rock has been investigated previously (Gunter et al., 2000; Knauss et al., 2005; Palandri and Kharaka, 2005; Palandri et al., 2005; Xu et al., 2007; Murphy et al., 2010, 2011; Lammers et al., 2011), but with a focus on the geochemical evolution in the storage formation at relatively high temperatures and pressures ($T > 70\,°C$, $P > 150$ bar). The behaviour of co-injected H_2S along a leakage pathway has been shown to be affected by chromatographic partitioning between CO_2 and water (Bachu and Bennion, 2009; Bachu et al., 2009). The question that arises is how a shallow groundwater aquifer would respond if $CO_2 + H_2S$ mixtures intruded.

In this chapter, the results from reactive transport models are presented to illustrate the potential impact of CO_2 and brine leakage on shallow groundwater, in terms of the fate of trace metals and organics, and the effect of co-injected H_2S. We present two sets of simulations. The first set is based on a hypothetical leakage of CO_2 and brine into the High Plains Aquifer (McMahon et al., 2004), focusing on the concentration changes in trace metals and organic compounds. The second set is developed assuming a generic sandstone aquifer intruded by a $CO_2 + H_2S$ mixture, to examine the interaction of $CO_2 + H_2S$ with metal-bearing sediments in the aquifer. For this second set, the focus is on the impact of leakage on dissolved Pb and As concentrations in groundwater.

7.2 The Fate of Trace Metals and Organics in a Shallow Aquifer in Response to a Hypothetical CO_2 and Brine Leakage Scenario

It has been shown in laboratory experiments (Smyth et al., 2009; Lu et al., 2010; Little and Jackson, 2010; Wei et al., 2011; Viswanathan et al., 2012; Humez et al., 2013; Varadharajan et al., 2013; Wunsch et al., 2014) – which typically involve the release of CO_2 into a pre-equilibrated water–rock environment, followed by monitoring of the geochemical changes in the aqueous phase – that the concentrations of many dissolved major and trace elements increase noticeably upon reaction of the water with CO_2. Because the range of experimental conditions and sediment types covered by these experiments is wide, results vary significantly between different experiments in terms of the types of elements being released and their dissolved concentrations. However, one commonality amongst these experiments is the rapid increase of alkali and alkaline earth metals in solution. Trace metals of environmental concern that have been found to be released include As, Pb, Ba, and Cd. Observations from field tests (Kharaka et al., 2010; Peter et al., 2012; Yang et al., 2013; Cahill and Jakobsen, 2013; Trautz et al., 2013) are largely consistent with observations from laboratory experiments, in that both field and laboratory experiments show a general increase in the dissolved concentrations of major and trace elements upon CO_2 release.

However, there are two noticeable differences between results of field and laboratory tests. Firstly, the rise in concentration observed during field tests is typically much less pronounced than that in laboratory experiments – field tests show increases of less than about one order of magnitude (20% to 700% in the studies cited above), in comparison with orders-of-magnitude increases observed in laboratory tests. Secondly, increases in trace elements, especially those of environmental concern such as As, Pb, Ba and Cd, are more frequently observed in laboratory tests than in field tests. As pointed out by Apps et al. (2010), how an aquifer responds to the leakage of CO_2 and/or brine depends on the site-specific conditions, including metal–sediment associations and brine characteristics. Here, we develop a model based on the characterization of a specific site, the High Plains Aquifer (McMahon et al., 2004, 2007) in Heskell County, Kansas, US. The model setup and results are presented in the following sections.

7.2.1 Simulator

The multiphase flow and reactive transport modeling is performed with TOUGHREACT v.2 (Xu et al., 2011), using the flow module ECO2N (Spycher and Pruess, 2005). This simulator was developed by introducing reactive chemistry into the existing framework of a non-isothermal, multiphase, multicomponent fluid and heat flow code, TOUGH (Pruess et al., 1999). The ECO2N fluid property module (Pruess, 2005) includes a comprehensive description of the thermodynamics and thermophysical properties of a H_2O–NaCl–CO_2 system (Spycher and Pruess, 2005). Flow processes can be modeled isothermally (as done in this study) or non-isothermally, and phase conditions may include a single (aqueous or CO_2-rich) phase, as well as two-phase mixtures. Depending on the temperature and pressure conditions, CO_2 may be in a supercritical or gaseous form. Aqueous chemical complexation and gas (CO_2) dissolution/exsolution are assumed to be at local equilibrium. For the conditions assumed in this study, the intrusion of gaseous CO_2 into a freshwater aquifer is modeled as a two-phase flow system, with dissolution of CO_2 into water occurring instantaneously using a local equilibrium assumption. Excess gaseous CO_2 migrates within the aquifer, driven mainly by buoyancy forces.

7.2.2 Model Setup

The simulation domain is 10,000 m long, 5000 m wide and 250 m deep. A nested mesh is used, with refined space discretization in the area where CO_2 and brine leakage occurs and a coarser mesh for the area where the CO_2 and brine plumes are less likely to migrate (Figure 7.1). A hydraulic gradient of 0.005 is applied by fixing the pressure at gridblocks on the left and right model boundaries. The modeled domain is assumed to be isothermal with a temperature of 17 °C, and under pressure values ranging from about 27–29 bar at the bottom model boundary to 7–9 bar at the top.

A single leaking point is assigned at $x = 2000$ m, $y = 2500$ m, and $z = 250$ m, with CO_2 and brine leakage rates shown in Figure 7.2. These rates represent a hypothetical leakage pathway related to a deep, leaky well connecting a deep geological reservoir for CO_2 storage with a shallow groundwater resource. The temporal evolution and magnitude of leakage is based on a generalized leakage scenario (Carroll et al., 2014a).

The aquifer is assumed to be confined, with a uniform thickness of 240 m, which corresponds to the average thickness of the High Plains Aquifer. The following two lithological facies are considered within the aquifer: a 'sand lithofacies', mainly composed of medium and coarse sand, and a 'clay lithofacies', mainly composed of very fine sand and silty clay. The 'sand lithofacy' has a porosity of 0.25 and a permeability of 10^{-12} m^2, whereas the 'clay lithofacy' has a porosity of 0.35 and a permeability of 10^{-18} m^2. The mineralogical compositions of these two facies are listed in Table 7.1, and fall largely into the range reported by McMahon et al. (2004) and Lawter et al. (2016).

The initial water chemistry used in the model, listed in Table 7.2, represents the arithmetic average of major and trace inorganic element data collected in two surveys in 1999 and 2010, as part of the High Plains Regional Ground-Water Study conducted by the USGS's National Water Quality Assessment Program (Becker et al., 2002). The surveys carried out analyses of 30 to 74 randomly selected wells. We assign a very small value to the initial concentration of benzene, phenol, and naphthalene, because their concentrations were all below the detection limits in the 2010 survey.

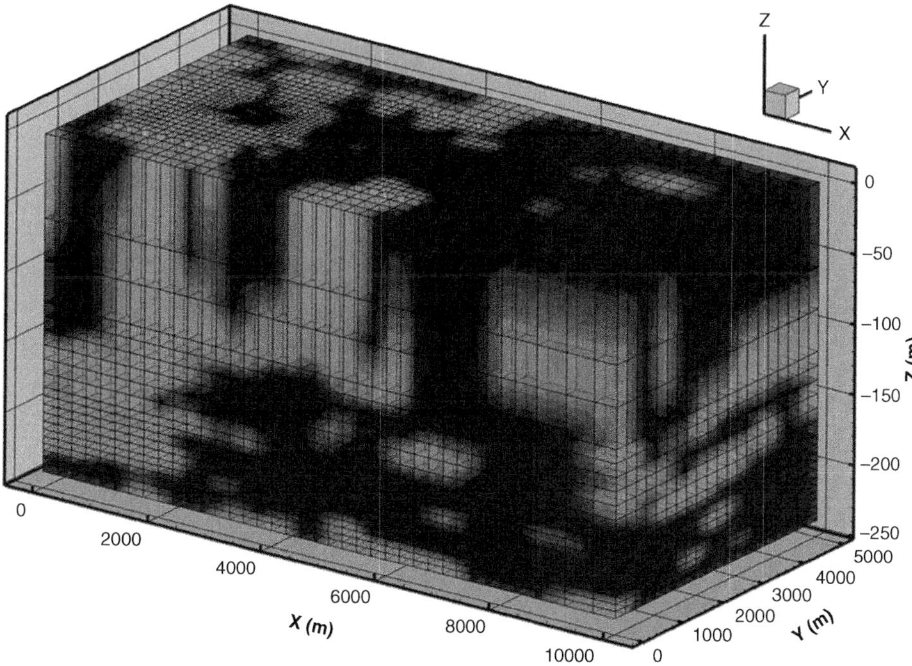

Figure 7.1 Numerical mesh used in the model and the lithological facies in the aquifer; grey represents sand facies and black represents clay facies.

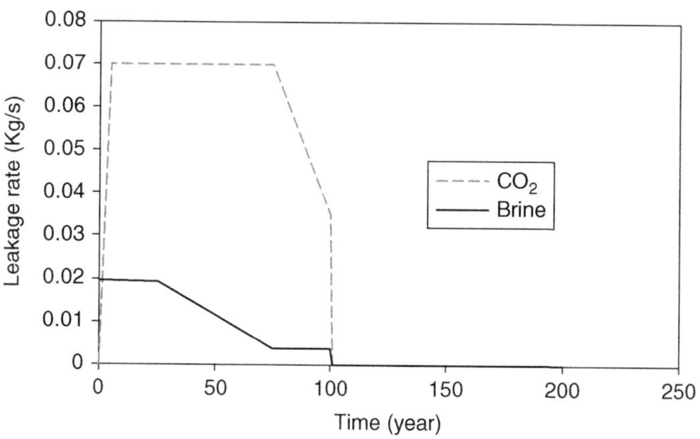

Figure 7.2 The leakage rate of CO_2 and brine as a function of time. *Source:* Data from Carroll et al. (2014a).

The composition of brine that migrates from the storage reservoir varies a great deal from site to site. In this model, to simplify the simulation and emphasize the impact on the trace metals, the brine is assumed to have the same concentrations as groundwater for most elements except for Na, Cl, Pb, Cd, As, benzene, phenol, and naphthalene, as listed in Table 7.2.

Table 7.1 The mass fraction (%) of primary minerals in the two modeled lithofacies: sand and clay.

Primary mineral	Sand facies	Clay facies
Quartz	73.8	34
Albite	2	3
K-feldspar	4.2	3
Kaolinite	2.5	10.5
Smectite-Na	1.5	15
Illite	1.5	20
Goethite	10	10
Calcite	4.5	4.5

Table 7.2 Chemical composition of groundwater in the aquifer and leaking brine. Concentrations of trace metals in the brine are based on experiments (Karamalidis et al., 2013), and organic concentrations in brine are based on Zheng et al. (2010).

Primary species	Concentration in the groundwater (molal)	Concentration in the leaking brine (molal)	Primary species	Concentration in the groundwater (molal)	Concentration in the leaking brine (molal)
Ca^{2+}	1.37×10^{-3}	1.37×10^{-3}	$Sb(OH)_3(aq)$	6.91×10^{-10}	5.59×10^{-10}
Mg^{2+}	8.74×10^{-4}	8.74×10^{-4}	$H_3AsO_3(aq)$	2.33×10^{-8}	2.33×10^{-8}
Na^{+}	1.60×10^{-3}	4.30×10^{-2}	Ba^{2+}	8.10×10^{-7}	1.99×10^{-4}
K^{+}	1.13×10^{-4}	1.13×10^{-4}	Cd^{2+}	6.91×10^{-10}	3.17×10^{-8}
Cl^{-}	1.85×10^{-3}	4.30×10^{-2}	Pb^{2+}	1.57×10^{-9}	1.78×10^{-7}
HS^{-}	6.48×10^{-4}	6.48×10^{-4}	Benzene	10^{-15}	2.51×10^{-7}
SO_4^{2-}	5.77×10^{-5}	5.77×10^{-5}	Phenol	10^{-15}	1.41×10^{-7}
HCO_3^{-}	3.95×10^{-3}	*	Naphthalene	10^{-15}	8.91×10^{-8}
$H_4SiO_4(aq)$	5.33×10^{-4}	5.33×10^{-4}	HSe^{-}	5.34×10^{-8}	5.34×10^{-8}
Fe^{2+}	1.01×10^{-6}	1.01×10^{-6}	Eh	−0.26 volts	−0.26 volts
Fe^{3+}	8.97×10^{-16}	8.97×10^{-16}	pH	7.4	*
AlO_2^{-}	1.57×10^{-7}	1.57×10^{-7}			

Source: Adapted from Karamalidis et al. (2013) and Zheng et al. (2010).
* The pH and the total dissolved carbonate concentration (as HCO_3^-) were calculated assuming equilibrium of the solution with CO_2 at the leaking point.

The CO_2 and brine leakage rates are based on the assumption that CO_2 and brine could be transported from a pressurized reservoir, through a borehole, and into protected groundwater (Zheng et al., 2014). Conceptually, CO_2 fluxes increase to a plateau during the onset of CO_2 injection into the storage reservoir, maintain a constant flux during injection, and then decrease over time to some minimal level. Brine leakage profiles are different from that of CO_2: brine leakage is characterized by a maximum and constant flux during injection, which falls to a final flux after injection stops. The CO_2 and brine leakage rates are shown in Figure 7.2. The simulations assume that leakage occurs continuously for a period of 100 years. In addition, we simulate a 100-year post-leakage time period, during which the leakage is assumed to have been detected and stopped.

7.2.3 Geochemical Model

The geochemical model is extensive and includes over 70 aqueous species, dissolution and precipitation kinetics for 14 minerals, 5 cation exchange reactions, 36 sorption reactions, and oxidation kinetics for 3 organic species. Kinetic constants, minerals, cation exchange, and sorption mass-balance reactions and constants are summarized in Tables A1–A3 in Appendix A. Equilibrium constants for aqueous complexes and minerals are taken primarily from the THERMODDEM database version 1.0.0.8 (Blanc et al., 2012). For those reactions that are not listed in THERMODDEM, thermodynamic data are taken from database data0.ymp.r5 in the EQ3/6 v.8.1 geochemical modeling package (Wolery and Jove-Colon, 2007) and Birkholzer et al. (2008). Cation exchange reactions are included for Na, K, Ca, Mg, and Ba, as are surface complexation reactions for As, Cd and Pb on goethite, illite, kaolinite, and montmorillonite. Surface complexation reactions on calcite are not considered in the model, because calcite is a much weaker adsorbent in comparison with goethite, illite, kaolinite, and montmorillonite, and is thus unlikely to make a significant impact on the adsorption of metals. Kinetic-rate parameters for most rock-forming minerals are taken from Palandri and Kharaka (2004); these are based mainly on experimental studies conducted under far-from-equilibrium conditions. The overall rate equation is tied to mineral equilibria through the saturation ratio (Q/K) with a pH-dependent rate consisting of acid, neutral, and base mechanisms:

$$r = \pm k\,A\left[1-\left(\frac{Q}{K}\right)^{\theta}\right]^{\eta} \tag{7.1}$$

where r is the kinetic rate, k is the rate constant, A is the specific reactive surface area, K is the equilibrium constant for the mineral–water reaction, and Q is the activity product of that reaction. θ is the inverse of the Temkin coefficient (the stoichiometric number of the activated complex involved in the rate-limiting step) and η is an empirical constant. In this study, the parameters θ and η are assumed equal to unity. The mineral reactive-surface areas are calculated assuming a grain diameter of 0.1 mm for non-clay minerals and 0.01 mm for clay minerals (Table 7.3).

Organic compounds undergo adsorption and degradation. The distribution coefficient, K_d, and the degradation coefficient are key parameters that control the concentration of benzene, phenols, and naphthalene. K_d can be calculated from the partitioning of a specific compound between water and organic carbon and the amount of organic carbon in sediments, by the equation:

$$K_d = f_{oc} \times K_{oc} \tag{7.2}$$

where f_{oc} is the mass fraction of organic carbon in sediments, defined as the mass of organic carbon in sediments m_{oc}^{s} divided by the mass of sediments m^{s}. K_{oc} may be

Table 7.3 Kinetic properties for minerals considered in the model (see text for data sources): 'primary' minerals are initially present in the aquifer; 'secondary' minerals are formed by precipitation.

		Parameters for kinetic rate law							
		Neutral mechanism		Acid mechanism			Base mechanism		
Mineral	A (cm^2/g)	k_{25} mol/m^2/s	E_a kJ /mol	k_{25}	E_a	$n(H^+)$	k_{25}	E_a	$n(OH-)$
Primary:									
Calcite	221	2.2×10^{-8}	62.76						
Illite	2124	1.66×10^{-13}	35.00	1.05×10^{-11}	23.60	0.34	3.02×10^{-17}	58.9	-0.40
Kaolinite	2309	6.91×10^{-14}	22.20	4.89×10^{-12}	65.90	0.78	8.91×10^{-18}	17.9	-0.47
Smectite	2149	1.66×10^{-13}	35.00	1.05×10^{-11}	23.60	0.34	3.02×10^{-17}	58.9	-0.4
Goethite	141	2.51×10^{-15}	66.20	4.07×10^{-10}	66.20	1.00			
Albite	229	3.89×10^{-13}	38.00	8.71×10^{-11}	51.70	0.50	6.31×10^{-12}	94.1	-0.82
Quartz	227	3.98×10^{-14}	87.7						
K-feldspar	234	3.89×10^{-13}	38.00	8.71×10^{-11}	51.70	0.50	6.31×10^{-12}	94.1	-0.82
Secondary:									
Dolomite	209	2.95×10^{-8}	62.76	2.34×10^{-7}	43.54	1.00			
Magnesite	199	4.57×10^{-10}	23.50	4.17×10^{-7}	14.40	1.00			
Dawsonite	247	3.89×10^{-13}	38.00	8.71×10^{-11}	51.70	0.50	6.31×10^{-12}	94.1	-0.82
Muscovite	212	3.89×10^{-13}	38.00	8.71×10^{-11}	51.70	0.50	6.31×10^{-12}	94.1	-0.82
Gibbsite	246	3.89×10^{-13}	38.00	8.71×10^{-11}	51.70	0.50	6.31×10^{-12}	94.1	-0.82
Pyrite	120	2.52×10^{-12}	62.76						

thought of as the ratio of the amount of chemical adsorbed per unit weight of organic carbon in the soil, to the concentration of the chemical in solution at equilibrium. Although the value of K_d depends on the properties of sediments, the hydrophobic partitioning theory as reviewed by Karickhoff et al. (1979) implies that the partitioning of a specific compound between water and organic carbon, as expressed by K_{oc}, is largely independent of the organic content of the solid material.

Based on published K_{oc} values and a typical range of f_{oc} values for the High Plains Aquifer, Carroll et al. (2014a) obtained plausible K_d ranges for benzene, phenol, and naphthalene. Here, we take the median values of these data, which are 0.012 l/kg, 0.0012 l/kg and 0.28 l/kg, respectively. For comparison, Table 7.4 lists some published K_d values for benzene, which show a large variation. Also shown in Table 7.4 is the weight fraction of organic carbon in sediments, f_{oc}, calculated based on a K_{oc} of 79 ($\log K_{oc} = 1.9$) for benzene. The large range of measured K_d values primarily results from the dependence of K_d on the organic matter content of soils, which is quite variable. In other words, K_d is soil-specific, and the same K_d value is not necessarily applicable to a range of different soil types.

Table 7.4 Compilation of published K_d values for benzene between soil and water.

Reference	K_d (ml/g)	f_{oc}	Comments
Larsen et al. (1992)	0.05–0.65	0.00063–0.0082	Measured for a range of different aquifer samples at various locations in Denmark
Donahue et al. (1999)	0.1–1.0	0.0013–0.013	Measured for Regina clay in Canada
Jean et al. (2002)	0.16–0.5	0.0021–0.0063	Measured in a laboratory experiment for artificial medium-sized sand
Hawthorne and Miller (2003)	28–59	0.35–0.75	Measured for manufactured gas plant soils (contaminated by organic compounds)

Degradation of the organics is calculated with a first-order degradation kinetic rate:

$$\frac{dC}{dt} = -kC \tag{7.3}$$

where C is the concentration of the organic compound and k is the first-order rate constant. Organic compounds are much more degradable under aerobic conditions than anaerobic conditions. Because the aquifer is under reducing conditions in the current model, anaerobic degradation constants should be used. Values of such constants for anaerobic systems have been sparsely reported. Here, we select these constants based on work reported by the US Environmental Protection Agency (EPA, 1999), adopting values of 3.7×10^{-7} l/s for phenol and 0 (no degradation) for benzene and naphthalene.

7.2.4 Metal Release from CO_2 and/or Brine Leakage

The intrusion of CO_2 into the aquifer forms a separate phase, which is carried horizontally down the regional hydraulic gradient by groundwater flow, and at the same time driven upward by buoyancy forces (Figure 7.3), following the zones of higher permeability (sand facies). After the CO_2 leakage stops at 100 years, the free CO_2 phase gradually fades, as shown by the simulated CO_2 plume at 110 years, and almost completely vanishes after about 130 years (Figure 7.3).

The dissolution of CO_2 in water produces carbonic acid, which decreases pH and increases the dissolved carbonate content in impacted groundwater:

$$CO_2(g) + H_2O = H_2CO_3(aq) \tag{7.4}$$

$$H_2CO_3(aq) = HCO_3^- + H^+ \tag{7.5}$$

The simulated plume of dissolved CO_2 (expressed as total inorganic carbon, TIC; Figure 7.4) occupies a much larger area than the free CO_2 phase, because of diffusion and dispersion in groundwater. The highest concentration of TIC (slightly higher than 1 molal) appears in the vicinity of the leakage point. After the CO_2 leakage stops at 100 years, the TIC plume is carried downgradient and the TIC concentrations gradually decrease, from 0.8 molal at 110 years to 0.2 molal at 200 years (Figure 7.4).

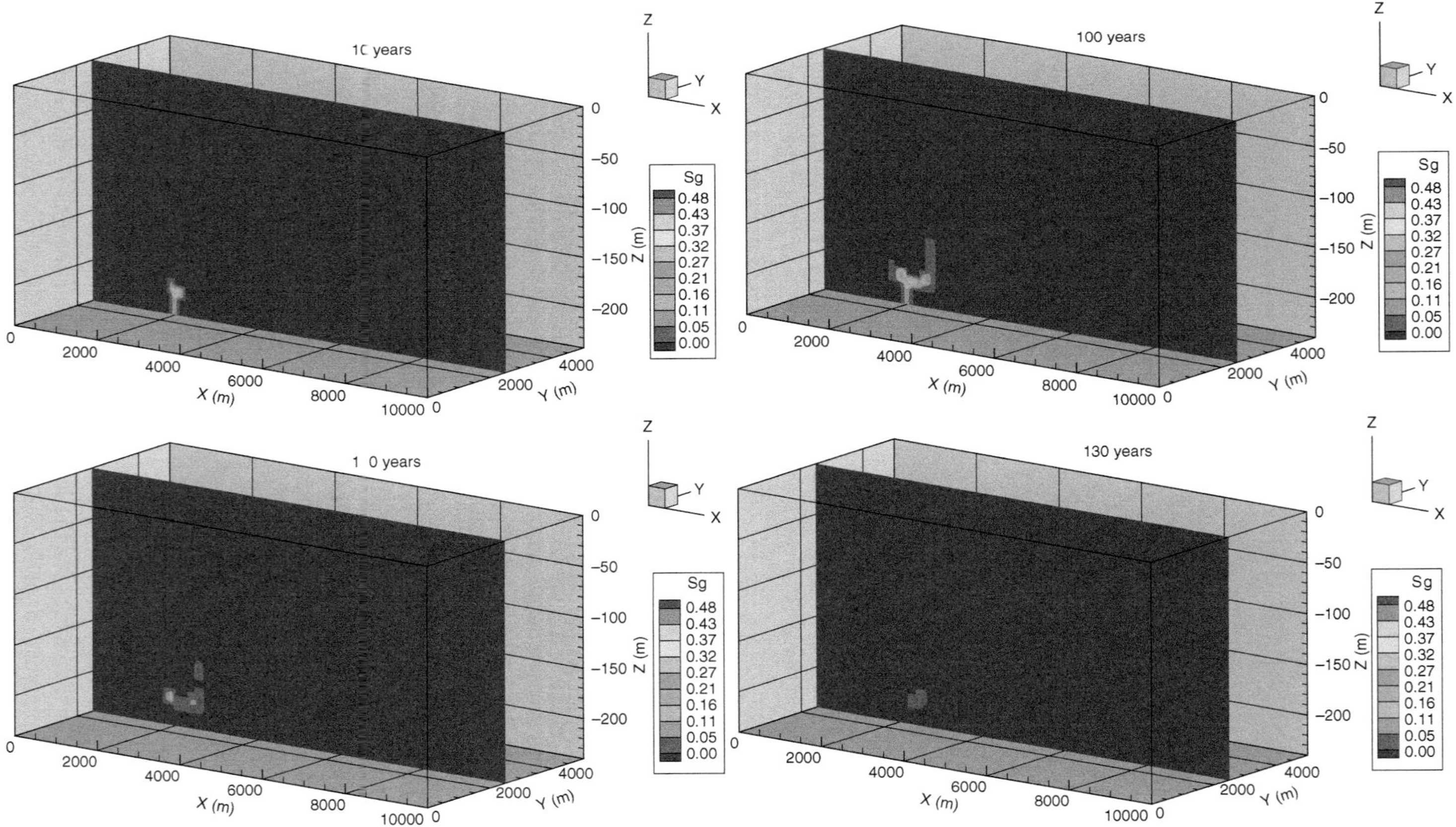

Figure 7.3 Simulated spatial distribut on of the free CO_2 phase (physical) saturation along a cross section at $y = 2500$ m at 10, 100, 110, and 130 years. (See colour plate section for the colour rep esentation of this figure.)

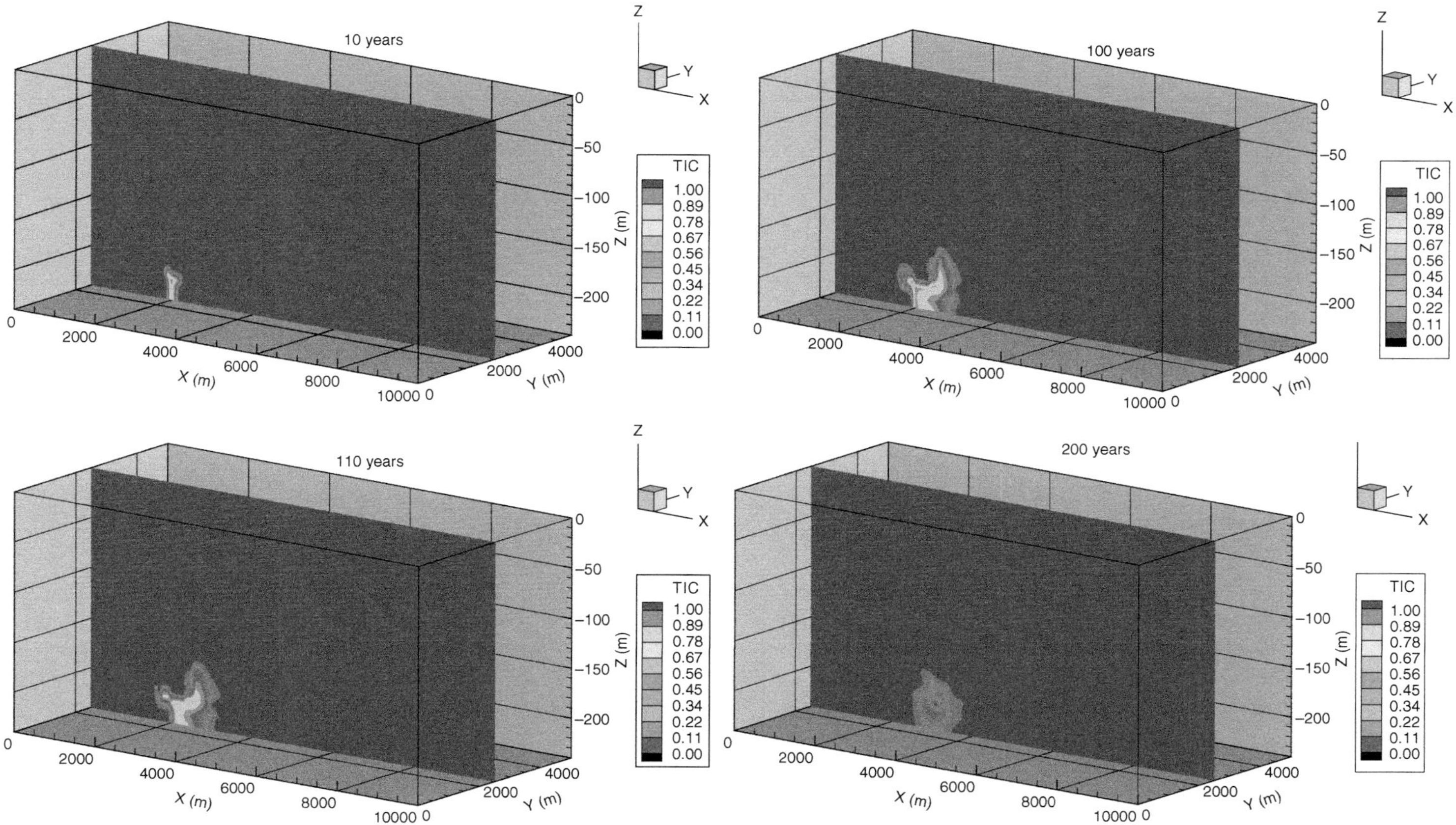

Figure 7.4 Simulated spatial distribution of total inorganic carbon (TIC, molal) along a cross-section at $y = 2500$ m at 10, 100, 110, and 200 years.

The evolution of pH closely follows the evolution of TIC (Figure 7.5), with pH decreasing as the concentration of TIC rises. In pure water, the dissolution of CO_2 at pressure > 5 bar yields a pH near 3.5; however, in natural systems, the pH drop is not as pronounced (values typically near or above 5) because of buffering by various reactions. For example, at the Frio deep CO_2 injection test site (Kharaka et al., 2009), injection of supercritical CO_2 at pressures ~150 bar into a deep sandstone aquifer resulted in pH dropping from about 6.5 to 5.7. Under totally different conditions in shallow sandy gravels, a few metres below the ground surface at the Montana State University Zero Emissions Research and Technology (MSU-ZERT) field test in Montana (Kharaka et al., 2010), the pH decreased from around 7 to 6 after injection of gaseous CO_2 through a 2–2.3 m deep horizontal pipe. A more significant pH decrease, from about 8 to 5.1, was observed in a field test in Mississippi (Trautz et al., 2013) upon injection of groundwater carbonated at a pressure near 4 bar into a confined, poorly buffered formation predominantly containing sandy sediments. By comparison, in the present model, the lowest pH is predicted to be around 5.2 within the two-phase zone (Figure 7.5). In this case, the pH buffering is caused primarily by the dissolution of calcite, with a slight contribution from protonation/deprotonation reactions on clay minerals and goethite surfaces. Once the CO_2 leakage stops, the pH is predicted to recover slowly from a value of 5.2 at 100 years to 6.8 at 200 years, which is primarily a result of mixing and dilution with groundwater, and secondarily a result of the continuous dissolution of calcite (albeit at a lower rate) and dissolution of siderite formed during the CO_2 leakage phase.

The decrease in groundwater pH upon carbonation can lead to the release of metals (M) by the dissolution of minerals, especially carbonates and sulfides:

$$MCO_3(s) + H^+ \rightarrow M^{2+} + HCO_3^- \tag{7.6}$$

$$MS(s) + H^+ \rightarrow M^{2+} + HS^- \tag{7.7}$$

Adsorption–desorption reactions are also expected to play an important role in metal mobilization, and also in buffering pH in the absence of significant amounts of fast-reacting carbonate minerals:

$$>SOM^+ + H^+ \rightarrow >SOH + M^{2+} \tag{7.8}$$

$$>SOH + H^+ \rightarrow >SOH_2^+ \tag{7.9}$$

Competitive sorption with bicarbonate ions (e.g. As, Se) could also mobilize metal oxyanions:

$$>SMO_4^{-n} + HCO_3^- \rightarrow >SHCO_3^- + MO_4^{-n} \tag{7.10}$$

In addition, metals released by the above reactions (e.g. Ca) may trigger exchange reactions that then further mobilize other metals into solution:

$$X-M + Ca^{2+} \rightarrow X-Ca + M^{2+} \tag{7.11}$$

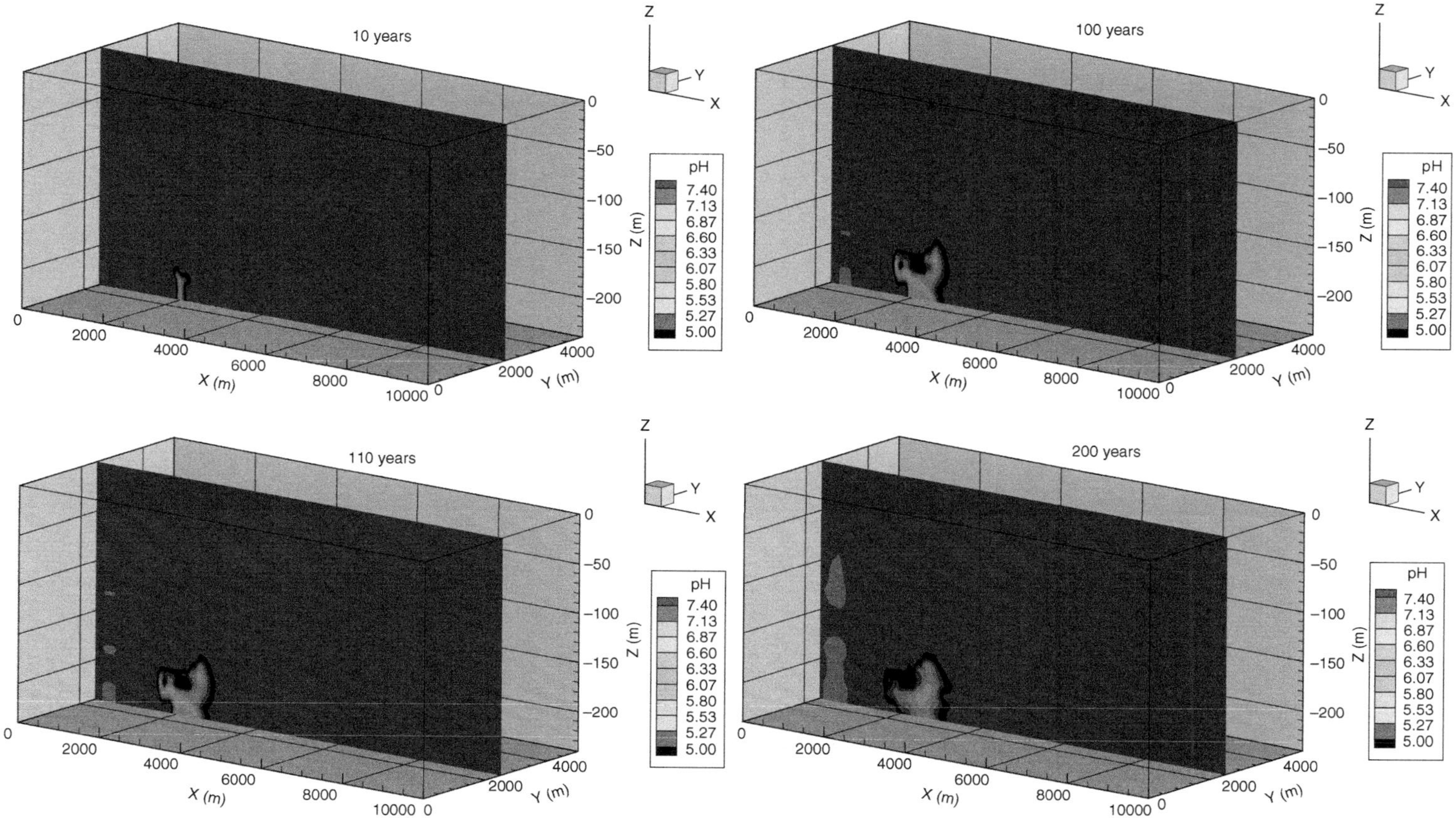

Figure 7.5 Simulated spatial distribution of pH along a cross section at Y = 2500 m at 10, 100, 110, and 200 years. (See colour plate section for the colour representation of this figure.)

All these reactions are plausible in controlling the release of trace metals from sediments upon exposure to CO_2. The dissolution of sulfide minerals was the first set of reactions that were proposed (using modeling studies) to be controlling the release of trace metals in groundwater after a CO_2 leak. For example, Wang and Jaffe (2004) simulated the release of Pb as a result of the dissolution of galena, and Zheng et al. (2009) and Apps et al. (2010) considered the control of galena on Pb, and arsenian pyrite on As. However, Zheng et al. (2009) suggested that adsorption/desorption reactions should more typically dominate the release of trace metals from sediments, because these reactions are typically much faster than mineral dissolution and precipitation. Moreover, the increases in concentrations of trace metals reported in laboratory experiments exposing sediments to CO_2 (e.g. Varadharajan et al., 2013; Lawter et al., 2016) are not elevated enough, and do not follow the trends required to suggest that the dissolution of sulfide minerals controls the concentrations of metals such as Pb and As in groundwater. For this reason, in the present model, the dissolution of sulfide minerals is not considered.

The dissolution of calcite with impurities of Sr and Ba has been proposed by Lu et al. (2010) and tested by Zheng et al. (2015) to explain increased concentrations of these metals in laboratory experiments. An alternative mechanism proposed by Zheng et al. (2015) involves calcite dissolution accompanied by Ca-driven cation exchange reactions. Simulations of these coupled processes give satisfactory results when compared with field measurements at two CO_2-release test sites (Zheng et al., 2015). For this reason, in the present model, we consider the dissolution of pure calcite and a series of cation exchange and desorption reactions (Tables A2 and A3), rather than the dissolution of impure calcite, as the reactions that control the release of metals.

The pH drop from CO_2 dissolution (reactions (7.4) and (7.5)) triggers a series of desorption reactions (following Reaction (7.11)) that lead to increases in Pb in solution. As seen in Figure 7.6, a plume with elevated Pb concentration develops and closely follows the plume with low pH surrounding the CO_2 intrusion. In our case, Pb^{2+} is initially adsorbed onto goethite, illite, kaolinite, and montmorillonite. Montmorillonite and illite are the most important adsorbents of Pb, with $(mon\_)_2Pb$ and $(ill\_)_2Pb$ (Table A3) being the dominant surface complexes. The concentration of dissolved Pb peaks at about 4.2×10^{-8} molal (~9 ppb) during the CO_2 leakage period. After CO_2 leakage stops at 100 years, the plume becomes diluted by mixing with groundwater, and the maximum Pb concentration decreases to about 2×10^{-8} molal (~4 ppb).

The case for Cd is very similar to Pb: an increase in dissolved concentration as the result of pH-driven desorption reactions, although the increase in concentration is less pronounced than with Pb. In this case, the dominant surface complexes are $(mon\_)_2Cd$ and $(ill\_)_2Cd$ (Table A3), and the maximum dissolved Cd concentration is $\sim 9 \times 10^{-9}$ molal (~1 ppb) during the CO_2 leakage period, decreasing to about 6×10^{-9} molal (~0.7 ppb) after 200 years (Figure 7.7).

Arsenian pyrite was initially postulated by Zheng et al. (2009) and Apps et al. (2010) as the mineral that thermodynamically controls the concentration of As in groundwater, based on statistical analyses of groundwater samples in a range of aquifers in the US (Birkholzer et al., 2008). However, laboratory and field tests since then did not support this hypothesis. First, not all laboratory tests published to date showed a release of As when sediments are exposed to CO_2. Even for those laboratory tests that did show a release of As (Little and Jackson, 2010; Viswanathan et al., 2012; Varadharajan

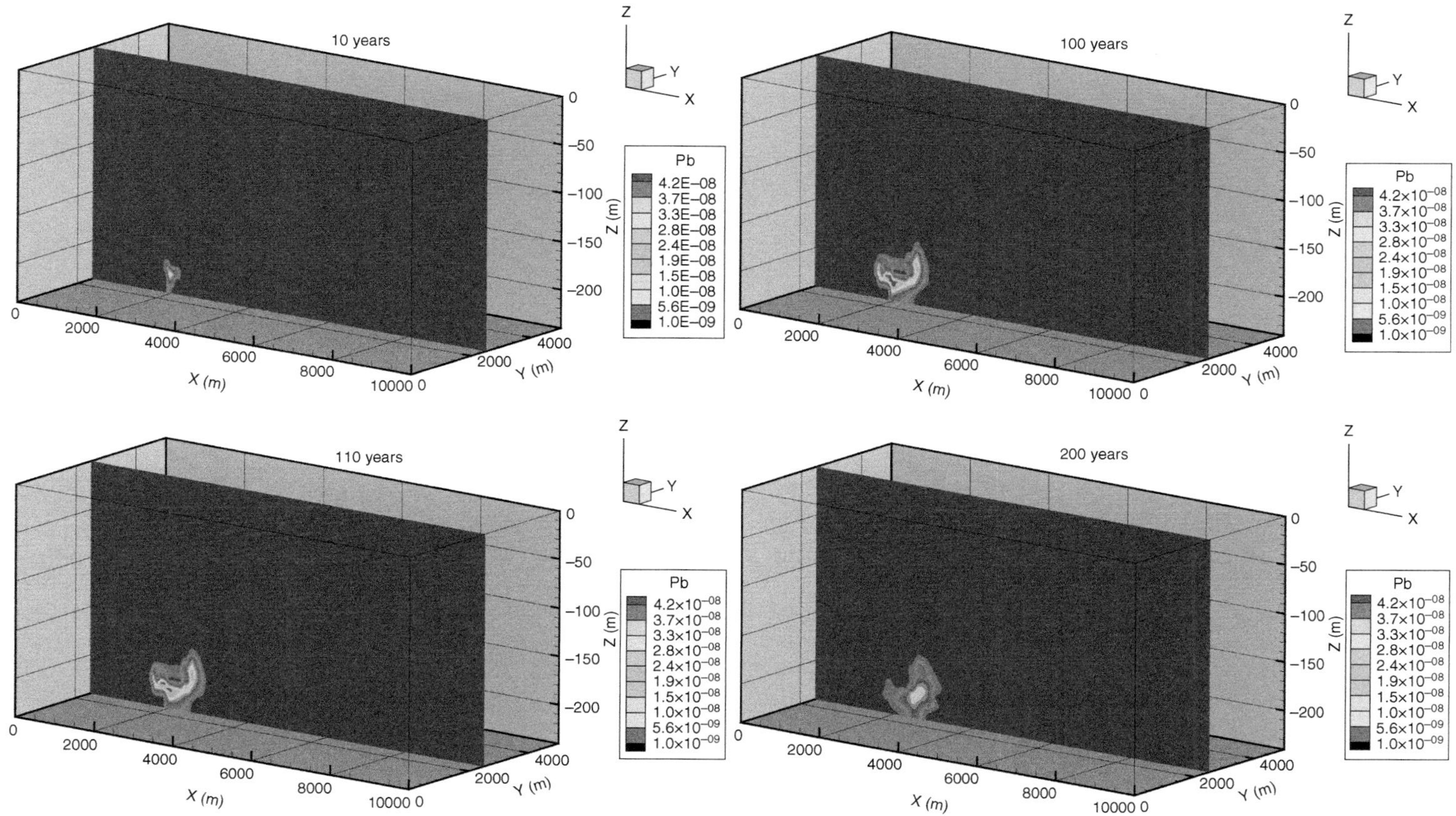

Figure 7.6 Simulated spatial distribution of dissolved Pb concentrations (molal) along a cross-section at $y = 2500$ m at 10, 100, 110, and 200 years.

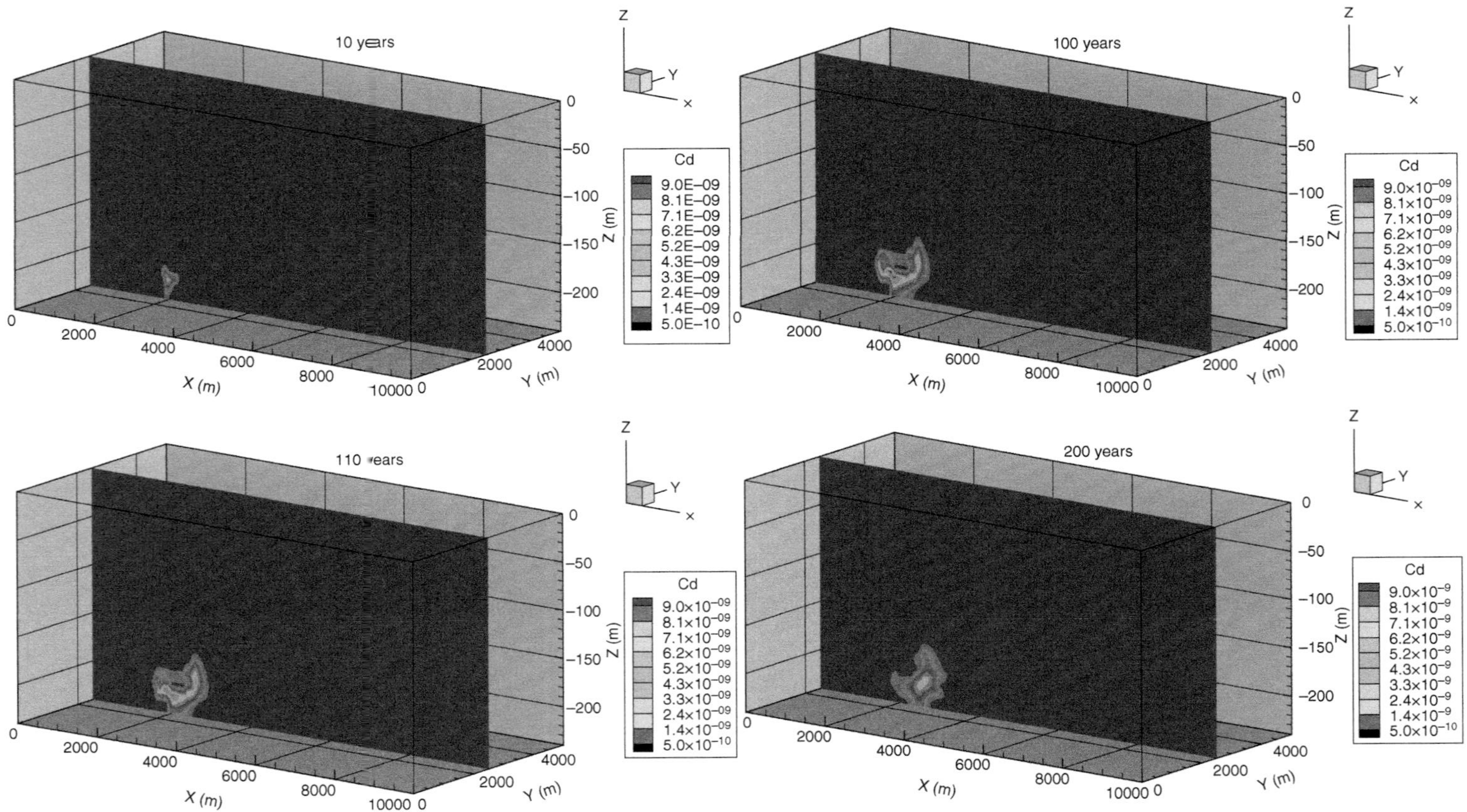

Figure 7.7 Simulated spatial distribution of dissolved Cd concentration (molal) along a cross section at $y = 2500$ m at 10, 100, 110, and 200 years.

et al., 2013), either the sediments were taken from aquifers known to have high background levels of As (Viswanathan et al., 2012) or the concentration of As observed in the laboratory was too low to be attributed to the dissolution of arsenian pyrite. Moreover, it is known that laboratory tests are more aggressive than field tests in leaching out trace metal with dissolved CO_2 (e.g. Yang et al., 2013; Zheng et al., 2015). Among field tests that have been reported in the literature so far (Kharaka et al., 2010; Peter et al., 2012; Yang et al., 2013; Trautz et al., 2013; Cahill and Jakobsen, 2013), only Yang et al. (2013) reported a release of As after the injection of CO_2 ($\sim 3 \times 10^{-9}$ molal As, or 0.2 ppb), which is orders of magnitude lower than if As concentrations were controlled by equilibrium with arsenian pyrite. For this reason, arsenian pyrite is not considered in the present model, and the geochemical behaviour of As is controlled solely by surface and aqueous complexes, with goe2_H_2AsO_4, goe2_$HAsO_4^-$, and mon_$HAsO_3^-$ (Table A3) dominating the surface complexation of As. Figure 7.8 shows the As plume at four different times. The maximum dissolved As concentration appears at the leakage point, where it is up to 1.3×10^{-7} molal ($\sim$10 ppb) during the leaking period. After the CO_2 leakage stops, the dissolved As concentration decreases somewhat to about 1×10^{-7} molal ($\sim$7 ppb).

Pb, Cd, and As are included in the model because they are elements of environmental concern and were observed in laboratory studies being leached out from sediments by carbonated water. However, as mentioned earlier, laboratory leaching experiments appear too aggressive compared with conditions expected for an actual CO_2 leak in the field. This seems to be supported by field experiments (e.g. Peter et al., 2012; Cahill and Jakobsen, 2013; Trautz et al., 2013), which show no release of Pb, Cd, or As in groundwater. In contrast, the mobilization of Ba has been typically observed in both laboratory and field studies. Modeling interpretations of CO_2 leakage, laboratory (Zheng et al., 2015) and field experiments (Trautz et al., 2013; Zheng et al., 2013) showed that Ca-driven cation exchange provides a robust explanation for the release of Ba in groundwater upon carbonation. The drop of pH leads to the dissolution of calcite, and then the increase in the concentration of Ca triggers a series of cation-exchange reactions that cause an increase in the concentrations of alkaline earth metals in groundwater. This explains the parallel temporal and spatial distribution of dissolved Ba (Figure 7.9) and Ca, and pH (Figure 7.5), which is also observed in the field experiments mentioned above. In the present case, the model predicts a maximum Ba concentration of 1.2×10^{-3} molal ($\sim$180 ppm) during the 100-year leakage phase, which then decreases to 8×10^{-5} molal ($\sim$11 ppm) after 200 years.

Trace metals released into the aquifer come from two sources: (1) the leaking brine and (2) the aquifer sediments resulting from the pH (and/or CO_2) driven chemical reactions. However, model results in Figures 7.5 to 7.9 show that the spatial distribution of the contaminant plume is largely dictated by the transport of CO_2, as manifested by the similarity between the plumes of TIC and trace metals. To distinguish the effect of the leaking brine on the contamination of the aquifer (from the CO_2-related mobilization), we conducted a simulation without CO_2 leakage. The simulated plume of trace metals obtained in this manner at 100 years is shown in Figure 7.10. Because brine is denser, the saline plume always stay at the bottom of the aquifer and only partially overlaps with the TIC plume (compare Figure 7.10 with Figure 7.4). The concentrations of trace metals in the saline plume during the injection phase are very low, typically about 1.5-fold higher than background concentrations, because of the dilution by clean groundwater and

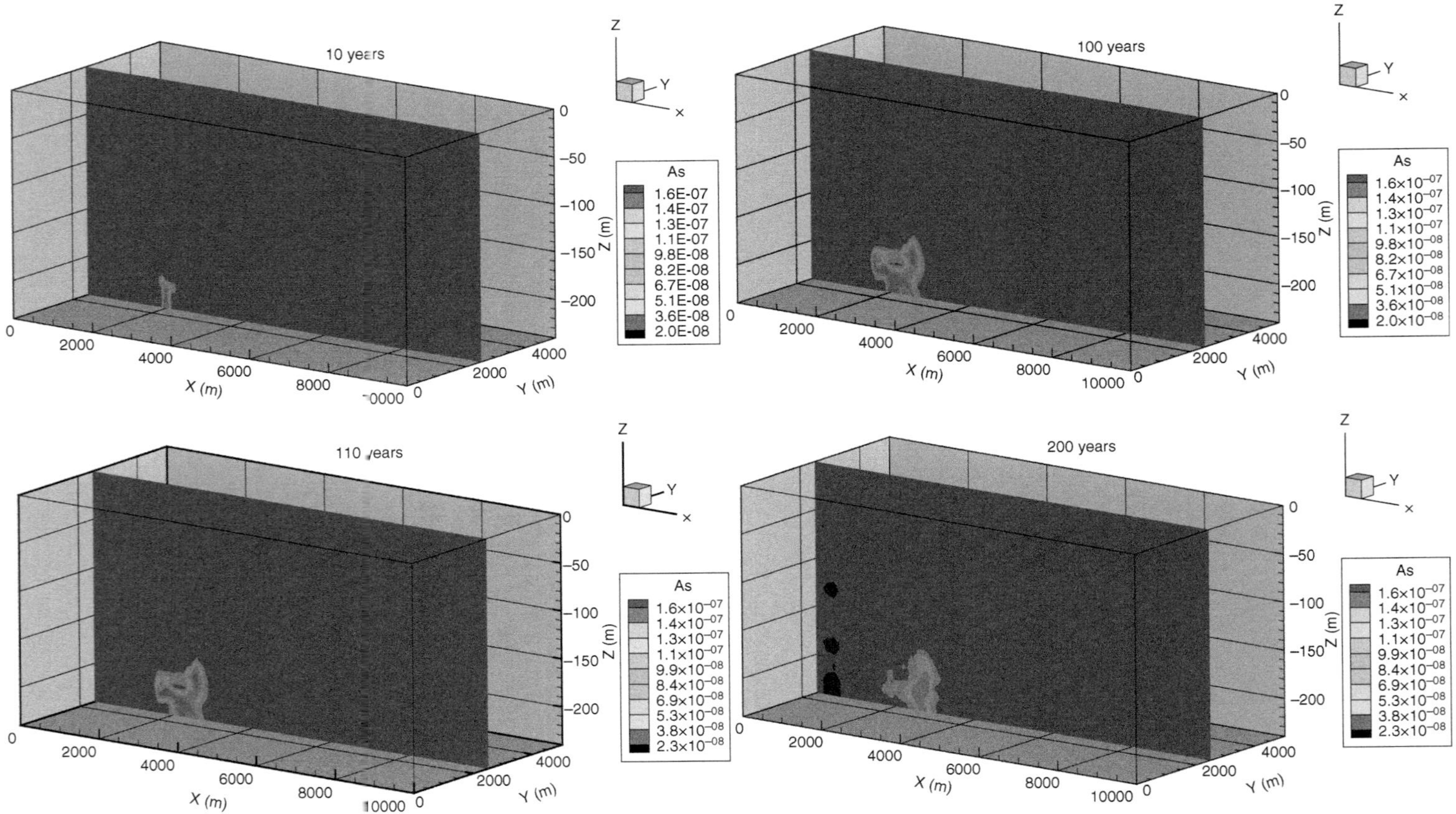

Figure 7.8 Simulated spatial distribution of As concentration (molal) along a cross-section at $y = 2500$ m at 10, 100, 110, and 200 years.

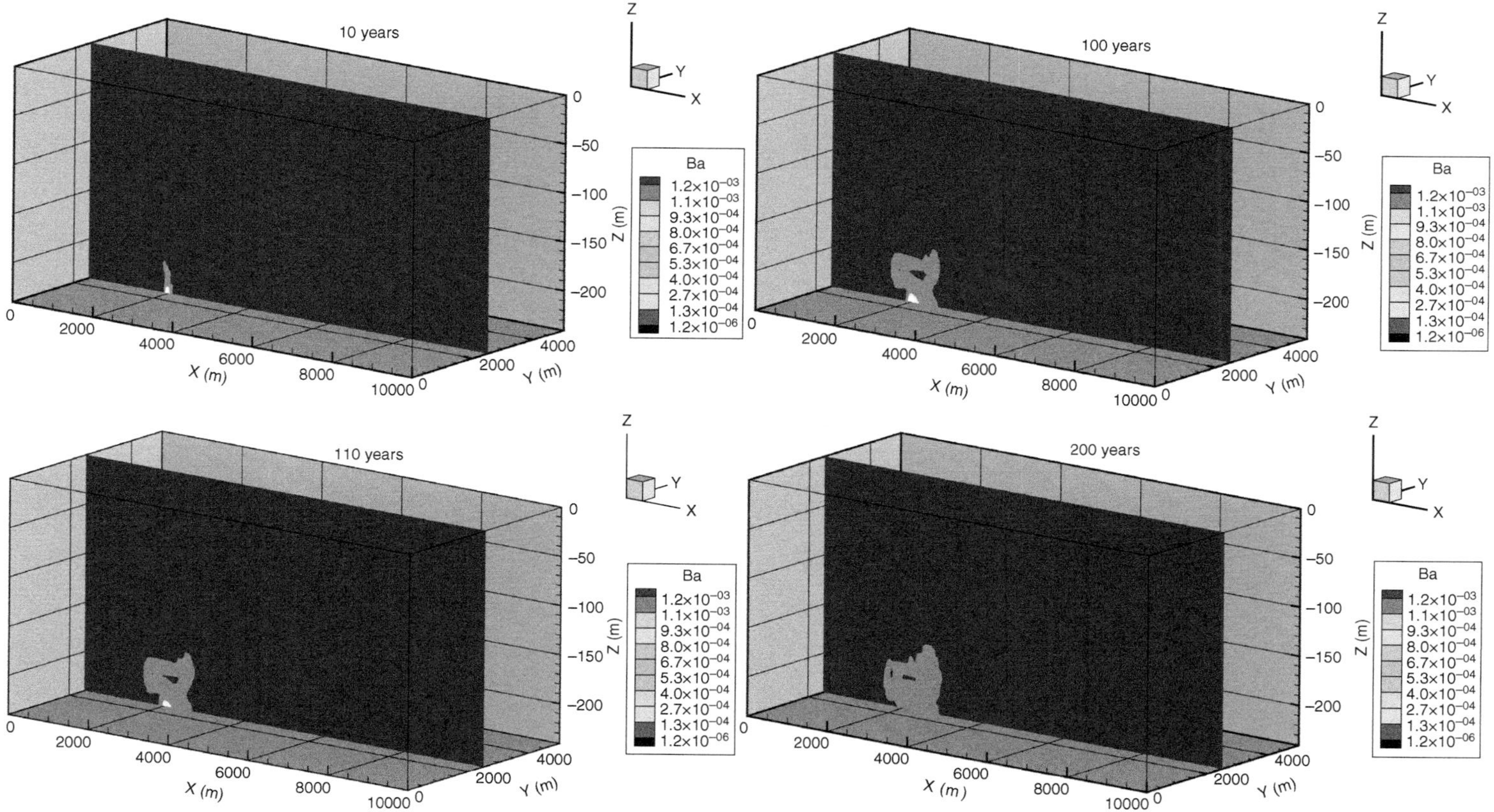

Figure 7.9 Simulated spatial distribution of dissolved Ba concentrations (molal) along a cross section at $y = 2500$ m at 10, 100, 110, and 200 years.

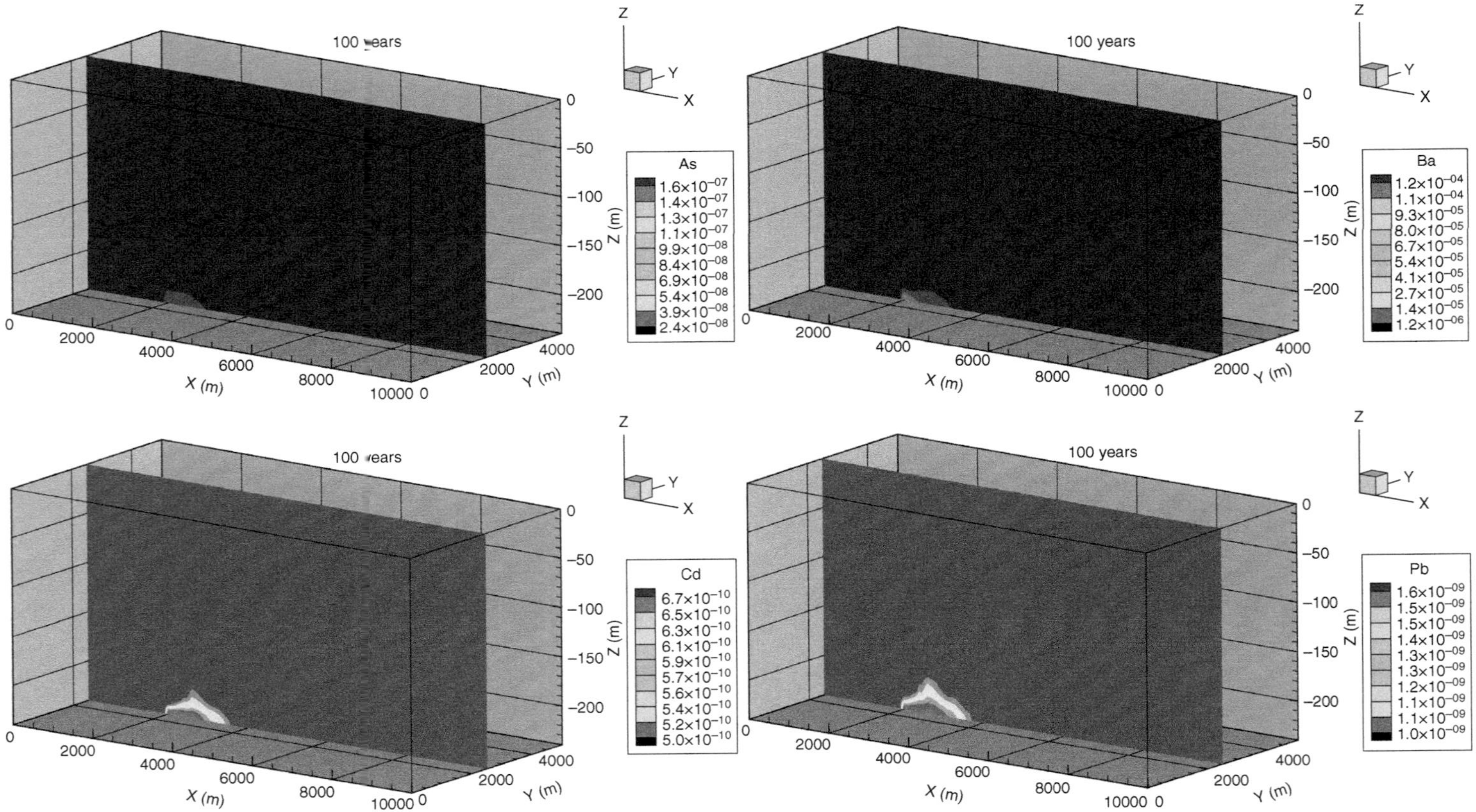

Figure 7.10 Simulated spatial distribution of As, Ba, Cd and Pb concentrations (molal) along a cross-section at $y = 2500$ m at 100 years, without CO_2 leakage.

adsorption to mineral surfaces. For this reason, the plume of trace metals caused by brine leakage is overshadowed by the effect of CO_2 leakage. This finding suggests that if leakage of both CO_2 and brine occurred, the CO_2 leakage would most likely dictate the spatial distribution of trace-metal contamination.

In the case of CO_2 geological storage in deep saline aquifers, the type of organic contaminants that could be transported to the shallow aquifer once leakage occurs depends on the type of organic compounds present in the brine. Organic compounds present in deep saline aquifers are seldom reported. However, their occurrence and concentrations in brines produced in the vicinity of gas and oil fields have been documented (e.g. RøeUtvik, 1999; Witter and Jones, 1999; Kharaka et al., 2006). From such studies, benzene, toluene, ethylbenzene, and xylenes (BTEX), as well as polycyclic aromatic hydrocarbons (PAHs) and phenols, were identified as potentially relevant candidates for our study. Obviously, many other organics could also warrant attention. Such compounds include carboxylic acids, which are commonly present at much higher concentrations (up to 5000 mg/l) in oil field brines (Kharaka and Hanor, 2007), and are also expected to be present in deep saline aquifers. Although these compounds are essentially nontoxic, they are good ligands for metal complexation, and their mobilization could exacerbate the release of trace metals upon CO_2 leakage into fresh water.

In the present model, we select benzene as a representative for BTEX, phenol for phenols, and naphthalene for PAHs. Because the aquifer is assumed free of these organic contaminants, the contamination of organic compounds comes exclusively from the leaking brine. Note that these organics included in the brine could conceptually also come from the CO_2 (CO_2 under supercritical conditions is a good solvent for these compounds; Anitescu and Tavlarides, 2006), and therefore could leach out organic compounds from the reservoir, carry them to shallower depths, and then dissolve into brine (Zheng et al., 2013)). Once reaching the aquifer, degradation and adsorption is predicted to essentially immobilize these organic compounds. Phenol, as shown in Figure 7.11, undergoes both degradation and adsorption, and its plume is very small. After the brine leakage stops at 100 years, degradation quickly consumes the remaining phenol, and its plume vanishes. Although naphthalene undergoes stronger adsorption, its dissolved plume (Figure 7.12) is much larger than that of phenol (Figure 7.11) because it suffers no degradation. After the brine leakage stops at 100 years, the concentration of naphthalene decreases owing to dilution, but the size of the plume increases slightly. Benzene (Figure 7.13) forms the largest plume of the three organic compounds included in the model, because it undergoes less adsorption than naphthalene and no degradation. These modeling results show that organic compounds in the leaking brine spread out at the bottom of the aquifer, but can be quickly consumed by degradation. If no degradation takes place, they can linger within the aquifer for a long time, even after the leakage stops.

7.3 Impact of Co-injected H_2S on the Quality of a Freshwater Aquifer

In order to understand the impact of co-injection of H_2S with CO_2 on the quality of groundwater, one important question to consider is how H_2S is transported together with CO_2 along the leakage pathway. Laboratory work (e.g. Bachu and Bennion, 2009) has shown that H_2S contained in an injected CO_2 stream chromatographically

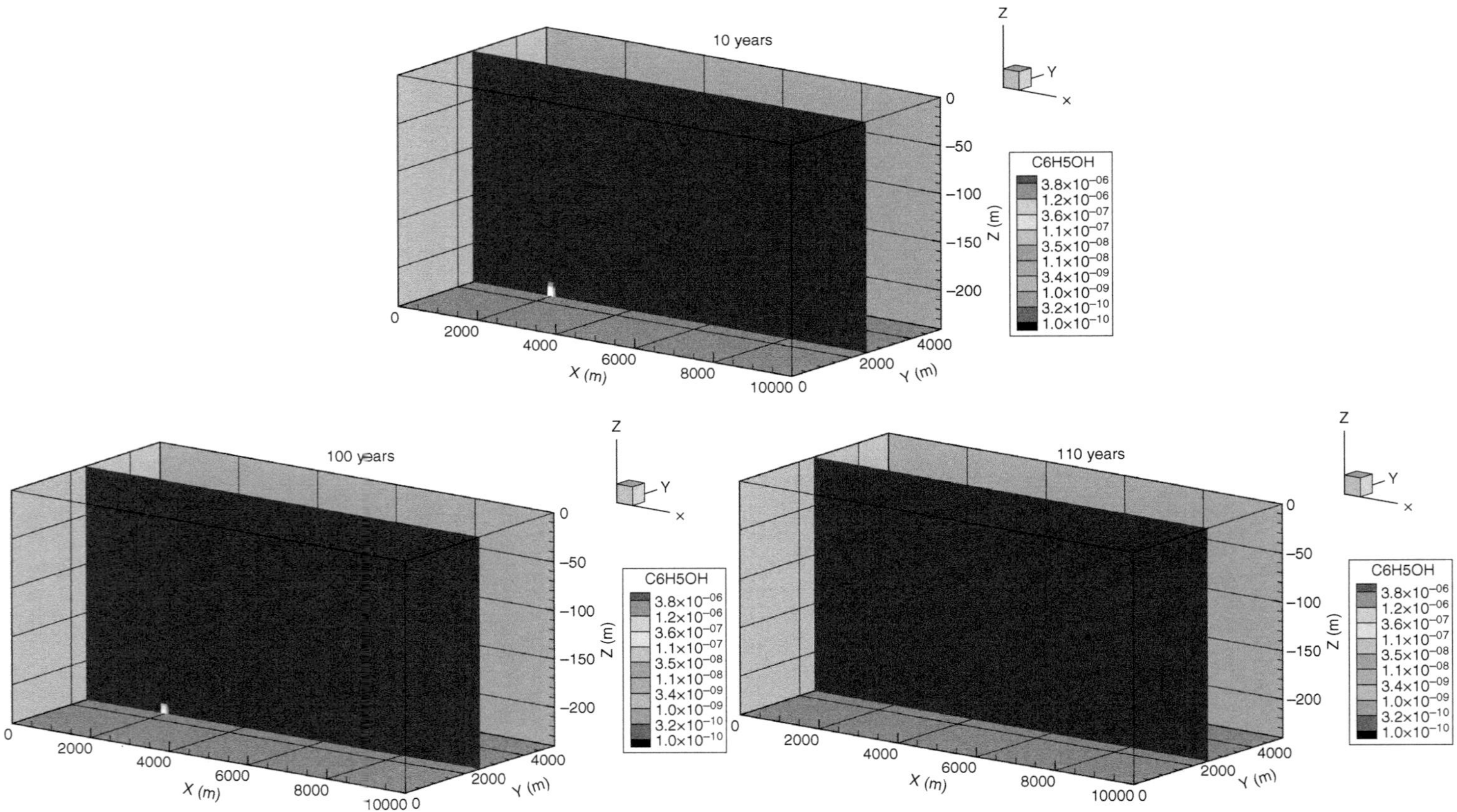

Figure 7.11 Simulated spatial distribution of phenol concentration (molal) along a cross-section at $y = 2500$ m at 10, 100, and 110 years.

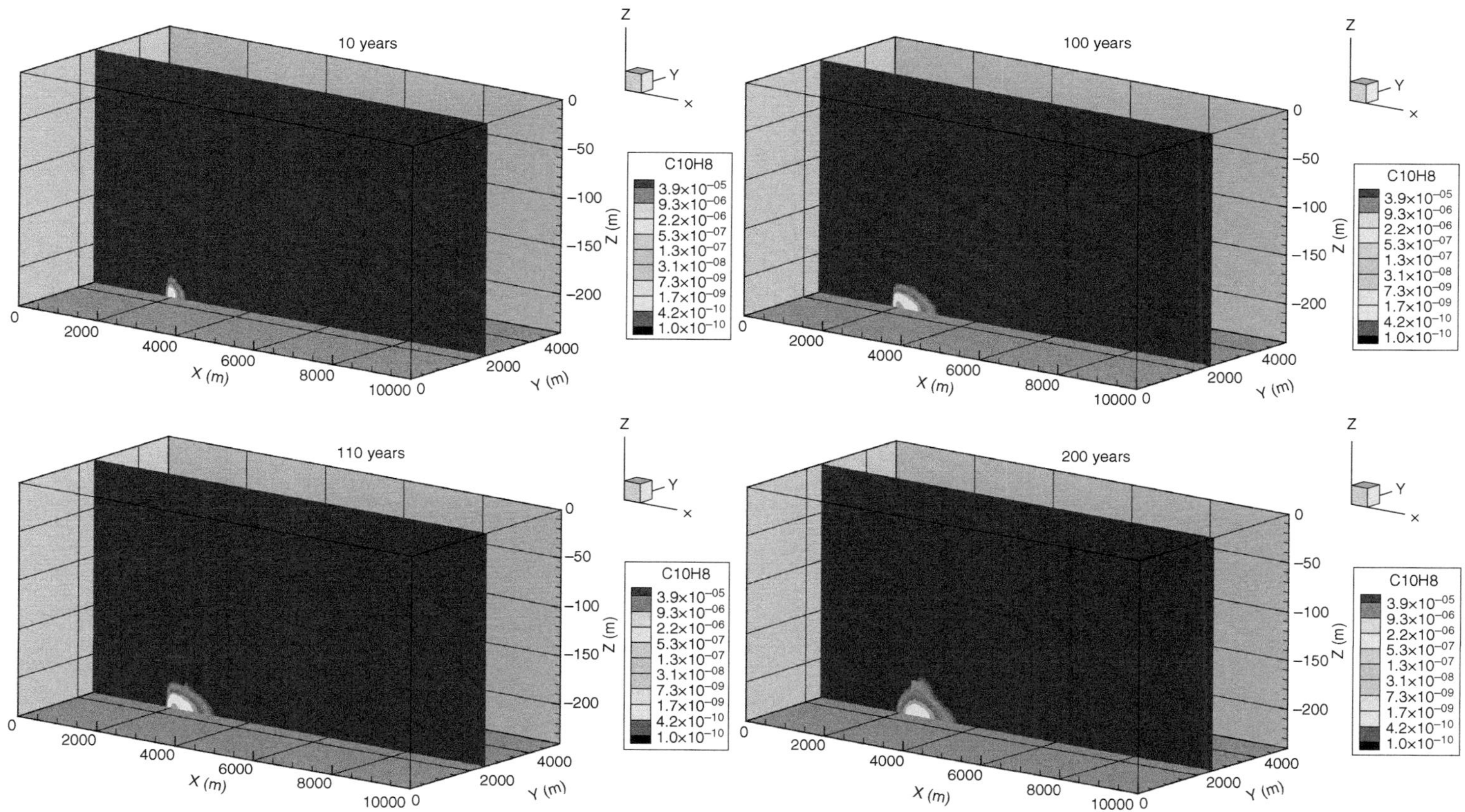

Figure 7.12 Simulated spatial distribution of naphthalene concentration (molal) along a cross-section at $y = 2500$ m at 10, 100, 110, and 200 years.

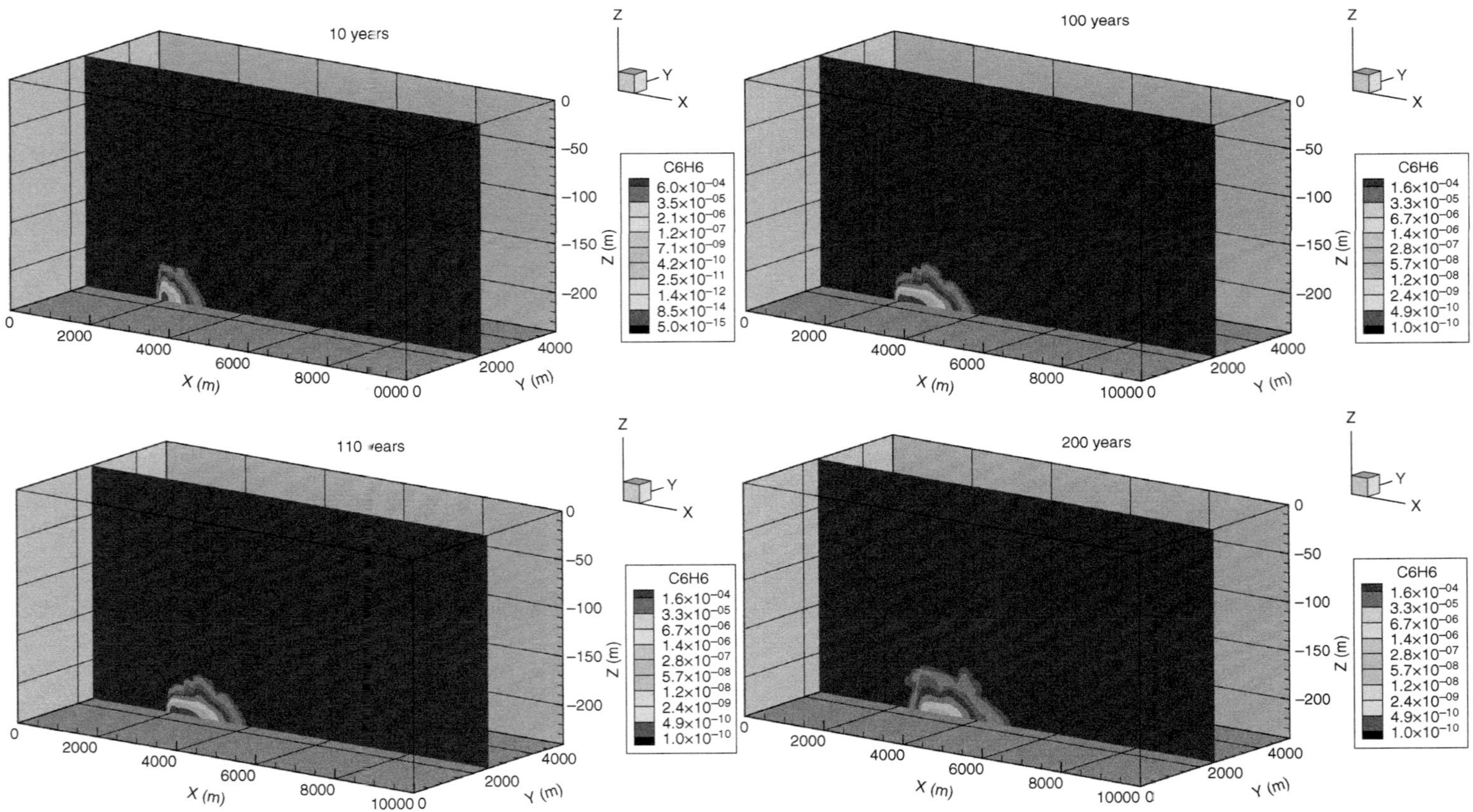

Figure 7.13 Simulated spatial distribution of benzene concentration (molal) along a cross-section at $y = 2500$ m at 10, 100, 110, and 200 years.

partitions when in contact with brines in deep saline aquifers. As discussed by Bachu et al. (2009), chromatographic partitioning of various gases, when in contact with subsurface fluids, is not a new phenomenon. Actually, this phenomenon has been used extensively in the petroleum industry for inter-well tracer tests to determine the oil saturation and reservoir properties between injection and observation wells (e.g. Illiassov and Datta-Gupta, 2002; Tang, 2005), and also for the detection and estimation of nonaqueous-phase liquids (NAPLs) present in the groundwater and in the vadose zone at contaminated sites (e.g. Deeds et al., 2000). Numerical models (Bachu et al., 2009) have shown that the preferential H_2S solubility in brine compared with that of CO_2 is the dominant effect on the chromatographic partitioning of these two gases, resulting in H_2S being stripped off at the leading edge of the advancing front of the $CO_2 + H_2S$ supercritical mixture. Because the experimental and modeling work conducted by Bachu and Bennion (2009) and Bachu et al. (2009) was for column-scale systems under isothermal conditions, and did not specifically address depth-dependent changes in pressure and temperature along a leakage pathway (e.g. a fracture zone or a fault), Zheng et al. (2013) developed a model to simulate a hypothetical scenario of leakage along a preferential pathway (including declining pressures and temperatures along the flow path) to investigate how the breakthrough of H_2S into an aquifer may be delayed compared with the breakthrough of CO_2.

The model simulations (Zheng et al., 2013) indicated that the breakthrough of H_2S co-migrating with CO_2 along a preferential pathway may be delayed compared with the arrival of the bulk free $CO_2 + H_2S$ phase. At some point, however, the ratio of H_2S to CO_2 in this leaking free 'gas' phase is predicted to approach the same value as in the injected gas mixture. Therefore, the extent to which a leak of $CO_2 + H_2S$ (compared with a leak of pure CO_2) could affect groundwater warrants further investigation.

Obviously, the presence of H_2S in water itself is a water-quality issue, because of its high toxicity at elevated concentrations, 'rotten egg' odour at low concentrations, and potential for oxidation to sulfuric acid, resulting in acidification and further degradation of water quality. A concentration limit of dissolved H_2S is not included in the EPA National Primary Drinking Water Regulations (the legally enforceable standards). However, a standard is specified in the National Secondary Drinking Water Regulations (the non-enforceable guidelines regulating contaminants) as three times the threshold odour number (TON), which is variable but typically in the ppb range for groundwater (e.g. Amoore and Hautala, 1983). Therefore, the leakage of H_2S with CO_2 could easily lead to concentrations exceeding this low limit.

Zheng et al. (2013) presented simulations to investigate the potential effects of trace H_2S leaking together with CO_2, focusing on thiophilic elements, lead and arsenic, in an anoxic environment. In this section, we summarize some of the results from Zheng et al. (2013) that illustrate how an aquifer could respond to the leakage of a $CO_2 + H_2S$ mixture. The simulator and model setup used for that study are also briefly described below.

7.3.1 The Simulator

Because of the need to simulate the leakage of CO_2 and H_2S, our modeling work here is conducted with a TOUGHREACT version (TMVOC_REACT) that allows the simulation of gases other than CO_2. TMVOC_REACT links TOUGHREACT (Xu et al., 2006) with TMVOC (Pruess and Battistelli, 2002), which simulates the three-phase

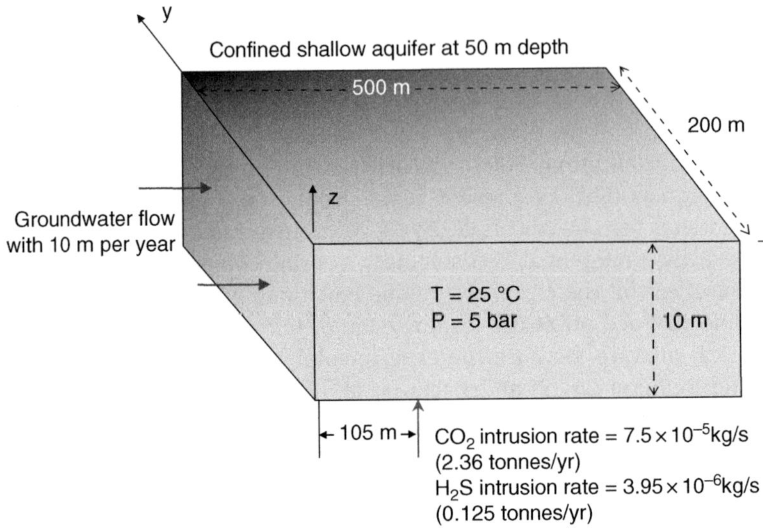

Figure 7.14 Schematic representation of the reactive transport model setup for simulations of CO_2+H_2S leakage.

non-isothermal flow of multicomponent hydrocarbon mixtures in variably saturated heterogeneous media. TMVOC was initially designed for studying subsurface contamination by volatile organic compounds (VOCs), such as hydrocarbon fuels and industrial solvents. Non-condensable gases that can be simulated with TMVOC include O_2, N_2, CO_2, CH_4, ethane, ethylene, acetylene, and air (a pseudo-component treated with properties averaged from N_2 and O_2). The capability of modeling H_2S was introduced into TMVOC_REACT specifically for this study.

7.3.2 Model Setup

A generic homogeneous sandstone aquifer is simulated as shown in Figure 7.14, with hydrogeological properties, initial chemical composition of groundwater, and host mineralogy of the aquifer listed in Appendix B. The model setup is similar to that in Zheng et al. (2009) except that the latter simulates the leakage of pure CO_2, whereas the present model (Zheng et al., 2013) assumes a hypothetical leakage of CO_2 + H_2S mixture from the bottom of the aquifer. Consequently, the results of these two models can be compared to distinguish the effect of co-injected H_2S from that of pure CO_2 leakage.

7.3.3 Metal Mobilization under CO_2+H_2S Leakage

The simulations from Zheng et al. (2013) evaluated whether the presence of H_2S in the gas mixture could exacerbate previously predicted CO_2 impacts on shallow groundwater, by possibly causing metal release from sediments. The dissolution of H_2S in water increases the acidity of the solution because of deprotonation, as shown by the following reaction:

$$H_2S(g) = H^+ + HS^- \tag{7.12}$$

However, the reductive dissolution of Fe(III) (hydr)oxides by sulfide (Poulton, 2003; Poulton et al., 2004), as expressed by the following reaction:

$$FeOOH(goethite) + 1.875H^+ + 0.125HS^- = Fe^{2+} + 0.125SO_4^{2-} + 1.5H_2O \quad (7.13)$$

consumes more protons than those being released by the dissolution of H_2S, and subsequently tends to counter the pH decrease from H_2S deprotonation. As a result, the pH of the groundwater in the case of $H_2S + CO_2$ leakage is slightly higher than the case of a pure CO_2 leakage presented in Zheng et al. (2009) (Figure 7.15). Xu et al. (2007) and Spycher and Oldenburg (2015) also report that co-injection of H_2S does not affect pH significantly.

As would be expected, elevated aqueous sulfide concentrations (Figure 7.16) are constrained by the precipitation of metal sulfides, in this case primarily pyrite (Figure 7.17), indicating that the precipitation of this mineral could sequester a significant amount of sulfide, if enough Fe is present in solution. The total aqueous sulfide concentration consistently peaks at the leakage source, and the magnitude of the peak concentration is proportional to the assumed rate of pyrite precipitation.

The reductive dissolution of Fe(III) minerals (such as goethite or hematite) by sulfide (thus with oxidation of sulfide to sulfate; Reaction 7.13) is a more likely sink for leaking H_2S, because of the ubiquitous presence of Fe(III) minerals in the subsurface. In the context of CO_2 sequestration, it has been shown experimentally and by numerical modeling (Palandri and Kharaka, 2005; Palandri et al., 2005; Xu et al., 2007; Spycher and Oldenburg, 2015) to supply ferrous iron in solution, which can then lead to the precipitation of Fe carbonates (such as siderite and/or ankerite). In the model discussed here, pyrite precipitation competes with siderite precipitation, with siderite forming in areas where pyrite precipitation is impeded by low aqueous sulfide concentrations (Figure 7.18). Note that increasing the dissolution rate of goethite can yield overlapping precipitation of both pyrite and siderite (Zheng et al., 2010). Batch experiments reacting hematite and CO_2 with SO_2 (Palandri et al., 2005) and sodium sulfide (Murphy et al., 2011) revealed the precipitation of both siderite and Fe sulfides, while similar experiments using ferrihydrite instead of hematite (Murphy et al., 2010) led to the precipitation of elemental sulfur instead of Fe sulfides. These findings point to the effects of competing rates of Fe(II) and sulfide supply and consumption. Whether Fe(II) is predicted to precipitate as pyrite preferentially to siderite (or ankerite) depends on the relative precipitation rates of these minerals, the rate of precipitation of these minerals relative to the goethite dissolution rate, and whether elemental sulfur is allowed to precipitate or not.

Previous reactive transport simulations have shown that dissolved Pb concentrations in groundwater could increase following a CO_2 leak, with or without H_2S (Zheng et al., 2009, 2013; Apps et al., 2010; and Section 7.2). In these simulations, the maximum aqueous lead concentrations mobilized in the case of $CO_2 + H_2S$ leakage (Zheng et al., 2013) is similar to that for the pure CO_2 leakage case (Zheng et al., 2009). Although both adsorbed Pb on mineral surfaces and galena could be the source of Pb release, the dissolution of galena is impeded by the desorption of Pb from mineral surfaces in the case of pure leakage (Zheng et al., 2009), and even more so for the case of $CO_2 + H_2S$ leakage (Zheng et al., 2013). Because in both cases adsorbed Pb on mineral surfaces is the

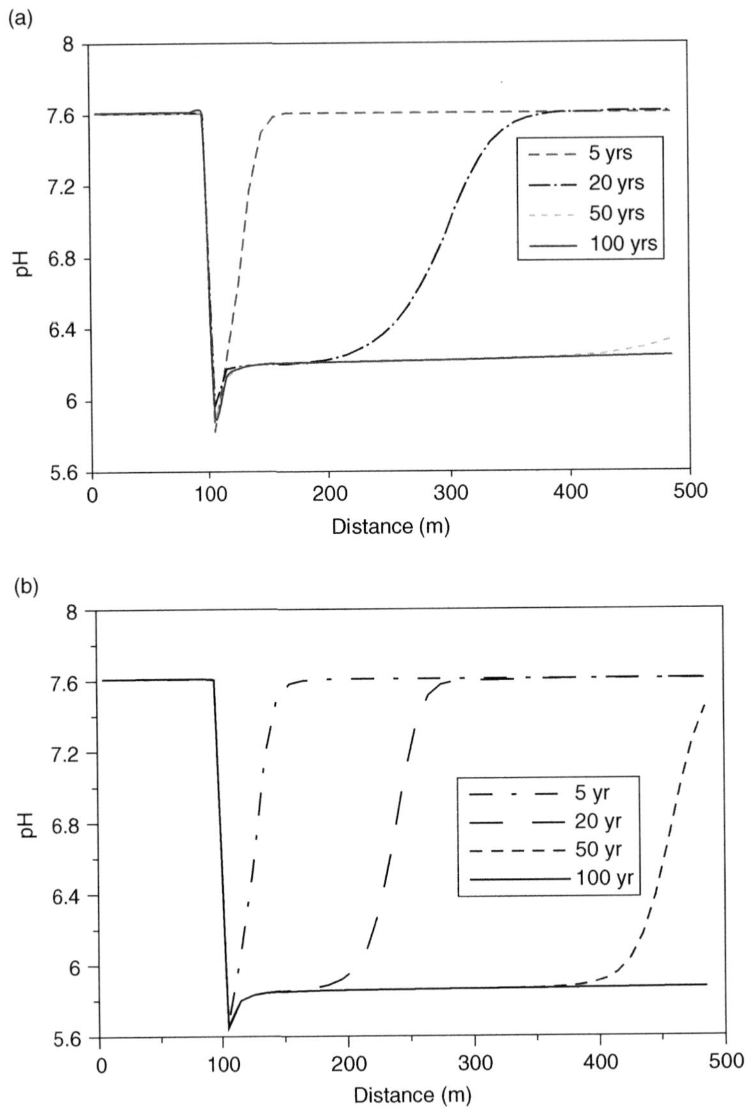

Figure 7.15 Computed pH profile along *x* at *y* = 0 at different times (the CO_2+H_2S mixture intrudes at *x* = 105 m). (a) CO_2+H_2S leak; (b) pure CO_2 leak (from Zheng et al., 2009).

dominant source for Pb, and desorption is not affected by the presence of high sulfide concentrations, the maximum aqueous Pb concentrations mobilized are quite similar in both cases. The leakage of CO_2+H_2S mixtures (Zheng et al., 2013), however, showed much higher As concentration than the pure CO_2 leakage case, because of the assumed presence of As-pyrite in the sediments and the strong aqueous complexation of As with sulfide (e.g. Helz and Tossell, 2008).

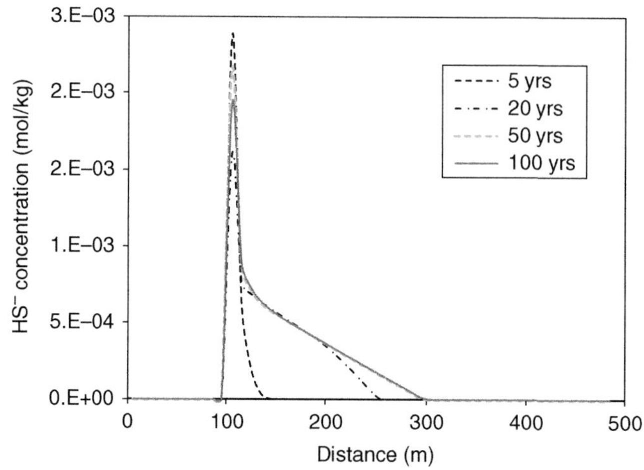

Figure 7.16 Predicted total aqueous sulfide concentration profile (as HS^-) along x at $y = 0$ at different times (the $CO_2 + H_2S$ mixture intrudes at $x = 105$ m).

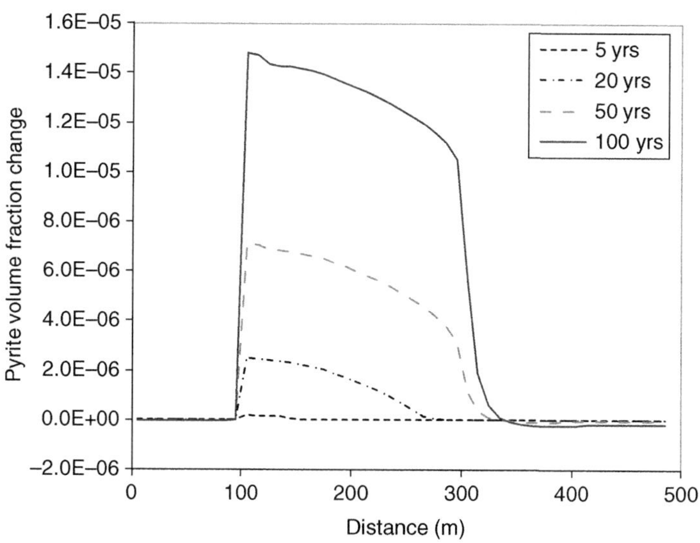

Figure 7.17 Profile of the pyrite volume fraction change (dimensionless) along x at $y = 0$ at different times.

7.4 Summary and Conclusion

Large-scale deployment of CO_2 geological storage operations requires assessment of the risks associated with each storage site. One of these risks is that a leak of CO_2 and/or brine from the storage reservoir could affect overlying shallow, potable groundwater aquifers. In the last decade, this issue has been studied via laboratory experiments, field tests, and numerical modeling studies to understand (1) to what degree

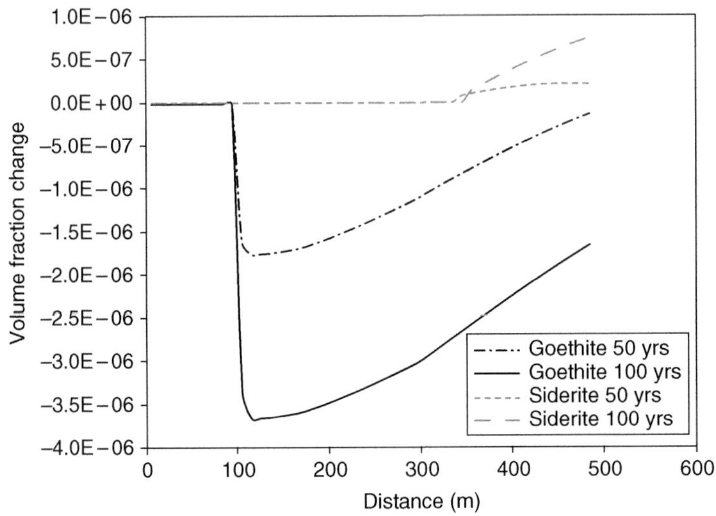

Figure 7.18 Computed profile of the goethite and siderite volume fraction change (dimensionless) along x at $y = 0$ at 50 and 100 years.

trace-metal and organic contaminants could be released into shallow aquifers as a result of CO_2 intrusion, and (2) which physical and geochemical processes control such a release. In this chapter, results of previous and new reactive transport simulations are presented to address this issue. Two sets of simulations are presented. In the first set, the leaking fluid comprises CO_2, brine-containing metals, and organic contaminants. In this case, reactive transport simulations are used to model the behaviour of trace metals and organic compounds in response to the leakage of CO_2 and brine into a shallow aquifer, using a model setup and properties aligned with the High Plains Aquifer in Kansas. The second set of simulations cover those presented by Zheng et al. (2013), who simulated the response of a generic sandstone aquifer to the intrusion of $CO_2 + H_2S$ mixtures, to examine whether the leakage of co-injected H_2S with CO_2 could exacerbate the mobilization of trace metals compared with leakage of pure CO_2.

The results from the first set of simulations show that CO_2 leakage leads to a two-phase zone (free CO_2 and aqueous) within the aquifer, shaped by the regional groundwater flow and buoyancy forces. A plume of lower pH and higher dissolved carbonate concentrations develops around, and extends away from, this two-phase zone. In this plume, As, Pb, Cd, and Ba are released as a result of the pH decrease. Desorption is the major reaction that leads to the increase in concentrations of dissolved As, Pb, and Cd. The dissolution of calcite and the subsequent Ca release, followed by Ca-driven cation exchange, leads to a rise in dissolved Ba concentrations, following a trend that parallels that of Ca. Once leakage stops and the pH rebounds, the dissolved concentrations of these elements decrease significantly, as a result of resorption and reversal of processes leading to Ca-driven cation exchange. Whether these processes are actually reversible in real systems needs further investigation.

Although the brine from the modeled deep storage reservoir contains elevated concentrations of As, Pb, Cd, and Ba, it does not contribute significantly to contamination of the modeled shallow aquifer. This is because the trace-metal concentrations in the brine, while higher than background concentrations, are not high enough to dominate the mobilization of adsorbed and exchangeable metals in aquifer sediments. In addition, the displaced brine, given its higher density, remains at the bottom of the aquifer. The organic compounds brought by the leaking brine (and/or CO_2) form a plume at the bottom of the aquifer that is carried downgradient. The concentration of these organic contaminants is strongly attenuated by adsorption and degradation, with degradation playing the major role in the modeled scenarios. Once the leakage stops, mobilized organic compounds that undergo degradation vanish, while less degradable compounds linger for a longer time.

Assessing the impact of leakage of $CO_2 + H_2S$ into shallow aquifers, in addition to the potentially adverse effects of H_2S itself (i.e. odour and taste), Zheng et al. (2013) in their simulations suggested that the leakage of $CO_2 + H_2S$ into an anoxic aquifer could lead to higher arsenic release in groundwater than cases of pure CO_2 leakage (e.g. Zheng et al., 2009), because of the formation of a stable As-sulfide aqueous species. In contrast, compared with cases of pure CO_2 leakage, the addition of H_2S to the leak has little effect on dissolved Pb concentrations, because dissolution of galena played an insignificant role in releasing Pb.

Obviously, the use of models alone to investigate an aquifer's response to intrusion of $CO_2 + H_2S$ mixtures (or even just CO_2) is seriously limited, because such a response depends on site-specific conditions, such as metal–sediment associations, groundwater composition and redox conditions, leakage rate versus regional groundwater flow rates, and many other hydrological and geochemical parameters. The leakage impact on groundwater is expected to be quite sensitive to the complex (and coupled) interaction between hydrological and chemical processes, as well as to the degree of mixing between the leaking fluid and groundwater. Using reactive transport models to investigate these complex coupled processes yields useful insights. However, the results of such modeling should be considered qualitative or semi-quantitative at best, because simulations require many input parameters, most of which are site-specific and often unknown or poorly constrained. The models presented in this chapter are used to illustrate the types of reaction mechanisms that could be expected in typical leakage scenarios. When conducting a risk assessment of specific CO_2 geological storage sites, it is recommended that extensive site characterization be conducted first, including the determination of hydrogeological and geochemical conditions in the storage reservoir, sediment–metal associations in overlying aquifers, and types of potential leakage pathways (e.g. wellbore, faults). Geochemical modeling integrated with laboratory experiments should also be carried out, to establish a robust conceptual model of contaminant mobilization. Only then can reactive transport modeling be applied with some confidence to predict the long-term impact on specific aquifers. These investigations should include uncertainty quantification to assess the effect of parameter and concept uncertainties on the model results, such as in the approach developed by the CO_2 geological storage National Risk Assessment Partnership (Carroll et al., 2014b; Dai et al., 2014; Pawar et al., 2014).

Appendix A

Table A1 Equilibrium constants of the major rock-forming minerals.

Minerals	Log *K* (25 °C)
Calcite + H^+ = Ca^{2+} + HCO_3^-	1.853
Illite + $6.3H_2O$ = H^+ + 0.25 Mg^{2+} + 0.85 K^+ + $3.4H_4SiO_4(aq)$ + $2.35AlO_2^-$	−43.490
Kaolinite + $3H_2O$ = $2H^+$ + $2H_4SiO_4(aq)$ + $2AlO_2^-$	−39.262
Smectite + $7.32H_2O$ = $0.68H^+$ + $0.17Ca^{2+}$ + 0.335 Mg^{2+} + $3.99H_4SiO_4(aq)$ + $1.68AlO_2^-$	−32.834
Goethite + $3H^+$ = $2H_2O$ + Fe^{3+}	0.363
Albite + $6H_2O$ = Na^+ + $3H_4SiO_4(aq)$ + AlO_2^-	−20.126
Quartz + $2H_2O$ = $H_4SiO_4(aq)$	−3.740
K-feldspar + $6H_2O$ = K^+ + $3H_4SiO_4(aq)$ + AlO_2^-	−22.394
Dolomite + $2H^+$ = Ca^{2+} + Mg^{2+} + $2HCO_3^-$	3.545
Magnesite + H^+ = Mg^{2+} + HCO_3^-	1.420
Dawsonite = H^+ + Na^+ + HCO_3^- + $2.35AlO_2^-$	−18.535
Muscovite + $6H_2O$ = $2H^+$ + K^+ + $3H_4SiO_4(aq)$ + $3AlO_2^-$	−57.264
Ferrihydrite + $3H^+$ = $3H_2O$ + Fe^{3+}	3.404
Gibbsite = H_2O + H^+ + AlO_2^-	−15.129

Table A2 Cation exchange reactions and selectivity coefficients, using the Gaines-Thomas convention (from Appelo and Postma, 1994).

Cation exchange reaction	$K_{Na/M}$
Na^+ + X-H = X-Na + H^+	1
Na^+ + X-K = X-Na + K^+	0.2
Na^+ + 0.5X-Ca = X-Na + $0.5Ca^{2+}$	0.4
Na^+ + 0.5X-Mg = X-Na + 0.5 Mg^{2+}	0.6
Na^+ + 0.5X-Ba = X-Na + $0.5Ba^{2+}$	0.2

Source: From Appelo 1994

Table A3 Surface sorption reactions for As, Cd, and Pb on goethite, illite, kaolinite, and montmorillonite.

Reactions	Log k_{int}	Site density mol/m^2	Surface area m^2/g	Type of SCM model (and capacitance as needed)	Reference
Goethite					
$goe^1\_OH_2^+ = goe^1\_OH + H^+$	−7.38	3.9×10^{-8}	80	DLM	1
$goe^1\_O^- + H^+ = goe^1\_OH$	10.74	3.9×10^{-8}	80	DLM	
$goe^2\_OH_2^+ = goe^2\_OH + H^+$	−7.38	3.8×10^{-6}	80	DLM	1
$goe^2\_O^- + H^+ = goe^2\_OH$	10.74	3.8×10^{-6}	80	DLM	
$goe^1\_OCd^+ + H^+ = goe^1\_OH + Cd^{2+}$	−1.29	3.9×10^{-8}	80	DLM	1
$goe^2\_OCd^+ + H^+ = goe^2\_OH + Cd^{2+}$	1.83	3.8×10^{-6}	80	DLM	
$goe^1\_OPb^+ + H^+ = goe^1\_OH + Pb^{2+}$	−4.78	3.9×10^{-8}	80	DLM	1
$goe^2\_OPb^+ + H^+ = goe^2\_OH + Pb^{2+}$	−1.52	3.8×10^{-6}	80	DLM	
$goe^2\_OH_2AsO_3 + H_2O = goe^2\_OH + H_3AsO_3(aq)$	−5.19	3.32×10^{-6}	80	DLM	2
$goe^2\_OHAsO_3^- + H_2O + H^+ = goe^2\_OH + H_3AsO_3(aq)$	2.34	3.32×10^{-6}	80	DLM	
Illite					
$ill\_OH_2^+ = ill\_OH + H^+$	−8.02	2.27×10^{-6}	66.8	CCM, = 2.0 F/m^2	3
$ill\_O^- + H^+ = ill\_OH$	8.93	2.27×10^{-6}	66.8	CCM, = 2.0 F/m^2	
$ill\_OCd^+ + H^+ = ill\_OH + Cd^{2+}$	3.62	2.27×10^{-6}	66.8	CCM, = 2.0 F/m^2	3
$(ill\_)_2Cd + 2H^+ = 2ill\_H + Cd^{2+}$	−0.63	1.3×10^{-6}	66.8	CCM, = 2.0 F/m^2	
$ill\_CdOH + 2H^+ = ill\_H + Cd^{2+} + H_2O$	6.49	1.3×10^{-6}	66.8	CCM, = 2.0 F/m^2	
$ill\_OPb^+ + H^+ = ill\_OH + Pb^{2+}$	0.70	2.27×10^{-6}	66.8	CCM, = 2.0 F/m^2	3
$(ill\_)_2Pb + 2H^+ = 2ill\_H + Pb^{2+}$	−1.37	1.3×10^{-6}	66.8	CCM, = 2.0 F/m^2	
$ill\_PbOH + 2H^+ = ill\_H + Pb^{2+} + H_2O$	3.65	1.3×10^{-6}	66.8	CCM, = 2.0 F/m^2	
$ill\_H_2AsO_3 + H_2O = ill\_OH + H_3AsO_3$	−2.12	3.83×10^{-6}	66.8	CCM, = 1.06 F/m^2	4
$ill\_HAsO_3^- + H_2O + H^+ = ill\_OH + H_3AsO_3$	5.66	3.83×10^{-6}	66.8	CCM, = 1.06 F/m^2	

(Continued)

Table A3 (Continued)

Reactions	Log k_{int}	Site density mol/m^2	Surface area m^2/g	Type of SCM model (and capacitance as needed)	Reference
Kaolinite					
$kao\_OH_2^+ = kao\_OH + H^+$	−4.63	2.24×10^{-6}	22.42	CCM, = 1.2 F/m^2	5
$kao\_O^- + H^+ = kao\_OH$	7.54	2.24×10^{-6}	22.42	CCM, = 1.2 F/m^2	
$Kao\_OCd^+ + H^+ = Kao\_OH + Cd^{2+}$	3.23	2.24×10^{-6}	22.42	CCM, = 1.2 F/m^2	5
$(Kao\_)_2Cd + 2H^+ = 2Kao\_H + Cd^{2+}$	−1.22	3.57×10^{-7}	22.42	CCM, = 1.2 F/m^2	
$Kao\_OPb^+ + H^+ = Kao\_OH + Pb^{2+}$	0.64	2.24×10^{-6}	22.42	CCM, = 1.2 F/m^2	5
$(Kao\_)_2Pb + 2H^+ = 2Kao\_H + Pb^{2+}$	−2.36	3.57×10^{-7}	22.42	CCM, = 1.2 F/m^2	
$kao\_HAsO_3^- + H_2O + H^+ = kao\_OH + H_3AsO_3$	5.43	3.83×10^{-6}	22.42	CCM, = 1.06 F/m^2	4
$kao\_AsO_4^{2-} + H_2O + 2H^+ = kao\_OH + H_3AsO_4$	4.69	3.83×10^{-6}	22.42	CCM, = 1.06 F/m^2	4
Montmorillonite					
$mon\_OH_2^+ = mon\_OH + H^+$	−6.04	4.41×10^{-6}	46	CCM, = 3.2 F/m^2	6
$mon\_O^- + H^+ = mon\_OH$	6.63	4.41×10^{-6}	46	CCM, = 3.2 F/m^2	
$mon\_Na^+ + H^+ = mon\_H + Na^+$	−0.18	1.53×10^{-5}	46	CCM, = 3.2 F/m^2	
$mon\_OCd^+ + H^+ = mon\_OH + Cd^{2+}$	2.93	4.41×10^{-6}	46	CCM, = 3.2 F/m^2	6
$(mon\_)_2Cd + 2H^+ = 2mon\_H + Cd^{2+}$	−2.37	1.53×10^{-5}	46	CCM, = 3.2 F/m^2	
$mon\_OPb^+ + H^+ = mon\_OH + Pb^{2+}$	−0.49	4.41×10^{-6}	46	CCM, = 3.2 F/m^2	6
$(mon\_)_2Pb + 2H^+ = 2mon\_H + Pb^{2+}$	−2.56	1.53×10^{-5}	46	CCM, = 3.2 F/m^2	
$mon\_H_2AsO_3 + H_2O = mon\_OH + H_3AsO_3$	−1.19	3.83×10^{-6}	46	CCM, = 1.06 F/m^2	4
$mon\_HAsO_3^- + H_2O + H^+ = mon\_OH - H_3AsO_3$	3.92	3.83×10^{-6}	46	CCM, = 1.06 F/m^2	4

1) Swedlund et al. (2009).
2) Dixit and Hering (2003).
3) Gu and Evans (2007).
4) Gu and Evans (2008).
5) Gu et al. (2010).
6) Goldberg (2002).
DLM, double layer model; CCM, constant capacitance model.

Appendix B

Table B1 Model hydrodynamic parameters (from Zheng et al., 2013).

Parameter	Value
Porosity	0.3
Intrinsic permeability	$10^{-12}\,m^2$ (hydraulic conductivity of $\approx 0.86\,m/day$)
Relative permeability (van Genuchten-Mualem model)	$k_{rl} = \sqrt{S^*}\left\{1-\left(1-[S^*]^{1/0.457}\right)^{0.457}\right\}^2$
	with $S^* = (S_l - 0.3)/(1.0-0.3)$ where S_l is the liquid saturation degree
Gas relative permeability	$k_{rg} = (1-\hat{S})^2(1-\hat{S}^2)$ with $\hat{S} = (S_l - 0.3)/(1-0.3-0.05)$
Capillary pressure function	$P_{cap} = -\frac{1}{5.1\times10^{-5}}\left([S^*]^{-1/0.457}-1\right)^{1-0.457}$
	with $S^* = S_l/0.999$ and subject to restriction $-10^7 \le P_{cap} \le 0$
Molecular diffusion coefficient	$10^{-9}\,m^2/s$
Tortuosity	0.3

Table B2 Initial total aqueous concentrations in porewater of the modeled aquifer (Zheng et al., 2013).

Species	Concentration (mol/l)	Species	Concentration (mol/l)
Ca	9×10^{-4}	TIC	3.3×10^{-3}
Mg	2.2×10^{-5}	SO_4^{2-}	1.9×10^{-4}
Na	2×10^{-3}	Cl	2.1×10^{-4}
K	2.7×10^{-4}	Pb	1.3×10^{-9}
Fe	5.6×10^{-6}	As	4.4×10^{-8}
HS^-	1.4×10^{-8}	HSe^-	1.7×10^{-9}
Si	9.3×10^{-4}	Eh	−0.23 V
pH	7.6	Ionic strength	0.0051

Table B3 Modal mineralogy of aquifer host rocks used in reactive transport simulations.

Mineral	Chemical composition	Volume fraction
Quartz	SiO_2	0.689
K-feldspar	$KAlSi_3O_8$	0.061
Oligoclase	$CaNa_4Al_6Si_{14}O_{40}$	0.099
Kaolinite	$Al_2Si_2O_5(OH)_4$	0.017
Smectite-Ca	$Ca_{0.14}Mg_{0.26}Si_{3.97}Al_{1.77}O_{10}(OH)_2$	0.013
Illite	$K_{0.6}Mg_{0.25}Si_{3.5}Al_{2.3}O_{10}(OH)_2$	0.028
Chlorite	$Fe_{2.5}Mg_{2.5}Si_3Al_2O_{10}(OH)_8$	0.010
Kerogen-OS	$C_{64}H_{102}O_{40}S_{10}$	0.0026
Dolomite	$CaMg(CO_3)_2$	0.000
Calcite	$CaCO_3$	0.015
Goethite	FeOOH	0.06
Arsenian pyrite	$FeAs_{0.05}S_{1.95}$	0.0051
Galena	PbS	8×10^{-6}

Acknowledgements

This work was funded by the Assistant Secretary for Fossil Energy, National Energy Technology Laboratory, National Risk Assessment Partnership, of the US Department of Energy, under Contract No. DEAC02-05CH11231.

References

Aiuppa, A., Federico, C., Allard, P., et al. (2005). Trace metal modeling of groundwater-gas-rock interactions in a volcanic aquifer: Mount Vesuvius, Southern Italy. *Chemical Geology* 216: 289–311.

Amoore, J.E. and Hautala, E. (1983). Odor as an aid to chemical safety: Odor thresholds compared with threshold limit values and volatilities for 214 chemicals in air and water dilution. *Journal of Applied Toxicology* 3 (6): 272–290.

Anitescu, G. and Tavlarides, L.L. (2006). Supercritical extraction of contaminants from soils and sediments. *Journal of Supercritical Fluids* 38: 167–180.

Appelo, C.J.A. and Postma, D. (1994). *Geochemistry, Groundwater and Pollution.* Rotterdam, Netherlands: A.A. Balkema.

Apps, J.A., Zheng, L., Zhang, Y., et al. (2010). Evaluation of groundwater quality changes in response to CO_2 leakage from deep geological storage. *Transport in Porous Media* 82 (1): 215–246.

Arnórsson, S., Hurtig, N., Gysi, A.P., et al. (2010). Carbon dioxide waters in Iceland: A natural analogue to CO_2 sequestration in basaltic aquifers. *Conference Proceedings in Water–Rock Interaction, Guanajuato, Mexico* (ed. P. Birkle and I. Torres-Alvarado). London: Taylor & Francis.

Bachu, S. and Bennion, D.B. (2009). Chromatographic partitioning of impurities contained in a CO_2 stream injected into a deep saline aquifer: Part 1. Effects of gas composition and in situ conditions. *International Journal of Greenhouse Gas Control* 3 (4): 458–467.

Bachu, S., Pooladi-Darvish, M., and Hong, H.F. (2009). Chromatographic partitioning of impurities (H_2S) contained in a CO_2 stream injected into a deep saline aquifer: Part 2. Effects of flow conditions. *International Journal of Greenhouse Gas Control* 3 (4): 468–473.

Becker, M.F., Bruce, B.W., Pope, L.M., and Andrews, W.J. (2002). Ground–water quality in the Central High Plains Aquifer, Colorado, Kansas, New Mexico, Oklahoma, and Texas, 1999. *Water Resources Investigations Report, 02-4112*. US Geological Survey.

Birkholzer, J.T., Apps, J.A., Zheng, L., et al. (2008). Research Project on CO_2 Geological Storage and Groundwater Resources: Water Quality Effects Caused by CO_2 Intrusion into Shallow Groundwater. Lawrence Berkeley National Laboratory, Berkeley, CA.

Blanc, P., Lassin, A., Piantone, P., et al. (2012). Thermoddem: A geochemical database focused on low temperature water/rock interactions and waste materials. *Applied Geochemistry* 27: 2107–2116.

Cahill, A.G. and Jakobsen, R. (2013). Hydro-geochemical impact of CO_2 leakage from geological storage on shallow potable aquifers: A field scale pilot experiment. *International Journal of Greenhouse Gas Control* 19: 678–688.

Carroll, S.A., Hao, Y., and Aines, R. (2009). Geochemical detection of carbon dioxide in dilute aquifers. *Geochemical Transactions* 10 (4): 1–18.

Carroll, S.A., Mansoor, K., and Sun, Y. (2014a). Second-generation reduced-order model for calculating groundwater impacts as a function of pH, total dissolved solids, and trace metal concentration. *NRAP-TRS-III-002*, NRAP Technical Report Series. US Department of Energy, National Energy Technology Laboratory, Morgantown, WV.

Carroll, S.A., Keating, E., Mansoor, K., et al. (2014b). Key factors for determining groundwater impacts due to leakage from geologic carbon sequestration reservoirs. *International Journal of Greenhouse Gas Control* 29: 153–168.

Dai, Z., Keating, E., Bacon, D., et al. (2014). Probabilistic evaluation of shallow groundwater resources at a hypothetical carbon sequestration site. *Scientific Reports* 4: 4006.

Deeds, N.E., McKinney, D.C., and Pope, G.A. (2000). Laboratory characterization of non-aqueous phase liquid/tracer interaction in support of a vadose zone partitioning interwell tracer test. *Journal of Contaminant Hydrology* 41: 193–204.

Dixit, S. and Hering, J.G. (2003). Comparison of arsenic(V) and arsenic(III) sorption onto iron oxide minerals: Implications for arsenic mobility. *Environmental Science and Technology* 37: 4182–4189.

Donahue, R.B., Barhour, S.L., and Headley, J.V. (1999). Diffusion and adsorption of benzene in Regina clay. *Canadian Geotechnical Journal* 36 (3): 430–442.

EPA (1999). *Anaerobic Biodegradation Rates of Organic Chemical in Groundwater: a Summary of Field and Laboratory Studies*. Washington, DC: US Environmental Protection Agency, Office of Solid Waste.

Flaathen, T.K., Gislason, S.R., Oelkers, E.H., and Sveinbjörnsdóttir, Á.E. (2009). Chemical evolution of the Mt. Hekla, Iceland, groundwaters: A natural analogue for CO_2 sequestration in basaltic rocks. *Applied Geochemistry* 24 (3): 463–474.

Goldberg, S. (2002). Competitive adsorption of arsenate and arsenite on oxides and clay minerals. *Soil Science Society of America Journal* 66: 413–421.

Gu, X. and Evans, L.J. (2007). Modelling the adsorption of Cd(II), Cu(II), Ni(II), Pb(II), and Zn(II) onto Fithian illite. *Journal of Colloid and Interface Science* 307: 317–325.

Gu, X. and Evans, L.J. (2008). Surface complexation modelling of Cd(II), Cu(II), Ni(II), Pb(II) and Zn(II) adsorption onto kaolinite. *Geochimica et Cosmochimica Acta* 72: 267–276.

Gu, X., Evans, L.J., and Barabash, S.J. (2010). Modeling the adsorption of Cd (II), Cu (II), Ni (II), Pb (II) and Zn (II) onto montmorillonite. *Geochimica et Cosmochimica Acta* 74, 5718–5728.

Gunter, W.D., Perkins, E.H., and Hutcheon, I. (2000). Aquifer disposal of acid gases: modeling of water-rock reactions for trapping of acid wastes. *Applied Geochemistry* 15 (8): 1085–1095.

Harvey, O.R., Qafoku, N.P., Cantrell, K.J., et al. (2012). Geochemical implications of gas leakage associated with geologic CO_2 storage, a qualitative review. *Environmental Science and Technology* 47: 23–36.

Hawthorne, S.B. and Miller, D.J. (2003). Evidence for very tight sequestration of BTEX compounds in manufactured gas plant soils based on selective supercritical fluid extraction and soil/water partitioning. *Environmental Science and Technology* 37: 3578–3594.

Helz, G.R. and Tossell, J.A. (2008). Thermodynamic model for arsenic speciation in sulfidic waters: a novel use of ab initio computations. *Geochimica et Cosmochimica Acta* 72 (18): 4457–4468.

Humez, P., Lagneau, V., Lions, J., and Negrel, P. (2013). Assessing the potential consequences of CO_2 leakage to freshwater resources: A batch-reaction experiment towards an isotopic tracing tool. *Applied Geochemistry* 30: 178–190.

Illiassov, P.A. and Datta-Gupta, A. (2002). Field-scale characterization of permeability and saturation distribution using partitioning tracer tests: the Ranger Field, Texas. *SPE Journal* 7 (4): 409–420.

Jean, J.-S., Tsai, C.-L., Ju, S.-H., et al. (2002). Biodegradation and transport of benzene, toluene, and xylenes in a simulated aquifer: comparison of modelled and experimental results. *Hydrological Processes* 16: 3151–3168.

Karamalidis, A.K., Torres, S.G., Hakala, J.A., et al. (2013). Trace metal source terms in carbon sequestration environments. *Environmental Science and Technology* 47 (1): 322–329.

Karickhoff, S.W., Brown, D.S., and Scott, T.A. (1979). Sorption of hydrophobic pollutants on natural sediments. *Water Research* 13 (3): 241–248.

Keating, E.H., Fessenden, J., Kanjorski, N., et al. (2010). The impact of CO_2 on shallow groundwater chemistry: observations at a natural analog site and implications for carbon sequestration. *Environmental Earth Sciences* 60 (3): 521–536.

Kharaka, Y.K. and J.S. Hanor, 2007. Deep fluids in the continents: I. sedimentary basins. In: J.I. Drever (Editor), Surface and Ground Water, Weathering and Soils. Elsevier, Treatise on Geochemistry, pp. 1–48.

Kharaka, Y.K., Cole, D.R., Hovorka, S.D., et al. (2006). Gas-water-rock interaction in Frio Formation following CO_2 injection: implications for the storage of greenhouse gases in sedimentary basins. *Geology* 34: 577–580.

Kharaka, Y.K., Thordsen, J.J., Hovorka, S.D., et al. (2009). Potential environmental issues of CO_2 storage in deep saline aquifers: geochemical results from the Frio-I Brine Pilot test, Texas, USA. *Applied Geochemistry* 24: 1106–1112.

Kharaka, Y.K., Thordsen, J.J., Kakouros, E., et al. (2010). Changes in the chemistry of shallow groundwater related to the 2008 injection of CO_2 at the ZERT Field Site, Bozeman, Montana. *Environmental Earth Sciences* 60 (2): 273–284.

Kirsch, K., Navarre-Sitchler, A.K., Wunsch, A., and McCray, J.E. (2014). Metal release from sandstones under experimentally and numerically simulated CO_2 leakage conditions. *Environmental Science and Technology* 48: 1436–1442.

Knauss, K.G., Johnson, J.W., and Steefel, C.I. (2005). Evaluation of the impact of CO_2, co-contaminant gas, aqueous fluid and reservoir rock interactions on the geologic sequestration of CO_2. *Chemical Geology* 217: 339–350.

Lammers, K., Murphy, R., Riendeau, A., et al. (2011). CO_2 sequestration through mineral carbonation of iron oxyhydroxides. *Environmental Science and Technology* 45 (24): 10422–10428.

Larsen, T., Kjeldsen, P., and Christensen, T.H. (1992). Correlation of benzene, 1,1,1-trichloroethane, and naphthalene distribution coefficient to the characteristics of aquifer materials with low organic carbon content. *Chemosphere* 24 (8): 979–991.

Lawter, A., Qafoku, N.P., Wang, G., et al. (2016). Evaluating impacts of CO_2 intrusion into an unconsolidated aquifer: I. Experimental data. *International Journal of Greenhouse Gas Control* 44: 323–333.

Lemieux, J.-M. (2011). Review: The potential impact of underground geological storage of carbon dioxide in deep saline aquifers on shallow groundwater resources. *Hydrogeology Journal* 19 (4): 757–778.

Little, M.G. and Jackson, R.B. (2010). Potential impacts of leakage from deep CO_2 geosequestration on overlying freshwater aquifers. *Environmental Science and Technology* 44 (23): 9225–9232.

Lu, J.M., Partin, J.W., Hovorka, S.D., and Wong, C. (2010). Potential risks to freshwater resources as a result of leakage from CO_2 geological storage: A batch-reaction experiment. *Environmental Earth Sciences* 60 (2): 335–348.

McMahon, P.B., Böhlke, J.K., and Christenson, S.C. (2004). Geochemistry, radiocarbon ages, and paleorecharge conditions along a transect in the central High Plains aquifer, southwestern Kansas, USA. *Applied Geochemistry*, 19 (11): 1655–1686.

McMahon, P.B., Böhlke, J.K., and Carney, C.P. (2007). Vertical gradients in water chemistry and age in the northern High Plains Aquifer, Nebraska, 2003. *US Geological Survey Scientific Investigations Report 2006–5294.*

Murphy, R., Lammers, K., Smirnov, A., et al. (2010). Ferrihydrite phase transformation in the presence of aqueous sulfide and supercritical CO_2. *Chemical Geology* 271 (1–2): 26–30.

Murphy, R., Lammers, K., Smirnov, A., et al. (2011). Hematite reactivity with supercritical CO_2 and aqueous sulfide. *Chemical Geology* 283 (3–4): 210–217.

Palandri, J. and Kharaka, Y.K. (2004). A compilation of rate parameters of water-mineral interaction kinetics for application to geochemical modeling. Open File Report 2004-1068, US Geological Survey.

Palandri, J.L. and Kharaka, Y.K. (2005). Ferric iron-bearing sediments as a mineral trap for CO_2 sequestration: Iron reduction using sulfur-bearing waste gas. *Chemical Geology* 217: 351–364.

Palandri, J.L., Rosenbauer, R.J., and Kharaka, Y.K. (2005). Ferric iron in sediments as a novel CO_2 mineral trap: CO_2–SO_2 reaction with hematite. *Applied Geochemistry* 20: 2038–2048.

Pawar, R., Bromhal, G., Carroll, S., et al. (2014). Quantification of key long-term risks at CO_2 sequestration sites: Latest results from US DOE's National Risk Assessment Partnership (NRAP) Project. *Energy Procedia* 63: 4816–4823.

Peter, A., Lamert, H., Beyer, M., et al. (2012). Investigation of the geochemical impact of CO_2 on shallow groundwater: design and implementation of a CO_2 injection test in Northeast Germany. *Environmental Earth Sciences* 67 (2): 335–349.

Poulton, S.W. (2003). Sulfide oxidation and iron dissolution kinetics during the reaction of dissolved sulfide with ferrihydrite. *Chemical Geology* 202 (1–2): 79–94.

Poulton, S.W., Krom, M.D., and Raiswell, R. (2004). A revised scheme for the reactivity of iron (oxyhydr)oxide minerals towards dissolved sulfide. *Geochimica et Cosmochimica Acta* 68 (18): 3703–3715.

Pruess, K. (2005). *ECO2N: A TOUGH2 Fluid Property Module for Mixture of Water, NaCl, and CO_2*. Berkeley, CA: Lawrence Berkeley National Laboratory.

Pruess, K. and Battistelli, A. (2002). *TMVOC, a Numerical Simulator for Three-Phase Non-Isothermal Flows of Multicomponent Hydrocarbon Mixtures in Saturated–Unsaturated Heterogeneous Media*. Berkeley, CA: Lawrence Berkeley National Laboratory.

Pruess, K., Oldenburg, C., and Moridis, G. (1999). *TOUGH2 User's Guide, Version 2.0*. Berkeley, CA: Lawrence Berkeley National Laboratory.

RøeUtvik, T.I. (1999). Chemical characterization of produced water from four offshore oil production platforms in the North Sea. *Chemosphere* 39: 2593–2606.

Smyth, R.C., Hovorka, S.D., Lu, J., et al. (2009). Assessing risk to fresh water resources from long term CO_2 injection-laboratory and field studies. *Energy Procedia* 1 (1): 1957–1964.

Spycher, N. and Oldenburg, C. (2015). Will mercury impurities impact CO_2 injectivity in deep sedimentary formations? II. Mineral dissolution and precipitation. *Greenhouse Gases: Science and Technology* 5 (1): 72–90.

Spycher, N. and Pruess, K. (2005). CO_2-H_2O mixtures in the geological sequestration of CO_2. II. Partitioning in chloride brines at 12–100°C and up to 600 bar. *Geochimica et Cosmochimica Acta* 69 (13): 3309–3320. doi:3310.1016/j.gca.2005.3301.3015

Swedlund, P.J., Webster, J.G., and Miskelly, G.M. (2009). Goethite adsorption of Cu(II), Pb(II), Cd(II), and Zn(II) in the presence of sulfate: Properties of the ternary complex. *Geochimica et Cosmochimica Acta* 73: 1548–1562.

Tang, J.S. (2005). Extended Brigham model for residual oil saturation measurement by partitioning tracer tests. *SPE Journal* 10 (2): 175–183.

Trautz, R.C., Pugh, J.D., Varadharajan, C., et al. (2013). Effect of dissolved CO_2 on a shallow groundwater system: A controlled release field experiment. *Environmental Science and Technology* 47 (1): 298–305.

Varadharajan, C., Tinnacher, R., Pugh, J., et al. (2013). A laboratory study of the initial effects of dissolved carbon dioxide (CO_2) on metal release from shallow sediments. *International Greenhouse Gas Control* 19: 183–211.

Viswanathan, H., Dai, Z., Lopano, C., et al. (2012). Developing a robust geochemical and reactive transport model to evaluate possible sources of arsenic at the CO_2 sequestration natural analog site in Chimayo, New Mexico. *International Journal of Greenhouse Gas Control* 10: 199–214.

Wang, S. and Jaffe, P.R. (2004). Dissolution of a mineral phase in potable aquifers due to CO_2 releases from deep formations; Effect of dissolution kinetics. *Energy Conversion and Management* 45: 2833–2848.

Wei, Y., Maroto-Valer, M., and Steven, M.D. (2011). Environmental consequences of potential leaks of CO_2 in soil. *Energy Procedia* 4: 3224–3230.

Wilkin, R.T. and Digiulio, D.C. (2010). Geochemical impacts to groundwater from geologic carbon sequestration: Controls on pH and inorganic carbon concentrations from reaction path and kinetic modeling. *Environmental Science and Technology* 44 (12): 4821–4827.

Witter, A.E. and Jones, A.D. (1999). Chemical characterization of organic constituents from sulfide-rich produced water using gas chromatography/mass spectrometry. *Environmental Toxicology and Chemistry* 18: 1920–1926.

Wolery, T. and Jove-Colon, C. (2007). *Qualification of Thermodynamic Data for Geochemical Modeling of Mineral-Water Interactions in Dilute Systems.* ANL-WIS-GS-000003 REV 01. Las Vegas, Nevada: Sandia National Laboratories. ACC: DOC.20070619.0007.

Wunsch, A., Navarre-Sitchler, A.K., Moore, J., and McCray, J.E. (2014). Metal release from limestones at high partial-pressures of CO_2. *Chemical Geology* 363: 40–55.

Xu, T., Sonnenthal, E., Spycher, N., and Pruess, K. (2006). TOUGHREACT: A simulation program for non-isothermal multiphase reactive geochemical transport in variably saturated geologic media. *Computers and Geosciences* 32: 145–165.

Xu, T., Apps, J.A., Pruess, K., and Yamamoto, H. (2007). Numerical modeling of injection and mineral trapping of CO_2 with H_2S and SO_2 in a sandstone formation. *Chemical Geology* 242: 319–346.

Xu, T., Spycher, N., Sonnenthal, E., et al. (2011). TOUGHREACT Version 2.0: A simulator for subsurface reactive transport under non-isothermal multiphase flow conditions. *Computers and Geosciences* 37 (6): 763–774.

Yang, C., Mickler, P.J., Reedy, R., et al. (2013). Single-well push–pull test for assessing potential impacts of CO_2 leakage on groundwater quality in a shallow Gulf Coast aquifer in Cranfield, Mississippi. *International Journal of Greenhouse Gas Control* 18: 375–387.

Zheng, L., Apps, J.A., Zhang, Y., et al. (2009). On mobilization of lead and arsenic in groundwater in response to CO_2 leakage from deep geological storage. *Chemical Geology* 268 (3–4): 281–297.

Zheng, L., Spycher, N., Birkholzer, J.T., et al. (2010). *Modeling Studies on the Transport of Benzene and H_2S in CO_2-Water Systems.* Technical Report LBNL-4339E. Berkeley, CA: Lawrence Berkeley National Laboratory.

Zheng, L., Apps, J.A., Spycher, N., et al. (2012). Geochemical modeling of changes in shallow groundwater chemistry observed during the MSU-ZERT CO_2 injection experiment. *International Journal of Greenhouse Gas Control* 7: 202–217.

Zheng, L., Spycher, N., Birkholzer, J., et al. (2013). On modeling the potential impacts of CO_2 sequestration on shallow groundwater: Transport of organics and co-injected H_2S by supercritical CO_2 to shallow aquifers. *International Journal of Greenhouse Gas Control* 14: 113–127.

Zheng, L., Carroll, S., Bianchi, M., et al. (2014). Reduced order models for prediction of groundwater quality impacts from CO_2 and brine leakage. *Energy Procedia* 63: 4875–4883.

Zheng, L., Spycher, N., Varadharajan, C., et al. (2015). On the mobilization of metals by CO_2 leakage into shallow aquifers: Exploring release mechanisms by modeling field and laboratory experiments. *Greenhouse Gases: Science and Technology* 5: 1–16.

Zheng, L., Qafoku, N.P., Lawter, A., et al. (2016). Evaluating impacts of CO_2 intrusion into an unconsolidated aquifer: II. Modeling results. *International Journal of Greenhouse Gas Control* 44: 300–309.

Zhong, L., Cantrell, K.J., Bacon, D.H., and Shewell, J. (2014). Transport of organic contaminants mobilized from coal through sandstone overlying a geological carbon sequestration reservoir. *International Journal of Greenhouse Gas Control* 21: 158–164.

8

Modeling the Long-term Stability of Multi-barrier Systems for Nuclear Waste Disposal in Geological Clay Formations

Francis Claret[1], Nicolas Marty[1] and Christophe Tournassat[1,2,3]

[1] BRGM, 3 Avenue Guillemin, 45060, Orléans Cedex 2, FRANCE
[2] Université d'Orléans - CNRS/INSU-BRGM, UMR 7327 Institut des Sciences de la Terre d'Orléans (ISTO), 45071, Orléans, FRANCE
[3] Energy Geoscience Division, Lawrence Berkeley National Laboratory, 1 Cyclotron Rd., Berkeley, CA 94720, USA

8.1 Introduction

8.1.1 Geological Final Disposal of Radioactive Waste

Large amounts of nuclear waste await final disposal worldwide. Repository facilities that will host this waste must be capable of ensuring very long term isolation to protect the environment and ensure the safety of the future generations (Linsley and Fattah, 1994; Hummel and Schneider, 2005). Because of this need, the concept of geological disposal of radioactive waste emerged in the late 1950s (Hess, 1957; de Marsily et al., 1977; Apted and Ahn, 2010; Chapman and Hooper, 2012). Ambitious research programmes have since been launched to study deep argillaceous formations, granitic rocks and salt formations as potential host rocks for radioactive waste disposal (Landais and Aranyossy, 2011). Geological disposal is now considered, in many countries, to be the best option to ensure safe, long-term containment of radioactive waste (National Research Council, 2001).

A combination of waste overpacks (e.g. metal canisters, concrete), engineered barriers such as bentonite (Sellin and Leupin, 2013), and natural barriers such as clay rocks, constitute the elements of the so-called 'multiple-barrier system' between the waste matrix and the biosphere (Apted and Ahn, 2010; Chapman and Hooper, 2012). The number, types and assigned safety functions of these various barriers depend on the chosen repository concept, the waste form, the radionuclide inventory in the waste, the selected host rock, the hydrogeological and geochemical settings of the repository site, and so on (Apted and Ahn, 2010). These barrier properties will evolve with time in response to the physical and chemical interactions between the various constituents of the barriers and the surrounding environment. Consequently, predicting how these properties evolve is of prime importance for performance and safety evaluations of the repository concepts. This chapter deals with the way the evolution of barrier properties with time has been predicted until now, with a focus on repository concepts in clay formations.

Reactive Transport Modeling: Applications in Subsurface Energy and Environmental Problems, First Edition. Edited by Yitian Xiao, Fiona Whitaker, Tianfu Xu and Carl Steefel.

8.1.2 The 'Clay Concept'

The water and oil industries consider sedimentary clay layers to be impermeable in the engineering sense. Consequently, argillaceous sedimentary rocks have been identified as potential host formations for 'radwaste' repositories, for which their low transmissivity should provide at least part of the basis for isolation of radionuclide contaminants from the biosphere. Clay minerals, the main constituents of these rocks, also have other properties of interest, namely their very high adsorption capacity. Clay minerals strongly adsorb and therefore slow many contaminants, so are ideal for use in barrier systems, particularly where there is a desire to improve confidence in the safety case beyond reliance on slower transport rates alone (Tournassat et al., 2015a). In addition, clay materials have interesting mechanical properties, such as fracture self-sealing (Bastiaens et al., 2007; Van Geet et al., 2008). For all the reasons mentioned, clay-based materials are foreseen as buffer and backfill materials in almost all repository concepts (Pusch, 2010).

8.1.3 How a Repository System Evolves in Time and Space

In repository facilities, the construction and exploitation phases and the phase when foreign materials like silicate glasses (Gin et al., 2015; Poinssot et al., 2010), metallic canisters (King and Shoesmith, 2010), and concrete (Alonso et al., 2010) are introduced will induce chemical gradients across the repository constituents (Andra, 2005, 2009; Nagra, 2002). Because of these chemical gradients, perturbations such as pH and redox changes may alter the performance of the barriers over time (Bildstein and Claret, 2015). Therefore, in the last two decades many in the scientific community have extensively studied how properties of clay barriers and of waste overpacks evolve during the construction and exploitation phases in repository facilities (see, for example, Ahn and Apted, 2010, and Tournassat et al., 2015a, for a review). Although exact phenomenological analysis of repository situations depends on the repository concept under investigation, some general statements can be made about the most relevant and identified phenomena that will occur during a repository system's lifetime (Plas and Landais, 2008). These phenomena can be segmented in space and time. To understand how a repository system will evolve, as a minimum, thermal, hydraulic, mechanical, chemical and radiological processes must be taken into account, especially for its near-field. The near-field refers to the portion of the host rock that is significantly affected by repository construction and waste emplacement, in opposition to the far-field, which describes the unaffected portion separating the near-field from the biosphere (Zhou and Arthur, 2010). A non-exhaustive list of these processes is given below, to show how complex and intimately connected some of them are.

During the operation period (repository construction and waste emplacement), which will last for approximately a hundred years in current scenarios (Plas and Landais, 2008), the ventilation systems used to ensure adequate gallery aeration for the workers and for the infrastructure will create heat and vapour exchanges with walls and storage packages, pressure variations and evaporation, and sometimes vapour condensation (Benet et al., 2014b). The heat released from the waste packages will be partly evacuated by air ventilation and partly dissipated in the over-packs and the geological medium (Benet et al., 2014a). During construction, excavation of underground drifts will cause both tunnel convergence (Lisjak et al., 2015) and damage to the rock in the vicinity of the opening, with the formation of an associated excavated disturbed zone (EDZ) (Armand et al., 2014). The change in rock microstructure in the EDZ compared with pristine rock might impact radionuclide transport parameters. As a result of excavation and ventilation, oxidation and desaturation processes will also occur in the storage's near-field (Matray et al., 2007; Craen et al., 2008; Vinsot et al., 2013, 2014; De Windt et al., 2014). Redox conditions in the rock will indeed be modified by this oxidation event. During the operation period, the disposal cells, once filled with waste packages, could be backfilled with cement-based materials and/or bentonite depending on the nature of the waste and the concept chosen (Nagra, 2002; Andra, 2009). Rehydration of the near-field rocks and the backfill materials will start after the repository closes and could last for 100 000 years (Andra, 2005). Meanwhile, gas production inside the repository (e.g. H_2 production by anaerobic corrosion of metals or through water radiolysis) will drive pressure changes in the near-field, and may delay material rehydration (Benet et al., 2014c). In addition, it is expected that processes such as corroding metals present in the containers and infrastructure material (e.g. reinforced concrete), and degrading organic matter from bitumen waste, for example, and also microbial activity, could affect the how redox properties evolve long-term (Duro et al., 2014a,b). These combined processes may be detrimental or favourable to the overall performance of a repository system over time. Since the combinations will occur on timescales that are not accessible to experimentation, we must develop modeling approaches that can help to predict how barriers will evolve in time and space.

8.1.4 Modeling How a Repository System Evolves

The containment, transport, and mechanical properties of barriers depend in large part on the mineralogy and chemistry of their constituent materials. For example, the swelling properties of clay barriers are linked to the nature and abundance of their constituent clay minerals, with smectite being favourable and illite unfavourable to swelling (Taylor and Smith, 1986). In turn, the presence of illite facilitates the retention of radioactive cesium (Cs) (Poinssot et al., 1999). The abundance and nature of clay minerals also impacts the permeability and diffusivity of clay materials (Jacquier et al., 2013; Bourg, 2015; Bourg and Tournassat, 2015; Gonçalvès et al., 2015). For these reasons, a great deal of attention has been paid to the stability of clay materials in response to the predicted perturbations arising from how a repository system evolves.

In recent decades, reactive transport modeling codes have become more and more efficient. Nowadays they can handle both complex geometries and complex chemistry while taking into account spatial heterogeneities and transient conditions (Steefel et al., 2015a). For these reasons, reactive transport modelling has established itself as a very useful tool in assessing the performance and safety of the various radioactive waste disposal concepts. A broad range of physical and chemical interactions between the various stored materials has already been modeled and published (Table 8.1). In what follows, we will describe first the basic principles that have guided how these models are constructed for establishing and evaluating the long-term geochemical behaviour of repository systems. We will especially focus on the generic conceptualization of these calculations, on how materials are described and represented, and on the importance of databases for the outcome of the simulations. In the second section, we will review the

Table 8.1 List of the large-scale reactive transport modeling studies published in the framework of clay repository systems.

Reference	Code	Database	Geometry	Dimensions	Conditions	Porosity feedback
Cement-based and clay-based material interactions						
Savage et al. (2002)	PRECIP (code description within the publication)	Mostly EQ3/6 (Wolery, 1983)	1D Cartesian grid (1 cm)	1 m	Water-saturated conditions	Yes
Soler (2003)	GIMRT (Steefel and Yabusaki, 1996)	Mostly EQ3/6 (Wolery, 1983)	2D Cartesian grid (10 cm by 5 cm)	6 × 1 m	Water-saturated conditions	Yes
De Windt et al. (2004a)	HYTEC (van der Lee et al., 2003)	Mostly EQ3/6 (Wolery, 1983)	1D Cartesian grid (5 cm)	20 m	Water-saturated conditions	Yes
De Windt et al. (2004b)	HYTEC (van der Lee et al., 2003)	Mostly EQ3/6 (Wolery, 1983)	2D with cylindrical coordinates	35 m length, 15 m height	Water-saturated conditions	Not defined
Gaucher et al. (2004b)	PHREEQC (Parkhurst and Appelo, 1999)	Mostly EQ3/6 (Wolery, 1983)	1D Cartesian grid (2.5 cm)	8 m	Water-saturated conditions	No

Temp. (°C)	Simulated time	Reaction processing	Surface reactions	Materials	Transport properties De in $m^2 s^{-1}$ Permeability unit depends on the authors
25 and 70	320 000 yr	Kinetic assumption	Not taken into account	Hyperalkaline solutions (pH 13.2, pH 12.7 and pH 11.2 at 25°C)	
				Bentonite	De = 10^{-10} and Ø = 0.4
25	5000–20 000 yr	Kinetic assumption	Not taken into account	Hyperalkaline solutions (pH 13.5 and pH 12.5 at 25°C)	
				Fractured marl	Advection ($3.3 \times 10^{-13} - 4.0 \times 10^{-10}$ m s^{-1}) and molecular diffusion 10^{-9} $m^2 s^{-1}$ Ø = 0.01 in the marl zone and 0.1 in the fractured region Permeabilities are updated according to Kozeny's equation
20	100 000 yr	Local equilibrium assumption	Cation exchange and surface complexation	Portland cement (pH 13.3 at 20°C)	De = 2×10^{-11} $\phi = 0.2$
				Argillaceous formation	De = 1.5×10^{-11} and Ø = 0.15 for pristine host-rock whereas De = 3.5×10^{-10} and Ø = 0.35 in the EDZ
70	100 000 yr	Local equilibrium assumption	Surface complexation and cation exchange for clay		ϕ ; $K_{\parallel} = K_{\perp}$ ms^{-1} ; D_e
				Waste package	0.4 ; 10^{-11} ; 2×10^{-10}
				MX80 bentonite	0.36 ; 10^{-13} ; 1.8×10^{-10}
				Portland cement liner (pH 13.2 at 25°C)	0.3 ; 10^{-11} ; 1.5×10^{-10}
				Portland cement wall (pH 13.2 at 25°C)	0.2 ; 10^{-11} ; 2×10^{-11}
				Clay-rock	0.12 ; 10^{-13} ; $1.2 \times 10^{-}$
				EDZ-I case A	0.35 ; 5×10^{-11} ; 3.5×10^{-10}
				EDZ-I case B	0.35 ; 5×10^{-10} ; 1.5×10^{-10}
				EDZ-I case C	0.35 ; 5×10^{-9} ; 1.5×10^{-10}
				EDZ-II case A	0.2 ; 5×10^{-12} ; 5×10^{-11}
				EDZ-II case B	0.2 ; 5×10^{-11} ; 5×10^{-11}
				EDZ-II case C	0.2 ; 5×10^{-10} ; 5×10^{-11}
				Handling drift	0.35 ; 5×10^{-10} ; 3.5×10^{-10}
				Major fracture	0.8 ; 10^{-8} ; 8×10^{-10}
25	100 000 yr	Local equilibrium assumption	Cation exchange	Hyperalkaline solutions (pH 12.5 at 25°C)	
				MX80 bentonite	De = 10^{-11} and Ø = 0.4

(Continued)

Table 8.1 (Continued)

Reference	Code	Database	Geometry	Dimensions	Conditions	Porosity feedback
Burnol et al. (2006)	TOUGHREACT (Xu et al., 2006)	Mostly EQ3/6 (Wolery, 1983)	1D radial grid	100 m	Unsaturated conditions (saturation state evolves)	No
Trotignon et al. (2006)	HYTEC (van der Lee et al., 2003)	Derived from EQ3/6 (Wolery, 1983)	1D radial grid (10 cm)	5.5 m	Water-saturated conditions	Yes
Trotignon et al. (2006)	HYTEC (van der Lee et al., 2003)	Derived from EQ3/6 (Wolery, 1983)	1D radial grid (10 cm)	17.5 m	Water-saturated conditions	Yes
Montarnal et al. (2007)	ALLIANCES (PHREEDC/ MT3D) code description within the publication	Not given	2D cartesian	60 m height, length is not given	Saturated condition	Not mentioned
Trotignon et al. (2007)	HYTEC (van der Lee et al., 2003)	Derived from EQ3/6 (Wolery, 1983)	1D radial grid (10 cm)	5.5 m	Water-saturated conditions	Yes (depending on the simulated case)
Trotignon et al (2007)	HYTEC (van der Lee et al., 2003)	Derived from EQ3/6 (Wolery, 1983)	1D radial grid (10 cm)	17.5 m	Water-saturated conditions	Yes (depending on the simulated case)
Yang et al. (2008)	CORE^{2D} V4 (Samper et al., 2003)	EQ3/6 (Wolery, 1983)	1D axisymmetric (16.6 cm)	25 m	Unsaturated conditions (saturation state evolves)	Not defined

Temp. (°C)	Simulated time	Reaction processing	Surface reactions	Materials	Transport properties De in $m^2 s^{-1}$ Permeability unit depends on the authors
Non-isothermal (max. 70)	2000 yr	Local equilibrium assumption	Not taken into account	Portland cement	$De = 10^{-10}$ and $\varnothing = 0.144$ base-case scenario $1/\alpha = 2 \times 10^{6}$ Pa $n = 1.46$
				Callovo-Oxfordian argillite	$De = 10^{-10}$ and $\varnothing = 0.1$ base-case scenario $1/\alpha = 15 \times 10^{6}$ Pa $n = 1.46$
25	500 000 yr	Local equilibrium assumption	Not taken into account	Portland cement paste (CEM I) and blended Portland cement paste (CEM V) (pH 13.24 at 25°C)	$De = 2.5 \times 10^{-12}$, $\varnothing = 0.27$ for the CEM I $De = 1.3 \times 10^{-13}$, $\varnothing = 0.33$ for the CEM V
				Interstitial Callovo-Oxfordian porewater	
25	200 000 yr	Local equilibrium assumption	Not taken into account	Portland cement (CEM I) and blended Portland (CEM V) (pH 13.24 at 25°C)	$De = 2.5 \times 10^{-1}$, $\varnothing = 0.27$ for the CEM I $De = 1.3 \times 10^{-13}$, $\varnothing = 0.33$ for the CEM V
				Callovo-Oxfordian argillite	$De = 1.0 \times 10^{-11}$ and $\varnothing = 0.33$
25	100 000 yr	Local equilibrium	Cation exchange	Bentonite	$De = 10^{-11}$. and $\varnothing$ is not given
				Concrete	$De = 10^{-11}$. and $\varnothing$ is not given
				Clay-rock	$De = 2.6 \times 10^{-11}$ and $\varnothing$ is not given
				EDZ	$De = 5.2 \times 10^{-11}$ and $\varnothing$ is not given
25	1 000 000 yr	Local equilibrium and kinetic approaches (depending on the simulated case)	Not taken into account	Portland cement (pH 13.24 at 25°C)	$De = 1.26 \times 10^{-12}$ (waste package)–2.52×10^{-12} $m^2 s^{-1}$ (Engineered barrier concrete) and $\varnothing = 0.09$
				Interstitial Callovo-Oxfordian porewater	
25	300 000 yr (only results at 100 000 yr are presented)	Local equilibrium and kinetic approaches (depending on the simulated case)	Cation exchange (depending on the simulated case)	Portland cement (pH 13.24 at 25°C)	$De = 1.26 \times 10^{-12}$ (waste package) $De = 2.52 \times 10^{-12}$ (Engineered barrier concrete) and $\varnothing = 0.09$
				Callovo-Oxfordian argillite	$De = 10^{-11}$ (normal case) $De = 10^{-12}$ (sensitivity case) $\varnothing = 0.15$ in both cases
Time-varying temperature function imposed upon the inner boundary	1 000 000 yr	Local equilibrium assumption	Cation exchange		ϕ K (ms^{-1}) D_e
				Bentonite	0.407 2.75×10^{-14} 1.66×10^{-11}
				Concrete (pH 13.25)	0.085 3.39×10^{-12} 1.83×10^{-10}
				Clay-rock	0.37 4.19×10^{-12} 5.01×10^{-11}

(Continued)

Table 8.1 (Continued)

Reference	Code	Database	Geometry	Dimensions	Conditions	Porosity feedback
Fernandez et al. (2009)	CRUNCH (Steefel et al., 2015a)	Mostly EQ3/6 (Wolery, 1983)	1D Cartesian grid	1.8 m	Water-saturated conditions	No
Marty et al. (2009)	TOUGHREACT (Xu et al., 2004)	THERMODDEM (Wolery, 1983)	1D Cartesian grid (1 mm to 10 cm)	10 m	Water-saturated conditions	Yes
Trotignon et al. (2011)	TOUGHREACT (Xu et al., 2004)	THERMODDEM (Wolery, 1983)	1D radial (from 5 mm close to air concrete boundary to 1 m)	40 m	Unsaturated conditions (saturation state evolves)	Yes
Berner et al. (2013)	OpenGeoSys-GEM (Kolditz et al., 2012)	CEMDATA2007 (Lothenbach and Wieland, 2006; Lothenbach and Winnefeld, 2006) and GEMS-PSI version by of the reference Nagra/PSI (Hummel et al., 2002) database and supplemented with newly evaluated thermodynamic data on montmorillonite	Axisymmetric 1D model (10 mm and enlarged toward the boundaries)	5.6 m	Water-saturated conditions	Yes

Temp. (°C)	Simulated time	Reaction processing	Surface reactions	Materials	Transport properties De in $m^2 s^{-1}$ Permeability unit depends on the authors
80	100 000 yr	Kinetic assumption	Cation exchange	Portland cement (CEM I)	$De = 7 \times 10^{-11}$ and $\varnothing = 0.1$
				FEBEX bentonite	$De = 7 \times 10^{-11}$ and $\varnothing = 0.4$
25	9600 to 389 120 yr	Local equilibrium and kinetic approaches	Not taken into account	Portland cement (pH 12.5 at 25°C)	$De = 1.5 \times 10^{-12}$ and $\varnothing = 0.15$
				Callovo-Oxfordian argillite	$De = 1.5 \times 10^{-11}$ and $\varnothing = 0.15$
Initially at 25°C and then submitted to hot (40°C) dry air at the concrete boundary. Non-isothermal	10 yr	Kinetic assumption	No	CEM I cement	$De = 9 \times 10^{-12}$ and $\varnothing = 0.13$ $1/\alpha = 2 \times 10^6$ Pa $n = 1.54$ $K_{rw}\,(m^2) = 9.5 \times 10^{-20}$ $K_{rg}\,(m^2) = 8 \times 10^{-17}$
				Fracturated EDZ	$De = 3.6 \times 10^{-11}$ and $\varnothing = 0.2$ $1/\alpha = 2.10^6$ Pa $n = 1.5$ $K_{rw}\,(m^2) = 9.2 \times 10^{-17}$ $K_{rg}\,(m^2) = 9.2 \times 10^{-15}$
				Microfissured EDZ	$De = 2.8 \times 10^{-11}$ and $\varnothing = 0.18$ $1/\alpha = 8 \times 10^6$ Pa $n = 1.5$ $K_{rw}\,(m^2) = 9.2 \times 10^{-19}$ $K_{rg}\,(m^2) = 9.2 \times 10^{-17}$
				Callovo-Oxfordian	$De = 2.8 \times 10^{-11}$ and $\varnothing = 0.18$ $1/\alpha = 1.5 \times 10^7$ Pa $n = 1.49$ $K_{rw}\,(m^2) = 4.6 \times 10^{-20}$ $K_{rg}\,(m^2) = 4.6 \times 10^{-18}$
25	30 000 yr	Local equilibrium or kinetic assumption	Cation exchange and neutral 'surface ions pairs'	Opalinus Clay	$\phi = 0.06$ $De = D_0\,\phi^2$ with $D_0 = 1.5 \times 10^{-9}$
				Bentonite	$\phi = 0.05$
				ESRED 'low pH' concrete (pH 11.07)	$\phi = 0.179$

(*Continued*)

Table 8.1 (Continued)

Reference	Code	Database	Geometry	Dimensions	Conditions	Porosity feedback
Kosakowski and Berner (2013)	OpenGeoSys-GEM (Kolditz et al., 2012)	CEMDATA2007 (Lothenbach and Wieland, 2006; Lothenbach and Winnefeld, 2006) and GEMS-PSI version by of the reference Nagra/PSI (Hummel et al., 2002) database and supplemented with newly evaluated thermodynamic data on montmorillonite	Linear 1D setup (10 mm and enlarged toward the boundaries)	46 m in the diffusive case, 51.5 m in the advective case (but not reported here)	Water-saturated conditions	Yes except for the advective case
Liu et al. (2014c)	A diffusive transport solver coupled with PHREEQC	CEMDATA2007 (Lothenbach and Wieland, 2006; Lothenbach and Winnefeld, 2006)	1D radial Grid (mesh size in the range 200 μm to 2 cm depending on the case)	1.01 m	Water-saturated conditions	Yes
Marty et al. (2014)	PHREEQC (Parkhurst and Appelo, 1999)	THERMOCHIMIE (Blanc et al., 2015; Giffaut et al., 2014)	1D Cartesian grid (5 cm)	40 m (case 1)	Water-saturated conditions	No (case 1)
Marty et al. (2014)	PHREEQC (Parkhurst and Appelo, 1999)	THERMOCHIMIE (Blanc et al., 2015; Giffaut et al., 2014)	1D Cartesian grid (5 cm) and 1D radial grid (5 cm)	43 and 45.4 m (depending on the simulated case)	Water-saturated conditions	Yes for the last simulated case
Marty et al. (2015a)	TOUGHREACT, PHREEQC, CRUNCH, HYTEC, ORCHESTRA, MIN3P-THCm (Steefel et al., 2015a)	THERMODDEM (Blanc et al., 2012)	1D radial grid (5 cm)	43 m	Water-saturated conditions	No
Iron (canister) clay-based materials interactions						
Montes-H et al. (2005)	KIRMAT (Gérard et al., 1998)	KIRMAT and EQ3/6 (Wolery, 1983)	1D Cartesian (5 cm)	1 m	Saturated condition	No
Bildstein et al. (2006)	CRUNCH (Steefel et al., 2015a)	EQ3/6 (Wolery, 1983)	1D cartesian grid	~10 m	Saturated condition	Yes

Temp. (°C)	Simulated time	Reaction processing	Surface reactions	Materials	Transport properties De in $m^2 s^{-1}$ Permeability unit depends on the authors
2	Depending on the case (~2500 to 4000 yr)	Local equilibrium assumption	Cation exchange and neutral 'surface ions pairs'	Opalinus Clay	ϕ = 0.06 De = $D_0 \phi^2$ with $D_0 = 1.5 \times 10^{-9}$
				Effingen Member	ϕ = 0.04
				Palfris formation	ϕ = 0.01
				CEM I concrete (pH 13.1)	ϕ = 0.2
16	Depending on the case. Up to clogging time (~10 to 100 000 yr)	Local equilibrium or kinetic assumption	Not taken into account	Portland cement (pH 13.57)	De is not explicitly given but based on the used Archie's law and a Ø = 0.1 should be close to 10^{-11} in pristine concrete
				Interstitial Boom clay porewater	
25	100 000 yr	Local equilibrium assumption	Cation exchange	Hyperalkaline solutions (pH 12.5 at 25°C)	
				Callovo-Oxfordian argillite	De = 2.6×10^{-11} and Ø = 0.18
25	100 000 yr	Local equilibrium and kinetic approaches (depending on the simulated case)	Cation exchange	Portland cement (pH 12.5 and pH 13.2 at 25°C)	De = 9.0×10^{-12} and Ø = 0.13
				Callovo-Oxfordian argillite	De = 2.6×10^{-11} and Ø = 0.18 in pristine clay-rock De = 5.2×10^{-11} and Ø = 0.2 in EDZ
25	10 000 yr	Local equilibrium and kinetic approaches	Cation exchange	Portland cement (CEM I) (pH 13.2 at 25°C)	De = 9.0×10^{-12} and Ø = 0.13
				Callovo-Oxfordian argillite	De = 2.6×10^{-11} and Ø = 0.18
100	10 000 yr	Corrosion rate of 5 μm yr^{-1} Kinetic assumption	Cation exchange	Clay-rock	Fix boundaries condition (fluid)
				MX80 bentonite	De = 10^{-11} and Ø = 0.3
				Container	Fix boundaries condition (fluid)
50	16 000 yr	Corrosion rate of 4.3 μm yr^{-1} Kinetic assumption	Cation exchange	Iron steel + bentonite	A 17 cm thick cell is considered which contains the steel canister, the technological gap and a fraction of clay
				Bentonite	De = 10^{-11} and Ø = 0.15
				Callovo-Oxfordian	De = 10^{-11} and Ø = 0.15

(*Continued*)

Table 8.1 (Continued)

Reference	Code	Database	Geometry	Dimensions	Conditions	Porosity feedback
Samper et al. (2008a)	CORE^{2D} V4 (Samper et al., 2003)	EQ3/6 (Wolery, 1983)	1D and 2D axisymmetric	25 m in radial direction, 6.5 in longitudinal direction	Saturated condition	No
Xu et al. (2008)	TOUGHREACT (Samper et al., 2003)	EQ3/6 (Wolery, 1983)	1D radial	150 m	Non-saturated condition (saturation state evolves)	
Marty et al. (2010)	KIRMAT (Gérard et al., 1998)	KIRMAT additionally completed by data as described in the publication	1D Cartesian (grid size 5 m)	1 m	Saturated condition	Yes
Savage et al (2010)	QPAC (Maul et al., 2008)	EQ3/6 (Wolery, 1983)	1D (most probably) Cartesian (grid size 10 cm)	80 cm	Saturated condition	Not mentioned
Lu et al. (2011)	CORE^{2D} V4 (Samper et al., 2003)	EQ3/6 (Wolery, 1983)	1D axisymmetric	25 m	Saturated condition	No

Temp. (°C)	Simulated time	Reaction processing	Surface reactions	Materials	Transport properties De in $m^2 s^{-1}$ Permeability unit depends on the authors
25	300 000 yr	Corrosion rate of 0.2 μm yr^{-1-}	Cation exchange and surface complexation	Canister	A dummy material zone is defined for the canister to model canister corrosion in a manner similar to mineral dissolution
				Bentonite	$De = 4.07 \times 10^{-11}$, $\varnothing = 0.407$ and $K = 5.71 \times 10^{-14}$
				EDZ	$De = 5.02 \times 10^{-14}$, $\varnothing = 0.05$ and $K = 8.72 \times 10^{-11}$
				Granite	$De = 5.02 \times 10^{-14}$, $\varnothing = 0.05$ and $K = 8.72 \times 10^{-12}$
40	200 000 yr	Corrosion rate depends on saturation and porosity. Simulations with constant rate are also performed Local equilibrium assumption except for iron corrosion and magnetite precipitation, which were under kinetic constraints Test of two constant rates for hydrogen production	Cation exchange	Waste canister	De is not given. and $\varnothing = 0.1$ $1/\alpha = 1$ Pa n = 2 $K_{rw} = K_{rg} (m^2) = 10^{-19}$
				Bentonite	De is not given. and $\varnothing = 0.4$ $1/\alpha = 1.8 \times 10^7$ Pa n = 1.82 $K_{rw} = K_{rg} (m^2) = 10^{-19}$
				Opalinus Clay	De is not given. and $\varnothing = 0.12$ $1/\alpha = 1.8 \times 10^7$ Pa n = 1.82 $K_{rw} = K_{rg} (m^2) = 10^{-20}$
100	100 000 yr	Corrosion rate decreases progressively from 5 to 0.21 μm yr^{-1} Kinetic	No	Canister	Corrosion cell (iron + water)
				MX-80 Bentonite	$De = 1.14 \times 10^{-11}$, $\varnothing = 0.39$
				Callovo-Oxfordian porewater	
Not mentioned for the modeling of iron-bentonite	1 000 000 yr	Kinetic, processes of nucleation, growth, precursor cannibalization and Ostwald ripening were included	No	MX-80 Bentonite	$\varnothing = 0.42$, De is not mentioned
25	300 000 yr	Corrosion rate (0.1 μm yr^{-1}) is either constant or kinetically controlled	Cation exchange and surface complexation	Canister	A dummy material zone is defined for the canister to model canister corrosion in a manner similar to mineral dissolution
				Bentonite	$De = 4.07 \times 10^{-11}$, $\varnothing = 0.407$ and $K = 5.71 \times 10^{-14}$
				EDZ	$De = 5.02 \times 10^{-14}$, $\varnothing = 0.05$ and $K = 8.72 \times 10^{-11}$
				Granite	$De = 5.02 \times 10^{-14}$, $\varnothing = 0.05$ and $K = 8.72 \times 10^{-12}$

(*Continued*)

Table 8.1 (Continued)

Reference	Code	Database	Geometry	Dimensions	Conditions	Porosity feedback
De Windt et al. (2014)	HYTEC (van der Lee et al., 2003)	EQ3/6 (Wolery, 1983)	2D cylindrical grid	Length ~24 m Height ~ 8 m	Unsaturated conditions (water saturation state does not evolve)	
Sedighi et al. (2014)	COMPASS-PHREEQC	PHREEQC (Parkhurst and Appelo, 1999)	2D Cartesian discretized in 80 equally-sized quadrilateral elements	50*35 cm	Unsaturated condition (saturation does not evolve)	No
Wersin and Birgersson (2014)	CRUNCH (Steefel et al., 2015a)	THERMODDEM (Blanc et al., 2012)	1D Cartesian (2 cm wide cells)	42 cm	Saturated condition	Yes
Wilson et al. (2015)	QPAC (Maul et al., 2008)	Geochemist's Workbench, THERMODDEM and SUPCRT92	1D Cartesian (grid size 1.75 cm in case 1)	70 cm	Saturated condition	Yes
Samper et al. (2016)	CORE^{2D} V4 (Samper et al., 2003)	EQ3/6 (Wolery, 1983)	1D axisymmetric discretized in 92 elements	85 cm	Saturated conditon	Not specified
Glass + overpack (iron) clay-based material interactions						
Bildstein et al. (2007)	CRUNCH (Steefel et al., 2015a)	EQ3/6 (Wolery, 1983)	1D cartesian	~10 m	Saturated condition	Yes

Temp. (°C)	Simulated time	Reaction processing	Surface reactions	Materials	Transport properties De in $m^2 s^{-1}$ Permeability unit depends on the authors
25	100 yr	Local equilibrium assumption except for corrosion of C-Steel and pyrite that were under kinetic constraints	Cation exchange and surface complexation	Clay-rock Fissured EDZ Fractured EDZ Bentonite	$De = 1.5 \times 10^{-11}$ and $\varnothing = 0.15$ $De = 1.5 \times 10^{-11}$ and $\varnothing = 0.15$ $De = 2 \times 10^{-10}$ and $\varnothing = 0.2$ $De = 1.5 \times 10^{-10}$ and $\varnothing = 0.36$
25	10 000 yr	Local equilibrium	Cation exchange and surface complexation	Hydrogen MX-80 Bentonite	Constant rate influx $\varnothing = 0.4$, diffusion coefficient in water is not given, gas diffusion coefficient according to equation 7 intrinsic gas permeability $(m^2) = 10^{12}$
25	50 000 yr	Corrosion rate ($1\ \mu m\ yr^{-1}$) modelled as first-order kinetic Kinetic assumption	No	Canister MX-80 Bentonite	A cell with iron and water. Amount of iron adjusted to give an iron/bentonite mass ratio which agrees with the KBS-3H design $De = 2.14 \times 10^{-11}$ and $\varnothing = 0.476$
70	Depends on the case, up to 100 000 yr	Model 1 includes a fixed steel corrosion rate; in Model 2, the corrosion rate of steel is diffusion-limited. In Model 3, a 1 mm thick, water-filled 'corrosion cell' was included.	Cation exchange	Canister Kunigel VI bentonite	Porewater boundary conditions for model 1 and 2, a water cell for model 3 in which precipitation of all minerals except magnetite, siderite and Fe(OH)2 was suppressed $De = 2.14 \times 10^{-10}$ and $\varnothing = 0.4$
Reference case at 25°C and sensitivity run in non-isothermal conditions ($T_{max} = 97°C$)	1 000 000 yr	Corrosion rate ($2\ \mu m\ yr^{1}$) for reference case between 0.1 and $5\ \mu m\ yr^{-1}$ for sensitivity Kinetic assumption	Cation exchange and surface complexation	Canister Bentonite Granite	$De = 4.07 \times 10^{-11}$ and $\varnothing = 0.407$ (sensitivity case with $\varnothing = 0.2$) $De = 4.07 \times 10^{-11}$ and $\varnothing = 0.407$ (sensitivity case with 0.5xDe or 2xDe) Constant flux ranging from 0.01 (reference case) to $0.1\ l yr^{-1}$ parallel to the axis of the gallery at the bentonite-granite interface
50	100 000 yr	Corrosion rate of $4.3\ \mu m\ yr^{-1}$ Glass dissolution according to equation 12 Kinetic assumption		Glass Iron canister Callovo-Oxfordian clay-rocks	A 3 cm thick cell is considered which contains the glass and the technological gap A 17 cm thick cell is considered which contains the steel canister, the technological gap and a fraction of clay $De = 10^{-11}$ and $\varnothing = 0.15$

studies published for large-scale simulation to demonstrate agreement, disagreement and current gaps in our understanding and quantification of the processes among the results. In the last section, we will give insights on recent improvements and perspectives for reactive transport modeling in the framework of studies on predicting how radioactive waste disposal will evolve.

8.2 Modeling Physical and Chemical Processes on Repository Scales

8.2.1 Reactive Transport Modeling Principles

8.2.1.1 Reactive Transport Constitutive Equations

Contaminants can be transported in the form of aqueous, gaseous or colloidal species in the pore space of the materials under investigation. The pore space is the voids between aggregates and mineral grains. Reactive transport models for porous media rely on constitutive equations that make it possible to describe the porous medium of interest at the continuum scale, with respect to its macroscopically measurable properties such as permeability, dispersivity, diffusibility, and so on (Steefel et al., 2015a). In this framework, a generic equation for advective/dispersive transport in the liquid phase coupled to biogeochemical reactions can be written:

$$\frac{\partial(\phi S_L \Psi_i)}{\partial t} = \nabla\left(\phi S_L D_i^* \nabla \Psi_i\right) - \nabla(q\Psi_i) - \sum_{r=1}^{Nr} \nu_{ir} R_r - \sum_{m=1}^{Nm} \nu_{im} R_m - \sum_{l=1}^{Nl} \nu_{il} R_l \quad (8.1)$$

$$\Psi_i = C_i + \sum_{j=1}^{Nx} \nu_{ji} \cdot C_j \quad (8.2)$$

where ϕ is the porosity (m^3_{void}/m^3_{medium}), S_L is the liquid saturation (unitless), Ψ_i (mol/m^3_{water}) is the total concentration term that integrates the portioning between primary species (with concentration C_i) and secondary species ($\Sigma_{j=1}^{Nx} \nu_{ji} \cdot C_j$), D_i^* is the diffusion coefficient for species i in the porous media (m^2/s), q is the volumetric (or Darcy flux) of water ($m^3_{H_2O} \cdot m^{-2} \cdot medium/s$), and R_r, R_m and R_l are the aqueous phase, mineral and gas reactions ($mol \cdot m^3 \cdot medium/s$) respectively. ν_{ji} is the number of moles of component i in one mole of secondary species j. ν_{ik} with $k = r$, m or l is the number of moles of component i in one mole of phase k. These equations can be expanded to consider (i) transport in phases other than the liquid (e.g. gases), (ii) a dispersion term, or (iii) the Nernst-Planck equation, instead of Fick's equation, to model diffusional transport. A comprehensive description of these equations can be found in Steefel et al. (2015a).

8.2.1.2 Geometry and Space Discretization

The values of the parameters that are considered in reactive transport modeling equations are the results of coupling many fundamental processes at the pore scale. In principle, it is possible to model a system at the scale of individual pores with a reactive transport approach, as long as the parameters of the continuous model remain physically meaningful (Steefel et al., 2013; Trebotich et al., 2014), but such an approach cannot currently be used for very large systems, such as a repository gallery, because the computational cost is too large. In any case, the reactive transport modeling approach stands on the hypothesis that the solid, aqueous and gaseous phases can be modeled with homogeneous properties at the scale of the system's spatial discretization (i.e. the numerical cell), and that it is therefore possible to define a representative elementary volume for each domain in the modeled system.

A repository design is given in Figure 8.1, together with examples of simplified representations for reactive transport model applications. Its geometry and the resolution of the spatial discretization (i.e. the number of cells) are the results of a compromise between (1) minimizing code execution time, (2) how accurate the system description is in the most reactive zones, (3) how representative it is of the symmetry of the real system, and (4) positioning the boundary far enough away to prevent the boundary conditions from influencing system evolution near the interfaces. The mesh size is in the millimetre–centimetre range for most of the reported studies (Table 8.1). 1D Cartesian and 1D radial geometries are the representations that have been used most commonly. 2D representations have also been used. Thermo-hydraulic two-phase transport calculations using 3D geometries at the scale of a repository object are reported (Enssle et al., 2011a, 2014; Brommundt et al., 2014), but to our knowledge, based on our literature survey (which might not be exhaustive), no full 3D reactive transport calculations have been published in the peer-reviewed literature at the time of writing. The simulation with geometry that is closest to a 3D representation is the one published by De Windt et al. (2004b), which makes use of 2D with cylindrical coordinates.

8.2.1.3 Where Everything Takes Place: the Pore Space

Porosity (ϕ) is one of the main parameters that controls mass transfer in Equation (8.1), because advection and diffusion fluxes are scaled to the porosity value, and because all reactive processes (dissolution, precipitation, surface processes, etc.) are mediated by the fluid phases. The fluid phase of interest for chemical reactions is usually the aqueous phase alone, since no reactivity is taken into account between the components of the gas phase or between gas components and the solid.

Defining physical porosity is straightforward: it is simply the ratio of void volume to total volume, that is, the volume not occupied by mineral grains. In most RTM publications, the porosity value matches this definition. However, defining the porosity of a clay material or a cement-based material is not so obvious as soon as we take into account one of the hypotheses of RTM calculations – the homogeneity hypothesis at the numerical cell scale. For an iron overpack, in which there is initially no physical connected porosity, defining a numerical porosity term is also challenging. These problems will be explored in more detail below.

8.2.1.4 Kinetic and Thermodynamic Databases

Databases used in the RTM studies published in the framework of clay repository systems are listed in Table 8.1. Obviously, a range of databases have been used in these studies, and their differences in tabulated thermodynamic and kinetic values may impact the result of the predictions made with RTM. The completeness and internal consistency of a database with regards to the investigated processes must be the guideline for choosing one database over another for a given application. In particular, supplying the chemical/transport code with too short a list of minerals can bias model results, because it forces the system to converge towards a given mineralogical assemblage without alternative choices (Gaucher and Blanc, 2006).

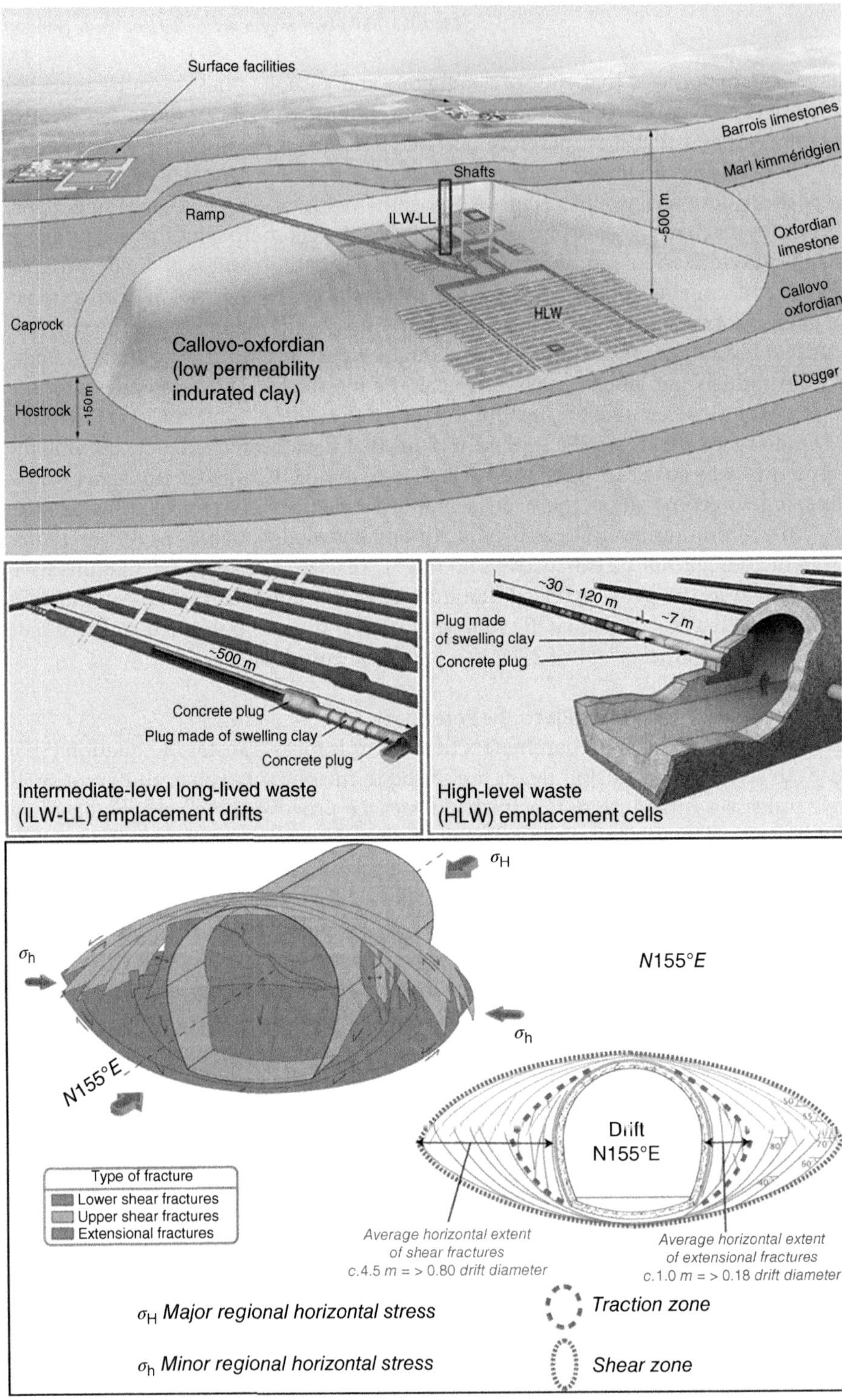

Figure 8.1 Left, from top to bottom: full-scale design of a repository concept in clay-rocks, zoom on the design of the emplacement drifts, representation of the EDZ. Right: schemes of simplified geometries modeled with reactive transport approaches. *Source*: Adapted from Brommundt et al. (2014) and Vinsot et al. (2014). (See colour plate section for the colour representation of this figure.)

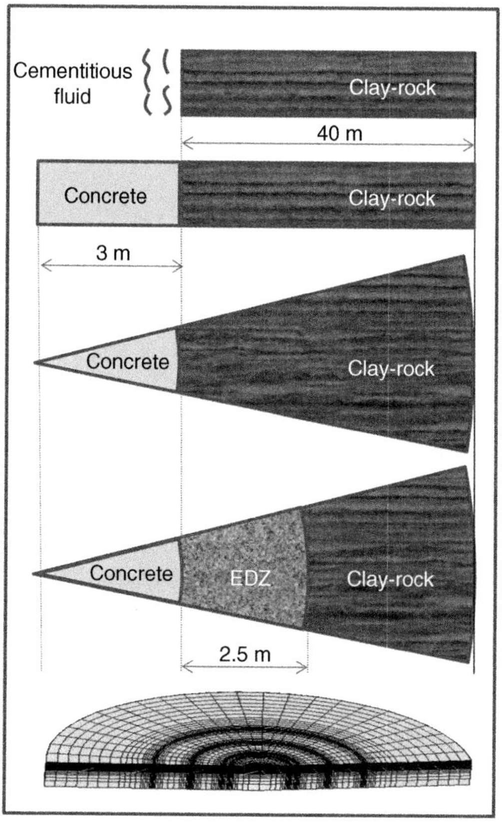

Figure 8.1 (Cont'd)

These important considerations for RTM calculation have recently been the focus of a special issue of *Applied Geochemistry* (Kulik et al., 2015). We see that in the framework of national radioactive waste disposal projects, database developments have been pushed forward (Blanc et al., 2012; Giffaut et al., 2014). This research effort combines new thermodynamic data acquisition, especially for clay minerals (Gailhanou et al., 2007, 2009, 2012, 2013), critical data selection for repository materials such as cement phases (Blanc et al., 2010a,b) and how to evaluate their capacity to describe interactions between materials (Blanc et al., 2015). In addition, kinetic databases with kinetic rate parameters are being built (Marty et al., 2015b). What is important for them is consistency with the associated thermodynamic database for stoichiometry and solubility values, so that they can calculate how rates depend on saturation state.

8.2.1.5 Initial Conditions

It is not possible to model how a part of the repository system evolves from first principles, taking into account all chemical and physical processes in space and time from excavation to its post-closure evolution. Therefore we must consider *a priori* initial conditions for the calculation. These initial conditions include temperature, water saturation, pressure gradients, porewater composition, and mineralogy. In addition, reactivity and transport properties must be assigned to each of the domains defined in the calculation. The way that these parameters are most frequently ascribed is defined in the following section.

8.2.2 Repository Material Properties

8.2.2.1 Generalities

Here we define the main calculation parameters for clay, cement, iron and glass materials, and include those related to mineralogy, porosity, initial porewater composition, saturation and transport properties.

8.2.2.2 Clay Materials

Mineralogy

Chemical conditions prevailing in clay-rocks have recently been reviewed by Tournassat et al. (2015b). As stated by the authors, the properties of clay-rock result from the individual properties of each of the mineral and organic constituents and the way they interact. The mineralogy of clay-rocks is complex and difficult to quantify. It is often challenging to characterize the mineralogy of its clay fraction, which contains not only pure clay mineral end-members, such as kaolinite, smectite and illite, but also mixed layer minerals (Claret et al., 2004). In addition, carbonate minerals with a range of chemical compositions and structures are present (see for example Lerouge et al., 2013), and pyrite, along with organic matter and quartz (Gaucher et al., 2004a; Lerouge et al., 2014b; Zeelmaekers et al., 2015). Proportions of total phyllosilicates, carbonates, quartz, pyrite and organic matter can be found in the literature for various clay formations (e.g. Opalinus clay, Boom clay, Callovian-Oxfordian formation, and Boda clay, in Altmann et al. (2012)). These proportions are used in reactive transport modeling calculations, for which homogeneous mineralogical compositions are considered for large spatial domains. These proportions can vary across a given clay formation (Claret et al., 2004; Gaucher et al., 2004a; Jacquier et al., 2013; Chen et al., 2014), but the variations in mineralogy are *a priori* considered to have little effect on the outcome of the simulations. Currently, available thermodynamic and kinetic databases are neither accurate nor comprehensive enough to allow for the introduction of the complexity of the actual chemical compositions of clay minerals and carbonate minerals in the simulations. For carbonate minerals, generic compositions corresponding to pure phases (e.g. calcite, dolomite and siderite) are used instead. For clay minerals, ideal compositions and/or compositions of well-characterized model samples are used (e.g. De Windt et al., 2004a; Bildstein et al., 2006; Trotignon et al., 2007; Marty et al., 2015a). We note that the solid solution description of mineral solubility has been tested in a very limited number of studies (e.g. Berner et al., 2013; Kosakowski and Berner, 2013).

Porosity

Depending on the mineralogy, the burial diagenesis history of the considered host rock formation's physical porosity lies in the approximate range of 5–40% (Mazurek et al., 2008; Altmann et al., 2012; Van Loon and Mibus, 2015). For bentonite in and engineered barrier system (EBS), the porosity can be calculated according to:

$$\phi = 1 - \frac{\rho_d}{\rho_s} \tag{8.3}$$

where ρ_d is the bulk dry density (g/cm^3) and ρ_s is the density of solid particle (g/cm^3) (grain density).

Cation exchange and surface complexation

Smectite, illite, and illite/smectite mixed-layer minerals are built from layers of oxygen atoms and cations (mainly Si^{4+}, Al^{3+}, Mg^{2+}, Fe^{2+}, Fe^{3+} and H^+) in tetrahedral–octahedral–tetrahedral (TOT) coordination. Their TOT layer structure has excess negative charge, due to Al^{3+} for Si^{4+} substitutions in tetrahedral layers and (Mg^{2+}, Fe^{2+}) for (Al^{3+}, Fe^{3+}) substitutions in octahedral layers. The TOT layer negative charge is compensated by hydrated cations (e.g. Na^+, Ca^{2+}, Mg^{2+}, K^+) in the vicinity of the charged surface (Figure 8.2). These cations are readily exchangeable with the neighbouring cations in solution. The sum of the charges of these exchangeable cations is termed cation exchange capacity (CEC, in mol of charge – mol$_c$ – per kg of material). In compacted clay materials and clay-rocks, the solid to liquid ratio is high, so the ratio of abundance of cations in exchangeable positions versus those in solution is also high. For example, if we consider a bentonite compacted at 1.6 kg/dm^3, corresponding to a total porosity of ~0.4, and a CEC of 0.8 mol$_c$/kg, which is typical for a bentonite (see Table 8.1), the total cation exchange capacity is ~1.3 mol$_c$/dm^3, i.e. ~3.25 mol$_c$/kg$_{water}$. This concentration is far higher than the total cation concentration in the porewater, and consequently cation exchange properties of clay minerals efficiently buffer the cation composition of the porewater (Tournassat et al., 2015b).

At the edges of layers, the clay mineral structure is cleaved in a manner that gives rise to surface functional groups analogous to those that exist on Al and Si oxides, so clay mineral edges carry a pH-dependent surface charge that arises from the acid–base

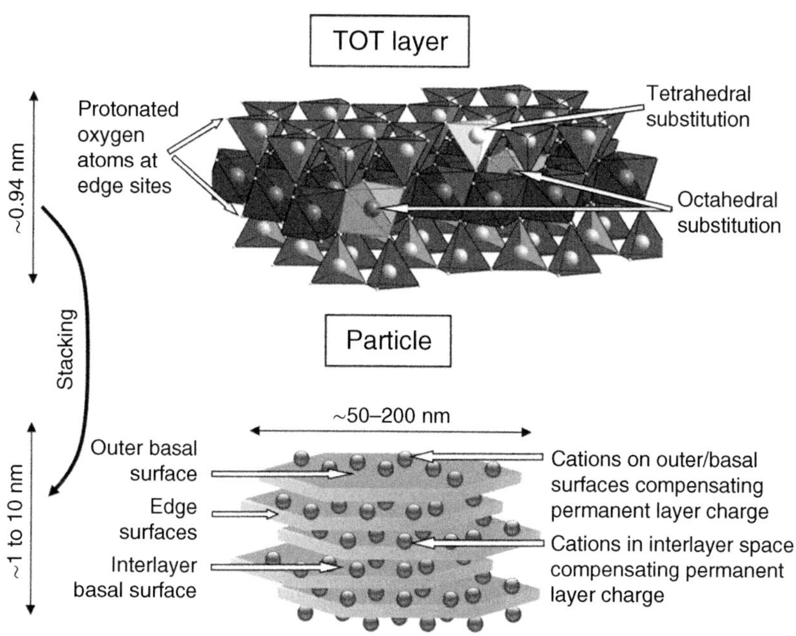

Figure 8.2 Structure of a TOT layer, and scheme of a TOT layer and compensating cations in a clay mineral particle. *Source*: Courtesy of Tournassat and Steefel (2015).

reactivity of edge surface functional groups (Figure 8.2). This property is responsible in part for the pH buffering properties of compacted clay materials, especially bentonite (see Wersin, 2003; Tournassat et al., 2011; among others).

Porewater composition

In the porewater of the three most studied clay rocks in Europe (Boom Clay, Callovian–Oxfordian clay and Opalinus clay), the pH lies in the range 7 to 8.5 (see Table 8.2), corresponding to partial CO_2 pressures (p_{CO2}) ranging from ~10^{-3} to ~10^{-2} bars – p_{CO2} values that are higher than the atmospheric pressure. Ionic strength can vary widely among clay-rocks, from 0.008 to 0.4, from fresh water to brackish water, with Na^+ being the main contributor to the cationic charge. Bicarbonate and carbonate anions are the main contributors to the anionic charge in the porewaters with low ionic strength (in Boom clay, for example) while chloride and sulfate are the main contributors in porewater with ionic strength higher than ~0.1 (the Callovian-Oxfordian and Opalinus clays, for example). The fact that values of ionic strength remain lower than ~0.5 implies that the use of the extended Debye-Hückel activity model for solute species, which is one of the most commonly used activity models implemented in RTM (Steefel et al., 2015a), is relevant for clay materials as long as the perturbation under investigation is not a high salinity plume. In addition, porewater composition models at equilibrium that rely on the Gibbs' phase rule, and that have been validated by comparison with *in situ* seepage water composition, are available for various clay-rocks, providing reliable models for far-field initial conditions (Bradbury and Baeyens, 1998; Gaucher et al., 2006, 2009; Kars et al., 2015; Pearson et al., 2003; Tournassat et al., 2015b; Tremosa et al., 2012; Vinsot et al., 2008). In these models, redox potential calculated at equilibrium with the pyrite/sulfate system is in the range −180 mV to −250 mV (Tournassat et al., 2015b). The porewater composition in highly compacted bentonite used in engineered barrier systems (EBS) can also be modeled using similar approaches (Wersin, 2003). The main phenomena that must be captured by the model are rapid total or partial dissolution of accessory minerals such as gypsum or calcite, ion exchange between the solution and the clay mineral surfaces, and protonation/deprotonation of the amphoteric sites on the edge of montmorillonite, which provide a powerful pH buffering effect (Muurinen and Lehikoinen, 1999; Bradbury and Baeyens, 2003, 2009). Fewer reactive minerals can be used to constrain the porewater composition in bentonite than in clay-rocks (Bildstein et al., 2006), so we must also note that the porewater composition in these systems will depend on (1) the composition of the fluid that will rehydrate the bentonite, and (2) the kinetics of mineral dissolution/precipitation. The porewater composition of bentonite material is thus dependent on the events taking place in the repository before closure and rehydration. The same remark holds true for clay-rocks in the EDZ where dehydration and oxidation take place before rehydration. These processes increase the salinity of the porewater later on (Vinsot et al., 2013). The impact of transient conditions on clay porewater composition is not taken into account in long-term predictions modeled with a water-saturated initial condition.

Transport properties

In saturated conditions, molecular diffusion is considered to be the main mass transport mechanism (Mazurek, 2010; Altmann et al., 2012; Bourg and Tournassat, 2015), so Darcy flux is often neglected because clay-rocks and smectite-based engineered

Table 8.2 Comparison between porewater chemistries of cements at different ages and clay-rocks.

Porewater type	Na (mmol/l)	K (mmol/l)	Ca (mmol/l)	Mg (mmol/l)	Cl (mmol/l)	S* (mmol/l)	C (mmol/l)	pH	Reference
Clay-rock type									
Meuse Haute Marne URL	45.7	0.6	5	4.3	35	12.5	4.2	7.4	Tournassat et al. (2015b) and references therein
Mont-Terri URL	235–281	1.4	16.8–19.3	22	276–325	17	3.6	7.2–7.7	
Hades URL	7.1	0.11	0.11	0.03	0.56	0.44	5.79	8.4	
Cement-based material type									
OPC for 69 days, extraction between 60 and 330 MPa	46–50	450–480				8.2–13		13.6–13.6	Lothenbach (2010)
OPC after 0.5 and 2 years of curing time	26.2–8.7	54.6–14.4	11.8–17.7			0.1		12.9–12.8	García Calvo et al. (2010)
OPC after 20 years of curing time	36–79	105–154	1.05–3.33	n.g.**	n.g.	n.g.	n.g.	n.g.	Luke and Lachowski (2008)
CEMI between 0.04 and 150 days	25–160	34–130	28–3			22–1.8		12.7–13.3	Lothenbach and Wieland (2006)
Low alkaline concrete B40, 12 months of curing time	1.4	0.8	4	n.g.	n.g.	2.5	n.g.	11.5	Bach et al. (2012)
Low alkaline concrete T1, 12 months of curing time	1.9	1.1	4.2	n.g.	n.g.	4.4	n.g.	11.2	
Low alkaline concrete between 0 and ≈ 3.5 years months of curing time	18–23	26–17	20–2.7	0.0018–0.0011	n.g.	25–22	≈0.05	12.4–12.2	Lothenbach et al. (2012)
ESRED formulation between 0 and ≈ 3.5 years of curing time	79–22	186–11	25–29	0.0063–0.0045	n.g.	45–1.8	n.g.	12.4–11.7	Lothenbach et al. (2014)

(*Continued*)

Table 8.2 (Continued)

Porewater type	Na (mmol/l)	K (mmol/l)	Ca (mmol/l)	Mg (mmol/l)	Cl (mmol/l)	S* (mmol/l)	C (mmol/l)	pH	Reference
60% OPC 40% SF after 0.5 and 2 years of curing time	5.2–1.1	8.3–0.8	15.3–8.2			0.9–0.7		12.2–12.2	García Calvo et al. (2010)
50% OPC 50% SF after 0.5 and 2 years of curing time	7.2–2.5	9.5–1.3	16.2–11.5			22.7–11.2		11.2–11.3	
80% OPC 10% SF 10% SF after 0.5 and 2 years of curing time	5.1–3.7	11.8–7.8	17.2–17.1			0.3–0.3		12.6–12.7	
35% OPC 35% SF 30% SF after 0.5 and 2 years of curing time	7.6–3.5	9.9–2.0	12.1–7.4			1.9–6.3		11.2–11.6	
20% OPC 50% SF 30% SF after 0.5 and 2 years of curing time	9.6–4.3	5.6–6.4	30.2–12.6			32.3–17.5		10.5–10.4	

* S(6) for clay-rock type and not defined otherwise.

** n.g., not given

barriers exhibit very low permeability values (Enssle et al., 2011b; Yu et al., 2013). In contrast, in the presence of non-saturated conditions, two-phase flow models must be considered.

In most RTM studies, molecular diffusion is treated in its simplest way by applying Fick's laws, which state that the diffusive flux (J_i) of a species in solution is proportional to the concentration gradient (C_i):

$$J_i = -D_{e,i} \nabla C_i \tag{8.4}$$

where $D_{e,i}$ is the effective diffusion coefficient, which is specific to each chemical species as indicated by the subscript i. The diffusion coefficient includes a correction for the tortuosity (τ) and the porosity (ϕ) of the porous media:

$$D_{e,i} = \phi \cdot D_i^* = \phi \cdot \tau \cdot D_{0,i} \tag{8.5}$$

where $D_{0,i}$ is the diffusion coefficient of species i in water (or self-diffusion coefficient), and D_i^* is the pore diffusion coefficient ($D_i^* = \tau \cdot D_{0,i}$) in Equation (8.1).

In the case of tritiated water (HTO) that accesses all the porosity, the effective diffusion coefficient is commensurable with 10^{-11} m^2/s in clay-rock with low porosity values (between 5 % and 17 %) (Altmann et al., 2012). D_e(HTO) increases with increasing accessible water porosity, approximately following Archie's law relationship (see Van Loon and Mibus, 2015, for more details):

$$\frac{D_{e,\mathrm{HTO}}}{D_{0,\mathrm{HTO}}} = \phi^m \tag{8.6}$$

where m is an empirical factor, called the cementation factor, and $D_{0,\mathrm{HTO}}$ is water's self-diffusion coefficient. Archie's law is certainly the most common relationship implemented in reactive transport codes that relate the diffusion coefficient to the porosity of the material. This kind of relationship is used to update the tortuosity and the pore diffusion coefficient as a function of the porosity changes that are a consequence of geochemical processes such as mineral dissolution or precipitation (Xie et al., 2015). However, in many studies described in Table 8.1, the coupling between the porosity, transport parameters and geochemistry was neglected, and the porosity value was kept constant in each numerical cell.

In most of the studies dealing with complex geochemical interactions and large systems (Table 8.1), a unique D_e (taken as equal to HTO) was taken for all the chemical species. Taking into account different diffusion coefficients for different chemical species (multi-component diffusion or MCD) necessitates the use of the Nernst-Planck equation instead of Fick's equation (Steefel et al., 2015a). MCD is computationally expensive, and few reactive transport codes are capable of handling it. Moreover, the parameterization of such a model would be under-constrained for most of the chemical species under consideration. This explains why the MCD formalism is seldom used in large-scale simulations (no study using the MCD formalism is reported in Table 8.1).

In the case of non-saturated conditions, reactive transport codes often include an exponential dependency of pore aqueous and gas diffusion coefficients (D_{ea} and D_{eg}) as a function of saturation and porosity using the Millington relationship (Steefel et al., 2015a):

$$D_{ea} = \phi^{\delta} S_a^{\delta+2} D_{0a}$$
$$D_{eg} = \phi^{\delta} S_g^{\delta+2} D_{0g} \quad (8.7)$$

where S_a and S_g are the saturations of the aqueous and the gas phases, D_{0g} the free phase diffusion in the gas phase and δ is equal to 4/3 according to Millington (1959).

Variably saturated flow can be described either with the full set of multiphase flow and conservation equations or with the Richards equation (Steefel et al., 2015a). Depending on the RT code used, the relative permeability and aqueous phase saturation can be approximated using different formalisms (e.g. Brooks and Corey, 1964; Mualem, 1976; van Genuchten, 1980; Parker et al., 1987; Luckner et al., 1989). In most of the RTM studies we reviewed, a capillary pressure-relative permeability relationship that is derived from the van Genuchten-Mualem approach (Mualem, 1976; Van Genuchten, 1980) is used to describe two-phase flow:

$$P_c = \frac{1}{\alpha}\left(S_{ec}^{n/1-n} - 1\right)^{1/n}$$
$$\text{with} \quad S_{ec} = \frac{S_w - S_{wr}}{S_{wm} - S_{wr}} \quad (8.8)$$

where P_c is the capillary pressure (Pa), $1/\alpha$ is the pseudo air entry pressure for the gas phase (Pa), S_{ec} is the effective saturation in the capillary pressure–saturation relationship (dimensionless), S_w is the water saturation (dimensionless), S_{wr} is the minimum (or irreducible) value of the water saturation achievable when gas displaces water (dimensionless), and S_{wm} is the maximum value of the water saturation (dimensionless), which often has a value of 1. The parameter n is a pore size distribution index.

$$k_{r,w} = S_e^{\eta}\left[1 - \left(1 - S_e^{n/(n-1)}\right)^{(n-1)/n}\right]^2 \quad \text{if } S_w < 1 - S_{gr}$$
$$\text{or} \quad 1 \quad \text{if } S_w \geq 1 - S_{gr}$$
$$k_{r,g} = (1 - S_e)^{\gamma}\left[\left(1 - S_e^{n/(n-1)}\right)^{(n-1)/n}\right]^2 \quad (8.9)$$
$$\text{with} \quad S_e = \frac{S_w - S_{wr}}{1 - S_{gr} - S_{wr}}$$

where $k_{r,w}$ and $k_{r,g}$ are the relative permeabilities of the water and gas phases (dimensionless) respectively, S_e is the effective saturation in the relative permeability saturation relationship (dimensionless), S_{gr} is the residual gas saturation (dimensionless), and η and γ are exponents describing the pore connectivity for the wetting and non-wetting phases. These two exponents are estimated to be equal to ½ for many soils. That is also the value considered for clay materials in the simulations reported in Table 8.1.

8.2.2.3 Cement Materials

Mineralogy and porewater composition

Concrete integrates cement (binder), aggregates and water. A cement material without aggregates is called a cement paste. If only fine aggregates (below 4–5 mm) are used, the cement material is a mortar. Otherwise, it is a concrete. In large-scale simulations, the presence of concrete is frequently taken into account, and since aggregates are often calcareous, the mineralogical compositions of the cement materials are dominated by calcite (Marty et al., 2015a). Ordinary Portland cement (OPC) is the cement formulation that is most frequently considered. When it is brought into contact with water, its calcium silicates (C_3S or C_2S) and calcium aluminates (C_3A and C_4AF) start to react, and various hydration products form, such as portlandite, C-S-H (calcium silicate hydrate), ettringite, monosulfate or monocarbonate. These hydration products are the mineral phases that are integrated in large-scale RTM. As evidenced in Table 8.2, hydration of cement-based material is kinetically driven, and thus the composition of its porewater evolves with the curing time. High pH values are characteristic of ordinary Portland-based cement material. In some studies, an early-age concrete (pH > 13.2) is considered (Marty et al., 2015a), whereas in other cases a concrete that has partly lost its alkali content and whose pH is maintained by equilibrium with portlandite (pH ~12.5) is considered. CEM I, which is made of finely ground Portland clinker, and CEM V, which is constituted of blended Portland, fly ash and blast furnace slag cement (Chen et al., 2012), are the most studied cement materials (Tables 8.1 and 8.2). CEM I and CEM V differ in their portlandite content being lower in CEM V than in CEM I (Trotignon et al., 2006). Recently, 'low pH' cement formulations have been developed that contain additives like pozzolan and blast furnace slag (see Codina et al., 2008, and Table 8.2). Advantages of 'low pH' cements compared with CEM V cements are twofold: (1) they have a low-heat hydration temperature, which minimizes the microcracking that can have negative consequences on cement's long-term durability, and (1), their alkali content is low, which may reduce changes in clay in contact with this material compared with CEM I or CEM V cement based materials. Both of these properties improve compatibility with the repository environment, yet it remains a high-strength concrete. These formulations are seldom considered in large-scale RTM (only the study of Berner et al. (2013) was referenced on this matter in Table 8.1).

In cement pastes, hydration of Ca_2SiO_4 and Ca_3SiO_5, two anhydrous calcium silicates accounting for more than two-thirds of Portland cement's weight, leads to cryptocrystalline synthetic products named calcium silicate hydrates (C-S-H). Their chemical composition is extremely variable, with the structural Ca/Si ratio varying from ~0.6 to ~2.3 depending on the formulation and structure, similar to nano-crystalline turbostratic tobermorite (Grangeon et al., 2013a,b). One of the main difficulties in modeling any cement material is deciding how to represent the range of C-S-H phase chemical compositions. The C-S-H representation depends on the authors and is made either by considering a discrete calcium to silica ratio (e.g. 1.6, 1.2 and 0.8), or by considering a jennite–tobermorite solid solution (see, for example, Blanc et al., 2010a, for a discussion and comparison of these two approaches). Some authors use only jennite and tobermorite end-members without solid solution for intermediate compositions.

Porosity

The physical porosity of concrete used in RTM studies is about 10% (Chen et al., 2012).

Transport properties

For cement-based materials, a unique effective diffusion coefficient for all chemical species is considered in the same way as for clay materials. The chloride diffusion coefficient is often taken as a reference value, although chloride is known to interact with

C-S-H (Beaudoin et al., 1990; Nguyen et al., 2006). Chloride diffusion in cement has been studied widely (Yang and Chiang, 2008) because the presence of chloride induces corrosion in reinforcing steel, so it alters the key performance requirement for the design, construction and maintenance of concrete structures. The effective diffusion coefficient of chloride in concrete is around 10^{-12} m^2/s (see, for example, Elfmarkova et al., 2015), a value that is often considered in RTM studies (Table 8.1).

When dealing with two-phase flow transport, capillary curves and gas and water relative humidities can be modeled using Equation (8.9) (Carlier et al., 2012; Chen et al., 2012). The same laws can be applied to both clay and cement-based materials, but the pseudo air entry $1/\alpha$ (see Equation (8.8)) pressure and the pore size distribution index are not the same, although in both cases the pseudo air entry pressure is quite high (>MPa).

8.2.2.4 Iron (Metals)

A porosity problem

Steel and other metals are not porous materials, and, unfortunately, reactive transport models are, by definition, not fitted to include non-porous media (Bildstein et al., 2007). To unravel this problem, the volume clearance that exists at the interface between the canister and the surrounding material, also known as the 'technological gap', is included in the numerical cells that represent the canister. Most often, the hydraulic conductivity and the diffusion coefficient of the adjacent material is used for the numerical cells that represent the metallic structure.

Mineralogy

Nuclear waste canister materials can be made of carbon steel and cast iron, stainless steels, copper and copper alloys, titanium alloys and ceramics (King and Shoesmith, 2010). In the reviewed RTM studies, only iron-based materials were considered (Table 8.1). In repository conditions, metallic iron is not thermodynamically stable in the presence of water, and corrosion of metallic iron occurs both in oxic (operation phase) and reduced conditions (post-closure period). In the first case, corrosion leads to the release of ferric iron and hydroxide in solution (Equation (8.10)), whereas in the second case corrosion produces aqueous ferrous iron, hydrogen and hydroxide (Equation (8.11)). It leads either to a pH increase and an increase in the redox potential, or to a pH increase and a decrease in redox potential. Oxic corrosion is seldom taken into account, at least according to our literature review (only in De Windt et al. (2014) in Table 8.1).

$$4Fe^0 + 6H_2O + 3O_2(aq) \Rightarrow 4Fe^{3+} + 12OH^- \tag{8.10}$$

$$Fe^0 + 2H_2O \Leftrightarrow Fe^{2+} + 2OH^- + H_2(aq) \tag{8.11}$$

Bildstein and Claret (2015) have recently reviewed the reported experimental corrosion rates in reducing conditions. As reported in Table 8.1, in large-scale simulations, corrosion rates in the range 0.1–5 μm/year (at 25°C) have been considered by the different authors. The kinetic rate for Equation (8.11) is often assumed to be constant or to be a first-order rate–time dependency (see Table 8.1). This kinetic rate depending on temperature has been also taken into account in a few studies using the Arrhenius relationship with activation energy of 11 kJ/mol (Foct and Gras, 2003). The dependence of the kinetic rate on water saturation has also been introduced in one study (Xu et al., 2008), following experimental evidence of the phenomenon given by Lagha et al., (2007). However, the aforementioned approaches are based on simple bulk chemistry models and do not integrate a detailed corrosion mechanism model such as the one proposed by Bataillon et al. (2010, 2012). Steel canister corrosion induces two-phase flow fluid transport due to water consumption by the corrosion process and gas pressure buildup following hydrogen production (Xu et al., 2008).

8.2.2.5 Glass

Borosilicate glass is the host matrix for reprocessed high-level waste. RT models have the same problem for this material as for the metallic component, how to define porosity. In addition to the waste itself, borosilicate glasses contain different cations that maximize their durability and maintain a low melting temperature. Silicon, boron and sodium oxide are the main constituents of glasses all over the world, but glasses also contain lithium, calcium, titanium, manganese, aluminium, zinc or zirconium oxide (Vance and Begg, 2010). In contact with water, these glasses will dissolve. Modeling the long-term degradation of borosilicate nuclear waste glasses is challenging because, even if thermodynamic calculations predict that glass in contact with water will irreversibly transform into more stable phases, the rate of this transformation remains an open question (Vienna et al., 2013). Although advanced geochemical models exist for this process, the few existing RTM large-scale studies that are devoted to glass alteration (only one study has been referenced in Table 8.1) make use of a simple model with a two-stage alteration (r_0 and r_r) (Bildstein et al., 2007):

$$
\begin{aligned}
r_0 &= r_{0T_0} \exp\left[-\frac{E_{a0}}{R}\left(\frac{1}{T}-\frac{1}{T_0}\right)\right]10^{N_0 f(\mathrm{pH})} \\
r_r &= r_{rT_r} \exp\left[-\frac{E_{ar}}{R}\left(\frac{1}{T}-\frac{1}{T_r}\right)\right]
\end{aligned}
\tag{8.12}
$$

with $T_0 = 373\,\mathrm{K}$, $E_{a0} = 76\,\mathrm{kJ/mol}$, $N_0 = 0.4$, $r_{0T_0} = 1.7\,\mathrm{g/m^2/day}$ (subscript standing for 'initial stage') and $f(\mathrm{pH}) = \max(\mathrm{pH} - 7;0)$ specifying that the glass alteration rate is affected by pH only when pH is greater than 7. When the alteration rate has reached a low residual rate (r_r), with solely temperature dependency, $T_r = 323\,\mathrm{K}$, $E_{ar} = 53\,\mathrm{kJ/mol}$ and $r_{0T_r} = 6 \times 10^{-5}\,\mathrm{g/m^2/day}$.

8.3 Literature Review

Thirty RTM publications are referenced in Table 8.1. This includes all that we found for clay repository concepts and excludes most RTM studies that deal with granitic repository concepts. Among those, clay-based materials are also used as engineered barriers (Sellin and Leupin, 2013), so studies dealing with that concept but focusing on the engineered barrier were added to Table 8.1.

8.3.1 Clay/Concrete Interactions

The expected gradients between clay and concrete materials are the following (Table 8.2):

- pH varies widely in the porewaters at the sides of the interface. The pH of porewater in clay materials is generally in the range of 7 to 9, whereas it may rise above 13.5 for cement-based materials, with actual pH depending on the concrete type and ageing. Even for low pH concrete material, the pH difference at the interface may be as large as two pH units.
- Partial CO_2 pressures are far higher in the porewater of clay-rocks than in cement materials (in relationship with the pH values).
- Redox conditions are reducing in claystone, whereas they may be oxidizing or reducing in cement materials.
- The alkali (Na^+, K^+) concentrations in porewater are higher in ordinary Portland-based cement material than in clay-rocks. The reverse is true for K^+ when the cement material is made of low-pH cement.

Two-thirds of the reactive transport studies reported in Table 8.1 deal with the interactions between cement-based materials and clay. In the 1990s, early calculations based on mass balance assumptions only, and reported in Gaucher and Blanc (2006) and Savage et al. (2007), led to the conclusion that 0.2–1 m^3 of bentonite are needed to buffer the chemical perturbation created by 1 m^3 of concrete. If true, this conclusion would have been problematic for the storage concepts that rely on the properties of unaltered clay materials, and this explains why so much effort has been put into reactive transport modeling studies of the interaction at the clay/concrete interface. In fact, reactive transport modeling results from the studies reported in Table 8.1 are all consistent, with very limited spatial extension of the cement perturbation in clay materials. Various reactive transport codes have been used to tackle this issue over the two past decades. During this period, the way the system is represented, capabilities of codes, and the quality and completeness of the databases used to carry out the calculations have evolved. Therefore, in order to evaluate the impact of different modeling assumptions among available studies, Marty et al. (2014) carried out calculations with a consistent set of data and input parameters arranged with increasing order of complexity. This standardized approach allowed for a proper comparison of numerical results and showed that modeled reaction pathways were mostly independent of the modeling assumptions for simulations carried out in the presence of water saturated conditions. As an additional indication of the robustness of the results obtained with RTM studies, the accuracy and numerical stability of the RTM codes have been successfully benchmarked (see Figures 8.3 and 8.4) recently for simulations carried out to describe clay/cement interactions in saturated conditions (Marty et al., 2015a).

On the other hand, simulations accounting for non-saturated and non-isothermal conditions are rare, so their results are difficult to evaluate. In these simulations, in addition to brines forming and sulfate/carbonate salts being deposited where drying takes place, mineralogical transformation pathways in concrete were found to be similar to those simulated in saturated conditions (e.g. portlandite degradation, carbonation, Ca to Si ratio decrease in C-S-H). The period necessary to reach complete saturation inside the concrete (considering an initial saturation of 30 %) is estimated to

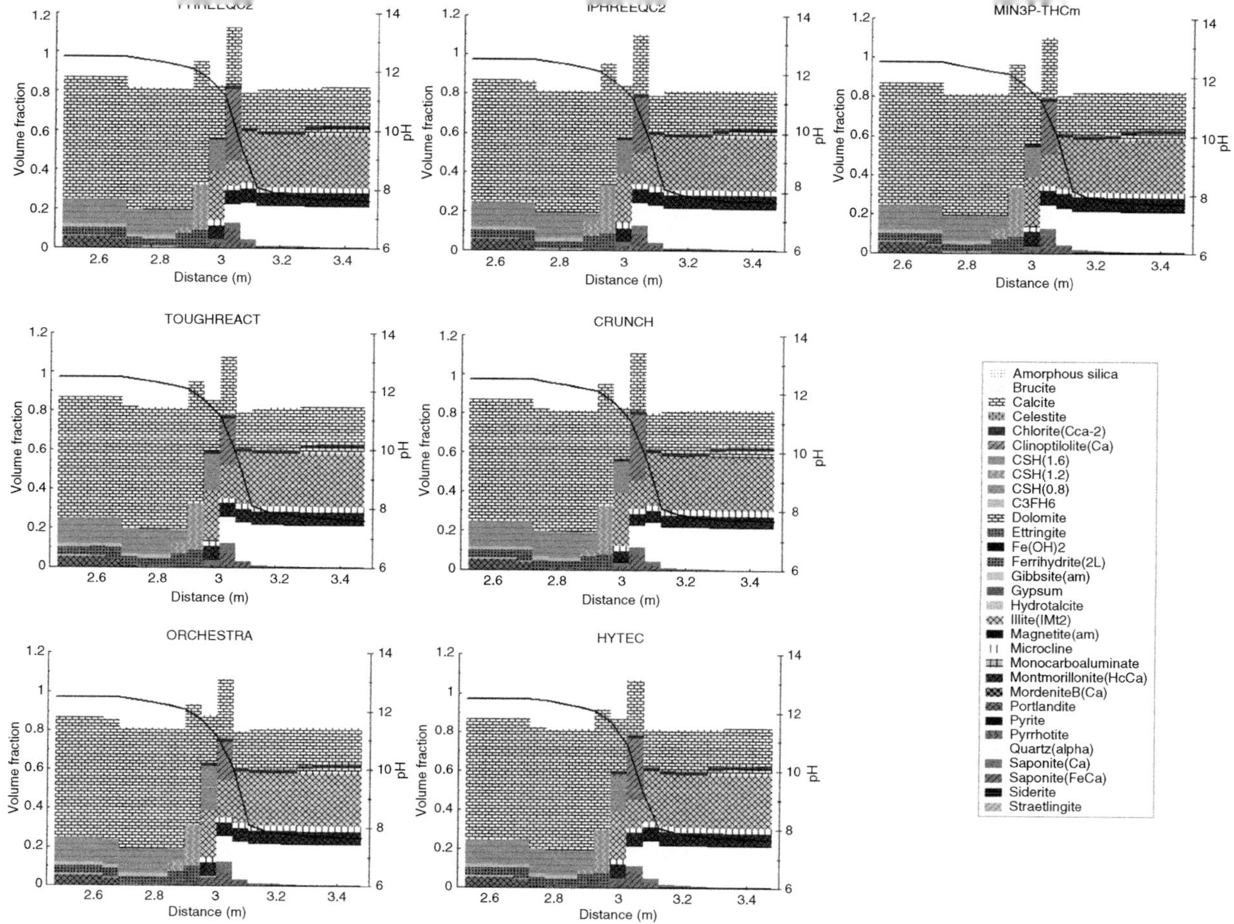

Figure 8.3 Mineralogical and pH changes obtained with TOUGHREACT, PHREEQC2, iPHREEQC2 (external transport module), CRUNCH, HYTEC, ORCHESTRA, and MIN3P-THCm after 10 000 years of concrete/claystone interaction (case 3, see Marty et al. (2015a) for details). The initial volume fraction value for $t = 0$ is the value far from the interface (2.5 m for concrete and 3.5 m for claystone) with main mineralogical changes modeled at the clay/concrete interface. Reproduced with permission of Springer. (See colour plate section for the colour representation of this figure.)

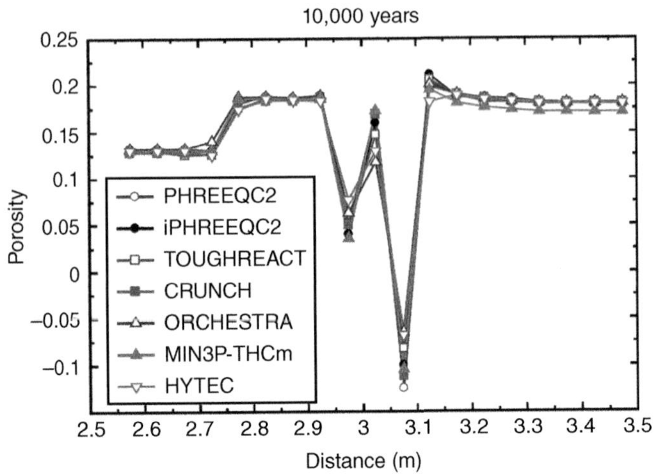

Figure 8.4 Porosity profiles obtained with TOUGHREACT, PHREEQC2, iPHREEQC2 (external transport module), CRUNCH, HYTEC, ORCHESTRA, and MIN3P-THCm after 10 000 years of concrete/claystone interaction (see case 3, Marty et al. (2015a) for details). Reproduced with permission of Springer.

be approximately 2000 years (Burnol et al., 2006), but it may depend on the grid size resolution and on the power exponent δ in Equation (8.7) (Trotignon et al., 2011), which is not a well-defined parameter for cement.

8.3.2 Iron/Clay Interactions

During the rehydration of the system in the post-closure period, metallic iron will not be thermodynamically stable in water, and it will dissolve:

$$Fe^0 + 2H_2O \rightarrow Fe^{2+} + 2OH^- + H_2 \tag{8.13}$$

Part of the ferrous iron produced by Fe^0 corrosion transforms into corrosion byproducts such as magnetite:

$$3Fe^{2+} + 4H_2O \rightarrow Fe_3O_4 + 6H^+ + H_2 \tag{8.14}$$

These reactions are anticipated to play a key role in H_2 gas pressure buildup in repository systems, and hence in the development of transient unsaturated conditions in the storage's near-field. In addition, we foresee a pH increase in the vicinity of the metallic component. Its magnitude depends on the relative value of corrosion rates versus magnetite precipitation rates. In the outer corrosion layer, other corrosion products may also form by incorporating chemical elements, such as carbonates (e.g. siderite and chukanovite) and silicates (e.g. greenalite (Rivard et al., 2015) and cronstedtite (Pignatelli et al. 2013)), which come from primary minerals dissolving from the clay materials, and diffusing from the clay-rock porewater (Bildstein and Claret, 2015). The reactivity of H_2 with clay-rock components is currently not well constrained. From a mineralogical point of view the most reactive mineral is probably pyrite, which is transformed into pyrrhotite through a dissolution/precipitation mechanism, releasing sulfides in solution but with a very limited effect on clay barrier stability (Truche et al., 2009, 2010, 2013).

Almost a third of the studies reported in Table 8.1 focus on iron/clay interactions. In agreement with Equation (8.13), the reported studies predict a pH increase at the iron–clay mineral interface up to 8–9 (Montes-H et al., 2005; Lu et al., 2011; Wersin and Birgersson, 2014), or even higher up to 10–11 (Bildstein et al., 2006; Samper et al., 2008a; Marty et al., 2010), depending on magnetite precipitation rate versus corrosion rate. Indeed, combining equations (8.13) and (8.14) shows that if the magnetite precipitation consumes all Fe^{2+} created by the Fe^0 corrosion, no pH change should take place ($3Fe^0 + 4H_2O \rightarrow Fe_3O_4 + 4H_2$). Consequently, the lowest pH values are usually associated with high precipitation rates (or local equilibrium assumption) or low corrosion rates ($\leq 1\ \mu m$/year). Irrespective of the hypotheses concerning the kinetics of mineral reactions or the presence of ion exchange in the simulation, and in agreement with the buffer capacity of clay-based materials, the high pH plume is always buffered in the clay material close to the iron/clay interface. The corrosion products predicted by the models are dominated by magnetite, but also include Fe-carbonates (Ca-siderite, chukanovite) and Fe-silicates (greenalite, cronstedtite, berthierine) in agreement with observations made on experimental systems. The extent and nature of the alteration front predicted by simulation seems to depend on the kinetic steel corrosion rate (e.g. fixed steel corrosion rate versus diffusion-limited corrosion rate) (Wilson et al., 2015). Similarly to clay/concrete interfaces, porosity reduction is predicted at the iron/clay interface in the corrosion product layer. It is also usually accompanied by a porosity increase in the clay material next to the interface (Montes-H et al., 2005; Bildstein et al., 2006; Samper et al., 2008b).

Among the available studies on iron/clay interactions, only two deal with non-saturated conditions (Table 8.1). One of them uses a two-phase flow model driven by gas-pressure buildup, originating from H_2 generation, and water consumption in equations (8.13) and (8.14) (Xu et al., 2008). In this simulation, H_2 generation rate varies with time, depending on factors such as amount of Fe^{2+}, water availability, contact area, and aqueous and solid chemistry. In these simulations, the predicted degree and extent of gas-pressure buildup is much smaller than predictions from models that neglect the coupling between flow and reactive transport mechanisms. H_2 reactivity with minerals in clay-rock was not considered in this study. If taken into account, this reactivity might further decrease H_2 pressure. Using a modeling approach where the level of saturation does not evolve over time, Sedighi et al. (2014) examined the chemical processes that may buffer excess H_2 in bentonite-engineered barriers. The reactivity of accessory minerals was shown to have an important role in mitigating excess hydrogen ions. However, some of the reactions considered, such as sulfate reduction, may not actually take place unless bacterial activity mediates the reaction.

8.3.3 Clay/Iron/Atmosphere (O_2) Interactions

Oxidation will occur in the anaerobic host-rock during construction (excavation, drilling operations) and operations (gallery ventilation) in a geological repository. Under these conditions the prevailing reducing condition will be perturbed and redox-sensitive minerals will react. Amongst the redox reactive phases that are present in clay-rocks, pyrite has been the most studied, and its oxidation is the priority mechanism to be considered (Charpentier et al., 2004; Vinsot et al., 2014):

$$FeS_2 + H_2O + 3.5O_2(aq) \rightarrow Fe^{2+} + 2H^+ + 2SO_4^{2-} \tag{8.15}$$

Pyrite oxidation by atmospheric O_2 increases sulfate concentration in the porewater of clay-rocks as well as decreasing pH. While the former can easily be tracked using porewater extraction techniques (De Craen et al., 2004, 2008), the latter is limited by the pH buffers in clay-rocks, such as calcite dissolution and reactivity of amphoteric edge sites on clay mineral surfaces (Tournassat et al., 2011). Because calcite dissolution follows pyrite oxidation, Ca^{2+} is released into the porewater and triggers ion exchange reactions that increase Na^+, K^+, and Mg^{2+} concentrations (De Craen et al., 2004). In addition, the increasing Ca^{2+} and SO_4^{2-} concentrations can induce gypsum precipitation (Charpentier et al., 2001; Lerouge et al., 2014a), as well as precipitation of other sulfate minerals such as celestite, bassanite and natrojarosite (Charpentier et al., 2004; De Craen et al., 2008; Vinsot et al., 2014). These precipitations can also be enhanced by desaturation in the rock. Fe released by pyrite oxidative dissolution can be oxidized by O_2 into Fe^{3+} and can precipitate in the form of various (hydr)oxides such as hematite, goethite or ferrihydrite depending on local conditions (Vinsot et al., 2014). Structural iron may be present in the clay minerals in the layer structure and in exchangeable positions (Hadi et al., 2013; Didier et al., 2014). These forms of iron are prone to redox reactions (Sander et al., 2015) and could constitute a major redox buffer (Didier et al., 2014). However, the reactivity of this iron pool has not been considered in RTM studies yet. Only one published RTM study deals with the oxic transient period (De Windt et al., 2014) and tackles the corrosion of the container under oxic conditions. In a closed system, simulations predict fast and complete O_2 consumption, whereas in an open system that takes into consideration the ventilation of the handling drifts, a redox contrast occurs between reducing conditions at the back of the disposal cell, with anoxic corrosion of steel and H_2 production, and oxidizing conditions at the front of the cell where oxic steel corrosion takes place. The extension of the position of the oxidizing/reducing front in the disposal cell depends on the transport gas property in the partially saturated zones.

8.3.4 Glass Corrosion and its Interaction with Clay

We could only reference one publication dealing with glass (and iron) corrosion (Bildstein et al., 2007). The calculations were performed with a glass alteration rate switching from a high initial rate to a residual rate according to the sorption capacity of corrosion products and following the mathematical treatment given by Equation (8.12). In these simulations, released silica was adsorbed on iron corrosion byproducts like magnetite and siderite or on mineral surfaces in the clay-rock, in agreement with the experimental findings of Philippini et al. (2007). The simulations indicated a pH increase at the interface up to 9.2 compared to the initial value of 7.3 in the clay-rock. As for concrete/clay and iron/clay interactions described previously, porosity reduction at the interface between glass and clay is predicted by the model due to the massive precipitation of iron corrosion products and glass alteration products.

8.4 Recent Improvements and Future Challenges in the RTM Approach to Repository Systems

8.4.1 Necessary Simplifications in the RTM Approach

Reactive transport models, like any other model, are only simplified representations of reality. We have described the simplifications that are inherent to the numerical approach and that are due to the incompleteness of available databases. Additional simplifications are associated with the conceptual model of the system under investigation. For example, the mineralogy of clay-rocks or cement materials that we consider is simpler than the actual mineralogy. The geometry chosen, the transport parameters selected, the fact that calculations are carried out using the local equilibrium hypothesis, and so on, also affect the necessary simplifications of the numerical problem. Most of these simplifications are a matter of choice for the modeler and should depend ultimately on issues that the RT simulations aim to shed light on. Modelling cation and anion diffusion in clay-rocks on short timescales may rely on conceptualizing the clay-rock as an exchanger and a carbonate phase only (Appelo and Wersin, 2007; Appelo et al., 2008, 2010), but modelling clay–concrete interactions for periods lasting thousands of years requires the introduction of a more complex mineralogical description that, in turn, leads to the need to rely on a simplified representation of the transport parameters of solute species, for computational cost reasons. By contrast, some simplifications are made by default because a given functionality simply does not exist in codes currently available. Next, we explore some recent improvements and challenges for future studies in the field.

8.4.2 Modeling Diffusion in Porous Systems with Consideration of Electrostatic Effects

The diffusion coefficients of water and solutes in clayey media have been extensively studied in conditions relevant to repository systems, particularly in the case of water tracers, anions, and alkali and alkaline earth metals (Appelo et al., 2010; Savoye et al., 2010; Gimmi and Kosakowski, 2011; Loomer et al., 2013; Tachi and Yotsuji, 2014; Bourg and Tournassat, 2015). These studies have repeatedly demonstrated differences in diffusion coefficients and apparent porosity values for anionic, neutral and cationic species. For example, while the porosity available for water diffusion is equal to the physical porosity defined in Equation (8.1), the porosity available for chloride diffusion is approximately half of the physical porosity in most clay-rocks (Pearson, 1999; Pearson et al., 2003; Van Loon et al., 2003; Fernandez et al., 2014; Tournassat et al., 2015b). This apparent inconsistency between diffusion transport parameters led to the appearance in the literature of a range of different kinds of porosity definitions: total porosity, geochemical porosity, anion accessible porosity, and so on (Pearson, 1999). However, in a RTM approach, it is not possible to specify porosity values that depend on the nature of the chemical species, so, except in two studies where anion porosity was considered (Berner et al., 2013; Kosakowski and Berner, 2013), total physical porosity has been used as a reference value for all chemical species in the studies reviewed in Table 8.2.

The reason for the observed differences in diffusive transport parameters of chemical species can be attributed, at least partly, to the presence of negatively charged surfaces on clay mineral particles. These repel anions and attract cations in the solution at the clay/water interface. This phenomenon gives rise to the presence of a solution with electric charge imbalance at the interface: the diffuse layer. Although this phenomenon has been known for more than a century (Gouy, 1910; Schofield, 1947), it has only recently been used to quantify the difference in diffusion properties of anionic, neutral and cationic species within a reactive transport modeling approach using the full Nernst-Planck equation to model diffusion instead of Fick's law (Appelo and Wersin, 2007; Appelo et al., 2010). More details on this model can be found in the review by Tournassat and Steefel (2015). This approach, based on a MCD mathematical treatment of the diffusion equations, makes it possible to consistently model ion and water tracer diffusion in a charged porous medium, and it has been applied successfully to a range of diffusion data in clay media (Appelo and Wersin, 2007; Appelo et al., 2010; Alt-Epping et al., 2015; Tinnacher et al., 2016). The same kind of approach has also been applied to cement materials where C-S-H phases carry charged surfaces (Elakneswaran et al., 2010). An additional advantage of this model is that non-Fickian diffusion behaviour, such as apparent uphill diffusion phenomena that have been evinced by Glaus et al. (2013) in the presence of salinity gradients, can be explained and quantified (Tournassat and Steefel, 2015). To the authors' knowledge, this modeling capability, which is made available in PHREEQC and CRUNCHFLOW-MC (Steefel et al., 2015a), has not been applied to large-scale simulations of repository systems yet.

8.4.3 Diffusion in Non-saturated Conditions

Although measurements conducted on clay-rocks confirm the exponential dependency of the diffusion coefficient of water on porosity and saturation as described in Equation (8.7), it seems necessary to introduce a percolation threshold reflecting that, for a given critical water saturation (S_{a_c}), the water phase is no longer connected at the scale of a representative volume in the clay-rock. Below this critical value, solutes can no longer be transported through the porewater. An empirical modified Archie's second law has been proposed by Savoye et al. (2010, 2012, 2014) to tackle this issue:

$$D_{ea,i} = \left(\frac{S_a - S_{a\_c}}{1 - S_{a_c}} \right)^m S_a^n \phi^m \tag{8.16}$$

with n equal to 2, and $m = 2.7$. The value of S_{a_c} is species dependent, pointing again to the need to rely on multi-component diffusion equations to model diffusion processes accurately in clay materials.

For concrete, a saturation dependency has also been reported for the effective chloride diffusion coefficient (Olsson et al., 2013):

$$D_{Cl^-,S_w} = S_w^{\lambda} D_{Cl^-,S_{w=1}} \tag{8.17}$$

where D_{Cl^-,S_w} is the aqueous effective diffusion for chloride at the value of S_w saturation, $D_{Cl^-,S_{w=1}}$ is the aqueous effective diffusion for chloride in saturated conditions, and λ is the saturation exponent in the range 4.5 to 6 (Olsson et al., 2013). Other authors (Zhang et al., 2015, and references therein) discussed the use of Millington formalism (Equation

(8.7)) for cement-based materials. Instead of $\delta = 4/3$ as given in Equation (8.7), a higher value (2.74) has been found, but one should keep in mind that the Millington model has been calibrated on soils whose pore structure is different from that of cement-based materials.

Again, like for clay-rocks, the existence of a threshold under which the liquid phase is disconnected and for which ionic diffusion becomes negligible has also been discussed (Baroghel-Bouny et al., 2011; de Vera et al., 2007; Martys, 1999). The value of the threshold might depend on the type of concrete considered.

8.4.4 Two-Phase Flow Models

Gas transport through low-permeability formations is complex (see Amann-Hildenbrand et al., 2015, for a review) and involves different transport mechanisms (Marschall et al., 2005; Xu et al., 2013), such as (1) advective-diffusive transport of gas dissolved in the porewater, (2) visco-capillary two-phase flow, (3) dilatancy-controlled gas flow when the gas pressure exceeds the minimal *in situ* principal stress, and (4) gas transport in tensile fractures (in this case the gas transport through macro-fractures becomes a single-phase flow process). Diffusion of the dissolved gas phase in the porewater is a ubiquitous process but with low efficiency from the transport point of view. In comparison with the water's effective diffusion coefficient, the effective diffusion coefficient of gas dissolved in the aqueous phase is between one to three orders of magnitude lower (Amann-Hildenbrand et al., 2015). Similarly to water permeability, gas permeability is very low in clay-rocks as well as engineered barriers, and the gas pressure entry is driven by capillary-scale force (Liu et al., 2014a,b). Gas transport properties of Opalinus clay and Callovo-Oxfordian clay-rocks have been determined experimentally (Marschall et al., 2005; Cariou et al., 2012). Both studies insist on the influence of the clay-rock microstructure on gas migration. A classical two-phase flow capillary pressure–relative permeability relationship based on the van Genuchten-Mualem approach (Mualem, 1976; Van Genuchten, 1980) seems to be valid for Opalinus clay samples (Croisé et al., 2006), but not for Callovian-Oxfordian clay-rock samples (Cariou et al., 2012). There the use of inverse modelling seems to show that it is not valid, as it was possible to properly reproduce the early stages (<3 months) of gas injection tests in the Callovo-Oxfordian claystone with a van Genuchten-Mualem model (De La Vaissiere et al., 2014). For the last stages of the gas injection experiment, Parker's gas permeability relationship (Parker et al., 1987) was a better alternative but did not contradict the use of a two-phase flow modelling approach. In addition, the parameterization used in this study for $k_{r,g}$ was very close to that given in Equation (8.9), with $S_e = S_w$ and $\gamma = 0.5$. These apparent discrepancies in models are in line with the statement of Amann-Hildenbrand et al. (2015) that 'standard viscocapillary models of two-phase flow (based on simple concepts of phase displacement) provide a poor representation of the underlying physics governing multiphase flow'. However, because standard two-phase flow viscocapillary models give the correct order of magnitude and are easy to implement in reactive transport codes for large-scale simulation on long timescales (a million years), they are widely used in modeling hydrogen gas transport through clay-rocks (Enssle et al., 2011a; Poller et al., 2011; Treille et al., 2011; Yu et al., 2011; Zhang et al., 2011; Brommundt et al., 2014).

For concrete, experimental evidence supports the use of Equation (8.8) (Carlier et al., 2012; Chen et al., 2012). Relative permeability curves for gas and water (difficult to measure for concrete because of the reactivity) can also be described using Equation (8.9) (Leech et al., 2008; Chen et al., 2012; Kameche et al., 2014), although some refinements have been proposed (Zhou, 2014; Zhou et al., 2015).

8.4.5 Water Consumption and Non-saturated Conditions

In relation to the saturation problem, for example, equations (8.13) and (8.14) indicate that water will be consumed during corrosion reactions. From a numerical point of view, ionic strength will increase in the cells where corrosion is considered. This leads to at least two numerical problems: (1) the activity model might be outside its validity domain as the ionic strength increases, and (2) the water fluxes should be updated in response to this water consumption. Some authors have tried to tackle this issue by associating water consumption induced by corrosion to an equivalent water withdrawal rate (Xu et al., 2008). However, this approach of kinetic dependence toward water content has not been systematized to iron corrosion or to other materials. Apart from this specific question of water consumption, we need more RTM modeling studies that account for non-saturated conditions. The fact that few RTM studies simulate non-saturated conditions is probably due to the long computing times required for a short simulated time. Indeed, Trotignon et al. (2011) reported run durations of 30 to 180 days to simulate 100 years. This long computational time is linked to the fact that in the version of TOUGHREACT used for this study, gas-phase transport is not iteratively coupled to reaction processes. It was therefore necessary to use a sequential non-iterative approach (SNIA) for reactive transport. However, as gas diffusion in the gas phase is approximatively four orders of magnitude higher than in the water phase, very small time-steps must be used in order to respect the Van Neumann (Sousa, 2003) convergence rule. Continuous code development and improvement (e.g. parallelization) will probably help to solve issues like this in the future.

8.4.6 Reducing Porosity and Coupling with Transport Parameters

As described above, at the interface, mineral dissolution and precipitation will occur. Depending on the conditions, porosity can either rise or fall. In the continuum approach, the effect of this mineralogy modification on the porosity and therefore on the transport is integrated using Archie's law as described in Equation (8.6). However, its applicability to complex and heterogeneous systems is questionable (Navarre-Sitchler et al., 2009). In particular, Archie's law failed to describe the effect of a precipitation front on the time-dependent transport properties of a porous medium (Chagneau et al., 2015). Using this law, one-dimensional continuum models always underestimated the remaining connected porosity in the precipitation zone, or overestimated the amount of precipitate. The results suggest that 1D single continuum models might fail in the long-term prediction of hazardous material transport in geochemically disturbed systems, and argue for more complex 2D/3D pore-scale reactive transport models, including complex precipitation kinetics (e.g. heterogeneous nucleation and inhibition processes) together with a modified diffusivity/porosity relationship.

8.4.7 Accounting for Material Heterogeneities

As described previously, most reactive transport studies only consider very simple geometries and homogeneous media, thus disregarding spatial heterogeneities at the repository and/or geological formation scale. One may argue that selected clay-rock formations are thick homogeneous clay layers (Delay, 2010; Delay et al., 2007), so that using an average mineralogy, porosity, transport properties, and so on, in each cell of the grid is satisfactory to a first approximation. The effect of the variability of parameters describing the materials can be evaluated by a sensitivity analysis in which the parameter of interest can be varied between its experimentally determined lower and upper limits. Nevertheless, for some heterogeneity, this approach may not be sufficient. In the case of how to describe the EDZ, instead of considering a homogeneous domain with degraded properties compared with the pristine rock (i.e. considering a higher diffusion coefficient and higher permeability), a multiple continua approach may be more appropriate. The dual-continuum approach is commonly and extensively applied with subsurface environment applications, but more sophisticated models are envisaged (Steefel et al., 2015a). This approach has already been tested to model natural analogues (Steefel and Lichtner, 1998). An approach considering the EDZ as a fractured medium has already been tested (De Windt et al., 2004b), but not in a multiple continua approach and not in a two-phase flow model.

8.4.8 Kinetics versus Local Equilibrium Calculations

For mineral dissolution precipitation reactions, simulations reported in Table 8.1 make use of the local chemical equilibrium assumptions, of kinetic controls, or of a mixture of both. For the investigated systems, in which transport is dominated by diffusion, reactions are controlled by diffusive mass transfer rather than by kinetic constraints, and the spatial extent of mineral alterations does not change when using thermodynamic equilibrium instead of kinetics (Kosakowski and Berner, 2013). However, differences in dissolution/precipitation rates and in grid size resolution affect the time at which the porosity is reduced to zero (Marty et al., 2009; Hayek et al., 2011). Using a kinetics model that treats portlandite dissolution and calcite precipitation based on the so-called shrinking core model (basically it is a diffusion-controlled model with precipitation of a calcite layer around portlandite particles), only a slight difference can be found for clogging time (Liu et al., 2014c). Therefore, the accuracy of predictions made for how a repository system evolves over time relies on the quality of the kinetic description of the reactions taking place at the interface.

The reaction rates taken into consideration by most RTM codes originate from the transition state theory (TST) described in detail by Aagaard and Helgeson (1982) and Lasaga (1981, 1984). The use of TST kinetic models appears to be unavoidable for large-scale simulation currently (Marty et al., 2015b), even if the use of a mean or rate-limiting value describing the reaction rate appears to be questionable, notably regarding the value to be considered for surface area (Lüttge et al., 2013). Furthermore, mineral dissolution and precipitation occur at the pore scale (Guarracino et al., 2014), and the continuum descriptions of these processes using parameters such as reactive surface area and intrinsic reaction rate as averaged properties of the porous medium have many limitations (Molins et al., 2014). In particular, mineral precipitation is dependent on pore size, degree of solution saturation, and surface chemistry, where the latter leads to the presence of domains either favourable or unfavourable to precipitation in a heterogeneous porous medium (Stack et al., 2014). In turn, such heterogeneous precipitation regimes lead to changes in transport parameters that are difficult to capture in simple single-continuum models. Theoretical models used to explain the behaviour of saturated solutions in nanoscale pore sizes (<100 nm) are especially controversial. The inhibition of mineral nucleation in nanopores has traditionally been ascribed to the pore-size controlled solubility (PCS) effect, which predicts a higher solubility in nanopores. This has been successfully applied to explain the lack of mineralization in nanopores in sedimentary formations or the lack of mineralization in some nanopores associated with stylolites (Putnis and Mauthe, 2001; Emmanuel and Berkowitz, 2007; Emmanuel and Ague, 2009). However, recent observations suggest that the PCS effect is not dominant in cases where surface chemistry offers a preferential template for mineral nucleation (Stack et al., 2014). This model has been recently reinforced by statistical physics models of heterogeneous nucleation (Hedges and Whitelam, 2013). These results highlight the need for further investigation to develop consistent thermodynamic models of precipitation phenomena at the nanoscale. Such models will also help to better describe porosity reduction and how it interacts with transport parameters. Another important aspect regarding kinetics and a repository situation is the integration of saturation dependency terms in the kinetic law, as, for example, described in Xu et al. (2008). To do so, a water contact area factor affecting the effective surface area must also be introduced.

8.4.9 Modeling Glass Alteration in Clay-rock Environments

Glass alteration in RTM studies is very rare; we have only referenced one study in Table 8.1, and its glass alteration model is very simple. Advances in modeling interfacial glass–water reactions have been reviewed recently (Pierce et al., 2014). The existing geochemical models usually describe evolving bulk solution chemistry by capturing the thermodynamic constraints associated with secondary phase formation. Such calculations might only provide a first approximation of the continuum scale processes controlling the evolution of the glass–water reaction, and one of the limitations of such a model is the need to adjust log K dissolution/precipitation constants to reproduce solubility data (Pierce et al., 2014). With a combination of diffusion and dissolution processes in the glass within a single set of equations, the GRAAL model (Glass Reactivity with Allowance for the Alteration Layer) could be a significant improvement in the RTM representation of glass alteration (Frugier et al., 2008). In this model, the glass–water interface is constituted from a pristine glass, a gel, and water. The main mechanisms implemented in GRAAL are:

- fast ion exchange and hydrolysis reactions between the mobile glass constituents (alkalis, boron, etc.) and the contacting solution;
- slow hydrolysis of the silica network to form a hydrated silica-rich amorphous layer by polymerization and/or condensation;
- dissolution of the amorphous layer itself as long as the external solution is undersaturated with respect to silica;

- transformation of the amorphous layer into a diffusive barrier (referred to as a passive reactive interface – PRI);
- secondary phase precipitation on the glass surface (e.g. phyllosilicates, calcium and rare earth element-bearing phosphates, zeolites and hydrated calcium silicates).

The main simplifying hypotheses in the GRAAL model are:

- water diffusion within the PRI being a rate-limiting step for glass corrosion;
- neglecting water diffusion in the glass and proton/alkali ion exchange;
- describing PRI reactivity with surrounding solutions using a local equilibrium approach.

The GRAAL model has been implemented in the HYTEC reactive transport code (Debure et al., 2013). Recently, a more general model called the Kinetic Micro-Continuum (KμC) and based on the reactive transport code CrunchFlow has been developed and successfully used to model glass corrosion (Steefel et al., 2015b). It makes no *a priori* hypotheses about the rate-limiting step in the overall glass process. In comparison with the GRAAL model, KμC includes the following processes:

- water diffusion through the pristine glass and corrosion byproducts;
- ion exchange between water and the cations borne by the glass;
- introduction of kinetics to model hydrolysis reactions, formation of amorphous/crystalline phases;
- multicomponent ion diffusion through the glass corrosion products;
- updated transport properties (e.g. tortuosity, diffusivity) as a function of porosity changes induced by geochemical processes like densification reactions or formation of new crystalline phases.

This model better represents the nanosized chemical fronts that have been observed experimentally (Hellmann et al., 2015). The integration of glass alteration advanced models in RTM calculations is thus possible in principle. However, they are yet to be implemented in modeling large repository systems, and the size of the reactive interface that must be taken into account in these approaches remains a challenge.

8.4.10 Coupling Mechanics and Chemistry

Geochemical changes, including, among others, changes in ion concentration, clay exchanger composition and abundance of swelling materials (e.g. smectite illitization), will modify bentonite's swelling behaviour and the associated multi-scale porosity (Massat et al., 2016). Swelling/shrinkage has being coupled to reactive transport models either implementing the Barcelona Basic Model (Alonso et al., 1990; Gens et al., 2006) into TOUHGREACT–FLAC3D (Rutqvist et al., 2014; Rutqvist, 2015) or linking TOUGHREACT with a swelling model based on diffuse double-layer theory (Zheng et al., 2014). Such couplings are promising but are still challenging in terms of integrating complex chemistry. In addition, both numerical modeling and investigations of laboratory experiments on aged sample interfaces retrieved from the underground research laboratory (see Bildstein and Claret, 2015, for a review) clearly indicated localized but significant porosity modifications (clogging or opening) induced by the geochemical reactions. These porosity changes are poorly understood in terms of mechanical behaviour and must be integrated in reactive transport modelling.

Acknowledgements

Concepts and research topics discussed in this chapter have been investigated mostly within the framework of the Andra-BRGM scientific partnership. Fruitful discussions with colleagues at Andra and BRGM are warmly acknowledged. We would like to dedicate this chapter to our colleague and friend Eric Giffaut, who sadly passed away in December 2015. His scientific vision has inspired us a great deal in our scientific career.

References

Aagaard, P. and Helgeson, H.C. (1982). Thermodynamic and kinetic constraints on reaction rates among minerals and aqueous solutions; I, Theoretical considerations. *American Journal of Science* 282: 237–285.

Ahn, J. and Apted, M.J. (2010). Preface. In: *Geological Repository Systems for Safe Disposal of Spent Nuclear Fuels and Radioactive Waste* (ed. J. Ahn and M.J. Apted), xxiii–xxvi. Woodhead Publishing.

Alonso, E.E., Gens, A., and Josa, A. (1990). A constitutive model for partially saturated soils. *Géotechnique* 40: 405–430.

Alonso, M.C., García Calvo, J.L., Hidalgo, A., and Fernández Luco, L. (2010). Development and application of low-pH concretes for structural purposes in geological repository systems, In: *Geological Repository Systems for Safe Disposal of Spent Nuclear Fuels and Radioactive Waste* (ed. J. Ahn and M.J. Apted), 286–322. Woodhead Publishing.

Alt-Epping, P., Tournassat, C., Rasouli, P., et al. (2015). Benchmark reactive transport simulations of a column experiment in compacted bentonite with multispecies diffusion and explicit treatment of electrostatic effects. *Computational Geosciences* 19 *(3)*: 535–550.

Altmann, S., Tournassat, C., Goutelard, F., et al. (2012). Diffusion-driven transport in clayrock formations. *Applied Geochemistry* 27: 463–478.

Amann-Hildenbrand, A., Krooss, B.M., Harrington, J., et al. (2015). Gas transfer through clay barriers. In: *Developments in Clay Science* Vol 6 Natural and Engineered Clay Barriers, 1st Edition (ed. C. Tournassat, C. Steefel, I.C. Bourg, F. Bergaya), 227–267. Amsterdam: Elsevier.

Andra. (2005). Dossier 2005 Argile. Tome évolution phénoménologique du stockage géologique. Available at http://www.andra.fr/download/site-principal/document/dossier2005/D05A_269_TEP.pdf

Andra. (2009). JALON 2009 HA-MAVL - Options de conception du stockage en formation géologique profonde. Available at https://www.andra.fr/download/site-principal/document/editions/394.pdf

Appelo, J. and Wersin, P. (2007). Multicomponent diffusion modeling in clay systems with application to the diffusion of tritium, iodide, and sodium in opalinus clay. *Environmental Science and Technology* 41: 5002–5007.

Appelo, C.A.J., Vinsot, A., Mettler, S., and Wechner, S. (2008). Obtaining the porewater composition of a clay rock by modeling the in- and out-diffusion of anions and cations from an in-situ experiment. *Journal of Contaminant Hydrology* 101: 67–76.

Appelo, C.A.J., Van Loon, L.R., and Wersin, P. (2010). Multicomponent diffusion of a suite of tracers (HTO, Cl, Br, I, Na, Sr, Cs) in a single sample of Opalinus Clay. *Geochimica et Cosmochimica Acta* 74: 1201–1219.

Apted, M. and Ahn, J. (2010). Multiple-barrier geological repository design and operation strategies for safe disposal of radioactive materials. In: *Geological Repository Systems for Safe Disposal of Spent Nuclear Fuels and Radioactive Waste* (ed. J. Ahn and M.J. Apted), 3–28. Woodhead Publishing.

Armand, G., Leveau, F., Nussbaum, C., et al. (2014). Geometry and properties of the excavation-induced fractures at the Meuse/Haute-Marne URL Drifts. *Rock Mechanics and Rock Engineering* 47: 21–41.

Bach, T.T.H., Coumes, C.C.D., Pochard, I., et al. (2012). Influence of temperature on the hydration products of low pH cements. *Cement and Concrete Research* 42: 805–817.

Baroghel-Bouny, V., Thiéry, M., and Wang, X. (2011). Modelling of isothermal coupled moisture–ion transport in cementitious materials. *Cement and Concrete Research* 41: 828–841.

Bastiaens, W., Bernier, F., and Li, X.L. (2007). SELFRAC: Experiments and conclusions on fracturing, self-healing and self-sealing processes in clays. *Physics and Chemistry of the Earth, Parts A/B/C* 32: 600–615.

Bataillon, C., Bouchon, F., Chainais-Hillairet, C., et al. (2010). Corrosion modelling of iron based alloy in nuclear waste repository. *Electrochimica Acta* 55: 4451–4467.

Bataillon, C., Bouchon, F., Chainais-Hillairet, C., et al. (2012). Numerical methods for the simulation of a corrosion model with moving oxide layer. *Journal of Computational Physics* 231: 6213–6231.

Beaudoin, J.J., Ramachandran, V.S., and Feldman, R.F. (1990). Interaction of chloride and C-S-H. *Cement and Concrete Research* 20: 875–883.

Benet, L.-V., Bouillet, C., and Wendling, J. (2014a). Analysis of the ambient conditions in an IL-LLW storage cell in a deep clay repository during the waiting closure period. In: *Clays in Natural and Engineered Barriers for Radioactive Waste Confinement* (ed. S. Norris, J. Bruno, M. Cathelineau, et al.), 145–161. Geological Society London Special Publication.

Benet, L.-V., Tulita, C., Calsyn, L., and Wendling, J. (2014b). Evolution of temperature and humidity in an underground repository over the operation period. In: *Clays in Natural and Engineered Barriers for Radioactive Waste Confinement* (ed. S. Norris, J. Bruno, M. Cathelineau, et al.), 413–426. Geological Society London Special Publication.

Benet, L.-V., Tulita, C., Pasteau, A., and Wendling, J. (2014c). Analysis of the long-term hydraulic-gas transient in the central zone of a deep clay repository. In: *Clays in Natural and Engineered Barriers for Radioactive Waste Confinement* (ed. S. Norris, J. Bruno, M. Cathelineau, et al.), 483–496. Geological Society London Special Publication.

Berner, U., Kulik, D.A., and Kosakowski, G. (2013). Geochemical impact of a low-pH cement liner on the near field of a repository for spent fuel and high-level radioactive waste. *Physics and Chemistry of the Earth, Parts A/B/C* 64: 46–56.

Bildstein, O. and Claret, F. (2015). Stability of clay barriers under chemical perturbations. In: *Developments in Clay Science* Vol 6 Natural and Engineered Clay Barriers, 1st Edition (ed. C. Tournassat, C. Steefel, I.C. Bourg, F. Bergaya), 155–188. Amsterdam: Elsevier.

Bildstein, O., Trotignon, L., Perronnet, M., and Jullien, M. (2006). Modelling iron-clay interactions in deep geological disposal conditions. *Physics and Chemistry of the Earth* 31: 618–625.

Bildstein, O., Trotignon, L., Pozo, C., and Jullien, M. (2007). Modelling glass alteration in an altered argillaceous environment. *Journal of Nuclear Materials* 362: 493–501.

Blanc, P., Bourbon, X., Lassin, A., and Gaucher, E.C. (2010a). Chemical model for cement-based materials: Temperature dependence of thermodynamic functions for nanocrystalline and crystalline C-S-H phases. *Cement and Concrete Research* 40: 851–866.

Blanc, P., Bourbon, X., Lassin, A., and Gaucher, E.C. (2010b). Chemical model for cement-based materials: Thermodynamic data assessment for phases other than C-S-H. *Cement and Concrete Research* 40: 1360–1374.

Blanc, P., Lassin, A., Piantone, P., et al. (2012). Thermoddem: A geochemical database focused on low temperature water/rock interactions and waste materials. *Applied Geochemistry* 27: 2107–2116.

Blanc, P., Vieillard, P., Gailhanou, H., et al. (2015). ThermoChimie database developments in the framework of cement/clay interactions. *Applied Geochemistry* 55: 95–107.

Bourg, I.C. (2015). Sealing shales versus brittle shales: A sharp threshold in the material properties and energy technology uses of fine-grained sedimentary rocks. *Environmental Science & Technology Letters* 2: 255–259.

Bourg, I.C. and Tournassat, C. (2015). Self-diffusion of water and ions in clay barriers. In: *Developments in Clay Science* Vol 6 Natural and Engineered Clay Barriers, 1st Edition (ed. C. Tournassat, C. Steefel, I.C. Bourg, F. Bergaya), 189–226. Amsterdam: Elsevier.

Bradbury, M.H. and Baeyens, B. (1998). A physicochemical characterisation and geochemical modelling approach for determining porewater chemistries in argillaceous rocks. *Geochimica et Cosmochimica Acta* 62: 783–795.

Bradbury, M.H. and Baeyens, B. (2003). Porewater chemistry in compacted re-saturated MX-80 bentonite. *Journal of Contaminant Hydrology* 61: 329–338.

Bradbury, M.H. and Baeyens, B. (2009). Experimental and modelling studies on the pH buffering of MX-80 bentonite porewater. *Applied Geochemistry* 24: 419–425.

Brommundt, J., Kaempfer, T.U., Enssle, C.P., et al. (2014). Full-scale 3D modelling of a nuclear waste repository in the Callovo-Oxfordian clay. Part 1: Thermo-hydraulic two-phase transport of water and hydrogen. In: *Clays in Natural and Engineered Barriers for Radioactive Waste Confinement* (ed. S. Norris, J. Bruno, M. Cathelineau, et al.), 443–467. Geological Society London Special Publication.

Brooks, R. and Corey, A. (1964). Hydraulic properties of porous media. *Hydrology Papers No. 3*, Colorado State University, Ft. Collins, CO.

Burnol, A., Blanc, P., Xu, T., et al. (2006). Uncertainty in the reactive transport model response to an alkaline perturbation in a clay formation. *TOUGH Symposium 2006*, Berkeley, CA.

Cariou, S., Skoczylas, F., and Dormieux, L. (2012). Experimental measurements and water transfer models for the drying of argillite. *International Journal of Rock Mechanics and Mining Sciences* 54: 56–69.

Carlier, J.P., Rougelot, T., and Burlion, N. (2012). Performance evaluation of models describing sorption isotherm in cementitious materials between saturation and oven dryness. *Construction and Building Materials* 37: 58–66.

Chagneau, A., Claret, F., Enzmann, F., et al. (2015). Mineral precipitation-induced porosity reduction and its effect on transport parameters in diffusion-controlled porous media. *Geochemical Transactions* 16: 13.

Chapman, N. and Hooper, A. (2012). The disposal of radioactive wastes underground. *Proceedings of the Geologists' Association* 123: 46–63.

Charpentier, D., Cathelineau, M., Mosser-Ruck, R., and Bruno, G. (2001). Mineralogical evolution of argillites in dehydrated-oxidised zones: the example of the argillitic walls from Tournemire tunnel. *Comptes Rendus De l' Academie Des Sciences Serie II Fascicule A-Sciences de la Terre et des Planete* 332: 601–607.

Charpentier, D., Mosser-Ruck, R., Cathelineau, M., and Guillaume, D. (2004). Oxidation of mudstone in a tunnel (Tournemire, France): consequences for the mineralogy and crystal chemistry of clay minerals. *Clay Minerals* 39: 135–149.

Chen, W., Liu, J., Brue, F., et al. (2012). Water retention and gas relative permeability of two industrial concretes. *Cement and Concrete Research* 42: 1001–1013.

Chen, Z., Montavon, G., Ribet, S., et al. (2014). Key factors to understand in-situ behavior of Cs in Callovo-Oxfordian clay-rock (France). *Chemical Geology* 387: 47–58.

Claret, F., Sakharov, B.A., Drits, V.A., et al. (2004). Clay minerals in the Meuse-Haute marne underground laboratory (France): Possible influence of organic matter on clay mineral evolution. *Clays and Clay Minerals* 52: 515–532.

Codina, M., Cau-dit-Coumes, C., Le Bescop, P., et al. (2008). Design and characterization of low-heat and low-alkalinity cements. *Cement and Concrete Research* 38: 437–448.

Croisé, J., Mayer, G., Marschall, P., et al. (2006). Gas threshold pressure test performed at the Mont Terri Rock Laboratory (Switzerland): Experimental data and data analysis. *Oil & Gas Science and Technology – Rev. IFP* 61: 631–645.

Debure, M., De Windt, L., Frugier, P., and Gin, S. (2013). HLW glass dissolution in the presence of magnesium carbonate: Diffusion cell experiment and coupled modeling of diffusion and geochemical interactions. *Journal of Nuclear Materials* 443: 507–521.

De Craen, M., Van Geet, M., Wang, L., and Put, M. (2004). High sulphate concentrations in squeezed Boom Clay pore water: evidence of oxidation of clay cores. *Physics and Chemistry of the Earth, Parts A/B/C* 29: 91–103.

De Craen, M., Geet, M.V., Honty, M., et al. (2008). Extent of oxidation in Boom Clay as a result of excavation and ventilation of the HADES URF: Experimental and modelling assessments. *Physics and Chemistry of the Earth, Parts A/B/C* 33, Supplement 1: S350–S362.

De La Vaissiere, R., Gerard, P., Radu, J.-P., et al. (2014). Gas injection test in the Callovo-Oxfordian claystone: data analysis and numerical modelling. In: *Clays in Natural and Engineered Barriers for Radioactive Waste Confinement* (ed. S. Norris, J. Bruno, M. Cathelineau, et al.), 427–441. Geological Society London Special Publication.

Delay, J. (2010). Clay geological repository systems: characterisation and site surveying technologies and techniques. In: *Geological Repository Systems for Safe Disposal of Spent Nuclear Fuels and Radioactive Waste* (ed. J. Ahn and M.J. Apted), 153–187. Woodhead Publishing.

Delay, J., Rebours, H., Vinsot, A., and Robin, P. (2007). Scientific investigation in deep wells for nuclear waste disposal studies at the Meuse/Haute Marne underground research laboratory, Northeastern France. *Physics and Chemistry of the Earth* 32: 42–57.

de Marsily, G., Ledoux, E., Barbreau, A., and Margat, J. (1977). Nuclear waste disposal: can the geologist guarantee isolation? *Science* 197: 519–527.

de Vera, G., Climent, M.A., Viqueira, E., et al. (2007). A test method for measuring chloride diffusion coefficients through partially saturated concrete. Part II: The instantaneous plane source diffusion case with chloride binding consideration. *Cement and Concrete Research* 37: 714–724.

De Windt, L., Pellegrini, D., and van der Lee, J. (2004a). Coupled modeling of cement/claystone interactions and radionuclide migration. *Journal of Contaminant Hydrology* 68: 165–182.

De Windt, L., Pellegrini, D., and van der Lee, J. (2004b). Reactive transport modelling of a spent fuel repository in a stiff clay formation considering excavation damaged zones. *Radiochimica Acta* 92: 841–848.

De Windt, L., Marsal, F., Corvisier, J., and Pellegrini, D. (2014). Modeling of oxygen gas diffusion and consumption during the oxic transient in a disposal cell of radioactive waste. *Applied Geochemistry* 41: 115–127.

Didier, M., Géhin, A., Grenèche, J.-M., et al. (2014). Method development for evaluating the redox state of Callovo-Oxfordian clayrock and synthetic montmorillonite for nuclear waste management. *Applied Geochemistry* 49: 184–191.

Duro, L., Bruno, J., Grivé, M., et al. (2014a). Redox processes in the safety case of deep geological repositories of radioactive wastes. Contribution of the European RECOSY Collaborative Project. *Applied Geochemistry* 49: 206–217.

Duro, L., Domenech, C., Grive, M., et al. (2014b). Assessment of the evolution of the redox conditions in a low and intermediate level nuclear waste repository (SFR1, Sweden). *Applied Geochemistry* 49: 192–205.

Elakneswaran, Y., Iwasa, A., Nawa, T., et al. (2010). Ion-cement hydrate interactions govern multi-ionic transport model for cementitious materials. *Cement and Concrete Research* 40: 1756–1765.

Elfmarkova, V., Spiesz, P., and Brouwers, H.J.H. (2015). Determination of the chloride diffusion coefficient in blended cement mortars. *Cement and Concrete Research* 78 Part B: 190–199.

Emmanuel, S. and Ague, J.J. (2009). Modeling the impact of nano-pores on mineralization in sedimentary rocks. *Water Resources Research* 45: W04406.

Emmanuel, S. and Berkowitz, B. (2007). Effects of pore-size controlled solubility on reactive transport in heterogeneous rock. *Geophysical Research Letters* 34: 1–5.

Enssle, C.P., Croisé, J., Poller, A., et al. (2011a). Full scale 3D-modelling of the coupled gas migration and heat dissipation in a planned repository for radioactive waste in the Callovo-Oxfordian clay. *Physics and Chemistry of the Earth, Parts A/B/C* 36: 1754–1769.

Enssle, C.P., Cruchaudet, M., Croisé, J., and Brommundt, J. (2011b). Determination of the permeability of the Callovo-Oxfordian clay at the metre to decametre scale. *Physics and Chemistry of the Earth, Parts A/B/C* 36: 1669–1678.

Enssle, C.P., Brommundt, J., Kaempfer, T.U., et al. (2014). Full-scale 3D modelling of a nuclear waste repository in the Callovo-Oxfordian clay. Part 2: Thermo-hydraulic two-phase transport of water, hydrogen, 14C and 129I. *In: Clays in Natural and Engineered Barriers for Radioactive Waste Confinement* (ed. S. Norris, J. Bruno, M. Cathelineau, et al.). Geological Society London Special Publication.

Fernandez, R., Cuevas, J., and Maeder, U.K. (2009). Modelling concrete interaction with a bentonite barrier. *European Journal of Mineralogy* 21: 177–191.

Fernandez, A.M., Sanchez-Ledesma, D.M., Tournassat, C., et al. (2014). Applying the squeezing technique to highly consolidated clayrocks for pore water characterisation: Lessons learned from experiments at the Mont Terri Rock Laboratory. *Applied Geochemistry* 49: 2–21.

Foct, F. and Gras, J. (2003). Semi-empirical model for carbon steel corrosion in long term geological nuclear waste disposal. Prediction of long term corrosion behaviour in nuclear waste systems. *Proceedings of an International Workshop*, Cadarache, France, 2002.

Frugier, P., Gin, S., Minet, Y., et al. (2008). SON68 nuclear glass dissolution kinetics: Current state of knowledge and basis of the new GRAAL model. *Journal of Nuclear Materials* 380: 8–21.

Gailhanou, H., van Miltenburg, J.C., Rogez, J., et al. (2007). Thermodynamic properties of anhydrous smectite MX-80, illite IMt-2 and mixed-layer illite-smectite ISCz-1 as determined by calorimetric methods. Part I: Heat capacities, heat contents and entropies. *Geochimica et Cosmochimica Acta* 71: 5463–5473.

Gailhanou, H., Rogez, J., van Miltenburg, J.C., et al. (2009). Thermodynamic properties of chlorite CCa-2. Heat capacities, heat contents and entropies. *Geochimica et Cosmochimica Acta* 73: 4738–4749.

Gailhanou, H., Blanc, P., Rogez, J., et al. (2012). Thermodynamic properties of illite, smectite and beidellite by calorimetric methods: Enthalpies of formation, heat capacities, entropies and Gibbs free energies of formation. *Geochimica et Cosmochimica Acta* 89: 279–301.

Gailhanou, H., Blanc, P., Rogez, J., et al. (2013). Thermodynamic properties of saponite, nontronite, and vermiculite derived from calorimetric measurements. *American Mineralogist* 98: 1834–1847.

García Calvo, J.L., Hidalgo, A., Alonso, C., and Fernández Luco, L. (2010). Development of low-pH cementitious materials for HLRW repositories: Resistance against ground waters aggression. *Cement and Concrete Research* 40: 1290–1297.

Gaucher, E.C. and Blanc, P. (2006). Cement/clay interactions – a review: Experiments, natural analogues, and modeling. *Waste Management* 26: 776–788.

Gaucher, E., Robelin, C., Matray, J.M., et al. (2004a). ANDRA underground research laboratory: interpretation of the mineralogical and geochemical data acquired in the Callovian-Oxfordian formation by investigative drilling. *Physics and Chemistry of the Earth* 29: 55–77.

Gaucher, E.C., Blanc, P., Matray, J.M., and Michau, N. (2004b). Modeling diffusion of an alkaline plume in a clay barrier. *Applied Geochemistry* 19: 1505–1515.

Gaucher, E.C., Blanc, P., Bardot, F., et al. (2006). Modelling the porewater chemistry of the Callovian-Oxfordian formation at a regional scale. *Comptes Rendus Geoscience* 338: 917–930.

Gaucher, E.C., Tournassat, C., Pearson, F.J., et al. (2009). A robust model for pore-water chemistry of clayrock. *Geochimica et Cosmochimica Acta* 73: 6470–6487.

Gens, A., Sánchez, M., and Sheng, D. (2006). On constitutive modelling of unsaturated soils. *Acta Geotechnica* 1: 137–147.

Gérard, F., Clément, A., and Fritz, B. (1998). Numerical validation of a Eulerian hydrochemical code using a 1D multisolute mass transport system involving heterogeneous kinetically controlled reactions. *Journal of Contaminant Hydrology* 30: 201–216.

Giffaut, E., Grivé, M., Blanc, P., et al. (2014). Andra thermodynamic database for performance assessment: ThermoChimie. *Applied Geochemistry* 49: 225–236.

Gimmi, T. and Kosakowski, G. (2011). How mobile are sorbed cations in clays and clay rocks? *Environmental Science and Technology* 45: 1443–1449.

Gin, S., Jollivet, P., Fournier, M., et al. (2015). Origin and consequences of silicate glass passivation by surface layers. *Nature Communication* 6.

Glaus, M.A., Birgersson, M., Karnland, O., and Van Loon, L.R. (2013). Seeming steady-state uphill diffusion of Na-22(+) in compacted montmorillonite. *Environmental Science and Technology* 47: 11522–11527.

Gonçalvès, J., Adler, P.M., Cosenza, P., et al. (2015). Semipermeable membrane properties and chemomechanical coupling in clay barriers. In: *Developments in Clay Science* Vol 6

Natural and Engineered Clay Barriers, 1st Edition (ed. C. Tournassat, C. Steefel, I.C. Bourg, F. Bergaya), 269–327. Amsterdam: Elsevier.
Gouy, M. (1910). Sur la constitution de la charge électrique à la surface d'un électrolyte. *Journal de Physique Théorique et Appliquée* 9: 457–468.
Grangeon, S., Claret, F., Lerouge, C., et al. (2013a). On the nature of structural disorder in calcium silicate hydrates with a calcium/silicon ratio similar to tobermorite. *Cement and Concrete Research* 52: 31–37.
Grangeon, S., Claret, F., Linard, Y., and Chiaberge, C. (2013b). X-ray diffraction: a powerful tool to probe and understand the structure of nanocrystalline calcium silicate hydrates. *Acta Crystallographica B - Structural Science* B69: 465–473.
Guarracino, L., Roetting, T., and Carrera, J. (2014). A fractal model to describe the evolution of multiphase flow properties during mineral dissolution. *Advances in Water Resources* 67: 78–86.
Hadi, J., Tournassat, C., Ignatiadis, I., et al. (2013). Modelling CEC variations versus structural iron reduction levels in dioctahedral smectites. Existing approaches, new data and model refinements. *Journal of Colloid and Interface Science* 407: 397–409.
Hayek, M., Kosakowski, G., and Churakov, S. (2011). Exact analytical solutions for a diffusion problem coupled with a precipitation-dissolution reaction and feedback of porosity change. *Water Resources Research* 47: 1–13.
Hedges, L.O. and Whitelam, S. (2013). Selective nucleation in porous media. *Soft Matter* 9: 9763–9766.
Hellmann, R., Cotte, S., Cadel, E., et al. (2015). Nanometre-scale evidence for interfacial dissolution–reprecipitation control of silicate glass corrosion. *Nature Materials* 14: 307–311.
Hess, H.H. (1957). *The Disposal of Radioactive Waste on Land: Report of the Committee on Waste Disposal of the Division of Earth Sciences* [chairman Harry H. Hess]. National Academy of Sciences, National Research Council.
Hummel, W. and Schneider, J.W. (2005). Safety of nuclear waste repositories. *Chimia* 59: 909–915.
Hummel, W., Berner, U., Curti, E., et al. (2002). Nagra/PSI chemical thermodynamic data base 01/01. *Radiochimica Acta* 90: 805–813.
Jacquier, P., Hainos, D., Robinet, J.C., et al. (2013). The influence of mineral variability of Callovo-Oxfordian clay rocks on radionuclide transfer properties. *Applied Clay Science* 83–84: 129–136.
Kameche, Z.A., Ghomari, F., Choinska, M., and Khelidj, A. (2014). Assessment of liquid water and gas permeabilities of partially saturated ordinary concrete. *Construction and Building Materials* 65: 551–565.
Kars, M., Lerouge, C., Grangeon, S., et al. (2015). Identification of nanocrystalline goethite in reduced clay formations: Application to the Callovian-Oxfordian formation of Bure (France). *American Mineralogist* 100: 1544–1553.
King, F. and Shoesmith, D.W. (2010). Nuclear waste canister materials, corrosion behaviour and long-term performance in geological repository systems. In: *Geological Repository Systems for Safe Disposal of Spent Nuclear Fuels and Radioactive Waste* (ed. J. Ahn and M.J. Apted), 379–420. Woodhead Publishing.
Kolditz, O., Bauer, S., Bilke, L., et al. (2012). OpenGeoSys: an open-source initiative for numerical simulation of thermo-hydro-mechanical/chemical (THM/C) processes in porous media. *Environmental Earth Sciences* 67: 589–599.

Kosakowski, G. and Berner, U. (2013). The evolution of clay rock/cement interfaces in a cementitious repository for low- and intermediate level radioactive waste. *Physics and Chemistry of the Earth, Parts A/B/C* 64: 65–86.

Kulik, D.A., Hummel, W., Lützenkirchen, J., and Lefèvre, G. (2015). Preface: SI: Geochemical speciation codes and databases. *Applied Geochemistry* 55: 1–2.

Lagha, S.B., Crusset, D., Mabille, I., et al. (2007). Corrosion of iron: A study for radioactive waste canisters. *Journal of Nuclear Materials* 362: 485–492.

Landais, P. and Aranyossy, J.-F. (2011). Clays in natural and engineered barriers for radioactive waste confinement. *Physics and Chemistry of the Earth, Parts A/B/C* 36: 1437.

Lasaga, A.C. (1981). Transition state theory. *Reviews in Mineralogy and Geochemistry* 8: 135–168.

Lasaga, A.C. (1984). Chemical kinetics of water–rock interactions. *Journal of Geophysical Research: Solid Earth* 89: 4009–4025.

Leech, C., Lockington, D., Hooton, R.D., et al. (2008). Validation of Mualem's conductivity model and prediction of saturated permeability from sorptivity. *ACI Materials Journal* 105: 44–51.

Lerouge, C., Vinsot, A., Grangeon, S., et al. (2013). Controls of Ca/Mg/Fe activity ratios in pore water chemistry models of the Callovian-Oxfordian Clay Formation. *Procedia Earth and Planetary Science* 7: 475–478.

Lerouge, C., Claret, F., Tournassat, C., et al. (2014a). Constraints from sulfur isotopes on the origin of gypsum at concrete/claystone interfaces. *Physics and Chemistry of the Earth, Parts A/B/C* 70–71: 84–95.

Lerouge, C., Grangeon, S., Claret, F., et al. (2014b). Mineralogical and isotopic record of diagenesis from the Opalinus clay formation at Benken, Switzerland: Implications for the modeling of pore-water chemistry in a clay formation. *Clays and Clay Minerals* 62: 286–312.

Linsley, G. and Fattah, A. (1994). The interface between nuclear safeguards and radioactive waste disposal: Emerging issues. *IAEA Bulletin* 36: 22–26.

Lisjak, A., Garitte, B., Grasselli, G., et al. (2015). The excavation of a circular tunnel in a bedded argillaceous rock (Opalinus Clay): Short-term rock mass response and FDEM numerical analysis. *Tunnelling and Underground Space Technology* 45: 227–248.

Liu, J.F., Davy, C.A., Talandier, J., and Skoczylas, F. (2014a). Effect of gas pressure on the sealing efficiency of compacted bentonite–sand plugs. *Journal of Contaminant Hydrology* 170: 10–27.

Liu, J.F., Skoczylas, F., and Liu, J. (2014b). Experimental research on water retention and gas permeability of compacted bentonite/sand mixtures. *Soils and Foundations* 54: 1027–1038.

Liu, S., Jacques, D., Govaerts, J., and Wang, L. (2014c). Conceptual model analysis of interaction at a concrete–Boom Clay interface. *Physics and Chemistry of the Earth, Parts A/B/C* 70–71: 150–159.

Loomer, D.B., Scott, L., Al, T.A., et al. (2013). Diffusion–reaction studies in low permeability shale using X-ray radiography with cesium. *Applied Geochemistry* 39: 49–58.

Lothenbach, B. (2010). Thermodynamic equilibrium calculations in cementitious systems. *Materials and Structures* 43: 1413–1433.

Lothenbach, B. and Wieland, E. (2006). A thermodynamic approach to the hydration of sulphate-resisting Portland cement. *Waste Management* 26: 706–719.

Lothenbach, B. and Winnefeld, F. (2006). Thermodynamic modelling of the hydration of Portland cement. *Cement and Concrete Research* 36: 209–226.

Lothenbach, B., Le Saout, G., Ben Haha, M., et al. (2012). Hydration of a low-alkali CEM III/B–SiO_2 cement (LAC). *Cement and Concrete Research* 42: 410–423.

Lothenbach, B., Rentsch, D., and Wieland, E. (2014). Hydration of a silica fume blended low-alkali shotcrete cement. *Physics and Chemistry of the Earth* 70–71: 3–16.

Lu, C., Samper, J., Fritz, B., et al. (2011). Interactions of corrosion products and bentonite: An extended multicomponent reactive transport model. *Physics and Chemistry of the Earth* 36: 1661–1668.

Luckner, L., Vangenuchten, M.T., and Nielsen, D.R. (1989). A consistent set of parametric models for the 2-phase flow of immiscible fluids in the subsurface. *Water Resources Research* 25: 2187–2193.

Luke, K. and Lachowski, E. (2008). Internal composition of 20-year-old fly ash and slag-blended ordinary Portland cement pastes. *Journal of the American Ceramic Society* 91: 4084–4092.

Lüttge, A., Arvidson, R.S., and Fischer, C. (2013). A stochastic treatment of crystal dissolution kinetics. *Elements* 9: 183–188.

Marschall, P., Horseman, S., and Gimmi, T. (2005). Characterisation of gas transport properties of the Opalinus Clay, a potential host rock formation for radioactive waste disposal. *Oil & Gas Science and Technology - Rev. IFP* 60: 121–139.

Marty, N.C.M., Tournassat, C., Burnol, A., et al. (2009). Influence of reaction kinetics and mesh refinement on the numerical modelling of concrete/clay interactions. *Journal of Hydrology* 364: 58–72.

Marty, N.C.M., Fritz, B., Clement, A., and Michau, N. (2010). Modelling the long term alteration of the engineered bentonite barrier in an underground radioactive waste repository. *Applied Clay Science* 47: 82–90.

Marty, N.M., Munier, I., Gaucher, E., et al. (2014). Simulation of cement/clay interactions: feedback on the increasing complexity of modelling strategies. *Transport in Porous Media* 104: 385–405.

Marty, N.C.M., Bildstein, O., Blanc, P., et al. (2015a). Benchmarks for multicomponent reactive transport across a cement/clay interface. *Computational Geosciences* 19: 635–653.

Marty, N.C.M., Claret, F., Lassin, A., et al. (2015b). A database of dissolution and precipitation rates for clay-rocks minerals. *Applied Geochemistry* 55: 108–118.

Martys, N.S. (1999). Diffusion in partially-saturated porous materials. *Materials and Structures* 32: 555–562.

Massat, L., Cuisinier, O., Bihannic, I., et al. (2016). Swelling pressure development and inter-aggregate porosity evolution upon hydration of a compacted swelling clay. *Applied Clay Science* 124–125: 197–210.

Matray, J.M., Savoye, S., and Cabrera, J. (2007). Desaturation and structure relationships around drifts excavated in the well-compacted Tournemire's argillite (Aveyron, France). *Engineering Geology* 90: 1–16.

Maul, P.R., Benbow, S.J., Bond, A.E., et al. (2008). The *Quintessa Multiphysics General-Purpose Code QPAC*. Quintessa Report QRS-3000B-8 version 1.0.

Mazurek, M. (2010). Far-field process analysis and radionuclide transport modelling in geological repository systems. In: *Geological Repository Systems for Safe Disposal of*

Spent Nuclear Fuels and Radioactive Waste (ed. J. Ahn and M.J. Apted), 222–257. Woodhead Publishing.
Mazurek, M., Gautschi, A., Marschall, P., et al. (2008). Transferability of geoscientific information from various sources (study sites, underground rock laboratories, natural analogues) to support safety cases for radioactive waste repositories in argillaceous formations. *Physics and Chemistry of the Earth, Parts A/B/C* 33, Supplement 1: S95–S105.
Millington, R.J. (1959). Gas diffusion in porous media. *Science* 130: 100–102.
Molins, S., Trebotich, D., Yang, L., et al. (2014). Pore-scale controls on calcite dissolution rates from flow-through laboratory and numerical experiments. *Environmental Science and Technology* 48: 7453–7460.
Montarnal, P., Mugler, C., Colin, J., et al. (2007). Presentation and use of a reactive transport code in porous media. *Physics and Chemistry of the Earth* 32: 507–517.
Montes-H, G., Marty, N., Fritz, B., et al. (2005). Modelling of long-term diffusion-reaction in a bentonite barrier for radioactive waste confinement. *Applied Clay Science* 30: 181–198.
Mualem, Y. (1976). New model for predicting hydraulic conductivity of unsaturated porous-media. *Water Resources Research* 12: 513–522.
Muurinen, A. and Lehikoinen, J. (1999). Porewater chemistry in compacted bentonite. *Engineering Geology* 54: 207–214.
Nagra. (2002). Project Opalinus Clay. Models, Codes and Data for Safety Assessment. Demonstration of disposal feasibility for spent fuel, vitrified high-level waste and long-lived intermediate-level waste. Available at http://www.nagra.ch/display.cfm/id/100188
National Research Council (2001). *Disposition of High-Level Waste and Spent Nuclear Fuel: The Continuing Societal and Technical Challenges*. Washington, DC: The National Academies Press. doi:10.17226/10119
Navarre-Sitchler, A., Steefel, C.I., Yang, L., et al. (2009). Evolution of porosity and diffusivity associated with chemical weathering of a basalt clast. *Journal of Geophysical Research: Earth Surface* 114: F02016.
Nguyen, T.Q., Baroghel-Bouny, V., and Dangla, P. (2006). Prediction of chloride ingress into saturated concrete on the basis of a multi-species model by numerical calculations. *Computers and Concrete* 3: 401–421.
Olsson, N., Baroghel-Bouny, V., Nilsson, L.-O., and Thiery, M. (2013). Non-saturated ion diffusion in concrete – A new approach to evaluate conductivity measurements. *Cement and Concrete Composites* 40: 40–47.
Parker, J.C., Lenhard, R.J., and Kuppusamy, T. (1987). A parametric model for constitutive properties governing multiphase flow in porous-media. *Water Resources Research* 23: 618–624.
Parkhurst, D.L. and Appelo, C. (1999). *User's Guide to PHREEQC (Version 2): A computer program for speciation, batch-reaction, one-dimensional transport, and inverse geochemical calculations*. US Geological Survey Water-Resources Investigations Report 99-4259. Available at https://pubs.er.usgs.gov/publication/wri994259
Pearson, F.J. (1999). What is the porosity of a mudrock? In: *Muds and Mudstones: Physical and Fluid Flow Properties* (ed. A.C. Aplin, A.J. Fleet, and J.H.S. Macquaker), 9–21. Geological Society London Special Publication.
Pearson, F.J., Arcos, D., Bath, A., et al. (2003). *Mont Terri Project – Geochemistry of Water in the Opalinus Clay Formation at the Mont Terri Rock Laboratory*. Bern: Federal Office for Water and Geology (FOWG). Geology Series 5.

Philippini, V., Naveau, A., Catalette, H., and Leclercq, S. (2007). Sorption of silicon on magnetite and other corrosion products of iron (vol. 348, pg. 60, 2006). *Journal of Nuclear Materials* 362: 139–139.

Pierce, E.M., Frugier, P., Criscenti, L.J., et al. (2014). Modeling Interfacial Glass-Water Reactions: Recent Advances and Current Limitations. *International Journal of Applied Glass Science* 5: 421–435.

Pignatelli, I., Mugnaioli, E., Hybler, J., et al. (2013). A multi-technique characterization of cronstedtite synthesized by iron-clay interaction in a step-by-step cooling procedure. *Clays and Clay Minerals* 61: 277–289.

Plas, F. and Landais, P. (2008). The Phenomenological Analysis of Repository Situations (PARS) – Application within the dossier 2005 ARGILE (Meuse/Haute-Marne site). Organisation for Economic Cooperation and Development - Nuclear Energy Agency, *Symposium on Safety Cases for Deep Geological Disposal of Radioactive Waste: Where Do We Stand?* Available at https://inis.iaea.org/search/search.aspx?orig_q=RN:39116159

Poinssot, C., Baeyens, B., and Bradbury, M.H. (1999). Experimental and modelling studies of caesium sorption on illite. *Geochimica et Cosmochimica Acta* 63: 3217–3227.

Poinssot, C., Fillet, C., and Gras, J.M. (2010). Post-containment performance of geological repository systems: source-term release and radionuclide migration in the near- and far-field environments. In: *Geological Repository Systems for Safe Disposal of Spent Nuclear Fuels and Radioactive Waste* (ed. J. Ahn and M.J. Apted), 421–493. Woodhead Publishing.

Poller, A., Enssle, C., Mayer, G., et al. (2011). Repository-scale modeling of the long-term hydraulic perturbation induced by gas and heat generation in a geological repository for high-and intermediate-level radioactive waste: Methodology and example of application. *Transport in Porous Media* 90: 77–94.

Pusch, R. (2010). Development and application of smectitic buffer and backfill materials in geological repository systems. In: *Geological Repository Systems for Safe Disposal of Spent Nuclear Fuels and Radioactive Waste* (ed. J. Ahn and M.J. Apted), 323–352. Woodhead Publishing.

Putnis, A. and Mauthe, G. (2001). The effect of pore size on cementation in porous rocks. *Geofluids* 1: 37–41.

Rivard, C., Pelletier, M., Michau, N., et al. (2015). Reactivity of Callovo-Oxfordian claystone and its clay fraction with metallic iron: role of non-clay minerals in the interaction mechanism. *Clays and Clay Minerals* 63: 290–310.

Rutqvist, J. (2015). Coupled thermo-hydro-mechanical behavior of natural and engineered clay barriers. In: *Developments in Clay Science* Vol 6 Natural and Engineered Clay Barriers, 1st Edition (ed. C. Tournassat, C. Steefel, I.C. Bourg, F. Bergaya), 329–355. Amsterdam: Elsevier.

Rutqvist, J., Zheng, L., Chen, F., et al. (2014). Modeling of coupled thermo-hydro-mechanical processes with links to geochemistry associated with bentonite-backfilled repository tunnels in clay formations. *Rock Mechanics and Rock Engineering* 47: 167–186.

Samper, J., Yang, C., and Montenegro, L, (2003). *User's Manual of CORE2D Version 4: A code for ground-water flow and reactive solute transport*. Universidad de A Coruña, Spain.

Samper, J., Lu, C., and Montenegro, L. (2008a). Reactive transport model of interactions of corrosion products and bentonite. *Physics and Chemistry of the Earth* 33: S306–S316.

Samper, J., Zheng, L., Maria Fernandez, A., and Montenegro, L. (2008b). Inverse modeling of multicomponent reactive transport through single and dual porosity media. *Journal of Contaminant Hydrology* 98: 115–127.

Samper, J., Naves, A., Montenegro, L., and Mon, A. (2016). Reactive transport modelling of the long-term interactions of corrosion products and compacted bentonite in a HLW repository in granite: Uncertainties and relevance for performance assessment. *Applied Geochemistry* 67: 42–51.

Sander, M., Hofstetter, T.B., and Gorski, C.A. (2015). Electrochemical analyses of redox-active iron minerals: a review of nonmediated and mediated approaches. *Environmental Science and Technology* 49: 5862–5878.

Savage, D., Noy, D., and Mihara, M. (2002). Modelling the interaction of bentonite with hyperalkaline fluids. *Applied Geochemistry* 17: 207–223.

Savage, D., Walker, C., Arthur, R., et al. (2007). Alteration of bentonite by hyperalkaline fluids: A review of the role of secondary minerals. *Physics and Chemistry of the Earth* 32: 287–297.

Savage, D., Watson, C., Benbow, S., and Wilson, J. (2010). Modelling iron-bentonite interactions. *Applied Clay Science* 47: 91–98.

Savoye, S., Page, J., Puente, C., et al. (2010). New experimental approach for studying diffusion through an intact and unsaturated medium: A case study with Callovo-Oxfordian Argillite. *Environmental Science and Technology* 44: 3698–3704.

Savoye, S., Beaucaire, C., Fayette, A., et al. (2012). Mobility of cesium through the Callovo-Oxfordian claystones under partially saturated conditions. *Environmental Science and Technology* 46: 2633–2641.

Savoye, S., Imbert, C., Fayette, A., and Coelho, D. (2014). Experimental study on diffusion of tritiated water and anions under variable water-saturation and clay mineral content: comparison with the Callovo-Oxfordian claystones. In: *Clays in Natural and Engineered Barriers for Radioactive Waste Confinement* (ed. S. Norris, J. Bruno, M. Cathelineau, et al.), 579–588. Geological Society London Special Publication.

Schofield, R.K. (1947). Calculation of surface areas from measurements of negative adsorption. *Nature* 160: 408.

Sedighi, M., Thomas, H.R., Masum, S.A., et al. (2014). Geochemical modelling of hydrogen gas migration in an unsaturated bentonite buffer. In: Gas Generation and Migration in Deep Geological Radioactive Waste Repositories (ed. R.P. Shaw) *Geological Society London Special Publication 415*.

Sellin, P. and Leupin, O.X. (2013). The use of clay as an engineered barrier in radioactive-waste management – a review. *Clays and Clay Minerals* 61: 477–498.

Soler, J.M. (2003). Reactive transport modeling of the interaction between a high-pH plume and a fractured marl: the case of Wellenberg. *Applied Geochemistry* 18: 1555–1571.

Sousa, E. (2003). The controversial stability analysis. *Applied Mathematics and Computation* 145: 777–794.

Stack, A.G., Fernandez-Martinez, A., Allard, L.F., et al. (2014). Pore-size-dependent calcium carbonate precipitation controlled by surface chemistry. *Environmental Science and Technology* 48: 6177–6183.

Steefel, C.I. and Lichtner, P.C. (1998). Multicomponent reactive transport in discrete fractures – II: Infiltration of hyperalkaline groundwater at Maqarin, Jordan, a natural analogue site. *Journal of Hydrology* 209: 200–224.

Steefel, C. and Yabusaki, S. (1996). *OS3D/GIMRT, software for multicomponent-multidimensional reactive transport, user manual and programmer's guide.* PNL-11166. Pacific Northwest National Laboratory, Richland, Washington.

Steefel, C.I., Molins, S., and Trebotich, D. (2013). Pore scale processes associated with subsurface CO_2 injection and sequestration. In: *Geochemistry of Geologic CO_2 Sequestration* (ed. D.J. DePaolo, D.R. Cole, A. Navrotsky, and I.C. Bourg), 259–303. Reviews in Mineralogy and Geochemistry 77. Mineralogical Society of America and the Geochemical Society. doi:10.2138/rmg.2013.77.8

Steefel, C.I., Appelo, C.A.J., Arora, B., et al. (2015a). Reactive transport codes for subsurface environmental simulation. *Computational Geosciences* 19 (3): 445–478.

Steefel, C.I., Beckingham, L.E., and Landrot, G. (2015b). Micro-continuum approaches for modeling pore-scale geochemical processes. In: *Pore-Scale Geochemical Processes* (ed. C.I. Steefel, S. Emmanuel, and L.M. Anovitz), 217–246. Reviews in Mineralogy and Geochemistry 80. Mineralogical Society of America and the Geochemical Society. doi:10.2138/rmg.2015.80.07

Tachi, Y. and Yotsuji, K. (2014). Diffusion and sorption of Cs^+, Na^+, I^- and HTO in compacted sodium montmorillonite as a function of porewater salinity: Integrated sorption and diffusion model. *Geochimica et Cosmochimica Acta* 132: 75–93.

Taylor, R. and Smith, T. (1986). The engineering geology of clay minerals; swelling, shrinking and mudrock breakdown. *Clay Minerals* 21: 235–260.

Tinnacher, R.M., Holmboe, M., Tournassat, C., et al. (2016). Ion adsorption and diffusion in smectite: molecular, pore, and continuum scale views. *Geochimica et Cosmochimica Acta* 177: 130–149.

Tournassat, C. and Steefel, C.I. (2015). Ionic transport in nano-porous clays with consideration of electrostatic effects. In: *Pore-Scale Geochemical Processes* (ed. C.I. Steefel, S. Emmanuel, and L.M. Anovitz), 287–329. Reviews in Mineralogy and Geochemistry 80. Mineralogical Society of America and the Geochemical Society. doi:10.2138/rmg.2015.80.09

Tournassat, C., Alt-Epping, P., Gaucher, E.C., et al. (2011). Biogeochemical processes in a clay formation in situ experiment: Part F – Reactive transport modelling. *Applied Geochemistry* 26: 1009–1022.

Tournassat, C., Steefel, C.I., Bourg, I.C., and Bergaya, F. (2015a). Introduction. In: *Developments in Clay Science* Vol 6 Natural and Engineered Clay Barriers, 1st Edition (ed. C. Tournassat, C. Steefel, I.C. Bourg, F. Bergaya), 1–4. Amsterdam: Elsevier.

Tournassat, C., Vinsot, A., Gaucher, E.C., and Altmann, S. (2015b). Chemical conditions in clay-rocks. In: *Developments in Clay Science* Vol 6 Natural and Engineered Clay Barriers, 1st Edition (ed. C. Tournassat, C. Steefel, I.C. Bourg, F. Bergaya), 71–100. Amsterdam: Elsevier.

Trebotich, D., Adams, M.F., Molins, S., et al. (2014). High-resolution simulation of pore-scale reactive transport processes associated with carbon sequestration. *Computing in Science & Engineering* 16: 22–31.

Treille, E., Wendling, J., and Plas, F. (2011). Simulation of a gas transfer at the scale of a disposal cell in the context of High-Level and Long-Lived radwaste disposal in a deep clay formation during the operating phase. *Nuclear Technology* 174: 353–363.

Tremosa, J., Arcos, D., Matray, J.M., et al. (2012). Geochemical characterization and modelling of the Toarcian/Domerian porewater at the Tournemire underground research laboratory. *Applied Geochemistry* 27: 1417–1431.

Trotignon, L., Peycelon, H., and Bourbon, X. (2006). Comparison of performance of concrete barriers in a clayey geological medium. *Physics and Chemistry of the Earth* 31: 610–617.

Trotignon, L., Devallois, V., Peycelon, H., et al. (2007). Predicting the long term durability of concrete engineered barriers in a geological repository for radioactive waste. *Physics and Chemistry of the Earth* 32: 259–274.

Trotignon, L., Thouvenot, P., Munier, I., et al. (2011). Numerical simulation of atmpspheric carbonation of concrete components in a deep geological radwaste disposal during operating period. *Nuclear Technology* 174: 424–437.

Truche, L., Berger, G., Destrigneville, C., et al. (2009). Experimental reduction of aqueous sulphate by hydrogen under hydrothermal conditions: Implication for the nuclear waste storage. *Geochimica et Cosmochimica Acta* 73: 4824–4835.

Truche, L., Berger, G., Destrigneville, C., et al. (2010). Kinetics of pyrite to pyrrhotite reduction by hydrogen in calcite buffered solutions between 90 and 180 °C: Implications for nuclear waste disposal. *Geochimica et Cosmochimica Acta* 74: 2894–2914.

Truche, L., Jodin-Caumon, M.-C., Lerouge, C., et al. (2013). Sulphide mineral reactions in clay-rich rock induced by high hydrogen pressure. Application to disturbed or natural settings up to 250 degrees C and 30 bar. *Chemical Geology* 351: 217–228.

Vance, E.R. and Begg, B.D. (2010). Immobilisation of spent nuclear fuel and high-level radioactive waste for safe disposal in geological repository systems, In: *Geological Repository Systems for Safe Disposal of Spent Nuclear Fuels and Radioactive Waste* (ed. J. Ahn and M.J. Apted), 261–285. Woodhead Publishing.

van der Lee, J., De Windt, L., Lagneau, V., and Goblet, P. (2003). Module-oriented modeling of reactive transport with HYTEC. *Computers and Geosciences* 29: 265–275.

Van Geet, M., Bastiaens, W., and Ortiz, L. (2008). Self-sealing capacity of argillaceous rocks: Review of laboratory results obtained from the SELFRAC project. *Physics and Chemistry of the Earth, Parts A/B/C* 33, Supplement 1: S396–S406.

Van Genuchten, M.T. (1980). A closed-form equation for predicting the hydraulic conductivity of unsaturated soils. *Soil Science Society of America Journal* 44: 892–898.

Van Loon, L.R. and Mibus, J. (2015). A modified version of Archie's law to estimate effective diffusion coefficients of radionuclides in argillaceous rocks and its application in safety analysis studies. *Applied Geochemistry* 59: 85–94.

Van Loon, L.R., Soler, J.M., and Bradbury, M.H. (2003). Diffusion of HTO, Cl-36(-) and I-125(-) in Opalinus Clay samples from Mont Terri – Effect of confining pressure. *Journal of Contaminant Hydrology* 61: 73–83.

Vienna, J.D., Ryan, J.V., Gin, S., and Inagaki, Y. (2013). Current understanding and remaining challenges in modeling long-term degradation of borosilicate nuclear waste glasses. *International Journal of Applied Glass Science* 4: 283–294.

Vinsot, A., Mettler, S., and Wechner, S. (2008). In situ characterization of the Callovo-Oxfordian pore water composition. *Physics and Chemistry of the Earth* 33: S75–S86.

Vinsot, A., Linard, Y., Lundy, M., et al. (2013). Insights on desaturation processes based on the chemistry of seepage water from boreholes in the Callovo-Oxfordian argillaceous rock. *Procedia Earth and Planetary Science* 7: 871–874.

Vinsot, A., Leveau, F., Bouchet, A., and Arnould, A. (2014). Oxidation front and oxygen transfer in the fractured zone surrounding the Meuse/Haute-Marne URL drifts in the Callovian–Oxfordian argillaceous rock. In: *Clays in Natural and Engineered Barriers for Radioactive Waste Confinement* (ed. S. Norris, J. Bruno, M. Cathelineau, et al.), 207. Geological Society London Special Publication.

Wersin, P. (2003). Geochemical modelling of bentonite porewater in high-level waste repositories. *Journal of Contaminant Hydrology* 61: 405–422.

Wersin, P. and Birgersson, M. (2014). Reactive transport modelling of iron-bentonite interaction within the KBS-3H disposal concept: the Olkiluoto site as a case study. In: *Clays in Natural and Engineered Barriers for Radioactive Waste Confinement* (ed. S. Norris, J. Bruno, M. Cathelineau, et al.), 237–250. Geological Society London Special Publication.

Wilson, J.C., Benbow, S., Sasamoto, H., et al. (2015). Thermodynamic and fully-coupled reactive transport models of a steel-bentonite interface. *Applied Geochemistry* 61: 10–28.

Wolery, T.J. (1983). *EQ3NR a computer program for geochemical aqueous speciation-solubility calculations: user's guide and documentation*. Lawrence Livermore National Laboratory, Livermore, CA.

Xie, M., Mayer, K.U., Claret, F., et al. (2015). Implementation and evaluation of permeability-porosity and tortuosity-porosity relationships linked to mineral dissolution-precipitation. *Computational Geosciences* 19: 655–671.

Xu, T., Sonnenthal, E., Spycher, N., and Pruess, K. (2004). *TOUGHREACT User's Guide: a simulation program for non-isothermal multiphase reactive geochemical transport in variable saturated geologic media*. LBNL-55460. Berkeley, CA: Lawrence Berkeley National Laboratory.

Xu, T., Sonnenthal, E., Spycher, N., and Pruess, K. (2006). TOUGHREACT – a simulation program for non-isothermal multiphase reactive geochemical transport in variably saturated geologic media: applications to geothermal injectivity and CO_2 geological sequestration. *Computers and Geosciences* 32: 145–165.

Xu, T., Senger, R., and Finsterle, S. (2008). Corrosion-induced gas generation in a nuclear waste repository: Reactive geochemistry and multiphase flow effects. *Applied Geochemistry* 23: 3423–3433.

Xu, W.J., Shao, H., Hesser, J., et al. (2013). Coupled multiphase flow and elasto-plastic modelling of in-situ gas injection experiments in saturated claystone (Mont Terri Rock Laboratory). *Engineering Geology* 157: 55–68.

Yang, C. and Chiang, S. (2008). A rapid method for determination of the chloride migration coefficient in concrete using electrical field. *Journal of ASTM International* 5 (4): 1–13.

Yang, C., Samper, J., and Montenegro, L. (2008). A coupled non-isothermal reactive transport model for long-term geochemical evolution of a HLW repository in clay. *Environmental Geology* 53: 1627–1638.

Yu, L., Weetjens, E., Perko, J., and Mallants, D. (2011). Comparison of numerical tools through Thermo-Hydro-Gas transport modeling for a geological repository in Boom clay. *Nuclear Technology* 174: 411–423.

Yu, L., Rogiers, B., Gedeon, M., et al. (2013). A critical review of laboratory and in-situ hydraulic conductivity measurements for the Boom Clay in Belgium. *Applied Clay Science* 75–76: 1–12.

Zeelmaekers, E., Honty, M., Derkowski, A., et al. (2015). Qualitative and quantitative mineralogical composition of the Rupelian Boom Clay in Belgium. *Clay Minerals* 50: 249–272.

Zhang, K., Croise, J., and Mayer, G. (2011). Computation of the Couplex-Gaz exercise with TOUGH2-MP: Hydrogen flow and transport in the pore water of a low-permeability clay rock hosting a nuclear waste repository. *Nuclear Technology* 174: 364–374.

Zhang, Z., Thiery, M., and Baroghel-Bouny, V. (2015). Numerical modelling of moisture transfers with hysteresis within cementitious materials: Verification and investigation of the effects of repeated wetting–drying boundary conditions. *Cement and Concrete Research* 68: 10–23.

Zheng, L., Rutqvist, J., Liu, H.-H., et al. (2014). Model evaluation of geochemically induced swelling/shrinkage in argillaceous formations for nuclear waste disposal. *Applied Clay Science* 97–98: 24–32.

Zhou, C. (2014). Predicting water permeability and relative gas permeability of unsaturated cement-based material from hydraulic diffusivity. *Cement and Concrete Research* 58: 143–151.

Zhou, C.S., Chen, W., Wang, W., and Skoczylas, F. (2015). Unified determination of relative molecular diffusivity and fluid permeability for partially saturated cement-based materials. *Cement and Concrete Research* 67: 300–309.

Zhou, W. and Arthur, R. (2010). Near-field processes, evolution and performance assessment in geological repository systems. In: *Geological Repository Systems for Safe Disposal of Spent Nuclear Fuels and Radioactive Waste* (ed. J. Ahn and M.J. Apted), 353–378. Woodhead Publishing.

9

Modeling Variably Saturated Water Flow and Multicomponent Reactive Transport in Constructed Wetlands

Günter Langergraber[1] *and Jirka Šimůnek*[2]

[1] *Institute of Sanitary Engineering and Water Pollution Control, University of Natural Resources and Life Sciences Vienna (BOKU), A-1190, Vienna, Austria*
[2] *Department of Environmental Sciences, University of California Riverside, Riverside, CA, 92521, USA*

9.1 Introduction

Constructed wetlands (CWs) are engineered treatment systems that optimize treatment processes found in natural environments and are therefore considered to be sustainable, environmentally friendly solutions to engineering problems. Processes occurring in CWs are very complex, and include a large number of simultaneously active physical, chemical, and biological processes that mutually influence each other. CWs are systems that efficiently treat different types of polluted water (e.g. Kadlec and Wallace, 2009; García et al., 2010).

For a long time, CWs have been considered 'black boxes'. Little effort has been made to understand the main processes leading to contaminant removal. Still today, most models for wetlands are using a 'black box' approach – they do not consider processes in wetlands in detail. For 'black box' models, data from experiments are needed to derive model equations. In process-based models the mathematical equations are based on processes in wetlands and include balance equations for energy, mass, charge, and so on. Data from experiments are used for calibration and validation of models. Better predictions should be possible using these models (Langergraber, 2011).

When developing a wetland model, a number of different processes have to be considered (Langergraber et al., 2009b):

- the flow model (describing water flow in the porous media);
- the transport model (describing transport of constituents as well as adsorption and desorption processes);
- the biokinetic model (describing biochemical transformation and degradation processes of the constituents);

Reactive Transport Modeling: Applications in Subsurface Energy and Environmental Problems, First Edition. Edited by Yitian Xiao, Fiona Whitaker, Tianfu Xu and Carl Steefel.

- the plant model (growth, decay, decomposition, nutrient uptake, root oxygen release, etc);
- the clogging model (the transport and deposition of suspended particulate matter, and bacterial and plant growth that may reduce the hydraulic capacity/conductivity of the filter medium).

Different types of CWs require different models to describe water flow. No free water level is visible in subsurface flow (SSF) CWs, and water flows, either horizontally or vertically, through the porous filter media. Horizontal flow (HF) systems can be simulated when only saturated water flow conditions are considered. For modeling vertical flow (VF) CWs with intermittent loading, transient variably-saturated flow models are required. Due to the intermittent loading, these systems are highly dynamic, adding to the complexity needed to model the overall system. Models applicable to VF CWs use either the Richards equation or various simplified approaches to describe variably-saturated flow (Langergraber, 2008).

Wastewaters contain particulate and solute compounds that are transported with the flowing water. Transport equations generally consider convective–dispersive transport in the liquid phase, diffusion in the gaseous phase, as well as adsorption and desorption processes between the solid and liquid phases.

Biokinetic models describe the transformation and degradation processes of the pollutants. The complexity of biokinetic models ranges from considering a single process affecting the concentration of one compound to multiple processes affecting multiple compounds. The reaction rates are often assumed to be constant, that is, independent of environmental conditions. However, reaction rates are dependent on a number of environmental conditions (e.g. concentrations of oxygen, substrate, nutrients). These dependencies are often modeled with Monod-type equations.

Plants are an essential part of wetland treatment systems, and thus plant-related processes have to be included in wetland models. Root growth influences the water flow pattern in the subsurface, and via evapotranspiration the water balance. During growth and decay, plants take up nutrients and release organic matter and nutrients, respectively. Additionally, some plants release substances via roots (e.g. oxygen and/or specific organic substances). Plant models describe uptake and release of substances either associated with water uptake or as a process dependent on concentration gradients.

Clogging models need to be able to describe (1) the transport and deposition of suspended particulate matter, and (2) the deposition of particulate matter, bacterial and plant growth that may reduce the hydraulic capacity/conductivity of the filter medium. This is important for simulations of the long-term CW performance and predictions of potential failure due to clogging (Langergraber and Morvannou, 2014).

During the last two decades a couple of mechanistic models have been developed for simulating CWs (Langergraber, 2008, Langergraber et al., 2009a; Kumar and Zhao, 2011; Meyer et al., 2015; Samsó et al., 2015b). Recent developments, especially towards implementation of the CWM1 biokinetic model (Langergraber et al., 2009b), include the works of Langergraber and Šimůnek (2012), Samsó and Garcia (2013), and Mburu et al. (2012). Table 9.1 compares three tools with respect to the considered submodels required for a wetland model.

In this chapter, we briefly describe version 2 of the HYDRUS Wetland module (Langergraber and Šimůnek, 2012) and discuss existing experience using wetland

Table 9.1 Comparison of recent simulation tools for constructed wetlands.

Simulation tool	HYDRUS Wetland module	BIO_PORE	AQUASIM
Reference	Langergraber and Šimůnek (2012)	Samsó and Garcia (2013)	Mburu et al. (2012)
Flow model	Richards equation (variably saturated flow)	Variable water table (saturated flow)	No flow considered
Transport model	Advection, dispersion, adsorption	Advection, dispersion, adsorption	No transport model
Biokinetic model	CW2D + CWM1	CWM1	CWM1
Influence of plants	Evapotranspiration, uptake and release of substances	Evapotranspiration, uptake and release of substances	Evapotranspiration, uptake and release of substances
Clogging model	Not considered	Included	Not considered

models in general and the HYDRUS Wetland module in particular. We further describe remaining challenges in using these tools and summarize needs for further development and research.

9.2 The HYDRUS Wetland Module

The HYDRUS software numerically solves the Richards equation for saturated/unsaturated water flow and the convection–dispersion equation for heat and solute transport. The flow equation incorporates a sink term to account for water uptake by plant roots. The solute transport equations consider convective–dispersive transport in the liquid phase, diffusion in the gaseous phase, as well as nonlinear nonequilibrium reactions between the solid and liquid phases (Šimůnek et al., 2008, 2011).

Version 2 of the HYDRUS wetland module (Langergraber and Šimůnek, 2012) includes two biokinetic model formulations: (1) the CW2D (Langergraber and Šimůnek, 2005), and (2) the CWM1 (Constructed Wetland Model 1) biokinetic models (Langergraber et al., 2009b). In CW2D, aerobic and anoxic transformation and degradation processes for organic matter, nitrogen and phosphorus are taken into account, whereas in CWM1, aerobic, anoxic and anaerobic processes for organic matter, nitrogen and sulfur are considered. CWM1 has been developed with the main goal of providing a widely accepted model formulation for biochemical transformation and degradation processes in SSF CWs. Table 9.2 summarizes processes and components described by, and applications of, the biokinetic models CW2D and CWM1, which are implemented in the HYDRUS Wetland module.

Both CW2D and CWM1 have been developed to describe processes in constructed wetlands treating domestic wastewater. A number of applications of the HYDRUS wetland module shows that a good match between measured data and simulation results can be achieved (e.g. Langergraber, 2003, 2007; Langergraber et al., 2007). In addition to

Table 9.2 Processes and components described by (number of components/processes in brackets), and applications of, the biokinetic models CW2D and CWM1 (adapted from Langergraber and Šimůnek, 2012).

Biokinetic model	CW2D	CWM1
Processes	Aerobic and anoxic (9)	Aerobic, anoxic, and anaerobic (17)
Components	Oxygen, organic matter, nitrogen, and phosphorus (12)	Oxygen, organic matter, nitrogen, and sulfur (16)
Type of CW	VF CWs Low-loaded HF beds	VF and HF CWs

applications involving treatment of municipal wastewater in CWs, the HYDRUS Wetland module has been also used to model:

- CWs treating combined sewer overflow (Dittmer et al., 2005; Henrichs et al. 2007, 2009; Meyer, 2011)
- CWs treating effluents of the wastewater treatment plant for irrigation purposes (Toscano et al., 2009; Pálfy et al., 2015a)
- CWs treating artificial wastewater (e.g. Pálfy and Langergraber, 2014; Rizzo et al., 2014)
- CWs treating greywater (Karlsson et al., 2015)
- Runoff from agricultural sites and the effects of streamside management zones (Smethurst et al., 2014).

Besides the Wetland module, the HYDRUS software also includes the HP1/2/3 multi-component solute transport modules, the colloid-facilitated solute transport C-Ride module, a module for fumigants, and the major ion UnsatChem module. All of these modules simulate water flow and solute/heat transport processes in two-dimensional transport domains and are supported by the HYDRUS (2D/3D) graphical user interface. Many processes of these specialized modules are also available as part of the public domain HYDRUS-1D software (Šimůnek et al., 2013).

9.3 The CW2D and CWM1 Biokinetic Models

In this section, the CW2D and CWM1 biokinetic models (i.e., the components and processes defined in the two models) are presented in detail. The information is derived from the original papers of Langergraber and Šimůnek (2005) and Langergraber et al. (2009b), respectively.

Table 9.3 compares the components defined in the CW2D and CWM1 model formulations. As described above, both biokinetic models consider processes affecting organic matter and nitrogen. Additionally, CW2D considers processes affecting phosphorus, whereas CWM1 considers processes affecting sulfur. All components except bacteria are assumed to be soluble (including the particulate Chemical Oxygen Demand (COD) fraction) and bacteria are immobile. Organic nitrogen (in CW2D and CWM1) and organic phosphorus (in CW2D) are modeled as part of the COD. Since wastewater

Table 9.3 Comparison of CW2D and CWM1 components in the HYDRUS implementation.

CW2D (Langergraber and Šimůnek, 2005) Organic matter, nitrogen, phosphorus	**CWM1 (Langergraber et al., 2009b)** Organic matter, nitrogen, sulfur
CW2D components 1) SO: Dissolved oxygen, O_2 2) CR: Readily biodegradable soluble COD 3) CS: Slowly biodegradable soluble COD 4) CI: Inert soluble COD 5) XH: Heterotrophic bacteria 6) XANs: Autotrophic ammonia oxidizing bacteria (Nitrosomonas spp.) 7) XANb: Autotrophic nitrite oxidizing bacteria (Nitrobacter spp.) 8) NH4N: Ammonium and ammonia nitrogen 9) NO2N: Nitrite nitrogen 10) NO3N: Nitrate nitrogen 11) N2: Elemental nitrogen 12) PO4P: Phosphate phosphorus *Additional component (in the HYDRUS implementation)* 13) Non-reactive tracer Nitrification is modeled as a two-step process	*Soluble components* 1) SO: Dissolved oxygen, O_2 2) SF: Fermentable, readily biodegradable soluble COD 3) SA: Fermentation products as acetate 4) SI: Inert soluble COD 5) SNH: Ammonium and ammonia nitrogen 6) SNO: Nitrate and nitrite nitrogen 7) SSO4: Sulfate sulfur 8) SH2S: Dihydrogensulfide sulfur *Particulate components* 9) XS: Slowly biodegradable particulate COD 10) XI: Inert particulate COD 11) XH: Heterotrophic bacteria 12) XA: Autotrophic nitrifying bacteria 13) XFB: Fermenting bacteria 14) XAMB: Acetotrophic methanogenic bacteria 15) XASRB: Acetotrophic sulfate reducing bacteria 16) XSOB: Sulfide oxidizing bacteria *Additional component (in the HYDRUS implementation)* 17) Non-reactive tracer

Table 9.4 Definitions of CW2D and CWM1 components in the liquid and solid phases (Langergraber and Šimůnek, 2011).

Component	**1**	**2**	**3**	**4**	**5**	**6**	**7**	**8**	**9**	**10**	**11**	**12**	**13**	**14**	**15**	**16**	**17**
CW2D	L	L+S	L+S	L+S	S	S	S	L+S	L	L	L	L+S	L+S	–	–	–	–
CWM1	L	L+S	L+S	L+S	L+S	L	L	L	L+S	L+S	S	S	S	S	S	S	L+S

L, defined in the liquid phase only; S, defined in the solid phase only; L+S, defined in both liquid and solid phases.

constituents considered in the CW2D and CWM1 biokinetic models are different, there is no direct conversion between two model components.

Table 9.4 summarizes in what phase (i.e. liquid and/or solid) the CW2D and CWM1 components are defined in the HYDRUS Wetland module. For components defined in both phases (L+S), adsorption and desorption processes can be considered. As mentioned above, suspended particulate organic matter compounds (XS and XI in CWM1) are considered as solute compounds. Note that the number of components in Table 9.4 is increased by one compared with that given in Table 9.2 for both CW2D and CWM1. In both models, a non-reactive tracer that is independent of other components, and that can be used to derive the hydraulic retention time, is added in the HYDRUS wetland module. This non-reactive tracer is defined in both liquid and solid phases (Langergraber and Šimůnek, 2011).

Table 9.5 compares the processes defined in the CW2D and CWM1 model formulations. Only aerobic and anoxic processes are defined in CW2D. Two main types of bacteria, heterotrophic and autotrophic bacteria, are considered. One special feature of CW2D is that nitrification is modeled as a two-step process, from ammonia via nitrite to nitrate. Since anaerobic processes are additionally defined in CWM1, six different types of bacteria need to be considered in this model. In addition to heterotrophic and

Table 9.5 Comparison of CW2D and CWM1 processes (Langergraber and Šimůnek, 2011).

CW2D (Langergraber and Šimůnek, 2005)	CWM1 (Langergraber et al., 2009b)
Heterotrophic bacteria: 1) Hydrolysis: conversion of CS into CR 2) Aerobic growth of XH on CR (mineralization of organic matter) 3) Anoxic growth of XH on CR (denitrification on NO2N) 4) Anoxic growth of XH on CR (denitrification on NO3N) 5) Lysis of XH	*Heterotrophic bacteria:* 1) Hydrolysis: conversion of XS into SF 2) Aerobic growth of XH on SF (mineralization of organic matter) 3) Aerobic growth of XH on SA (mineralization of organic matter) 4) Anoxic growth of XH on SF (denitrification) 5) Anoxic growth of XH on SA (denitrification) 6) Lysis of XH
Autotrophic bacteria: 6) Aerobic growth of XANs on SNH (ammonium oxidation) 7) Lysis of XANs 8) Aerobic growth of XANb on SNH (nitrite oxidation) 9) Lysis of XANb	*Autotrophic bacteria:* 7) Aerobic growth of XA on SNH (nitrification) 8) Lysis of XA
	Fermenting bacteria: 9) Growth of XFB (fermentation) 10) Lysis of XFB
	Acetotrophic methanogenic bacteria: 11) Growth of XAMB: Anaerobic growth of acetotrophic, methanogenic bacteria XAMB on acetate SA 12) Lysis of XAMB
	Acetotrophic sulfate reducing bacteria: 13) Growth of XASRB: Anaerobic growth of acetotrophic, sulfate reducing bacteria 14) Lysis of XASRB
	Sulfide oxidizing bacteria: 15) Aerobic growth of XSOB on SH2S: The opposite process to 13, the oxidation of SH2S to SSO4 16) Anoxic growth of XSOB on SH2S: Similar to process 15 but under anoxic conditions 17) Lysis of XSOB

autotrophic bacteria, fermenting, acetotrophic methanogenic, acetotrophic sulfate reducing, and sulfide oxidizing bacteria are also considered, in order to account for the main anaerobic processes.

9.3.1 CW2D Biokinetic Model

9.3.1.1 Stoichiometric Matrix and Reaction Rates

Table 9.6 and Table 9.7 show stoichiometric coefficients for ammonium nitrogen and inorganic phosphorus, respectively. Table 9.8 shows the stoichiometric matrix of reactions in CW2D, whereas Table 9.9 shows the reaction rates.

9.3.1.2 Model Parameters

Table 9.10 shows the kinetic parameters, and Table 9.11 the temperature dependences, stoichiometric parameters, composition parameters and parameters describing oxygen transfer for the CW2D biokinetic model, as described in Langergraber and Šimůnek (2005).

Table 9.6 Stoichiometric coefficients for ammonium nitrogen (Langergraber and Šimůnek, 2011; see Table 9.11 for definitions of the composition and stoichiometric parameters).

$\nu_{1,N} =$	$i_{N,CS} - (1 - f_{Hyd,CI}).i_{N,CR} - f_{Hyd,CI}.i_{N,CI}$
$\nu_{2,N} =$	$1/Y_H.i_{N,CR} - i_{N,BM}$
$\nu_{3,N} =$	$1/Y_H.i_{N,CR} - i_{N,BM}$
$\nu_{4,N} =$	$1/Y_H.i_{N,CR} - i_{N,BM}$
$\nu_{5,N} =$	$i_{N,BM} - (1 - f_{BM,CR} - f_{BM,CI}).i_{N,CS} - f_{BM,CR}.i_{N,CR} - f_{BM,CI}.i_{N,CI}$
$\nu_{6,N} =$	$-1/Y_{ANs} - i_{N,BM}$
$\nu_{7,N} =$	$i_{N,BM} - (1 - f_{BM,CR} - f_{BM,CI}).i_{N,CS} - f_{BM,CR}.i_{N,CR} - f_{BM,CI}.i_{N,CI}$
$\nu_{8,N} =$	$-i_{N,BM}$
$\nu_{9,N} =$	$i_{N,BM} - (1 - f_{BM,CR} - f_{BM,CI}).i_{N,CS} - f_{BM,CR}.i_{N,CR} - f_{BM,CI}.i_{N,CI}$

Table 9.7 Stoichiometric coefficients for inorganic phosphorus (Langergraber and Šimůnek, 2011; see Table 9.11 for definitions of the composition and stoichiometric parameters).

$\nu_{1,P} =$	$i_{P,CS} - (1 - f_{Hyd,CI}).i_{P,CR} - f_{Hyd,CI}.i_{P,CI}$
$\nu_{2,P} =$	$1/Y_H.i_{P,CR} - i_{P,BM}$
$\nu_{3,P} =$	$1/Y_H.i_{P,CR} - i_{P,BM}$
$\nu_{4,P} =$	$1/Y_H.i_{P,CR} - i_{P,BM}$
$\nu_{5,P} =$	$i_{P,BM} - (1 - f_{BM,CR} - f_{BM,CI}).i_{P,CS} - f_{BM,CR}.i_{P,CR} - f_{BM,CI}.i_{P,CI}$
$\nu_{6,P} =$	$-i_{P,BM}$
$\nu_{7,P} =$	$i_{P,BM} - (1 - f_{BM,CR} - f_{BM,CI}).i_{P,CS} - f_{BM,CR}.i_{P,CR} - f_{BM,CI}.i_{P,CI}$
$\nu_{8,P} =$	$-i_{P,BM}$
$\nu_{9,P} =$	$i_{P,BM} - (1 - f_{BM,CR} - f_{BM,CI}).i_{P,CS} - f_{BM,CR}.i_{P,CR} - f_{BM,CI}.i_{P,CI}$

Table 9.8 CW2D stoichiometric matrix (Langergraber and Šimůnek, 2005; see Table 9.11 for definitions of the stoichiometric coefficients).

	i →		1	2	3	4	5	6	7	8	9	10	11	12
j	Process	Component	O2	CR	CS	CI	XH	XANs	XANb	NH4N	NO2N	NO3N	N2N	IP
↓		expressed as →	O_2	COD	COD	COD	COD	COD	COD	N	N	N	N	P
1	Hydrolysis			$1-f_{\mathrm{Hyd,CR}}$	−1	$1-f_{\mathrm{Hyd,CI}}$				$\nu_{1,N}$				$\nu_{1,P}$
2	Aerobic growth of heterotrophs on readily biodegradable COD		$1-\frac{1}{Y_H}$	$-1/Y_{\mathrm{H}}$			1			$\nu_{2,N}$				$\nu_{2,P}$
3	NO_3-growth of heterotrophs on readily biodegradable COD			$-1/Y_{\mathrm{H}}$			1			$\nu_{3,N}$		$-\frac{1-Y_H}{2.86\cdot Y_H}$	$\frac{1-Y_H}{2.86\cdot Y_H}$	$\nu_{3,P}$
4	NO_2-growth of heterotrophs on readily biodegradable COD			$-1/Y_{\mathrm{H}}$			1			$\nu_{4,N}$	$-\frac{1-Y_H}{1.71\cdot Y_H}$		$\frac{1-Y_H}{1.71\cdot Y_H}$	$\nu_{4,P}$
5	Lysis of heterotrophs			$f_{\mathrm{BM,CR}}$	ν_{Lysis}	$f_{\mathrm{BM,CI}}$	−1			$\nu_{5,N}$				$\nu_{5,P}$
6	Aerobic growth of *Nitrosomonas* on NH_4		$-\frac{3.43-Y_A}{Y_A}$					1		$i_{\mathrm{N,BM}}-1/Y_{\mathrm{ANs}}$	$1/Y_{\mathrm{ANs}}$			$\nu_{6,P}$
7	Lysis of *Nitrosomonas*			$f_{\mathrm{BM,CR}}$	ν_{Lysis}	$f_{\mathrm{BM,CI}}$		−1		$\nu_{7,N}$				$\nu_{7,P}$
8	Aerobic growth of *Nitrobacter* on NO_2		$-\frac{1.14-Y_A}{Y_A}$						1	$\nu_{8,N}$	$-1/Y_{\mathrm{ANb}}$	$1/Y_{\mathrm{ANb}}$		$\nu_{8,P}$
9	Lysis of *Nitrobacter*			$f_{\mathrm{BM,CR}}$	ν_{Lysis}	$f_{\mathrm{BM,CI}}$			−1	$\nu_{9,N}$				$\nu_{9,P}$

$\nu_{\mathrm{Lysis}} = 1 - f_{BM,CR} - f_{BM,CI}$

Table 9.9 Reaction rates in CW2D (Langergraber and Šimůnek, 2005; see Table 9.10 for definitions of rate coefficients).

R	Process/reaction rate rc_j
	Heterotrophic organisms
1)	Hydrolysis
	$K_h \cdot \frac{c_{CS}/c_{XH}}{K_X + c_{CS}/c_{XH}} \cdot c_{XH}$
2)	Aerobic growth of heterotrophs on readily biodegradable COD
	$\mu_H \cdot \frac{c_{O2}}{K_{Het,O2} + c_{O2}} \cdot \frac{c_{CR}}{K_{Het,CR} + c_{CR}} \cdot f_{N,Het} \cdot c_{XH}$
3)	NO_3-growth of heterotrophs on readily biodegradable COD
	$\mu_{DN} \cdot \frac{K_{DN,O2}}{K_{DN,O2} + c_{O2}} \cdot \frac{c_{NO3}}{K_{DN,NO3} + c_{NO3}} \cdot \frac{K_{DN,NO2}}{K_{DN,NO2} + c_{NO2}} \cdot \frac{c_{CR}}{K_{DN,CR} + c_{CR}} \cdot f_{N,DN} \cdot c_{XH}$
4)	NO_2-growth of heterotrophs on readily biodegradable COD
	$\mu_{DN} \cdot \frac{K_{DN,O2}}{K_{DN,O2} + c_{O2}} \cdot \frac{c_{NO2}}{K_{DN,NO2} + c_{NO2}} \cdot \frac{c_{CR}}{K_{DN,CR} + c_{CR}} \cdot f_{N,DN} \cdot c_{XH}$
5)	Lysis of heterotrophs
	$b_H \cdot c_{XH}$
	Autotrophic organisms 1 – Nitrosomonas
6)	Aerobic growth of *Nitrosomonas* on NH_4
	$\mu_{ANs} \cdot \frac{c_{O2}}{K_{ANs,O2} + c_{O2}} \cdot \frac{c_{NH4}}{K_{ANs,NH4} + c_{NH4}} \cdot \frac{c_{IP}}{K_{ANs,IP} + c_{IP}} \cdot c_{XANs}$
7)	Lysis of *Nitrosomonas*
	$b_{HANs} \cdot c_{XANs}$
	Autotrophic organisms 2 – Nitrobacter
8)	Aerobic growth of *Nitrobacter* on NO_2
	$\mu_{ANb} \cdot \frac{c_{O2}}{K_{ANb,O2} + c_{O2}} \cdot \frac{c_{NO2}}{K_{ANb,NO2} + c_{NO2}} \cdot f_{N,ANb} \cdot c_{XANb}$
9)	Lysis of *Nitrobacter*
	$b_{HANb} \cdot c_{XANb}$
	Conversion of solid and liquid phase concentrations
	$c_{XY} = \frac{\rho}{\theta} \cdot s_{XY}, \text{where} Y = \text{H, ANs, ANb}$
	Factor for nutrients
	$f_{N,x} = \frac{c_{NH4}}{K_{x,NH4} + c_{NH4}} \cdot \frac{c_{IP}}{K_{x,IP} + c_{IP}}, \text{ where } x = \text{Het, DN, ANb}$

Table 9.10 Kinetic parameters in the CW2D biokinetic model (Langergraber and Šimůnek, 2005).

Parameter	Description [unit]	Value
Hydrolysis		For 20 °C (10 °C)
K_h	Hydrolysis rate constant [1/d]	3 (2)
K_X	Saturation/inhibition coefficient for hydrolysis [g COD_{CS}/g COD_{BM}]	0.1 (0.22)*
Heterotrophic bacteria (aerobic growth)		
μ_H	Maximum aerobic growth rate on CR [1/d]	6 (3)
b_H	Rate constant for lysis [1/d]	0.4 (0.2)
$K_{\text{het,O2}}$	Saturation/inhibition coefficient for S_O [mg O_2/l]	0.2
$K_{\text{het,CR}}$	Saturation/inhibition coefficient for substrate [mg COD_{CR}/l]	2
$K_{\text{het,NH4N}}$	Saturation/inhibition coefficient for NH_4 (nutrient) [mg N/l]	0.05
$K_{\text{het,IP}}$	Saturation/inhibition coefficient for P [mg N/l]	0.01
Heterotrophic bacteria (denitrification)		
μ_{DN}	Maximum aerobic growth rate on CR [1/d]	4.8 (2.4)
$K_{\text{DN,O2}}$	Saturation/inhibition coefficient for S_O [mg O_2/l]	0.2
$K_{\text{DN,NO3N}}$	Saturation/inhibition coefficient for NO_3 [mg N/l]	0.5
$K_{\text{DN,NO2N}}$	Saturation/inhibition coefficient for NO_2 [mg N/l]	0.5
$K_{\text{DN,CR}}$	Saturation/inhibition coefficient for substrate [mg COD_{CR}/l]	4
$K_{\text{DN,NH4N}}$	Saturation/inhibition coefficient for NH_4 (nutrient) [mg N/l]	0.05
$K_{\text{DN,IP}}$	Saturation/inhibition coefficient for P [mg N/l]	0.01
Ammonia oxidizing bacteria (Nitrosomonas spp.)		
μ_{ANs}	Maximum aerobic growth rate on S_{NH} [1/d]	0.9 (0.3)
b_{ANs}	Rate constant for lysis [1/d]	0.15 (0.05)
$K_{\text{ANs,O2}}$	Saturation/inhibition coefficient for S_O [mg O_2/l]	1
$K_{\text{ANs,NH4N}}$	Saturation/inhibition coefficient for NH_4 [mg N/l]	0.5 (5.5)*
$K_{\text{ANs,IP}}$	Saturation/inhibition coefficient for P [mg N/l]	0.01
Nitrite oxidizing bacteria (Nitrobacter spp.)		
μ_{ANb}	Maximum aerobic growth rate on S_{NH} [1/d]	1 (0.35)
b_{ANb}	Rate constant for lysis [1/d]	0.15 (0.05)
$K_{\text{ANb,O2}}$	Saturation/inhibition coefficient for S_O [mg O_2/l]	0.1
$K_{\text{ANb,NO2N}}$	Saturation/inhibition coefficient for NO_2 [mg N/l]	0.1
$K_{\text{ANb,NH4N}}$	Saturation/inhibition coefficient for NH_4 (nutrient) [mg N/l]	0.05
$K_{\text{ANb,IP}}$	Saturation/inhibition coefficient for P [mg N/l]	0.01

*Langergraber (2007).

Table 9.11 Temperature dependences, stoichiometric parameters, composition parameters and parameters describing oxygen transfer in the CW2D biokinetic model (Langergraber and Šimůnek, 2005).

Parameter	Description [unit]	Value
Temperature dependencies (activation energy [J/mol] for Arrhenius equation)		
Tdep_het	Activation energy for processes caused by XH [J/mol]	47,800
Tdep_aut	Activation energy for processes caused by XA [J/mol]	69,000
Tdep_Kh	Activation energy hydrolyses [J/mol]	28,000
Tdep_KX	Activation energy factor KX for hydrolyses [J/mol]	–53,000*
Tdep_KNHA	Activation energy for factor KNHA for nitrification [J/mol]	–160,000*
Stoichiometric parameters		
$f_{Hyd,CI}$	Production of CI in hydrolysis	0.0
$f_{BM,CR}$	Fraction of CR generated in biomass lysis	0.1
$f_{BM,CI}$	Fraction of CI generated in biomass lysis	0.02
Y_{Het}	Yield coefficient for XH	0.63
Y_{ANs}	Yield coefficient for XANs	0.24
Y_{ANb}	Yield coefficient for XANb	0.24
Composition parameters		
$i_{N,CR}$	N content of CR [$g N/g COD_{CR}$]	0.03
$i_{N,CS}$	N content of CS [$g N/g COD_{CS}$]	0.04
$i_{N,CI}$	N content of CI [$g N/g COD_{CI}$]	0.01
$i_{N,BM}$	N content of biomass [$g N/g COD_{BM}$]	0.07
$i_{P,CR}$	P content of CR [$g P/g COD_{CR}$]	0.01
$i_{P,CS}$	P content of CS [$g P/g COD_{CS}$]	0.01
$i_{P,CI}$	P content of CI [$g P/g COD_{CI}$]	0.01
$i_{P,BM}$	P content of biomass [$g P/g COD_{BM}$]	0.02
Oxygen		
cO2_sat_20	Saturation concentration of oxygen [g/m^3]	9.18
Tdep_cO2_sat	Activation energy for saturation concentration of oxygen [J/mol]	–15,000
rate_O2	Re-aeration rate [1/d]	240

*Langergraber (2007).

9.3.2 CWM1 Biokinetic Model

9.3.2.1 Stoichiometric Matrix and Reaction Rates

Table 9.12 shows the stoichiometric matrix of reactions in CWM1, Table 9.13 the stoichiometric coefficients for ammonium nitrogen, and Table 9.14 the reaction rates.

Table 9.12 CWM1 stoichiometric matrix (Langergraber et al., 2009b; see Table 9.17 for definitions of the stoichiometric coefficients).

	i →	1	2	3	4	5	6	7	8	9	10	11	12	13	14	15	16
j	**Process** / **Component**	**SO**	**SF**	**SA**	**SI**	**SNH**	**SNO**	**SSO4**	**SH2S**	**XS**	**XI**	**XH**	**XA**	**XFB**	**XAMB**	**XASRB**	**XSOB**
↓	expressed as →	O_2	COD	COD	COD	N	N	S	S	COD	COD	COD	COD	COD	COD	COD	COD
1	Hydrolysis		$1-f_{Hyd,SI}$		$f_{Hyd,SI}$	$\nu_{1,5}$				-1							
2	Aerobic growth of X_H on S_F	$1-\frac{1}{Y_H}$	$-1/Y_H$			$\nu_{2,5}$						1					
3	Anoxic growth of X_H on S_F		$-1/Y_H$			$\nu_{3,5}$	$-\frac{1-Y_H}{2.86 \cdot Y_H}$					1					
4	Aerobic growth of X_H on S_A	$1-\frac{1}{Y_H}$		$-1/Y_H$		$\nu_{4,5}$						1					
5	Anoxic growth of X_H on S_A			$-1/Y_H$		$\nu_{5,5}$	$-\frac{1-Y_H}{2.86 \cdot Y_H}$					1					
6	Lysis of X_H		$f_{BM,SF}$			$\nu_{6,5}$				ν_{Lysis}	$f_{BM,XI}$	-1					
7	Aerobic growth of X_A on S_{NH}	$-\frac{4.57-Y_A}{Y_A}$				$-i_{N,BM}-\frac{1}{Y_A}$	$1/Y_A$						1				
8	Lysis of X_A		$f_{BM,SF}$			$\nu_{8,5}$				ν_{Lysis}	$f_{BM,XI}$		-1				
9	Growth of X_{FB}		$-1/Y_{FB}$	$\frac{1-Y_{FB}}{Y_{FB}}$		$\nu_{9,5}$								1			
10	Lysis of X_{FB}		$f_{BM,SF}$			$\nu_{10,5}$				ν_{Lysis}	$f_{BM,XI}$			-1			

11 Growth of X_{AMB}			$-1/Y_{AMB}$	$\nu_{11,5}$						1		
12 Lysis of X_{AMB}		$f_{BM,SF}$		$\nu_{12,5}$				ν_{Lysis}	$f_{BM,XI}$	−1		
13 Growth of X_{ASRB}			$-1/Y_{ASRB}$	$\nu_{13,5}$		$-\frac{1-Y_{ASRB}}{2\cdot Y_{ASRB}}$	$\frac{1-Y_{ASRB}}{2\cdot Y_{ASRB}}$				1	
14 Lysis of X_{ASRB}		$f_{BM,SF}$		$\nu_{14,5}$				ν_{Lysis}	$f_{BM,XI}$		−1	
15 Aerobic growth of X_{SOB} on S_{H2S}	$-\frac{2-Y_{SOB}}{Y_{SOB}}$			$\nu_{15,5}$		$1/Y_{SOB}$	$-1/Y_{SOB}$					1
16 Anoxic growth of X_{SOB} on S_{H2S}				$\nu_{16,5}$	$-\frac{1-Y_{SOB}}{0.875\cdot Y_{SOB}}$	$1/Y_{SOB}$	$-1/Y_{SOB}$					1
17 Lysis of X_{SOB}		$f_{BM,SF}$		$\nu_{17,5}$				ν_{Lysis}	$f_{BM,XI}$			−1

$\nu_{Lysis} = 1 - f_{BM,SF} - f_{BM,XI}$

Table 9.13 Stoichiometric coefficients for ammonia nitrogen (see Table 9.17 for definitions of the composition and stoichiometric parameters).

$\nu_{1,5} =$	$i_{N,XS} - (1 - f_{HYD,SI}) * i_{N,SF} - f_{HYD,SI} * i_{N,SI}$
$\nu_{2,5} = \nu_{3,5} =$	$i_{N,SF}/Y_H - i_{N,BM}$
$\nu_{4,5} = \nu_{5,5} = \nu_{11,5} = \nu_{13,5} = \nu_{15,5} = \nu_{16,5} =$	$-i_{N,BM}$
$\nu_{6,5} = \nu_{8,5} = \nu_{10,5} = \nu_{12,5} = \nu_{14,5} = \nu_{17,5} =$	$i_{N,BM} - f_{BM,SF} * i_{N,SF} - (1 - f_{BM,SF} - f_{BM,XI}) * i_{N,XS} - f_{BM,XI} * i_{N,XI}$
$\nu_{7,5} =$	$-i_{N,BM} - \frac{1}{Y_A}$
$\nu_{9,5} =$	$i_{N,SF}/Y_{FB} - i_{N,BM}$

9.3.2.2 Model Parameters

Table 9.15 shows the kinetic parameters in the CWM1 biokinetic model; Table 9.16 temperature dependences, stoichiometric parameters, composition parameters and parameters describing oxygen transfer as described in Langergraber et al. (2009b).

9.4 Simulation Results for Vertical Flow Constructed Wetlands Treating Domestic Wastewater

This section shows typical simulation results for VF CWs. As described by Langergraber (2008), VF pilot-scale CWs (Figure 9.1) have been operated in the technical laboratory hall of the Institute of Sanitary Engineering at BOKU University in Vienna, Austria. The systems had a surface area of 1 m^2 and were loaded four times per day with mechanically pre-treated municipal wastewater. The 60 cm main layer consisted of sand with a grain size of 0.06–4 mm ($d_{10} = 0.2$ mm; $d_{60} = 0.8$ mm). The intermediate layer (gravel 4–8 mm) prevents fine particles from being washed out into the drainage layer (gravel 16–32 mm).

For the numerical simulations, only the main layer was considered. The transport domain (i.e. a vertical cross-section of the main layer) was discretized into 11 columns and 40 rows. This resulted in a two-dimensional finite element mesh consisting of 440 nodes and 780 finite elements. An atmospheric boundary condition is assigned to the top of the system, a constant pressure head boundary condition with constant head of −2 cm (Langergraber, 2008).

Table 9.17 shows three parameter sets of the van Genuchten-Mualem model (van Genuchten, 1980) of the soil hydraulic properties used in simulations of water flow. The "HYDRUS parameter set for 'Sand'" uses default soil hydraulic properties for sand provided by HYDRUS, the '*Measured K_s and porosity*' parameter set measured values of porosity and saturated hydraulic conductivity K_s, and the default residual water content and shape parameters for sand. The third parameter set was obtained by inverse simulation. Measured effluent flow rates were used to estimate the soil hydraulic parameters (i.e. residual water content and shape parameters) using the inverse simulation routine provided by HYDRUS (Šimůnek et al., 2011). Measured values of porosity and K_s were kept constant. Figure 9.2 compares the measured and simulated effluent flow rates using these three parameter sets. The simulated effluent flow rates are rather constant when using the default parameter set for sand due to the low K_s value. When using

Table 9.14 Reaction rates in CWM1 part 1 (Langergraber et al., 2009b; see Tables 9.16 and 9.17 for definitions of rate coefficients).

R	Process/reaction rate rc_j
Heterotrophic organisms	
1)	Hydrolysis $k_h * \left[\frac{X_S/(X_H+X_{FB})}{K_X+(X_S/(X_H+X_{FB}))}\right] * (X_H+\eta_h * X_{FB})$
2)	Aerobic growth of XH on SF (mineralization) $\mu_H * \left(\frac{S_F}{K_{SF}+S_F}\right) * \left(\frac{S_F}{S_F+S_A}\right) * \left(\frac{S_O}{K_{OH}+S_o}\right) * \left(\frac{S_{NH}}{K_{NHH}+S_{NH}}\right) * \left(\frac{K_{H2SH}}{K_{H2SH}+S_{H2S^*}}\right) * X_H$
3)	Aerobic growth of XH on SA (mineralization) $\eta_g * \mu_H * \left(\frac{S_F}{K_{SF}+S_F}\right) * \left(\frac{S_F}{S_F+S_A}\right) * \left(\frac{K_{OH}}{K_{OH}+S_o}\right) * \left(\frac{S_{NO}}{K_{NOH}+S_{NO}}\right) * \left(\frac{S_{NH}}{K_{NHH}+S_{NH}}\right) * \left(\frac{K_{H2SH}}{K_{H2SH}+S_{H2S^*}}\right) * X_H$
4)	Anoxic growth of XH on SF (denitrification) $\mu_H * \left(\frac{S_A}{K_{SA}+S_A}\right) * \left(\frac{S_A}{S_F+S_A}\right) * \left(\frac{S_O}{K_{OH}+S_o}\right) * \left(\frac{S_{NH}}{K_{NHH}+S_{NH}}\right) * \left(\frac{K_{H2SH}}{K_{H2SH}+S_{H2S^*}}\right) * X_H$
5)	Anoxic growth of XH on SA (denitrification) $\eta_g * \mu_H * \left(\frac{S_A}{K_{SA}+S_A}\right) * \left(\frac{S_A}{S_F+S_A}\right) * \left(\frac{K_{OH}}{K_{OH}+S_o}\right) * \left(\frac{S_{NO}}{K_{NOH}+S_{NO}}\right) * \left(\frac{S_{NH}}{K_{NHH}+S_{NH}}\right) * \left(\frac{K_{H2SH}}{K_{H2SH}+S_{H2S^*}}\right) * X_H$
6)	Lysis of XH $b_H * X_H$
Autotrophic bacteria	
7)	Aerobic growth of XA on SNH (nitrification) $\mu_A * \left(\frac{S_{NH}}{K_{NHA}+S_{NH}}\right) * \left(\frac{S_O}{K_{OA}+S_o}\right) * \left(\frac{K_{H2SA}}{K_{H2SA}+S_{H2S^*}}\right) * X_A$
8)	Lysis of XA $b_A * X_A$
Fermenting bacteria	
9)	Growth of XFB (fermentation) $\mu_{FB} * \left(\frac{S_F}{K_{SFB}+S_F}\right) * \left(\frac{K_{H2SFB}}{K_{H2SFB}+S_{H2S^*}}\right) * \left(\frac{K_{OFB}}{K_{OFB}+S_O}\right) * \left(\frac{K_{NOFB}}{K_{NOFB}+S_{NO}}\right) * \left(\frac{S_{NH}}{K_{NHFB}+S_{NH}}\right) * X_{FB}$

(Continued)

Table 9.14 (Continued)

R	Process/reaction rate rc_j
10)	Lysis of XFB $b_{FB} * X_{FB}$
	Acetotrophic methanogenic bacteria
11)	Growth of XAMB $\mu_{AMB} * \left(\frac{S_A}{K_{SAMB}+S_A}\right) * \left(\frac{K_{H2SAMB}}{K_{H2SAMB}+S_{H2S*}}\right) * \left(\frac{K_{OAMB}}{K_{OAMB}+S_O}\right) * \left(\frac{K_{NOAMB}}{K_{NOAMB}+S_{NO}}\right) * \left(\frac{S_{NH}}{K_{NHAMB}+S_{NH}}\right) * X_{AMB}$
12)	Lysis of XAMB $b_{AMB} * X_{AMB}$
	Acetotrophic sulfate reducing bacteria
13)	Growth of XASRB $\mu_{ASRB} * \left(\frac{S_A}{K_{SASRB}+S_A}\right) * \left(\frac{S_{SO4}}{K_{SOASRB}+S_{SO4}}\right) * \left(\frac{K_{H2SASRB}}{K_{H2SASRB}+S_{H2S*}}\right) * \left(\frac{K_{OASRB}}{K_{OASRB}+S_O}\right) * \left(\frac{K_{NOASRB}}{K_{NOASRB}+S_{NO}}\right) * \left(\frac{S_{NH}}{K_{NHASRB}+S_{NH}}\right) * X_{ASRB}$
14)	Lysis of XASRB $b_{ASRB} * X_{ASRB}$
	Sulfide oxidizing bacteria
15)	Aerobic growth of XSOB on SH2S $\mu_{SOB} * \left(\frac{S_{H2S}}{K_{SSOB}+S_{H2S}}\right) * \left(\frac{S_O}{K_{OSOB}+S_O}\right) * \left(\frac{S_{NH}}{K_{NHSOB}+S_{NH}}\right) * X_{SOB}$
16)	Anoxic growth of XSOB on SH2S $\mu_{SOB} * \eta_{SOB} * \left(\frac{S_{H2S}}{K_{SSOB}+S_{H2S}}\right) * \left(\frac{S_{NO}}{K_{NOSOB}+S_{NO}}\right) * \left(\frac{K_{OSOB}}{K_{OSOB}+S_O}\right) * \left(\frac{S_{NH}}{K_{NHSOB}+S_{NH}}\right) * X_{SOB}$
17)	Lysis of XSOB $b_{SOB} * X_{SOB}$

measured values of porosity and saturated hydraulic conductivity (the '*Measured* K_s *+porosity*' parameter set), measured and simulated effluent flow rates match better; however, the peak of the effluent flow rate is overpredicted. Using the '*Fitted parameter set*', a good match between simulated and measured data can be achieved. This indicates that it is necessary to measure at least the porosity and the saturated hydraulic conductivity of the filter material to obtain reasonable simulation results for water flow.

Table 9.18 shows measured and simulated concentrations of ammonia and nitrate nitrogen (NH_4-N and NO_3-N, respectively) for the VF pilot-scale CWs with two

Table 9.15 Kinetic parameters in the CWM1 biokinetic model (Langergraber et al., 2009b).

Parameter	Description [unit]	Value
Hydrolysis		For 20 °C (10 °C)
K_h	Hydrolysis rate constant [1/d]	3 (2)
K_X	Saturation/inhibition coefficient for hydrolysis [g COD_{SF}/g COD_{BM}]	0.1 (0.22)
K_H	Correction factor for hydrolysis by fermenting bacteria [-]	0.1
Heterotrophic bacteria (aerobic growth and denitrification)		
μ_H	Maximum aerobic growth rate on SF and SA [1/d]	6 (3)
K_g	Correction factor for denitrification by XH [-]	0.8
b_H	Rate constant for lysis [1/d]	0.4 (0.2)
K_{OH}	Saturation/inhibition coefficient for SO [mg O_2/l]	0.2
K_{SF}	Saturation/inhibition coefficient for SF [mg COD_{SF}/l]	2
K_{SA}	Saturation/inhibition coefficient for SA [mg COD_{SA}/l]	4
K_{NOH}	Saturation/inhibition coefficient for SNO [mg N/l]	0.5
K_{NHH}	Saturation/inhibition coefficient for SNH (nutrient) [mg N/l]	0.05
K_{H2SH}	Saturation/inhibition coefficient for SH2S [mg S/l]	140
Autotrophic bacteria		
μ_A	Maximum aerobic growth rate on SNH [1/d]	1 (0.35)
b_A	Rate constant for lysis [1/d]	0.15 (0.05)
K_{OA}	Saturation/inhibition coefficient for SO [mg O_2/l]	1
K_{NHA}	Saturation/inhibition coefficient for SNH [mg N/l]	0.5 (5)
K_{H2SA}	Saturation/inhibition coefficient for SH2S [mg S/l]	140
Fermenting bacteria		
μ_{FB}	Maximum aerobic growth rate for XFB [1/d]	3 (1.5)
b_{FB}	Rate constant for lysis [1/d]	0.02
K_{OFB}	Saturation/inhibition coefficient for SO [mg O_2/l]	0.2
K_{SFB}	Saturation/inhibition coefficient for SF [mg COD_{SF}/l]	28
K_{NOFB}	Saturation/inhibition coefficient for SNO [mg N/l]	0.5
K_{NHFB}	Saturation/inhibition coefficient for SNH (nutrient) [mg N/l]	0.01
K_{H2SFB}	Saturation/inhibition coefficient for SH2S [mg S/l]	140
Acetotrophic methanogenic bacteria		
μ_{AMB}	Maximum aerobic growth rate on for XAMB [1/d]	0.085
b_{AMB}	Rate constant for lysis [1/d]	0.008
K_{OAMB}	Saturation/inhibition coefficient for SO [mg O_2/l]	0.0002
K_{SAMB}	Saturation/inhibition coefficient for SA [mg COD_{SA}/l]	56
K_{NOAMB}	Saturation/inhibition coefficient for SNO [mg N/l]	0.0005
K_{NHAMB}	Saturation/inhibition coefficient for SNH (nutrient) [mg N/l]	0.01
K_{H2SAMB}	Saturation/inhibition coefficient for SH2S [mg S/l]	140

(Continued)

Table 9.15 (Continued)

Parameter	Description [unit]	Value
Acetotrophic sulfate reducing bacteria		
μ_{ASRB}	Maximum aerobic growth rate for XASRB [1/d]	0.18
b_{ASRB}	Rate constant for lysis [1/d]	0.012
K_{OASRB}	Saturation/inhibition coefficient for SO [mg O_2/l]	0.0002
K_{SASRB}	Saturation/inhibition coefficient for SA [mg COD_{SA}/l]	24
K_{NOASRB}	Saturation/inhibition coefficient for SNO [mg N/l]	0.0005
K_{NHASRB}	Saturation/inhibition coefficient for SNH (nutrient) [mg N/l]	0.01
K_{SOASRB}	Saturation/inhibition coefficient for SSO4 [mg S/l]	19
$K_{H2SASRB}$	Saturation/inhibition coefficient for SH2S [mg S/l]	140
Sulfide oxidizing bacteria		
μ_{SOB}	Maximum aerobic growth rate for XSOB [1/d]	5.28
K_{SOB}*	Correction factor for anoxic growth of XSOB [-]	0.8
b_{SOB}	Rate constant for lysis [1/d]	0.15
K_{OSOB}	Saturation/inhibition coefficient for SO [mg O_2/l]	0.2
K_{NOSOB}	Saturation/inhibition coefficient for SNO [mg N/l]	0.5
K_{NHSOB}	Saturation/inhibition coefficient for SNH (nutrient) [mg N/l]	0.05
K_{SSOB}	Saturation/inhibition coefficient for SH2S [mg S/l]	0.24

*Typing error in the original CWM1 publication.

different sandy filter materials used for the main layer. Besides the sand with a grain size of 0.06–4 mm, sand with a grain size of 1–4 mm was also tested. The different levels of the NO_3-N effluent concentrations for the two filter materials could be simulated. The effluent concentrations of NO_3-N for the 1-4 mm substrate have been higher, because less readily degradable organic matter is produced by hydrolysis that could be utilized for denitrification in deeper layers (Langergraber, 2003).

For the VF pilot-scale CWs with main layer of 0.06–4 mm sand, the distribution of microbial and bacterial biomass in different depths of the filter was also measured (Tietz et al., 2007). Converting the measured data into biomass COD results in mean values of 3400 to 5100 μg COD/g DW (dry weight of the substrate, i.e. sand) for the first centimetre of the main layer, 1100 to 2600 μg COD/g DW from 1–5 cm, and 640 to 1400 μg COD/g DW from 5–10 cm, respectively. Most of the biomass could be found in the upper 10 cm of the VF bed (Langergraber et al., 2007).

Simulated microbial biomass COD in the first centimetre of the main layer is between 5600 and 3400 μg COD/g DW (the range of the measured values) when using heterotrophic lysis rates of between 0.25 and 0.35 d^{-1}. Figure 9.3 compares calculated (from substrate-induced respiration) and simulated microbial biomass COD in different depths of the main layer for a heterotrophic lysis rate of 0.30 d^{-1}. When comparing measured and simulated biomass COD in different depths of the main layer, simulations seem to overpredict biomass COD in 1–5 cm depth and underpredict biomass COD in 5–10 cm depth. This could be an indication that the

Table 9.16 Temperature dependences, stoichiometric parameters, composition parameters and parameters describing oxygen transfer in the CW2D biokinetic model (Langergraber et al., 2009b).

Parameter	Description [unit]	Value
Temperature dependences (activation energy [J/mol] for Arrhenius equation)		
Tdep_HyKh	Activation energy hydrolyses [J/mol]	28,000
Tdep_HyKX	Activation energy factor KX for hydrolyses [J/mol]	−54,400
Tdep_H	Activation energy for processes caused by XH [J/mol]	47,800
Tdep_A	Activation energy for processes caused by XA [J/mol]	75,800
Tdep_KNHA	Activation energy for factor KNHA for nitrification [J/mol]	−160,000
Tdep_mueFB	Activation energy for XFB growth [J/mol]	47,800
Tdep_bFB	Activation energy for XFB lysis [J/mol]	0
Tdep_AMB	Activation energy for processes caused by XAMB [J/mol]	0
Tdep_ASRB	Activation energy for processes caused by XASRB [J/mol]	0
Tdep_SOB	Activation energy for processes caused by XSOB [J/mol]	0
Stoichiometric parameters		
$f_{Hyd,SI}$	Production of S_I in hydrolysis	0.0
$f_{BM,SF}$	Fraction of S_F generated in biomass lysis	0.05
$f_{BM,XI}$	Fraction of X_I generated in biomass lysis	0.1
Y_H	Yield coefficient for XH	0.63
Y_A	Yield coefficient for XA	0.24
Y_{FB}	Yield coefficient for XFB	0.053
Y_{AMB}	Yield coefficient for XAMB	0.032
Y_{ASRB}	Yield coefficient for XASRB	0.05
Y_{SOB}	Yield coefficient for XSOB	0.12
Composition parameters		
$i_{N,SF}$	N content of S_F [g N/g COD_{SF}]	0.03
$i_{N,SI}$	N content of S_I [g N/g COD_{SI}]	0.01
$i_{N,XS}$	N content of X_S [g N/g COD_{XS}]	0.04
$i_{N,XI}$	N content of X_I [g N/g COD_{XI}]	0.03
$i_{N,BM}$	N content of biomass [g N/g COD_{BM}]	0.07
Oxygen		
cO2_sat_20	Saturation concentration of oxygen [g/m^3]	9.18
Tdep_cO2_sat	Activation energy for saturation concentration of oxygen [J/mol]	−15,000
rate_O2	Re-aeration rate [1/d]	240

influence of biomass growth on the hydraulic properties has to be considered in the model (Langergraber et al., 2007).

Additionally, outdoor experiments were carried out at experimental CWs located at the wastewater treatment plant at Ernsthofen (Lower Austria). The experimental plant consisted of three parallel operating VF beds with a surface area of about 20 m^2 each.

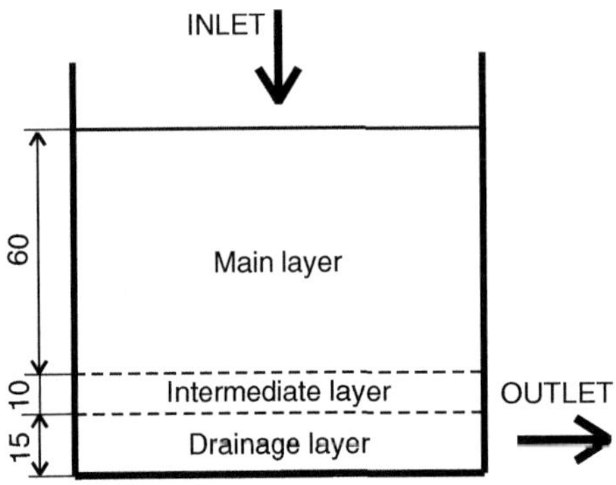

Figure 9.1 Schematic representation of the indoor pilot-scale CW (values are in cm).

Table 9.17 Soil hydraulic parameters of the van Genuchten-Mualem model for the 0.06–4 mm main layer (shape parameters α, N and L: van Genuchten, 1980).

Parameter	Residual water content ffj$_r$ (–)	Saturated water content ffj$_s$ (–)	Shape parameters α (cm^{-1})	N (–)	L (–)	Saturated hydraulic conductivity K_s (cm/h)
HYDRUS parameter set for 'Sand'	0.045	0.43	0.145	2.68	0.5	29.7
Measured K_s and porosity	0.045	0.30	0.145	2.68	0.5	117
Fitted parameter set	0.013	0.30	0.147	2.42	0.636	117

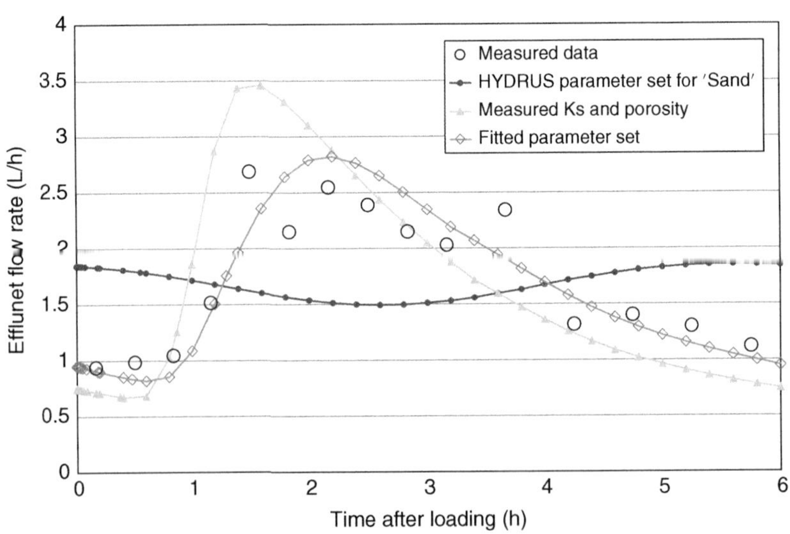

Figure 9.2 Measured and simulated effluent flow rates for a single feeding of 10 litres ('HYDRUS parameter set or Sand', 'Measured Ks and porosity' and 'Fitted parameter set' represent the parameter sets from Table 9.17).

Table 9.18 Median values of measured and simulated influent and effluent concentrations for NH_4-N and NO_3-N in mg/l (adapted from Langergraber, 2003).

	Influent	**Effluent (0.06–4 mm sand)**		**Effluent (1–4 mm sand)**	
Parameter	**Measured**	**Measured**	**Simulated**	**Measured**	**Simulated**
NH_4-N	60.0	0.15	0.01	1.20	0.16
NO_3-N	3.0	38.5	41.1	50.0	63.0

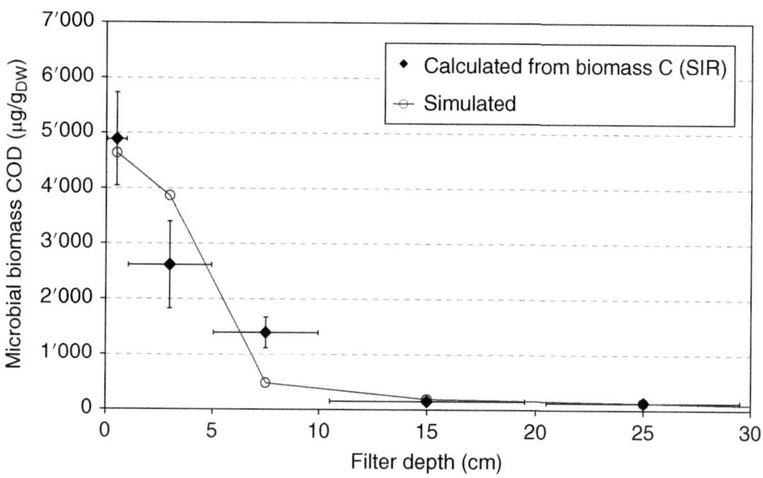

Figure 9.3 Calculated and simulated microbial biomass COD in different depths of the main layer (*Source:* adapted from Langergraber et al., 2007; Langergraber, 2015).

The organic loads for bed 1 to 3 were 20, 27 and 40 g COD/m^2/d, which correspond to hydraulic loading rates of 32.2, 43.0, and 64.7 mm/d. The main layer of the filter consisted of 50 cm sandy substrate (gravel size 0.06–4 mm, $d_{10} = 0.2$ mm; $d_{60} = 0.8$ mm). The beds are planted with common reed (*Phragmites australis*). The data gained from this experiment have been used to verify the temperature model incorporated into CW2D (Langergraber, 2007).

The width of the transport domain in the numerical simulations was 4 m, and its depth 0.8 m, while the transport domain itself was discretized into 31 columns and 33 rows. This resulted in a two-dimensional finite element mesh consisting of 1023 nodes and 1920 triangular finite elements. An atmospheric boundary condition was assigned to the top of the system representing the influent distribution system, and a constant pressure head boundary condition (constant head of −4 cm) was assigned to one side of the drainage layer (Langergraber, 2007).

Results for water flow comparing measured and simulated effluent flow rates and the cumulative effluent of the experimental CW are shown in Langergraber (2007). Simulated effluent flow rate and cumulative effluent flow match the measured data well.

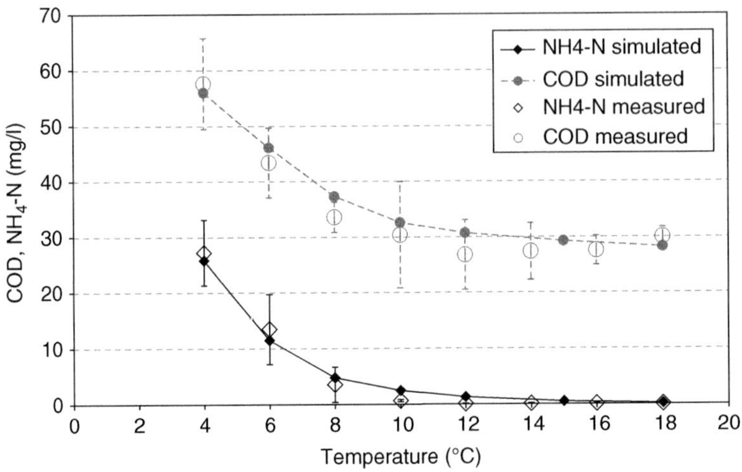

Figure 9.4 Measured and simulated COD and NH4-N effluent concentrations using the modified parameter set (adapted from Langergraber, 2007).

Using the calibrated flow model, the effluent concentrations during summer could be simulated using the standard CW2D parameter set (Langergraber and Šimůnek, 2005). However, measured COD and NH_4-N effluent concentrations at low temperatures could not be simulated because hydrolysis and nitrification at low temperatures were overpredicted. The standard CW2D parameter set considers temperature dependencies only for maximum growth, decay and hydrolysis rates. Figure 9.4 shows measured and simulated COD and NH_4-N effluent concentrations using a modified parameter set that includes temperature dependencies for the half-saturation constants of hydrolysis and nitrification. Using this modified parameter set, it was possible to simulate the COD and NH_4-N effluent concentrations at low temperatures (Langergraber, 2007).

9.5 Experiences and Challenges using Wetland Models

9.5.1 Description of Water Flow

It is generally agreed that the hydraulics have a high impact on the treatment performance of CWs. When simulating SSF CWs, the first step thus should be a good calibration of the water flow model. Experience shows that simulated effluent concentrations can match the measured data only when the hydraulic behaviour of the system is well described.

For VF CWs, the influence of the parameters describing the hydraulic properties of the filter material (such as for the van Genuchten-Mualem function in HYDRUS; Šimůnek et al., 2011) is much higher than the influence of the parameters of the biokinetic model. To achieve a good calibration of the water flow model, it is advised to measure at least the porosity and the saturated hydraulic conductivity of the filter material and, if possible, the volumetric effluent flow rate between two intermittent loadings (Langergraber, 2011). With these measurements, it is possible to estimate the remaining parameters describing the hydraulic properties of the filter material using the inverse simulation implemented in HYDRUS (Šimůnek et al., 2011).

Morvannou et al. (2012) showed that macropores/cracks in some layers of a VF CW might serve as preferential flow paths through which water can bypass most of the soil porous matrix in a largely unpredictable way. This is especially true for the sludge layer in French-type VF CWs (Molle et al., 2005; Troesch and Esser, 2012). Water flow in such systems cannot be modeled with uniform flow models (such as using the Richards equation combined with the van Genuchten-Mualem function in HYDRUS; Šimůnek et al., 2011). The comparison between measured and simulated tracer breakthrough curves indicates that the nonequilibrium approach (i.e. using a model that separately describes flow and transport in preferred flow paths and slow or stagnant pore regions) seems to be more appropriate for simulating preferential flow. Such a dual-porosity model therefore also needs to be incorporated into the software tools to describe water flow and solute transport accurately in French VF CWs.

For HF CWs, in which the filter material is usually gravel and the saturated hydraulic conductivity is much larger than the actual water flow in the filter, tracer experiments are usually used to calibrate the water flow and single solute transport models.

9.5.2 Values of the Biokinetic Model Parameters and Influent Fractionation

Bacteria in CWs behave similarly to those in activated sludge systems. Based on this assumption, Langergraber (2001), when developing the first biokinetic model for CWs, proposed that parameters of the biokinetic models developed for activated sludge systems should be applicable to describe processes in CWs as well. This assumption was confirmed when it was shown that a good match between measured and simulated concentrations can be achieved when the hydraulic behaviour of the system is well described (see above). Additionally, Morvannou et al. (2011) found good agreement between measured and calculated volumetric nitrification rates (Table 9.19).

When using the biokinetic models such as CW2D or CWM1, it is therefore advised not to change the default parameters of the biokinetic model, except for good reasons. However, parameters characterizing the wastewater are related to the chosen biokinetic model. Influent fractionation has a high impact on the simulation results and needs to be adapted for each simulation study. These parameters include (1) fractionation of influent COD (i.e. estimation of different COD model fractions from measured total COD) and (2) organic N content of different COD fractions.

Table 9.19 Comparison of measured and calculated volumetric nitrification rates (adapted from Morvannou et al., 2011).

Method	Measured with solid respirometry	Calculated from simulation results*
Results [mg $O_2/l_{sample}/h$]	32–50 (mean = 41, SD = 9; 2 values)	30.5

*From simulations using parameters for the biokinetic model from activated sludge systems.

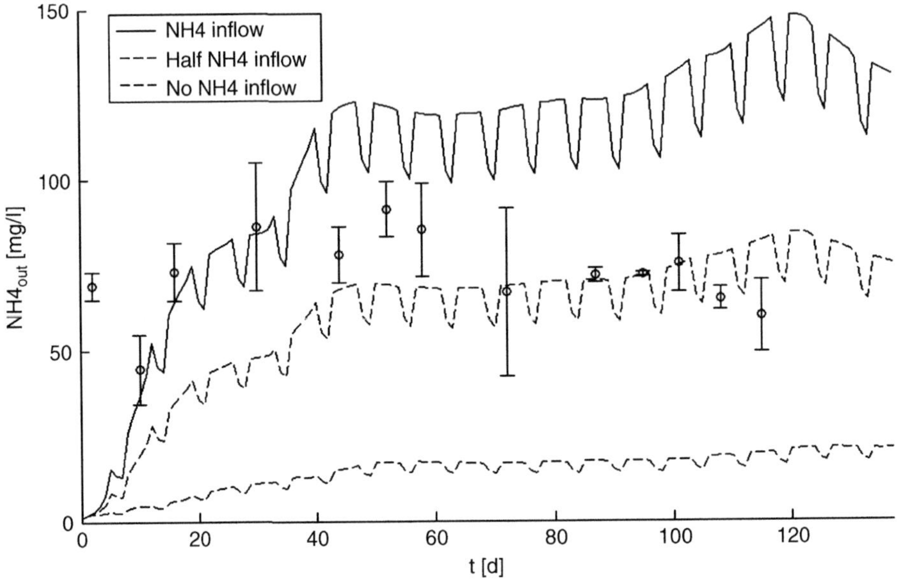

Figure 9.5 Influence of inflow NH_4-N concentrations on effluent NH_4-N concentrations (adapted from Langergraber and Morvannou, 2014).

Rizzo et al. (2014) described the setup of a model to simulate experimental data from a HF CW fed with artificial wastewater. During the experiments, the only nitrogen parameter measured in influent was TKN (Total Kjehldal Nitrogen). As the model requires influent concentrations of ammonia nitrogen, the organic N content of different COD fractions had to be adapted from standard values for the type of used artificial wastewater. This results in different ratios of ammonia nitrogen to TKN. Figure 9.5 shows the impact of different influent concentrations of ammonia nitrogen on simulated effluent concentrations. When all influent TKN is assumed to be ammonia (i.e. no organic nitrogen is present), simulated effluent concentrations of ammonia nitrogen are higher than measured effluent concentrations. In contrast, when all influent TKN is assumed to be organic nitrogen (i.e. no ammonia is present), simulated effluent concentrations are too low.

Pucher and Langergraber (2015) simulated vertical flow filters treating domestic wastewater using experimental data from Dal Santo et al. (2010). The main layer of the filters consisted of sand with a grain size of 1–4 mm or zeolite with a grain size of 2–5 mm. Both filters had an impounded drainage layer. After calibrating the flow model for the sand and zeolite filters (Table 9.20), the procedure for calibrating the biokinetic model was as follows: the standard parameter set of the CW2D biokinetic model (Langergraber and Šimůnek, 2012) was used in the first simulation of the sand filter ('simulated sand 1' in Table 9.21). This resulted in overprediction of nitrification, that is, simulated ammonia nitrogen concentrations were much lower than measured data. After reducing the maximum nitrification rate from 0.9 to 0.175 d^{-1}, the simulated effluent concentrations of ammonia nitrogen matched measured data well ('simulated sand 2' in Table 9.21). After adding the measured adsorption isotherms according to Dal Santo et al. (2010) and a first-order rate coefficient for nonequilibrium adsorption

Table 9.20 Estimated soil hydraulic parameters of the van Genuchten-Mualem model (shape parameters α, N and L: van Genuchten, 1980).

Parameter	Residual water content θ_r (-)	Saturated water content θ_s (-)	Shape parameters α (cm^{-1})	N (-)	L (-)	Saturated hydraulic conductivity K_s (cm/min)
Sand (1–4 mm)	0.063	0.37	0.124	3.2	0.49	55
Zeolite (2–5 mm)	0.045	0.42	0.096	4.6	1.27	154

Table 9.21 Measured and simulated influent and effluent concentrations (in mg/l, measured data: average values and standard deviations in brackets, simulated data: average values over 1 day).

Parameter		CR	CS	CI	COD	NH_4-N	NO_2-N	NO_3-N
Influent	Measured	–	–	–	294	93	0.008	0.59
	Simulated	274	100	20	–	93	0.008	0.59
Effluent (sand)	Measured	–	–	–	**35 (5)**	**14.5 (4.6)**	0.026 (0.040)	**70.8 (8.4)**
	Simulated sand 1	0.17	0.08	31.2	31.5	0.42	<0.003	85.7
	Simulated sand 2	0.14	0.01	29.4	**30.0**	**15.3**	0.016	**76.2**
Effluent (zeolite)	Measured	–	–	–	**29 (9)**	**0.06 (0.05)**	0.52 (1.10)	**51.8 (4.6)**
	Simulated zeolite	0.16	0.01	29.1	**29.3**	**0.27**	<0.003	**43**

of 0.01 d^{-1}, a good match between simulated and measured data could be obtained for the zeolite filter ('simulated zeolite' in Table 9.21).

Pálfy and Langergraber (2014) simulated experimental results from batch-fed column experiments using the CWM1 biokinetic model. They described the need to adjust some parameters of the biokinetic model to be able to simulate anaerobic, anoxic, and aerobic processes to occur in parallel. These phenomena occurred in practice and can be explained by the local effect of root zone re-aeration. Figure 9.6 shows measured and simulated sulfate concentrations before and after the adjustment of parameters of the biokinetic model. Batch experiments can be used to calibrate the biokinetic model parameters as there is no impact of water flow on the treatment performance.

9.5.3 Clogging Model

Modeling of clogging processes is not possible using the HYDRUS Wetland module. Langergraber and Šimůnek (2009) coupled the colloids attachment/detachment model according to Bradford et al. (2003) (implemented in the standard HYDRUS module) with the CW2D biokinetic model. It was assumed that CS, the slowly biodegradable organic matter (as defined by Langergraber and Šimůnek, 2005), consists of a particulate matter XS that can be divided into five different classes according to different

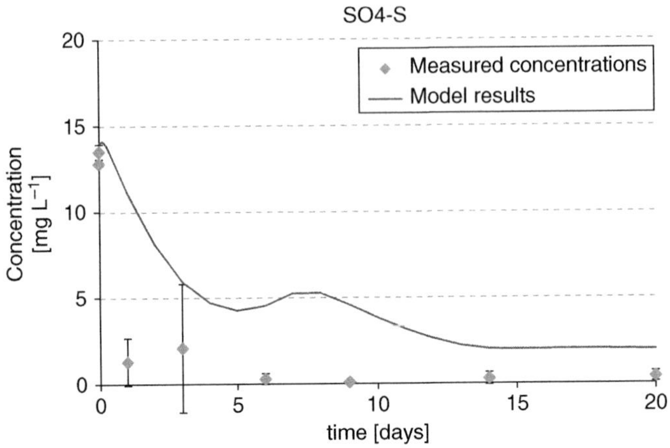

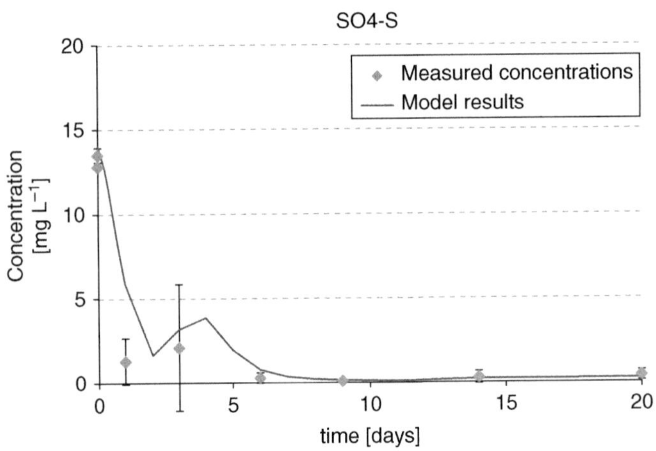

Figure 9.6 Measured and simulated sulfate concentrations for a batch-fed column at 24 °C planted with *Carex* (left: with the standard parameter set of the biokinetic model; right: after adjusting inhibition and half-saturation coefficients to allow anaerobic, anoxic, and aerobic processes to occur in parallel). Adapted from Pálfy (2013).

particle sizes (<1, 1–3, 3–10, 10–30, and 30–100 μm, respectively). It was further assumed that particles are transported in the liquid phase until they are attached. Once attached the particles undergo hydrolysis and cCR (readily biodegradable organic matter; Langergraber and Šimůnek, 2005) in the liquid phase is produced. An additional assumption was that particles with different sizes have the same density and degradability.

The model was tested using measurements of particle concentrations in different depths within the 50-cm main layer of the VF pilot-scale CW, as described earlier. Particle measurements at the inlet were used to calculate fractionation of COD influent. Measured and simulated soluble particulate COD concentrations for the 10–30 μm particle class before and after loading of the VF filter along the filter depth are shown in Figure 9.7. 'Measured' COD was derived from particle measurements. In general, the

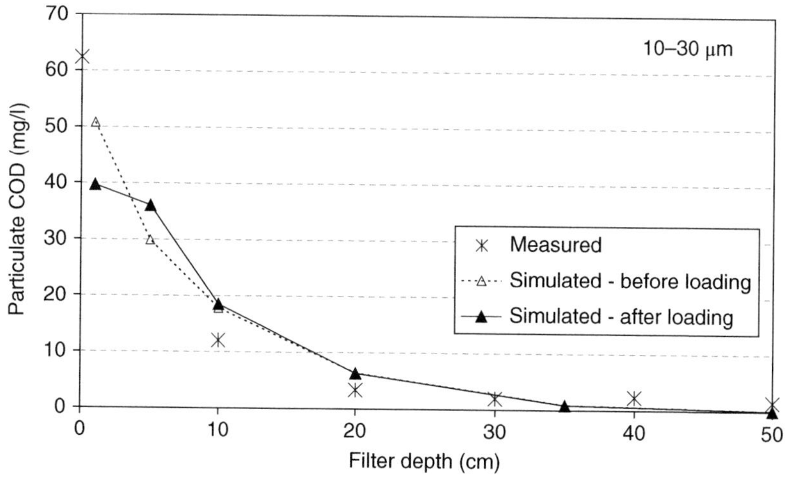

Figure 9.7 Comparison of measured and simulated particulate COD for the 10–30 μm particle class (adapted from Langergraber and Šimůnek, 2009).

first test showed a good agreement between measured and simulated particulate COD. However, there is still a great need for further testing of the particle transport model and for finding an optimal parameter set for various conditions present in constructed wetlands.

Several clogging models have been developed for BIO_PORE (Samsó and García, 2013). Samsó and García (2014) described the relationship between bacterial communities and accumulated solids, leading to clogging in a HF CW. The so-called 'Cartridge Theory' describes how the active bacterial zone moves in the filter. In this model, the effect of the inert organic matter accumulated in pores upon the hydraulic conductivity is not considered. Description of limited growth of bacteria was implemented by adding two parameters. The first parameter provides a negative feedback term to the growth of all bacteria groups to prevent their unlimited growth in areas where substrate concentrations are high (which is especially relevant for the influent zones of HF CWs). The second parameter provides a negative feedback term to the growth equations, but in this case, it decreases the growth rate of bacteria due to progressive accumulation of inert solids in the pore space of the granular media. Samsó et al. (2015a) described in their simulation study the effect of bacteria density and accumulated inert solids on effluent pollutant concentrations, but again only without considering the effect of organic matter accumulation on water flow. Samsó et al. (2016) additionally described the effects of a pore size reduction on the hydraulic conductivity using the biological clogging model proposed by Mostafa and Van Geel (2007). Using this approach, Samsó et al. (2016) were able to simulate realistic behaviour of HF CWs, including overland flow and re-infiltration of water in the unused filter material.

9.5.4 Models as CW Design Tools

Langergraber (2011) concluded that further developments of existing models are needed to make them a useful and applicable tool for a CW design. A simplified computer-based CW design tool, based on process-based numerical models, should be developed that:

- could be used with knowledge about a CW design, but without requiring special knowledge about numerical modeling;
- allows designing CWs for different boundary conditions (such as different climatic conditions, wastewater characterization, filter material, etc.); and
- makes the description of the dynamic behaviour of the designed CW possible, thus allowing demonstration of the greater robustness of CW treatment systems, for example, for fluctuating inflows and peak loads.

Several advances have been made to develop such a simplified design tool. Meyer and Dittmer (2015) developed RSF_Sim, which supports the design of CWs for a combined sewer overflow treatment: retention soil filters (RSFs). This simplified, yet robust and reliable, model was developed for design purposes based on experiences with the HYDRUS Wetland module. RSF_Sim has only a few parameters and is therefore simple to use. RSF_Sim was developed specifically for RSFs for German conditions. Pálfy et al. (2015b) adapted RSF_Sim for CWs for a combined sewer overflow treatment for French conditions (e.g. for different design rules and effluent requirements).

A similar approach was taken by Morvannou et al. (2015). The aim of their work was to develop a simple model to support the design of French-type VF CWs. Numerical simulation using the HYDRUS Wetland module were first performed, varying a number of design parameters. The design parameters included the depth of the main layer of the first-stage filter, the hydraulic loading rate, and the inlet COD concentration and temperature. Effluent concentration of the first-stage filter was then verified using measured data and a surrogate model was developed to estimate COD and NH_4-N removal in relation to the design parameters. Although the first results were promising, Morvannou et al. (2015) concluded that more work needs to be done to develop a robust design model.

9.6 Summary and Conclusions

Interest in modeling processes in CWs has increased over the last 15 years. Several mechanistic models to describe processes in subsurface flow CWs have been developed. However, the HYDRUS Wetland module is the only one that is commercially available and thus available for wider use.

Experiences with applying the HYDRUS Wetland module and other existing simulation tools can be summarised as follows:

- Simulation studies often suffer from a lack of data. When the use of dynamic models is anticipated, experiments have to be planned in a way that data for model calibration and validation are available. Data requirements are different for different models, but always include careful planning of sampling frequency and analysed parameters (Meyer et al., 2015). The amount and nature of data needed for calibration depends on the objectives of the simulation study. In general, for calibration of the water flow model, more data are required for vertical flow CWs compared with horizontal flow CWs. The amount of data required for calibrating reactive transport simulations depends on biokinetic model parameters (e.g. influent fractionations), as well as the dynamics of influent in the system.

- Good calibration of the water flow model is a prerequisite for achieving a good match between measured and simulated pollutant concentrations. If the water flow model is calibrated, good results can be obtained in most applications when using the standard parameter sets of the biokinetic CW2D and CWM1 models (Langergraber and Šimůnek, 2012). Influent fractionation (i.e. fractionation of influent COD and the N and P contents of different COD fractions) has a high impact on simulation results and thus is an essential part of calibrating reactive transport models. This is especially true when simulating CWs treating other than domestic wastewater.
- One of the main obstacles for the wider use of available simulation tools is that they are rather complicated and difficult to run. Simplified, yet robust and reliable, models for the design of CWs need to be developed (such as RSF_Sim, developed to support the design of CWs treating combined sewer overflow; Meyer and Dittmer, 2015).

References

Bradford, S.A., Šimůnek, J., Bettahar, M., et al. (2003). Modeling colloid attachment, straining, and exclusion in saturated porous media. *Environmental Science and Technology* 37 (10): 2242–2250.

Dal Santo, S., Canga, E., Pressl, A., et al. (2010). Investigation of nitrogen removal in a two-stage subsurface vertical flow constructed wetland system using natural zeolite. In: *Proceedings of the 12th IWA Specialized Group Conference on Wetland Systems for Water Pollution Control* (ed. F. Masi and J, Nivala), 263–270. 3–8 October 2010, San Servolo, Venice, Italy.

Dittmer, U., Meyer, D., and Langergraber, G. (2005). Simulation of a subsurface vertical flow constructed wetland for CSO treatment. *Water Science and Technology* 51 (9): 225–232.

García, J., Rousseau, D.P.L., Morató, J., et al. (2010). Contaminant removal processes in subsurface-flow constructed wetlands: a review. *Critical Reviews in Environmental Science and Technology* 40 (7): 561–661.

Henrichs, M., Langergraber, G., and Uhl, M. (2007). Modelling of organic matter degradation in constructed wetlands for treatment of combined sewer overflow. *Science of the Total Environment* 380 (1–3): 196–209.

Henrichs, M., Welker, A., Uhl, M. (2009). Modelling of biofilters for ammonium reduction in combined sewer overflow. *Water Science and Technology* 60 (3), 825–831.

Kadlec, R.H. and Wallace, S. (2009). *Treatment Wetlands*, 2nd ed. Boca Raton, FL: CRC Press.

Karlsson, S.C., Langergraber, G., Pell, M., et al. (2015). Simulation and verification of hydraulic properties and organic matter degradation in sand filters for greywater treatment. *Water Science and Technology* 71 (3): 426–433.

Kumar, J.L.G. and Zhao, Y.Q. (2011). A review on numerous modeling approaches for effective, economical and ecological treatment wetlands. *Journal of Environmental Management* 92: 400–406.

Langergraber, G. (2001). *Development of a simulation tool for subsurface flow constructed wetlands*. PhD dissertation, BOKU University, Vienna, Austria.

Langergraber, G. (2003). Simulation of subsurface flow constructed wetlands – Results and further research needs. *Water Science and Technology* 48 (5): 157–166.

Langergraber, G. (2007). Simulation of the treatment performance of outdoor subsurface flow constructed wetlands in temperate climates. *Science of the Total Environment* 380 (1–3): 210–219.

Langergraber, G. (2008). Modeling of processes in subsurface flow constructed wetlands – A review. *Vadose Zone Journal* 7 (2): 830–842.

Langergraber, G. (2011). Numerical modelling: A tool for better constructed wetland design? *Water Science and Technology* 64 (1): 14–21.

Langergraber, G. (2015). *Erratum: Water Science and Technology* 56 (3), 2007, 233–240: Comparison of measured and simulated distribution of microbial biomass in subsurface vertical flow constructed wetlands. *Water Science and Technology* 71 (1): 157–158.

Langergraber, G. and Šimůnek, J. (2005). Modeling variably-saturated water flow and multi-component reactive transport in constructed wetlands. *Vadose Zone Journal* 4 (4): 924–938.

Langergraber, G. and Šimůnek, J. (2009). Simulating particle transport in subsurface flow constructed wetlands with CW2D/HYDRUS. In: *3rd International Symposium on Wetland Pollutant Dynamics and Control – WETPOL 2009 Book of Abstracts* (ed. J.M. Bayona and J. García), 135–136. Barcelona, 20–24 September 2009.

Langergraber, G. and Šimůnek, J. (2011). *HYDRUS Wetland Module, Version 2, Manual.* Hydrus Software Series 4, Department of Environmental Sciences, University of California Riverside, Riverside, CA.

Langergraber, G. and Šimůnek, J. (2012). Reactive transport modeling of subsurface flow constructed wetlands using the HYDRUS Wetland module. *Vadose Zone Journal* 11 (2): Special Issue on Reactive Transport Modeling. doi:10.2136/vzj2011.0104

Langergraber, G. and Morvannou, A. (2014). Modelling of treatment wetlands. *Sustainable Sanitation Practice* 18 (January 2014): 31–36. Available at www.ecosan.at/ssp

Langergraber, G., Tietz, A., and Haberl, R. (2007). Comparison of measured and simulated distribution of microbial biomass in subsurface vertical flow constructed wetlands. *Water Science and Technology* 56 (3): 233–240.

Langergraber, G., Giraldi, D., Mena, J., et al. (2009a). Recent developments in numerical modelling of subsurface flow constructed wetlands. *Science of the Total Environment* 407 (13): 3931–3943.

Langergraber, G., Rousseau, D., García, J., and Mena, J. (2009b). CWM1 – A general model to describe biokinetic processes in subsurface flow constructed wetlands. *Water Science and Technology* 59 (9): 1687–1697.

Mburu, N., Sanchez-Ramos, D., Rousseau, D.P.L., et al. (2012). Simulation of carbon, nitrogen and sulphur conversion in a batch-operated experimental wetland mesocosm. *Ecological Engineering* 42: 304–315.

Meyer, D. (2011). Modellierung und Simulation von Retentionsbodenfiltern zur weitergehenden Mischwasserbehandlung [Modelling and simulation of constructed wetlands for treatment of combined sewer overflow]. PhD thesis, Technical University of Kaiserslautern, Germany. Available at https://kluedo.ub.uni-kl.de/frontdoor/index/index/docId/2843

Meyer D. and Dittmer U. (2015). RSF_Sim – A simulation tool to support the design of constructed wetlands for combined sewer overflow treatment. *Ecological Engineering* 80: 198–204.

Meyer, D., Chazarenc, F., Claveau-Mallet, D., et al. (2015). Modelling constructed wetlands: Scopes and aims – A comparative review. *Ecological Engineering* 80: 205–213.

Molle, P., Liénard, A., Boutin, C., et al. (2005). How to treat raw sewage with constructed wetlands: an overview of the French systems. *Water Science and Technology* 51 (9): 11–21.

Morvannou, A., Choubert, J.-M., Vanclooster, M., and Molle, P. (2011). Solid respirometry to characterize nitrification kinetics: A better insight for modelling nitrogen conversion in vertical flow constructed wetlands. *Water Research* 45 (16): 4995–5004.

Morvannou, A., Forquet, N., Vanclooster, M., and Molle, P. (2012). Which hydraulic model to use in vertical flow constructed wetlands? *Proceedings of the 13th IWA Specialized Group Conference on Wetland Systems for Water Pollution Control* – Conference Papers Volume 2, 145–153. 25–29 November 2012, Perth.

Morvannou, A., Froquet, N., Troesch, S., and Molle, P. (2015). Impact of design and operating conditions on vertical flow constructed wetland performances: modelling and global assessment confrontation. *6th International Symposium on Wetland Pollution Dynamics and Annual Conference of the Constructed Wetland Association* – Book of Abstracts, 196–197. 13–18 September 2015, York, UK.

Mostafa, M. and Van Geel, P.J. (2007). Conceptual models and simulations for biological clogging in unsaturated soils. *Vadose Zone Journal* 6: 175–185.

Pálfy, T. (2013). *Verification of the implementation of CWM1 in the HYDRUS wetland module.* MSc thesis, BOKU University, Vienna, Austria.

Pálfy, T.G. and Langergraber, G. (2014). The verification of the Constructed Wetland Model No. 1 implementation in HYDRUS using column experiment data. *Ecological Engineering* 68: 105–115.

Pálfy, T.G., Gribovszki, Z., and Langergraber, G. (2015a). Design-support and performance estimation using HYDRUS/CW2D: A horizontal flow constructed wetland for polishing SBR effluent. *Water Science and Technology* 71 (7): 965–970.

Pálfy, T.G., Meyer, D., and Molle, P. (2015b). Orage: Simulation of planted detentive filters treating urban storm water. *6th International Symposium on Wetland Pollution Dynamics and Annual Conference of the Constructed Wetland Association* – Book of Abstracts, 216–217. 13–18 September 2015, York, UK.

Pucher, B. and Langergraber, G. (2015). Simulation of two-stage vertical flow sand and zeolite filters treating domestic wastewater using the HYDRUS wetland module. *6th International Symposium on Wetland Pollution Dynamics and Control and Annual Conference of the Constructed Wetland Association* – Book of Abstracts, 198–199. 13–18 September 2015, York, UK.

Rizzo, A., Langergraber, G., Galváo, A., et al. (2014). Modelling the response of horizontal flow constructed wetlands to unsteady organic loads with HYDRUS-CWM1. *Ecological Engineering* 68: 209–213.

Samsó, R. and Garcia, J. (2013). BIO PORE, a mathematical model to simulate biofilm growth and water quality improvement in porous media: Application and calibration for constructed wetlands. *Ecological Engineering* 54: 116–127.

Samsó, R. and García, J. (2014). The Cartridge Theory: A description of the functioning of horizontal subsurface flow constructed wetlands for wastewater treatment, based on modelling results. *Science of The Total Environment* 473–474: 651–658.

Samsó, R., Blázquez, J., Agulló, N., et al. (2015a). Effect of bacteria density and accumulated inert solids on the effluent pollutant concentrations predicted by the constructed wetlands model BIO_PORE. *Ecological Engineering* 80: 172–180.

Samsó, R., Meyer, D., and García, J. (2015b). Subsurface flow constructed wetland models: review and prospects. In: *The Role of Natural and Constructed Wetlands in Nutrient Cycling and Retention on the Landscape* (ed. J. Vymazal), 149–174. Springer International Publishing.

Samsó, R., García, J., Molle, P., and Forquet, N. (2016). Modelling bioclogging in variably saturated porous media and the interactions between surface/subsurface flows: application to Constructed Wetlands. *Journal of Environmental Management* 165: 271–279.

Šimůnek, J., van Genuchten, M.Th., and Šejna, M. (2008). Development and applications of the HYDRUS and STANMOD software packages and related codes. *Vadose Zone Journal* 7 (2): 587–600.

Šimůnek J., Šejna, M., and van Genuchten, M.Th. (2011). *The HYDRUS Software Package for Simulating the Two- and Three-Dimensional Movement of Water, Heat, and Multiple Solutes in Variably-Saturated Media.* Technical Manual, Version 2.0. PC-Progress, Prague.

Šimůnek, J., Jacques, D., Langergraber, G., et al. (2013). Numerical modeling of contaminant transport using HYDRUS and its specialized modules. *Journal of the Indian Institute of Science* 93 (2): 265–284.

Smethurst, P.J, Petrone, K., Langergraber, G., et al. (2014). Nitrate dynamics in a rural headwater catchment: measurements and modeling. *Hydrological Processes* 28 (4): 1820–1834.

Tietz, A., Langergraber, G., Sleytr, K., et al. (2007). Characterization of microbial biocoenosis in vertical subsurface flow constructed wetlands. *Science of the Total Environment* 380 (1–3): 163–172.

Toscano, A., Langergraber, G., Consoli, S., and Cirelli, G.L. (2009). Modelling pollutant removal in a pilot-scale two-stage subsurface flow constructed wetlands. *Ecological Engineering* 35 (2): 281–289.

Troesch, S. and Esser, D. (2012). Constructed wetlands for the treatment of raw wastewater: The French experience. *Sustainable Sanitation Practice* 12 (July 2012): 9–15. Available at www.ecosan.at/ssp

van Genuchten, M.Th. (1980). A closed form equation for predicting the hydraulic conductivity of unsaturated soils. *Soil Science Society of America Journal* 44: 892–898.

10

Reactive Transport Modeling and Biogeochemical Cycling

Christof Meile[1] and Timothy D. Scheibe[2]

[1] *Department of Marine Sciences, University of Georgia, Athens, GA, 30602, USA*
[2] *Pacific Northwest National Laboratory, Richland, WA, 99352, USA*

10.1 Introduction

Models that describe the spatial and temporal distribution of chemical substances in the environment have a rich tradition in modern Earth Sciences. They integrate knowledge of what controls biogeochemical transformations with a description of how constituents are transported, and they express this knowledge in a quantitative mathematical framework. This allows testing of whether the understanding of biogeochemical processes codified in the model is consistent with observations and helps to identify knowledge gaps. Models also allow for a knowledge-based interpolation between sparse observations, which can be used to gain insight into how a system responds to gradual changes in external forcings, or to predict system behaviour following a sudden perturbation.

Multicomponent reactive transport modeling has become prominent over the last decades for the study of biogeochemical cycling in many near-surface environments where water acts as a solvent for chemicals, provides a habitat for microorganisms that catalyse many of the key biogeochemical transformations, and furthers the transport of dissolved entities. Reactive transport modeling has been used in porous media such as sediments, groundwater, the riparian and hyporheic zones, soils and the rhizosphere, fractured media, biofilms, as well as in both fresh and marine surface waters. Examples include the quantification of interacting C and S cycles in marine sediments (Hensen et al., 2003), the assessment of the distribution of arsenic in groundwater and its connection to iron oxide reduction (Postma et al., 2007), aquifer pollution (Brun and Engesgaard, 2002), the study of elemental cycling driven by the breakdown of organic matter (Hunter et al., 1998), or the migration of radionuclides in sparsely fractured rock (MacQuarrie and Mayer, 2005). A description of the state and impact of reactive transport modeling in Earth Sciences can be found in Steefel et al. (2005).

The increasing computational resources, the availability of new data streams, and novel experimental capabilities probing environmental systems at a wide range of

Reactive Transport Modeling: Applications in Subsurface Energy and Environmental Problems,
First Edition. Edited by Yitian Xiao, Fiona Whitaker, Tianfu Xu and Carl Steefel.

temporal and spatial scales, are providing opportunities for new developments in reactive transport modeling. Li et al. (2017) identified a number of research areas that are likely to benefit from the development and application of novel modeling approaches, including the study of chemical weathering in the critical zone and the role of roots or fungi, the controls on organic matter mineralization in the subsurface under varying external forcings, the integration of (semi-)mechanistic biogeochemical models in larger modeling frameworks that quantify the effect of climate change, and the improved representation of microorganisms.

Microbes are responsible for a large portion of the chemical transformations in Earth's near-surface environments and exhibit a tremendous diversity in function. In subsurface environments, this includes the breakdown of organic matter by heterotrophic bacteria and archaea with the use of a range of solutes, dissolved gases or minerals as electron acceptors, the redox reactions carried out by chemolithotrophs, or the assimilation of C, N and P during growth. As the partitioning between aqueous, solid and gaseous phases – and hence an element's mobility in the environment – depends on the chemical form, microbial activity drives biogeochemical cycling (Burgin et al., 2011), and lies at the heart of biogeochemical reactive transport models. Thus, in this chapter we focus on some of the challenges arising in the modeling of microbially-driven systems. Building on a brief discussion of common modeling approaches, we discuss the representation of microbes in reactive transport models and how microbial community interactions can be described. We then discuss the challenges associated with the integration of data, and ways to tackle the vast range of scales relevant to models of biogeochemical cycling.

10.2 Reactive Transport Model Formulations

Historically, multicomponent reactive transport models build upon geochemical modeling calculations of the equilibrium speciation in aqueous systems (e.g. Jenne, 1979). This approach was then expanded by considering a sequence of speciation calculations that describe the trajectory of a water parcel, tracking the chemical composition as it follows a 'reaction path.' Several widely used models including EQ6 (Wolery and Daveler, 1992) and GWB (Bethke, 2007) implement this approach. However, reaction path modeling falls short in fully capturing the open nature of biogeochemical cycling, dealing with nonequilibrium conditions, and the potentially bidirectional coupling between reaction and transport processes that affect the fate and distribution of chemicals. Thus, dynamic formulations have become common. Such reactive transport models rely on conservation equations to describe the spatio-temporal evolution of state variables (Steefel et al., 2005). The conservation equations are used at the continuum scale (Lichtner, 1985), so that conservation of momentum can be expressed through the Navier-Stokes equations, or the Darcy equation can be applied in porous media under laminar flow conditions. The flow field is then used in the mass conservation equation that describes a concentration field, which for a saturated porous medium takes the form of:

$$\frac{\partial \zeta(x) C_j(x,t)}{\partial t} = \nabla \cdot \left(\mathbf{D} \zeta \nabla C_j - \mathbf{v} \zeta C_j \right) + \sum_{m=1}^{N_r} \upsilon_{mj} R_m$$

where ζ is porosity (ϕ) for solutes and $1 - \phi$ for solids, **D** is an effective diffusion tensor describing bioturbation, hydrodynamic dispersion and/or molecular diffusion, **v** is an effective advection velocity, C_j is the concentration of substance j, R_m the source/sink term reflecting the rate of reaction m, and υ_{mj} is the corresponding stoichiometric coefficient. These mass conservation equations are commonly complemented by constitutive equations expressing medium properties such as permeability or fluid density as functions of porosity or solution composition, respectively. These properties may vary over time and space and can constitute a feedback of reactions on flow dynamics.

The reaction rates connect the mass conservation equations for individual chemicals that are substrate or products of a given reaction, so that, mathematically, one has to solve a system of coupled partial differential equations. The numerical solution (see e.g. Steefel and MacQuarrie, 1996) requires discretization, with finite difference, finite volume and finite element approximations being common. The mass conservation equations thus can be recast into a system of linear equations of the form:

$$\mathbf{A}\cdot\vec{C}=\vec{d}$$

which can computationally be solved efficiently (e.g. Saad, 2003).

The main challenge for the solution of biogeochemical reactive transport models lies in the reaction operator, which couples the mass conservation equations. This coupling formally precludes their sequential solution, resulting in a large, sparse matrix **A** of the size (ns*nx) * (ns*nx), where ns is the number of state variables, and nx is the number of spatial nodes, and C is of size (ns*nx) * 1. Furthermore, expressions of reaction rates in a set of linear equations requires the reaction rates to be estimated from known concentrations (or use of linear rate laws). For fast kinetics, this can cause severe constraints on the model timestep and lead to long simulation times. Also, most reactive transport models across a broad range of environments account for diffusion/dispersion and advection, and their key difference lies in the chemicals of interest (i.e. the state variables) and the reactions they consider. Hence, R varies widely between models.

To address these issues, approaches that separate the transport and reaction operators have been developed (e.g. Steefel and MacQuarrie, 1996; Geiser, 2011). As the transport of chemical species in many applications is independent of each other (or only weakly coupled; exceptions may include settings in which coulombic interactions lead to co-migration of charged species), their transport can be solved separately and simultaneously. When separating transport from reaction, one can further recast the reaction problem as a system of ordinary differential equations at each spatial node, which can be solved using well-established numerical techniques (e.g. Gear 1971). The selection of the relevant reaction processes and the corresponding rate laws, which are non-unique and can be arbitrary, impacts the model results (Curtis, 2003; King et al., 2010). The chosen expressions may be adapted over the course of iterative model improvement. Thus, modeling frameworks have been developed to facilitate the implementation of new processes or modification of existing rate expressions, through for example the automated generation of Jacobian matrices that can be part of solving the systems of ordinary differential equations (Regnier et al., 2002) or have included kinetic considerations in the selection of reaction components (Fang et al., 2003; Molins et al., 2004).

10.3 The Representation of Microbes

The goal of reactive transport model applications commonly is to predict macroscopic phenomena, such as the fate of a contaminant in a spill. As a consequence of this interest in aggregate responses, explicit representation of processes taking place at the micro-scale is rare. However, the ability to trace and characterize microbial communities has grown tremendously over the past decade, both with respect to phylogeny and activity through genomics, transcriptomics, and proteomics. Combined with the ability to trace chemicals through metabolomics, this has provided a much more refined view of the microbial community and revealed a dynamic microbial world with functional responses that to date are not yet reflected in reactive transport models. This motivates the development and application of a diverse range of novel modeling approaches, as reviewed by Song et al. (2014).

10.3.1 Implicit Presence of Microbes

In settings with abundant resources, bacteria can exhibit doubling times on the order of an hour. Microbial populations thus have the intrinsic ability to adapt rapidly to environmental conditions. In addition, the adaptation of metabolic capacity and capability through the expression of enzymes is commonly observed at shorter timescales. However, the timescales of interest are often much longer, or changes in environmental conditions occur much more slowly, so the size of the microbial community as well as its functional profile is assumed to be adapted to environmental conditions. As a consequence, a process carried out by a microorganism, such as the oxidation of sulfide to sulfate with molecular oxygen as electron acceptor:

$$\mathrm{HS^-} + 2\mathrm{O_2} \rightarrow \mathrm{HSO_4}^-$$

can then be described by a simple rate law (e.g. Hunter et al., 1998):

$$R = k * \left[\mathrm{HS^-}\right] * \left[\mathrm{O_2}\right]$$

Such a rate expression recognizes the importance of the availability of substrates, and mimics chemical kinetics, although does not necessarily follow it exactly; here, the rate dependence on O_2 is linear, not quadratic. This reflects that such transformations are not elementary reactions but rather complex processes, involving the uptake of substrate and catalysed intracellular enzymatic reactions. The rate-controlling step in such an overall reaction may follow different kinetics, and commonly a Monod-type dependency (Monod, 1942) on substrate concentration S is observed:

$$R = k * \frac{[S]}{K_m + [S]}$$

where the half saturation constant K_m depicts the substrate concentration at which the reaction proceeds at half of its maximum rate. Theoretical underpinnings for such a rate expression are summarized in Liu (2007) and Ramkrishna and Song (2008).

A cell's metabolic activity depends not only on the availability of substrates, but also on whether the use of that substrate is energetically beneficial. This is particularly true for a process that is central to the cell's catabolism, such as organic matter oxidation by heterotrophic microorganisms, because if it does not yield sufficient energy then it is not a viable option to sustain an organism. In reactive transport modeling, such focus on the key metabolic processes of a particular community is indeed common, because they are responsible for most of the elemental throughput. Differences in energy gains may explain commonly observed sequential use of terminal electron acceptors (Froelich et al., 1979), and have been modeled using inhibition terms. Numerous descriptions exist that reflect different mechanisms (Mulchandani and Luong, 1989), including actual inhibition of an enzyme, or an indirect effect through competitive exclusion. The latter can be tied to energetics – if one microbe can perform a process that yields more energy than another, and everything else being equal, it will be able to grow faster and outperform its competitor for resources. This can be accounted for by including an energetic constraint (F_T) in the rate expression:

$$R = F_k * F_T$$

where F_k represents the kinetic dependence on substrate availability. Jin and Bethke (2003) defined the thermodynamic factor $F_T = 1 - \exp(-f/(\chi R_g T))$, where f denotes the driving force, expressed as the difference between the Gibbs free energy of reaction and the energy utilized for converting ADP to ATP, χ is a stoichiometric coefficient, R_g is the gas constant and T is temperature (see Jin and Bethke, 2003, for a derivation and detailed explanation). This reflects that a minimum energy gain is required for an organism to persist (Pirt, 1965; Carlson et al., 2007; Hoehler and Jørgensen, 2013), and that under energy-replete conditions, the energy gain no longer controls the rate of reaction.

10.3.2 Explicit Representations

The assumption of a microbial community that is adapted and static simplifies the modeling tremendously, yet there are many instances where this may not be justified. Firstly, a system perturbation will prompt a response in the microbial community with regard to abundance and composition, for example in response to an oil spill (Joye et al., 2014) or aquifer CO_2 injection (Morozova et al., 2011). Secondly, the above-mentioned rapid doubling times of bacteria and archaea can be 4–6 orders of magnitude slower in oligotrophic settings than under nutrient-replete conditions (Hoehler and Jørgensen, 2013). Thirdly, models that do not reflect microbial dynamics tend to ignore the ability of the microbial community (or even of individual organisms with metabolic plasticity) to adapt their function to changing environmental conditions.

A key distinction between different approaches to model microbial dynamics lies in the choice of interacting units (Song et al., 2014), reflecting that microbial functional capabilities can be compartmentalized at different scales, ranging from cells to ecosystems (Figure 10.1).

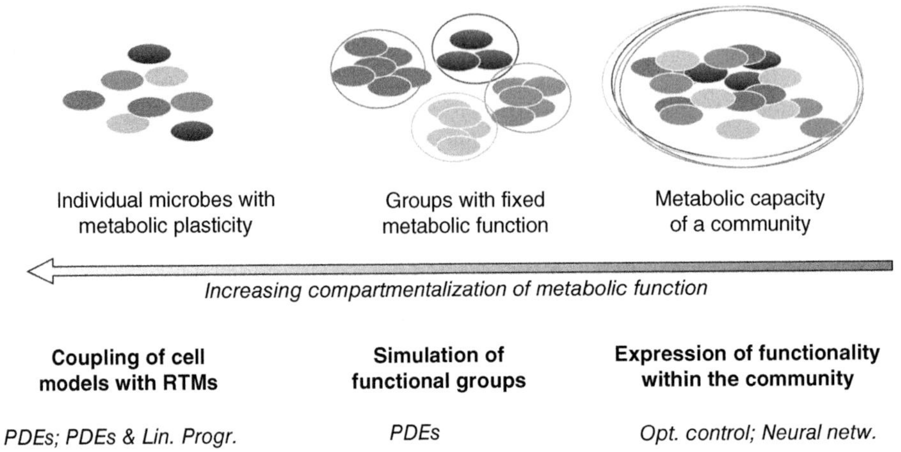

Figure 10.1 Sketch of microbial modeling approaches with varying extent of compartmentalization (denoted by open ellipses) of function (indicated by differing colours). Typical mathematical approaches are noted at the bottom: PDE, partial differential equation; Lin. Progr., linear programming; Opt. Control, optimal control problem; Neural Netw., neural networks and related statistical approaches. Examples for formulations using individual microbial cell models include King et al. (2009), Fang et al. (2011) and Zhuang et al. (2011); models with functional groups include Thullner et al. (2005) and King et al. (2010); descriptions with whole community approaches are presented in Vallino (2003) and Algar and Vallino (2014). (See colour plate section for the colour representation of this figure.)

10.3.2.1 Functional Populations

Early attempts to explicitly incorporate microbes in models (e.g. Molz et al., 1986; Wang and Papenguth, 2001) built upon the description of the spatio-temporal evolution of chemical species:

$$\frac{\partial \vec{C}}{\partial t} = T + R$$

where T and R are the transport and reaction operator, respectively, and $\vec{C}$ represents the concentrations of chemicals. This set of mass conservation equations was expanded to include the biomass of functional microbial groups:

$$\frac{\partial \vec{C}}{\partial t} = T + R_{cell}\vec{B}$$

$$\frac{\partial \vec{B}}{\partial t} = T + \gamma R_{cell}\vec{B} - d$$

where $\vec{B}$ represents the abundance of distinct functional microbial groups, T represents the transport operator, R and R_{cell} are bulk and cell specific reaction rates, and γ and d are growth efficiencies and loss terms, respectively.

An example of this approach can be found in Bethke et al. (2010). Considering iron, sulfate and methane dynamics driven by acetate produced in the breakdown of natural organic matter, they show the development of microbial and biogeochemical zonations

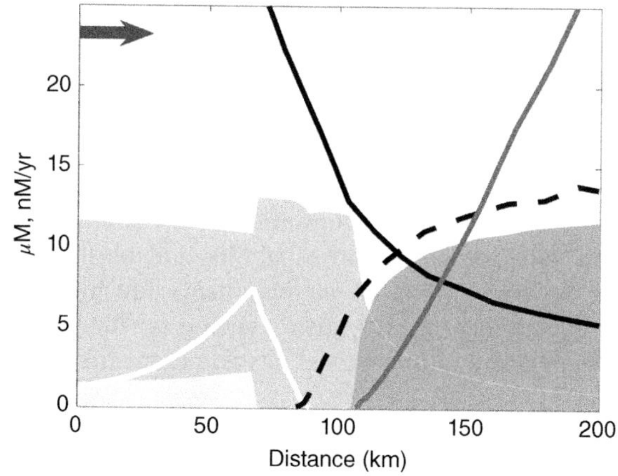

Figure 10.2 Dissolved iron, sulfate, sulfide and methane profiles along an aquifer flow path, where iron, sulfate reduction and methanogenesis are fueled by the breakdown of natural organic matter. Flow of sulfate-containing fluid is from left to right as denoted by the arrow, with iron oxides present in the upstream third of the domain. Metabolic rates are indicated as shaded areas (light grey: iron reduction, intermediate grey: sulfate reduction, dark grey: methanogenesis). Lines indicate concentrations (solid white, dissolved Fe^{II}; black, sulfate; black dashed, hydrogen sulfide; solid grey, methane). Note that sulfide is only present after the resupply of dissolved iron from iron reduction is exhausted and no longer titrates out the sulfide produced during sulfate reduction, active throughout the domain. This sketch is based on results presented in Bethke et al. (2010). *Source:* Data from Bethke 2010.

observed in aquifers, and discuss the underlying causes. They also point out that the buildup of metabolites alone can be a poor indicator of the major metabolic processes taking place in the subsurface (Figure 10.2), as is the case when metabolites are rapidly removed or recycled (e.g. Holmkvist et al., 2011).

Such a formulation connects cell specific rates and cell abundance to chemical transformations, and the use of growth efficiencies relates microbial activity to population growth. However, although critical, this connection is highly simplified. For example, growth efficiencies may vary depending on cell state, (micro-)nutrient conditions, or antagonistic interactions with other microorganisms (Goldman et al., 1987; Cajal-Medrano and Maske, 2005; Bernier and Surette, 2013), which are typically not explicitly accounted for. A central aspect for biogeochemical reactive transport models is how microbes interact with their surroundings. In the absence of an explicit representation of microorganisms, this is commonly reflected in Monod kinetics (see above), with reaction rates decreasing with substrate concentrations dropping below levels represented by the half saturation constant. Theoretical underpinnings have been developed that relate the widely used Monod kinetics to transport limitations or energetic constraints (Liu, 2007; Ramkrishna and Song, 2008). Alternative descriptions such as the Droop model (Droop, 1973) relate microbial growth to cell-internal nutrient content, rather than ambient substrate availability alone. Subsequent work has shown that this cell-quota model is equivalent to an optimization of net growth (Pahlow and Oschlies, 2013). As such, it relates the development of a microbial population to the optimal resource allocation, a strategy also highlighted in the work by Vallino et al. (1996).

10.3.2.2 Trait-based Models

Recent developments in reactive transport modeling have focused on trait-based models, in which microorganisms are characterized by traits that can provide them with advantages allowing them to thrive in one environmental setting, while in other settings they get outcompeted. Such characteristics or traits can encompass physiological, behavioural and morphological characteristics, and may include substrate affinity, with a high affinity providing an advantage in low nutrient environments, maximum growth rate, nutrient requirements, metabolic capability and plasticity, or motility. These traits can further be constrained by an organism's life history (Litchman and Klausmeier, 2008) and they affect an organism's ecological function, which in concert with trade-offs between traits, ultimately defines an organism's success in occupying an ecological niche.

Trait-based modeling has been very successful in describing the spatial distribution of phytoplankton communities in the ocean and predicting primary productivity (Litchman et al., 2007; Follows et al., 2007; Dutkiewicz et al., 2009; Follows and Dutkiewicz, 2011). This reflects that in a setting in which a myriad of traits exist, narrow constraints on metabolic function can no longer be pre-imposed but have to be relaxed. Randomly seeding such functional groups, and replacing them if not successful in the simulated environment, allows the selection of competitive traits. The activity of dominant organisms then provides a feedback on the environment. This gives rise to emerging organism distributions, which in the surface ocean compare favourably to observed patterns (Dutkiewicz et al., 2009; Follows and Dutkiewicz, 2011). In soils, simulations considering the trade-off between different strategies to acquire resources have produced testable predictions of microbial traits during litter breakdown (Allison 2012), and Bouskill et al. (2012) presented a trait-based modeling approach that simulated the community composition (ammonium oxidizing archaea vs. bacteria). These are noteworthy achievements, because this approach inherently can simulate the evolution of system properties from a (potentially probabilistic) distribution of unique traits for which there is a direct correspondence between the mathematical formulation and biological features. Furthermore, many microbial traits are observable through genomic, transcriptomic or proteomic data, and hence are observable.

10.3.2.3 Bottom-up Approaches

The number of microorganisms for which *in silico* representations of cell metabolism exist is increasing rapidly (Kim et al., 2012). Cell metabolism is established from a sequenced and annotated genome, and can be validated through comparison of data such as ^{13}C label in metabolites reflecting intracellular carbon fluxes (Sauer, 2006). Most such models of cell metabolism rely on a steady-state formulation, so that $0 = \mathbf{S}f$, where $\mathbf{S}$ is a stoichiometric matrix describing the metabolic network and f are the metabolic fluxes. This system is then solved as a linear programming problem, fulfilling constraints on lower (lb) and upper (ub) bounds, $lb \le f \le ub$, and optimizing an objective function $g(f)$. This function is commonly chosen as maximizing growth, although Schuetz et al. (2012) showed that other factors, such as costs of adaptation to changing environmental conditions, may be important.

Genome-scale metabolic reconstructions of microbial species have successfully predicted microbial metabolic function and allocation of resources (Durot et al., 2009).

For batch systems, *in silico* representations of two organism communities have been able to predict observed interactions and competitive advantages (Stolyar et al., 2007; Zhuang et al., 2011), and the joint analysis of stoichiometric cell models have shed light on their interactions and mass and energy exchanges (Taffs et al., 2009; Klitgord and Segrè, 2010).

Models of individual cells have been successfully integrated into environmental reactive transport models (King et al., 2009; Scheibe et al., 2009; Fang et al., 2011), albeit for only a single species. This reflects three fundamental challenges. Firstly, cell models are typically flux-based, rather than employing kinetic rate expressions that can require extensive parameterization. This includes the exchange fluxes between the cell and its environment. As a consequence, in order to relate the intracellular fluxes to the extracellular environmental conditions, the exchange fluxes need to be parameterized, which requires experimental data. Secondly, these models may be formulated to compete for resources, but they typically do not account for complex community interactions (Zomorrodi and Maranas, 2012; Zomorrodi et al., 2014), including the use of antibiotics as a means to establish a niche (Zengler and Palsson, 2012) or mutualistic interactions (Amin et al., 2015). Thirdly, only in extremely rare cases is the microbial community simple enough to be reflected by a select few organisms. While the number of cell models is increasing rapidly, this may leave out metabolic capabilities, and hence underestimate the adaptive capacity of a microbial community.

10.3.2.4 Metabolic Activity as Ecosystem Response

On the other end of the spectrum, the ecosystem's metabolic function can be described by the set of half-reactions, which are then combined to maximize a goal function (Vallino, 2003). In such a formulation, the functional capability results from the combination of two half-reactions to form a redox process that can be harvested for microbial maintenance and growth. It is only restricted by the presence of metabolites, and the microbial community is assumed to follow the paradigm 'everything is everywhere, but the environment selects' (Baas Becking, 1934) and modeled as a distributed metabolic network. Building on these ideas, a self-organizing metabolic network approach has been implemented by Algar and Vallino (2014) and Vallino (2010) as an optimal control problem, in which the system maximizes entropy production over a finite time interval:

$$\max\left(\dot{\sigma}_i\right)=\frac{1}{\delta_t}\int_{t+\delta_t}^{t} f(t)\sum_i \dot{\sigma}_i\, d\tau$$

where $f(t)$ is a discount function that weights entropy production over the time interval δ_t (Vallino et al., 2014), and the entropy production of reaction i $(\dot{\sigma}_i)$ is defined by the negative of a reaction rate times the Gibbs free energy of the reaction divided by temperature (Vallino, 2010). The model is formulated with optimal control variables that are part of the reaction rate formulations and differentiate between allocation to growth vs. energy production, and the use of alternative nitrogen substrates. Optimal control theory also is at the centre of the cybernetic approach (Ramkrishna and Song, 2012), which has been implemented to study the multi-step process of denitrification (Song and Liu, 2015).

10.3.2.5 Emerging Patterns

The goal of reactive transport models that include microbial dynamics is to reproduce and predict observations at the macro-scale, ideally as patterns naturally emerging from the coded interactions at the micro-scale. Intrinsically, these models include the effect of both transport (of chemicals and microorganisms) and reactions, which depend on (and affect) ecological niches. Both of these aspects can be important in shaping the community composition (Stegen et al., 2013).

The simulation of a whole microbial community faces numerous challenges, which depend upon the modeling approach used. Röling and van Bodegom (2014) argued that a *bottom-up* approach can be powerful, as flux balance approaches (FBA) integrate knowledge of cell metabolism, and the analysis of intracellular fluxes for a given metabolic network has developed into a mature research area (Lewis et al., 2012). They have the advantage that they do not require extensive parameterization and can build on a metagenome from which individual organisms can be identified (Iverson et al., 2012; Luo et al., 2012; Nielsen et al., 2014). Challenges include that the microbial community, in particular in soils, can be very complex (Torsvik and Øvreås, 2002). The number of microbial species exceeds the number of *in silico* models by orders of magnitude, and will continue to do so in the foreseeable future. From a modeling perspective, however, this might be addressed by introducing a metabolic 'superbug' that complements the metabolism of the microbes considered individually and includes the remaining pathways. Yet such an approach requires flux constraints on these additional reaction rates, which requires quantitative rate data, and constrains the ability of the ecosystem to naturally evolve under changing external forcings. Furthermore, the uptake of nutrients and the interactions with other organisms can be difficult to constrain, and may require extensive experimental data.

Trait-based approaches have yielded promising results; they have been successfully applied to explain phytoplankton distributions in the marine environment (Follows and Dutkiewicz, 2011). For soil and subsurface environments, similar work is in active development. The approach extends the modeling of functional microbial groups into a more flexible framework (Figure 10.3). Furthermore, the use of functional traits allows the comparison with genomic data on functional genes, as well as the corresponding transcriptomic or proteomic data. Challenges include the alignment of emerging model organisms with biological observations at the organism level.

Model descriptions that focus on the wide distribution of functionality that can emerge under the right environmental conditions, such as the distributed metabolic network approach, generally do not face the same challenge of imposed and limited metabolic functions. Rather, the difficulty lies in identifying the appropriate time horizon and the correct goal function that governs the self-organization of the ecosystem. Similar to the issue encountered in FBA, there is still limited data that unequivocally show that maximizing entropy production (Vallino, 2010) governs the evolution of the microbial community. Indeed, there is a rich literature in ecological theory suggesting a number of governing principles (e.g. Ulanowicz and Abarca-Arenas, 1997; Fath et al., 2001; Jørgensen and Svirezhev, 2004).

These mechanistic process- and organism-based modeling approaches are complemented by statistical approaches in environments with sufficient data density. For example, at the sea surface, extensive remote sensing data-sets have been connected with large microbial sequencing efforts through statistical analysis to predict the

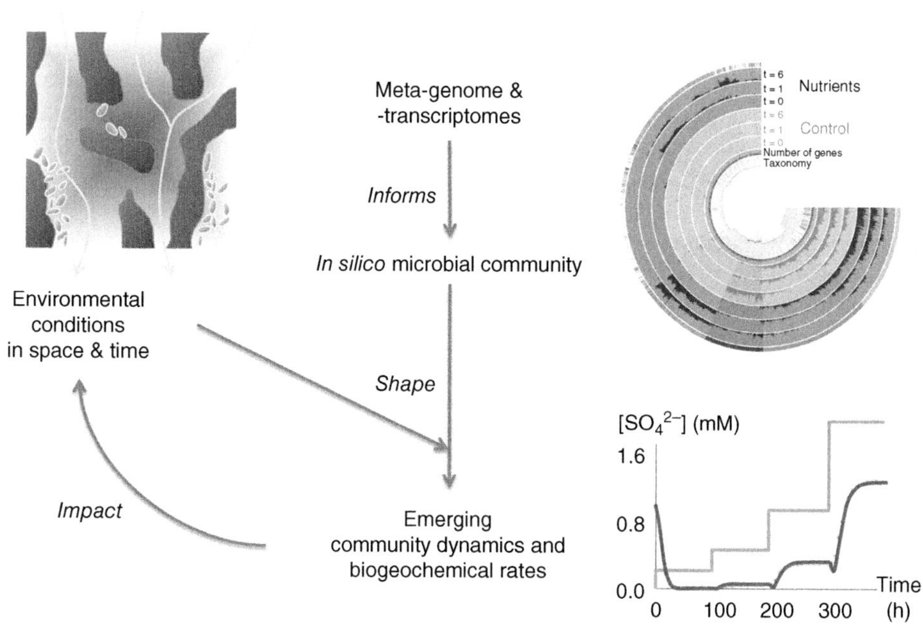

Figure 10.3 Trait-based approach of simulating microbially-driven biogeochemical processes. The microbial community is established as a large number of organisms whose metabolic traits may be assigned probabilistically. Traits may reflect hypothesized functions and processes, or are based on the interpretation of meta-genomes, -transcriptomes or other -omic data that constrain metabolic capabilities in the environment (top right: the image depicts the transcriptional response of a microbial community over time in the presence and absence of nutrient enrichments. *Source*: courtesy of T. Montenegro). Combining the functional capabilities with the environmental conditions determines the expression of metabolic functions and the modeled process rates (bottom right: the graph depicts the sulfate concentration in the inflow (pink) and outflow (red) in a column experiment over time). These rates then impact the composition of the local environment across the model domain (top left: illustrating fluid flow and the distribution of microbes in a porous medium). It also informs microbial growth dynamics, which in turn shapes the community composition. *Source:* Courtesy of T. Montenegro (See colour plate section for the colour representation of this figure.)

functioning and possibly the evolution of the microbial ecosystem (Larsen et al., 2011). The power of this approach is the combination of neural networks to predict the microbial community composition (Larsen et al., 2012; Larsen et al., 2015), and linking that with turnover rates of metabolites. This is a hugely promising approach mining big data, but as is always the case with approaches that lack strong mechanistic underpinnings, the usefulness is restricted to the range of conditions sampled. This may limit prediction of the system behaviour beyond these bounds, for example, in conditions that may be encountered in a changing climate.

10.4 Data Integration

To ground mathematical models of biogeochemical processes in reality, inclusion of fundamental knowledge and fusion with observational data are central. The former is commonly reflected in the governing equations. They typically include diffusive and advective transport, whose formulations are well established. In contrast, the descriptions of microbially mediated reactions vary, yet reflect rich background knowledge and generally observed patterns, such as a dependency on substrate concentrations. In addition, macroscopic rates can also rely on the consumption or production of chemicals predicted by *in silico* cell models whose metabolic networks are assembled from sequenced genomes.

Model formulations are informed by general prior knowledge, complemented by the integration of site-specific data. This can include the presence or absence of chemical compounds, which can point to whether a particular process may be active and hence part of the model or not (e.g. methane as an indicator of methanogenesis). Phylogenetic information indicates the presence of particular microorganisms, or the presence and expression of functional genes and proteins show potential or realized metabolic function. Such observations can be used to establish a reaction network. In models that explicitly resolve enzyme dynamics, their presence can also be used to validate the modeled process dynamics (Reed et al., 2014). This is particularly powerful because omic data can detect individual processes and reveal the pathways rather than just the net increase or decrease in the concentration of a chemical substance involved in multiple reactions (e.g. Bowen et al., 2014; Hawley et al., 2014; Morales-Castilla et al., 2015). This is critical for predicting the system response to perturbations. Another example is the bioremediation study of Fang et al. (2012), who used environmental proteomic data to evaluate the metabolic functions predicted by a genome-scale *in silico* cell model and to identify active pathways that were not predicted by the model. As shown by Vieira-Silva and Rocha (2010), bias in codon (nucleotide sequences coding amino acids) usage correlates with bacterial growth efficiencies. Thus, genome sequence data may also contain information on critical dynamic model parameters such as growth efficiencies, which govern the temporal dynamics of microbial populations.

The incorporation of new insights from systems biology into reactive transport modeling increases model complexity and the number of – sometimes poorly constrained – parameters. This raises the question of whether the increase in complexity leads to an improved capability to predict system behaviour. Comparison between models of different complexity suggests that it can, and importantly, more complex descriptions allow the use of the same model across a range of environments (Friedrichs et al., 2007). However, such application requires the integration of site-specific data and their objective integration through data assimilation (Hood et al., 2007). While common in weather forecast and ocean circulation modeling, its use in biogeochemistry to date is limited (Dowd et al., 2014). This is in part because, unlike the aforementioned fields, which use well-established governing equations and observational data-sets, the model structure of reactive transport models is variable, in particular with regard to the reactions considered and their corresponding rate laws. Furthermore, the observations – including omics data – are often indirect measurements of the quantity of interest. Both these aspects make data assimilation less routine and hence less widespread. Nevertheless, recent developments include the use of Ensemble Kalman filters as one approach to improve predictions of reactive transport models (Tong et al., 2014). Using time series measurements to estimate unknown model parameters, such methods have also been applied to estimate biogeochemical parameters in atmospheric land surface flux models (Trudinger et al., 2008; Zhang et al., 2015). In addition, Bayesian methods and Ensemble Kalman filtering (EnKF) methods have been applied to infer flow and transport parameters needed for reactive transport modeling (e.g. Murakami et al., 2010; Chen et al., 2013), and increases in computational capabilities offer promise for incorporating biogeochemical reaction observations as well. Because many codes used for hydrological parameter estimation also have reactive transport capabilities and offer the computational performance needed (e.g. Chen et al. 2013), the extension of EnKF to include estimation of biogeochemical parameters such as reaction rate coefficients is relatively straightforward. Figure 10.4 (modified from Chen et al., 2013) shows an example of the application of EnKF to estimate hydrogeological parameters for improvement of model predictions of tracer transport, and a logic chart for extension of the method used therein to incorporate biogeochemical parameters.

10.5 Linking Models Across Scales

Biogeochemical models deal with phenomena that cover a wide range of temporal and spatial scales. One may separate the timescales, so that slow processes (e.g. tied to geological processes) can be prescribed as environmental conditions/steady external forcing, intermediate ones can be explicitly modeled (e.g. the dynamics in an NPZD model), while fast processes (e.g. the adaptation of the microbial community) can be included in the model parameterization. Traditionally, this has been the most common approach, although accounting for variability in fast processes through a stochastic approach as noise may be more realistic (Berglund and Gentz, 2002). Separation of timescales is common, and it has also prompted the use of differential-algebraic equations in the description of biogeochemical dynamics, and has benefited from the development of efficient numerical algorithms for efficiently solving such problems (e.g. Brenan et al., 1996; Hindmarsh et al., 2005).

A primary obstacle to the integration of microbial processes into reactive transport models is the large disparity in physical length scales between (1) the physical environment that humans observe and impact (scales of metres to kilometres and larger), and (2) the physical environment that subsurface microorganisms directly sense and modify (scale of tens of microns and smaller). An unavoidable fact is that the natural environment is strongly heterogeneous across this full range of length scales: heterogeneous in processes (e.g. flow velocities), states (e.g. chemical concentrations), physical properties (e.g. porosity, permeability), geochemical properties (e.g. mineral phases, organic matter content), and microbial abundance and community composition. A ubiquitous impact of this heterogeneity is the creation of local microenvironments in which conditions vary strongly over scales that are very small relative to scales of interest in modeling applications, and small relative to scales that can be practically resolved in most numerical simulations. The grid resolution of a reactive transport model can also vary greatly depending on the application, and is limited by the domain size of interest, computational demands of model solution, and sparsity of available characterization data. For groundwater contaminant plume fate and transport, grid elements are typically at least one metre in size in the horizontal plane, while in watershed simulations, grid elements of hundreds of metres to kilometres in size would not be unusual. Models that resolves the geometry of, and processes within, individual pores of a porous medium (pore-scale models) can achieve resolution approaching relevance to the world that microbes experience, but are currently limited to domain sizes on the order of a cubic

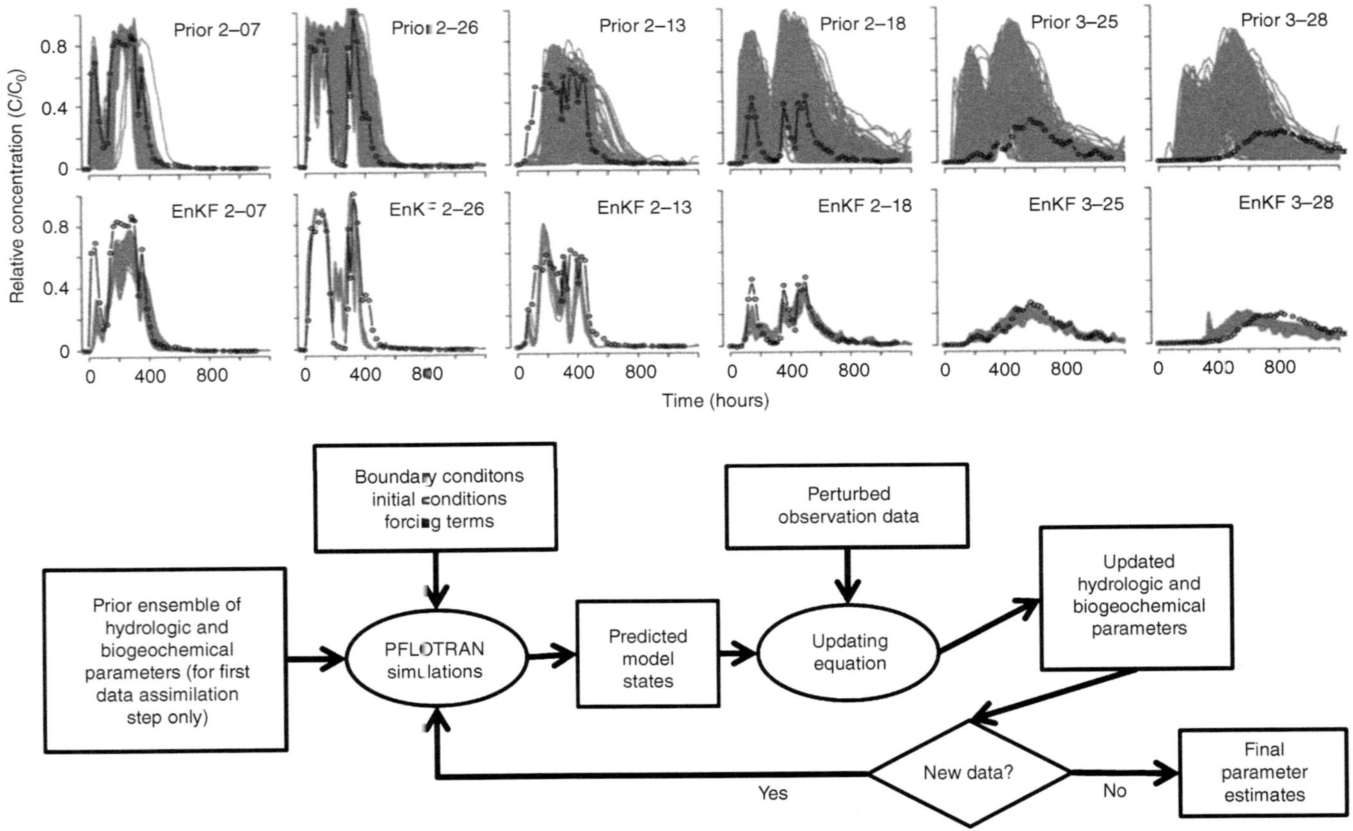

Figure 10.4 Top: Comparison of prior model predictions (top row) with updated model predictions following application of the Ensemble Kalman Filtering (EnKF) method to estimate hydraulic conductivity fields, modified from Chen et al. (2013). Black symbols are observed tracer breakthrough curves at several wells (denoted 2-07, 2-26, etc.), and shaded curves are the ensemble of model predictions. Vertical axis scales on all plots are tracer concentration normalized to the injection concentration (maximum value of 1.0); horizontal axis scales on all plots are time in hours from 0 to 1200. The plots clearly demonstrate the dramatic improvement in model reproduction of observations and reduction of model uncertainty following parameter updating using EnKF. Bottom: Logic diagram for parameter estimation using EnKF, modified from Chen et al. (2013) to include biogeochemical data and parameters. Note that the PFLOTRAN code referenced in the diagram is a high-performance code capable of simulating large ensembles of complex flow and reactive transport in a time-efficient manner required for application of EnKF. *Source*: Adapted from Chen et al. (2013).

decimetre (Scheibe et al., 2015b) – very small relative to typical application model domains. Within a single grid element, models typically assume uniformity (or at least smooth variation) of properties, states and processes. Thus any variations below the resolved scales are usually treated only in an average sense. If a function (such as a fundamental reaction rate law) is applied to average concentrations at some scale, it is accurate only if the function applied to average concentrations is equal to the average of the function applied to local concentrations. Stated in mathematical terms, this constraint is:

$$\frac{1}{V}\int_V f\big(C(v)\big)dv = f\left[\frac{1}{V}\int_V C(v)dv\right]$$

where C is a vector of reactant concentrations over some local incremental volume dv, V is the total volume of the averaging domain (e.g. a numerical grid cell), and $f()$ is a rate law function. This equality will generally hold only in the simple cases that (1) the function $f()$ is linear in C (violated if rates depend on reaction concentration products or are nonlinear functions of single constituent concentrations such as Monod functions), or (2) the concentration vector C is spatially uniform (complete mixing at all scales smaller than those explicitly resolved). In heterogeneous systems involving microbial reactions, these conditions do not usually hold. Commonly, the result is that observed biogeochemical reaction rates in heterogeneous systems are lower than rates that would be predicted based on average concentrations and rate law parameters derived from well-mixed batch studies (Li et al., 2006; Meile and Tuncay, 2006). This can easily be understood as the effect of incomplete mixing at the subgrid scale; biogeochemical reactions require direct contact among reacting agents (including microorganisms), and if they are poorly mixed due to small-scale heterogeneity they will not react as effectively as if they were uniformly mixed (Gramling et al., 2002). This problem may be particularly challenging in the case of microorganisms, as they can actively adapt their metabolic pathways and rates to their geochemical environment. Clearly, an individual bacterial cell (on the order of one micron in size) cannot respond to averages of chemical and mineral concentrations over volumes much larger than a single pore (tens of microns in size), and within a single model grid cell there may exist a wide range of microbial microenvironments that one would not imagine existed, given the average concentrations of key constituents. Accordingly, one may observe methanogenesis (a strictly anaerobic process) taking place within a porous medium volume which is on average well oxygenated (aerobic); this apparent paradox is easily explained by the existence of small pockets of anaerobic water within the volume of interest, but cannot be predicted based on average conditions.

At least four approaches, varying widely in their complexity and breadth of use, have been applied to address this challenge.

1) *Empirical fitting/adjustment of rate law parameters*: A simple solution is to treat rate law parameters as 'tuning' knobs of a model. Rather than specifying them independently based on laboratory measurements or theoretical considerations, one can perform an experiment or make observations at the scale of interest and use the results to estimate or fit reaction model parameters. Although this is a very common approach and may be useful for constructing a quantitative description of system

behaviour, it has limited utility in a predictive sense. Obviously, one must observe the process of interest over the time and length scales of interest before model parameters can be estimated, whereas it would be preferable to be able to predict the outcome *a priori* based on laboratory measurements and theory. Perhaps more importantly, the parameters so estimated strictly apply only to the conditions under which the observations were made and should not be extrapolated to predict response to system perturbations or alternative environmental management strategies.

2) *Rigorous upscaling methods*: Another common approach is to derive effective reaction parameters mathematically at the scale of interest which properly quantify the impacts of heterogeneity at smaller scales (Cushman et al., 2002). As pointed out by Wood (2009), this process requires the specification of a scaling law: a mathematical statement of assumptions about the (e.g. statistical) nature of microscale variability. As long as these assumptions are valid (the scaling law is applicable), the upscaled rate law and associated effective parameters will provide a predictive model capability. In many problems, however, it is well known that upscaling conditions are not applicable and the upscaled models break down (Battiato et al., 2009). Recent research has developed and demonstrated an approach for quantifying applicability conditions for upscaled models, typically expressed as a function of dimensionless variables such as the Peclet (Pe) and Damkohler (Da) numbers (Battiato and Tartakovsky, 2011; Boso and Battiato, 2013).
3) *Quantification of the degree of mixing as a process variable*: Since effective reaction rates depend strongly on the degree of local mixing of reactants, another approach is to independently model the degree of mixing and define a variable reaction rate formulation that depends on the mixing quantifier. Kitanidis (1994) clarified the distinction between mechanical spreading and diffusive mixing, which are typically combined in effective dispersion models, and introduced an entropy-based macroscopic metric called the dilution index to represent the degree of actual mixing in heterogeneous systems. Many subsequent researchers have used the dilution index concept to quantify the impact of heterogeneity on mixing and associated effects on reaction rates (e.g. Kapoor and Kitanidis, 1998; Tartakovsky et al., 2009; Chiogna et al., 2011; de Barros et al., 2012; see literature review in de Barros et al., 2015). A related measure of dilution, the scalar dissipation rate, has also been used to quantify the impacts of heterogeneity on mixing (e.g. Le Borgne et al., 2010; de Dreuzy et al., 2012; de Barros et al., 2015). De Simoni et al. (2007) introduced the concept of a mixing ratio applied to conservative and reactive solutes, and used it to quantify impacts of mixing on reactions. Le Borgne et al. (2014) introduced an approach for representing fluid mixing based on quantifying the advective stretching and folding of interfaces between reacting fluids and its resulting impact on concentration gradients, diffusive mass transfer, and effective reaction rates.
4) *Hybrid multiscale methods*: Another recent approach is to model processes directly at multiple scales of resolution and couple multiple models together within a single simulation framework. These methods (hybrid multiscale methods) have many diverse forms (e.g. Kevrekidis et al., 2004; E et al., 2007; Fish, 2009) and have been extensively applied to many scientific domains but only recently to reactive transport in porous media. Scheibe et al. (2015a) reviewed a broad spectrum of multiscale methods with emphasis on hybrid multiscale methods; the reader is referred there for additional detail and example applications. In the context of modeling

microbially-mediated biogeochemical reactions, a critical scale is the pore scale at which geometric and geochemical features are resolved down to the micrometre scale. Tartakovsky et al. (2013) coupled a genome-scale FBA model of a metal-reducing organism with a pore-scale flow and reactive transport model and demonstrated that diffusion-limited pore-scale mixing processes significantly impacted macroscopic reaction rates. This result has significant implications for the application of mechanism-based microbial metabolism models to reactive transport in natural porous media. The application of genome-scale metabolic models is attractive because of its mechanistic nature and ability to reflect organismal adaptation to changing environmental conditions. However, the validity of this approach depends on the ability to characterize the actual environment experienced by the organism (not some average environmental conditions over a much larger volume). While we are conceptually and numerically able to model the behaviour and interactions of microorganisms in heterogeneous environments even at the individual cell level (e.g. Resat et al., 2012), properly representing the manifestation of these processes at relatively large scales remains a daunting challenge. Explicit coupling of pore- and continuum-scale models in a computationally efficient manner is a promising avenue to address this challenge, and is also likely to lead to improvements in understanding that will support better upscaling approaches and mixing models.

10.6 Summary and Outlook

Biogeochemical cycling in Earth (near-)surface environments is driven by microbes. This recognition, combined with new capabilities to study microbial communities in the field, prompts the development of new and more realistic modeling approaches. This includes the explicit representation of microbes, as well as an improved resolution of smaller scales critical for biogeochemical transformations. The increase in computational power, advances in *in situ* characterization, software infrastructure and the development of multiscale modeling frameworks are all critical for progress. New omics data-sets provide novel challenges and opportunities to the modeling community.

Regardless of how microbes are incorporated into reaction transport models, all new developments face the task of connecting the models with novel omic data-sets. This can improve models at three levels. Firstly, it can inform the model formulation. At the cellular level, this means that genome sequencing and annotation informs the metabolic cell model (e.g. Mahadevan et al. 2006), while at the system level, knowledge of the presence and expression of functional genes can inform the processes to be included into the model (Mahadevan et al., 2011; Reed et al., 2014). Secondly, rather than inferring statistical connections between omic and process data, genomic or transcript data may help to parameterize key model parameters, so genomic data can be mined to infer dynamic system properties. Thirdly, omic data may be integrated into model validation and become part of the data assimilation process. Such data are complementary to concentration and rate measurements, and have the potential to constrain metabolic potential (genomics), active processes (transcriptomics), and enzyme and substrate levels (proteomics and metabolomics). As it does not *directly* match the modeled properties, its use requires either establishing these connections for integration in data assimilation, or adaptation of the models to represent properties more closely related to observational data.

One of the biggest challenges in assimilation of observational data for model validation and updating is the wide range of support scales for model and observational data. All measurements have an associated physical scale of support, which may not match the associated model resolution scale. This is one of the factors motivating the linkage of models across scales. We anticipate that the use of data assimilation techniques in conjunction with the use of omic data will transform reactive transport modeling of biogeochemical cycling significantly. Effective use of field observations and omic information provides the opportunity to increase the mechanistic fidelity of reactive transport models, which can be expected to improve predictive capability significantly. However, adoption of a more mechanistic approach to simulation requires significant advances in connecting models across physical and temporal scales, since most mechanisms of interest are well-defined at small scales relative to scales of practical interest in applications.

Model coupling is also critical in a multiprocess (or multi-physics) sense. Biogeochemical systems (as the word implies) involve several different types of processes (biological, hydrogeological, and chemical). For example, while we have focused here on microbial processes in the subsurface environment, plants obviously play a major role in biogeochemical cycling in the near-surface environment as well, and the interaction of plants and microbes in the rhizosphere is a critical element of natural nutrient cycles and productivity. Accordingly, future simulation efforts will become increasingly interdisciplinary in nature, requiring a community-based approach to model development and curation with an emphasis on code modularity and interoperability (US Department of Energy, 2015). The dramatic increases in omics information, process knowledge, property characterization, and computational capacity present the opportunity to explore ever-increasing levels of system complexity. As we commence this exploration, it is clear that natural biogeochemical systems are far more complex in nature than we can currently understand or simulate. Modeling reactive transport in biogeochemically active systems remains a critical challenge for the future and a fertile ground for transformative research.

Acknowledgements

Christof Meile acknowledges that this research was made possible in part by a grant from the Gulf of Mexico Research Initiative to support the 'Ecosystem Impacts of Oil and Gas in the Gulf' (ECOGIG) research consortium, and in part by the US Department of Energy, Office of Biological Environmental Research (award number DE-SC0010574). This is ECOGIG contribution no. 368. Tim Scheibe acknowledges support of the US Department of Energy (DOE) Office of Biological and Environmental Research (BER), Subsurface Biogeochemical Research Program, through the Pacific Northwest National Laboratory Subsurface Science Scientific Focus Area project.

References

Algar, C.K. and Vallino, J.J. (2014). Predicting microbial nitrate reduction pathways in coastal sediments. *Aquatic Microbial Ecology: International Journal* 71 (3): 223–238.

Allison, S.D. (2012). A trait-based approach for modelling microbial litter decomposition. *Ecology Letters* 15 (9): 1058–1070.

Amin, S.A., Hmelo, L.R., van Tol, H.M., et al. (2015). Interaction and signalling between a cosmopolitan phytoplankton and associated bacteria. *Nature* 522 (7554): 98–101.

Baas Becking, L.G.M. (1934). *Geobiologie of Inleiding Tot de Milieukunde*. Den Haag: W.P. Van Stockum & Zoon.

Battiato, I. and Tartakovsky, D.M. (2011). Applicability regimes for macroscopic models of reactive transport in porous media. *Journal of Contaminant Hydrology* 120–121: 18–26.

Battiato, I., Tartakovsky, D.M., Tartakovsky, A.M., and Scheibe, T. (2009). On breakdown of macroscopic models of mixing-controlled heterogeneous reactions in porous media. *Advances in Water Resources* 32 (11): 1664–1673.

Berglund, N. and Gentz, B. (2002). Metastability in simple climate models: Pathwise analysis of slowly driven Langevin equations. *Stochastics and Dynamics* 2 (3): 327–356.

Bernier, S.P. and Surette, M.G. (2013). Concentration-dependent activity of antibiotics in natural environments. *Frontiers in Microbiology* 4 (February): 20.

Bethke, C.M. (2007). *Geochemical and Biogeochemical Reaction Modeling*. New York: Cambridge University Press.

Bethke, C.M., Ding, D., Jin, Q., and Sanford, R.A. (2010). Origin of microbiological zoning in groundwater flows. *Geology* 36 (9): 739–742.

Boso, F. and Battiato, I. (2013). Homogenizability conditions for multicomponent reactive transport. *Advances in Water Resources* 62: 254–265.

Bouskill, N.J., Tang, J., Riley, W.J., and Brodie, E.L. (2012). Trait-based representation of biological nitrification: Model development, testing, and predicted community composition. *Frontiers in Microbiology* 3 (October): 364.

Bowen, J.L., Babbin, A.R., Kearns, P.J., and Ward, B.B. (2014). Connecting the dots: Linking nitrogen cycle gene expression to nitrogen fluxes in marine sediment mesocosms. *Frontiers in Microbiology* 5 (August): 429.

Brenan, K.E., Campbell, S.L., and Petzold, L.R. (1996). *Numerical Solution of Initial-Value Problems in Differential-Algebraic Equations*. Philadelphia, PA: Society for Industrial and Applied Mathematics (SIAM).

Brun, A. and Engesgaard, P. (2002). Modelling of transport and biogeochemical processes in pollution plumes: Literature review and model development. *Journal of Hydrology* 256 (3–4): 211–227.

Burgin, A.J., Yang, W.H., Hamilton, S.K., and Silver, W.L. (2011). Beyond carbon and nitrogen: How the microbial energy economy couples elemental cycles in diverse ecosystems. *Frontiers in Ecology and the Environment* 9 (1): 44–52.

Cajal-Medrano, R. and Maske, H. (2005). Growth efficiency and respiration at different growth rates in glucose-limited chemostats with natural marine bacteria populations. *Aquatic Microbial Ecology: International Journal* 38: 125–133.

Carlson, C., del Giorgio, P., and Herndl, G. (2007). Microbes and the dissipation of energy and respiration: From cells to ecosystems. *Oceanography* 20 (2): 89–100.

Chen, X., Hammond, G.E., Murray, C.J., et al. (2013). Application of ensemble-based data assimilation techniques for aquifer characterization using tracer data at Hanford 300 Area. *Water Resources Research* 49 (10): 7064–7076.

Chiogna, G., Cirpka, O.A., Grathwohl, P., and Rolle, M. (2011). Transverse mixing of conservative and reactive tracers in porous media: quantification through the concepts of flux-related and critical dilution indices. *Water Resources Research* 47 (2). doi:10.1029/2010wr009608

Curtis, G.P. (2003). Comparison of approaches for simulating reactive solute transport involving organic degradation reactions by multiple terminal electron acceptors. *Computers and Geosciences* 29 (3): 319–329.

Cushman, J.H., Bennethum, L.S., and Hu, B.X. (2002). A primer on upscaling tools for porous media. *Advances in Water Resources* 25 (8–12): 1043–1067.

de Barros, F.P.J., Dentz, M., Koch, J., and Wolfgang, N. (2012). Flow topology and scalar mixing in spatially heterogeneous flow fields. *Geophysical Research Letters* 39 (8). doi:10.1029/2012gl051302

de Barros, F.P.J., Fiori, A., Boso, F., and Bellin, A. (2015). A theoretical framework for modeling dilution enhancement of non-reactive solutes in heterogeneous porous media. *Journal of Contaminant Hydrology* 175–176 (April): 72–83.

de Dreuzy, J.-R., Carrera, J., Dentz, M., and Le Borgne, T. (2012). Time evolution of mixing in heterogeneous porous media. *Water Resources Research* 48 (6). doi:10.1029/2011wr011360

De Simoni, M., Sanchez-Vila, X., Carrera, J., and Saaltink, M.W. (2007). A mixing ratios-based formulation for multicomponent reactive transport. *Water Resources Research* 43 (7). doi:10.1029/2006wr005256

Dowd, M., Jones, E., and Parslow, J. (2014). A statistical overview and perspectives on data assimilation for marine biogeochemical models. *Environmetrics* 25 (4): 203–213.

Droop, M.R. (1973). Some thoughts on nutrient limitation in algae. *Journal of Phycology* 9 (3): 264–272.

Durot, M., Bourguignon, P.-Y., and Schachter, V. (2009). Genome-scale models of bacterial metabolism: Reconstruction and applications. *FEMS Microbiology Reviews* 33 (1): 164–190.

Dutkiewicz, S., Follows, M.J., and Bragg, J.G. (2009). Modeling the coupling of ocean ecology and biogeochemistry. *Global Biogeochemical Cycles* 23 (4). doi:10.1029/2008gb003405

E, W., Engquist, B., Li, X.T., et al. (2007). Heterogeneous multiscale methods: A review. *Communications in Computational Physics* 2 (3): 367–450. Available at http://www.global-sci.com/openaccess/v2_367.pdf

Fang, Y., Yeh, G.-T., and Burgos, W.D. (2003). A general paradigm to model reaction-based biogeochemical processes in batch systems. *Water Resources Research* 39 (4): 1083.

Fang, Y., Scheibe, T.D., Mahadevan, R., et al. (2011). Direct coupling of a genome-scale microbial in silico model and a groundwater reactive transport model. *Journal of Contaminant Hydrology* 122 (1–4): 96–103.

Fang, Y., Wilkins, M.J., Yabusaki, S.B., et al. (2012). Evaluation of a genome-scale in silico metabolic model for *Geobacter metallireducens* by using proteomic data from a field biostimulation experiment. *Applied and Environmental Microbiology* 78 (24): 8735–8742.

Fath, B.D., Patten, B.C., and Choi, J.S. (2001). Complementarity of ecological goal functions. *Journal of Theoretical Biology* 208 (4): 493–506.

Fish, J. (2009). *Multiscale Methods: Bridging the Scales in Science and Engineering*. Oxford: Oxford University Press.

Follows, M.J. and Dutkiewicz, S. (2011). Modeling diverse communities of marine microbes. *Annual Review of Marine Science* 3: 427–451.

Follows, M.J., Dutkiewicz, S., Grant, S., and Chisholm, S.W. (2007). Emergent biogeography of microbial communities in a model ocean. *Science* 315 (5820): 1843–1846.

Friedrichs, M.A.M., Dusenberry, J.A., Anderson, L.A., et al. (2007). Assessment of skill and portability in regional marine biogeochemical models: Role of multiple planktonic groups. *Journal of Geophysical Research* 112, C08001. doi:10.1029/2006JC003852

Froelich, P.N., Klinkhammer, G.P., Bender, M.L., et al. (1979). Early oxidation of organic matter in pelagic sediments of the Eastern Equatorial Atlantic: Suboxic diagenesis. *Geochimica et Cosmochimica Acta* 43 (7): 1075–1090.

Gear, C.W. (1971). *Numerical Initial Value Problems in Ordinary Differential Equations.* Prentice Hall.

Geiser, J. (2011). *Iterative Splitting Methods for Differential Equations.* CRC Press.

Goldman, J.C., Caron, D.A., and Dennett, M.R. (1987). Regulation of gross growth efficiency and ammonium regeneration in bacteria by substrate C:N ratio. *Limnology and Oceanography* 32 (6): 1239–1252.

Gramling, C.M., Harvey, C.F., and Meigs, L.C. (2002). Reactive transport in porous media: a comparison of model prediction with laboratory visualization. *Environmental Science & Technology* 36: 2508–2514.

Hawley, A.K., Brewer, H.M., Norbeck, A.D., et al. (2014). Metaproteomics reveals differential modes of metabolic coupling among ubiquitous oxygen minimum zone microbes. *Proceedings of the National Academy of Sciences of the United States of America* 111 (31): 11395–11400.

Hensen, C., Zabel, M., Pfeifer, K., et al. (2003). Control of sulfate pore-water profiles by sedimentary events and the significance of anaerobic oxidation of methane for the burial of sulfur in marine sediments. *Geochimica et Cosmochimica Acta* 67 (14): 2631–2647.

Hindmarsh, A.C., Brown, P.N., Grant, K.E., et al. (2005). SUNDIALS. ACM Transactions on Mathematical Software. *Association for Computing Machinery* 31 (3): 363–396.

Hoehler, T.M. and Jørgensen, B.B. (2013). Microbial life under extreme energy limitation. Nature Reviews. *Microbiology* 11 (2): 83–94.

Holmkvist, L., Ferdelman, T.G., and Jørgensen, B.B. (2011). A cryptic sulfur cycle driven by iron in the methane zone of marine sediment (Aarhus Bay, Denmark). *Geochimica et Cosmochimica Acta* 75: 3581–3599.

Hood, R., Laws, E., Follows, M., and Siegel, D. (2007). Modeling and prediction of marine microbial populations in the genomic era. *Oceanography* 20 (2): 155–165.

Hunter, K.S., Wang, Y., and Van Cappellen, P. (1998). Kinetic modeling of microbially-driven redox chemistry of subsurface environments: Coupling transport, microbial metabolism and geochemistry. *Journal of Hydrology* 209 (1–4): 53–80.

Iverson, V., Morris, R.M., Frazar, C.D., et al. (2012). Untangling genomes from metagenomes: Revealing an uncultured class of marine *Euryarchaeota*. *Science* 335 (6068): 587–590.

Jenne, E.A. (1979). Chemical modeling in aqueous systems: Speciation, sorption, solubility, and kinetics. Symposium Cosponsored by the Society of Environmental Geochemistry and Health and the ACS Division of Environmental Chemistry. *176th Meeting of the American Chemical Society*, Miami Beach, Florida, September 11–13, 1978.

Jin, Q. and Bethke, C.M. (2003). A new rate law describing microbial respiration. *Applied and Environmental Microbiology* 69 (4): 2340–2348.

Jørgensen, S.E. and Svirezhev, Y.M. (2004). *Towards a Thermodynamic Theory for Ecological Systems.* Amsterdam: Elsevier.

Joye, S.B., Teske, A.P., and Kostka, J.E. (2014). Microbial dynamics following the Macondo Oil Well blowout across Gulf of Mexico environments. *Bioscience* 64 (9): 766–777.

Kapoor, V. and Kitanidis, P.K. (1998). Concentration fluctuations and dilution in aquifers. *Water Resources Research* 34 (5): 1181–1193.

Kevrekidis, I.G., Gear, C.W., and Hummer, G. (2004). Equation-free: The computer-aided analysis of complex multiscale systems. *AIChE Journal* 50 (7): 1346–1355.

Kim, T.Y., Sohn, S.B., Kim, Y.B., et al. (2012). Recent advances in reconstruction and applications of genome-scale metabolic models. *Current Opinion in Biotechnology* 23 (4): 617–623.

King, E.L., Tuncay, K., Ortoleva, P., and Meile, C. (2009). In silico Geobacter sulfurreducens metabolism and its representation in reactive transport models. *Applied and Environmental Microbiology* 75 (1): 83–92.

King, E.L., Tuncay, K., Ortoleva, P., and Meile, C. (2010). Modeling biogeochemical dynamics in porous media: Practical considerations of pore scale variability, reaction networks, and microbial population dynamics in a sandy aquifer. *Journal of Contaminant Hydrology* 112 (1-4): 130–140.

Kitanidis, P.K. (1994). The concept of the Dilution Index. *Water Resources Research* 30 (7): 2011–2026.

Klitgord, N. and Segrè, D. (2010). Environments that induce synthetic microbial ecosystems. *PLoS Computational Biology* 6 (11): e1001002.

Larsen, P.E., Collart, F.R., Field, D., et al. (2011). Predicted Relative Metabolomic Turnover (PRMT): Determining metabolic turnover from a coastal marine metagenomic dataset. *Microbial Informatics and Experimentation* 1 (1): 4.

Larsen, P.E., Field, D., and Gilbert, J.A. (2012). Predicting bacterial community assemblages using an artificial neural network approach. *Nature Methods* 9 (6): 621–625.

Larsen, P., Dai, Y., and Collart, F.R. (2015). Predicting Bacterial Community Assemblages Using an Artificial Neural Network Approach. In: Cartwright H. (eds) Artificial Neural Networks. Methods in Molecular Biology, vol 1260. Springer, New York, NY **DOI** https://doi.org/10.1007/978-1-4939-2239-0_3 **Publisher Name** Springer, New York, NY **Print ISBN** 978-1-4939-2238-3 https://link.springer.com/protocol/10.1007/978-1-4939-2239-0_3#citeas.

Le Borgne, T., Dentz, M., Bolster, D., et al. (2010). Non-Fickian mixing: Temporal evolution of the scalar dissipation rate in heterogeneous porous media. *Advances in Water Resources* 33 (12): 1468–1475.

Le Borgne, T., Ginn, T.R., and Dentz, M. (2014). Impact of fluid deformation on mixing-induced chemical reactions in heterogeneous flows. *Geophysical Research Letters* 41 (22): 7898–7906.

Lewis, N.E., Nagarajan, H., and Palsson, B.O. (2012). Constraining the metabolic genotype-phenotype relationship using a phylogeny of in silico methods. *Nature Reviews Microbiology* 10 (4): 291–305.

Li, L., Peters, C.A., and Celia, M.A. (2006). Upscaling geochemical reaction rates using pore-scale network modeling. *Advances in Water Resources* 29 (9): 1351–1370.

Li L., Maher, K., Navarre-Sitchler, A., Druhan, J., Meile, C., Lawrence, C., Moore, J., Perdrial, J., Sullivan, P., Thompson, A., Jin, L., Bolton, E.W., Brantley, S.L., Dietrich, W.E., Mayer, K.U., Steefel, C.I., Valocchi, A., Zachara, J., Kocar, B., Mcintosh, J., Tutolo, B.M., Kumar, M., Sonnenthal, E., Bao, C., Beisman, J. (2017). Expanding the role of reactive transport models in critical zone processes, Earth-Science Reviews 165: 280-301. ISSN 0012-8252, https://doi.org/10.1016/j.earscirev.2016.09.001.

Lichtner, P.C. (1985). Continuum model for simultaneous chemical reactions and mass transport in hydrothermal systems. *Geochimica et Cosmochimica Acta* 49 (3): 779–800.

Litchman, E. and Klausmeier, C.A. (2008). Trait-based community ecology of phytoplankton. *Annual Review of Ecology, Evolution, and Systematics* 39 (1): 615–639.

Litchman, E., Klausmeier, C.A., Schofield, O.M., and Falkowski, P.G. (2007). The role of functional traits and trade-offs in structuring phytoplankton communities: Scaling from cellular to ecosystem level. *Ecology Letters* 10 (12): 1170–1181.

Liu, Y. (2007). Overview of some theoretical approaches for derivation of the Monod equation. *Applied Microbiology and Biotechnology* 73 (6): 1241–1250.

Luo, C., Tsementzi, D., Kyrpides, N.C., and Konstantinidis, K.T. (2012). Individual genome assembly from complex community short-read metagenomic Datasets. *ISME Journal* 6 (4): 898–901.

MacQuarrie, K.T.B. and Mayer, K.U. (2005). Reactive transport modeling in fractured rock: A state-of-the-science review. *Earth Science Reviews* 72 (3–4): 189–227.

Mahadevan, R., Bond, D.R., Butler, J.E., et al. (2006). Characterization of metabolism in the Fe(III)-reducing organism *Geobacter sulfurreducens* by constraint-based modeling. *Applied and Environmental Microbiology* 72 (2): 1558–1568.

Mahadevan, R., Palsson, B., and Lovley, D. (2011). In situ to in silico and back: elucidating the physiology and ecology of Geobacter spp. using genome-scale modeling. *Nature Reviews Microbiology* 9 (1): 39–50.

Meile, C. and Tuncay, K. (2006). Scale dependence of reaction rates in porous media. *Advances in Water Resources* 29 (1): 62–71.

Molins, S., Carrera, J., Ayora, C., and Saaltink, M.W. (2004). A formulation for decoupling components in reactive transport problems. *Water Resources Research* 40 (10): W10301.

Molz, F.J., Widdowson, M.A., and Benefield, L.D. (1986). Simulation of microbial growth dynamics coupled to nutrient and oxygen transport in porous media. *Water Resources Research* 22 (8): 1207–1216.

Monod, J. (1942). *Recherches Sur La Croissance Des Cultures Bactériennes*. Hermann et cie.

Morales-Castilla, I., Matias, M.G., Gravel, D., and Araújo, M.B. (2015). Inferring biotic interactions from proxies. *Trends in Ecology & Evolution* 30 (6): 347–356.

Morozova, D., Zettlitzer, M., Let, D., and Würdemann, H. (2011). Monitoring of the microbial community composition in deep subsurface saline aquifers during CO_2 storage in Ketzin, Germany. *Energy Procedia* 4: 4362–4370.

Mulchandani, A. and Luong, J.H.T. (1989). Microbial inhibition kinetics revisited. *Enzyme and Microbial Technology* 11 (2): 66–73.

Murakami, H., Chen, X., Hahn, M.S., et al. (2010). Bayesian approach for three-dimensional aquifer characterization at the Hanford 300 Area. *Hydrology and Earth System Sciences* 14 (10): 1989–2001.

Nielsen, H.B., Almeida, M., Juncker, A.S., et al. (2014). Identification and assembly of genomes and genetic elements in complex metagenomic samples without using reference genomes. *Nature Biotechnology* 32 (8): 822–828.

Pahlow, M. and Oschlies, A. (2013). Optimal allocation backs Droop's cell-quota model. *Marine Ecology Progress Series* 473: 1–5.

Pirt, S.J. (1965). The maintenance energy of bacteria in growing cultures. *Proceedings of the Royal Society B: Biological Sciences* 163 (991): 224–231.

Postma, D., Flemming, L., Hue, N.T.M., et al. (2007). Arsenic in groundwater of the Red River Floodplain, Vietnam: Controlling geochemical processes and reactive transport modeling. *Geochimica et Cosmochimica Acta* 71 (21): 5054–5071.

Ramkrishna, D. and Song, H.-S. (2008). A rationale for Monod's biochemical growth kinetics. *Industrial & Engineering Chemistry Research* 47 (23): 9090–9098.

Ramkrishna, D. and Song, H.-S. (2012). Dynamic models of metabolism: Review of the cybernetic approach. *AIChE Journal* 58 (4): 986–997.

Reed, D.C., Algar, C.K., Huber, J.A., and Dick, G.J. (2014). Gene-centric approach to integrating environmental genomics and biogeochemical models. *Proceedings of the National Academy of Sciences of the United States of America* 111 (5): 1879–1884.

Regnier, P., O'Kane, J.P., Steefel, C.I., and Vanderborght, J.P. (2002). Modeling complex multi-component reactive-transport systems: Towards a simulation environment based on the concept of a knowledge base. *Applied Mathematical Modelling* 26 (9): 913–927.

Resat, H., Bailey, V., McCue, L.A., and Konopka, A. (2012). Modeling microbial dynamics in heterogeneous environments: Growth on soil carbon sources. *Microbial Ecology* 63 (4): 883–897.

Röling, W.F.M. and van Bodegom, P.M. (2014). Toward quantitative understanding on microbial community structure and functioning: A modeling-centered approach using degradation of marine oil spills as example. *Frontiers in Microbiology* 5 (March): 125.

Saad, Y. (2003). *Iterative Methods for Sparse Linear Systems*. Philadelphia, PA: Society for Industrial and Applied Mathematics (SIAM). Available at http://epubs.siam.org/doi/book/10.1137/1.9780898718003

Sauer, U. (2006). Metabolic networks in motion: 13C-based flux analysis. *Molecular Systems Biology* 2 (November): 62.

Scheibe, T.D., Mahadevan, R., Fang, Y., et al. (2009). Coupling a genome-scale metabolic model with a reactive transport model to describe in situ uranium bioremediation. *Microbial Biotechnology* 2 (2): 274–286.

Scheibe, T.D., Murphy, E.M., Chen, X., et al. (2015a). An analysis platform for multiscale hydrogeologic modeling with emphasis on hybrid multiscale methods. *Ground Water* 53 (1): 38–56.

Scheibe, T.D., Perkins, W.A., Richmond, M.C., et al. (2015b). Pore-scale and multiscale numerical simulation of flow and transport in a laboratory-scale column. *Water Resources Research* 51 (2): 1023–1035.

Schuetz, R., Zamboni, N., Zampieri, M., et al. (2012). Multidimensional optimality of microbial metabolism. *Science* 336 (6081): 601–604.

Song, H.-S. and Liu, C. (2015). Dynamic metabolic modeling of denitrifying bacterial growth: The cybernetic approach. *Industrial & Engineering Chemistry Research* 54 (42): 10221–10227. doi:10.1021/acs.iecr.5b01615

Song, H.-S., Cannon, W., Beliaev, A., and Konopka, A. (2014). Mathematical modeling of microbial community dynamics: A methodological review. *Processes* 2 (4): 711–752.

Steefel, C.I. and MacQuarrie, K.T.B. (1996). Approaches to modeling of reactive transport in porous media. In: *Reactive Transport in Porous Media* (ed. C.I. Steefel, E.H. Oelkers and P.C. Lichtner), 797–798. Reviews in Mineralogy 34. Washington, DC: Mineralogical Society of America.

Steefel, C., Depaolo, D., and Lichtner, P. (2005). Reactive transport modeling: An essential tool and a new research approach for the earth sciences. *Earth and Planetary Science Letters* 240 (3–4): 539–558.

Stegen, J.C., Lin, X., Fredrickson, J.K., et al. (2013). Quantifying community assembly processes and identifying features that impose them. *ISME Journal* 7 (11): 2069–2079.

Stolyar, S., Van Dien, S., Hillesland, K.L., et al. (2007). Metabolic modeling of a mutualistic microbial community. *Molecular Systems Biology* 3 (March): 92.

Taffs, R., Aston, J.E., Kristen, B., et al. (2009). In silico approaches to study mass and energy flows in microbial consortia: A syntrophic case study. *BMC Systems Biology* 3 (1): 114.

Tartakovsky, A.M., Tartakovsky, G.D., and Scheibe, T.D. (2009). Effects of incomplete mixing on multicomponent reactive transport. *Advances in Water Resources* 32 (11): 1674–1679.

Tartakovsky, G.D., Tartakovsky, A.M., Scheibe, T.D., et al. (2013). Pore-scale simulation of microbial growth using a genome-scale metabolic model: Implications for Darcy-scale reactive transport. *Advances in Water Resources* 59: 256–270.

Thullner, M., Van Cappellen, P., and Regnier, P. (2005). Modeling the impact of microbial activity on redox dynamics in porous media. *Geochimica et Cosmochimica Acta* 69 (21): 5005–5019.

Tong, J., Hu, B.X., Huang, H., et al. (2014). Application of a data assimilation method via an ensemble Kalman filter to reactive urea hydrolysis transport modeling. *Stochastic Environmental Research and Risk Assessment: Research Journal* 28 (3): 729–741.

Torsvik, V. and Øvreås, L. (2002). Microbial diversity and function in soil: From genes to ecosystems. *Current Opinion in Microbiology* 5 (3): 240–245.

Trudinger, C.M., Raupach, M.R., Rayner, P.J., and Enting, I.G. (2008). Using the Kalman filter for parameter estimation in biogeochemical models. *Environmetrics* 19: 849–870.

Ulanowicz, R.E. and Abarca-Arenas, L.G. (1997). An informational synthesis of ecosystem structure and function. *Ecological Modelling* 95 (1): 1–10.

US Department of Energy. (2015). Building virtual ecosystems: Computational challenges for mechanistic modeling of terrestrial environments. Workshop report, DOE/SC-0171, US Department of Energy Office of Science. Available at http://doesbr.org/VirtualEcosystems/

Vallino, J.J. (2003). Modeling microbial consortiums as distributed metabolic networks. *Biological Bulletin* 204 (2): 174–179.

Vallino, J.J. (2010). Ecosystem biogeochemistry considered as a distributed metabolic network ordered by maximum entropy production. *Philosophical Transactions of the Royal Society of London, Series B: Biological Sciences* 365 (1545): 1417–1427.

Vallino, J.J., Hopkinson, C.S., and Hobbie, J E. (1996). Modeling bacterial utilization of dissolved organic matter: Optimization replaces Monod growth kinetics. *Limnology and Oceanography* 41 (8): 1591–1609.

Vallino, J.J., Algar, C.K., González, N.F., and Huber, J.A. (2014). Use of receding horizon optimal control to solve MaxEP-based biogeochemistry problems. In: *Beyond the Second Law: Understanding Complex Systems* (ed. R. Dewar, C. Lineweaver, R. Niven and K. Regenauer-Lieb), 337–359. Berlin: Springer.

Vieira-Silva, S. and Rocha, E.P.C. (2010). The systemic imprint of growth and its uses in ecological (meta)genomics. *PLoS Genetics* 6 (1): e1000808.

Wang, Y. and Papenguth, H.W. (2001). Kinetic modeling of microbially-driven redox chemistry of radionuclides in subsurface environments: Coupling transport, microbial metabolism and geochemistry. *Journal of Contaminant Hydrology* 47 (2–4): 297–309.

Wolery, T.J. and Daveler, S.A. (1992). *EQ6, a Computer Program for Reaction Path Modeling of Aqueous Geochemical Systems: Theoretical Manual, User's Guide, and Related Documentation (Version 7.0); Part 4*. Livermore, CA: Lawrence Livermore National Laboratory.

Wood, B.D. (2009). The role of scaling laws in upscaling. *Advances in Water Resources* 32 (5): 723–736.

Zengler, K. and Palsson, B.O. (2012). A road map for the development of Community Systems (CoSy) biology. *Nature Reviews Microbiology*. doi:10.1038/nrmicro2763

Zhang, S., Zheng, X., Chen, J.M., et al. (2015). A global carbon assimilation system using a modified ensemble Kalman filter. *Geoscientific Model Development* 8: 805–816.

Zhuang, K., Izallalen, M., Mouser, P., et al. (2011). Genome-scale dynamic modeling of the competition between Rhodoferax and Geobacter in anoxic subsurface environments. *ISME Journal* 5 (2): 305–316.

Zomorrodi, A.R. and Maranas, C.D. (2012). OptCom: A multi-level optimization framework for the metabolic modeling and analysis of microbial communities. *PLoS Computational Biology* 8 (2): e1002363.

Zomorrodi, A.R., Islam, M.M., and Maranas, C.D. (2014). D-OptCom: Dynamic multi-level and multi-objective metabolic modeling of microbial communities. *ACS Synthetic Biology* 3 (4): 247–257.

11

Effective Stochastic Model For Reactive Transport

Alexandre M. Tartakovsky

Pacific Northwest National Laboratory, Richland, WA, 99352, USA

11.1 Introduction

Transport and mixing-controlled reactions are abundant in many natural and engineered systems, which could be as small as biological cells or as large as an aquifer, oil reservoir, the oceans or Earth's atmosphere. Yet, despite being an active research area, many existing effective models have poor predictive ability. Specifically, effective models obtained from homogenization and stochastic or volume averaging tend to overestimate the extent of mixing-controlled reactions (Gramling et al., 2002; Tartakovsky et al., 2008a; Tartakovsky, 2010; Bolster et al., 2010). Solutes, involved in the reactions, mix by two mechanisms: (1) molecular diffusion; and (2) mechanical mixing caused by variations in the fluid velocities advecting the solutes. While these two mechanisms are not completely independent (i.e. mechanical mixing facilitates molecular mixing), only molecular mixing contributes directly to the mixing-controlled reactions. On the pore scale, characterized by the average pore size l (Figure 11.1), an equation for the general advection-diffusion-reaction process with N species arises naturally from mass conservation considerations (Bear, 1988):

$$\frac{dC_i}{dt} = \frac{D_m}{\rho} \nabla \cdot \left(\rho \nabla C_i \right) - R_i \left(C_1, \ldots, C_N \right), \tag{11.1}$$

where $d/dt = \partial/\partial t + \mathbf{v} \cdot \nabla$, $\mathbf{v}$ is the fluid velocity, ρ is the solvent density, C_i is the mass fraction of the ith component, D_m is the molecular diffusion constant (assumed here to be the same for all species), and R_i is the rate of chemical reaction generating (or consuming) the ith species. The fluid velocity $\mathbf{v}$ satisfies the Navier-Stokes equations, including the continuity equation,

$$\frac{d\rho}{dt} = -\rho \nabla \cdot \mathbf{v}, \tag{11.2}$$

Reactive Transport Modeling: Applications in Subsurface Energy and Environmental Problems,
First Edition. Edited by Yitian Xiao, Fiona Whitaker, Tianfu Xu and Carl Steefel.

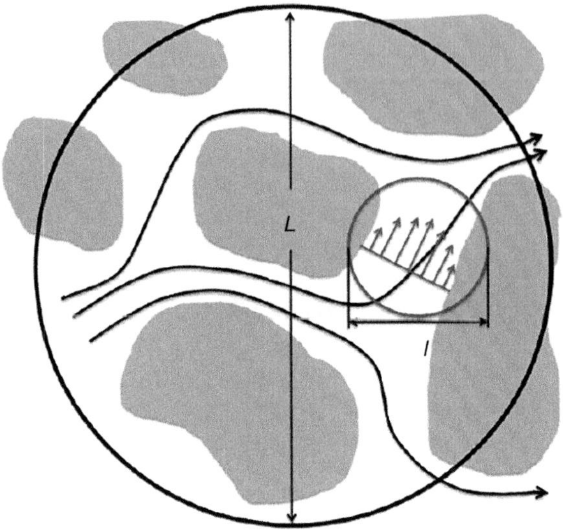

Figure 11.1 Mixing in porous media occurs because of velocity variations within the averaging volume *L* and the Taylor dispersion on the scale *l*.

and the momentum conservation equation,

$$\frac{dv}{dt} = -\frac{\nabla p}{\rho} + \mathbf{g} + \frac{\mu}{\rho}\nabla\cdot\left(\nabla\mathbf{v} + \nabla\mathbf{v}^T\right), \tag{11.3}$$

subject to the no-slip boundary condition at the surface of the soil grains (fluid-solid interface). Here, $\mathbf{g}$ is the body force per unit mass, and ρ and μ are the density and viscosity of the fluid.

Under most conditions, the direct calculations of $\mathbf{v}$ and C_i are not practical, and Equation (11.1) is commonly replaced with the advection-dispersion-reaction model (Bear, 1988; Li and Helm, 1998):

$$\frac{d\bar{C}_i}{dt} = \frac{1}{\rho}\nabla\cdot\left(\rho\mathbf{D}\nabla\bar{C}_i\right) - R_i\left(\bar{C}_1,\ldots,\bar{C}_N\right). \tag{11.4}$$

Here, $d/dt = \partial/\partial t + \mathbf{u}\cdot\nabla$, $\mathbf{u}$ and $\overline{C_i}$ are the volume averages of $\mathbf{v}$ and C_i over the representative volume with characteristic size L (Figure 11.1), and $\mathbf{D}$ is the dispersion tensor. The average pore velocity $\mathbf{u}$ satisfies the combination of the continuity equation,

$$\frac{d\rho}{dt} = -\rho\nabla\cdot\mathbf{u}, \tag{11.5}$$

and Darcy flow equation,

$$\frac{d\mathbf{u}}{dt} = -\frac{\nabla p}{\rho} + \mathbf{g} - \gamma\mathbf{u}. \tag{11.6}$$

The friction coefficient $\gamma = \phi\mu / (\rho k)$, porosity ϕ, permeability k, and hydraulic conductivity $\phi g/\gamma$ are some of the macroscopic parameters characterizing the continuum properties of the porous media. Under certain assumptions, the Darcy-scale transport equations can be obtained by volume-averaging the pore-scale equations (11.1)-(11.3) (Bear, 1988; Li and Helm, 1998).

In the case of transport in porous media, the averaging volume has size $L \gg l$ (Figure 11.1). In the advection-diffusion-reaction equation, the $\nabla \cdot (\mathbf{v}c_i)$ term models advection and mechanical mixing, and $D_m \nabla^2 C_i$ term represents diffusive mixing. On the other hand, in the advection-dispersion-reaction equation, both the mechanical and diffusive mixings are modeled by the $\mathbf{D}\nabla^2 \bar{C}_i$ term. We will demonstrate that this is one reason why the dispersion model often fails to accurately predict mixing-controlled reactions. In atmospheric and ocean modeling, subgrid models are commonly used to approximate $\mathbf{v}$ and its effect on the mechanical mixing (Frederiksen, 2012). In transport in porous media, stochastic Lagrangian particle models are commonly used to describe mechanical mixing and statistical models are employed for reactions (LeBorgne et al., 2008; Benson and Meerschaert, 2009; Bolster et al., 2010). While these models have been shown to accurately model transport processes, parameterization of these models is not trivial, and the results may be resolution dependent (Benson and Meerschaert, 2009).

Here, we describe the Langevin advection-diffusion-reaction (LADR) model, an effective stochastic model defined on the scale L, originally introduced in Tartakovsky et al. (2008a) and Tartakovsky (2010). In the dispersion model, Equation (11.4), it is assumed that solutes are well mixed on the scale L. This mixing is due to large-scale velocity variations within the averaging volume and the Taylor dispersion. The former is due to variations of pore sizes and complex pore geometry, and the latter is the result of velocity variations within individual pores and molecular diffusion. Chemical reactions, especially of the bimolecular type, create large concentration gradients on scales much smaller than L, which violates this assumption. In the LADR model, we assume that solutes are well mixed only on the scale l, mostly due to Taylor dispersion and molecular diffusion. Furthermore, we assume that the mixing at the scale l is a Fickian process with effective diffusion coefficient D_{eff}. Mechanical mixing in the LADR model is caused by random velocities with statistics determined from the pore-scale simulations or laboratory experiments. We formulate the LADR model as the stochastic partial differential equation:

$$\frac{d\chi_i}{dt} = \frac{\mathbf{D}_{eff}}{\rho} \nabla \cdot (\rho \nabla \chi_i) - R_i(\chi_1, \dots, \chi_N). \tag{11.7}$$

Here, $d/dt = \partial/\partial t + \mathbf{U} \cdot \nabla$ and $\mathbf{U}(\mathbf{x}, \mathbf{t}) = \langle \mathbf{U}(\mathbf{x}, \mathbf{t}) \rangle + \tilde{\mathbf{U}}(\mathbf{x}, \mathbf{t})$ is the random advective velocity with ensemble mean $\langle \mathbf{U} \rangle$ and random fluctuations $\mathbf{U}'$ ($\langle \cdot \rangle$ denotes an ensemble average). A stochastic equation for $\mathbf{U}$ will be introduced later in Section 11.3.

The velocity $\mathbf{U}$ is governed by a Langevin equation, obtained by adding a white noise into the Darcy flow equation (11.6). In combination with the continuity equation, the Langevin flow equation describes both the first and the second moments of the pore-scale velocity distribution. The velocity obtained from the Langevin flow equation not only advects the centre of mass but also mechanically disperses the solute. The separate treatment of the mechanical and diffusive mixing produces a mixing zone that is more realistic (in terms of solute dilution) than that obtained from the Darcy transport equations. The LADR model is parameterized such that, for a non-reactive transport, $\langle \chi_i(x, t) \rangle \equiv \overline{C_i}(x, t)$.

Multicomponent reactive transport with mixing-controlled reactions is one example where the Darcy description can be inadequate. Specifically, the Darcy flow and advection-dispersion equations overpredict the extent of mixing-controlled reactions (Cirpka et al., 1999; Knutson et al., 2007; Tartakovsky et al., 2009). Tartakovsky et al. (2009) found that the error in the Darcy-scale model increases with increasing Pe and is strongly correlated with a dilution index (Kitanidis, 1994) (the dilution index decreases with increasing Pe). Kitanidis (1994) demonstrated that a Darcy-scale transport model with constant coefficients overestimates the dilution index, which may lead to overpredicting the extent of the mixing-controlled reactions.

The LADR model is based on the idea that the dispersion tensor, $\mathbf{D}$, can be decomposed into two parts: $\mathbf{D} = \mathbf{D}_{mech} + \mathbf{D}_{eff}$, where the mechanical dispersion coefficient $\mathbf{D}_{mech}$ only depends on the pore-scale velocity field, and the effective diffusion coefficient $\mathbf{D}_{eff}$ (used in Equation (11.7)) depends on both molecular diffusion and velocity. The longitudinal and transverse components, D_L and D_T, of $\mathbf{D}$ can be determined numerically or experimentally from, for example, a conservative tracer experiment (Bear, 1988). A number of empirical relationships for the dispersion tensor have been proposed, including (Coelho et al., 1997; Carvalho and Delgado, 1999; Olsson and Grathwohl, 2007)

$$D_i = \mathcal{D} + \mathcal{D}\alpha_i Pe^{1+\beta_i}, \quad i = L,T \tag{11.8}$$

where $\mathcal{D}$ is the effective molecular diffusion coefficient of a porous medium, $Pe = |\mathbf{u}| l / \mathcal{D}$ is the Peclet number, and $0 < \alpha_i < 0.5$ and $0 < \beta_i < 0.5$ are experimentally determined constants (Coelho et al., 1997; Carvalho and Delgado, 1999; Olsson and Grathwohl, 2007). Expanding Equation (11.8) in Taylor series in powers of β yields:

$$D_i = \mathcal{D} + \mathcal{D}\alpha_i Pe\left[1 + \beta_i \log Pe + \frac{\beta_i^2 (\log Pe)^2}{2!} + o\left(\beta_i^3\right)\right], \quad i = L,T. \tag{11.9}$$

Only the zero-order (in β_i) term in Equation (11.9) is independent of molecular diffusion and, hence, contributes to the mechanical dispersion. Therefore, we define components of the $\mathbf{D}_{mech}$ and $\mathbf{D}_{eff}$ tensors as:

$$D_{mech,i} = \mathcal{D}\alpha_i Pe \quad i = L,T \tag{11.10}$$

and

$$D_{eff,i} = \mathcal{D} + \mathcal{D}\alpha_i Pe\left[\beta_i \log Pe + \frac{\beta_i^2 (\log Pe)^2}{2!} + o\left(\beta_i^3\right)\right], \quad i = L,T. \tag{11.11}$$

Another commonly used model for the dispersion coefficient is (Bear, 1988):

$$D_i = \mathcal{D} + \lambda_i \mathcal{D} Pe, \quad i = L,T. \tag{11.12}$$

where λ_i are the principal components of the dispersivity tensor. It is evident that Equation (11.12) is a special case of Equations (11.10)-(11.11), which assumes that the effective diffusion coefficient is a scalar constant independent of flow conditions. To the contrary, our results below show a strong directional dependence of $\mathbf{D}_{eff}$ on Pe.

11.2 Pore and Darcy Models for Transport with Bimolecular Reactions

To demonstrate that the traditional advection-dispersion-reaction equations may fail to accurately predict reactive transport even in perfectly homogeneous porous media, Tartakovsky et al. (2009) considers isothermal steady-state flow of an incompressible fluid and reactive transport of solutes A, B, and C ($A + B \rightarrow C$) in a two-dimensional periodic porous medium made of impermeable non-overlapping, randomly arranged circular grains of random size (see top panel in Figure 11.2). The boundaries of the periodic unit cell of size L are shown in Figure 11.2 by dashed lines. On the pore scale, fluid flow is governed by the incompressible Navier-Stokes equations, and the reactive transport of A, B, and C is governed by the advection-diffusion equations

$$\frac{DA}{Dt} = D_m \nabla^2 A - kAB \tag{11.13}$$

$$\frac{DB}{Dt} = D_m \nabla^2 B - kAB \tag{11.14}$$

$$\frac{DC}{Dt} = D_m \nabla^2 C + kAB, \tag{11.15}$$

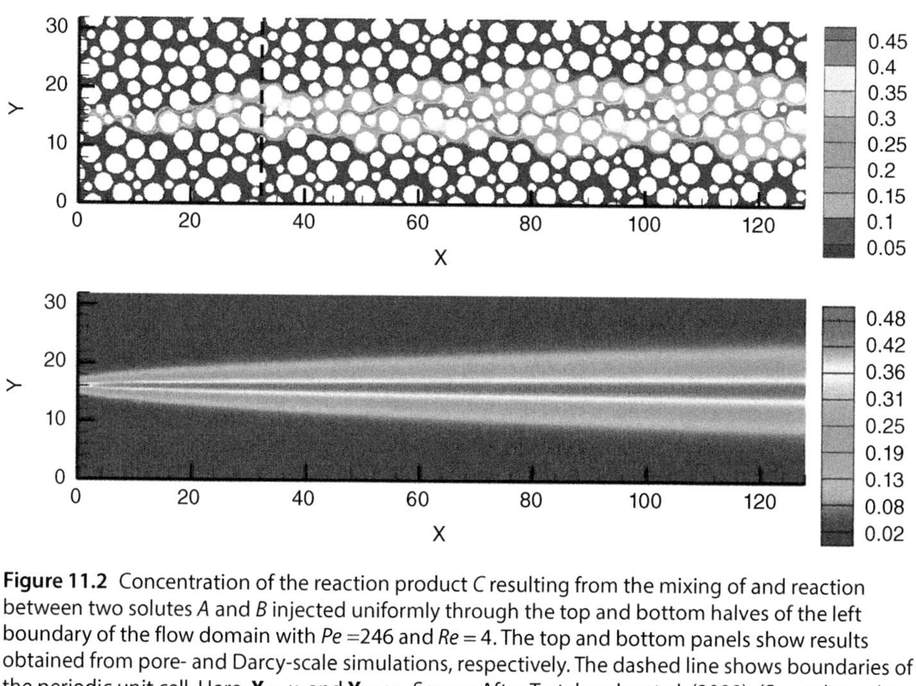

Figure 11.2 Concentration of the reaction product *C* resulting from the mixing of and reaction between two solutes *A* and *B* injected uniformly through the top and bottom halves of the left boundary of the flow domain with Pe =246 and Re = 4. The top and bottom panels show results obtained from pore- and Darcy-scale simulations, respectively. The dashed line shows boundaries of the periodic unit cell. Here, $\mathbf{X} = x_1$ and $\mathbf{Y} = x_2$. *Source:* After Tartakovsky et al. (2009). (See colour plate section for the colour representation of this figure.)

where $D/Dt = \partial/\partial t + \mathbf{v}\cdot\nabla$ and t is time. A and B are mass fractions of A and B normalized with the 'input' mass fractions, A_0 and B_0, C is the mass fraction of C normalized with $A_0 + B_0$, $\kappa = \kappa^* M^*$ is the normalized reaction rate with units of 1 over time, and M^* is the referenced molar concentration assumed to be equal to the input molar concentrations of A and B. The domain's external boundaries are located at $x_1 = 0$, $x_1 = L_1$, $x_2 = 0$, and $y_2 = L_2$. The flow is driven by a body force acting in the x_1 direction. At the horizontal boundaries located at $x_2 = 0$ and $x_2 = L_2$, normal diffusive transfer of A, B, and C is zero. At the vertical boundaries, the parallel injection of A and B into different halves of the domain is described by the boundary conditions:

$$A\left(x_2 < L_2/2\right) = A_0, \quad A\left(x_2 > L_2/2\right) = 0, \quad x_1 = 0, \tag{11.16}$$

$$B\left(x_2 < L_2/2\right) = 0, \quad B\left(x_2 > L_2/2\right) = B_0, \quad x_1 = 0, \tag{11.17}$$

$$C = 0, \quad x_1 = 0, \tag{11.18}$$

$$\partial A/\partial x_1 = 0, \ x_1 = L_1, \tag{11.19}$$

$$\partial B/\partial x_1 = 0, \ x_1 = L_1, \tag{11.20}$$

$$\partial C/\partial x_1 = 0, \ x_1 = L_1. \tag{11.21}$$

Initially, the domain is tracer-free:

$$A = 0 \ B = 0 \ C = 0, t = 0. \tag{11.22}$$

Pore-scale flow and transport equations are solved numerically using the Smoothed Particle Hydrodynamics (SPH) method (Tartakovsky et al., 2008b) for Reynolds number $Re = \rho v l/\mu = 4$ (v is the average pore velocity) and several Peclet and Damkohler numbers. Figure 11.2 (top panel) shows the distribution of the reaction product C resulting from a pore-scale simulation with Pe =246 and Da =11500. Figure 11.3 shows the mass of the reaction product C, normalized by the final (steady-state) mass of C as a function of time, obtained from the pore-scale simulations for several Pe and Da.

The results of the pore-scale simulations are used in Tartakovsky et al. (2009) to test the accuracy of the corresponding advection-dispersion-reaction equations:

$$\frac{D\bar{A}}{Dt} = \mathbf{D}\nabla^2 \bar{A} - k\bar{A}\times\bar{B}, \tag{11.23}$$

$$\frac{D\bar{B}}{Dt} = \mathbf{D}\nabla^2 \bar{B} - k\bar{A}\times\bar{B}, \tag{11.24}$$

$$\frac{D\bar{C}}{Dt} = \mathbf{D}\nabla^2 \bar{C} + k\bar{A}\times\bar{B}, \tag{11.25}$$

where $D/Dt = \partial/\partial t + \mathbf{u}\cdot\nabla$. Advective velocity $\mathbf{u} = \int_{\Omega_p} v dx/V_p$ (V_p is the total volume of pores), determined from the pore-scale simulations. The longitudinal and transverse components of the dispersion tensor $\mathbf{D}$ are calculated from the pore-scale conservative tracer simulations in the same porous medium. The advection-dispersion-reaction equations were solved for several Pe and Da numbers. Table 11.2 summarizes the parameters used in the simulations.

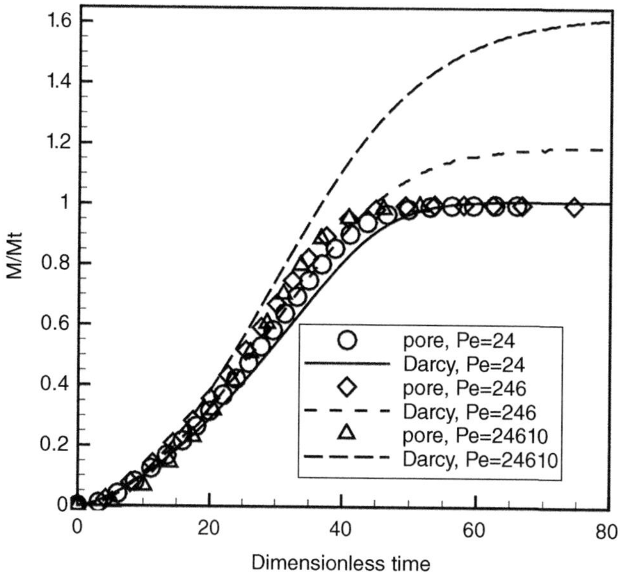

Figure 11.3 Dimensionless mass of reaction product C versus dimensionless time $t\langle v\rangle/L$ obtained from the pore- and macro-scale simulations for several Peclet numbers and $Da = 11500$. The mass is normalized with respect to the asymptotic (steady-state) value of mass obtained from the pore-scale simulations. *Source:* After Tartakovsky et al. (2009).

Figure 11.2 compares the distributions of C obtained from Darcy- and pore-scale simulations using Pe =246 and Da =11500. Figure 11.3 compares the normalized mass of C as a function of time obtained from the Darcy- and pore-scale simulations for several values of Pe and Da. For the simulated time, the Darcy-scale model overestimates the steady-state mass of C by 1% for small Peclet numbers (Pe = 24). The error increases to 18% for Pe =246 and to 52 for Pe =24610. However, varying the Damkohler number does not have a significant effect on the error for $Pe = 24$, but it increases the error by 10% for $Pe = 24610$. Table 11.2 summarizes the differences between the pore- and Darcy-scale solutions for different Peclet and Damkohler numbers. The degree of solute mixing increases with the distance from the $x = 0$ boundary where the solutes are injected. Figure 11.4 shows that error decreases with the domain length for all considered Pe and Da, as in the longer domains, solutes have more time to diffuse/dilute. On the other hand, we observe relatively large errors for $Pe = 246$ and 24610 even in the largest modeled domain. For a conservative tracer, Cao and Kitanidis (1998) showed that at the boundary of a plume, the solute never is completely diluted. Therefore, it should be expected that even in significantly larger domains the Darcy-scale model will overestimate the extent of reactions.

The degree of mixing can be measured by the so-called dilution index $E(t)$ (Kitanidis, 1994):

$$E(t) = \exp\left[-\int_{\Omega_p} P(x,t)\ln\big(P(x,t)\big)dx\right], \tag{11.26}$$

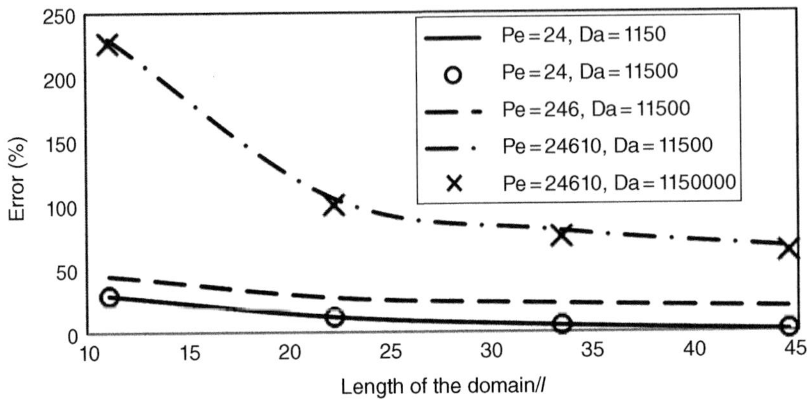

Figure 11.4 Error (the percentage of mass of the reaction product C overestimated by the Darcy-scale model relative to the prediction of the pore-scale model in the domain of size L_x) versus L_x normalized with the average soil grain diameter for several *Pe* and *Da* numbers. *Source:* After Tartakovsky et al. (2009).

where P is concentration, normalized by the total tracer mass, and Ω_p is the pore domain. To quantify mixing in the considered reactive flows, Tartakovsky et al. (2009) simulated transport of the conservative tracer in the same porous domain and for the same Pe as used in the reactive transport simulations. The domain size in these simulations is much larger than the the tracer plume size, so that the domain could be considered as infinite. Under such conditions, Kitanidis (1994) proved that the maximum value of $E(t)$ is given by

$$E_{max}(t) = 2\phi\pi \exp(1)(S_{xx}(t)S_{yy}(t) - S_{xy}^2(t))^{1/2}, \tag{11.27}$$

where ϕ is the porosity of the porous medium and $S_{xx}(t)$, $S_{yy}(t)$, and $S_{xy}(t)$ are the second plume moments computed from the pore-scale simulations of the tracer plume as $S_{ll}(t) = \int (S_l - x_l)^2 dM(\mathbf{x}, t)/M$. Here, M is the total solute mass and S_l is the first moment of the plume in the x_l direction ($l = 1,2$): $S_l(t) = \int x_l dM(\mathbf{x}, t)/M$. Then, the principal components of the dispersion tensor $D_{ll}, l = 1,2$, are given by

$$\lim_{t\to\infty} \frac{dS_{ll}}{dt} = 2D_{ll}. \tag{11.28}$$

Kitanidis (1994) also demonstrated that Equation (11.27) is satisfied by the solution of the advection-dispersion equation, with the components of the dispersion tensor found from Equation (11.28). The degree to which the continuum description overestimates the plume's dilution is given by the normalized dilution index $E(t)/E_{max}(t)$. It follows from the theoretical analysis in Kitanidis (1994) that $E(t)/E_{max}(t) \leq 1$, that is, the advection-dispersion equation generally overestimates the degree of pore-scale dilution.

Figure 11.5 shows the normalized dilution index as a function of time for various values of *Pe*. Initial concentration of solute within the plume is uniform, resulting in $E(t)/E_{max}(t) \approx 0.75$. This shows that a plume with the uniform concentration distribution has a high – but not maximum – degree of dilution as $E_{max}(t)$ corresponds to the

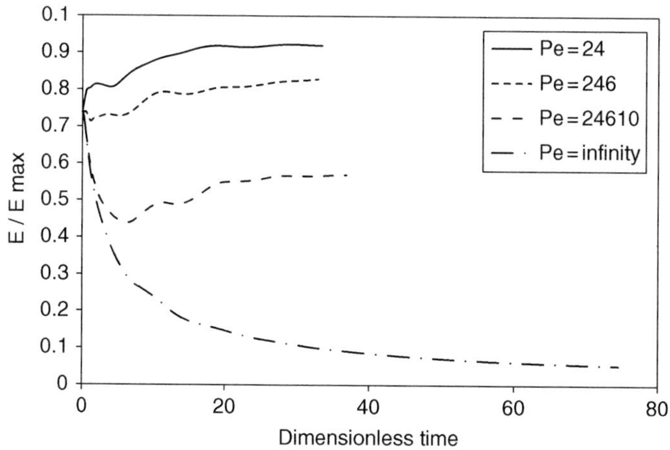

Figure 11.5 Normalized dilution index $E(t)/E_{max}(t)$ versus dimensionless time $t\langle v\rangle/L$ for different *Pe*. *Re* = 4. *Source:* After Tartakovsky et al. (2009).

concentration profile following the Gaussian distribution. Early in the process, advection distorts the plume, leading to the decrease in $E(t)/E_{max}(t)$. Later, the solute has more time to diffuse, and $E(t)/E_{max}(t)$ increases with time. At all times, $E(t)/E_{max}(t)$ decreases with increasing *Pe*. This means that the solute in the plume is less dilute for large *Pe*. For infinitely large *Pe*, $E(t)/E_{max}(t)$ monotonically decreases with time. Our results quantitatively agree with Kapoor and Kitanidis (1998), who studied Darcy-scale advection of a conservative plume in a randomly heterogeneous porous media.

For reactive transport, our results show that the error in the Darcy advection-dispersion-reaction model increases with the decreasing dilution index, where we define error as the ratio between masses of the reaction product obtained from the pore- and Darcy-scale simulations.

Table 11.1 shows the ratio τ_D^p/τ_A^D used in the simulations for different *Pe* numbers, where $\tau_D^p = l^2/d_m$ is the (pore-scale) diffusion time and $\tau_A^D = L/\langle v\rangle$ is the advection time. When the diffusion time τ_D^p is of the same order or smaller than τ_A^D, solute inside the pores is going to be well diluted on the Darcy advection timescale. Otherwise, the pore-scale mixing may not be complete, even on relevant Darcy time and length scales. This may affect the accuracy of the Darcy advection-dispersion model. The ratio τ_D^p/τ_A^D is of the order of unity for $Pe = 24$ and much larger than one for $Pe = 246$ and 24610.

Two primary conclusions can be drawn from the comparison of pore- and Darcy-scale reactive transport simulations: (1) classical advection-dispersion equations with constant dispersion tensor accurately describe reactive transport with mixing-controlled reactions for moderate Peclet numbers and large normalized times t/τ_D^p; for large *Pe* and/or small normalized times, the advection-dispersion equations overestimate the extent of reaction; and (2) the normalized dilution index computed from pore-scale simulations can be used to predict the accuracy of the advection-dispersion-reaction model – its accuracy increasing with an increasing dilution index.

In the following section, we describe the LADR model and demonstrate that separate treatment of advective and diffusion mixing is more accurate for modeling multicomponent reactive transport.

Table 11.1 Parameters *Pe*, *Da*, and a ratio of (pore-scale) diffusion time $\tau_D^P = l^2/d_m$ to Darcy-scale advection time $\tau_A^D = L/\langle v\rangle$ used in the simulations. Diagonal components of the dispersion tensor, D_{11} and D_{22}, are estimated from the pore-scale conservative transport simulations and used in the Darcy-scale simulations. The values of normalized dilution index E/E_{max} are found from the conservative tracer simulations at normalized time $t/\tau_D^P = 1.34$ for $Pe = 24$, $t/\tau_D^P = 0.13$ for $Pe = 246$, and $t/\tau_D^P = 0.0015$ for $Pe = 24610$. The error refers to the percentage of mass of the reaction product C overestimated by the Darcy-scale model relative to the prediction of the pore-scale model.

Pe	*Da*	$\frac{D_{11}}{\langle v\rangle d_m}$	$\frac{D_{22}}{\langle v\rangle d_m}$	$\frac{\tau_D^D}{\tau_A^P}$	Error (%)	E/E_{max}
24	1150	0.03	0.01	2.2	1.1	0.92
24	11500	0.03	0.01	2.2	1.1	0.92
246	11500	0.05	0.003	22.1	19.1	0.83
24610	11500	0.11	0.0001	2211.1	62.8	0.57
24610	1150000	0.11	0.0001	2211.1	66.6	0.57

Source: After Tartakovsky et al. (2009).

11.3 Langevin Advection-Diffusion-Reaction Model

The Langevin model is obtained by adding a random force (per unit mass) into the Darcy-scale momentum conservation equation (11.6):

$$\frac{d\mathbf{U}}{dt} = -\frac{\nabla P}{\rho} + g - \gamma\mathbf{U} + \sqrt{|\mathbf{U}|}\xi, \tag{11.29}$$

where $\mathbf{U}(\mathbf{x}, \mathbf{t}) = \langle \mathbf{U}(\mathbf{x}, \mathrm{t})\rangle + \tilde{\mathbf{U}}(\mathbf{x}, \mathrm{t})$ and $P(\mathbf{x}, t) = \langle P(\mathbf{x}, t)\rangle + \tilde{P}(\mathbf{x}, t)$ are the random velocity and pressure. Here, $|\ |$ denotes a magnitude of a vector, $\langle\ \rangle$ denotes the ensemble mean of the random variables, and $\tilde{}$ denotes random fluctuations around the mean.

The random vector, $\xi(\mathbf{x}, t) = (\xi_1(\mathbf{x}, t), \xi_2(\mathbf{x}, t), \xi_3(\mathbf{x}, t))^T$ has zero ensemble mean, $\langle \xi_l(t)\rangle = 0$, and the covariance function:

$$\langle \xi_l(\mathbf{x}, t)\xi_l(\mathbf{x}', t')\rangle = \Gamma_l^2 \delta(\mathbf{x} - \mathbf{x}')\delta(t - t'), \tag{11.30}$$

where Γ_l is a constant and δ is the Dirac delta function. The random velocity satisfies the continuity equation:

$$\frac{d\rho}{dt} = -\rho\nabla \cdot \mathbf{U}. \tag{11.31}$$

Solute transport is described by the stochastic advection-diffusion equation (11.7). The Lagrangian displacements are defined as:

$$\frac{d\mathbf{X}^L(t)}{dt} = \mathbf{U}\left(\mathbf{X}_0^L, t\right), \quad \mathbf{X}^L(0) = \mathbf{X}_0^L, \tag{11.32}$$

where,

$$\mathbf{X}^L(t) = \left\langle \mathbf{X}^L(t) \right\rangle + \mathbf{X}^L(t). \tag{11.33}$$

Assuming that average flow is in the x_1 direction, the longitudinal and transverse components, D_L^M and D_T^M, of the *mechanical* dispersion tensor could be found from the Einstein relationship (Van Kampen, 2001):

$$\lim_{t\to\infty} \frac{d\left\langle \tilde{X}_1^L(t)\tilde{X}_1^L \right\rangle(t)}{dt} = 2D_L^M, \tag{11.34}$$

$$\lim_{t\to\infty} \frac{d\left\langle \tilde{X}_l^L(t)\tilde{X}_l^L \right\rangle(t)}{dt} = 2D_T^M, \qquad l = 2,3. \tag{11.35}$$

For the dispersion tensor defined according to Equation (11.12), the variance of displacements can be related to the dispersivity tensor as:

$$\lim_{t\to\infty} \frac{d\left\langle \tilde{X}_1^L(t)\tilde{X}_1^L \right\rangle(t)}{dt} = 2\alpha_L \left|\left\langle \mathbf{U}^\infty \right\rangle\right|, \tag{11.36}$$

$$\lim_{t\to\infty} \frac{d\left\langle \tilde{X}_l^L(t)\tilde{X}_l^L \right\rangle(t)}{dt} = 2\alpha_T \left|\left\langle \mathbf{U}^\infty \right\rangle\right|, \qquad l = 2,3, \tag{11.37}$$

where $\left|\left\langle \mathbf{U}^\infty \right\rangle\right|$ is the steady-state value of the average velocity. For given values of Γ_l ($l = 1,2,3$), the dispersivity tensor, α, can be found numerically by solving Langevin equations (11.29)-(11.31) and using the relationships (11.36)–(11.37). Alternatively, for a given α, a model calibration is required to determine Γ_l. A purely mechanical dispersion is difficult to measure directly in laboratory or field experiments because of the ever-present molecular diffusion. In a porous medium with known pore geometry, the mechanical dispersion coefficient or dispersivity tensor can be found by solving the Navier-Stokes equations for the pore-scale fluid velocity. Here, the Langevin model is parameterized for the two-dimensional porous medium used in the pore- and Darcy-scale simulations described in the preceding section.

11.4 Parameterization of the Stochastic Model

In the pore-scale conservative and reactive flow simulations, the flow is assumed to be driven by the body force and is subject to periodic boundary conditions. The fluid is discretized with N_p Lagrangian particles, and the solution of the (two-dimensional) Navier-Stokes equations is obtained using the SPH method (Tartakovsky et al., 2007). The average pore velocity $\overline{v}$ is determined from the SPH simulations as:

$$\overline{v} = \frac{1}{N_p} \sum_i^{N_p} v_{1,i}, \tag{11.38}$$

where the x_1 direction coincides with the average flow direction.

Then, the friction coefficient γ is calculated as:

$$\gamma = \frac{g}{\bar{v}}, \tag{11.39}$$

where g is the magnitude of the body force per unit mass.

The coefficients of mechanical dispersion and dispersivities of the porous medium are found from the pore-scale simulation (Figure 11.6) using the relationships:

$$\lim_{t\to\infty} \frac{\partial \sigma_{x_1}^2(t)}{\partial t} = \lim_{t\to\infty} \frac{\partial}{\partial t}\left[\frac{1}{N_p}\sum_i^{N_p} \langle \tilde{x}_{1,i}(t)\tilde{x}_{1,i}(t)\rangle\right] = 2D_L^M = 2\alpha_L \bar{v}, \tag{11.40}$$

and

$$\lim_{t\to\infty} \frac{\partial \sigma_{x_2}^2(t)}{\partial t} = \lim_{t\to\infty} \frac{\partial}{\partial t}\left[\frac{1}{N_p}\sum_i^{N_p} \langle \tilde{x}_{2,i}(t)\tilde{x}_{2,i}(t)\rangle\right] = 2D_T^M = 2\alpha_T \bar{v}, \tag{11.41}$$

where $\sigma_{x_1}^2$ and $\sigma_{x_2}^2$ are the variances of particle displacements in the x_1 and x_2 directions and

$$\tilde{\mathbf{x}}_i(t) = \mathbf{x}_i(t) - \bar{\mathbf{x}}_i(t), \tag{11.42}$$

where $\mathbf{x}_i(t)$ is the position of the fluid particle i ($i = 1, N_p$) at the time t found from the SPH solution of the Navier-Stokes equations. The average particle position $\bar{\mathbf{x}}_i(t)$ is given by

$$\bar{\mathbf{x}}_i(t) = \mathbf{x}_i^0 + \frac{1}{N_p}\sum_j^{N_p}\left(\mathbf{x}_j(t) - \mathbf{x}_j^0\right), \tag{11.43}$$

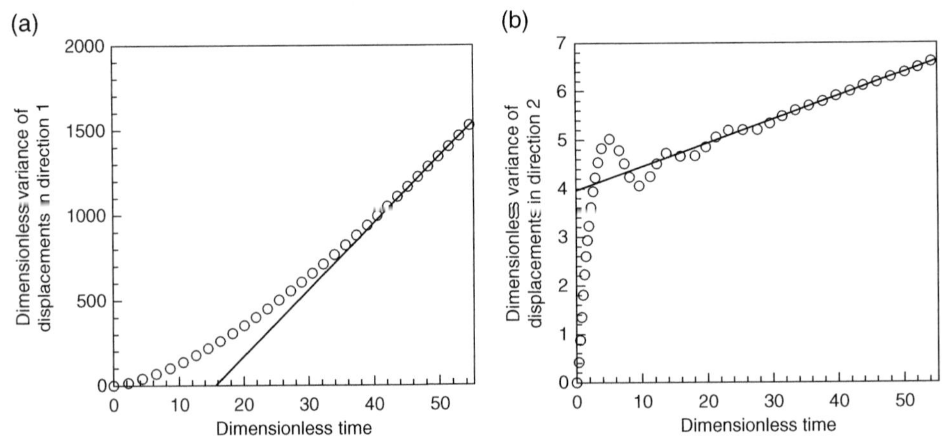

Figure 11.6 Open circles represent dimensionless variance of displacements, $\sigma_{x_i}^2/d^2$, versus dimensionless time, $t\bar{v}/d$, obtained from the pore-scale simulation: (a) the variance in the x_1 direction (a direction of average flow); and (b) variance in the x_2 direction (a direction perpendicular to the direction of average flow). Solid lines are obtained as a best fit to the linear parts of the curves. Components of the dimensionless mechanical dispersion tensor, $2D_L^M/(d\bar{v})$ and $2D_T^M/(d\bar{v})$, are given by the slopes of the solid lines. After Tartakovsky (2010).

where $\mathbf{x}_i^0$ is the position of particle i at time zero, and $\mathbf{x}_i(t)$ is the deviation of the position of particle i.

Parameters $\Gamma_1 = 13.3$ and $\Gamma_2 = 0.32$, satisfying conditions (11.36) and (11.37), are found from the Langevin model calibration.

The effective diffusion tensor in Equation (11.7) is found by matching the longitudinal and transverse components of the dispersion coefficient obtained from the pore-scale simulations of the conservative tracer. This is done by simulating advection of a conservative tracer with the stochastic model and matching the resulting total dispersion coefficient with the coefficient obtained from the corresponding pore-scale simulations.

11.5 The Langevin Model for Multicomponent Reactive Transport

The Langevin model's accuracy is tested by simulating the reactive transport of A, B, and C species in the homogeneous porous medium and comparing it with the pore- and Darcy-scale simulations. The stochastic advection-diffusion equations for A, B, and C take the form:

$$\frac{dA}{dt} = \frac{1}{\rho}\nabla\cdot\left(\rho D_{eff}\nabla A\right) - \kappa AB, \tag{11.44}$$

$$\frac{dB}{dt} = \frac{1}{\rho}\nabla\cdot\left(\rho D_{eff}\nabla B\right) - \kappa AB, \tag{11.45}$$

$$\frac{dC}{dt} = \frac{1}{\rho}\nabla\cdot\left(\rho D_{eff}\nabla C\right) + \kappa AB. \tag{11.46}$$

As before, A, B, and C are the normalized mass fractions of A, B, and C. These equations are solved subject to the same boundary and initial conditions as in the pore- and Darcy-scale simulations.

In Tartakovsky (2010), the Langevin flow and transport equations are solved using the SPH method. In SPH, a filed $A(\mathbf{X})$ can be approximated as:

$$A_I(\mathbf{X}) = \int A(\mathbf{X}) W\left(\left|\mathbf{G}\cdot(\mathbf{X}-\mathbf{X}')\right|\right) d\mathbf{X}', \tag{11.47}$$

where W is the SPH weighting function (a smooth, bell-shaped function) satisfying the normalization condition:

$$\int W\left(\left|\mathbf{G}\cdot(\mathbf{X}-\mathbf{X}')\right|\right) d\mathbf{X}' = 1. \tag{11.48}$$

The form of the second-rank tensor G depends on the dimensionality of the space. In two spatial dimensions, G is given by

$$\mathbf{G} = \begin{pmatrix} h_1^{-1}\cos^2\psi + h_2^{-1}\sin^2\psi & \left(h_1^{-1} - h_2^{-1}\right)\cos\psi\sin\psi \\ \left(h_1^{-1} - h_2^{-1}\right)\cos\psi\sin\psi & h_1^{-1}\sin^2\psi + h_2^{-1}\cos^2\psi \end{pmatrix}, \tag{11.49}$$

where h_1 and h_2 are the semi-major and semi-minor axes of the ellipse and ψ is the angle between the x_1 coordinate direction and semi-major axis. If the semi-major axis of the ellipse coincides with the x_1 direction, then **G** reduces to:

$$\mathbf{G}=\begin{pmatrix} h_1^{-1} & 0 \\ 0 & h_2^{-1} \end{pmatrix}. \tag{11.50}$$

In the $h_1 \to 0$ limit, W is required to reduce to the Dirac delta function. In this limit the SPH interpolation scheme becomes exact, i.e., $A_I \equiv A$. To simplify notation, in the following, we will drop the subscript I.

Next, fluids occupying a computational domain are represented by a set of N particles with positions X_i $(i=1,N)$, and the integral in Equation (11.47) is discretized as:

$$A_i=\sum_j A_j / n_j W\left(\left|G\cdot\left(\mathbf{X}_i-\mathbf{X}_j\right)\right|\right), \tag{11.51}$$

where $A_i = A(\mathbf{X}_i)$, $n_i = \rho_i / m_i$ is the particle number density, and m_i is the mass of particle i. Using this SPH interpolation scheme (11.51), the Langevin equation (11.29) can be discretized as (Welton and Pope, 1997):

$$\frac{d\mathbf{U}_i}{dt}=\mathbf{F}_i^A+\mathbf{F}_i^B, \tag{11.52}$$

where

$$\mathbf{F}_i^A=-\frac{1}{m_i}\sum_j\left(\frac{P_j}{n_j^2}+\frac{P_i}{n_i^2}\right)\nabla_i W\left(\left|\mathbf{G}\cdot\mathbf{X}_{ij}\right|\right)+\mathbf{g}-\gamma_i\mathbf{U}_i, \tag{11.53}$$

$$\mathbf{F}_i^B=\sqrt{\left|\langle\mathbf{U}\rangle\right|}\xi_i, \tag{11.54}$$

where $\mathbf{X}_{ij}=\mathbf{X}_i-\mathbf{X}_j$, $\mathbf{U}_i$ is the velocity of particle i, $\left|\langle\mathbf{U}\rangle\right|=$ is the magnitude of the average velocity, and m_i is the mass of particle i. The evolution of particle density, n_i, is from Equation (11.51) with $A=n$:

$$n_i-\sum_j W\left(\left|\mathbf{G}\cdot\mathbf{X}_{ij}\right|\right). \tag{11.55}$$

The system of equations is closed by the equation of state:

$$P_i=P_{eq}n_i / n_{eq}, \tag{11.56}$$

where n_{eq} is the average number of particles per unit volume and P_{eq} is the pressure of the particles in the equilibrium state (state where, in the absence of body and random forces, the total force acting on each particle is zero).

The SPH discretization of the diffusion equations (11.44)-(11.46) is given by

$$\frac{dA_i}{dt}=\frac{D_{eff}}{m_i}\sum_j\frac{\left(m_i n_i+m_j n_j\right)A_{ij}}{n_i n_j \mathbf{X}_{ij}^2}\mathbf{X}_{ij}\cdot\nabla_i W\left(\left|\mathbf{G}\cdot\mathbf{X}_{ij}\right|\right)-\kappa A_i B_i, \tag{11.57}$$

$$\frac{dB_i}{dt} = \frac{D_{eff}}{m_i} \sum_j \frac{(m_i n_i + m_j n_j) B_{ij}}{n_i n_j \mathbf{X}_{ij}^2} \mathbf{X}_{ij} \cdot \nabla_i W\left(\left|\mathbf{G} \cdot \mathbf{X}_{ij}\right|\right) - \kappa A_i B_i, \tag{11.58}$$

and

$$\frac{dC_i}{dt} = \frac{D_{eff}}{m_i} \sum_j \frac{(m_i n_i + m_j n_j) C_{ij}}{n_i n_j \mathbf{X}_{ij}^2} \mathbf{X}_{ij} \cdot \nabla_i W\left(\left|\mathbf{G} \cdot \mathbf{X}_{ij}\right|\right) + \kappa A_i B_i, \tag{11.59}$$

where $a_{ij} = a_i - a_j$ $(a = A, B, C)$ and A_i, B_i, and C_i are the mass fractions of solutes A, B, and C carried by particle i. For simplicity, all SPH particles have mass m_0 that is independent of the solute concentration. The assumption of a highly dilute solution can be easily relaxed (Tartakovsky et al., 2008a).

Different forms of $\mathbf{G} = \mathbf{G}(h_1, h_2)$ can be used for discretizing different terms in the SPH equations. In Tartakovsky (2010), an isotropic kernel W ($h_1 = h_2 = 1$ (in model units of length)) is used in the SPH diffusion equations (11.57)-(11.59). Due to the anisotropic white noise ($\Gamma_1 = 13.3$ and $\Gamma_2 = 0.32$) in the SPH Langevin equation (11.52)-(11.55), the anisotropic kernel W with $h_1 = 10$ and $h_2 = 1$ (in model length units) is employed.

Figure 11.7 shows the mass fraction of the solute C, a product of the irreversible reaction $A + B \rightarrow C$, obtained from the Langevin model for $Pe = 246$ and $Da = 11500$, while

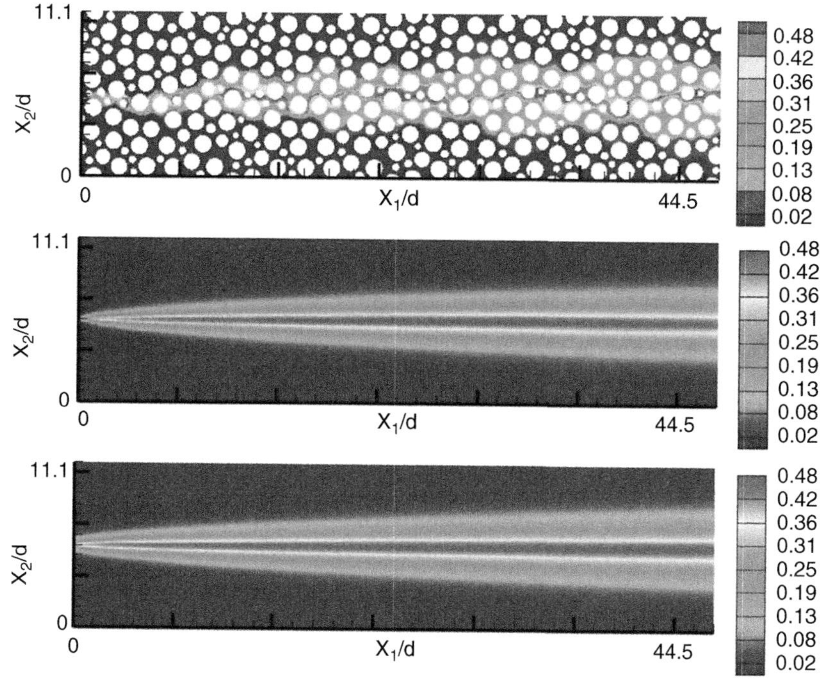

Figure 11.7 Dimensionless concentration of solute C, a product of the reaction between solutes A and B injected uniformly through the top and bottom halves of the left boundary of the flow domain, for $Pe = 246$. The top, middle, and bottom panels show results obtained from pore-scale, Langevin, and Darcy simulations, respectively. After Tartakovsky (2010).

Figure 11.8 shows a distribution of C for $Pe = 24610$ and $Da = 115000$. The Reynolds number was set to $Re = 4$. For comparison purposes, Figures 11.7 and 11.8 also show pore-scale and Darcy simulations of the same transport phenomena in the same porous medium. Table 11.2 provides a summary of all simulations conducted in this study.

The pore-scale simulations show a 'tortuous' mixing zone. The Darcy model predicts a mixing zone with a smoothly varying C in the x_2 (Y) direction. On the other hand, the stochastic simulations produced a mixing zone with highly heterogeneous distribution of the solute C, which is clearly seen in the simulation with large Pe (Figure 11.8).

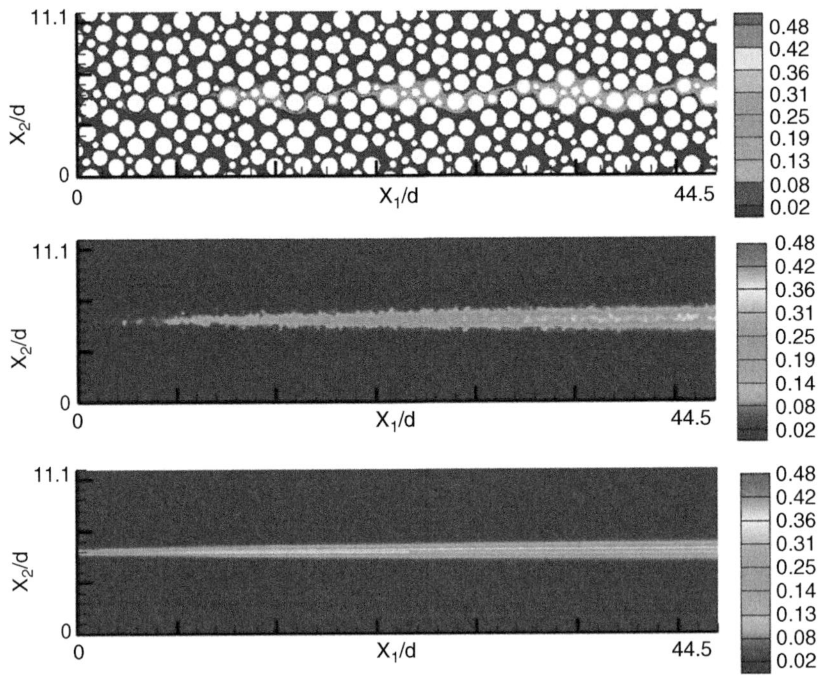

Figure 11.8 Concentration of solute C, a product of the reaction between solutes A and B injected uniformly through the top and bottom halves of the left boundary of the flow domain, for $Pe = 24610$. The top, middle, and bottom panels show results obtained from pore-scale, Langevin, and Darcy simulations, respectively. After Tartakovsky (2010)

Table 11.2 Parameters for the Langevin and Darcy transport equations. After Tartakovsky (2010).

Pe	Da	D_m	D_L	D_T	D_{eff}	Langevin Error (%)	Darcy Error (%)
24.6	1150	6.25×10^{-4}	0.008	0.003	0.0027	3	1
24.6	11500	6.25×10^{-4}	0.008	0.003	0.0027	3	1
246.1	11500	6.25×10^{-5}	0.01	0.0007	6.25×10^{-4}	4	19
24610	115000	6.25×10^{-7}	0.03	0.00003	6.25×10^{-6}	18	52

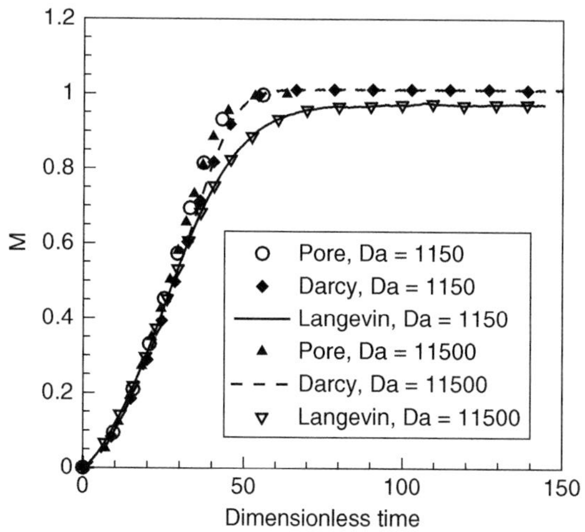

Figure 11.9 Dimensionless mass of *C* versus dimensionless time, $t\bar{v}/d$, obtained from the pore-scale, Langevin, and Darcy simulations for $Pe = 24$ and different *Da*. The mass is normalized with respect to the asymptotic value of mass obtained from the pore-scale simulation with corresponding *Da*. After Tartakovsky (2010).

Figures 11.9 and 11.10 show the dimensional mass of C obtained from pore-scale, Langevin and Darcy simulations for several *Pe* and *Da*:

$$M_I(t) = \frac{\int\limits_{\Omega_I} C_I(\mathbf{x}, t)\rho(\mathbf{x}, t)d\mathbf{x}}{\int\limits_{\Omega_P} C_P(\mathbf{x}, t^*)\rho(\mathbf{x}, t^*)d\mathbf{x}}, \tag{11.60}$$

where subscript $I = P, L, D$ denotes quantities obtained from pore-scale, Langevin, and Darcy models, respectively. Ω_I is a computational domain of the corresponding model, and t^* is the time when M_P reaches a steady-state value. Comparing three models shows that for small Peclet numbers (Pe = 24, Figure 11.9), both Langevin and Darcy models predict mass of *C* within 3% of the mass of *C* obtained from the pore-scale model – regardless of the magnitude of the Damkohler number. For small *Pe*, the Darcy-scale model appears to be slightly more accurate (1% error) than the Langevin model (3% error). Errors in the Darcy model increase rapidly with increasing *Pe* (Figure 11.10). For $Pe = 246$, the relative error in the Darcy model becomes 19%, while the error in the Langevin model stays within 4%. For $Pe = 24610$, the relative error in the Darcy model increases to 52%. For $Pe = 24610$, the Langevin model remains more accurate than the Darcy model with a relative error of 19%. Table 11.1 lists the errors in the Langevin and Darcy models for different Peclet and Damkohler numbers. These results show that the separate treatment of mechanical dispersion and molecular and Taylor-type mixing is superior to the effective diffusion (Darcy) model, especially for mixing-controlled reactions with high Peclet numbers.

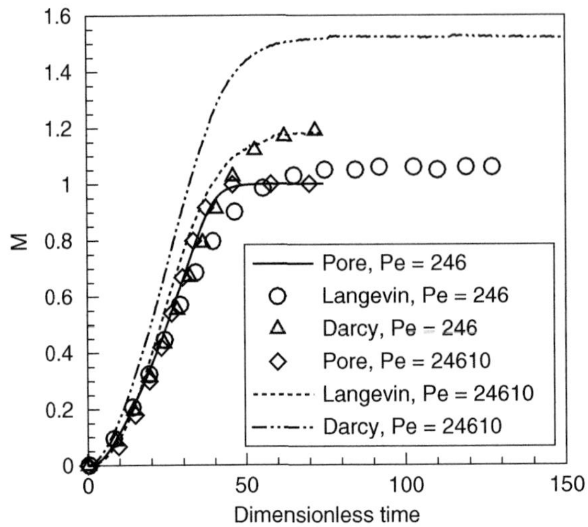

Figure 11.10 Dimensionless mass of *C* versus dimensionless time, $t\langle v\rangle/d$, obtained from the pore-scale, Langevin, and Darcy simulations for Peclet numbers *Pe* = 246 and 24610. The mass is normalized with respect to the asymptotic value of mass obtained from the pore-scale simulation with corresponding *Pe*. After Tartakovsky (2010).

11.6 Rayleigh-Taylor Instability

Tartakovsky et al. (2008a) considered the effect of the Langevin treatment of mixing on the Rayleigh-Taylor (RT) instability resulting from a more dense fluid displacing a less dense fluid under the action of gravity. For simplicity, they assumed that the density ρ and viscosity μ of a fluid depend linearly on the solute concentration: $\rho=\rho_0+k^{\rho}C$ and $\mu=\mu_0+k^{\mu}C$, where ρ_0 and μ_0 are the solvent's density and viscosity. The deterministic advection-dispersion model consists of equations (11.4)–(11.6) with $i=1$ and $R_1=0$. The dispersion tensor is taken as $D=D_{eff}+\alpha|\mathbf{u}|$ with $\alpha=0.02$. In the Langevin advection-diffusion model, transport is described by Equation (11.7) (with $i=1$ and $R_1=0$) and Equations (11.29)-(11.31) with $\Gamma=0.046$. Transport is simulated in a two-dimensional domain with impermeable boundaries. The initial solute concentration is set to C_0 in the upper half of the domain and zero in the bottom half. The flow is driven by the gravitational force acting downward. Snapshots of concentrations obtained from the advection-dispersion and Langevin advection-diffusion models are shown in Figure 11.11. Both models predict the front becoming unstable with a single perturbation growing with time. Figure 11.12 shows the position of the front (tip of the iso-surface $C/C_0=0.5$) as a function of time obtained from the two models. The front predicted by the Langevin model propagates significantly faster than the one obtained by the advection-dispersion model. This is because the front velocity is proportional to the concentration gradient across the front, and the concentration gradient is much smaller in the advection-dispersion model simulation, as evident from Figure 11.11.

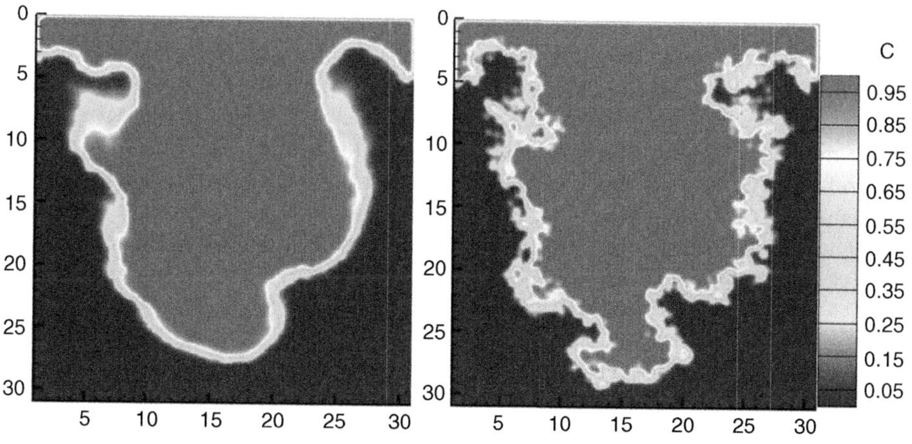

Figure 11.11 Deterministic (left) and stochastic (right) predictions of the Rayleigh-Taylor density-driven instability. (See colour plate section for the colour representation of this figure.)

Figure 11.12 Position of the front (tip of the iso-surface $C / C_0 = 0.5$) as a function of time predicted by the two models.

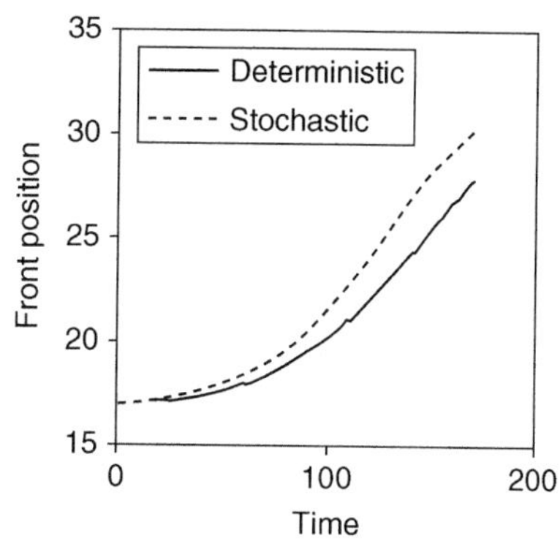

11.7 Summary and Conclusions

Multicomponent reactive transport in porous media is a challenging problem because of the complex interplay between diffusive and advective mixing and reactions. Under certain conditions, the traditional Darcy-scale advection-dispersion model with constant dispersion coefficient may significantly overestimate the extent of reactions, due to overestimating mixing by treating it as an effective Fickian process, leading to inaccurate predictions of concentration-gradient driven flows. Here, we reviewed an alternative Langevin advection-diffusion model, originally proposed in Tartakovsy et al. (2008a) and Tartakovsky (2010). In the model, fluid flow is governed by a Langevin equation, leading to the mechanical dispersion of a solute. Solute mixing due to molecular diffusion and velocity variations within individual pores and pore throats (Taylor dispersion) is governed by the advection-diffusion equation with the stochastic advective velocity found from the Langevin equation.

The comparison study has shown that separate treatment of the advective and diffusive mixing in the Langevin model results in a more accurate prediction of the multicomponent reactive transport. For the considered A + B→C system, the Darcy transport model was shown to overestimate the mass of C, with the error increasing from 1 % for $Pe = 24$ to 19 % for $Pe = 246$ and 52 % for $Pe = 24610$. The Langevin model remains accurate within 4 % for Pe values as large as 246. For $Pe = 24610$, the error in the Langevin model increases to 18 %, which is significantly smaller than the error in the Darcy model for the same Pe. In the case of the concentration-gradient-driven Rayleigh-Taylor instability, our results show that the advection-dispersion model underestimates the concentration gradient across the front separating two miscible fluids and the rate of the Rayleigh-Taylor instability growth.

Acknowledgement

This research was supported by the Advanced Scientific Computing Research Program of the U.S. Department of Energy (DOE) Office of Science. Pacific Northwest National Laboratory is operated by Battelle for the DOE under Contract DE-AC05-76RL01830.

References

Bear, J. (1988) *Dynamics of fluids in porous media*, New York: Dover.

Benson, D. and Meerschaert, M. (2009). Simulation of chemical reaction via particle tracking: Diffusion-limited versus thermodynamic rate-limited regimes. *Water Resources Research*, **44** (12): W12 201.

Bolster, D., Benson, D., Borgne, T.L., and Dentz, M. (2010). Anomalous mixing and reaction induced by superdiffusive nonlocal transport. *Physical Review E* **82**: 021 119.

Cao, J. and Kitanidis, P. (1998). Pore-scale dilution of conservative solutes: An example. *Water Resources Research*, **34** (8): 1941–1949.

Carvalho, J. and Delgado, J. (1999). Mass transfer from a large sphere buried in a packed bed along which liquid flows. *Chemical Engineering Sciences* **54**: 1121–1129.

Cirpka, O., Frind, E., and Helmig, R. (1999). Numerical simulation of biodegradation controlled by transverse mixing. *Journal of Contaminant Hydrology*, **40**: 159–182.

Coelho, D., Thovert, J.F., and Adler, P.M. (1997). Geometrical and transport properties of random packings of spheres and aspherical particles. *Physical Review E*, **55** (2): 1959–1978.

Frederiksen, J.S. (2012). Statistical dynamical closures and subgrid modeling for inhomogeneous qg and 3d turbulence. *Entropy* **14**: 32–57.

Gramling, C.M., Harvey, C.F., and Meigs, L.C. (2002). Reactive transport in porous media: a comparison of model prediction with laboratory visualization. *Environmental Science and Technology*, **36**: 2508–2514.

Kapoor, V. and Kitanidis, P. (1998). Concentration fluctuations and dilution in aquifers. *Water Resources Research*, **34** (5): 1181–1193.

Kitanidis, P. (1994). Mixing-induced precipitation: experimental study and multi-scale numerical analysis. *Water Resources Research*, **30** (7): 2011–2026.

Knutson, C., Valocchi, A., and Werth, C. (2007). Comparison of continuum and pore-scale models of nutrient biodegradation under transverse mixing conditions. *Environmental Science and Technology*, **30** (6-7): 1421–1431.

LeBorgne, T., Dentz, M., and Carrera, J. (2008). Lagrangian statistical model for transport in highly heterogeneous velocity fields. *Physical Review Letters*, **101** (9), 090 601.

Li, J. and Helm, D. (1998). Viscous drag, driving forces, and their reduction to Darcy's law. *Water Resources Research*, **34** (7): 1675–1684.

Olsson, A. and Grathwohl, P. (2007) Transverse dispersion of non-reactive tracers in porous media: A new nonlinear relationship to predict dispersion coefficients. *Journal of Contaminant Hydrology* **92**: 149–161.

Tartakovsky, A.M. (2010). Langevin model for reactive transport in porous media. *Physical Review E* **82** (2). doi:10.1103/PhysRevE.82.026302

Tartakovsky, A., Meakin, P., Scheibe, T., and Wood, B. (2007). A smoothed particle hydrodynamics model for reactive transport and mineral precipitation in porous and fractured porous media. *Water Resources Research*, **43**: W05 437.

Tartakovsky, A.M., Tartakovsky, D.M., and Meakin, P. (2008a) Stochastic langevin model for flow and transport in porous media. *Physical Review Letters* **101**: 044 502.

Tartakovsky, A., Redden, G., Lichtner, P., et al. (2008b) Mixing-induced precipitation: experimental study and multi-scale numerical analysis. *Resources Research*, **44**: W06S04. doi:10.1029/2006WR005 725

Tartakovsky, A.M., Tartakovsky, G.D., and Scheibe, T.D. (2009). Effects of incomplete mixing on multicomponent reactive transport. *Advances in Water Resources* **32**: 1674–1679. doi:10.1016/j.advwatres.2009.08.012

Van Kampen, N. (2001). *Stochastic Processes in Physics and Chemistry*. Elsevier.

Welton, W. and Pope, S. (1997). Pdf model calculations of compressible turbulent flows using smoothed particle hydrodynamics. *Journal of Computational Physics*, **134**: 150–168.

Index

Reactive Transport Modeling: Applications in Subsurface Energy and Environmental Problems,
First Edition. Edited by Yitian Xiao, Fiona Whitaker, Tianfu Xu and Carl Steefel.

c

d

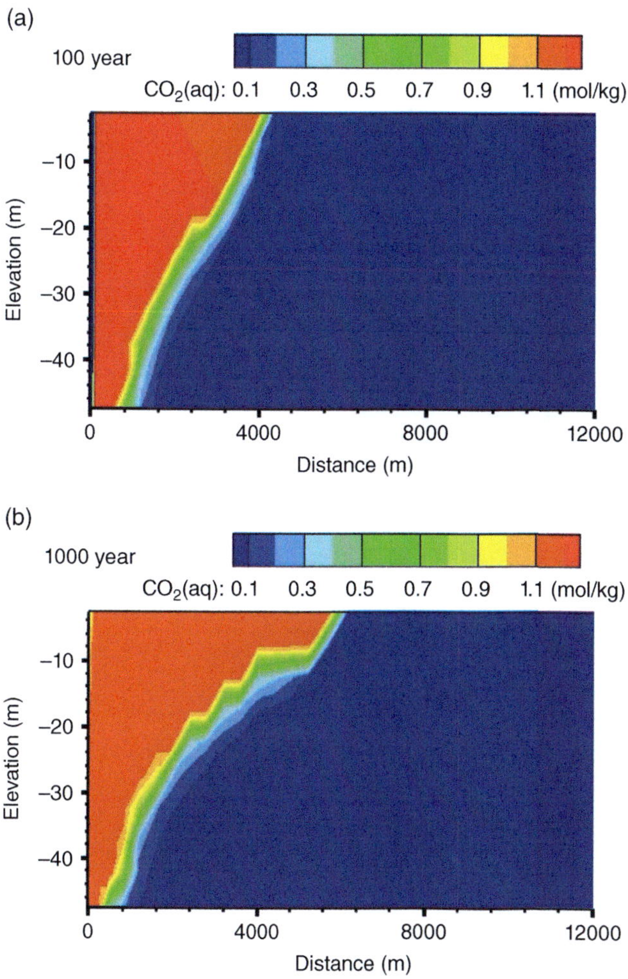

Figure 1.3 Spatial distribution of dissolved CO_2 at (a) 100, (b) 1000, (c) 5000 and (d) 10,000 years.

Reactive Transport Modeling: Applications in Subsurface Energy and Environmental Problems,
First Edition. Edited by Yitian Xiao, Fiona Whitaker, Tianfu Xu and Carl Steefel.

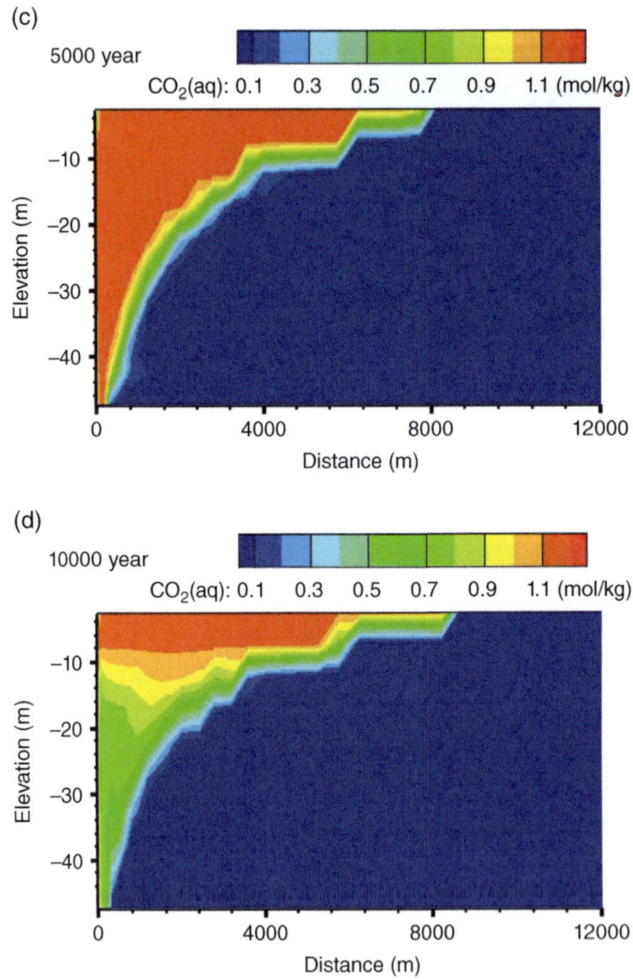

Figure 1.3 (Cont'd)

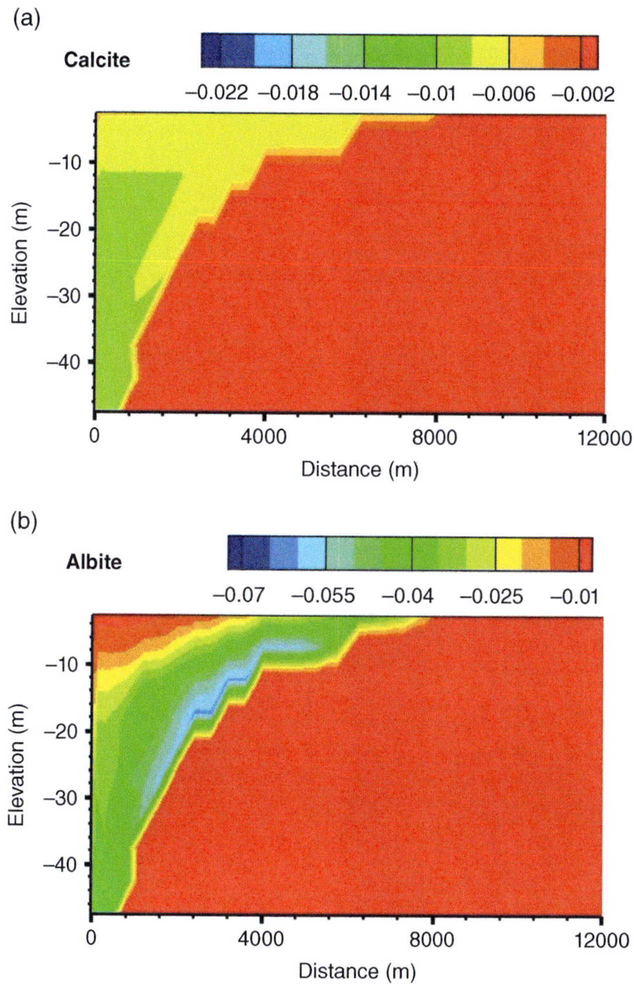

Figure 1.5 Spatial distribution of volume fraction change for dissolved minerals at 10,000 years (negative values indicate dissolution, positive precipitation).

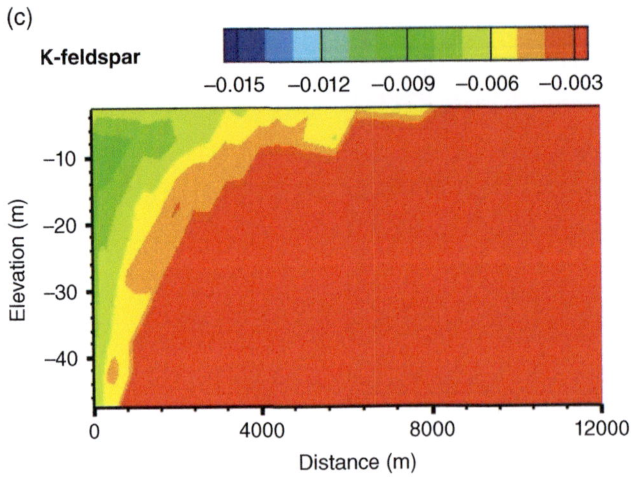

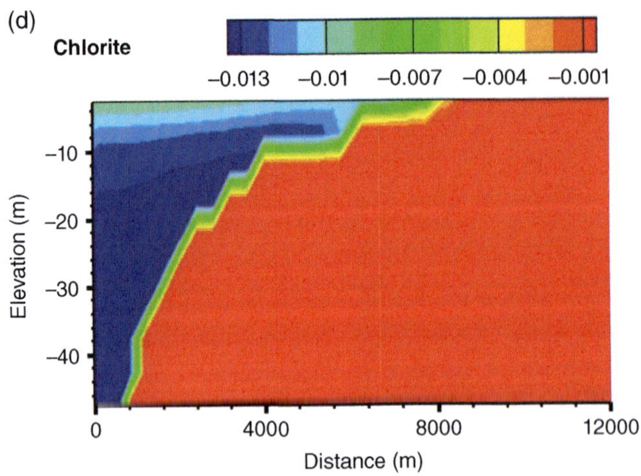

Figure 1.5 (Cont'd)

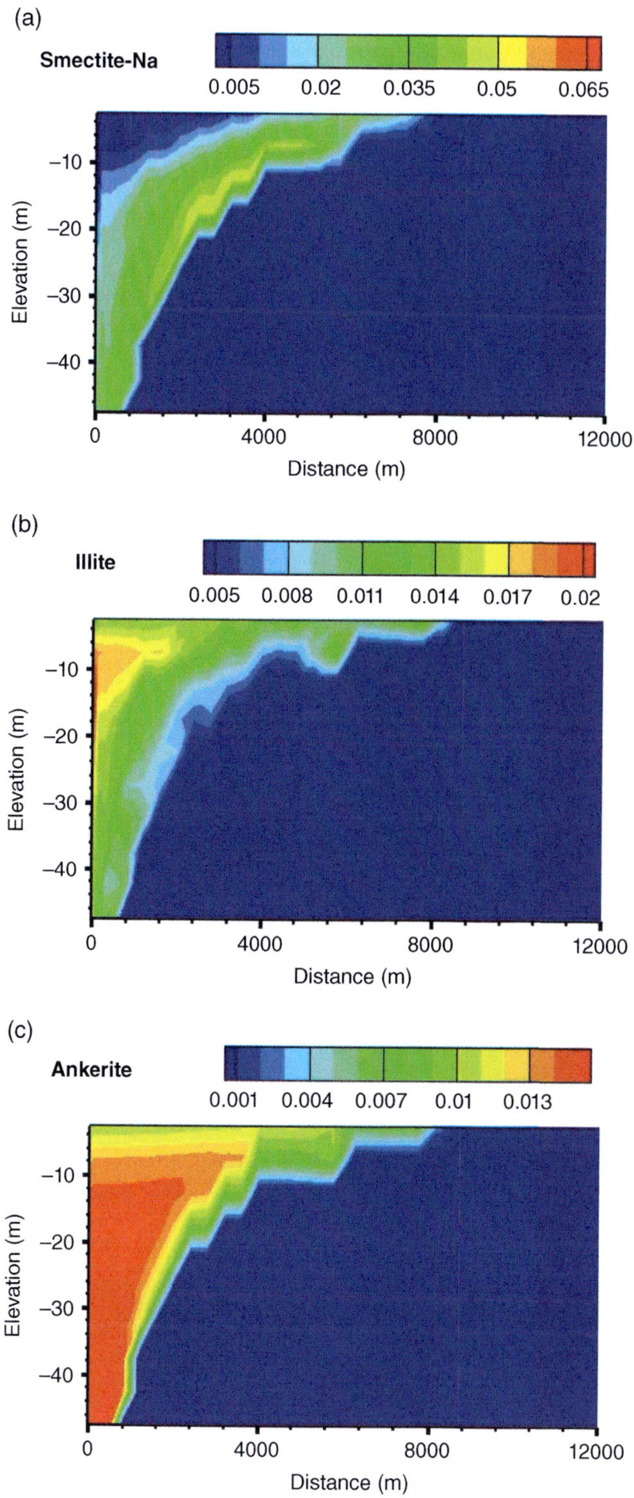

Figure 1.6 Spatial distribution of volume fraction change for precipitated minerals at 10,000 years.

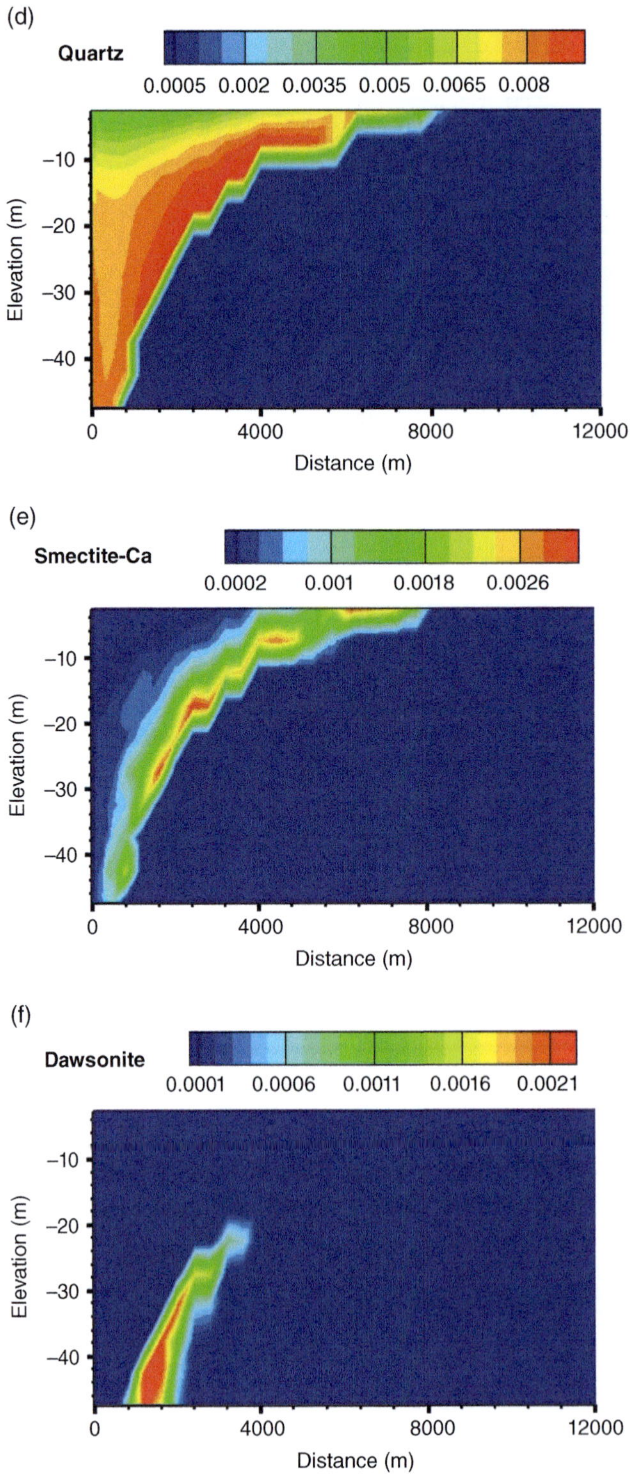

Figure 1.6 (Cont'd)

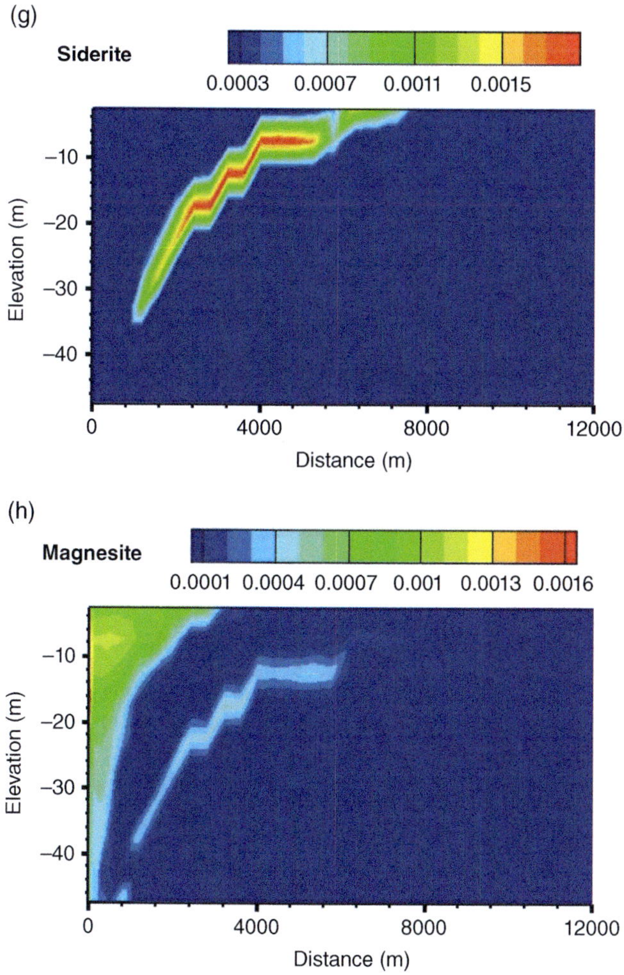

Figure 1.6 (Cont'd)

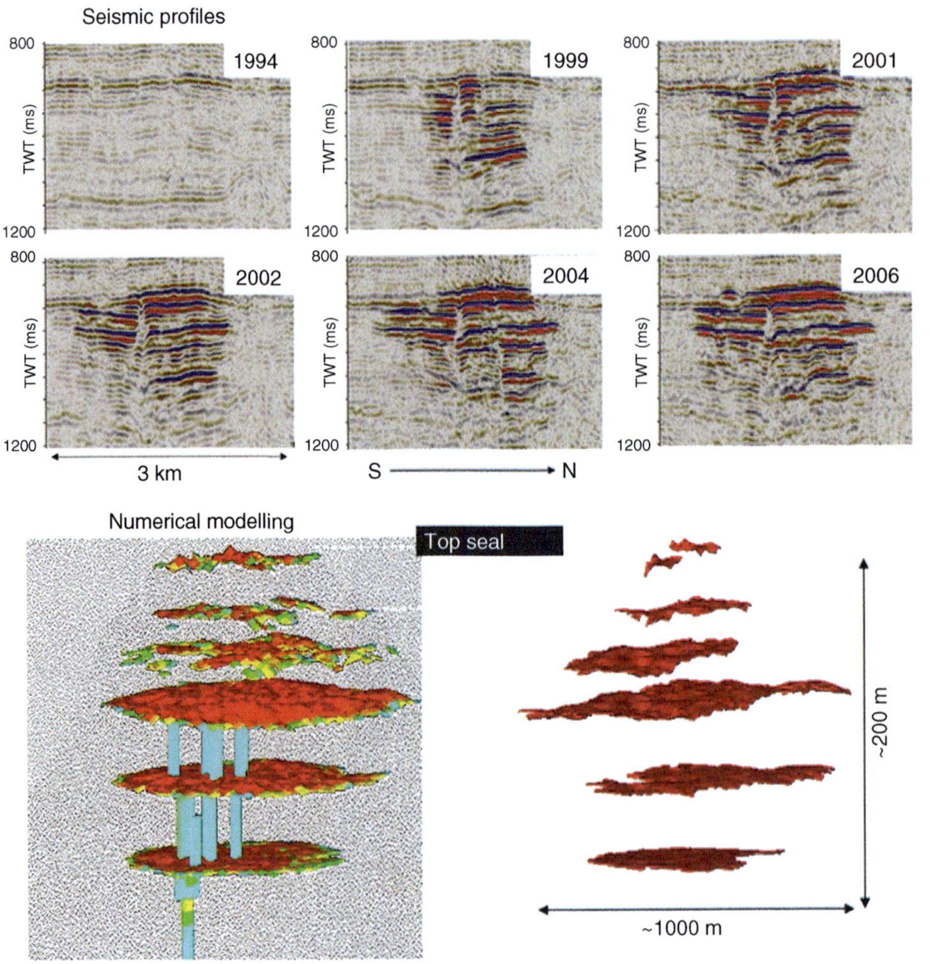

Figure 2.5 Seismic profiles before and during injection (Arts et al., 2008) and example of modeling of the related CO_2 plume illustrating CO_2 accumulation beneath semi-permeable shale lenses (Lindeberg et al., 2000). *Source:* Reproduced from Lindeberg et al. (2000), with permission from CSIRO Publishing.

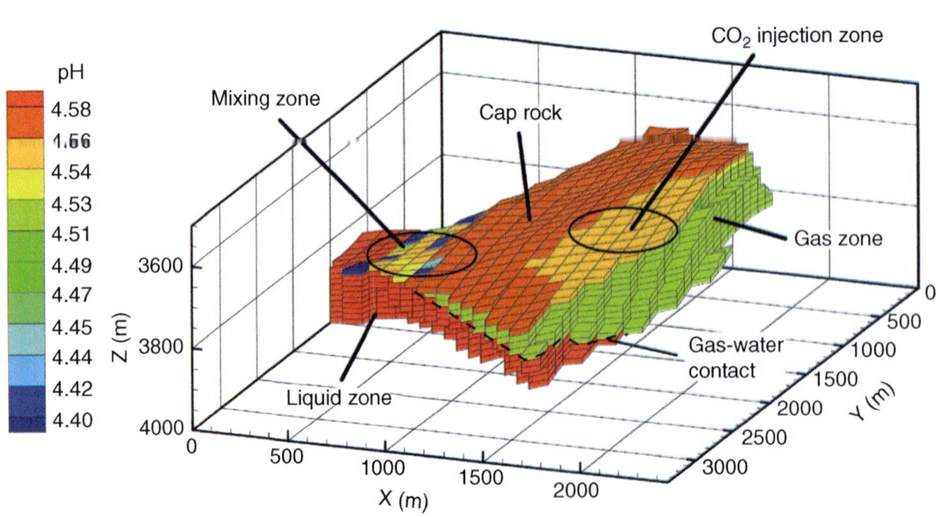

Figure 2.11 pH simulated in the K12B reservoir after 10 years of injection (Audigane et al., 2008, *Source:* used with permission of AAPG).

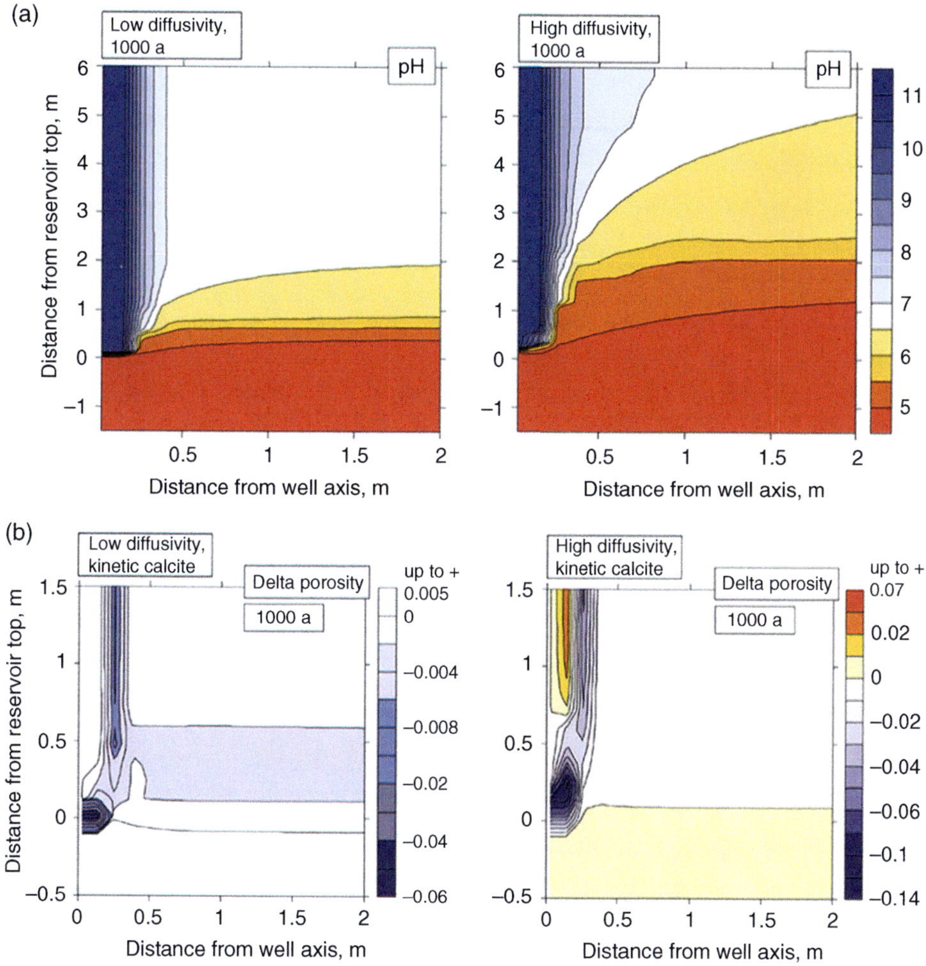

Figure 2.17 Simulated spatial distribution of (a) pH and (b) porosity after 1000 years (Gherardi et al., 2012).

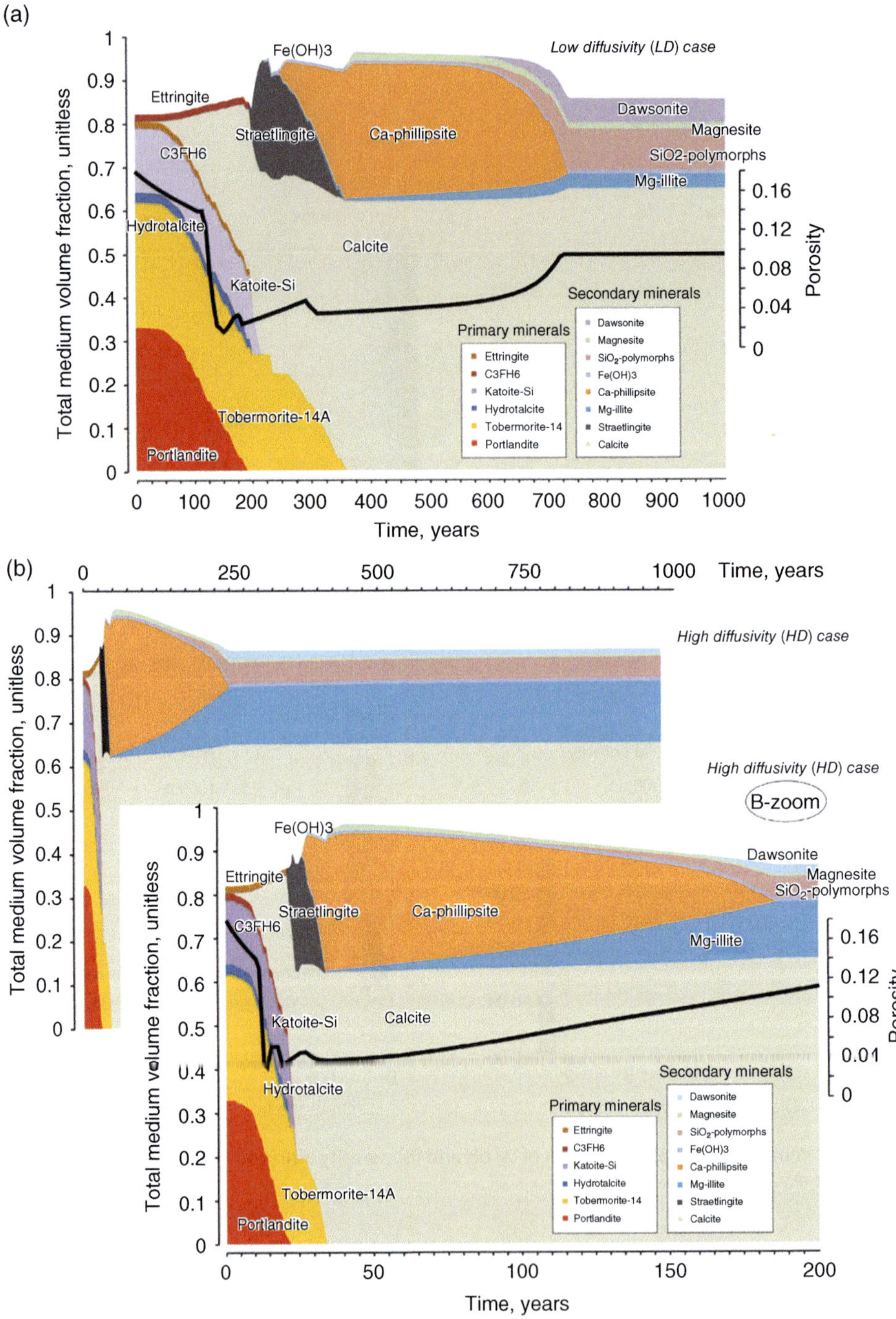

Figure 2.18 Sequence of dissolution and precipitation of the mineral phases of the cement at the interface with the reservoir and the caprock (Gherardi et al., 2012).

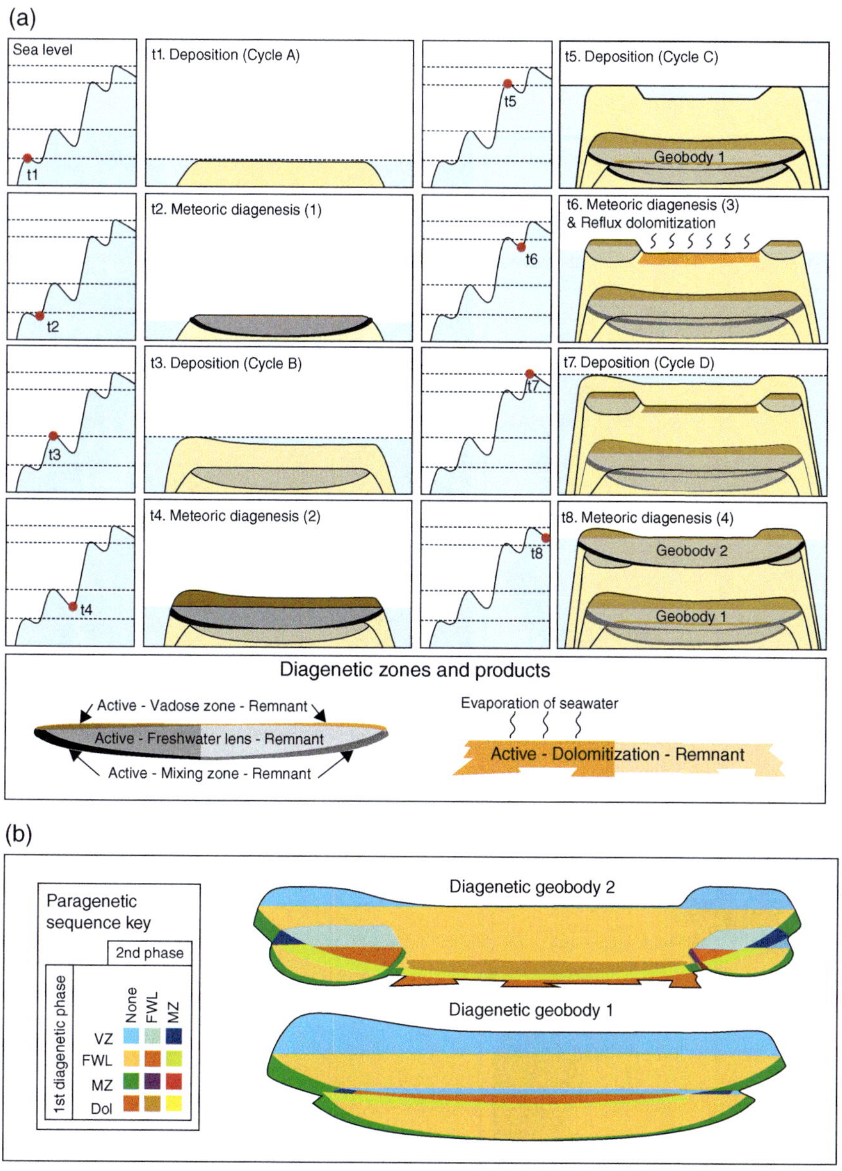

Figure 3.6 (a) A conceptual sequence stratigraphic scenario integrating syn-depositional diagenesis. **t1**: Cycle A is deposited as relative sea-level rises (transgressive sequence) and is then subjected to alteration by a meteoric diagenetic system as relative sea-level falls to **t2**. Cycle B is deposited during a second transgressive phase (ending at **t3**) but during the subsequent exposure event (**t4**) meteoric diagenesis cross-cuts the A/B cycle boundary. Cycle C is deposited during a significant rise in relative sea-level (**t5**), but the subsequent low stand (**t6**) is insufficient to expose the entire carbonate platform. Instead, meteoric alteration is restricted to two smaller hydrological systems beneath platform margin islands. These islands also restrict flow of ocean water to the platform interior, allowing evaporative concentration and dolomitization of platform interior sediments by brine reflux. These processes do not penetrate deep enough within the succession to overprint previous diagenetic phases. These processes cease upon the re-flooding of the platform and deposition of Cycle D (**t7**), which occurs before the final meteoric alteration phase in **t8**, during which the meteoric system overprints the diagenetic phases that record the **t6** diagenetic system. These periods of deposition and exposure lead to the production of two internally complex but spatially related diagenetic geobodies. The internal architecture of these geobodies formed because of the events detailed in (a) can be seen in (b). The diagenetic events that produced Geobody 1 leads to seven possible paragenetic sequences being recorded within 11 different regions within the geobody. The spatial complexities of diagenesis at t6 lead to a more complex internal architecture within Geobody 2, which contains 12 paragenetic sequences arranged within 27 paragenetic regions. Such a conceptual exercise helps us understand the complexities that could arise from even a relatively simplistic scenario. In reality, the 3D nature of these diagenetic process, as well as their temporally and spatially transitory nature, will lead to greater levels of complexity.

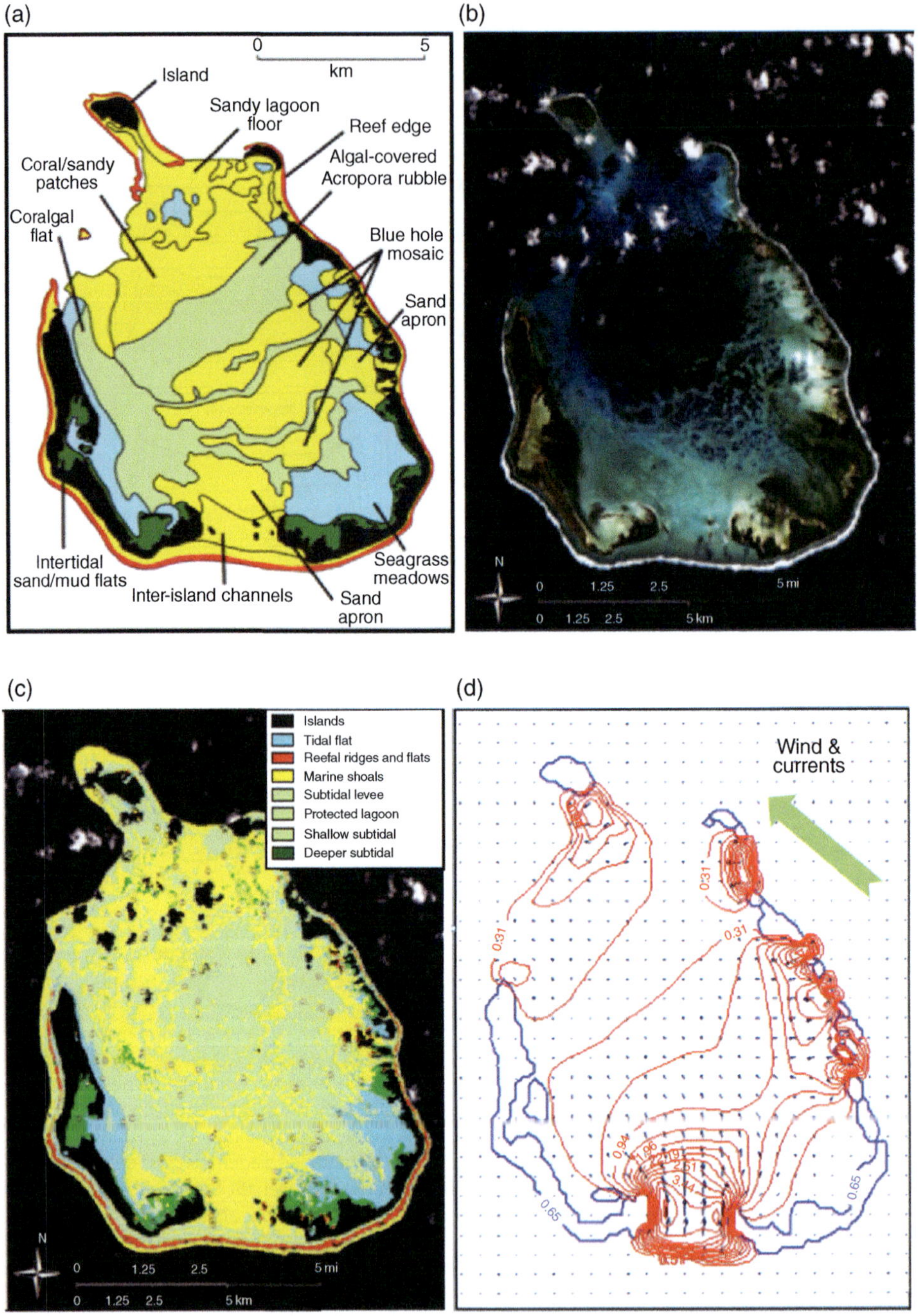

Figure 3.10 Evaluation of CARB3D$^+$ sediment modelling by comparison with the modern Cocos-Keeling platform (a–f) and the Caicos platform (g–i). Synthetic facies (f and i) are compared with direct observation of facies distribution based on sample characterization (a, after map of Smithers, 1994) and that inferred by an analysis of Landsat imagery (by Kaczmarek et al., 2010) (c and h). For Coco-Keeling we also show predicted current velocity (d), and the resulting redistribution of coarse and fine sediments (e). Paterson et al., unpublished data.

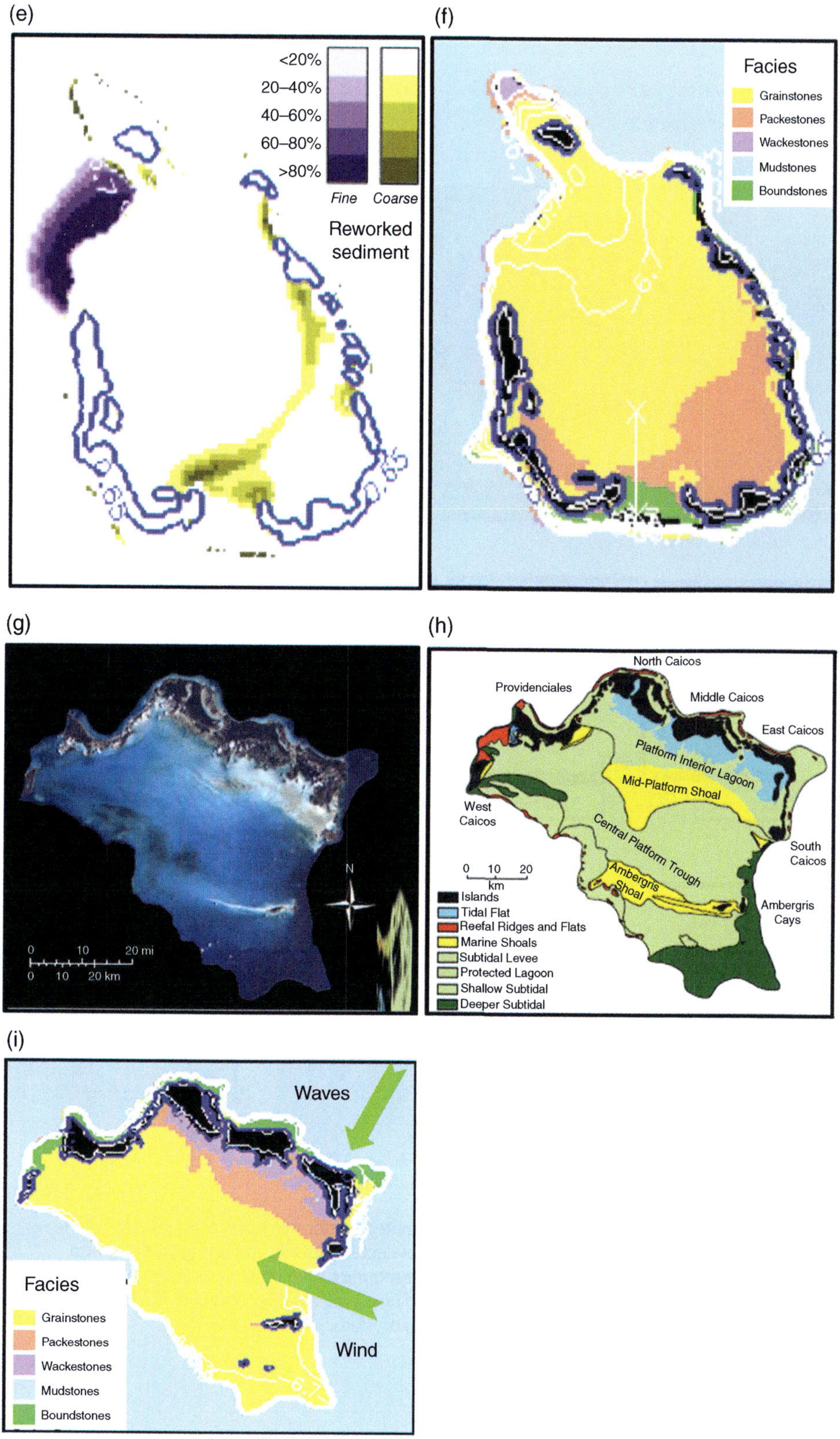

Figure 3.10 (Cont'd)

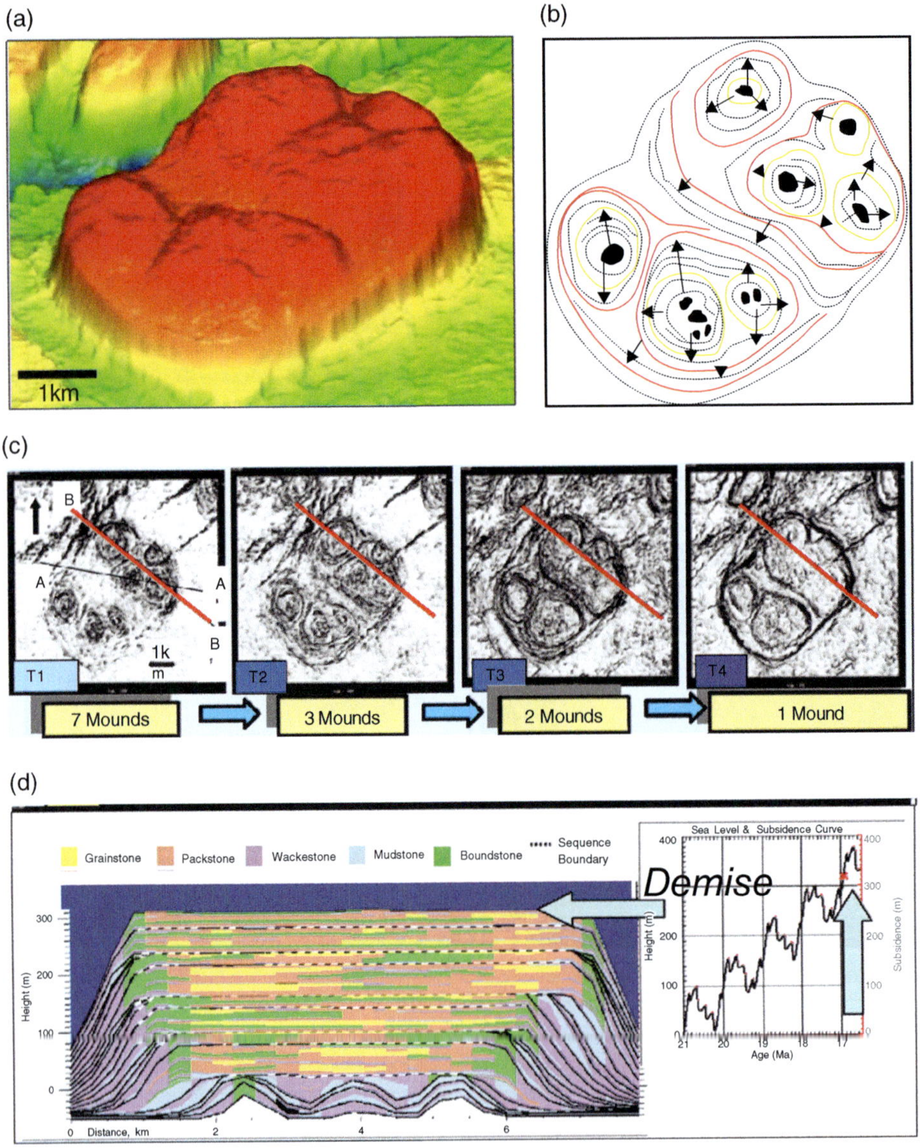

Figure 3.11 (a) PGS seismic data showing the amalgamated platform in North Madura with (b) the growth pattern interpreted from time slices. In (c), four selected slices illustrate stages in platform growth, with nucleation and amalgamation of separate mounds, then drowning caused by a reduction in production rate; (d) as simulated using CARB3D^{+} and displayed as a 2D section marked in red in (c). *Source:* After Hughes et al. (2008), used with permission of AAPG.

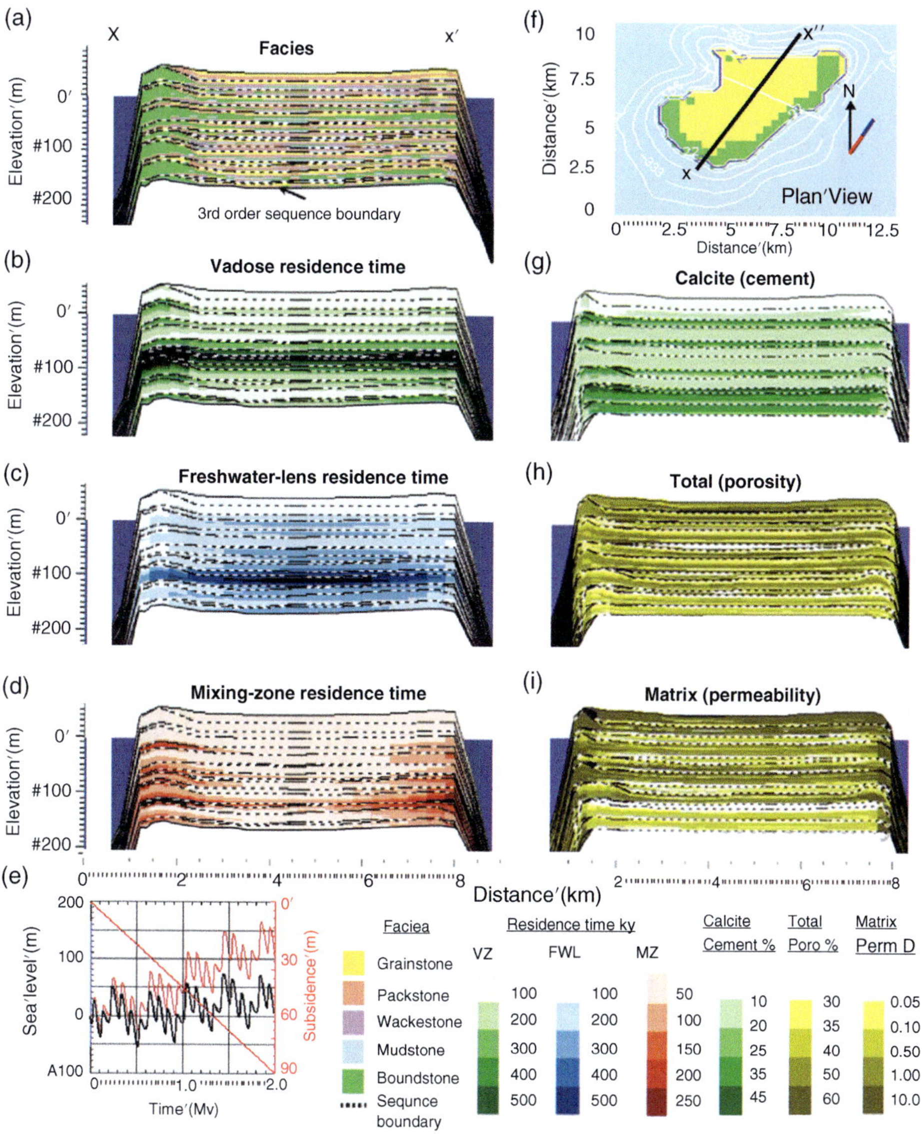

Figure 3.13 CARB3D[+] simulations for an isolated platform based on Miocene Natuna, Sarawak Basin, Indonesia. 2D sections through a 3D model show distribution show facies (a), residence time in the vadose zone (b), freshwater lens (c), and mixing zone (d), given the sea-level curve (e) with eustatic (black curve), subsidence (red line) and 'relative' sea-level variations (red curve). Cross-sections (x–x′) parallel to the principal wind direction (red/blue line) shown in map (g). In addition, we show one set of potential distributions of calcite cement (h), porosity (i) and matrix permeability (j). *Source:* Based on Paterson et al. (2008), used with permission of SEPM.

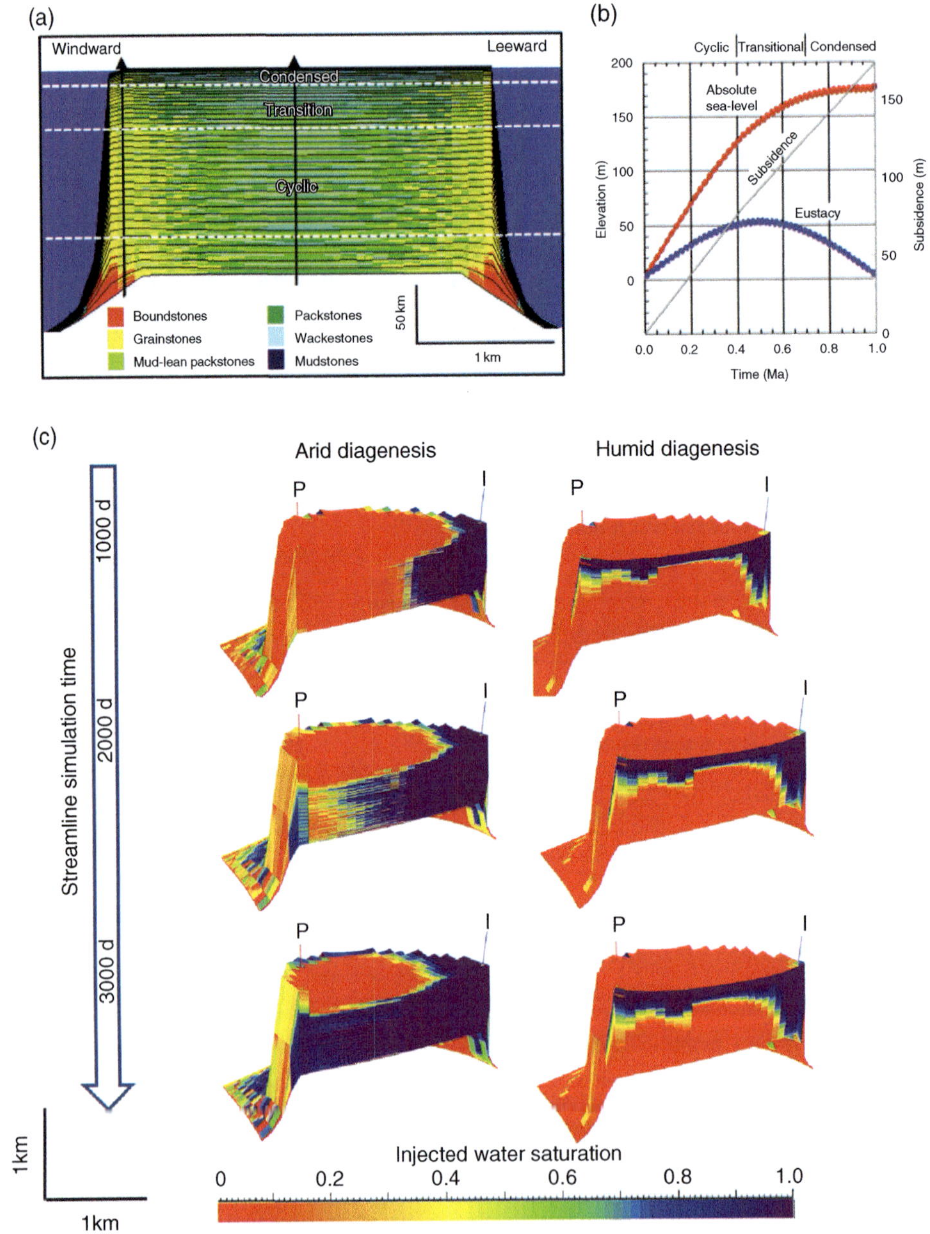

Figure 3.15 (a) 2D section oriented along principal wind direction through 3D simulation of the Latemar platform, showing platform architecture and facies distribution, arbitrary divisions into cyclic, transitional and condensed units (each 300 ky) reflecting an overall reduction in accommodation shown in (b) sea-level curve. (c) 3D distribution in half the platform (cut along line of 2D section) showing connate water saturation for arid and humid simulations at three time intervals during single-phase tracer experiment. Blue and red vertical lines mark the positions of injector (I) and producer (P) wells. *Source:* After Whitaker et al. (2014), used with the permission of The Geological Society.

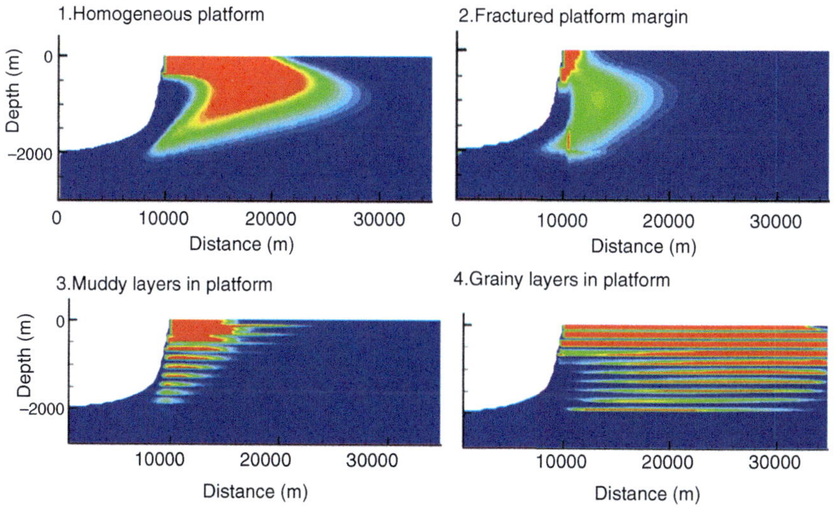

Figure 4.9 Comparison of dolomite (red) distribution as a fraction of rock volume from: (1) a homogeneous platform; (2) a fractured platform margin; (3) muddy layers in a platform; (4) grainy layers in a platform. From Whitaker and Xiao (2010).

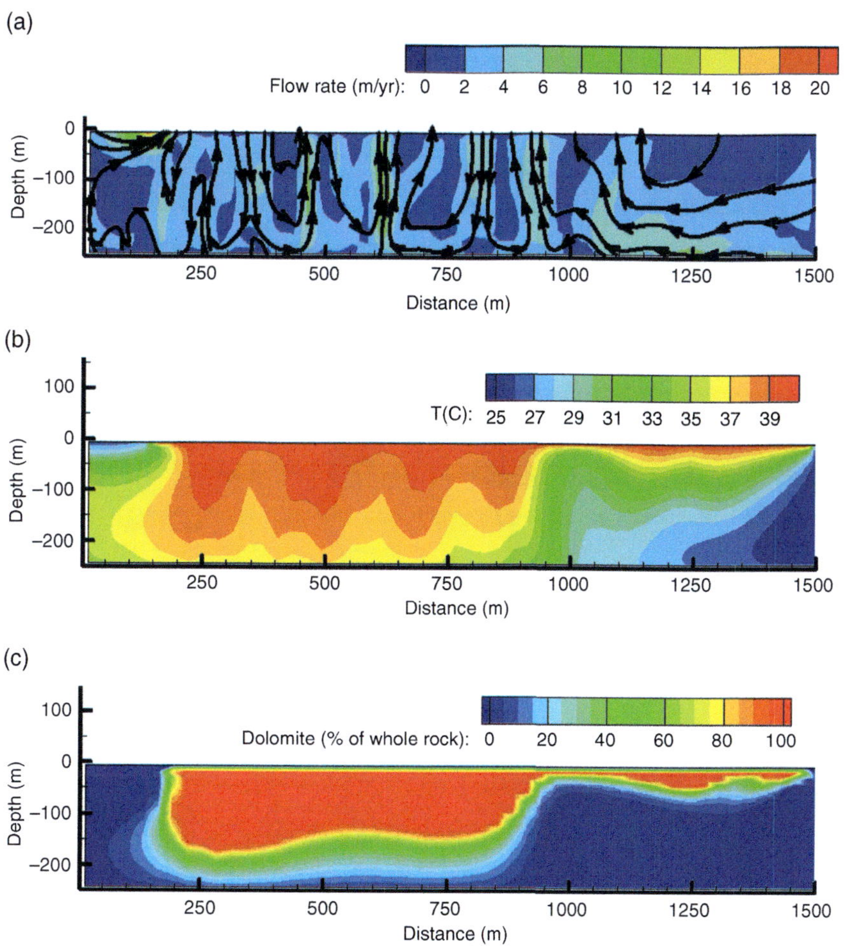

Figure 4.14 RTM results illustrating ascending freshwater–mesohaline water mixing mechanisms of observed extensive dolomitization at La Molata, Miocene. (a) Fluid flow rate; (b) temperature distribution; and (c) dolomite distribution. Arrows represent streamlines of fluid flow. Dolomite is in volume % of rock. From Cooper et al. (2014).

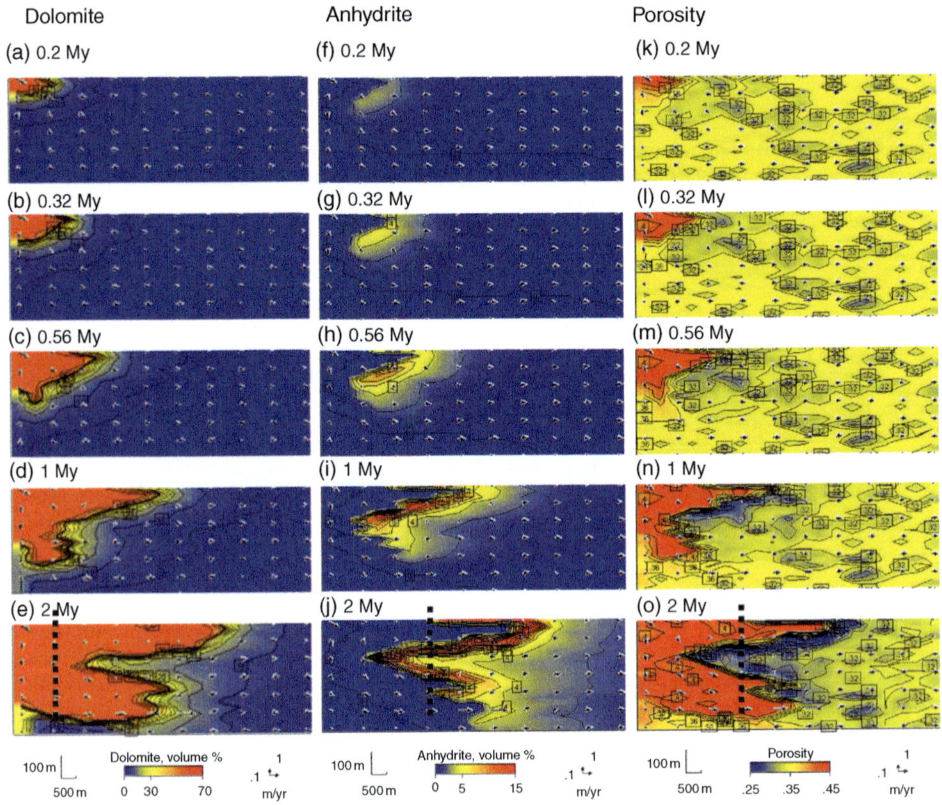

Figure 4.16 Effect of initial porosity and permeability heterogeneity on dolomite, anhydrite and porosity. (a–e) Dolomite (contours are percentage dolomite with a contour interval of 5); (f–j) anhydrite (contours are percentage anhydrite with a contour interval of 2); (k–o) porosity (contours are porosity [fraction] with a contour interval of 0.02). From Jones and Xiao (2005).

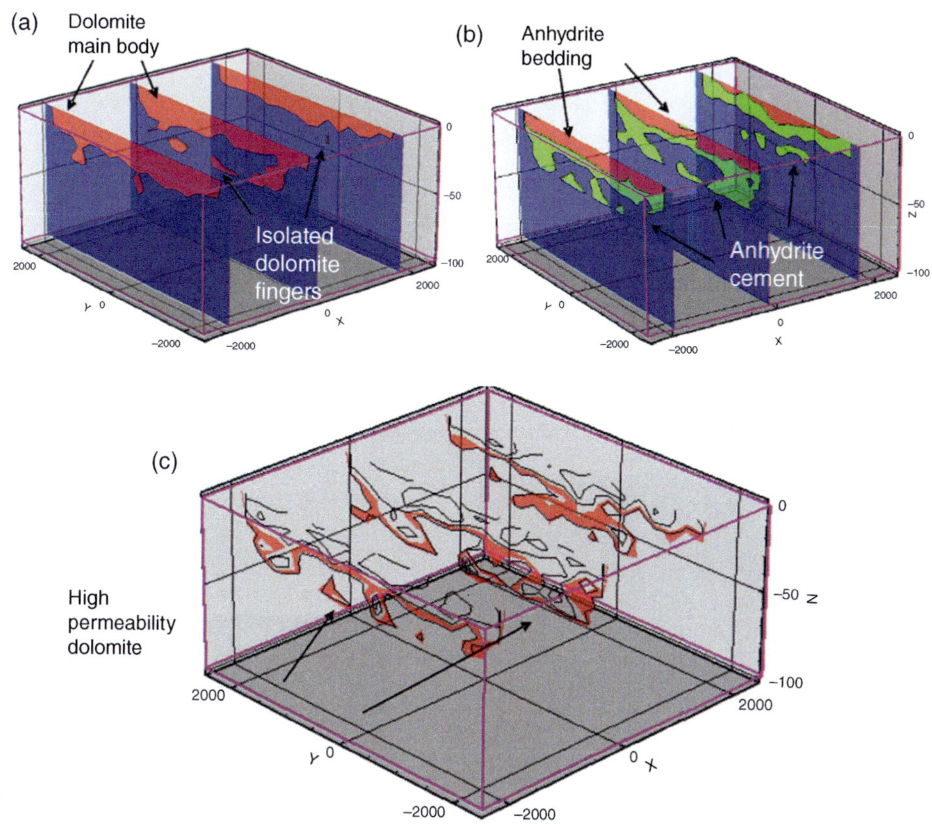

Figure 4.21 RTM 3D distribution of (a) dolomite; (b) anhydrite; and (c) high-permeability dolomite (dolomite with <5% anhydrite and >15% porosity). All properties are shown in 2D slices in *x*-direction. Porosity intervals are shown in contours and 'high-K' dolomite is shaded in red. From Xiao and Jones (2007).

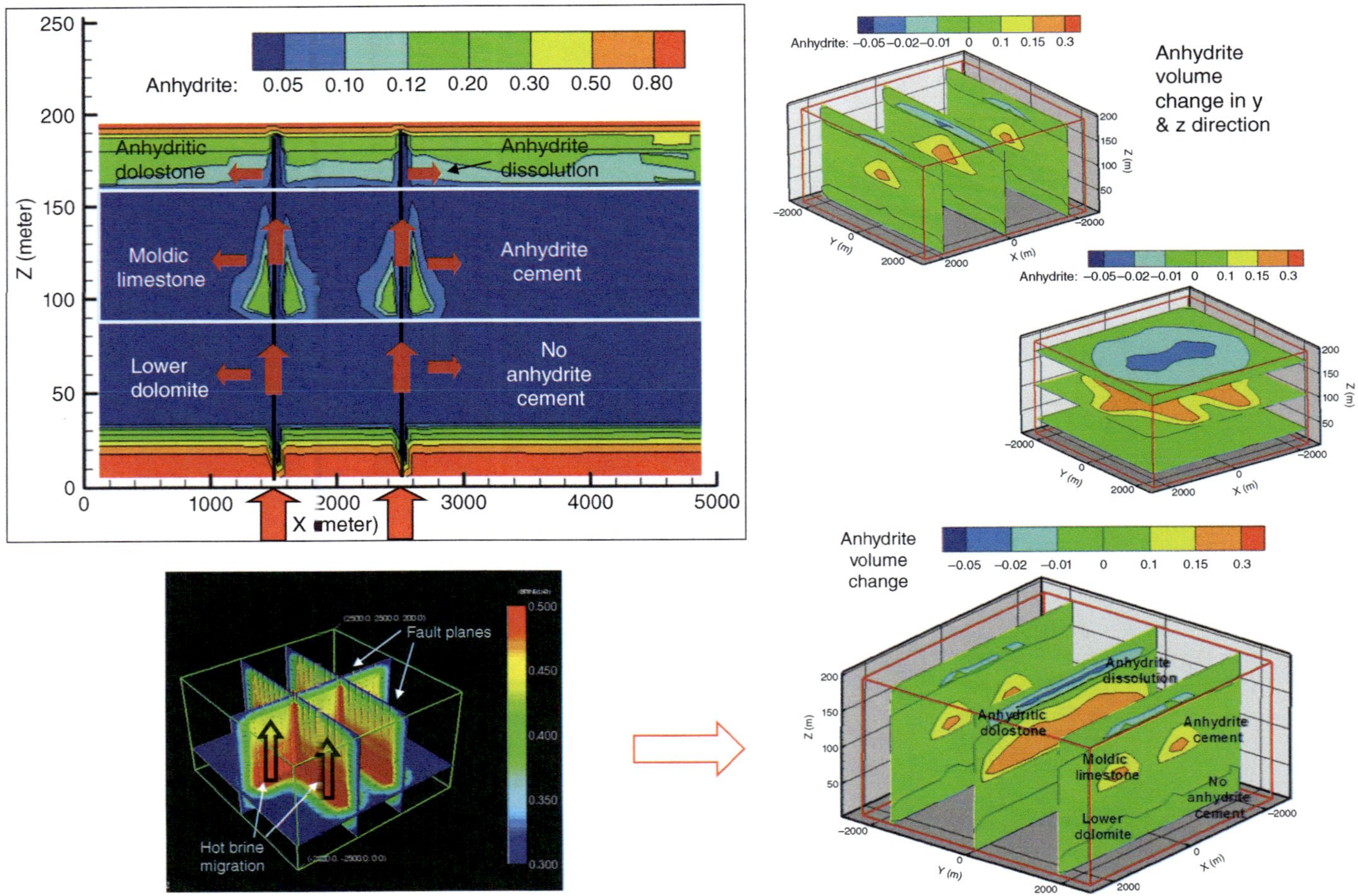

Figure 4.30 3D RTM results showing that HTD fluids entering a carbonate reservoir can lead to no anhydrite cement in the lower dolomite interval, anhydrite cement in the middle limestone interval, and anhydrite dissolution in the top anhydritic dolostone interval. From Xiao et al. (2013).

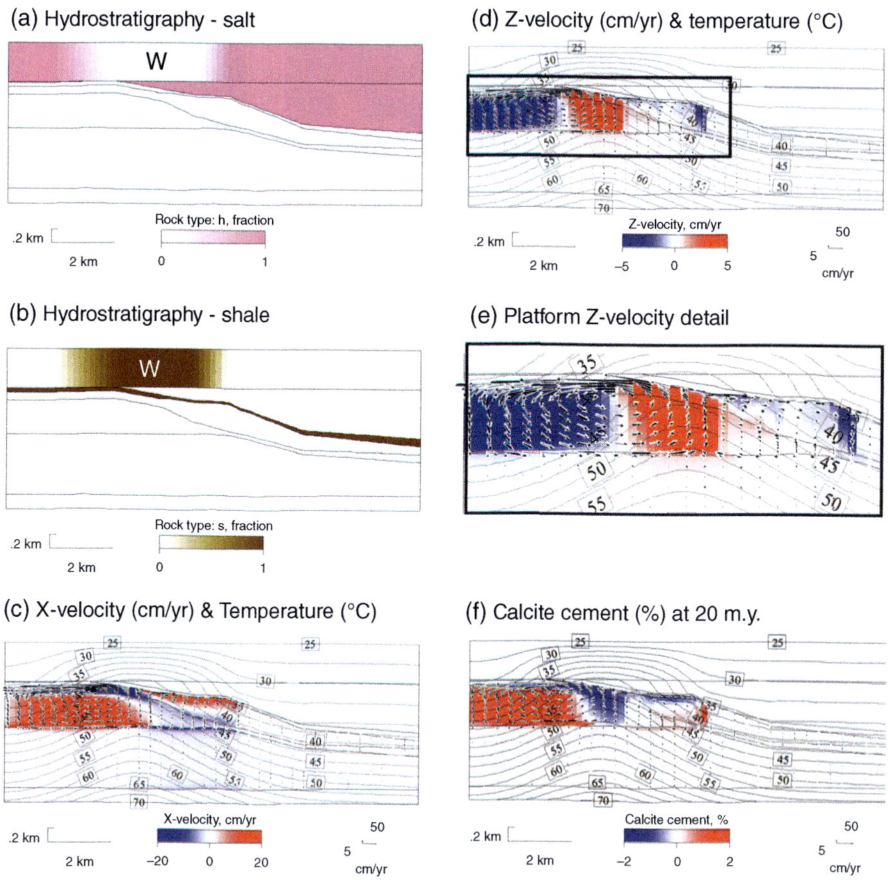

Figure 4.36 Geothermal circulation after platform burial and development of a 5.1 km shale-filled salt withdrawal basin that is 500 m deep. (a) Halite, 'W', withdrawal basin; (b) shale; (c) horizontal fluid velocity, velocity vectors and temperature (contour interval of 2.5 °C); (d) vertical fluid velocity, fluid velocity vectors, and temperature; (e) enlargement of vertical fluid velocity distribution in platform; (f) calcite cementation/dissolution at 20 My. Note: to enhance visualization of diagenetic patterns, the colour bar scale does not reflect the full range of the data, as discussed in the text. From Jones and Xiao (2006).

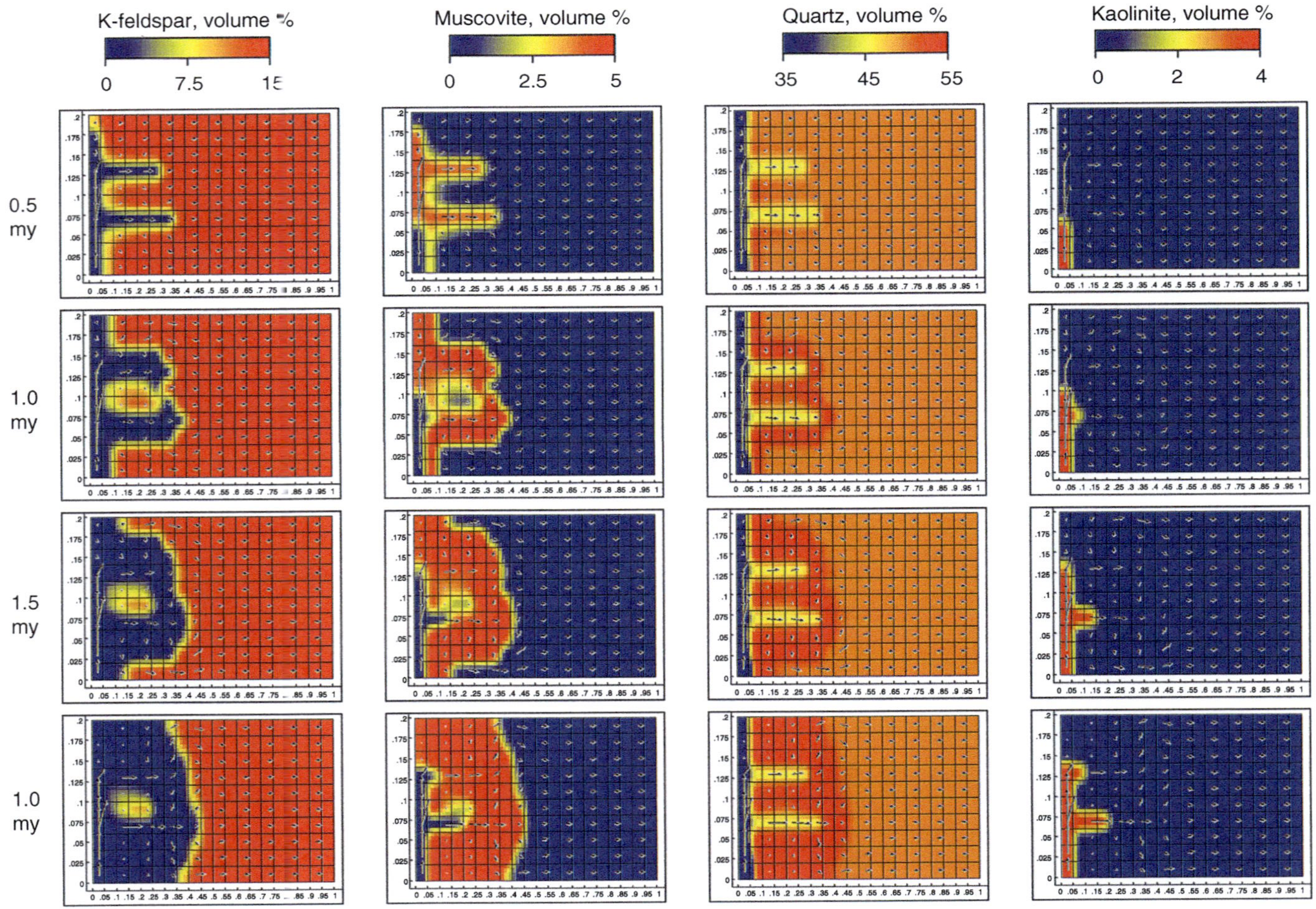

Figure 4.43 2D reactive transport modeling results showing the mineral zoning associated with fault-controlled burial diagenesis representing Rotliegendes after 2 My of illitization. From Xiao and Jones (2006).

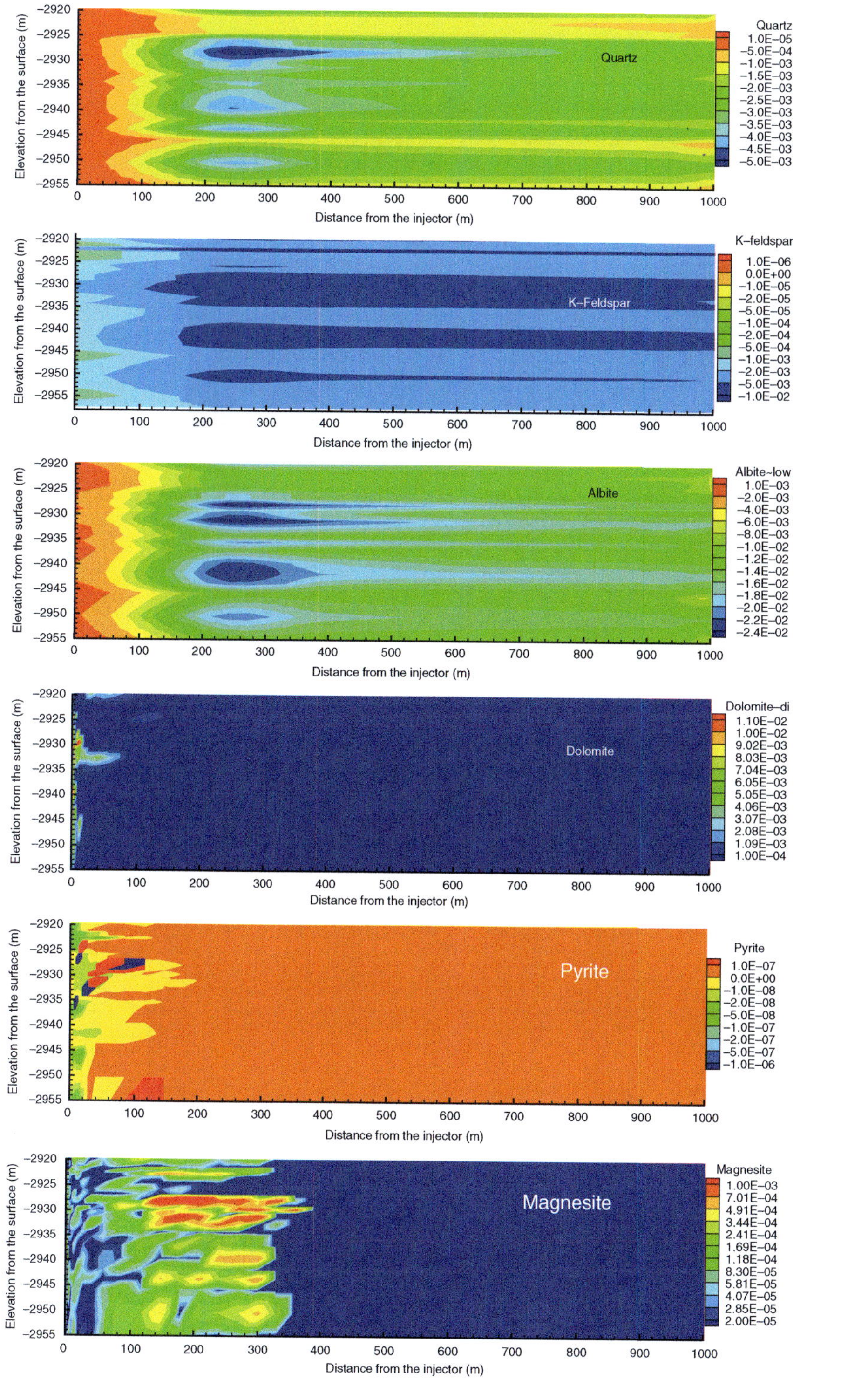

Figure 5.4 Model-predicted dissolution/precipitation of minerals (mol mineral/m^3 of rock) at the 20th year of the injection of untreated seawater. The colour scales represent the distribution of concentrations, from high (red) to low (blue). Note that precipitation or dissolution of most of these minerals is considered minor in terms of the amount of change except for calcite.

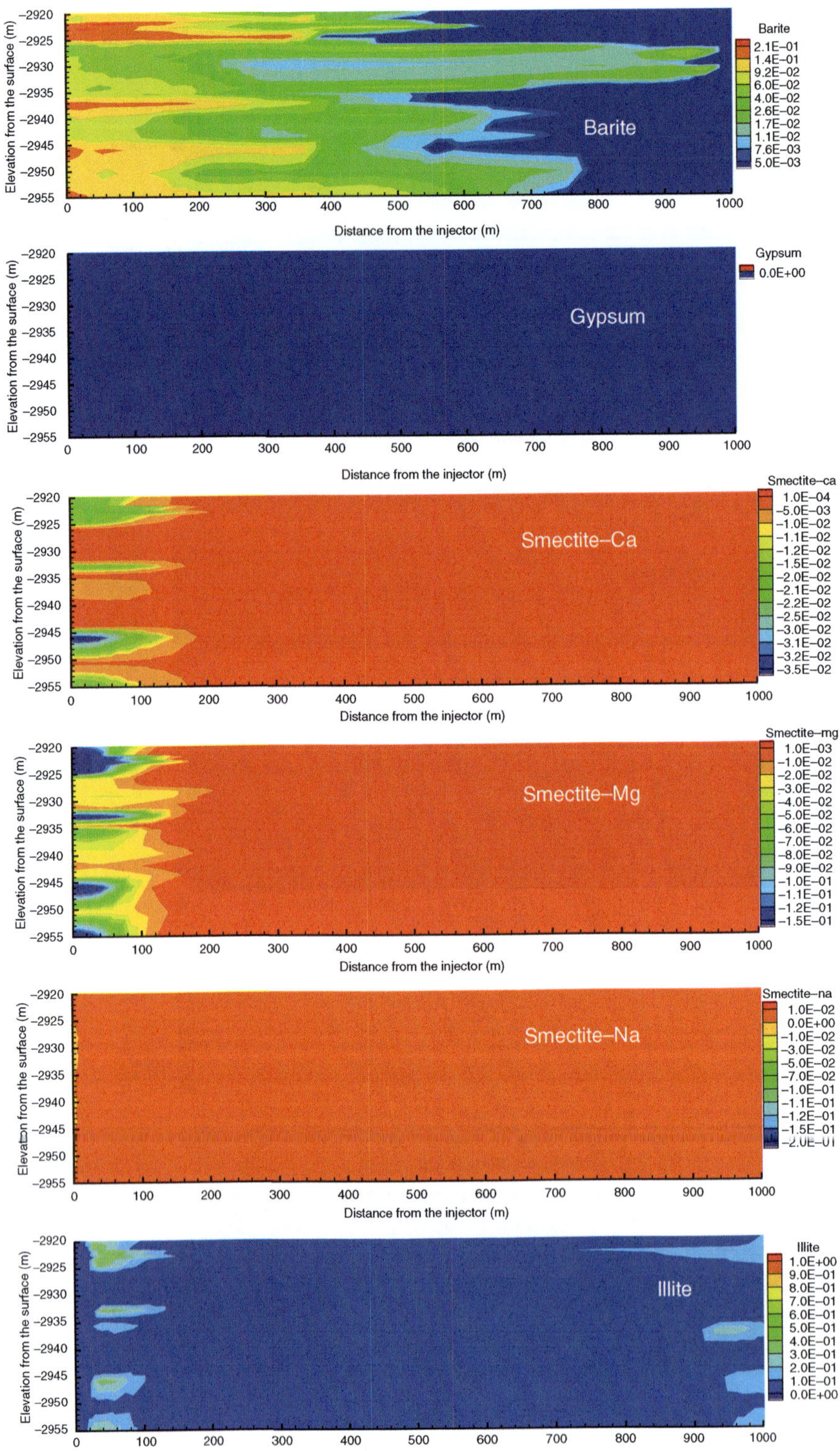

Figure 5.4 (Cont'd)

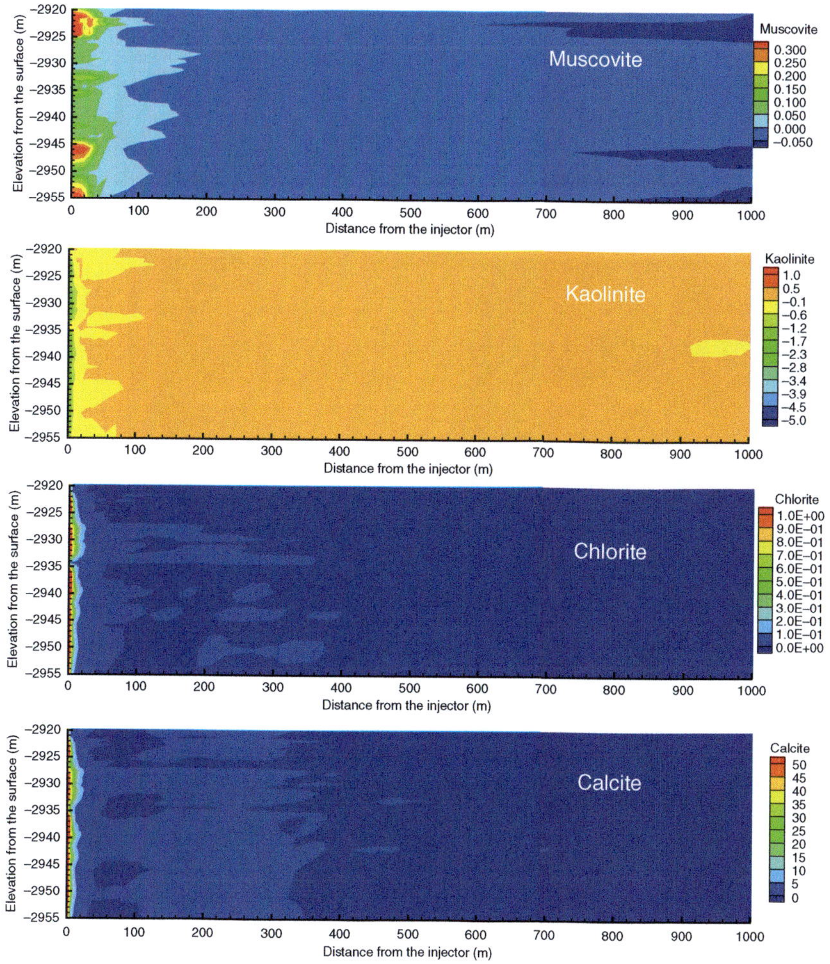

Figure 5.4 (Cont'd)

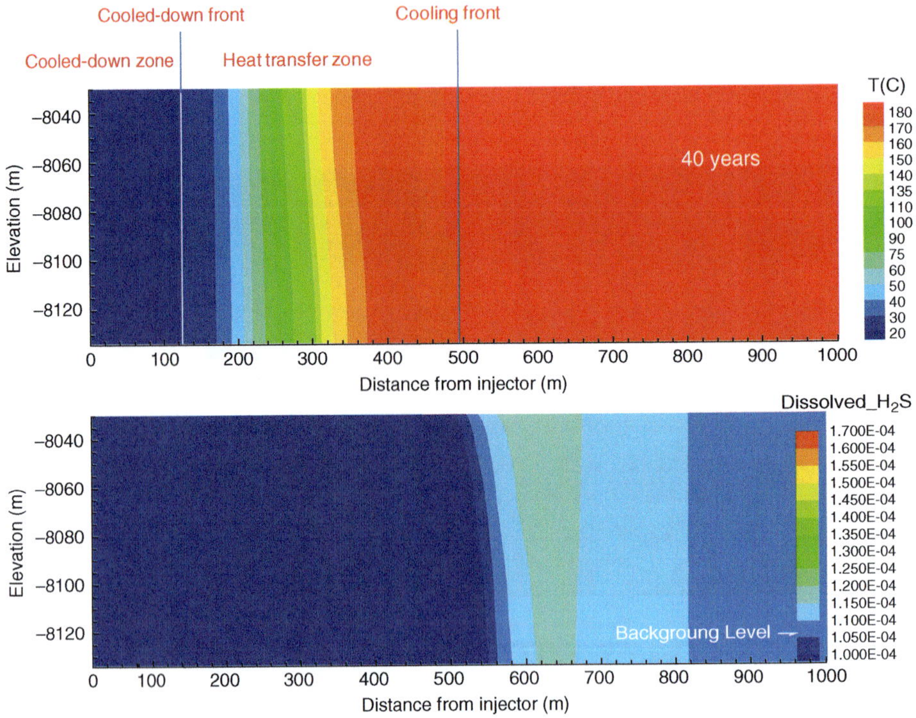

Figure 5.14 TOUGHREACT RTM (Phase 1, screening) predicted the H_2S (mol/kg of water) in the formation water during waterflood (the contour maps shows the concentration at 40 years).

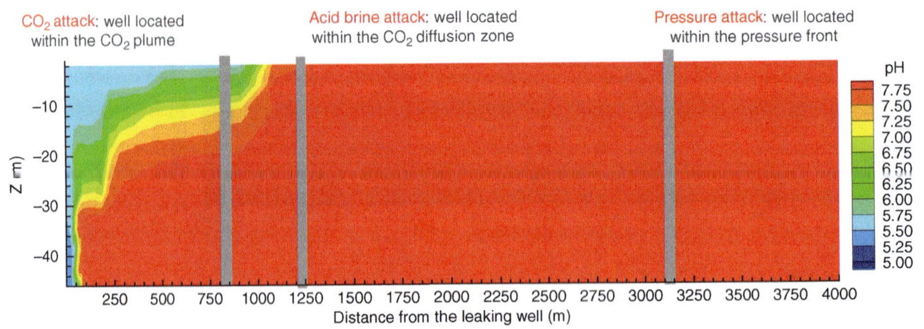

Figure 5.20 In a synthetic CO_2/acid gas injection scenario, a legacy well is located at a certain distance from the injection, and sequentially approached by a pressure front, CO_2 diffusion front (acid brine), and advection front (CO_2 plume) during CO_2 injection. Note that the injector is located at the left edge of the figure.

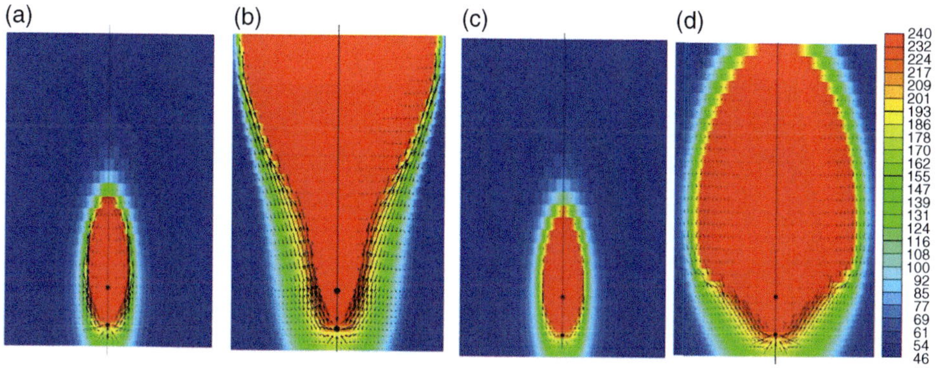

Figure 6.19 The temperature distributions of the SAGD and MTFs-AGD processes. (a) Initial stage of SAGD process; (b) steam chamber shape of SAGD process; (c) initial stage of MTFs-AGD process; (d) steam chamber shape of MTFs-AGD process.

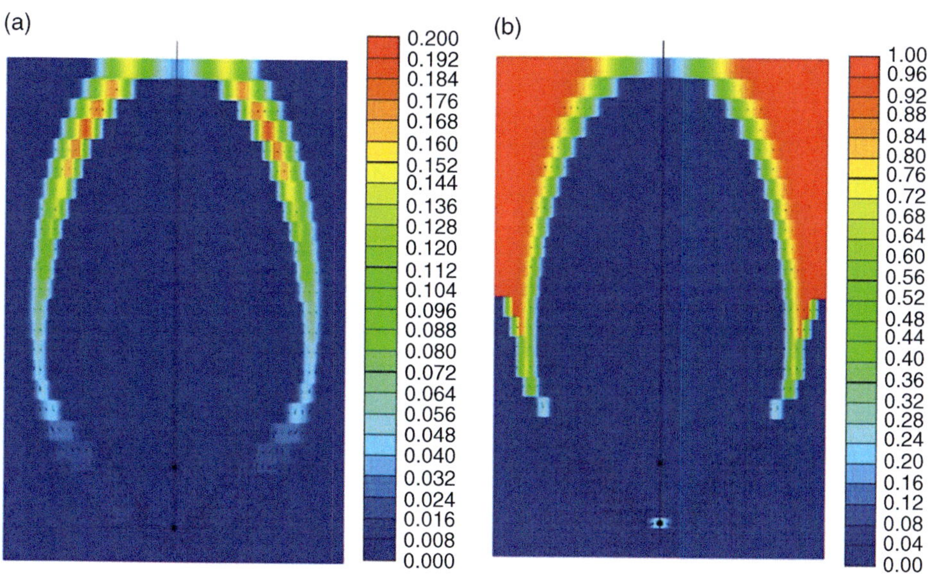

Figure 6.20 The NCG concentration during the MTFs-AGD process: (a) CO_2 and (b) N_2.

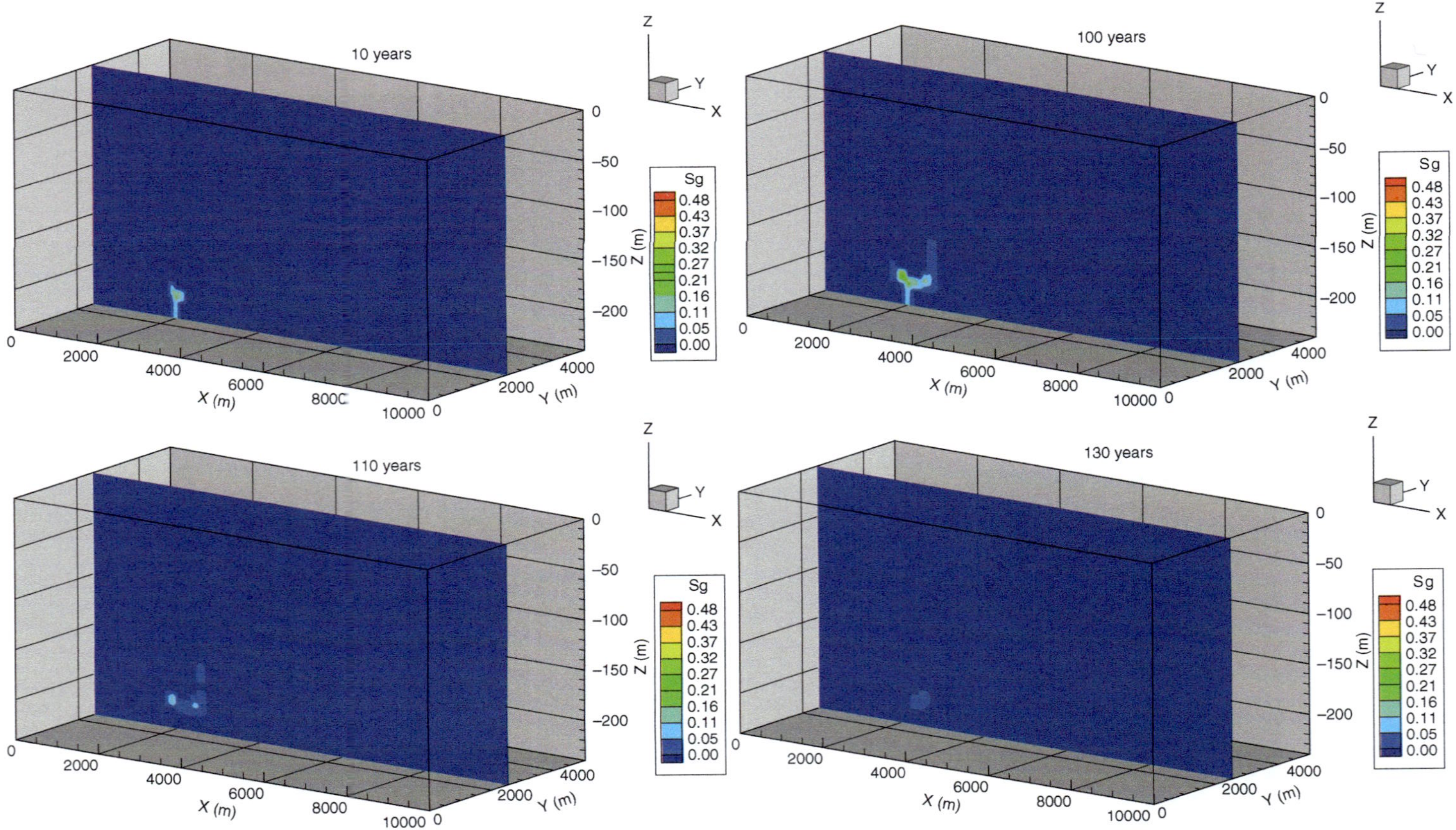

Figure 7.3 Simulated spatial distribution of the free CO2 phase (physical) saturation along a cross section at y = 2500 m at 10, 100, 110, and 130 years.

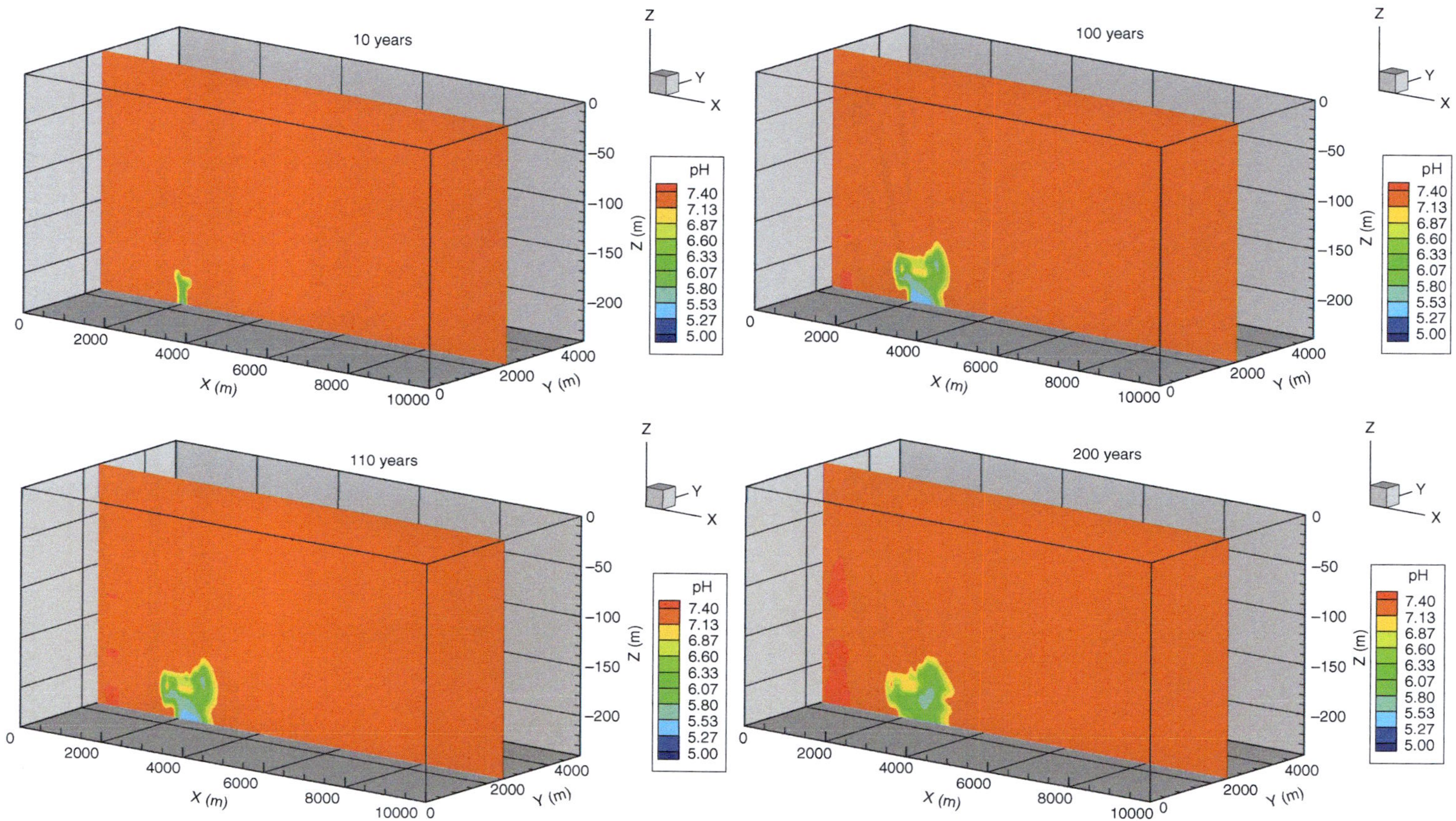

Figure 7.5 Simulated spatial distribution of pH along a cross section at Y = 2500 m at 10, 100, 110, and 200 years.

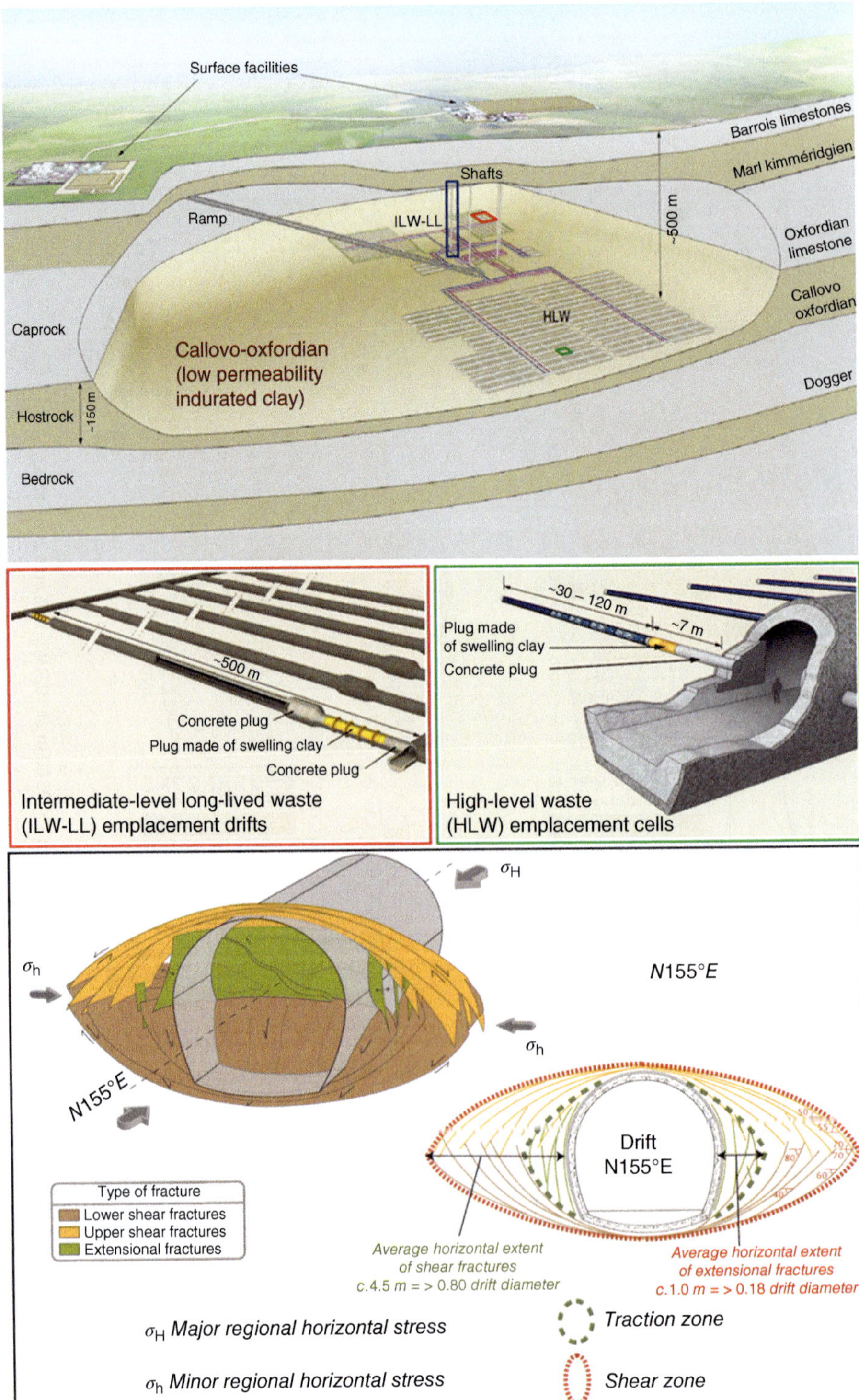

Figure 8.1 Left, from top to bottom: full-scale design of a repository concept in clay-rocks, zoom on the design of the emplacement drifts, representation of the EDZ. Right: schemes of simplified geometries modeled with reactive transport approaches. *Source:* Adapted from Brommundt et al. (2014) and Vinsot et al. (2014).

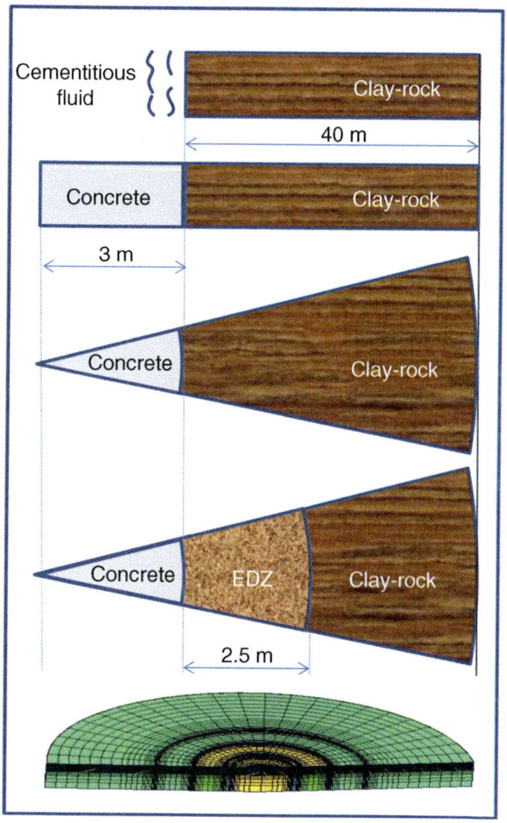

Figure 8.1 (Cont'd)

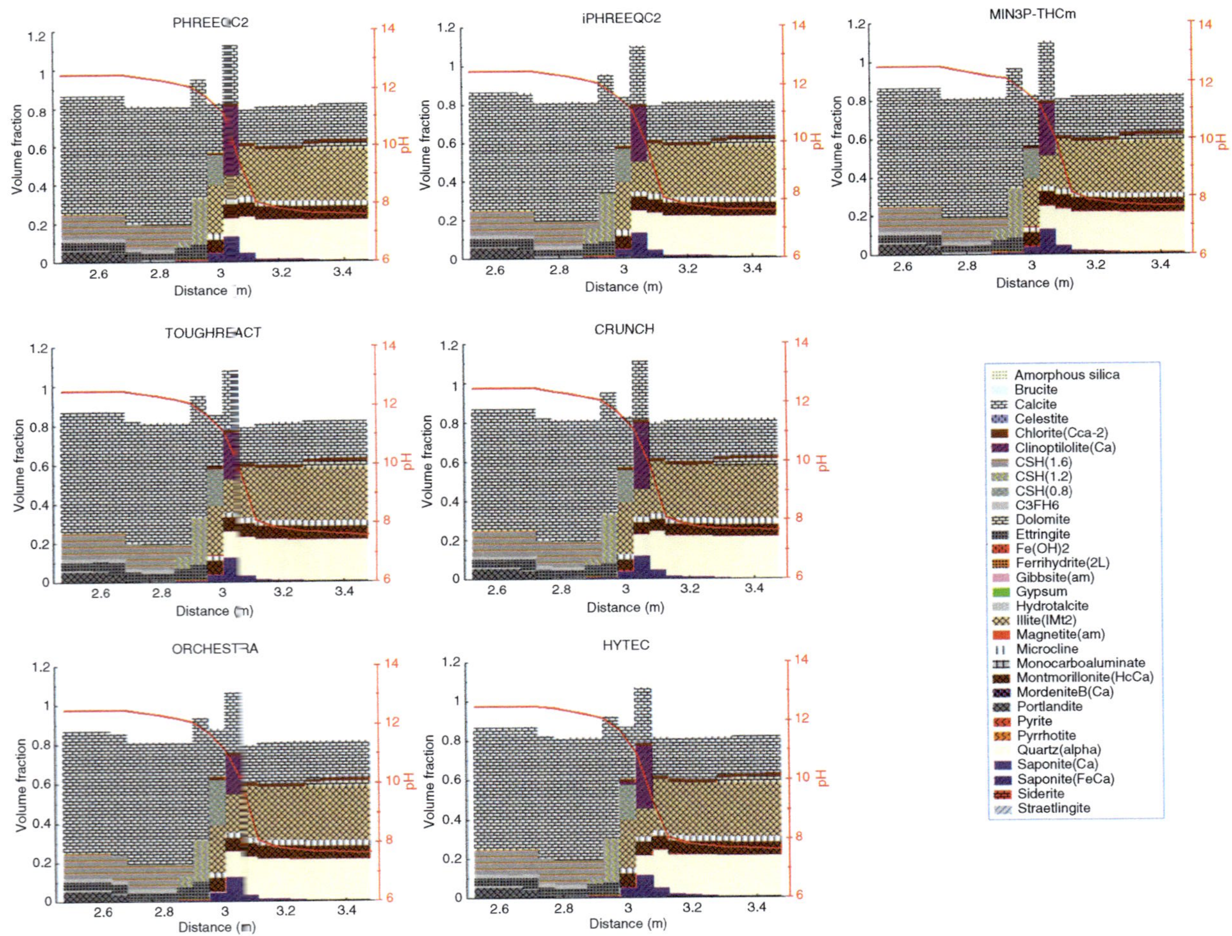

Figure 8.3 Mineralogical and pH changes obtained with TOUGHREACT, PHREEQC2, iPHREEQC2 (external transport module), CRUNCH, HYTEC, ORCHESTRA, and MIN3P-THCm after 10,000 years of concrete/claystone interaction (case 3, see Marty et al. (2015a) for details). The initial volume fraction value for $t = 0$ is the value far from the interface (2.5 m for concrete and 3.5 m for claystone) with main mineralogical changes modeled at the clay/concrete interface. Reproduced with permission of Springer.

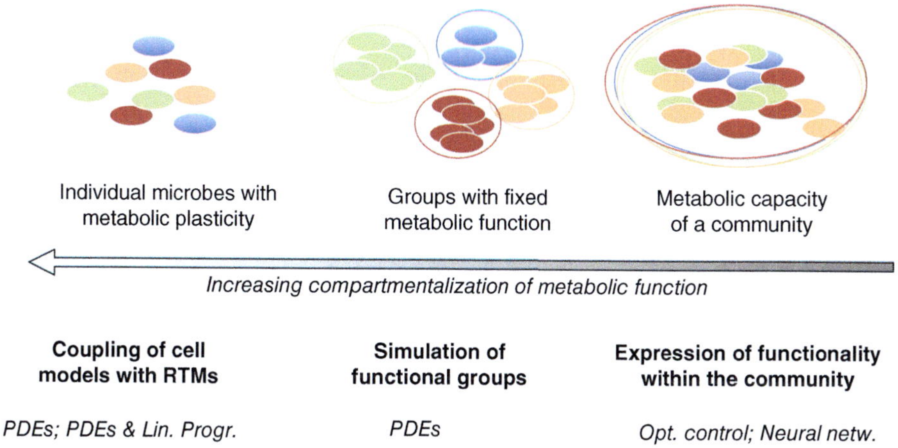

Figure 10.1 Sketch of microbial modeling approaches with varying extent of compartmentalization (denoted by open ellipses) of function (indicated by differing colours). Typical mathematical approaches are noted at the bottom: PDE, partial differential equation; Lin. Progr., linear programming; Opt. Control, optimal control problem; Neural Netw., neural networks and related statistical approaches. Examples for formulations using individual microbial cell models include King et al. (2009), Fang et al. (2011) and Zhuang et al. (2011); models with functional groups include Thullner et al. (2005) and King et al. (2010); descriptions with whole community approaches are presented in Vallino (2003) and Algar and Vallino (2014).

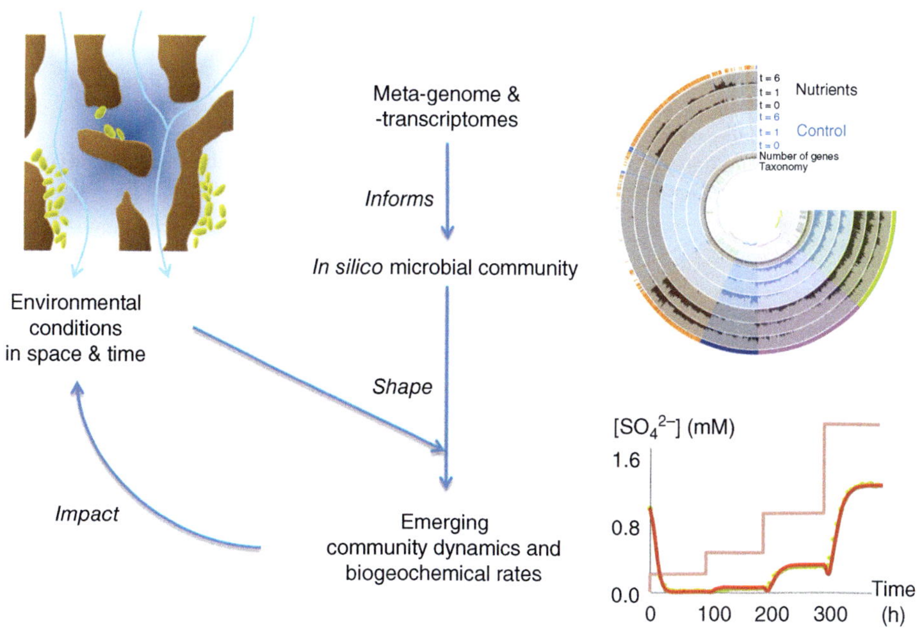

Figure 10.3 Trait-based approach of simulating microbially-driven biogeochemical processes. The microbial community is established as a large number of organisms whose metabolic traits may be assigned probabilistically. Traits may reflect hypothesized functions and processes, or are based on the interpretation of meta-genomes, -transcriptomes or other -omic data that constrain metabolic capabilities in the environment (top right: the image depicts the transcriptional response of a microbial community over time in the presence and absence of nutrient enrichments. *Source:* courtesy of T. Montenegro). Combining the functional capabilities with the environmental conditions determines the expression of metabolic functions and the modeled process rates (bottom right: the graph depicts the sulfate concentration in the inflow (pink) and outflow (red) in a column experiment over time). These rates then impact the composition of the local environment across the model domain (top left: illustrating fluid flow and the distribution of microbes in a porous medium). It also informs microbial growth dynamics, which in turn shapes the community composition. *Source:* Courtesy of T. Montenegro

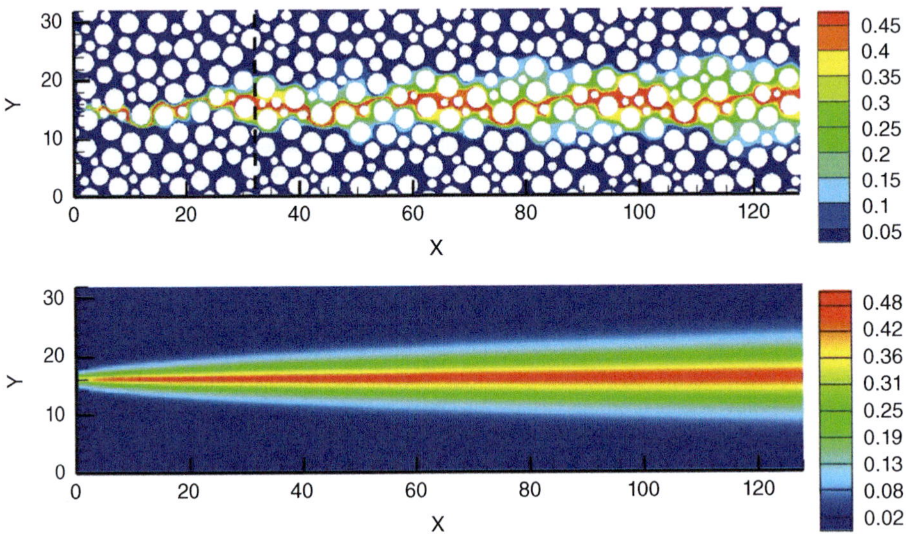

Figure 11.2 Concentration of the reaction product C resulting from the mixing of and reaction between two solutes A and B injected uniformly through the top and bottom halves of the left boundary of the flow domain with $Pe = 246$ and $Re = 4$. The top and bottom panels show results obtained from pore- and Darcy-scale simulations, respectively. The dashed line shows boundaries of the periodic unit cell. Here, $\mathbf{X} = x_1$ and $\mathbf{Y} = x_2$. *Source:* After Tartakovsky et al. (2009).

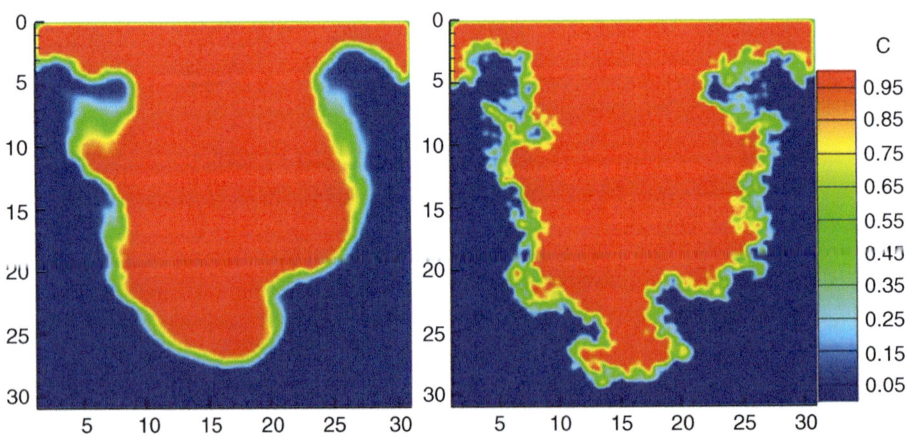

Figure 11.11 Deterministic (left) and stochastic (right) predictions of the Rayleigh-Taylor density-driven instability.

Printed and bound by CPI Group (UK) Ltd, Croydon, CR0 4YY

25/09/2026

14969414-0001